ELECTRONIC SYSTEMS MAINTENANCE HANDBOOK

Second Edition

ELECTRONICS HANDBOOK SERIES

Series Editor:
Jerry C. Whitaker
*Technical Press
Morgan Hill, California*

PUBLISHED TITLES

AC POWER SYSTEMS HANDBOOK, SECOND EDITION
Jerry C. Whitaker

THE COMMUNICATIONS FACILITY DESIGN HANDBOOK
Jerry C. Whitaker

THE ELECTRONIC PACKAGING HANDBOOK
Glenn R. Blackwell

POWER VACUUM TUBES HANDBOOK, SECOND EDITION
Jerry C. Whitaker

THERMAL DESIGN OF ELECTRONIC EQUIPMENT
Ralph Remsburg

THE RESOURCE HANDBOOK OF ELECTRONICS
Jerry C. Whitaker

MICROELECTRONICS
Jerry C. Whitaker

SEMICONDUCTOR DEVICES AND CIRCUITS
Jerry C. Whitaker

SIGNAL MEASUREMENT, ANALYSIS, AND TESTING
Jerry C. Whitaker

ELECTRONIC SYSTEMS MAINTENANCE HANDBOOK, SECOND EDITION
Jerry C. Whitaker

DESIGN FOR RELIABILITY
Dana Crowe and Alec Feinberg

FORTHCOMING TITLES

THE RF TRANSMISSION SYSTEMS HANDBOOK
Jerry C. Whitaker

ELECTRONIC SYSTEMS MAINTENANCE HANDBOOK

Second Edition

Edited by
Jerry C. Whitaker

CRC PRESS

Boca Raton   London   New York   Washington, D.C.

Library of Congress Cataloging-in-Publication Data

Electronic systems maintenance handbook / Jerry C. Whitaker, editor-in-chief.—2nd ed.
    p. cm. — (The Electronics handbook series)
  Rev. ed. of: Maintaining electronic systems. c1991.
  Includes bibliographical references and index.
  ISBN 0-8493-8354-4 (alk. paper)
   1. Electronic systems—Maintenance and repair—Handbooks, manuals, etc. 2. Electronic systems—Reliability—Handbooks, manuals, etc. I. Whitaker, Jerry C. II. Maintaining electronic systems. III. Series.

TK7870 .E212 2001
621.381'028'8—dc21
                                                2001043885
                                                              CIP

This book contains information obtained from authentic and highly regarded sources. Reprinted material is quoted with permission, and sources are indicated. A wide variety of references are listed. Reasonable efforts have been made to publish reliable data and information, but the authors and the publisher cannot assume responsibility for the validity of all materials or for the consequences of their use.

Neither this book nor any part may be reproduced or transmitted in any form or by any means, electronic or mechanical, including photocopying, microfilming, and recording, or by any information storage or retrieval system, without prior permission in writing from the publisher.

All rights reserved. Authorization to photocopy items for internal or personal use, or the personal or internal use of specific clients, may be granted by CRC Press LLC, provided that $1.50 per page photocopied is paid directly to Copyright clearance Center, 222 Rosewood Drive, Danvers, MA 01923 USA The fee code for users of the Transactional Reporting Service is ISBN 0-8493-8354-4/02/$0.00+$1.50. The fee is subject to change without notice. For organizations that have been granted a photocopy license by the CCC, a separate system of payment has been arranged.

The consent of CRC Press LLC does not extend to copying for general distribution, for promotion, for creating new works, or for resale. Specific permission must be obtained in writing from CRC Press LLC for such copying.

Direct all inquiries to CRC Press LLC, 2000 N.W. Corporate Blvd., Boca Raton, Florida 33431.

Trademark Notice: Product or corporate names may be trademarks or registered trademarks, and are used only for identification and explanation, without intent to infringe.

Visit the CRC Press Web site at www.crcpress.com

© 2002 by CRC Press LLC

No claim to original U.S. Government works
International Standard Book Number 0-8493-8354-4
Library of Congress Card Number 2001043885
Printed in the United States of America 1 2 3 4 5 6 7 8 9 0
Printed on acid-free paper

Preface

Technology is a moving target. Continuing advancements in hardware and software provide new features and increased performance for consumer, commercial, and industrial customers. Those same advancements, however, place new demands on the engineering and maintenance departments of the facility. Today—more than ever—the reliability of a system can have a direct and immediate impact on the profitability of an operation.

The days of troubleshooting a piece of gear armed only with a scope, multimeter, and general idea of how the hardware works are gone forever. Today, unless you have a detailed maintenance manual and the right test equipment, you are out of luck. The test bench of the 1980s—stocked with a VTVM, oscilloscope, signal generator, and signal analyzer—is a relic of the past. The work bench of today resembles more a small computer repair center than anything else.

It is true that some equipment problems can still be located with little more than a digital multimeter (DMM) and oscilloscope, given enough time and effort. But time costs money. Few technical managers are willing to make the trade. With current technology equipment, the proper test equipment is a must.

Two of the most important pieces of equipment for a maintenance technician servicing modern products are good lighting and a whopping big magnifier! While that is certainly an exaggeration, it points up a significant problem in equipment maintenance today: there are many tiny components, most of them jammed into a small amount of circuit board real estate. Tight component packaging makes printed wiring boards (PWBs) difficult to repair, at best. When complex and inter-related circuitry is added to the servicing equation, repair down to the component level may be virtually impossible. The equipment is just too complex, electrically and mechanically. The sophistication of hardware today has ushered in a new era in equipment maintenance—that of *repair by replacement*.

Some equipment manufacturers have built sophisticated test and diagnostic routines into their products. This trend is welcomed, and will likely accelerate as the *maintainability* of products becomes an important selling point. Still, however, specialized test equipment is often necessary to trace a problem to the board level.

Analytical Approach to Maintenance

Because of the requirement for maximum uptime and top performance, a *comprehensive preventive maintenance* (CPM) program should be considered for any facility. Priority-based considerations of reliability and economics are applied to identify the applicable and appropriate preventive maintenance tasks to be performed. CPM involves a realistic assessment of the vulnerable sections or components within the system, and a cause-and-effect analysis of the consequences of component failure. Basic to this analysis is the importance of keeping the system up and running at all times. Obvious applications of CPM include the stocking of critical spare parts used in stages of the equipment exposed to high temperatures and/or high voltages, or the installation of standby power/transient overvoltage protection equipment at the AC input point of critical equipment. Usually, the sections of a system most vulnerable to failure are those exposed to the outside world.

The primary goals of any CPM program are to prevent equipment deterioration and/or failure, and to detect impending failures. There are, logically, three broad categories into which preventive maintenance work can be classified:

- **Time-directed**: Tasks performed based upon a timetable established by the system manufacturer or user.
- **Condition-directed**: Maintenance functions undertaken because of feedback from the equipment itself (such as declining power output or frequency drift).
- **Failure-directed**: Maintenance performed first to return the system to operation, and second to prevent future failures through the addition of protection devices or component upgrades recommended by the manufacturer.

Regardless of whether such work is described as CPM or just plain common sense, the result is the same. Preventive maintenance is a requirement for reliability.

Training of Maintenance Personnel

The increasingly complex hardware used in industry today requires competent technical personnel to keep it running. The need for well-trained engineers has never been greater. Proper maintenance procedures are vital to top performance. A comprehensive training program can prevent equipment failures that impact productivity, worker morale, and income. Good maintenance is good business.

Maintenance personnel today must think in a "systems mode" to troubleshoot much of the hardware now in the field. New technologies and changing economic conditions have reshaped the way maintenance professionals view their jobs. As technology drives equipment design forward, maintenance difficulties will continue to increase. Such problems can be met only through improved test equipment and increased technician training.

The goal of every maintenance engineer is to ensure top quality performance from each piece of hardware. These objectives do not just happen. They are the result of a carefully planned maintenance effort.

It is easy to get into a rut and conclude that the old ways, tried and true, are best. Change for the sake of change does not make sense, but the electronics industry has gone through a revolution within the past 10 years. Every facility should re-evaluate its inventory of tools, supplies, and procedures.

Technology has altered the way electronic products are designed and constructed. The service bench needs to keep up as well. That is the goal of this book.

Jerry C. Whitaker
Editor-in-Chief

Editor-in-Chief

Jerry C. Whitaker is Technical Director of the Advanced Television Systems Committee, Washington D.C. He previously operated the consulting firm Technical Press. Mr. Whitaker has been involved in various aspects of the communications industry for more than 25 years. He is a Fellow of the Society of Broadcast Engineers and an SBE-certified Professional Broadcast Engineer. He is also a member and Fellow of the Society of Motion Picture and Television Engineers, and a member of the Institute of Electrical and Electronics Engineers. Mr. Whitaker has written and lectured extensively on the topic of electronic systems installation and maintenance.

Mr. Whitaker is the former editorial director and associate publisher of *Broadcast Engineering* and *Video Systems* magazines. He is also a former radio station chief engineer and TV news producer.

Mr. Whitaker is the author of a number of books, including:

- *The Resource Handbook of Electronics*, CRC Press, 2000.
- *The Communications Facility Design Handbook*, CRC Press, 2000.
- *Power Vacuum Tubes Handbook*, 2nd ed., CRC Press, 1999.
- *AC Power Systems*, 2nd ed., CRC Press, 1998.
- *DTV Handbook*, 3rd ed., McGraw-Hill, 2000.
- Editor-in-Chief, *NAB Engineering Handbook*, 9th ed., National Association of Broadcasters, 1999.
- Editor-in-Chief, *The Electronics Handbook*, CRC Press, 1996.
- Co-author, *Communications Receivers: Principles and Design*, 3rd ed., McGraw-Hill, 2000.
- *Electronic Display Engineering*, McGraw-Hill, 2000.
- Co-editor, *Standard Handbook of Video and Television Engineering*, 3rd ed., McGraw-Hill, 2000.
- Co-editor, *Information Age Dictionary*, Intertec/Bellcore, 1992.
- *Radio Frequency Transmission Systems: Design and Operation*, McGraw-Hill, 1990.

Mr. Whitaker has twice received a Jesse H. Neal Award Certificate of Merit from the Association of Business Publishers for editorial excellence. He also has been recognized as Educator of the Year by the Society of Broadcast Engineers. He resides in Morgan Hill, California.

Contributors

Samuel O. Agbo
California Polytechnic University
San Luis Obispo, California

Bashir Al-Hashimi
Staffordshire University
Stafford, England

David F. Besch
University of the Pacific
Stockton, California

Glenn R. Blackwell
Purdue University
West Lafayette, Indiana

Iuliana Bordelon
University of Maryland
College Park, Maryland

Gene DeSantis
DeSantis Associates
Little Falls, New Jersey

James E. Goldman
Purdue University
West Lafayette, Indiana

Jerry C. Hamann
University of Wyoming
Laramie, Wyoming

Dave Jernigan
National Instruments
Austin, Texas

Hagbae Kim
NASA Langely Research Center
Hampton, Virginia

Ravindranath Kollipara
LSI Logic Corporation
Milpitas, Calfornia

Edward McConnell
National Instruments
Austin, Texas

Michael Pecht
University of Maryland
College Park, Maryland

John W. Pierre
University of Wyoming
Laramie, Wyoming

Richard Rudman
KFWB Radio
Los Angeles, Calfornia

Jerry E. Sergent
BBS PowerMod, Incorporated
Victor, New York

Carol Smidts
University of Maryland
College Park, Maryland

Zbigniew J. Staszak
Technical University of Gdansk
Gdansk, Poland

Vijai Tripathi
Oregon State University
Corvallis, Oregon

Jerry C. Whitaker
ATSC
Morgan Hill, California

Allan White
NASA Langely Research Center
Hampton, Virginia

Don White
emf-emf control, Inc.
Gainesville, Virginia

Tsong-Ho Wu
Bellcore
Redbank, New Jersey

Rodger E. Ziemer
University of Colorado
Colorado Springs, Colorado

Contents

1 Probability and Statistics 1-1
Allan White and Hagbae Kim

2 Electronic Hardware Reliability 2-1
Michael Pecht and Iuliana Bordelon

3 Software Reliability 3-1
Carol Smidts

4 Thermal Properties 4-1
David F. Besch

5 Heat Management 5-1
Zbigniew J. Staszak

6 Shielding and EMI Considerations 6-1
Don White

7 Resistors and Resistive Materials 7-1
Jerry C. Whitaker

8 Capacitance and Capacitors 8-1
Jerry C. Whitaker

9 Inductors and Magnetic Properties 9-1
Jerry C. Whitaker

10 Printed Wiring Boards 10-1
Ravindranath Kollipara and Vijai Tripathi

11 Hybrid Microelectronics Technology 11-1
Jerry E. Sergent

12 Surface Mount Technology 12-1
Glenn R. Blackwell

13 Semiconductor Failure Modes 13-1
Jerry C. Whitaker

14 Power System Protection Alternatives 14-1
Jerry C. Whitaker

15 Facility Grounding 15-1
Jerry C. Whitaker

16 Network Switching Concepts 16-1
Tsong-Ho Wu

17 Network Communication 17-1
James E. Goldman

18 Data Acquisition 18-1
Edward McConnell and Dave Jernigan

19 Computer-Based Circuit Simulation 19-1
Bashir Al-Hashimi

20 Audio Frequency Distortion Mechanisms and Analysis 20-1
Jerry C. Whitaker

21 Video Display Distortion Mechanisms and Analysis 21-1
Jerry C. Whitaker

22 Radio Frequency Distortion Mechanisms and Analysis 22-1
Samuel O. Agbo

23 Digital Test Equipment and Measurement Systems 23-1
Jerry C. Whitaker

24 Fourier Waveform Analysis 24-1
    Jerry C. Hamann and John W. Pierre

25 Computer-Based Signal Analysis 25-1
    Rodger E. Ziemer

26 Systems Engineering Concepts 26-1
    Gene DeSantis

27 Disaster Planning and Recovery 27-1
    Richard Rudman

28 Safety and Protection Systems 28-1
    Jerry C. Whitaker

29 Conversion Tables 29-1

Index I-1

1
Probability and Statistics

Allan White
NASA Langely Research Center

Hagbae Kim
NASA Langely Research Center

1.1 Survey of Probability and Statistics 1-1
   General Introduction to Probability and Statistics
   • Probability • Statistics
1.2 Components and Systems 1-15
   Basics on Components • Systems • Modeling and Computation Methods
1.3 Markov Models 1-18
   Basic Constructions • Model for a Reconfigurable Fourplex • Correlated Faults • The Differential Equations for Markov Models • Computational Examples
1.4 Summary 1-24

1.1 Survey of Probability and Statistics

General Introduction to Probability and Statistics

The approach to probability and statistics used in this chapter is the pragmatic one that probability and statistics are methods of operating in the presence of incomplete knowledge.

The most common example is tossing a coin. It will land heads or tails, but the relative frequency is unknown. The two events are often assumed equally likely, but a skilled coin tosser may be able to get heads almost all of the time. The statistical problem in this case is to determine the relative frequency. Given the relative frequency, probability techniques answer such questions as how long can we expect to wait until three heads appear in a row.

Another example is light bulbs. The outcome is known: any bulb will eventually fail. If we had complete knowledge of the local universe, it is conceivable that we might compute the lifetime of any bulb. In reality, the manufacturing variations and future operating conditions for a lightbulb are unknown to us. It may be possible, however, to describe the failure history for the general population. In the absence of any data, we can propose a failure process that is due to the accumulated effect of small events (corrosion, metal evaporation, and cracks) where each of these small events is a random process. It is shown in the statistics section that a failure process that is the sum of many events is closely approximated by a distribution known as the normal (or Gaussian) distribution. This type of curve is displayed in Fig. 1.1(a).

A manufacturer can try to improve a product by using better materials and requiring closer tolerances in the assembly. If the light bulbs last longer and there are fewer differences between the bulbs, then the failure curve should move to the right and have less dispersion, as shown in Fig. 1.1(b).

There are three types of statistical problems in the light bulb example. One is identifying the shape of the failure distributions. Are they really the normal distribution as conjectured? Another is to

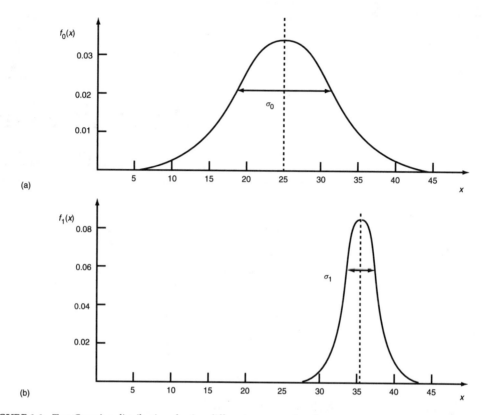

FIGURE 1.1 Two Gaussian distributions having different means and variances.

estimate such parameters as the average and the deviation from the average for the populations. Still another is to compare the old manufacturing process with the new. Is there a real difference and is the difference significant?

Suppose the failure curve (for light bulbs) and its parameters have been obtained. This information can then be used in probability models. As an example, suppose a city wishes to maintain sufficient illumination for its streets by constructing a certain number of lamp posts and by having a maintenance crew replace failed lights. Since there will be a time lag between light failure and the arrival of the repair crew, the city may construct more than the minimum number of lamp posts needed in order to maintain sufficient illumination. Determining the number of lamp posts depends on the failure curve for the bulbs and the time lag for repair. Furthermore, the city may consider having the repair crew replace all bulbs, whether failed or not, as the crew makes its circuit. This increases the cost for bulbs but decreases the cost for the repair crew. What is the optimum strategy for the number of lamp posts, the repair crew circuit, and the replacement policy? The objective is to maximize the probability of sufficient illumination while minimizing the cost.

Incomplete Knowledge vs Complete Lack of Knowledge

People unfamiliar with probability sometimes think that random means completely arbitrary, and there is no way to predict what will happen. Physical reasoning, however, can supply a fair amount of knowledge:

- It is reasonable to assume the coin will land heads or tails and the probability of heads remains constant. This gives a unique probability distribution with two possible outcomes.
- It is reasonable to assume that light bulb failure is the accumulation of small effects. This implies that the failure distribution for the light bulb population is approximately the normal distribution.

- It will be shown that the probability distribution for time to failure of a device is uniquely determined by the assumption that the device does not wear out.

Of course, it is possible (for random as for deterministic events) that our assumptions are incorrect. It is conceivable that the coin will land on its edge or disintegrate in the air. Power surges that burn out large numbers of bulbs are not incremental phenomena, and their presence will change the distribution for times to failure. In other words, using a probability distribution implies certain assumptions are being made. It is always good to check for a match between the assumptions underlying the probability model and the phenomenon being modeled.

Probability

This section presents enough technical material to support all our discussions on probability. It covers the (1) definition of a **probability space**, (2) Venn (set) diagrams, and (3) density functions.

Probability Space

The definition of a probability space is based on the intuitive ideas presented in the previous section. We begin with the set of all possible events, call this set X.

- If we are tossing a coin three times, X consists of all sequences of heads (H) and tails (T) of length three.
- If we are testing light bulbs until time to failure, X consists of the positive real numbers.

The next step in the definition of a probability space is to ensure that we can consider simple combinations of events. We want to consider the AND and OR combinations as intersections and unions. We also want to include complements.

In the coin-tossing space, getting at least two heads is the union of the four events, that is,

$$\{HHT, HTH, THH, HHH\}$$

In the light-bulb-failure space, a bulb lasting more than 10 h but less than 20 h is the intersection of the two events: (1) bulb lasts more than 10 h and (2) bulb lasts less than 20 h. In the coin-tossing space, getting less than two heads is the complement of the first set, that is,

$$\{HTT, THT, TTH, TTT\}$$

When it comes to assigning probabilities to events, the procedure arises from the idea of assigning areas to sets. This idea can be conveyed by pictures called Venn diagrams. For example, consider the disjoint sets of events A and B in the Venn diagram of Fig. 1.2.

We want the probability of being in either A or B to be the sum of their probabilities. Finally, when considering physical events it is sufficient to consider only finite or countably infinite combinations of simpler events. (It is also true that uncountably infinite combinations can produce mathematical pathologies.)

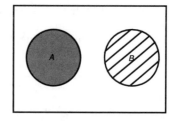

FIGURE 1.2 Venn diagram of two disjoint sets of events.

Example. It is necessary to include countably infinite combinations, not just finite combinations. The simplest example is tossing a coin until a head appears. There is no bound for the number of tosses. The sample space consists of all sequences of heads and tails. The combinatorial process in operation for this example is AND; the first toss is T, and the second toss is T, ... and the n-th toss is H.

Hence, with this in mind, a probability space is defined as follows:

1. It is a collection of subsets, called events, of a universal space X. This collection of subsets (events) has the following properties:
   - X is in the collection.
   - If A is in the collection, its set complement A' is in the collection.
   - If $\{A_i\}$ is a finite or countably infinite collection of events, then both $\bigcup A_i$ and $\bigcap A_i$ are in the collection.

2. To every event A there is a real number, $Pr[A]$, between 0 and 1 assigned to A such that:
   - $Pr[X] = 1$.
   - If $\{A_i\}$ is a finite or countably infinite collection of disjoint events, then $Pr[\bigcup A_i] = \sum Pr[A_i]$.

As the last axiom suggests, numerous properties about probability are derived by decomposing complex events into unions of disjoint events.

Example. $Pr[A \cup B] = Pr[A] + Pr[B] - Pr[A \cap B]$.

Let $A \backslash B$ be the elements in A that are not in B. Express the sets as the disjoint decompositions

$$A = (A \backslash B) \cup (A \cap B)$$
$$B = (B \backslash A) \cup (A \cap B)$$
$$A \cup B = (A \backslash B) \cup (B \backslash A) \cup (A \cap B)$$

which gives

$$Pr[A] = Pr[A \backslash B] + Pr[A \cap B]$$
$$Pr[B] = Pr[B \backslash A] + Pr[A \cap B]$$
$$Pr[A \cup B] = Pr[A \backslash B] + Pr[B \backslash A] + Pr[A \cap B]$$

and the result follows.

Probability from Density Functions and Integrals

The interpretation of probability as an area, when suitably generalized, gives a universal framework. The formulation given here, in terms of Riemann integrals, is sufficient for many applications. Consider a nonnegative function $f(x)$ defined on the real line with the property that its integral over the entire line is equal to one.

The probability that the values lie on the interval $[a, b]$ is defined to be

$$Pr[a \leq x \leq b] = \int_a^b f(x)dx$$

The function $f(x)$ is called the *density function*. The probability that the values are less than or equal to t is given by

$$F(t) = Pr[x \leq t] = \int_{-\infty}^{t} f(x)dx$$

This function $F(t)$ is called the *distribution function*.

In this formulation by density functions and (Riemann) integrals, the basic events are the intervals $[a, b]$ and countable unions of these disjoint intervals. It is easy to see that this approach satisfies all of the axioms for a probability space listed earlier. This formulation lets us apply all of the techniques of calculus and analysis to probability. This feature is demonstrated repeatedly in the sections to follow.

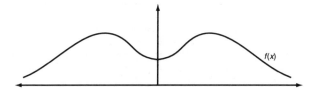

FIGURE 1.3 A nonnegative function $f(x)$ defined on the real line.

One last comment should be made. Riemann integration and continuous (or piecewise continuous) density functions are sufficient for a considerable amount of applied probability, but they cannot handle all of the topics in probability because of the highly discontinuous nature of some probability distributions. A generalization of Riemann integration (measure theory and general integration), however, does handle all aspects of probability.

Independence

Conditional probability, and the related topic of independence, give probability a unique character. As already mentioned, mathematical probability can be presented as a topic in measure theory and general integration. The idea of conditional probability distinguishes it from the rest of mathematical analysis. Logically, independence is a consequence of conditional probability, but the exposition to follow attempts an intuitive explanation of independence before introducing conditional events.

Informal Discussion of Independence. In probability, two events are said to be independent if information about one gives no information about the other. Independence is often assumed and used without any explicit discussion, but it is stringent condition. Furthermore, the probabilistic meaning is different from the usual physical meaning of two events having nothing in common.

Figure 1.4 gives three diagrams illustrating various degrees of correlation and independence. In the first diagram, event A is contained in event B. Hence, if A has occurred then B has occurred, and the events are not independent (since knowledge of A yields some knowledge of B). In the second diagram event A lies half in and half out of event B. Event B has the same relationship to event A. Hence, the two events in the third diagram are independent—knowledge of the occurrence of one gives no information about the occurrence of the other. This mutual relationship is essential. The third diagram illustrates disjoint events. If one event occurs, then the other cannot. Hence, the occurrence of one event yields information about the occurrence of the other event, which means the two events are not independent.

Conditional Probability and the Multiplicative Property. Conditional probability expresses the amount of information that the occurrence of one event gives about the occurrence of another event. The notation is $Pr[A \mid B]$ for the probability of A given B. To motivate the mathematical definition consider the events A and B in Fig. 1.5.

By relative area, $Pr[A] = 12/36$, $PrB = 6/36$, and $Pr[A \mid B] = 2/6$. The mathematical expression for conditional probability is

$$Pr[A \mid B] = Pr[A \text{ and } B]/Pr[B]$$

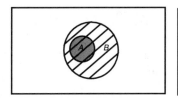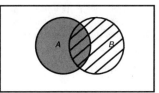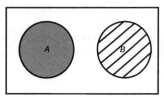

FIGURE 1.4 Various degrees of correlation and independence.

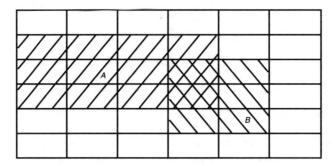

FIGURE 1.5 Pictorial example of conditional probability.

From this expression it is easy to derive the multiplication rule for independent events. Suppose A is independent of B. Then information about B gives no information about A. That is, the conditional probability of A given B is the same as the probability of A or (by formula)

$$Pr[A] = Pr[A \mid B] = PR[A \text{ and } B]/Pr[B]$$

A little algebra gives

$$Pr[A \text{ and } B] = Pr[A]Pr[B]$$

This multiplicative property means that independence is symmetrical. If A is independent of B, then B is independent of A. In its abstract formulation conditional probability is straightforward, but in applications it can be tricky.

Example (The Monte Hall Problem). A contestant on a quiz show can choose one of three doors, and there is a prize behind one of them. After the contestant has chosen, the host opens one of the doors not chosen and shows that there is no prize behind the door he has just opened. The contestant is offered the opportunity to change doors. Should the contestant do so? There is a one-third chance that the contestant has chosen the correct door. There is a two-thirds chance that the prize is behind one of the other doors, and it would be nice if the contestant could choose both of the other doors. By opening the other door that does not contain the prize, the host is offering the contestant an opportunity to choose both of the other doors by choosing the other door that has not been opened. Changing doors increases the chance of winning to two-thirds. There are several reasons why this result appears counterintuitive. One is that the information offered is negative information (no prize behind this door), and it is hard to interpret negative information. Another is the timing. If the host shows that there is no prize behind a door before the contestant chooses, then the chance of a correct choice is one-half. The timing imposes an additional condition on the host.

The (sometimes) subtle nature of conditional probability is important when conducting an experiment. It is almost trite to say that an event A can be measured only if A or its effect can be observed. It is not quite as trite to say that observability is a special property. Experiments involve the conditional probability:

$$Pr[A \text{ has some value} \mid A \text{ or its effect can be observed}]$$

A primary goal in experimental sampling is to make the observed value of A independent of its observability. One method that accomplishes this is to arrange the experiment so that all events are potentially observable. If X is the set of all events then the equations for this case are

$$Pr[A \mid X] = Pr[A \text{ and } X]/Pr[X] = Pr[A]/1 = Pr[A]$$

Probability and Statistics

For this approach to work it is only necessary that all events are potentially observable, not that all events are actually observed. The last section on statistics will show that a small fraction of the population can give a good estimate, provided it is possible to sample the entire population.

An experiment where part of the population cannot be observed can produce misleading results. Imagine a telephone survey where only half of the population has a telephone. The population with the geographic location and wealth to own a telephone may have opinions that differ from the rest of the population.

These comments apply immediately to fault injection experiments for reconfigurable systems. These experiments attempt to study the diagnostic and recovery properties of a system when a fault arrives. If it is not possible to inject faults into the entire system, then the results can be biased, because the unreachable part of the system may have different diagnostic and recovery characteristics than the reachable part.

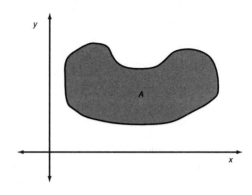

FIGURE 1.6 A set represented by two density functions $f(x)$ and $g(y)$.

Independence and Density Functions. The product formula for the probability of independent events translates into a product formula for the density of independent events. If x and y are random variables with densities $f(x)$ and $g(y)$, then the probability that (x, y) lies in the set A, displayed in Fig. 1.6, is given by

$$\iint_A f(x)g(y)dy\,dx$$

This property extends to any finite number of random variables. It is used extensively in the statistics section to follow because items in a **sample** are **independent events**.

The Probability Distribution for Memoryless Events

As an example of deriving the density function from the properties of the event, this section will show that memoryless events must have an exponential density function. The derivation uses the result from real analysis that the only continuous function $h(x)$ with the property, $h(a + b) = h(a)h(b)$ is the exponential function $h(t) = e^{\alpha t}$ for some α.

An event is memoryless if the probability of its occurrence in the next time increment does not depend on the amount of time that has already elapsed. That is,

Pr[occurrence by time $(s + t)$ given has not occurred at time s] = Pr[occurrence by time t]

Changing to distribution functions and using the formula for conditional probability gives

$$\frac{F(s+t) - F(s)}{1 - F(s)} = F(t)$$

A little algebra yields

$$1 - F(s + t) = [1 - F(s)][1 - F(t)]$$

Let $g(x) = 1 - F(x)$, then $g(s + t) = g(s)g(t)$, which means $g(x) = e^{\alpha t}$. Hence, the density function is

$$f(x) = F'(x) = -\alpha e^{\alpha t}$$

Since the density function must be nonnegative, $\alpha < 0$.

This derivation illustrates a key activity in probability. If a probability distribution has certain properties, does this imply that it is uniquely determined? If so, what is the density function?

Moments

A fair amount of information about a probability distribution is contained in its first two moments. The first moment is also called the **mean**, the **average**, the **expectation**, or the **expected value**. If the density function is $f(x)$ the first moment is defined as

$$\mu = E(x) = \int_{-\infty}^{+\infty} x f(x) dx$$

The variance is the second moment about the mean

$$\sigma^2 = E[(x - \mu)^2] = \int_{-\infty}^{+\infty} (x - \mu)^2 f(x) dx$$

The variance measures the dispersion of the distribution. If the variance is small then the distribution is concentrated about the mean. A mathematical version of this is given by Chebychev's inequality

$$Pr[\|x - \mu\| \geq \delta] \leq \frac{\sigma^2}{\delta^2}.$$

This inequality says it is unlikely for a member of the population to be a great distance from the mean.

Statistics

The purpose of this section is to present the basic ideas behind statistical techniques by means of concrete examples.

Sampling

Sampling consists of picking individual items from a population in order to measure the parameter in question. For the coin tossing experiment mentioned at the beginning of this chapter, an individual item is a toss of the coin and the parameter measured is heads or tails. If the matter of interest is the failure distribution for a class of devices, then a number of devices are selected and the parameter measured is the time until failure. Selecting and measuring an individual item is known as a trial.

The theoretical formulation arises from the assumptions that (1) selecting an individual from the population is a random event governed by the probability distribution for that population, (2) the results of one trial do not influence the results of another trial, and (3) the population (from which items are chosen) remains the same for each trial. If these assumptions hold, then the samples are said to be independent and identically distributed.

Suppose $f(x)$ is the density function for the item being sampled. The probability that a sample lies in the interval $[a, b]$ is $\int_a^b f(x) dx$. By the assumption of independence, the density function for a sample of size n is the n-fold product

$$f(x_1) f(x_2) \cdots f(x_n)$$

Probability and Statistics

The probability that a sample of size n, say, $x_1, x_2, \ldots, x_n$, is in the set A is

$$\int\int\cdots\int f(x_1)f(x_2)\cdots f(x_n)dx_n\cdots dx_2 dx_1$$

where the integral is taken over the (multidimensional) set A. This formulation for the probability of a sample (in terms of a multiple integral) provides a mathematical basis for statistics.

Estimators: Sample Average and Sample Variance

Estimators are functions of the sample values (and constants). Since samples are random variables, estimators are also random variables. For example, suppose g is a function of the sample $x_1, x_2, \ldots, x_n$, where $f(x)$ is the density function for each of the x. Then the expectation of g is

$$E(g) = \int\int\cdots\int g(x_1, \ldots, x_n) f(x_1) f(x_2) \ldots f(x_n) dx_n \ldots dx_2 dx_1$$

To be useful, of course, estimators must have a good relationship to the unknown parameter we are trying to determine. There are a variety of criteria for estimators. We will only consider the simplest. If g is an estimator for the (unknown) parameter θ, then g is an *unbiased* estimator for θ if the expected value of g equals θ. That is

$$E(g) = \theta$$

In addition, we would like the variance of the estimator to be small since this improves the accuracy. Statistical texts devote considerable time on the efficiency (the **variance**) of estimators.

The most common statistic used is the sample average as an estimator for the mean of the distribution. For a sample of size n it is defined to be

$$\bar{x} = \frac{(x_1 + \cdots + x_n)}{n}$$

Suppose the distribution has mean, variance, and density function of μ, σ^2 and $f(x)$. To illustrate the mathematical formulation of the estimator, we will do a detailed derivation that the sample mean is an unbiased estimator for the mean of the underlying distribution. We have:

$$E(\bar{x}) = \int_{-\infty}^{+\infty}\int_{-\infty}^{+\infty}\cdots\int_{-\infty}^{+\infty} \left(\frac{x_1 + \cdots + x_n}{n}\right) f(x_n)\cdots f(x_2) f(x_1) dx_n \cdots dx_2 dx_1$$

$$= \sum_{i=1}^{n} \frac{1}{n} \int_{-\infty}^{+\infty}\int_{-\infty}^{+\infty}\cdots\int_{-\infty}^{+\infty} x_i f(x_n)\cdots f(x_2) f(x_1) dx_n \cdots dx_2 dx_1$$

Writing the ith term in the last sum as an iterated integral and then as a product integral gives

$$\int_{-\infty}^{+\infty} 1f(x_1)dx_1 \cdots \int_{-\infty}^{+\infty} x_i f(x_i) dx_i \int_{-\infty}^{+\infty} 1f(x_{i+1})dx_{i+1} \cdots \int_{-\infty}^{+\infty} 1f(x_n)dx_n$$

The factors with 1 in the integrand are equal to 1. The factor with x_i in the integrand is equal to μ, the mean of the distribution. Hence,

$$E(\bar{x}) = \frac{1}{n}\sum_{i=1}^{n}\mu = \mu$$

The sample average also has a convenient expression for its variance. A derivation similar to, but more complicated than, the preceding one gives

$$\text{Var}(\bar{x}) = \frac{\sigma^2}{n}$$

where σ^2 is the variance of the underlying population. Hence, increasing the sample size n improves the estimate by decreasing the variance of the estimator. This is discussed at greater length in the section on confidence intervals.

We can also estimate the variance of the underlying distribution. Define the sample variance by

$$S^2 = \frac{\sum_{i=1}^{n}(x-\bar{x})^2}{n-1}$$

A similar derivation shows that

$$E(S^2) = \sigma^2$$

There is also an expression for $\text{Var}(S^2)$ in terms of the higher moments of the underlying distribution.

For emphasis, all of these results about estimators arrive from the formulation of a sample as a collection of independent and identically distributed random variables. Their joint distribution is derived from the n-fold product of the density function of the underlying distribution. Once these ideas are expressed as multiple integrals, deriving results in statistics are exercises in calculus and analysis.

The Density Function of an Estimator

This material motivates the central limit theorem presented in the next section, and it illustrates one of the major ideas in statistics. Since estimators are random variables, it is reasonable to ask about their density functions. If the density function can be computed we can determine the properties of the estimator.

This point of view permeates statistics, but it takes a while to get used to it. There is an underlying distribution. We take a random sample from this distribution. We consider a function of these sample points, which we call an estimator. This estimator is a random variable, and it has a distribution that is related to, but different from, the underlying distribution from which the samples were taken.

We will consider a very simple example for our favorite estimator—the sample average. Suppose the underlying distribution is a constant rate process with the density function $f(z) = e^{-z}$. Suppose the estimator is the average of a sample of size two $\bar{z} = (x + y)/2$. The distribution function for $\bar{z}$ is $G(t) = Pr[x + y \le 2t]$. The integration is carried out over the shaded area in Fig. 1.7 bounded by the line $y = -x + 2t$ and the axes.

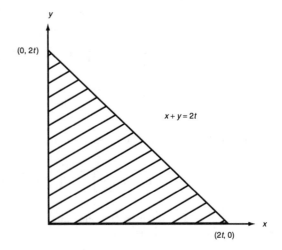

FIGURE 1.7 $x + y \le 2t$.

The probability that $x + y \leq 2t$ is given by the double integral

$$G(t) = \int_0^{2t} \int_0^{2t-x} e^{-x} e^{-y} dy\, dx = 1 - e^{-2t} - 2te^{-2t}$$

which implies the density function is

$$g(t) = G'(t) = 4te^{-2t}$$

Hence, if x and y are sample points from a constant rate process with rate one, then the probability that $a \leq (x+y)/2 \leq b$ is $\int_a^b 4te^{-2t}\, dt$.

Figure 1.8 displays the density functions for the underlying exponential distribution and for the sample average of size 2. Obviously, we want the density function for a sample of size n, we want the density function of the estimator for other underlying distributions, and we want the density function for estimators other than the sample average. Statisticians have expended considerable effort on this topic, and this material can be found in statistics tests and monographs.

Consider the two density functions in Fig. 1.8 from a different point of view. The first is the density function of the sample average when the sample size is 1. The second when the sample size is 2. As the sample size increases, the graph becomes more like the bell-shaped curve for the normal distribution. This is the content of the Central Limit Theorem in the next section. As the sample size increases, the distribution of the sample average becomes approximately normal.

The Central Limit Theorem

A remarkable result in probabihty is that for a large class of underlying distributions the sample average approaches one type of distribution as the sample size becomes large. The class is those distributions with finite mean and variance. The limiting distribution is the normal (Guassian) distribution. The density function for the normal distribution with mean μ and variance σ^2 is

$$\frac{1}{\sigma\sqrt{2\pi}} e^{-(x-\mu)^2/2\sigma^2},$$

where $\sigma = \sqrt{\sigma^2}$.

Theorem. Suppose a sample $x_1, \ldots, x_n$ is chosen from a distribution with mean μ and variance σ^2. As n becomes large, the distribution for the sample average $\bar{x}$ approaches the normal distribution with mean μ and variance σ^2/n.

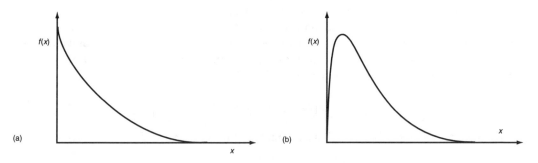

FIGURE 1.8 Density function: (a) underlying exponential distribution, (b) for the estimator of size 2.

This result is usually used in the following manner. As a consequence of the result, it can be shown that the function, $(\bar{x} - \mu)/\sqrt{\sigma^2/n}$, is (approximately) normal with mean zero and variance one (the standard normal).

Extensive tables are available that give the probability that $\alpha \leq Z \leq \beta$ if Z has the standard normal distribution. The probability that $a \leq \bar{x} \leq b$ can be computed from the inequalities

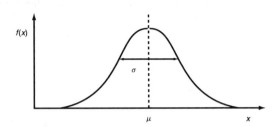

FIGURE 1.9 A Gaussian distribution with mean μ and variance σ^2.

$$\frac{a - \mu}{\sqrt{\sigma^2/n}} \leq \frac{\bar{x} - \mu}{\sqrt{\sigma^2/n}} \leq \frac{b - \mu}{\sqrt{\sigma^2/n}}$$

and a table lookup. This will be illustrated in the next section on confidence intervals.

Confidence Intervals

It is possible for an experiment to mislead us. For example, when estimating the average lifetime for a certain type of device, the sample could contain an unusual number of the better components (or of the poorer components).

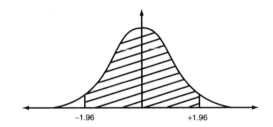

FIGURE 1.10 $-1.96 \leq t \leq 1.96$.

We need a way to indicate the quality of an experiment—the probability that the experiment has not misled us. A quantitative method of measuring the quality of an experiment is by using a confidence interval, which usually consists of three parts:

1. The value of the **estimator**: $\hat{x}$
2. An interval (called the **confidence interval**) about the estimator: $[\hat{x} - \alpha, \hat{x} + \beta]$
3. The probability (called the **confidence level**) that this interval contains the true (but unknown) value

The confidence level in part three is usually conservative, the probability that the interval contains the true value is greater than or equal to the stated confidence level.

The confidence interval has the following frequency interpretation. Suppose we obtain a 95% confidence interval from an experiment and statistical techniques. If we performed this experiment many times, then at least 95% of the time the interval will contain the true value of the parameter. This discussion of the confidence interval sounds similar to the previous discussion about estimators; they are random variables, and they have a certain probability of being within certain intervals.

To illustrate the general procedure, suppose we wish a 95% confidence interval for the mean μ of a random variable. Suppose the sample size is n. As described earlier, the sample mean $\bar{x}$ and the sample variance S^2 are unbiased estimators for the mean and variance of the underlying distribution. For a symmetric confidence interval the requirement is

$$Pr[\|\bar{x} - \mu\| \leq \alpha] \geq 0.95$$

It is necessary to solve for α. We will use the central limit theorem. Rewrite the preceding equation:

$$Pr\left[\frac{-\alpha}{\sqrt{S^2/n}} \leq \frac{\bar{x} - \mu}{\sqrt{S^2/n}} \leq \frac{\alpha}{\sqrt{S^2/n}}\right]$$

The term in the middle of the inequality has (approximately) the standard normal distribution. Here, 95% of the standard normal lies between -1.96 and $+1.96$. Hence, set $t = \alpha/\sqrt{S^2/n} = 1.96$ and solve for α.

As an example of what can be done with more information about the underlying distribution available, consider the exponential (constant rate) failure distribution with density function $f(t) = \lambda e^{-\lambda t}$. Suppose the requirement is to estimate the mean-time-to-failure within 10% with a 95% confidence level. We will use the central limit theorem to determine how many trials are needed. The underlying population has mean and variance $\mu = 1/\lambda$ and $\sigma^2 = (1)/\lambda^2$. For a sample of size n the sample average $\bar{x}$ has mean and variance:

$$\mu(\bar{x}) = \frac{1}{\lambda} \quad \text{and} \quad \sigma^2(\bar{x}) = \frac{1}{n\lambda^2}$$

The requirement is

$$Pr[\|\bar{x} - \mu\| \leq 0.1\mu] \geq 0.95$$

Since the estimator is unbiased, $\mu = \mu(\bar{x})$. A little more algebra gives

$$Pr\left[\frac{-0.1\mu(\bar{x})}{\sqrt{\sigma^2(\bar{x})}} \leq \frac{\bar{x} - \mu(\bar{x})}{\sqrt{\sigma^2(\bar{x})}} \leq \frac{0.1\mu(\bar{x})}{\sqrt{\sigma^2(\bar{x})}}\right] \geq 0.95$$

Once again, the middle term of the inequality has (approximately) the standard normal distribution, and 95% percent of the standard normal lies between -1.96 and $+1.96$. Set

$$1.96 = \frac{0.1\mu(\bar{x})}{\sqrt{\sigma^2(\bar{x})}} = 0.1\sqrt{n}$$

write $\mu(\bar{x})$ and $\sigma^2(\bar{x})$ in terms of λ; solve to get $n = 384$.

Opinion Surveys, Monte Carlo Computations, and Binomial Sampling

One of the sources for skepticism about statistical methods is one of the most common applications of statistics, the opinion survey where about 1000 people are polled to determine the preference of the entire population. How can a sample of size 1000 tell us anything about the preference of several hundred million people?

This question is more relevant to engineering applications of statistics than it might appear, because the answer lies in the general nature of binomial sampling. As a similar problem consider a **Monte Carlo** computation of the area enclosed in an irregular graph given in Fig. 1.11. Suppose the area of the square is A. Choose 1000 points at random in the square. If N of the chosen points lie within the enclosed graph, then the area of the graph is estimated to be $NA/1000$. This area estimation poses a greater problem than the opinion survey. Instead of a total population of several hundred million, the total population of points in the square is uncountably infinite. (How can 1000 points chosen at random tell us anything since 1000 is 0%

FIGURE 1.11 An area enclosed in an irregular graph.

of the population?) In both the opinion survey and the area problem we are after an unknown probability p, which reflects the global nature of the entire population. For the opinion survey:

$$p = \frac{\text{number of yeses}}{\text{total number of population}}$$

For the area problem

$$p = \frac{\text{area enclosed by graph}}{\text{area of square}}$$

Suppose the sampling technique makes it equally likely that any person or point is chosen when a random selection is made. Then p is the probability that the response is yes or inside the curve. Assign a numerical value of 1 to yes or inside the curve. Assign 0 to the complement. The average (the expected value) for the entire population is:

$$1 \text{ (probability of a 1)} + 0(\text{probability of a 0}) = 1p + 0(1 - p) = p$$

The population variance is

$$(\text{probability of a 1})(1 - \text{average})^2 + (\text{probability of a 0})(0 - \text{average})^2$$
$$= p(1-p)^2 + (1-p)(0-p)^2 = p(1-p)$$

Hence, for a sample of size n, the mean and variance of the sample average are $E(\bar{x}) = p$ and $\text{Var}(\bar{x}) = p(1-p)/n$.

The last equation answers our question about how a sample that is small compared to the entire population can give an accurate estimate for the entire population. The last equation says the variance of the estimator (which measures the accuracy of the estimator) depends on the sample size, not on the population size. The population size does not appear in the variance, which indicates the accuracy is the same whether the population is ten thousand, or several hundred million, or uncountably infinite.

The next step is to use the **central limit theorem**. Since the binomial distribution has a finite mean and variance, the adjusted sample average

$$\frac{\bar{x} - \mu}{\sqrt{\sigma^2/n}} = \frac{\bar{x} - p}{\sqrt{p(1-p)/n}}$$

converges to the standard normal distribution (which has mean zero and standard deviation one). The usual confidence level used for opinion surveys is 95%. A table with the normalized Gaussian distribution shows that 95% of the standard normal lies between -1.96 and $+1.96$. Hence,

$$Pr\left[-1.96 \leq \frac{\bar{x} - p}{\sqrt{p(1-p)/n}} \leq 1.96\right] \geq 0.95$$

or

$$Pr\left[\frac{-1.96\sqrt{p(1-p)}}{\sqrt{n}} \leq \bar{x} - p \leq \frac{1.96\sqrt{p(1-p)}}{\sqrt{n}}\right] \geq 0.95$$

Instead of replacing $\sqrt{p(1-p)}$ with an estimator, we will be more conservative and replace it with its maximum value, which is 1/2. Approximating 1.96 by 2 and taking n to be 1000 gives

$$Pr[-0.03 \leq \bar{x} - p \leq 0.03] \geq 0.95$$

We have arrived at the news announcement that the latest survey gives some percentage, plus or minus 3%. The 3% error comes from a 95% confidence level with a sample size of 1000.

1.2 Components and Systems

Basics on Components

Quantitative reliability (the probability of failure by a certain time) begins by considering how fast and in what manner components fail. The source for this information is field data, past experience with classes of devices, and consideration of the manufacturing process. Good data is hard to get. For this reason, reliability analysts tend to be conservative in their assumptions about device quality and to use worst-case analyses when considering failure modes and their effects.

Failure Rates

There is a great body of statistical literature devoted to identifying the failure rate of devices under test. We just present a conceptual survey of rates.

Increasing Rate. There is an increasing failure rate if the device is wearing out. As the device ages, it is more likely to fail in the next increment of time. Most mechanical items have this property. For long-term reliability of a system with such components, there is often a replace before failure strategy as in the streetlight example earlier.

Constant Rate. The device is not wearing out. The most prosaic example is the engineer's coffee mug. It fails by catastrophe (if we drop it), otherwise it is immortal. A constant failure rate model is often used for mathematical convenience when the wear out rate is small. This appears to be accurate for high-quality solid-state electronic devices during most of their useful life.

Decreasing Rate. The device is improving with age. Software is a (sometimes controversial) example, but this requires some discussion. Software does not break, but certain inputs will lead to incorrect output. The frequency of these critical inputs determines the frequency of failure. Debugging enables the software to handle more inputs correctly. Hence, the rate of failures decreases.

A major issue in software is the independence of different versions of the same program. Initially, software reliability was treated similarly to hardware reliability, but software is now recognized as having different characteristics. The main item is that hardware has components that are used over and over (tried and true), whereas most software is custom.

Mathematical Formulation of Rates. The occurrence rate $R(t)$ is defined to be the density function divided by the probability that the event has not yet occurred. Hence, if the density function and distribution functions are $f(t)$ and $F(t)$,

$$R(t) = \frac{f(t)}{1 - F(t)}$$

It appears reasonable that a device that is neither wearing out nor improving with age has the memoryless property. This can be shown mathematically. If the failure rate is constant, then

$$\frac{f(t)}{1 - F(t)} = c$$

Since the density function is the derivative of the distribution function, we have the differential equation

$$F'(t) + cF(t) = c$$

with the initial condition $F(0) = 0$. The solution is $F(t) = 1 - e^{-ct}$.

There are infinitely many distributions with increasing or decreasing rates. The Weibull distributions are popular and can be found in any reliability text.

Component Failure Modes

In addition to failing at different rates, devices can exhibit a variety of behaviors once they have failed. One variety is the on–off cycle of faulty behavior. A device, when it fails, can exhibit constant faulty behavior. In some ways, this is best for overall reliability. These devices can be identified and replaced more quickly than devices that exhibit intermittent faulty behavior. There are transient failures, usually due to external shocks, where the device is temporarily faulty, but will perform correctly once it has recovered. It can be difficult to distinguish between a device that is intermittently faulty (which should be replaced) and a device that is transiently faulty (which should not be replaced). In addition, a constantly faulty device in a complicated system may only occasionally cause a measurable performance error.

Devices can exhibit different faulty behavior, some of them malicious. The most famous is the lying clock problem, which can occur in system synchronization when correcting for clock drift.

- Clock A is good, but slow, and sends 11 am to the other two clocks.
- Clock B is good, but fast, and sends 1 pm to the other two clocks.
- Clock C is faulty. It sends 6 am to clock A and 6 pm to clock B.
- Clock A averages the three times and resets itself to 10 am.
- Clock B averages the three times and resets itself to 2 pm.

Even though a majority of the components are good, the system will lose synchronization and fail. The general version of this problem is known as *Byzantine* agreement.

Designing systems to tolerate and diagnose faulty components is an open research area.

Systems

Redundancy

There are two strands in reliability–fault avoidance and fault tolerance. Fault avoidance consists of building high-quality components and designing systems in a conservative manner. Fault tolerance takes the point of view that, despite all our efforts, components will fail and highly reliable systems must function in the presence of these failed components. These systems attempt achieving reliability beyond the reach of any single component by using redundancy.

There is a price in efficiency, however, for using redundancy, especially if the attempt is made to achieve reliability far beyond the reliability of the individual components. This is illustrated in Fig. 1.12, which plots the survival probability of a simple redundant system against the mean-time-to-failure (MTTF) of a single component. The system consists of n components, and the system works if a majority of the components are good.

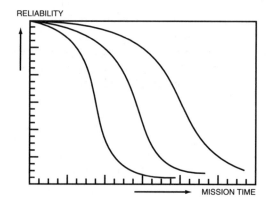

FIGURE 1.12 Reliability of n-modular redundant systems: (a) 3-MR, (b) 5-MR, (c) 7-MR.

There are several methods of improving the efficiency. One is by reconfiguration, which removes faulty components from the working group. This lets the good components remain in the majority for a longer period of time. Another is to use spares that can replace failed members of the working group. For long-term reliability, the spares can be unpowered, which usually reduces their failure rate. Although reconfiguration improves efficiency, it introduces several problems. One is a more complex system with an increased design cost and an increased possibility of design error. It also introduces a new failure mode. Even if reconfiguration works perfectly, it takes time to detect, identify, and remove a faulty component. Additional component failures during this time span can overwhelm the working group before the faulty components are removed. This failure mode is called a *coverage failure* or a *coincident-fault failure*. Assessing this failure mode requires modeling techniques that reflect system dynamics. One appropriate technique is **Markov models** explained in the next section.

Periodic Maintenance

In a sense, redundant systems have increasing failure rates. As individual components fail, the system becomes more and more vulnerable. Their reliability can be increased by periodic maintenance that replaces failed components. Figure 1.13 gives the reliability curve of a redundant system. The failure rate is small during the initial period of its life because it is unlikely that a significant number of its components fail in a short period of time. Figure 1.13 also gives the reliability curve of the same system with periodic maintenance. For simplicity, it is assumed that the maintenance restores the system as good as new. The reliability curve for the maintained system repeats the first part of the curve for the original system over and over.

Modeling and Computation Methods

If a system consists of individual components, it is natural to attempt to derive the failure characteristics of the system from the characteristics of the components and the system structure. There are three major modeling approaches.

Combinatorial and Fault Trees

Combinatorial techniques are appropriate if the system is static (nonreconfigurable). The approach is to construct a function that gives the probability of system failure (or survival) in terms of component failure (or survival) and the system structure. Anyone who remembers combinatorial analysis from high school or a course in elementary probability can imagine that the combinatorial expressions for complex systems can become very complicated. Fault trees have been developed as a tool to help express the combinatorial structure of a system. Once the user has described the system, a program computes the combinatorial probability. Fault trees often come with a pictorial input.

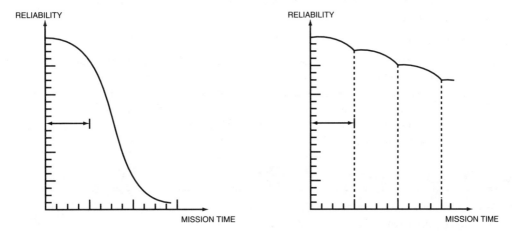

FIGURE 1.13 Reliability curves of an *unmaintained* (left) redundant system and a *maintained* (right) system.

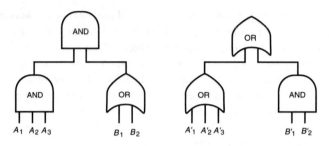

FIGURE 1.14 Success tree (left) and failure tree (right).

Example. A system works if both its subsystems work. Subsystem A works if all three of its components A_1, A_2, and A_3 work. Subsystem B works if either of its subsystems B_1 or B_2 work. Fault trees can be constructed to compute either the probability of success or the probability of failure. Figure 1.14 gives both the success tree and the failure tree for this example.

Markov and Semi-Markov Models

Markov and semi-Markov models are convenient tools for dynamic (reconfigurable) systems because the states in the model correspond to system states and the transition between states in the model correspond to physical processes (fault occurrence or system recovery). Because of this correspondence, they have become very popular, especially for electronic systems where the devices can be assumed to have a constant failure rate. Their disadvantages stem from their successes. Because of their convenience, they are applied to large and complex systems, and the models have become hard to generate and compute because of their size (state-space explosion). Markov models assume that transitions between states do not depend on the time spent in the state. (The transitions are memoryless.) Semi-Markov models are more general and let the transition distributions depend on the time spent in the state. This survey dedicates a special section to Markov models.

Monte Carlo Methods

Monte Carlo is essentially computation by random sampling. The major concern is using sampling techniques that are efficient techniques that yield a small confidence interval for a given sample size.

Monte Carlo is often the technique of last resort. It is used when the probability distributions are too arbitrary or when the system is too large and complex to be modeled and computed by other approaches. The Monte Carlo approach is usually used when failure rates are time varying. It can be used when maintenance is irregular or imperfect. The last subsection on statistics gave a short presentation of the Monte Carlo approach.

Model Verification

Here is an area for anyone looking for a research topic. Models are based on an inspection of the system and on assumptions about system behavior. There is a possibility that important features of the system have been overlooked and that some of the assumptions are incorrect. There are two remedies for these possibilities. One is a detailed examination of the system, a failure modes and effects analysis. The size and complexity of current electronic systems make this an arduous task. Experiments can estimate parameters, but conducting extensive experiments can be expensive.

1.3 Markov Models

Markov models have become popular in both reliability and performance analysis. This section explains this technique from an engineering point of view.

Basic Constructions

It only takes an intuitive understanding of three basic principles to construct Markov models. These are (1) constant rate processes, (2) independent competing events, and (3) independent sequential events. These three principles are described, and it is shown (by several examples) that an intuitive understanding of these principles is sufficient to construct Markov models. A Markov model consists of states and transitions. The states in the model correspond to identifiable states in the system, and the transitions in the model correspond to transitions between the states of the system.

Conversely, constructing a Markov model assumes these three principles, which means assuming these properties about a system. These properties may not be true for all systems. Hence, one reason for a detailed examination of these principles is to understand what properties are being assumed about the system. This is essential for knowing when and when not to use Markov models. The validity of the model depends on a match between the assumptions of the model and the properties of the system.

As discussed in a previous section, constant rate processes are random events with the memoryless property. This probability distribution appears well suited to modeling the failures of high-quality devices operating in a benign environment. The constant rate probability distribution, however, does not appear to be well suited to such events as system reconfiguration. Some of these procedures, for example, are fixed time programs. If the system is halfway through a 10-ms procedure, then it is known that the procedure will be completed in 5 ms. That is, this procedure has memory: how long it takes to complete the procedure depends on how long the procedure has been running. Despite this possible discrepancy, this exposition will describe Markov models because of their simplicity.

Competing events arise naturally in redundant and reconfigurable systems. An important example is the competition between the arrival of a second fault and system recovery from a first fault. Sequential events also arise naturally in redundant and reconfigurable systems. A component failure is followed by system recovery. Another component failure is followed by another system recovery.

The Markov model for an event that has rate λ is given by Fig. 1.15. In state S the event has not occurred. In state F the event has occurred.

Suppose A and B are independent events that occur at rates α and β, respectively. The model for A and B as competing events is given in Fig. 1.16. In state SS neither A nor B has occurred. In state SA, event A has occurred before event B, whereas event B may or may not have occurred. In state SB, event B has occurred before event A, whereas event A may or may not have occurred.

FIGURE 1.15 Model of a device failing at a constant rate λ.

If A and B are independent, then the probability that one or the other or both have occurred can be modeled by adding their rate of occurrence. This is a standard result in probability [Rohatgi 1976; Hoel, Port, and Stone 1972]. For the two events depicted in Fig. 1.16, the sum is $\alpha + \beta$, and the model (for either one or both) is given in Fig. 1.17. The probability of being in state C in Fig. 1.17 is equal to the sum of the probabilities of being in states SA or SB in Fig. 1.16.

The model in Fig. 1.16 tells us whether or not an event has occurred and which event has occurred first. The model does not tell us whether or not both events have occurred. Models with this additional information require considering sequential events.

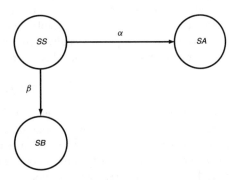

FIGURE 1.16 Model for independent competing events.

Figure 1.18 displays a model for two independent sequential events where the first event occurs at rate α and the second rate occurs at rate β. Such a model can arise when a device with failure rate α is operated until it fails, whereupon it is replaced by a device with failure rate β. State S represents the first device operating correctly; state A represents the first device having failed and the second device operating correctly; and state B represents the second device having failed.

Figure 1.19 displays a model with both competing and sequential events. Even though this model is simple, it uses all three of the basic principles. Suppose devices A and B with failure rates α and β are operating. In state S, there are two competing events, the failure of device A and the failure of device B. If device A fails first, the system is in state SA where device A has failed but device B has not failed.

In state SA, the memoryless property of a constant rate process is used to construct the transition to state SC. When it arrives in state SA, device B does not remember having spent some time in state S. Hence, for device B, the transition out of state SA is the same as the transition out of state S. If device B were a component that wore out, then the transition from state SA to SC in Fig. 1.19 would depend on the amount of time the system spent in state S. The modeling and computation would be more difficult. A similar discussion holds for state SB. In state SC, both devices have failed.

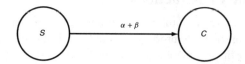

FIGURE 1.17 Model for occurrence of either independent event.

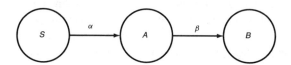

FIGURE 1.18 Model for sequential independent events.

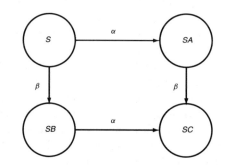

FIGURE 1.19 Model with both completing and sequential events.

Model for a Reconfigurable Fourplex

Having covered the three basic principles (constant rate processes, independent competing events, and independent sequential events), it is possible to construct Markov reliability models. This section describes the model for a reconfigurable fourplex. This fourplex uses majority voting to detect the presence of a faulty component. Faulty components are removed from the system.

The reconfiguration sequence for this fourplex is

fourplex → threeplex → twoplex → failure

This fourplex has two failure modes. The coincident-fault failure mode occurs when a second component fails before the system can remove a component that has previously failed. With two faulty components present, the majority voter cannot be guaranteed to return the correct answer. A coincident-fault failure is sometimes called a *coverage failure*. Since a failure due to lack of diagnostics is also called a coverage failure, this chapter will use the term coincident-fault failure for the system failure mode just described.

The exhaustion of parts failure mode occurs when too many components have failed for the system to operate correctly. In this reconfigurable fourplex, an exhaustion of parts failure occurs when one component of the final twoplex fails since this fourplex uses only majority voting to detect faulty components. A system using self-test diagnostics may be able to reconfigure to a simplex before failing.

In general, a system can have a variety of reconfiguration sequences depending on design considerations, the sophistication of the operating system, and the inclusion of diagnostic routines. This first example will use a simple reconfiguration sequence.

The Markov reliability model for this fourplex is given in Fig. 1.20. Each component fails at rate λ. The system recovers from a fault at rate δ. This system recovery includes the detection, identification, and removal of the faulty component. The mnemonics in Fig. 1.20 are S for a fault free state, R for a recovery mode state, C for a coincident-fault failure state, and E for the exhaustion of parts failure state.

The initial states and transitions for the model of the fourplex will be examined in detail. In state S_1 there are four good components each failing at rate λ. Since these components are assumed to fail independently, the failure rates can be summed to get the rate for the failure of one of the four. This sum is 4λ, which is the transition rate from state S_1 to R_1. State R_1 has a transition with rate 3λ to state C_1 which represents the failure of one of the three good components. Hence, in state R_1 precisely one component has failed because the 4λ transition (from S_1) has been taken, but the 3λ transition (out of R_1) has not been taken. There is also a δ transition out of state R_1 representing system recovery. System recovery transitions are usually regarded as independent of component failure transitions because system recovery depends on the architecture and software, whereas component failure depends on the quality of the devices. The system recovery procedure would not change if the devices were replaced by electronically equivalent devices with a higher or lower failure rate. Hence, state R_1 has the competing independent events of component failure and system recovery. If system recovery occurs first, the system goes to the fault free state S_2 where there are three good components in the system. The transition 3λ out of state S_2 represents the memoryless failure rate of these three good components. If, in state R_1, component failure occurs before system recovery, then the system is in the failed state C_1 where there is the possibility of the faulty components overwhelming the majority voter. The descriptions of the remaining states and transitions in the model are similar to the preceding descriptions.

There are several additional comments about the system failure state C_1. First, it is possible that the fourplex can survive a significant fraction of coincident faults. One possibility is that a fair amount of time elapses between the second fault occurrence and the faulty behavior of the second component. In this case, the system has time to remove the first faulty component before the majority is overwhelmed. Establishing this recovery mode, however, requires experiments with double fault injections, which may be expensive. It may be cheaper and easier to build a system that is more reliable than required instead of requiring an expensive set of experiments to establish the system's reliability, although more extensive experiments are a possible option. Models, such as the one in Fig. 1.20, that overestimate the probability of system failure are said to be conservative. Most models tend to be conservative because of the cost of obtaining detailed information about system performance, especially about system fault recovery.

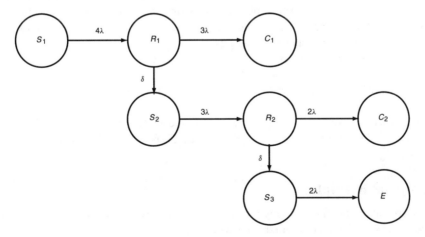

FIGURE 1.20 Reliability model of a reconfigurable fourplex.

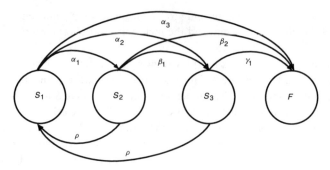

FIGURE 1.21 Fiveplex subjected to shocks.

Correlated Faults

This example demonstrates that Markov models can depict correlated faults. The technique used in this model is to have transitions represent the occurrence of several faults at once. Figure 1.21 displays a model for a fiveplex subjected to shocks, which produce transient faults. These shocks arrive at some constant rate. During a shock, faults appear in 0–5 components with some probability. In Fig. 1.21, the system is in state S_i if it has i faulty components for $i = 0, 1, 2$. The system is in the failure state F if there are three or more faulty components. The system removes all transient faults at rate ρ.

The Differential Equations for Markov Models

Markov processes are characterized by the manner in which the probabilities change in time. The rates give the flow for changes in probabilities. The flow for constant rate processes is smooth enough that the change in probabilities can be described by the (Chapman–Kolmogorov) differential equations. Solving the differential equations gives the probabilities of being in the states of the model.

One of the attractive features of Markov models is that numerical results can be obtained without an extensive knowledge of probability theory and techniques. The intuitive ideas of probability discussed in the previous section are sufficient for the construction of a Markov model. Once the model is constructed, it can be computed by differential equations. The solution does not require any additional knowledge of probability.

Writing the differential equations for a model can be done in a local manner. The equation for the derivative of a state depends only on the single step transitions into that state and the single step transitions out of that state. Suppose there are transitions from states $\{A_1,\ldots, A_m\}$ into state B with rates $\{\alpha_1,\ldots, \alpha_m\}$ respectively, and that these are the only transitions into state B. Suppose there are transitions from state B to states $\{C_1,\ldots,C_n\}$ with rates $\{\beta_1,\ldots, \beta_n\}$ respectively, and that these are the only transitions out of state B. This general condition is shown in Fig. 1.22.

The differential equation for state B is:

$$p'_B(t) = \alpha_1 p_{A_1}(t) + \alpha_2 p_{A_2}(t) + \cdots + \alpha_m p_{A_m}(t) - (\beta_1 + \cdots + \beta_n) p_B(t)$$

This formula is written for every state in the model, and the equations are solved subject to the initial conditions. While writing the formula for state B, it is possible to ignore all of the other states and transitions.

Computational Examples

This section presents a conceptual example that relates Markov models to combinatorial events and a computational example for a reconfigurable system.

Probability and Statistics

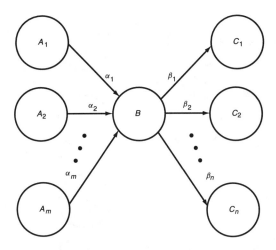

FIGURE 1.22 General diagram for a state in a Markov model.

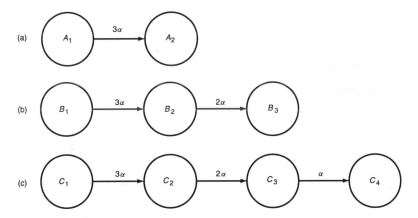

FIGURE 1.23 Markov models for combinatorial failures.

Combinatorial Failure of Components

The next example considers the failure of three identical components with failure rate α. This example uses the basic properties of competing events, sequential events, and memoryless processes. Figure 1.23 presents the model for the failure of at least one component. State A_2 represents the failure of one, two, or three components. In Fig. 1.23(b) state B_2 represents the failure of exactly one component, whereas state B_3 represents the failure of two or three components. Figure 1.23(c) gives all of the details.

Choosing the middle diagram in Fig. 1.23(b), the differential equations are

$$p_1'(t) = -3\alpha p_1(t)$$
$$p_2'(t) = 3\alpha p_1(t) - 2\alpha p_2(t)$$
$$p_3'(t) = 2\alpha p_2(t)$$

with the initial conditions $p_1(0) = 1$, $p_2(0) = 0$, and $p_3(0) = 0$.

Let Q be the probability that a single component has failed by time T. That is,

$$Q = 1 - e^{-\alpha T}$$

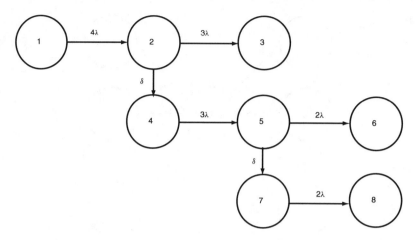

FIGURE 1.24 Original reliability model of a reconfigurable fourplex.

Solving the differential equations gives

$$Pr[B_3] = 3Q^2(1-Q) + Q^3$$

which is the combinatorial probability that two or three components out of three have failed. A similar result holds for the other models in Fig. 1.23.

Computing the Fourplex

The model of the reconfigurable fourplex using numbers instead of mnemonics to label the states is represented by Fig. 1.24. The differential equations are

$$p'_1(t) = -4\lambda p_1(t)$$
$$p'_2(t) = 4\lambda p_1(t) - (3\lambda + \delta)\lambda p_2(t)$$
$$p'_3(t) = 3\lambda p_2(t)$$
$$p'_4(t) = \delta p_2(t) - 3\lambda p_4(t)$$
$$p'_5(t) = 3\lambda p_4(t) - (2\lambda + \delta)\lambda p_5(t)$$
$$p'_6(t) = 2\lambda p_5(t)$$
$$p'_7(t) = 5\delta p_5(t) - 2\lambda p_7(t)$$
$$p'_8(t) = 2\lambda p_7(t)$$

Once the parameter values are known, this set of equations can be computed by numerical methods. Suppose the parameter values are $\lambda = 10^{-4}$/h and $\delta = 10^4$/h. Suppose the operating time is $T = 2$, and suppose the initial conditional is $p_1(0) = 1$. Then the probability of being in the failed states is:

$$p_3(2) = 2.4 \times 10^{11}, \qquad p_6(2) = 4.8 \times 10^{15}, \qquad p_8(2) = 3.2 \times 10^{11}$$

1.4 Summary

An attempt has been made to accomplish two things, to explain the ideas underlying probability and statistics and to demonstrate some of the most useful techniques. Probability is presented as a method of describing phenomena in the presence of incomplete knowledge and uncertainty. Elementary events are assigned a likelihood of occurring. Complex events comprise these elementary events, and this is used to compute the probability of complex events. Statistical methods are a way of determining the probability of elementary events by experiment. Sampling and observation are emphasized. These two items are more important than fancy statistical techniques.

The analysis of fault-tolerant systems is presented in light of the view of probability described. Component failure and component behavior when faulty are the elementary events. System failure and behavior is computed according to how the system is built from its components. The text discussed some of the uncertainties and pathologies associated with component failure. It is not easy to design a system that functions as specified when all of the components are working correctly. It is even more difficult to accomplish this when some of the components are faulty.

Markov models of redundant and reconfigurable systems are convenient because states in the model correspond to states in the system and transitions in the model correspond to system processes. The Markov section gives an engineering presentation.

Defining Terms

Confidence interval and confidence level: Indicates the quality of an experiment. A confidence interval is an interval around the estimator. A confidence level is a probability. The interval contains the unknown parameter with this probability.

Central limit theorem: As the sample size becomes large, the distibution of the sample average becomes approximately normal.

Density function of a random variable: Defined implicitly: the probability that the random variable lies between a and b is the area under the density function between a and b. This approach introduces the methods of calculus and analysis into probability.

Estimator: Strictly speaking, any function of the sample points. Hence, an estimator is a random variable. To be useful an estimator must have a good relationship to some unknown quality that we are trying to determine by experiment.

Independent event: Events with the property that the occurrence of one event gives no information about the occurrence of the other events.

Markov model: A modeling technique where the states of the model correspond to states of the system and transitions between the states in the model correspond to system processes.

Mean, average, expected value: The first moment of a distribution; the integral of x with respect to its density function $f(x)$.

Monte Carlo: Computation by statistical estimation.

Probability space: A set together with a distinguished collection of its subsets. A subset in this collection is called an event. Probabilities are assigned to events in a manner that preserves additive properties.

Sample: A set of items, each chosen independently from a distribution.

Variance: The second moment of a distribution about its mean; measures the dispersion of the distribution.

References

Hoel, Port, and Stone. 1972. *Introduction to Stochastic Processes*. Houghton Mifflin, Boston, MA.
Rohatgi. 1976. *An Introduction to Probability Theory and Mathematical Statistics*. Wiley, New York.
Ross. 1990. A Course in Simulation. Macmillan, New York.
Siewiorek and Swarz. 1982. *The Theory and Practice of Reliable System Design*. Digital Press, Maynard.
Trivedi. 1982. *Probability and Statistics with Reliability, Queing, and Computer Science Applications*. Prentice-Hall, Englewood Cliffs.

Further Information

Additional information on the topic of probability and statistics applied to reliability is available from the following journals:

Microelectronics Reliability, Pergamon Journals Ltd.
IEEE Transactions on Computers
IEEE Transactions on Reliability

and proceedings of the following conferences:

IEEE/AIAA Reliability and Maintainability Symposium
IEEE Fault-Tolerant Computing Symposium

2
Electronic Hardware Reliability

Michael Pecht
University of Maryland

Iuliana Bordelon
University of Maryland

2.1 Introduction 2-1
2.2 The Life Cycle Usage Environment 2-2
2.3 Characterization of Materials, Parts, and the Manufacturing Processes 2-3
2.4 Failure Mechanisms 2-3
2.5 Design Guidelines and Techniques 2-4
     Preferred Parts • Redundancy • Protective Architectures • Stress Margins • Derating
2.6 Qualification and Accelerated Testing 2-7
2.7 Manufacturing Issues 2-8
     Process Qualification • Manufacturability • Process Verification Testing
2.8 Summary 2-11

2.1 Introduction

Reliability is the ability of a product to perform as intended (i.e., without failure and within specified performance limits) for a specified life cycle. Reliability is a characteristic of a product, in the sense that reliability can be designed into a product, controlled in manufacture, measured during test, and sustained in the field.

To achieve **product performance** over time requires an approach that consists of tasks, each having total engineering and management commitment and enforcement. The tasks impact electronic hardware reliability through the selection of materials, structural geometries, design tolerances, manufacturing processes and tolerances, assembly techniques, shipping and handling methods, and maintenance and maintainability guidelines [Pecht 1994]. The tasks are as follows:

1. Define realistic system requirements determined by mission profile, required operating and storage life, performance expectations, size, weight, and cost. The design team must be aware of the environmental and operating conditions for the product. This includes all stress and loading conditions.

2. Characterize the materials and the manufacturing and assembly processes. Variabilities in material properties and manufacturing processes can induce failures. Although knowledge of variability may not be required for some engineering products, due to the inherent strength of the product compared to the stresses to which the product is subjected, concern with product weight, size, and cost often force the design team to consider such extreme margins as wasteful.

3. Identify the potential failure sites and **failure mechanisms**. Critical parts, part details, and the potential failure mechanisms and modes must be identified early in the design, and appropriate measures must be implemented to assure control. Potential architectural and stress interactions must be defined and assessed.

4. Design to the usage and process capability (i.e., the quality level that can be controlled in manufacturing and assembly) considering the potential failure sites and failure mechanisms. The design stress spectra, the part test spectra, and the full-scale test spectra must be based on the anticipated life cycle usage conditions. Modeling and analysis are steps toward assessment. Tests are conducted to verify the results for complex structures. The goal is to provide a physics-of-failure basis for design decisions with the assessment of possible failure mechanisms for the anticipated product. The proposed product must survive the life cycle profile while being cost effective and available to the market in a timely manner.
5. Qualify the product manufacturing and assembly processes. If all the processes are in control and the design is proper, then product testing is not warranted and is therefore not cost effective. This represents a transition from *product* test, analysis, and screening, to *process* test, analysis, and screening.
6. Control the manufacturing and assembly processes addressed in the design. During manufacturing and assembly, each process must be monitored and controlled so that process shifts do not arise. Each process may involve screens and tests to assess statistical process control.
7. Manage the life cycle usage of the product using closed loop management procedures. This includes realistic inspection and maintenance procedures.

2.2 The Life Cycle Usage Environment

The life cycle usage environment or scenario for use of a product goes hand in hand with the product requirements. The life cycle usage information describes the storage, handling, and operating stress profiles and thus contains the necessary load input information for failure assessment and the development of design guidelines, screens, and tests. The stress profile of a product is based on the application profile and the internal stress conditions of the product. Because the performance of a product over time is often highly dependent on the magnitude of the stress cycle, the rate of change of the stress, and even the time and spatial variation of stress, the interaction between the application profile and internal conditions must be specified. Specific information about the product environment includes absolute temperature, temperature ranges, temperature cycles, temperature gradients, vibrational loads and transfer functions, chemically aggressive or inert environments, and electromagnetic conditions. The life cycle usage environment can be divided into three parts: the application and life profile conditions, the external conditions in which the product must operate, and the internal product generated stress conditions.

The application and life profile conditions include the application length, the number of applications in the expected life of the product, the product life expectancy the product utilization or nonutilization (storage, testing, transportation) profile, the deployment operations, and the maintenance concept or plan. This information is used to group usage platforms (i.e., whether the product will be installed in a car, boat, satellite, underground), develop duty cycles (i.e., on–off cycles, storage cycles, transportation cycles, modes of operation, and repair cycles), determine design criteria, develop screens and test guidelines, and develop support requirements to sustain attainment of reliability and maintainability objectives.

The external operational conditions include the anticipated environment(s) and the associated stresses that the product will be required to survive. The stresses include temperature, vibrations, shock loads, humidity or moisture, chemical contamination, sand, dust and mold, electromagnetic disturbances, radiation, etc.

The internal operational conditions are associated with product generated stresses, such as power consumption and dissipation, internal radiation, and release or outgassing of potential contaminants. If the product is connected to other products or subsystems in a system, stresses associated with the interfaces (i.e., external power consumption, voltage transients, electronic noise, and heat dissipation) must also be included.

The life-cycle application profile is a time-sequential listing of all of the loads that can potentially cause failure. These loads constitute the parameters for quantifying the given application profile. For example, a flight application could be logged at a specified location and could involve engine warm-up, taxi, climb, cruising, high-speed maneuvers, gun firing, ballistic impact, rapid descent, and emergency landing. This information is of little use to a hardware designer unless it is associated with the appropriate application load histories, such as acceleration, vibration, impact force, temperature, humidity, and electrical power cycle.

2.3 Characterization of Materials, Parts, and the Manufacturing Processes

Design is intrinsically linked to the materials, parts, interfaces, and manufacturing processes used to establish and maintain functional and structural integrity. It is unrealistic and potentially dangerous to assume defect-free and perfect-tolerance materials, parts, and structures. Materials often have naturally occurring defects and manufacturing processes that can induce additional defects to the materials, parts, and structures. The design team must also recognize that the production lots or vendor sources for parts that comprise the design are subject to change. Even greater variability in parts characteristics is likely to occur during the fielded life of a product as compared to its design or production life cycle phases.

Design decisions involve the selection of components, materials, and controllable process techniques, using tooling and processes appropriate to the scheduled production quantity. Often, the goal is to maximize part and configuration standardization; increase package modularity for ease in fabrication, assembly, and modification; increase flexibility of design adaptation to alternate uses; and to utilize alternate fabrication processes. The design decisions also involve choosing the best materials interfaces and best geometric configurations, given the product requirements and constraints.

2.4 Failure Mechanisms

Failure mechanisms are the physical processes by which stresses can damage the materials included in the product. Investigating failure mechanisms aids in the development of failure-free, reliable designs. Numerous studies focusing on material failure mechanisms and physics-based damage models and their role in obtaining reliable electronic products have been extensively illustrated in a series of tutorials comprising all relevant **wearout** and **overstress failures** [Dasgupta and Pecht 1991; Dasgupta and Hu 1992a, 1992b, 1992c, 1992d; Dasgupta and Haslach 1993; Engel 1993; Li and Dasgupta 1993, 1994; Dasgupta 1993; Young and Christou 1994; Rudra and Jennings 1994; Al-Sheikhly and Christou 1994; Diaz, Kang, and Duvvury 1995; Tullmin and Roberge 1995].

Catastrophic failures that are due to a single occurrence of a stress event, when the intrinsic strength of the material is exceeded, are termed overstress failures. Failure mechanisms due to monotonic accumulation of incremental **damage** beyond the endurance of the material are termed wearout mechanisms. When the damage exceeds the endurance limit of the component, failure will occur. Unanticipated large stress events can either cause an overstress (catastrophic) failure or shorten life by causing the accumulation of wearout damage. Examples of such stresses are accidental abuse and acts of God. On the other hand, in well-designed and high-quality hardware, stresses should cause only uniform accumulation of wearout damage; the threshold of damage required to cause eventual failure should not occur within the usage life of the product.

The design team must be aware of all possible failure mechanisms in order to design hardware capable of withstanding loads without failing. Failure mechanisms and their related models are also important for planning tests and screens to audit the nominal design and manufacturing specifications, as well as the level of defects introduced by excessive variability in manufacturing and material parameters.

Electrical performance failures can be caused when individual components have incorrect resistance, impedance, voltage, current, capacitance, or dielectric properties or by inadequate shielding from electromagnetic interference (EMI) or particle radiation. The failure modes can be manifested as reversible drifts in electrical transient and steady-state responses such as delay time, rise time, attenuation, signal-to-noise ratio, and cross talk. Electrical failures, common in electronic hardware, include overstress mechanisms due to *electrical overstress* (EOS) and *electrostatic discharge* (ESD) such as dielectric breakdown, junction breakdown, hot electron injection, surface and bulk trapping, and surface breakdown and wearout mechanisms such as electromigration.

Thermal performance failures can arise due to incorrect design of thermal paths in an electronic assembly. This includes incorrect conductivity and surface emissivity of individual components, as well as incorrect convective and conductive paths for the heat transfer path. Thermal overstress failures are a result of heating a component beyond such critical temperatures as the glass-transition temperature, melting point, fictive point, or flash point. Some examples of thermal wearout failures are aging due to depolymerization, intermetallic growth, and interdiffusion. Failures due to inadequate thermal design may be manifested as components running too hot or too cold, causing operational parameters to drift beyond specifications, although the degradation is often reversible upon cooling. Such failures can be caused either by direct thermal loads or by electrical resistive loads, which in turn generate excessive localized thermal stresses. Adequate design checks require proper analysis for thermal stress and should include conductive, convective, and radiative heat paths.

Incorrect product response to mechanical overstress and wearout loads may compromise the product performance, without necessarily causing any irreversible material damage. Such failures include incorrect elastic deformation in response to mechanical static loads, incorrect transient response (such as natural frequency or damping) to dynamic loads, and incorrect time-dependent reversible (anelastic) response. Mechanical failure can also result from buckling, brittle and/or ductile fracture, interfacial separation, fatigue crack initiation and propagation, creep, and creep rupture. To take one example, excessive elastic deformations in slender structures in electronic packages due to overstress loads can sometimes constitute functional failure, such as excessive flexing of interconnection wires, package lids, or flex circuits in electronic devices, causing shorting and/or excessive cross talk. When the load is removed, however, the deformations disappear completely without any permanent damage. Examples of wearout failure mechanism include fatigue damage due to thermomechanical stresses during power cycling of electronic hardware, corrosion rate due to anticipated contaminants, and electromigration in high-power devices.

Radiation failures are principally caused by uranium and thorium contaminants and secondary cosmic rays. Radiation can cause wearout, aging, embrittlement of materials, or overstress soft errors in such electronic hardware as logic chips. Chemical failures occur in adverse chemical environments that result in corrosion, oxidation, or ionic surface dendritic growth. There may also be interactions between different types of stresses. For example, metal migration may be accelerated in the presence of chemical contaminants and composition gradients and a thermal load can accelerate the failure mechanism due to a thermal expansion mismatch.

2.5 Design Guidelines and Techniques

Generally, products replace other products. The replaced product can be used for comparisons (i.e., a baseline comparison product). Lessons learned from the baseline comparison product can be used to establish new product parameters, to identify areas of focus in the new product design, and to avoid the mistakes of the past.

Once the parts, materials, and processes are identified along with the stress conditions, the objective is to design a product using parts and materials that have been sufficiently characterized in terms of how they perform over time when subjected to the manufacturing and application profile conditions. Only through a methodical design approach using physics of failure and root cause analysis can a reliable and cost effective product be designed.

In using design guidelines, there may not be a unique path to follow. Instead, there is a general flow in the design process. Multiple branches may exist depending on the input design constraints. The design team should explore enough of the branches to gain confidence that the final design is the best for the prescribed input information. The design team should also assess the use of the guidelines for the complete design and not those limited to specific aspects of an existing design. This statement does not imply that guidelines cannot be used to address only a specific aspect of an existing design, but the design team may have to trace through the implications that a given guideline suggests.

Design guidelines that are based on physics of failure models can also be used to develop tests, screens, and derating factors. Tests can be designed from the physics of failure models to measure specific quantities and to detect the presence of unexpected flaws or manufacturing or maintenance problems. Screens can be designed to precipitate failures in the weak population while not cutting into the design life of the normal population. Derating or safety factors can be determined to lower the stresses for the dominant failure mechanisms.

Preferred Parts

In many cases, a part or a structure much like the required one has already been designed and tested. This "preferred part or structure" is typically mature in the sense that variabilities in manufacturing, assembly, and field operation that can cause problems have been identified and corrected. Many design groups maintain a list of preferred parts and structures of proven performance, cost, availability, and reliability.

Redundancy

Redundancy permits a product to operate even though certain parts and interconnections have failed, thus increasing its reliability and availability. Redundant configurations can be classified as either active or standby. Elements in active redundancy operate simultaneously in performing the same function. Elements in standby redundancy are designed so that an inactive one will or can be switched into service when an active element fails. The reliability of the associated function is increased with the number of standby elements (optimistically assuming that the sensing and switching devices of the redundant configuration are working perfectly, and failed redundant components are replaced before their companion component fails). One preferred design alternative is that a failed redundant component can be repaired without adversely impacting the operation of the product and without placing the maintenance person or the product at risk.

A design team may often find that redundancy is:

- The quickest way to improve product reliability if there is insufficient time to explore alternatives or if the part is already designed
- The cheapest solution, if the cost of redundancy is economical in comparison with the cost of redesign
- The only solution, if the reliability requirement is beyond the state of the art

On the other hand, in weighing its disadvantages, the design team may find that redundancy will:
- Prove too expensive, if the parts and redundant sensors and switching devices are costly
- Exceed the limitations on size and weight, particularly in avionics, missiles, and satellites
- Exceed the power limitations, particularly in active redundancy
- Attenuate the input signal, requiring additional amplifiers (which increase complexity)
- Require sensing and switching circuitry so complex as to offset the reliability advantage of redundancy.

Protective Architectures

It is generally desirable to include means in a design for preventing a part, structure, or interconnection from failing or from causing further damage if it fails. Protective architectures can be used to sense failure and protect against possible secondary effects. In some cases, self-healing techniques, in that they self-check and self-adjust to effect changes automatically to permit continued operation after a failure, are employed.

Fuses and circuit breakers are examples used in electronic products to sense excessive current drain and disconnect power from a failed part. Fuses within circuits safeguard parts against voltage transients or excessive power dissipation and protect power supplies from shorted parts. Thermostats may be used to sense critical temperature limiting conditions and shutting down the product or a component of the system until the temperature returns to normal. In some products, self-checking circuitry can also be incorporated to sense abnormal conditions and operate adjusting means to restore normal conditions or activate switching means to compensate for the malfunction.

In some instances, means can be provided for preventing a failed part or structure from completely disabling the product. For example, a fuse or circuit breaker can disconnect a failed part from a product in such a way that it is possible to permit partial operation of the product after a part failure, in preference to total product failure. By the same reasoning, degraded performance of a product after failure of a part is often preferable to complete stoppage. An example is the shutting down of a failed circuit whose design function is to provide precise trimming adjustment within a deadband of another control product. Acceptable performance may thus be permitted, perhaps under emergency conditions, with the deadband control product alone.

Sometimes the physical removal of a part from a product can harm or cause failure of another part of the product by removing load, drive, bias, or control. In these cases, the first part should be equipped with some form of interlock to shut down or otherwise protect the second part. The ultimate design, in addition to its ability to act after a failure, would be capable of sensing and adjusting for parametric drifts to avert failures.

In the use of protective techniques, the basic procedure is to take some form of action, after an initial failure or malfunction, to prevent additional or secondary failures. By reducing the number of failures, such techniques can be considered as enhancing product reliability, although they also affect availability and product effectiveness. Another major consideration is the impact of maintenance, repair, and part replacement. If a fuse protecting a circuit is replaced, what is the impact when the product is re-energized? What protective architectures are appropriate for postrepair operations? What maintenance guidance must be documented and followed when fail-safe protective architectures have or have not been included?

Stress Margins

A properly designed product should be capable of operating satisfactorily with parts that drift or change with time, temperature, humidity, and altitude, for as long as the parameters of the parts and the interconnects are within their rated tolerances. To guard against out-of-tolerance failures, the designer must consider the combined effects of tolerances on parts to be used in manufacture, subsequent changes due to the range of expected environmental conditions, drifts due to aging over the period of time specified in the reliability requirement, and tolerances in parts used in future repair or maintenance functions. Parts and structures should be designed to operate satisfactorily at the extremes of the parameter ranges, and allowable ranges must be included in the procurement or reprocurement specifications.

Methods of dealing with part and structural parameter variations are statistical analysis and worst-case analysis. In statistical design analysis, a functional relationship is established between the output characteristics of the structure and the parameters of one or more of its parts. In worst-case analysis, the effect that a part has on product output is evaluated on the basis of end-of-life performance values or out-of-specification replacement parts.

Derating

Derating is a technique by which either the operational stresses acting on a device or structure are reduced relative to rated strength or the strength is increased relative to allocated operating stress levels. Reducing the stress is achieved by specifying upper limits on the operating loads below the rated capacity of the hardware. For example, manufacturers of electronic hardware often specify limits for supply voltage, output current, power dissipation, junction temperature, and frequency. The equipment designer may

Electronic Hardware Reliability

decide to select an alternative component or make a design change that ensures that the operational condition for a particular parameter, such as temperature, is always below the rated level. The component is then said to have been derated for thermal stress.

The derating factor, typically defined as the ratio of the rated level of a given stress parameter to its actual operating level, is actually a margin of safety or margin of ignorance, determined by the criticality of any possible failures and by the amount of uncertainty inherent in the reliability model and its inputs. Ideally, this margin should be kept to a minimum to maintain the cost effectiveness of the design. This puts the responsibility on the reliability engineer to identify as unambiguously as possible the rated strength, the relevant operating stresses, and reliability.

To be effective, derating criteria must target the right stress parameter to address modeling of the relevant failure mechanisms. Field measurements may also be necessary, in conjunction with modeling simulations, to identify the actual operating stresses at the failure site. Once the failure models have been quantified, the impact of derating on the effective reliability of the component for a given load can be determined. Quantitative correlations between derating and reliability enable designers and users to effectively tailor the margin of safety to the level of criticality of the component, leading to better and more cost-effective utilization of the functional capacity of the component.

2.6 Qualification and Accelerated Testing

Qualification includes all activities that ensure that the nominal design and manufacturing specifications will meet or exceed the reliability targets. The purpose is to define the acceptable range of variabilities for all critical product parameters affected by design and manufacturing, such as geometric dimensions and material properties. Product attributes that are outside the acceptable ranges are termed defects because they have the potential to compromise the product reliability [Pecht et al. 1994].

Qualification validates the capacity of the design and manufacturing specifications of the product to meet customer's expectations. The goal of qualification testing is to verify whether the anticipated reliability is indeed achieved under actual life cycle loads. In other words, qualification tests are intended to assess the performance of survival of a product over the complete life cycle period of the product. Qualification testing thus audits the ability of the design specifications to meet reliability goals. A well-designed qualification procedure provides economic savings and quick turnaround during development of new products or mature products subject to manufacturing and process changes.

Investigating the failure mechanisms and assessing the reliability of products where long life is required may be a challenge because a very long test period under actual operating conditions is necessary to obtain sufficient data to determine actual failure characteristics. One approach to the problem of obtaining meaningful qualification test data for high-reliability devices is **accelerated testing** to achieve test-time compression, sometimes called *accelerated-stress life testing*. When qualifying the reliability for overstress mechanisms, however, a single cycle of the expected overstress load may be adequate, and acceleration of test parameters may not be necessary. This is sometimes called *proof-stress testing*.

Accelerated testing involves measuring the performance of the test product at accelerated conditions of load or stress that are more severe than the normal operating level, in order to induce failures within a reduced time period. The goal of such testing is to accelerate the time-dependent failure mechanisms and the accumulation of damage to reduce the time to failure. A requirement is that the failure mechanisms and modes in the accelerated environment are the same as (or can be quantitatively correlated with) those observed under usage conditions and it is possible to extrapolate quantitatively from the accelerated environment to the usage environment with some reasonable degree of assurance.

A scientific approach to accelerated testing starts with identifying the relevant wearout failure mechanism. The stress parameter that directly causes the time-dependent failure is selected as the acceleration parameter and is commonly called the accelerated stress. Common accelerated stresses include thermal stresses, such as temperature, temperature cycling, or rates of temperature change; chemical stresses, such as humidity, corrosives, acid, or salt; electrical stresses, such as voltage, current, or power; and mechanical stresses, such as vibration loading, mechanical stress cycles, strain cycles, and shock/impulse. The accelerated environment may include one or a combination of these stresses. Interpretation of results for combined stresses requires a very clear and quantitative understanding of their relative interactions and the contribution of each stress to the overall damage.

Once the failure mechanisms are identified, it is necessary to select the appropriate acceleration stress; determine the test procedures and the stress levels; determine the test method, such as constant stress acceleration or step-stress acceleration; perform the tests; and interpret the test data, which includes extrapolating the accelerated test results to normal operating conditions. The test results provide designers with qualitative failure information for improving the hardware through design and/or process changes.

Failure due to a particular mechanism can be induced by several acceleration parameters. For example, corrosion can be accelerated by both temperature and humidity; creep can be accelerated by both mechanical stress and temperature. Furthermore, a single acceleration stress can induce failure by several wearout mechanisms simultaneously. For example, temperature can accelerate wearout damage accumulation not only due to electromigration, but also due to corrosion, creep, and so forth. Failure mechanisms that dominate under usual operating conditions may lose their dominance as stress is elevated. Conversely, failure mechanisms that are dormant under normal use conditions may contribute to device failure under accelerated conditions. Thus, accelerated tests require careful planning in order to represent the actual usage environments and operating conditions without introducing extraneous failure mechanisms or nonrepresentative physical or material behavior. The degree of stress acceleration is usually controlled by an *acceleration factor*, defined as the ratio of the life under normal use conditions to that under the accelerated condition. The acceleration factor should be tailored to the hardware in question and should be estimated from an acceleration transform that gives a functional relationship between the accelerated stress and reduced life, in terms of all of the hardware parameters.

Detailed failure analysis of failed samples is a crucial step in the qualification and validation program. Without such analysis and feedback to designers for corrective action, the purpose of the qualification program is defeated. In other words, it is not adequate simply to collect failure data. The key is to use the test results to provide insights into, and consequent control over, relevant failure mechanisms and ways to prevent them, cost effectively.

2.7 Manufacturing Issues

Manufacturing and assembly processes can impact the quality and reliability of hardware. Improper assembly and manufacturing techniques can introduce defects, flaws, and residual stresses that act as potential failure sites or stress raisers later in the life of the product. The fact that the defects and stresses during the assembly and manufacturing process can affect the product reliability during operation necessitates the identification of these defects and stresses to help the design analyst account for them proactively during the design and development phase. The task of auditing the merits of the manufacturing process involves two crucial steps. First, qualification procedures are required, as in design qualification, to ensure that manufacturing specifications do not excessively compromise the long-term reliability of the hardware. Second, lot-to-lot screening is required to ensure that the variabilities of all manufacturing-related parameters are within specified tolerances [Pecht et al. 1994]. In other words, screening ensures the quality of the product and improves short-term reliability by precipitating latent defects before they reach the field.

Process Qualification

Like design qualification, process qualification should be conducted at the prototype development phase. The intent is to ensure that the nominal manufacturing specifications and tolerances produce acceptable reliability in the hardware. Once qualified, the process needs requalification only when process parameters, materials, manufacturing specifications, or human factors change.

Process qualification tests can be the same set of accelerated wearout tests used in design qualification. As in design qualification, overstress tests may be used to qualify a product for anticipated field overstress loads. Overstress tests may also be exploited to ensure that manufacturing processes do not degrade the intrinsic material strength of hardware beyond a specified limit. However, such tests should supplement, not replace, the accelerated wearout test program, unless explicit physics-based correlations are available between overstress test results and wearout field-failure data.

Manufacturability

The control and rectification of manufacturing defects has typically been the concern of production and process-control engineers, but not of the designer. In the spirit and context of concurrent product development, however, hardware designers must understand material limits, available processes, and manufacturing process capabilities in order to select materials and construct architectures that promote producibility and reduce the occurrence of defects, and consequently increase yield and quality. Therefore, no specification is complete without a clear discussion of manufacturing defects and acceptability limits. The reliability engineer must have clear definitions of the threshold for acceptable quality, and of what constitutes nonconformance. Nonconformance that compromises hardware performance and reliability is considered a defect. Failure mechanism models provide a convenient vehicle for developing such criteria. It is important for the reliability analyst to understand what deviations from specifications can compromise performance or reliability, and what deviations are benign and can hence be accepted.

A *defect* is any outcome of a process (manufacturing or assembly) that impairs or has the potential to impair the functionality of the product at any time. The defect may arise during a single process or may be the result of a sequence of processes. The *yield* of a process is the fraction of products that are acceptable for use in a subsequent process in the manufacturing sequence or product life cycle. The *cumulative yield* of the process is determined by multiplying the individual yields of each of the individual process steps. The source of defects is not always apparent, because defects resulting from a process can go undetected until the product reaches some downstream point in the process sequence, especially if screening is not employed.

It is often possible to simplify the manufacturing and assembly processes in order to reduce the probability of workmanship defects. As processes become more sophisticated, however, process monitoring and control are necessary to ensure a defect-free product. The bounds that specify whether the process is within tolerance limits, often referred to as the *process window*, are defined in terms of the independent variables to be controlled within the process and the effects of the process on the product, or the dependent product variables. The goal is to understand the effect of each process variable on each product parameter in order to formulate control limits for the process, that is, the points on the variable scale where the defect rate begins to possess a potential for causing failure. In defining the process window, the upper and lower limits of each process variable, beyond which it will produce defects, have to be determined. Manufacturing processes must be contained in the process window by defect testing, analysis of the causes of defects, and elimination of defects by process control, such as closed-loop corrective action systems. The establishment of an effective feedback path to report process-related defect data is critical. Once this is done and the process window is determined, the process window itself becomes a feedback system for the process operator.

Several process parameters may interact to produce a different defect than would have resulted from the individual effects of these parameters acting independently. This complex case may require that the interaction of various process parameters be evaluated in a matrix of experiments. In some cases, a defect cannot be detected until late in the process sequence. Thus, a defect can cause rejection, rework, or failure of the product after considerable value has been added to it. These cost items due to defects can reduce yield and return on investments by adding to hidden factory costs. All critical processes require special attention for defect elimination by process control.

Process Verification Testing

Process verification testing is often called *screening*. Screening involves 100% auditing of all manufactured products to detect or precipitate defects. The aim is to preempt potential quality problems before they reach the field. In principle, this should not be required for a well-controlled process. When uncertainties are likely in process controls, however, screening is often used as a safety net.

Some products exhibit a multimodal probability density function for failures, with a secondary peak during the early period of their service life due to the use of faulty materials, poorly controlled manufacture and assembly technologies, or mishandling. This type of early-life failure is often called *infant mortality*. Properly applied screening techniques can successfully detect or precipitate these failures, eliminating or reducing their occurrence in field use. Screening should only be considered for use during the early stages of production, if at all, and only when products are expected to exhibit infant mortality field failures. Screening will be ineffective and costly if there is only one main peak in the failure probability density function. Further, failures arising due to unanticipated events such as acts of God (lightning, earthquake) may be impossible to screen in a cost effective manner.

Since screening is done on a 100% basis it is important to develop screens that do not harm good components. The best screens, therefore, are nondestructive evaluation techniques, such as microscopic visual exams, X rays, acoustic scans, nuclear magnetic resonance (NMR), electronic paramagnetic resonance (EPR), and so on. Stress screening involves the application of stresses, possibly above the rated operational limits. If stress screens are unavoidable, overstress tests are preferred to accelerated wearout tests, since the latter are more likely to consume some useful life of good components. If damage to good components is unavoidable during stress screening, then quantitative estimates of the screening damage, based on failure mechanism models must be developed to allow the designer to account for this loss of usable life. The appropriate stress levels for screening must be tailored to the specific hardware. As in qualification testing, quantitative models of failure mechanisms can aid in determining screen parameters.

A stress screen need not necessarily simulate the field environment, or even utilize the same failure mechanism as the one likely to be triggered by this defect in field conditions. Instead, a screen should exploit the most convenient and effective failure mechanism to stimulate the defects that would show up in the field as infant mortality. Obviously, this requires an awareness of the possible defects that may occur in the hardware and extensive familiarity with the associated failure mechanisms.

Unlike qualification testing, the effectiveness of screens is maximized when screens are conducted immediately after the operation believed to be responsible for introducing the defect. Qualification testing is preferably conducted on the finished product or as close to the final operation as possible; on the other hand, screening only at the final stage, when all operations have been completed, is less effective, since failure analysis, defect diagnostics, and troubleshooting are difficult and impair corrective actions. Further, if a defect is introduced early in the manufacturing process, subsequent value added through new materials and processes is wasted, which additionally burdens operating costs and reduces productivity. Admittedly, there are also several disadvantages to such an approach. The cost of screening at every manufacturing station may be prohibitive, especially for small batch jobs. Further, components will experience repeated screening loads as they pass through several manufacturing steps, which increases the risk of accumulating wearout damage in good components due to screening stresses. To arrive at a screening matrix that addresses as many defects and failure mechanisms as feasible with each screen test, an optimum situation must be sought through analysis of cost-effectiveness, risk, and the criticality of the defects. All defects must be traced back to the root cause of the variability.

Any commitment to stress screening must include the necessary funding and staff to determine the root cause and appropriate corrective actions for all failed units. The type of stress screening chosen should be derived from the design, manufacturing, and quality teams. Although a stress screen may be necessary during the early stages of production, stress screening carries substantial penalties in capital, operating expense, and cycle time, and its benefits diminish as a product approaches maturity. If almost all of the products fail in a properly designed screen test, the design is probably incorrect. If many products fail, a revision of the manufacturing process is required. If the number of failures in a screen is small, the processes are likely to be within tolerances, and the observed faults may be beyond the resources of the design and production process.

2.8 Summary

Hardware reliability is not a matter of chance or good fortune; rather, it is a rational consequence of conscious, systematic, rigorous efforts at every stage of design, development, and manufacture. High product reliability can only be assured through robust product designs, capable processes that are known to be within tolerances, and qualified components and materials from vendors whose processes are also capable and within tolerances. Quantitative understanding and modeling of all relevant failure mechanisms provide a convenient vehicle for formulating effective design and processing specifications and tolerances for high reliability. Scientific reliability assessments may be supplemented by accelerated qualification testing.

The physics-of-failure approach is not only a tool to provide better and more effective designs, but it is also an aid for cost-effective approaches for improving the entire approach to building electronic systems. Proactive improvements can be implemented for defining more realistic performance requirements and environmental conditions, identifying and characterizing key material properties, developing new product architectures and technologies, developing more realistic and effective accelerated stress tests to audit reliability and quality, enhancing manufacturing-for-reliability through mechanistic process modeling and characterization to allow pro-active process optimization, and increasing first-pass yields and reducing hidden factory costs associated with inspection, rework, and scrap.

When utilized early in the concept development stage of a product's development, reliability serves as an aid to determine feasibility and risk. In the design stage of product development, reliability analysis involves methods to enhance performance over time through the selection of materials, design of structures, choice of design tolerance, manufacturing processes and tolerances, assembly techniques, shipping and handling methods, and maintenance and maintainability guidelines. Engineering concepts such as strength, fatigue, fracture, creep, tolerances, corrosion, and aging play a role in these design analyses. The use of physics-of-failure concepts coupled with mechanistic, as well as probabilistic, techniques is often required to understand the potential problems and tradeoffs and to take corrective actions when effective. The use of factors of safety and worst-case studies as part of the analysis is useful in determining stress screening and burn-in procedures, reliability growth, maintenance modifications, field testing procedures, and various logistics requirements.

Defining Terms

Accelerated testing: Tests conducted at higher stress levels than normal operation but in a shorter period of time for the specific purpose to induce failure faster.
Damage: The failure pattern of an electronic or mechanical product.
Failure mechanism: A physical or chemical defect that results in partial degradation or complete failure of a product.
Overstress failure: Failure mechanisms due to a single occurrence of a stress event when the intrinsic strength of an element of the product is exceeded.
Product performance: The ability of a product to perform as required according to specifications.
Qualification: All activities that ensure that the nominal design and manufacturing specifications will meet or exceed the reliability goals.
Reliability: The ability of a product to perform at required parameters for a specified period of time.
Wearout failure: Failure mechanisms caused by monotonic accumulation of incremental damage beyond the endurance of the product.

References

Al-Sheikhly, M. and Christou, A. 1994. How radiation affects polymeric materials. *IEEE Trans. on Reliability* 43(4):551–556.

Dasgupta, A. 1993. Failure mechanism models for cyclic fatigue. *IEEE Trans. on Reliability* 42(4):548–555.

Dasgupta, A. and Haslach, H.W., Jr. 1993. Mechanical design failure models for buckling. *IEEE Trans. on Reliability* 42(1):9–16.

Dasgupta, A. and Hu, J.M. 1992a. Failure mechanism models for brittle fracture. *IEEE Trans. on Reliability* 41(3):328–335.

Dasgupta, A. and Hu, J.M. 1992b. Failure mechanism models for ductile fracture. *IEEE Trans. on Reliability* 41(4):489–495.

Dasgupta, A. and Hu, J.M. 1992c. Failure mechanism models for excessive elastic deformation. *IEEE Trans. on Reliability* 41(1):149–154.

Dasgupta, A. and Hu, J.M. 1992d. Failure mechanism models for plastic deformation. *IEEE Trans. on Reliability* 41(2):168–174.

Dasgupta, A. and Pecht, M. 1991. Failure mechanisms and damage models. *IEEE Trans. on Reliability* 40(5):531–536.

Diaz, C., Kang, S.M., and Duvvury, C. 1995. Electrical overstress and electrostatic discharge. *IEEE Trans. on Reliability* 44(1):2–5.

Engel, P.A. 1993. Failure models for mechanical wear modes & mechanisms. *IEFE Trans. on Reliability* 42(2):262–267.

Ghanem and Spanos. 1991. *Stochastic Finite Element Methods*. Springer-Verlag, New York.

Haugen, E.B. 1980. *Probabilistic Mechanical Design*. Wiley, New York.

Hertzberg, R.W. 1989. *Deformation and Fracture Mechanics of Engineering Materials*. Wiley, New York.

Kapur, K.C. and Lamberson, L.R. 1977. *Reliability in Engineering Design*. Wiley, New York.

Lewis, E.E. 1987. *Introduction to Reliability Engineering*. Wiley, New York.

Li, J. and Dasgupta, A. 1994. Failure-mechanism models for creep and creep rupture. *IEEE Trans. on Reliability* 42(3):339–353.

Li, J. and Dasgupta, A. 1994. Failure mechanism models for material aging due to interdiffusion. *IEEE Trans. on Reliability* 43(1):2–10.

Pecht, M. 1994. Physical architecture of VLSI systems. In *Reliability Issues*. Wiley, New York.

Pecht, M. 1994. *Integrated Circuit, Hybrid, and Multichip Module Package Design Guidelines—A Focus on Reliability*. John Wiley and Sons, New York.

Pecht, M. 1995. *Product Reliability, Maintainability, and Supportability Handbook*. CRC Press, Boca Raton, FL.

Rudra, B. and Jennings, D. 1994. Failure-mechanism models for conductive-filament formation. *IEEE Trans. on Reliability* 43(3):354–360.

Sandor, B. 1972. *Fundamentals of Cyclic Stress and Strain*. City Univ. of Wisconsin Press, WI.

Tullmin, M. and Roberge, P.R. 1995. Corrosion of metallic material. *IEEE Trans. on Reliability* 44(2):271–278.

Young, D. and Christou, A. 1994. Failure mechanism models for electromigration. *IEEE Trans. on Reliability* 43(2):186–192.

Further Information

IEEE Transactions on Reliability: Corporate Office, New York, NY 10017-2394.

Pecht, M. 1995. *Product Reliability, Maintainability, and Supportability Handbook*. CRC Press, Boca Raton, FL.

Pecht, M., Dasgupta, A., Evans, J.W., and Evans, J.Y. 1994. *Quality Conformance and Qualification of Micro-electronic Packages and Interconnects*. Wiley, New York.

3
Software Reliability

Carol Smidts
University of Maryland

3.1 Introduction 3-1
3.2 Software Life Cycle Models 3-2
   The Waterfall Life Cycle Model • The Spiral Model
3.3 Software Reliability Models 3-5
   Definitions • Classification of Software Reliability Models
   • Predictive Models: The Rome Air Development Center
   Reliability Metric • Models for Software Reliability Assessment
   • Criticisms and Suggestion for New Approaches
3.4 Conclusions 3-21

3.1 Introduction

Software is without doubt an important element in our lives. Software systems are present in air traffic control systems, banking systems, manufacturing, nuclear power plants, medical systems, etc. Thus, software failures can have serious consequences: customer annoyance, loss of valuable data in information systems accidents, and so forth, which can result in millions of dollars in lawsuits. Software is invading many different technical and commercial fields. Our level of dependence is continuously growing. Hence, software development needs to be optimized, monitored, and controlled.

Models have been developed to characterize the quality of software development processes in industries. The capability maturity model [Paulk et al. 1993] is one such model. The CMM has five different levels, each of which distinguishes an organization's software process capability:

1. *Initial:* The software development process is characterized as ad hoc; few processes are defined and project outcomes are hard to predict.
2. *Repeatable:* Basic project management processes are established to track cost, schedule, and functionality; processes may vary from project to project, but management controls are standardized; current status can be determined across a project's life; with high probability, the organization can repeat its previous level of performance on similar projects. At level 2, the key process areas are as follows: requirements management, project planning, project tracking and oversight, subcontract management, quality assurance, and configuration management.
3. *Defined:* The software process for both management and engineering activities is documented, standardized, and integrated into an organization-wide software process; all projects use a documented and approved version of the organization's process for developing and maintaining software.
4. *Managed:* Detailed measures of the software process are collected; both the process and the product are quantitatively understood and controlled, using detailed measures.
5. *Optimizing:* Continuous process improvement is made possible by quantitative feedback from the process and from testing innovative ideas and technology.

Data reported to the Software Engineering Institute (SEI) through March 1994 from 261 assessments indicate that: 75% of the organizations were at level 1, 16% at level 2, 8% at level 3 and 0.5% at level 5.

In a letter dated September 25,1991, the Department of the Air Force, Rome Laboratory, Griffiss Air Force Base, notified selected computer software contractors who bid for and work on U.S. government contracts [Saiedian and Kuzara 1995] of the following:

> We wish to point out that at some point in the near future, all potential software developers will be required to demonstrate maturity level 3 before they can compete in ESD/RL [Electronics Systems Division/Rome Laboratory] major software development initiatives . . . . Now is the time to start preparing for this initiative.

Unfortunately, as already pointed out, most companies (75%) are still at level 1. Hence, preparing for most companies means to develop the capabilities to move from CMM level 1 to CMM level 2 and from CMM level 2 to CMM level 3. These capabilities include abilities to predict, assess, control, and improve software reliability, safety, and quality.

In addition to government related incentives to move software development processes to higher levels of maturity, industries have their own incentives for reaching for these higher levels. Studies [Saiedian and Kuzara 1995] have shown that the cost of improving development processes is largely counterbalanced by the resulting savings. As an example, Hughes estimates the assessment (i.e., determination of their software development maturity level) of their process at $45,000 and the subsequent two-year program of improvements cost at about $400,000; as a result Hughes estimates the annual savings to be about $2 million.

This chapter provides a description of the state of the art of software reliability prediction and assessment. The limitations to these models are examined, and a fruitful new approach is examined. Besides reliability models, many other techniques to improve software reliability exist: testing, formal specification, structured design, etc. A review of these techniques can be found in Smidts and Kowalski [1995].

3.2 Software Life Cycle Models

Knowledge of the software development life cycle is needed to understand **software reliability** concepts and techniques. Several life cycle models have been developed over the years; from the *code-and-fix* model to the *spiral model*. We will describe two such models, which have found a wide variety of applications: the *waterfall life cycle model* and the spiral model.

The Waterfall Life Cycle Model

The principal components of this software development process are presented in Fig. 3.1. Together they form the waterfall life cycle model [Royce 1970, 1987]. In actual implementations of the waterfall model variations in the number of phases may exist, naming of phases may differ, and who is responsible for what phase may change.

1. *Requirements definition and analysis* is the phase during which the analyst (usually an experimented software engineer) will determine the user's requirements for the particular piece of software being developed. The notion of requirements will be examined in a later section.
2. *The preliminary design* phase consists of the development of a high-level design of the code based on the requirements.
3. *The detailed design* is characterized by increasing levels of detail and refinement to the preliminary design from the subsystem level to the subroutine level of detail. This implies describing in detail the user's inputs, system outputs, input/output files, and the interfaces at the module level.
4. *The implementation phase* consists of three different activities: coding, testing at the module level (an activity called *unit testing*), and integration of the modules into subsystems. This integration is usually accompanied by integration testing, a testing activity which will ensure compatibility of interfaces between the modules being integrated.

Software Reliability   3-3

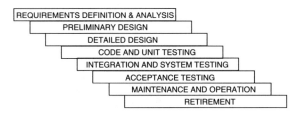

FIGURE 3.1 The waterfall life cycle.

   5. *System testing* is defined as when the development team is in charge of testing the entire system from the point of view of the functions the software is supposed to fulfill. The validation process is completed when all tests required in the system test plan are satisfied. The development team will then produce a system description document.
   6. *Acceptance testing* is defined as the point at which the development team relinquishes its testing responsibility to an independent acceptance test team. The independent acceptance test team is to determine whether or not the system now satisfies the original system requirements. An acceptance test plan is developed, and the phase ends with successful running of all tests in the acceptance test plan.
   7. *Maintenance and operations* is the next phase. If the software requirements are stable, activities will be focused on fixing errors appearing in the exercise of the software or on fine tuning the software to enhance its performance. On the other hand, if the requirements continue to change in response to changing user needs, this stage could resemble to a minilife cycle in itself as software is modified to keep pace with the operational needs.
   8. *Retirement occurs* when the user may decide at some point not to use the software anymore and to throw it away. Because of recurrent changes, the software might have become impossible to maintain.

The end of each phase of the life cycle should see the beginning of a new phase. Unfortunately, reality is often very different. Preliminary design will often start before the requirements have been completely analyzed. Actually, the real life cycle is characterized by a series of back-and-forth movements between the eight phases.

The Spiral Model

The waterfall model was extremely influential in the 1970s and still has widespread application today. However, the waterfall model has a number of limitations such as its heavy emphasis on fully developed documents as criteria of completion for early requirements and design phases. This constraint is not practical for many categories of software (for example, interactive end-user applications) and has led to development of large quantities of unusable software. Other process models were developed to remedy these weaknesses. The spiral model [Boehm 1988] (see Fig. 3.2) is one such model. It is a risk-driven software development model that encompasses most existing process models. The software lifecycle model is not known a priori, and it is the risk that will dynamically drive selection of a specific process model or of the steps of such process. The spiral ceases when the risk is considered acceptable. In Fig. 3.2, the radial dimension represents the cost of development at that particular stage. The angular dimension describes the progress made in achieving each cycle of the spiral. A new spiral starts with assessing:

   1. The objectives of the specific part of the product in development (functionality, required level of performance, etc.)
   2. The alternative means of implementing the product (design options, etc.)
   3. The constraints on the product development (schedule, cost, etc.)

Given the different objectives and constraints, one will select the optimal alternative. This selection process will usually uncover the principal sources of uncertainty in the development process. These sources of

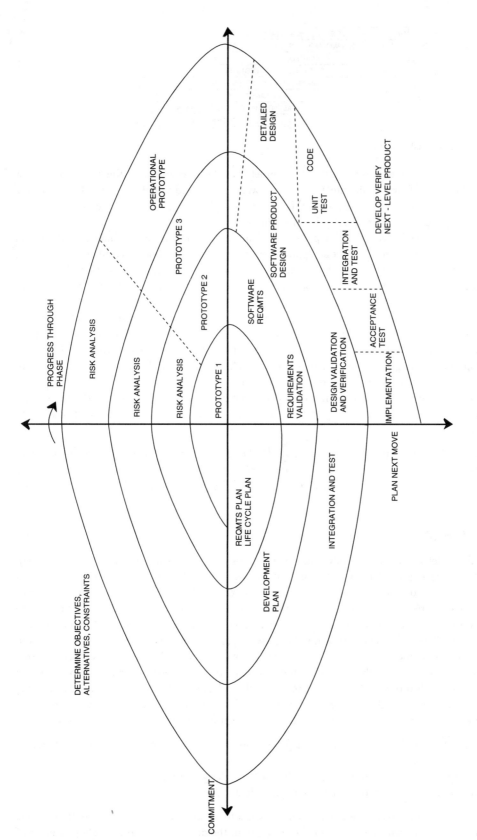

FIGURE 3.2 The spiral model.

uncertainty are prime contributors to risk, and a cost-effective strategy will be formulated for their prompt resolution (for example, by prototyping, simulating, developing analytic models, etc.).

Once the risks have been evaluated, the next step is to determine the relative remaining risks and resolve them with the appropriate strategy. For example, if previous efforts (such as prototyping) have resolved to issues related to user interface or software performance and remaining risks can be tied to software development, a waterfall life cycle development will be selected and implemented as part of the next spiral. Finally, let us note that each cycle of the spiral ends with a review of the documents that were generated in the cycle. This review ensures that all parties involved in the software development are committed to the phase of the spiral.

3.3 Software Reliability Models

Definitions

A number of definitions are needed to introduce the concept of software reliability. Following are IEEE [1983] definitions.

- **Errors** are human actions that result in the software containing a fault. Examples of such faults are the omission or misinterpretation of the user's requirements, a coding error, etc.
- **Faults** are manifestations of an error in the software. If encountered, a fault may cause a failure of the software.
- **Failure** is the inability of the software to perform its mission or function within specified limits. Failures are observed during testing and operation.
- **Software reliability** is the probability that software will not cause the failure of a product for a specified time under specified conditions; this probability is a function of the inputs to and use of the product, as well as a function of the existence of faults in the software; the inputs to the product will determine whether an existing fault is encountered or not.

The definition of software reliability is willingly similar to that of hardware reliability in order to compare and assess the reliability of systems composed of hardware and software components. One should note, however, that in some environments and for some applications such as scientific applications, time (and more precisely time between failures) is an irrelevant concept and should be replaced with a specified number of runs. The concept of software reliability is, for some, difficult to understand. For software engineers and developers, software has deterministic behavior, whereas hardware behavior is partly stochastic. Indeed, once a set of inputs to the software has been selected, once the computer and operating system with which the software will run is known, the software will either fail or execute correctly. However, our knowledge of the inputs selected, of the computer, of the operating system, and of the nature and position of the fault is uncertain. We can translate this uncertainty into probabilities, hence the notion of software reliability as a probability.

Classification of Software Reliability Models

Many software reliability models (SRMs) have been developed over the years. For a detailed description of most models see Musa, Iannino, and Okumoto [1987]. Within these models one can distinguish two main categories: predictive models and assessment models.

Predictive models typically address the reliability of the software early in the life cycle at the requirements or at the preliminary design level or even at the detailed design level in a waterfall life cycle process or in the first spiral of a spiral software development process (see next subsection). Predictive models could be used to assess the risk of developing software under a given set of requirements and for specified personnel before the project truly starts.

Assessment models evaluate present and project future software reliability from failure data gathered when the integration of the software starts (discussed subsequently).

Predictive software reliability models are few in number; most models can be categorized in the assessment category. Different classification schemes for assessment models have been proposed in the past.

A classification of assessment models generalized from Goel's classification [1985] (see also Smidts and Kowalski [1995]) is presented later in this section. It includes most existing SRMs and provides guidelines for the selection of a software reliability model fitting a specific application.

Predictive Models: The Rome Air Development Center Reliability Metric

As explained earlier, predictive models are few. The Air Force's Development Center (RADC) proposed one of the first predictive models [ASFC 1987]. A large range of software programs and related failure data were analyzed in order to identify the characteristics that would influence software reliability. The model identifies three characteristics: the application type (A), the development environment (D), and the software characteristics (S). A new software will be examined with reference to these different characteristics. Each characteristic will be quantified, and reliability R in faults per executable lines of code is obtained by multiplying these different metrics

$$R = A \times D \times S \tag{3.1}$$

A brief description of each characteristic follows:

The *application type* (A) is a basic characteristic of a software: it determines how a software is developed and run. Categories of applications considered by the Air Force were: airborne systems, strategic systems, tactical systems, process control systems, production systems (such as decision aids), and developmental systems (such as software development tools). An initial value for the reliability of the software to be developed is based only on the application type. This initial value is then modified when other factors characterizing the software development process and the product become available.

Different *development environments* (D) are at the origin of variations in software reliability. Boehm [1981] divides development environments into three categories:

- *Organic mode:* Small software teams develop software in a highly familiar, in-house environment. Most software personnel are extremely experienced and knowledgeable about the impact of this software development on the company's objectives.
- *Semidetached mode:* Team members have an intermediate level of expertise with related systems. The team is a mixture of experienced and inexperienced people. Members of the team have experience with some specific aspects of the project but not with others.
- *Embedded mode:* The software needs to operate under tight constraints. In other words, the software will function in a strongly coupled system involving software, hardware, regulations, and procedures. Because of the costs involved in altering other elements of this complex system, it is expected that the software developed will incur all required modifications due to unforeseen problems.

The *software characteristics* (S) metric includes all software characteristics that are likely to impact software reliability. This metric is subdivided into a *requirements and design representation metric* ($S1$) and a *software implementation metric* ($S2$) with

$$S = S1 \times S2 \tag{3.2}$$

The *requirements and design metric* ($S1$) is the product of the following subset of metrics.

The *anomaly management metric* (SA) degree to which fault tolerance exists in the software studied.

The *traceability metric* (ST) degree to which the software being implemented matches the user requirements.

The effect of traceability on the code is represented by a metric, *ST*,

$$ST = \frac{k_{tc}}{TC} \qquad (3.3)$$

where k_{tc} is a coefficient to be determined by regression and *TC* is the traceability metric calculated by

$$\frac{NR}{(NR - DR)} \qquad (3.4)$$

where *NR* is the number of user requirements, and *DR* is the number of user requirements that cannot be found in design or code.

The last characteristic is the *quality review results metric (SQ)*. During large software developments formal and informal reviews of software requirements and design are conducted. Any problems identified in the documentation or in the software design are recorded in discrepancy reports. Studies show a strong correlation between errors reported at these early phases and existence of unresolved problems that will be found later during test and operation. The quality review results metric is a measure of the effect of the number of discrepancy reports found on reliability

$$SQ = k_q \frac{NR}{(NR - NDR)} \qquad (3.5)$$

where k_q is a correlation coefficient to be determined, *NR* is the number of requirements and *NDR* is the number of discrepancy reports.

Finally,

$$S1 = SA \times ST \times SQ \qquad (3.6)$$

The *software implementation metric (S2)* is the product of *SL*, *SS*, *SM*, *SU*, *SX*, and *SR*, defined as follows.

A *language type metric (SL)*: a significant correlation seems to exist between fault density and language type. The software implementation metric identifies two categories of language types: assembly languages ($SL = 1$) and high-order languages, such as Fortran, Pascal and C ($SL = 1.4$).

Program size metric (SS): Although fault density does account directly for the size of the software (i.e., fault density is the number of faults divided by the number of executable lines of code), nonlinear effects of size on reliability due to factors such as complexity and inability of humans to deal with large systems still need to be assessed. Programs are divided in four broad categories: smaller than 10,000 lines of code, between 10,000 and 50,000 lines of code, between 50,000 and 100,000 lines of code, and larger than 100,000 lines of code.

Modularity metric (SM): It is considered that smaller modules are generally more tractable, can be more easily reviewed, and hence will ultimately be more reliable. Three categories of module size are considered: (1) modules smaller than 200 lines of code, (2) modules between 200 and 3,000 lines of code and (3) modules larger than 3,000 lines of code. For a software composed of α modules in category 1, β modules in category 2, and γ modules in category 3,

$$SM = (\alpha \times SM(1) + \beta \times SM(2) + \gamma \times SM(3))/(\alpha + \beta + \gamma) \qquad (3.7)$$

with $SM(i)$ the modularity metric for modules in category *i*.

Extent of reuse metric (SU): reused code is expected to have a positive impact on reliability because this reused code was part of an application that performed correctly in the field. Two cases need to be considered: (1) The reused code is to be used in an application of the same type as the application from which it originated. (2) The reused code is to be used in an application of a different type requesting different interfaces and so forth. In each case *SU* is a factor obtained empirically, which characterizes the impact of reuse on reliability.

Complexity metric (SX): it is generally considered that a correlation exists between a module's complexity and its reliability. Although complexity can be measured in different ways, the authors of the model define *SX* as

$$SX = k_x \cdot \frac{\sum_{i=1,n} SX_i}{n} \qquad (3.8)$$

where k_x is a correlation coefficient determined empirically, n is the number of modules, and SX_i is module i McCabe's cyclomatic complexity. (McCabe's cyclomatic complexity [IEEE 1989b] may be used to determine the structural complexity of a module from a graph of the module's operations. This measure is used to limit the complexity of a module in order to enhance understandability.)

TABLE 3.1 List of Criteria Determining Generation of a Discrepancy Report for a Code Module

- Design is not top down
- Module is not independent
- Module processing is dependent on prior processing
- Module description does not include input, output, processing or limitations
- Module has multiple entrances or multiple exits
- Database is too large
- Database is not compartmentalized
- Functions are duplicated
- Existence of global data

Standards review results metric (SR): Reviews and audits are also performed on the code. The standards review results metric accounts for existence of problem discrepancy reports at the code level. A discrepancy report is generated each time a module meets one of the criteria given in Table 3.1.

SR is given by

$$SR = kv \cdot \frac{n}{(n - PR)} \qquad (3.9)$$

where n is the number of modules, *PR* is the number of modules with severe discrepancies, and kv is a correlation coefficient.

Finally,

$$S2 = SL \cdot SS \cdot SM \cdot SU \cdot SX \cdot SR \qquad (3.10)$$

The RADC metric was developed from data taken from old projects much smaller than typical software developed today and in a lower level language. Hence its applicability to today's software is questionable.

Models for Software Reliability Assessment

Classification

Most existing SRMs may be grouped into four categories:

- *Times between failures category* includes models that provide estimates of the times between failures (see subsequent subsection, Jelinski and Moranda's model).

- *Failure count category* is interested in the number of faults or failure experienced in specific intervals of time (see subsequent subsection, Musa's basic execution time model and Musa–Okumoto logarithmic Poisson, execution time model).
- *Fault seeding category* includes models to assess the number of faults in the program at time 0 via seeding of extraneous faults (see subsequent subsection, Mills fault seeding model).
- *Input-domain based category* includes models that assess the reliability of a program when the test cases are sampled randomly from a well-known operational distribution of inputs to the program. The **clean room methodology** is an attempt to implement this approach in an industrial environment (the Software Engineering Laboratory, NASA) [Basili and Green 1993]. The reliability estimate is obtained from the number of observed failures during execution (see subsequent subsection, Nelson's model and Ramamoorthi and Bastani).

Table 3.2 lists the key assumptions on which each category of models is based, as well as representative

TABLE 3.2 Key Assumptions on Which Software Reliability Models Are Based

Key Assumptions	Specific Models
Times between failures models • Independent times between failures • Equal probability of esposure of each fault • Embedded faults are independent of each other • No new faults introduced during correction	• Jelinski-Moranda's de-eutrophication model [1972] • Schick and Wolverton's model [1973] • Goel and Okumoto's imperfect debugging model [1979] • Littlewood-Verrall's Bayesian model [1973]
Fault court models • Testing intervals are independent of each other • Testing during intervals is homogeneously distributed • Number of faults detected during nonoverlapping intervals are independent of each other	• Shooman's exponential model [1975] • Goel-Okumoto's nonhomogeneous Poisson process [1979] • Goel's generalized nonhomogeneous Poisson process model [1983] • Musa's execution time model [1975] • Musa-Okumoto's logarithmic Poisson execution time model [1983]
Fault seeding models • Seeded faults are randomly distributed in the program • Indigenous and seeded faults have equal probabilities of being detected	• Mills seeding model [1972]
Input-domain based models • Input profile distribution is known • Random testing is used (inputs are selected randomly) • Input domain can be partitioned into equivalence classes	• Nelson's model [1978] • Ramamoorthy and Bastani's model [1982]

models of the category. Additional assumptions specific to each model are listed in Table 3.3. Table 3.4 examines the validity of some of these assumptions (either the generic assumptions of a category or the additional assumptions of a specific model).

The software development process is environment dependent. Thus, even assumptions that would seem reasonable during the testing of one function or product may not hold true in subsequent testing. The ultimate decision about the appropriateness of the assumptions and the applicability of a model have to be made by the user.

Model Selection

To select an SRM for a specific application, the following practical approach can be used. First, determine to which category the software reliability model you are interested in belongs. Then, assess which specific model in a given category fits the application. Actually, a choice will only be necessary if the model is in

TABLE 3.3 Specific Assumptions Related to Each Specific Software Reliability Model

Specific Representatives of a Model Category	Specific Assumptions
Times between failures models • Jelinski and Moranda (JM) de-eutrophication model • Schick and Wolverton model • Goel and Okumoto imperfect debugging model • Littlewood-Verrall Bayesian model	• N faults at time 0; detected faults are immediately removed; the hazard rate[a] in the interval of time between two failures is proportional to the number of remaining faults • Same as above with the hazard rate a function of both the number of faults remaining and the time elapsed since last failure • Same as above, but the fault, even if detected, is not removed with certainty • Initial number of failures is unknown, times between failures are exponentially distributed, and the hazard rate is gamma distributed
Fault count models • Shooman's exponential model • Goel-Okumoto's nonhomogeneous Poisson process (GONHPP) • Goel's generalized nonhomogeneous Poisson process model • Musa's execution time model • Musa–Okumoto's logarithmic Poisson time model	• Same assumptions as in the JM model • The cumulative number of failures experienced follows a nonhomogeneous Poisson process (NHPP); the failure rate decreases exponentially with time • Same assumptions as in the GONHPP, but the failure rate attempts better to replicate testing experiment results that show that failure rate first increases then decreases with time • Same assumptions as in the JM model Same assumptions as in the GONHPP model, but the time element considered in the model is the execution time
Fault seeding models • Mills seeding model	
Input domain based models • Nelson's model • Ramamoorthy and Bastani's model	The outcome of a test case provides some stochastic information about the behavior of the program for other inputs close to the inputs used in the test

[a] Software hazard rate is $z(t) = f(t)/R(t)$, where $R(t)$ is the reliability at time t and $f(t) = -dR(t)/d(T)$. The software failure rate density is $\lambda(t) = d\mu(t)/d(t)$, where $\mu(t)$ is the mean value of the cumulative number of failures experienced by time t.

one of the two first categories, that is, times between failures or failure count models. If this is the case, select a software reliability model based on knowledge of the software development process and environment. Collect such failure data as the number of failures, their nature, the time at which the failure occurred, failure severity level, and the time needed to isolate the fault and to make corrections. Plot the cumulative number of failures and the failure intensity as a function of time. Derive the different parameters of the models from the data collected and use the model to predict future behavior. If the future behavior corresponds to the model prediction, keep the model.

Models

Jelinski and Moranda's Model [1972]. Jelinksi and Moranda [1972] developed one of the earliest reliability models. It assumes that:

- All faults in a program are equally likely to cause a failure during test.
- The hazard rate is proportional to the number of faults remaining.
- No new defects are incorporated into the software as testing and debugging occur.

Software Reliability

TABLE 3.4 Validity of Some Software Reliability Models Assumptions

Assumptions	Intrinsic Limitations of Such Assumptions
• Times between failures are independent	• Only true if derivation of test cases is purely random (never the case)
• A detected fault is immediately corrected	• Faults will not be removed immediately; they are usually corrected in batches; however, the assumption is valid as long as further testing avoids the path in which the fault is active
• No new faults are introduced during the fault removal process.	• In general, this is not true
• Failure rate decreases with test time	• Reasonable approximation in most cases
• Failure rate is proportional to the number of remaining faults for all faults	• A reasonable assumption if the test cases are chosen to ensure equal probability of testing different parts of code
• Time is used as a basis for failure rate	• Usually time is a good basis for failure rate; if this is not true, the models are valid for other units
• Failure rate increases between failures for a given failure interval	• Generally not the case unless testing intensity increases
• Testing is representative of the operational usage	• Usually not the case since testing usually selects error-prone situations

Originally, the model assumed only one fault was removed after each failure, but an extension of the model, credited to Lipow [1974], permits more than one fault to be removed.

In this model, the hazard rate for the software is constant between failures and is

$$Z_i(T) = \phi[N - n_{i-1}], \quad i = 1, 2, \ldots, m \tag{3.11}$$

for the interval between the $(i-1)$st ith failures. In this equation, N is the initial number of faults in the software, ϕ is a proportionality constant, and n_{i-1} is the cumulative number of faults removed in the first $(i - 1)$ intervals.

The maximum likelihood estimates of N and ϕ are given by the solution of the following equations:

$$\sum_{i=1}^{m} \frac{1}{N - n_{i-1}} - \sum_{i=1}^{m} \phi x_{li} = 0 \tag{3.12}$$

and

$$\frac{n}{\phi} - \sum_{i=1}^{m} [N - n_{i-1}] x_{li} = 0 \tag{3.13}$$

where x_{li} is the length of the interval between the $(i - 1)$st and ith failures, and n is the number of errors removed so far.

Once N and ϕ are estimated, the number of remaining errors is given by

$$N(\text{remaining}) = N - n_i \tag{3.14}$$

The mean time to the next software failure is

$$MTTF = \frac{1}{(N - n_i)\phi} \tag{3.15}$$

The reliability is

$$R_{i+1}(t|t_i) = \exp[-(N-n_i)\phi(t-t_i)], \quad t \geq t_i \tag{3.16}$$

where $R_{i+1}(t/t_i)$ is the reliability of the software at time t in the interval $[t_i, t_{i+1}]$, given that the ith failure occurred at time t_i.

Musa Basic Execution Time Model (BETM) [Musa 1975]. This model was first described by Musa in 1975. It assumes that failures occur as a nonhomogeneous Poisson process. The units of failure intensity are failures per central processing unit (CPU) time. This relates failure events to the processor time used by the software. In the BETM, the reduction in the failure intensity function remains constant, irrespective of whether the first or the Nth failure is being fixed.

The failure intensity is a function of failures experienced:

$$\lambda(\mu) = \lambda_0(1 - \mu/v_0) \tag{3.17}$$

where $\lambda(\mu)$ is the failure intensity (failures per CPU hour at μ failures), λ_0 is the initial failure intensity (at $\tau_e = 0$), 14 is the mean number of failures experienced at execution time and τ_e, and v_0 is the total number of failures expected to occur in infinite time.

Then, the number of failures that need to occur to move from a present failure intensity, λ_p, to a target intensity, λ_F, is given by

$$\Delta\mu = \frac{v_0}{\lambda_0}(\lambda_p - \lambda_F) \tag{3.18}$$

and the execution time required to reach this objective is

$$\Delta\tau = \frac{v_0}{\lambda_0}\ln(\lambda_p/\lambda_F) \tag{3.19}$$

In practice, v_0 and λ_0 can be estimated in three ways:

- Use previous experience with similar software. The model can then be applied prior to testing.
- Plot the actual test data to establish or update previous estimates. Plot failure intensity execution time; the y intercept of the straight line fit is an estimate of λ_0. Plot failure intensity failure number; the x intercept of the straight line fits is an estimate of v_0.
- Use the test data to develop a maximum-likelihood estimate. The details of this approach are described in Musa et at. [1987].

Musa [1987] also developed a method to convert execution time predictions to calendar time. The calendar time component is based on the fact that available resources limit the amount of execution time that is practical in a calendar day.

Musa–Okumoto Logarithmic Poisson Execution Time Model (LPETM). The logarithmic Poisson execution time model was first described by Musa and Okumoto [1983]. In the LPETM, the failure intensity is given by

$$\lambda(\mu) = \lambda_0\exp(-\theta\mu) \tag{3.20}$$

where θ is the failure intensity decay parameter and, λ, μ, and λ_0 are the same as for the BETM. The parameter θ represents the relative change of failure intensity per failure experienced. This model assumes that repair of the first failure has the greatest impact in reducing failure intensity and that the impact of each subsequent repair decreases exponentially.

Software Reliability

In the LPETM, no estimate of v_0 is needed. The expected number of failures that must occur to move from a present failure intensity of λ_p to a target intensity of λ_F is

$$\Delta\mu = (1/\theta)\ln(\lambda_p/\lambda_F) \tag{3.21}$$

and the execution time to reach this objective is given by

$$\Delta\tau = \frac{1}{\theta}\left[\frac{1}{\lambda_F} - \frac{1}{\lambda_p}\right] \tag{3.22}$$

In these equations, λ_0 and θ can be estimated based on previous experience by plotting the test data to make graphical estimates or by making a least-squares fit to the data.

Mills Fault Seeding Model [IEEE 1989a]. An estimate of the number of defects remaining in a program can be obtained by a seeding process that assumes a homogeneous distribution of a representative class of defects. The variables in this measure are: N_S the number of seeded faults, n_s the number of seeded faults found, and n_F the number of faults found that were not intentionally seeded.

Before seeding, a fault analysis is needed to determine the types of faults expected in the code and their relative frequency of occurrence. An independent monitor inserts into the code N_S faults that are representative of the expected indigenous faults. During reviews (or testing), both seeded and unseeded faults are identified. The number of seeded and indigenous faults discovered permits an estimate of the number of faults remaining for the fault type considered. The measure cannot be computed unless some seeded faults are found. The maximum likelihood estimate of the indigenous (unseeded) faults is given by

$$NF = n_F N_S / n_S \tag{3.23}$$

Example. Here, 20 faults of a given type are seeded and, subsequently, 40 faults of that type are uncovered: 16 seeded and 24 unseeded. Then, $NF = 30$, and the estimate of faults remaining is $NF(\text{remaining}) = NF - n_F = 6$.

Input Domain Models. Let us first define the notion of input domain. The input domain of a program is the set of all inputs to this program.

Nelson's Model. Nelson's model [TRW 1976] is typically used for systems with ultrahigh-reliability requirements, such as software used in nuclear power plants and limited to 1000 lines of code [Goel 1985, Ramamoorthy and Bastani 1982]. The model is applied to the validation phase of the software (acceptance testing) to estimate reliability. Typically during this phase, if any error is encountered, it will not be corrected (and usually no error will occur).

Nelson defines the reliability of a software run n times (for n test cases) and which failed n_f times as

$$R = 1 - n_f/n$$

where n is the total number of test cases and n_f is the number of failures experienced out of these test cases.

The Nelson model is the only model whose theoretical foundations are really sound. However, it suffers from a number of practical drawbacks.

- We need to run a huge number of test cases in order to reduce our uncertainty on the reliability R.
- Nelson's model assumes random sampling whereas testing strategies tend to use test cases that have a high probability of revealing errors.
- The model does not account for the characteristics of the program. One would expect that to reach an equal level of confidence in the reliability estimate, a logically complex program should be tested more often than a simple program.
- The model does not account for the notion of equivalence classes.

(Equivalence classes are subdomains of the input domain to the program. The expectation is that two sets of input selected from a same equivalence class will lead to similar output conditions.) The model was modified in order to account for these issues. The resulting model [Nelson 1978] defines the reliability of the program as follows:

$$R = \sum_{j=1,m} P_j(1 - \varepsilon_j)$$

where the program is divided in m equivalence classes, P_j is the probability that a set of user's inputs are selected from the equivalence class j and $(1 - \varepsilon_j)$ is the probability that the program will execute correctly for any set of inputs in class j knowing that it executed correctly for any input randomly selected in j. Here $(1 - \varepsilon_j)$ is the class j correctness probability.

The weakness of the model comes from the fact that values provided for ε_j are ad hoc and depend only on the functional structure of the program.

Empirical values for ε_j are:

- If more than one test case belongs to G_j where G_j is the set of inputs that execute a given logic path L_j, then $\varepsilon_j = 0.001$.
- If only one test case belongs to G_j, then $\varepsilon_j = 0.01$.
- If no test case belongs to G_j but all segments (i.e., a sequence of executable statements between two branch points) and segment pairs in L_j have been exercised in the testing, then $\varepsilon_j = 0.05$.
- If all segments L_j but not all segment pairs have been exercised in the testing, then $\varepsilon_j = 0.1$.
- If m segments ($1 < m < 4$) of L_j have not been exercised in testing, then $\varepsilon_j = 0.1 + 0.2m$.
- If more than 4 segments of L_j have not been exercised in the testing, then $\varepsilon_j = 1$.

Ramamoorthy and Bastani. Ramamoorthy and Bastani [1982] developed an improved version of Nelson's model. The model is an attempt to give an analytical expression of the correctness probability $1 - \varepsilon_j$. To achieve this objective the authors introduce the notion of probabilistic equivalence class.

E is a probabilistic equivalence class if $E \vee I$, where I is the input domain of the program P and P is correct for all elements in E with probability $P\{X_1, X_2, \ldots, X_d\}$ if P is correct for each $X_i \in E$, $i = 1, \ldots, d$. Then $P\{E \mid X\}$ is the correctness probability of P based on the set of test cases X. Knowing the correctness probability one derives the reliability of the program easily,

$$R = \sum_{i=1,m} P(E_i \mid X_i) P(E_i)$$

where E_i is a probabilistic equivalence class, $P(E_i)$ is the probability that the user will select inputs from the equivalence class, and $P(E_i \mid X_i)$ is the correctness probability of E_i given the set of test cases X_i.

It is possible to estimate the correctness probability of a program using the continuity assumption (i.e., that closely related points in the input domain are correlated). This assumption holds especially for algebraic equations. Furthermore, we make the assumption that a point in the input domain depends only on its closest neighbors. Figure 3.3 shows the correctness probability for an equivalence class $E_i = [a, a + V]$ where only one test case is selected [Fig. 3.3(a)] and where two test cases are selected [Fig. 3.3(b)].

The authors show that, in general,

$$P\{E_i \text{ is correct} \mid 1 \text{ test case is correct}\} = e^{-\lambda V}$$

$P(E_i \text{ is correct} \mid n \text{ test cases have successive distances } x_j \text{ are correct},$

$$j = 1, 2, \ldots, n-1) = e^{(-\lambda v)} \prod_{j=1}^{n-1} \left(\frac{2}{1 + e^{-\lambda x_j}} \right)$$

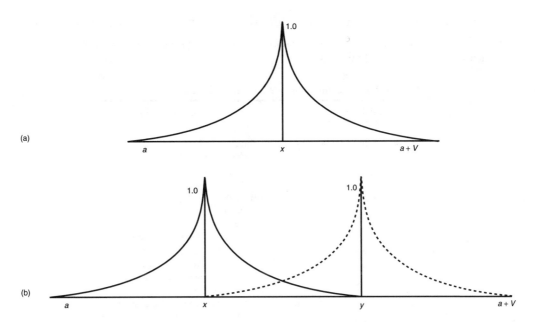

FIGURE 3.3 Interpretation of continuous equivalence classes: (a) single test case, (b) two test cases.

where λ is a parameter of the equivalence class [Ramamoorthy and Bastani 1982].

A good approximation of λV is found for

$$\lambda V \simeq (D-1)//N$$

where N is the number of elements in the class (due to the finite word of the computer) and D is the degree of the equivalence class (i.e., number of distinct test cases that completely validate the class).

Ramamoorthy and Bastani warn that application of the model has several disadvantages. It can be relatively expensive to determine the equivalence classes and their complexity and the probability distribution of the equivalence classes.

Derived Software Reliability Models. From the basic models presented previously, models can be built for applications involving more than one software, such as models to assess the reliability of fault-tolerant software designs [Scott, Gault, and McAllister 1987] or the reliability of a group of modules assembled during the integration phase.

Littlewood [1979] explicitly takes into account the structure (i.e., the modules) of the software, and models the exchange of control between modules (time of sojourn in a module and target of exchange) using a semi-Markovian process. The failure rates of a given module can be obtained from the basic reliability models applied to the module; interface failures are being modeled explicitly. This model can be used to study the integration process.

Data Collection

This subsection examines the data needed to perform a software reliability assessment with the models just described. In particular, we will define the concepts of criticality level and operational profiles. The complete list of data required by each model can be found in Table 3.5.

Criticality: The faults encountered in a software are usually placed in three or five categories called *severity levels* or *criticality levels*. This categorization helps distinguish those faults that are real contributors to software reliability from others such as mere enhancements. Such distinction is, of course, important if we wish to make a correct assessment of the reliability of our software. The following is a classification with five criticality levels [Neufelder 1993]:

- *Catastrophic.* This fault may cause unrecoverable mission failure, a safety hazard or loss of life.
- *Critical:* This fault may lead to mission failure, loss of data, system downtime with consequences such as lawsuits or loss of business.

TABLE 3.5 Data Requirements for Each Model Category

Model Type	Minimal Data Required
Times between failures category	• Criticality level • Operational profile • Failure count • Time between two failures
Failure count category	• Criticality • Operational profile • Failure count • Time at which failure occurred
Fault seeding category	• Types of faults • Number of seeded faults of a given type • Number of seeded faults of a given type that were discovered • Number of indigenous faults of a given type that were discovered.
Input-domain based category	• Equivalence classes • Probability of execution of an equivalence class • Distance between test cases in an equivalence class or correctness probability

- *Moderate:* This fault may lead to a partial loss of functionalities and undesired downtimes. However, workarounds exist, which can mitigate the effect of the fault.
- *Negligible:* This fault does generally not affect the software functionality. It is merely a nuisance to the user and may never be fixed.
- *All others:* This category includes all other faults.

It will be clear to the reader that the two first categories pertain without question to reliability assessment. But what is it about the three other categories?

Operational profile: A software's reliability depends inherently on its use. If faults exist in a region of the software never executed by its users and no faults exist in regions frequently executed, the software's perceived reliability will be high. On the other hand, if a software contains faults in regions frequently executed, the reliability will be perceived to be low. This remark leads to the concept of *input profile*. An input profile is the distribution obtained by putting the possible inputs to the software on the x-axis and the probability of selection of these inputs on the y-axis (see Fig. 3.4). As the reader will see the notion of input profile is extremely important: input profiles are needed for software reliability assessment but they also drive efficient software testing and even software design. Indeed, testing should start by exercising the software under sets of high probability inputs so that if the software has to be unexpectedly released early in the field, the reliability level achieved will be acceptable. It may even be that a reduced version of the software, the reduced operational software (ROS), constructed from the user's highly demanded functionalities, may be built first to satisfy the customer.

Because of the large number of possible inputs to the software, it would be an horrendous task to build a *pure input* profile (a pure input profile would spell out all combinations of input with their respective occurrence probabilities.) An approach to build a *tractable input profile* is proposed next. This approach assesses the operations that will be performed by the software and their respective occurrence probabilities. This profile is called an *operational profile*.

Software Reliability

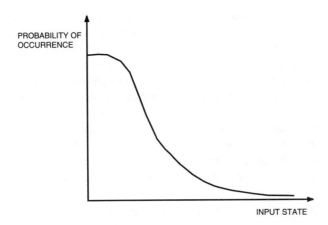

FIGURE 3.4 Input profile for a software program.

To build an operational profile five steps should be considered [Musa 1993]: establishing the customer's profile, and refinement to the user's profile, to the system-mode profile, to the functional profile, and finally to the operational profile itself.

Customer's profile: a customer is a person, group, or institution that acquires the software or, more generally, a system. A customer group is a set of customers who will use the software in an identical manner. The customer profile is the set of all customer groups with their occurrence probabilities.

User's profile: a user is a person, group, or institution that uses the software. The user group is then the set of users who use the system in an identical manner. The user profile is the set of user's groups (all user's groups through all customer's groups) and their probability of occurrence. If identical user's groups exist in two different customer's groups, they should be combined.

System mode profile: a system mode is a set of operations or functions grouped in analyzing system execution. The reliability analyst selects the level of granularity at which a system mode is defined. Some reference points for selecting system modes could be: *user group*, administration vs students in a university; *environmental condition*; overload vs normal load; *operational architectural structure*, network communications vs stand-alone computations; *criticality*, normal nuclear power plant operation vs after trip operations; *hardware components*, functionality is performed on hardware component 1 vs component 2. The system mode profile is the set of all system modes and their occurrence probabilities.

Functional profile: functions contributing to the execution of a system mode are identified. These functions can be found at the requirements definition and analysis stage (see Fig. 3.1). They may also be identifiable in a prototype of the final software product if a spiral software development model was used. However, in such cases care should be exercised: most likely the prototype will not possess all of the functionalities of the final product. The number of functions to be defined varies from one application to another. It could range from 50 functions to several hundreds. The main criteria governing the definition of a function is the extent to which processing may differ. Let us consider a command C with input parameters X and Y. Three admissible values of X exist, $x1$, $x2$, and $x3$, and five admissible values of parameter Y exist, namely, $y1$, $y2$, $y3$, $y4$, and $y5$. From these different combinations of input parameters we could build 15 different functionalities [i.e., combinations $(x1, y1)$, $(x1, y2)$, ..., $(x1, y5)$; $(x2, y1)$, ..., $(x2, y5)$, $(x3, y1)$, ..., $(x3, y5)$]. However, the processing may not differ significantly between these 15 different cases but only for cases that differ in variable X. Each variable defining a function is called a *key input variable*. In our example, the command C and the input X are two key input variables. Please note that even environmental variables could constitute candidate key input variables. The functional profile is the set of all functions identified (for all system modes) and their occurrence probability.

Operational profile: functions are implemented into code. The resulting implementation will lead to one or several operations that need to be executed for a function to be performed. The set of all operations obtained from all functions and their occurrence probability is the operational profile. To derive the operational profile, one will first divide execution into *runs*. A run can be seen as a set of tasks initiated by a given set of inputs. For example, sending e-mail to a specific address can be considered as a run. Note that a run will exercise different functionalities. Identical runs can be grouped into run types. The two runs sending e-mail to address @ a on June 1, 1995 and June 4, 1995 belong to the same run type. However, sending e-mail to address @ a and sending e-mail to address @ b belong to two different run types. Once the runs have been identified, the input state will be identified. The input state is nothing but the values of the input variables to the program at the initiation of the run. We could then build the input profile as the set of all input states and their occurrence probabilities and test the program for each such input state. However, this would be extremely expensive. Consequently, the run types are grouped into operations. The input space to an operation is called a *domain*. Run types that will be grouped share identical input variables. The set of operations should include all operations of high criticality.

Criticisms and Suggestion for New Approaches

Existing software reliability models will either base reliability predictions on historical data existing for similar ware (prediction models) or on failure data associated with the specific software being developed (software assessment models). A basic criticism held against these models is the fact that they are not based on a depth of knowledge and understanding of software behavior. The problem resides in the inherent complexity of software. Too many factors seem to influence the final performance of a software. Furthermore, the software engineering process itself is not well defined. For instance, two programmers handed the same set of requirements and following the same detailed design will still generate different software codes.

Given these limitations, how can we modify our approach to the precision, assessment, and measurement of software reliability? How can we integrate our existing knowledge on software behavior in new models? The main issues we need to resolve are:

- Why does software fail or how is a specific software failure mode generated?
- How does software fail or what are the different software failure modes?
- What is the likelihood of a given software failure mode?

Software failures are generated through the process of creating software. Consequently, it is in the software development process that lie the roots of software failures. More precisely, failures are generated either within a phase of the software development process or at the interface between two phase (see Fig. 3.5). To determine the ways in which software fail, we will then need to consider each phase of the software development process (for example, the requirements definition and analysis phase, the coding phase, etc.) and for each such phase identify the realm of failures that might occur. Then we will need to consider the interfaces between consequent phases and identify all of the failures that might occur at these interfaces: for example, at the interface between the requirements definition and analysis phase and the preliminary design phase.

To determine how software failure modes are created within a particular software development phase, we should first identify the activities and participants of a specific life cycle phase. For instance, let us

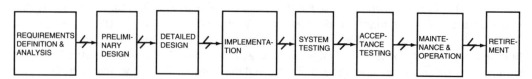

⚡: DISCONNECTION OF INTERACTION

FIGURE 3.5 Sources of failures.

consider the requirements definition and analysis phase of software S. Some of the activities involved in the software requirements definition phase are elicitation of the following:

1. Functional requirements: the different functionalities of the software.
2. Performance requirements: speed requirements.
3. Reliability and safety requirements: how reliable and how safe the software should be.
4. Portability requirements: the software should be portable from machine A to machine B.
5. Memory requirements.
6. Accuracy requirements.

An exhaustive list of the types of requirements can be found in Sage and Palmer [1990].

The participants of this phase are the user of the software and the software analyst, and generally a systems engineer working for a software development company. Note that the number of people involved in this phase may vary from one for each category (user category and analyst category) to several in each category. We will assume for the sake of simplicity that only one representative of each category is present.

We need now to identify the failure modes. These are the failure modes of the user, the failure modes of the analyst, and the failure modes of the communication between user and analyst (see Fig. 3.6). A preliminary list of user's failure modes are: omission of a known requirement (the user forgets), the user is not aware of a specific requirement, a requirement specified is incorrect (the user misunderstands his/her own needs), and the user specifies a requirement that conflicts with another requirement. Failure modes of the analyst are: the analyst omits (forgets) a specific user requirement, the analyst misunderstands (cognitive error) a specific user requirement, the analyst mistransposes (typographical error) a user requirement, the analyst fails to notice a conflict between two users' requirements (cognitive error). Communication failure is: requirement is misheard. Each such failure is applicable to all types of requirements.

As an example, let us consider the case of a software for which the user defined the following requirements:

- $R(f, 1)$ and $R(f, 2)$, two functional requirements
- A performance requirement $R(p, 1)$ on the whole software system
- A reliability requirement $R(r, 1)$ on the whole software
- One interoperability requirement $R(io, 1)$

Let us further assume for the sake of simplicity that there can only be four categories of requirements.

The failure modes in Table 3.6 are obtained by affecting each requirement category by the failure modes identified previously. Note that we might need to further refine this failure taxonomy in order to correctly assess the impact of each failure mode on the software system. Let us, for instance, consider

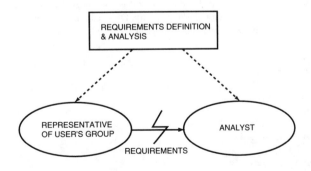

FIGURE 3.6 Sources of failures in the requirements definition and analysis phase.

the case of a reliability requirement. From previous definitions, reliability is defined as the probability that the system will perform its mission under given conditions over a given period of time. Then if the analyst omits a reliability requirement, it might be either too low or too high. If the reliability requirement is too low, the system may still be functional but over a shorter period of time, which may be acceptable or unacceptable. If the reliability requirement is too high, there is an economic impact but the mission itself will be fulfilled.

Once the failure modes are identified, factors that influence their generation should be determined. For example, in the requirements definition and analysis phase, a failure mode such as "user is unaware of a requirement" is influenced by the user's level of knowledge in the task domain.

Once all failure modes and influencing factors have been established, data collection can start. Weakness areas can be identified and corrected. The data collected is meaningful because it ties directly into the failure generation process. Actually, it is likely that a generic set of failure modes can be developed for specific software life cycles and that this set would only need to be tailored for a specific application.

The next step is to find the combination of failure modes that will either fail the software or degrade its operational capabilities in an unacceptable manner. These failure modes are product (software) dependent. Let us focus only on functional requirements. Let us assume that the software will function

TABLE 3.6 Preliminary List of Failure Modes of the Software Requirements Definition and Analysis Phase

Category: By:	Functional Requirements	Performance Requirements	Reliability Requirements	Interoperability Requirements
User	• Omission of known requirement • Unaware of requirement • $R(f, 1)$ is incorrect • $R(f, 2)$ is incorrect	• Omission of known requirement • Unaware of requirement • $R(p, 1)$ is incorrect • $R(p, 1)$ conflicts with other requirements	• Omission of known requirement • Unaware of requirement • $R(r, 1)$ is incorrect • $R(r, 1)$ conflicts with other requirements	• Omission of known requirement • Unaware of requirement • $R(po, 1)$ is incorrect • $R(po, 1)$ conflicts with other requirements
Analyst	• $R(f, 1)$ conflicts with other requirement • Analyst omits $R(f, 1)$ • Analyst omits $R(f, 2)$ • Analyst does not notice conflict between $R(f, 1)$ and other requirements • Analyst does not notice conflict between $R(f, 2)$ and other requirements • Analyst misunderstands $R(f, 1)$ • Analyst misunderstands $R(f, 2)$ • Analyst mistransposed $R(f, 1)$ • Analyst mistransposed $R(f, 2)$	• Analyst omits $R(p, 1)$ • Analyst misunderstands $R(p, 1)$ • Analyst mistransposed $R(p, 1)$ • Analyst does not notice conflict between $R(p, 1)$ and other requirements	• Analyst omits $R(r, 1)$ • Analyst misunderstands $R(r, 1)$ • Analyst mistransposed $R(r, 1)$ • Analyst does not notice conflict between $R(r, 1)$ and other requirements	• Analyst omits $R(po, 1)$ • Analyst misunderstands $R(po, 1)$ • Analyst does not notice conflict between $R(po, 1)$ and other requirements
Communication between User and Analyst	• User's requirement $R(f, 1)$ is misheard • User's requirement $R(f, 2)$ is misheard	• User's requirement $R(p, 1)$ is misheard	• User's requirement $R(r, 1)$ is misheard	• User's requirement $R(po, 1)$ is misheard

Software Reliability

satisfactorily if any of the two functional requirements $R(f, 1)$ and $R(f, 2)$ is carried out correctly. Hence, the probability that the software will fail P_f if we take only into account the requirements and definition analysis phase is given by

$$P_f = PE_{11} \cup (E_{12} \cap E_{13}) \cup E_{14} \cup E_{15} \cup E_{16} \cup E_{21} \cup (E_{22} \cap E_{23}) \cup E_{24} \cup E_{25} \cup E_{26}$$

where $E_{1i}, E_{2i}, i = 1, 6$ are defined in the fault tree given in Fig. 3.7. The actual fault tree will be more complex as failure modes generated by several phases of the software development process contribute to failure of functions 1 and 2. For instance, if we consider a single failure mode for coding such as "software developer fails to code $R(f, 1)$ correctly" and remove, for the sake of the example, contributions for all other phases, a new fault tree is obtained (see Fig. 3.7).

The last question is to determine if we could use such data for reliability prediction. Early in the process, for example, at the requirements definition and analysis phase, our knowledge of the attributes and characteristics of the software development process is limited: coding personnel and coding languages are undetermined, the design approach may not be defined, etc. Hence, our reliability assessment will be uncertain. The more we move into the life cycle process, the more information becomes available, and hence our uncertainty bound will be reduced. Let us call M the model of the process, i.e., the set of characteristics that define the process. The characteristics of such model may include:

- The process is a waterfall life cycle
- Top-down design is used
- Language selected is C++
- Number of programmers is 5, etc.

Our uncertainty is in M. Let $p(M)$ be the probability that the model followed is M. This probability is a function of the time into the life cycle, $p(M, t)$. A reasonable approximation of $p(M, t)$ is $p(M, k)$ where k is the kth life cycle phase. Let R_i be the reliability of the software at the end of the software development process if it was developed under model M_i. Initially, at the end of the requirements definition and analysis phase ($k = 1$), the expected reliability $\langle R \rangle$ is

$$\langle R \rangle = \sum_i p(M_i, 1) \cdot R_i$$

and the variance of R due to our uncertainty on the model is

$$\langle R^2 \rangle = \sum_i p(M_i, 1)(R_i - \langle R_i \rangle)^2$$

If information about the process becomes available, such as, for example, at the end of phase 2 (preliminary design), the process can be updated using Bayesian statistics,

$$p(M_i, m+1) = \frac{L(I|M_i)p(M_i, m)}{\sum_j p(M_j, m) \cdot L(I|M_i)}$$

where I is information obtained between phase m and $m + 1$ and where $L(I | M_i)$ is the likelihood of information I if the model is M_i.

3.4 Conclusions

The objective of this chapter was to establish a state of the art in software reliability. To achieve this objective, an overview of two software development life cycle processes, the waterfall life cycle process

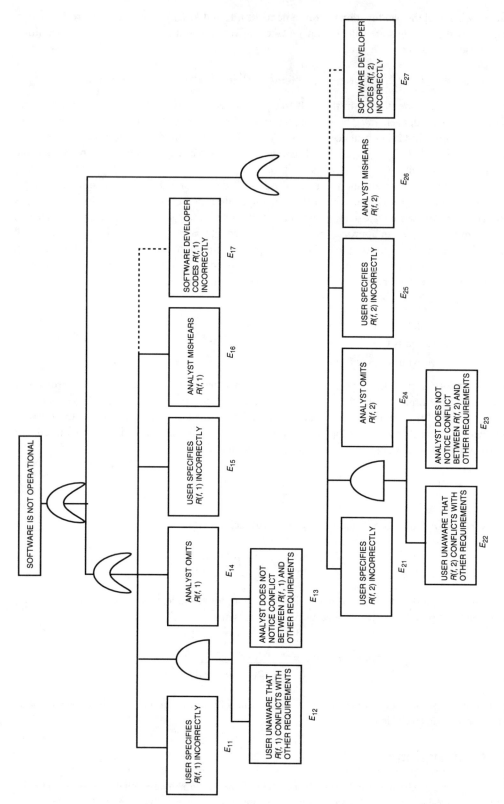

FIGURE 3.7 Software fault tree. The solid lines define the requirements definition and analysis phase fault tree. The requirements definition and analysis phase and coding phase fault tree is obtained by adding solid and dashed contributions.

and the spiral model was provided. These should help the reader understand the context in which the software is developed and how errors are actually generated.

Software reliability models belong to two different categories: the software prediction models that assess reliability early in the life cycle and software assessment models that assess reliability on the basis of failure data obtained during test and operation. Software reliability prediction models are rare and, unfortunately, based on older projects in lower level languages. Consequently, they cannot be easily extrapolated to current software development projects. A number of software reliability assessment models exist from which to choose depending on the nature of the software and of the development environment.

An approach to the prediction of software reliability which attempts to remove this basic criticism is briefly outlined. The approach is based on a systematic identification of software failure modes and of their influencing factors. Data can then be collected on these failure modes for different environment conditions (i.e., for different values of the influencing factors). A direct impact of such data collection is the identification of weakness areas that we can feed back immediately to the process. The data collected is meaningful because it ties directly into the failure generation process. The next step is to find the combination of failure modes that will either fail the software or degrade its operational capabilities in an unacceptable manner. Practicality, feasibility, and domain of application of such an approach will need to be determined. For instance, it is not clear if prediction is feasible very early in the life cycle. The uncertainty of the model to be used for development might be too large to warrant any valuable reliability prediction. Furthermore, such approach could be valid for a waterfall life cycle model but not for other software development models, such as the flexible spiral model.

Defining Terms

Errors: Human actions that result in the software containing a fault. Examples of such errors are the omission or misinterpretation of the user's requirements, coding error, etc.
Failure: The inability of the software to perform its mission or function within specified limits. Failures are observed during testing and operation.
Faults: Manifestations of an error in the software. If encountered, a fault may cause a failure of the software.
Software reliability: The probability that software will not cause the failure of a product for a specified time under specified conditions. This probability is a function of the inputs to and use of the product, as well as a function of the existence of faults in the software. The inputs to the product will determine whether an existing fault is encountered or not.

References

Air Force Systems Command. 1987. Methodology for software prediction. RADC-TR-87-171. Griffiss Air Force Base, New York.
Basili, V. and Green, S. 1993. The evolution of software processes based upon measurement in the SEL: The clean-room example. Univ. of Maryland and NASA/GSFC.
Boehm, B.W. 1988. A spiral model of software development and enhancement. *IEEE Computer* 21:61–72.
Boehm, B.W. 1981. *Software Engineering Economics*. Prentice-Hall, New Jersey.
Goel, A.L. 1983. A Guidebook for Software Reliability Assessment, Rept. RADC TR-83-176.
Goel, A.L. 1985. Software reliability models: Assumptions, limitations, and applicability. *IEEE Trans. Soft. Eng.* SE-11(12):1411.
Goel, A.L. and Okumoto, K. 1979. A Markovian model for reliability and other performance measures of software systems. In *Proceedings of the National Computing Conference* (New York), vol. 48.
Goel, A.L. and Okumoto, K. 1979. A time dependent error detection rate model for software reliability and other performance measures. *IEEE Trans. Rel.* R28:206.

Institute of Electrical and Electronics Engineers. 1983. *IEEE Standard Glossary of Software Engineering Terminology*, ANSI/IEEE Std. 729. IEEE.

Institute of Electrical and Electronics Engineers. 1989a. *IEEE Guide for the Use of IEEE Standard Dictionary of Measures to Produce Reliable Software*, ANSI/IEEE Std. 982.2-1988. IEEE.

Institute of Electrical and Electronics Engineers. 1989b. *IEEE Standard Dictionary of Measures to Produce Reliable Software*, IEEE Std. 982.1-1988. IEEE.

Jelinski, Z. and Moranda, P. 1972. Software reliability research. In *Statistical Computer Performance Evaluation*, ed. W. Freiberger. Academic Press, New York.

Leveson, N.G. and Harvey, P.R. 1983. Analyzing software safety. *IEEE Trans. Soft. Eng.* SE-9(5).

Lipow, M. 1974. Some variations of a model for software time-to-failure. TRW Systems Group, Correspondence ML-74-2260.1.9-21, Aug.

Littlewood, B. 1979. Software reliability model for modular program structure. *IEEE Trans. Reliability* R-28(3).

Littlewood, B. and Verrall, J.K. 1973. A Bayesian reliability growth model for computer software. *Appl. Statist.* 22:332.

Mills, H.D. 1972. On the statistical validation of computer programs. IBM Federal Systems Division, Rept. 72-6015, Gaithersburg, MD.

Musa, J.D. 1975. A theory of software reliability and its application. *IEEE Trans. Soft. Eng.* SE- 1:312.

Musa, J.D. 1993. Operational profiles in software reliability engineering. *IEEE Software* 10:14–32.

Musa, J.D., Iannino, A., and Okumoto, K. 1987. *Software Reliability*. McGraw-Hill, New York.

Musa, J.D. and Okumoto, K. 1983. A logarithmic poisson execution time model for software reliability measurement. In *Proceedings of the 7th International Conference on Software Engineering* (Orlando, FL), March.

Nelson, E. 1978. Estimating software reliability from test data. *Microelectronics Reliability* 17:67.

Neufelder, A.M. 1993. *Ensuring Software Reliability*. Marcel Dekker, New York.

Paulk, M.C., Curtis, B., Chrissis, M.B., and Weber, C.V. 1993. Capability maturity, model, Version 1.1. *IEEE Software* 10:18–27.

Ramamoorthy, C.V. and Bastani, F.B. 1982. Software reliability: Status and perspectives. *IEEE Trans. Soft. Eng.* SE-8:359.

Royce, W.W. 1970. Managing the development of large software systems: Concepts and techniques. Proceedings of Wescon, Aug.; also available in *Proceedings of ICSE9*, 1987. Computer Society Press.

Sage, A.P. and Palmer, J.D. 1990. *Software Systems Engineering*. Wiley Series in Systems Engineering. Wiley, New York.

Saiedian, H. and Kuzara, R. 1995. SEI capability maturity model's impact on contractors. *Computer* 28:16–26.

Schick, G.J. and Wolverton, R.W. 1973. Assessment of software reliability. 11th Annual Meeting German Oper. Res. Soc., DGOR, Hamburg, Germany; also in *Proc. Oper. Res.*, Physica-Verlag, Wirzberg-Wien.

Scott, R.K., Gault, J.W., and McAllister, D.G. 1987. Fault tolerant software reliability modeling. *IEEE Trans. Soft. Eng.* SE-13(5).

Shooman, M.L. 1975. Software reliability measurement and models. In *Proceedings of the Annual Reliability and Maintainability Symposium* (Washington, DC).

Smidts, C. and Kowalski, R. 1995. Software reliability. In *Product Reliability, Maintainability and Supportability Handbook*, ed. M. Pecht. CRC Press, Boca Raton, FL.

TRW Defense and Space System Group. 1976. Software reliability study. Rept. 76-2260, 1-9.5. TRW, Redondo Beach, CA.

Further Information

Additional information on the topic of software reliability can be found in the following sources:

Computer magazine is a monthly periodical dealing with computer software and hardware. The magazine is published by IEEE Computer Society. IEEE headquarters: 345 E. 47th St., New York, NY 10017–2394.

IBM System Journal is published four times a year by International Business Corporation, Old Orchard Road, Armonk, NY 10504.

IEEE Software is a bimonthly periodical published by the IEEE Computer Society, IEEE headquarters: 345 E. 47th St., New York, NY 10017-2394.

IEEE Transactions on Software is a monthly periodical by the IEEE Computer Society, IEEE headquarters: 345 E. 47th St. New York, NY l0017-2394.

In addition, the following books are recommended.

Friedman, M.A. and Voas, J.M. 1905. *Software Assessment: Reliability, Safety, Testability.* McGraw-Hill, New York.

Littlewood, B. 1987. *Software Reliability.* Blackwell Scientific, Oxford, UK.

Rook, P. 1990. *Software Reliability Handbook.* McGraw-Hill, New York.

4
Thermal Properties

4.1 Introduction 4-1
4.2 Fundamentals of Heat 4-1
Temperature • Heat Capacity • Specific Heat • Thermal Conductivity • Thermal Expansion • Solids • Liquids • Gases
4.3 Other Material Properties... 4-4
Insulators • Semiconductors • Conductors • Melting Point
4.4 Engineering Data... 4-6
Temperature Coefficient of Capacitance • Temperature Coefficient of Resistance • Temperature Compensation

David F. Besch
University of the Pacific

4.1 Introduction

The rating of an electronic or electrical device depends on the capability of the device to dissipate heat. As miniaturization continues, engineers are more concerned about heat dissipation and the change in properties of the device and its material makeup with respect to temperature.

The following section focuses on heat and its result. Materials may be categorized in a number of different ways. In this chapter, materials will be organized in the general classifications according to their resistivities:

- Insulators
- Semiconductors
- Conductors

It is understood that with this breakdown, some materials will fit naturally into more than one category. Ceramics, for example, are insulators, yet with alloying of various other elements, can be classified as semiconductors, resistors, a form of conductor, and even conductors. Although, in general, the change in resistivity with respect to temperature of a material is of interest to all, the design engineer is more concerned with how much a resistor changes with temperature and if the change will drive the circuit parameters out of specification.

4.2 Fundamentals of Heat

In the commonly used model for materials, heat is a form of energy associated with the position and motion of the material's molecules, atoms and ions. The position is analogous with the state of the material and is potential energy, whereas the motion of the molecules, atoms, and ions is kinetic energy. Heat added to a material makes it hotter and vice versa. Heat also can melt a solid into a liquid and convert liquids into gases, both changes of state. Heat energy is measured in calories (cal), British thermal units (Btu), or joules (J). A calorie is the amount of energy required to raise the temperature of one gram (1 g) of water one degree Celsius (1°C) (14.5 to 15.5°C). A Btu is a unit of energy necessary to raise the temperature of one pound (1 lb) of water by one degree Fahrenheit (1°F). A joule is an equivalent amount of energy equal to work done when a force of one newton (1 N) acts through a distance of one meter (1 m). Thus heat energy can be turned into mechanical energy to do work. The relationship among the three measures is: 1 Btu = 251.996 cal = 1054.8 J.

Temperature

Temperature is a measure of the average kinetic energy of a substance. It can also be considered a relative measure of the difference of the heat content between bodies. Temperature is measured on either the Fahrenheit scale or the Celsius scale. The Fahrenheit scale registers the freezing point of water as 32°F and the boiling point as 212°F. The Celsius scale or centigrade scale (old) registers the freezing point of water as 0°C and the boiling point as 100°C.

The Rankine scale is an absolute temperature scale based on the Fahrenheit scale. The Kelvin scale is an absolute temperature scale based on the Celsius scale. The absolute scales are those in which zero degree corresponds with zero pressure on the hydrogen thermometer. For the definition of temperature just given, zero °R and zero K register zero kinetic energy.

The four scales are related by the following:

°C = 5/9(°F – 32)
°F = 9/5(°C) + 32
K = °C + 273.16
°R = °F + 459.69

Heat Capacity

Heat capacity is defined as the amount of heat energy required to raise the temperature of one mole or atom of a material by 1°C without changing the state of the material. Thus it is the ratio of the change in heat energy of a unit mass of a substance to its change in temperature. The heat capacity, often called thermal capacity, is a characteristic of a material and is measured in cal/g per °C or Btu/lb per °F,

$$c_p = \frac{\partial H}{\partial T}$$

Specific Heat

Specific heat is the ratio of the heat capacity of a material to the heat capacity of a reference material, usually water. Since the heat capacity of water is 1 Btu/lb and 1 cal/g, the specific heat is numerically equal to the heat capacity.

Thermal Conductivity

Heat transfers through a material by conduction resulting when the energy of atomic and molecular vibrations is passed to atoms and molecules with lower energy. In addition, energy flows due to free electrons,

$$Q = kA\frac{\partial T}{\partial l}$$

where:

Q = heat flow per unit time
k = thermal conductivity
A = area of thermal path
l = length of thermal path
T = temperature

Thermal Properties

The coefficient of thermal conductivity k is temperature sensitive and decreases as the temperature is raised above room temperature.

Thermal Expansion

As heat is added to a substance the kinetic energy of the lattice atoms and molecules increases. This, in turn, causes an expansion of the material that is proportional to the temperature change, over normal temperature ranges. If a material is restrained from expanding or contracting during heating and cooling, internal stress is established in the material.

$$\frac{\partial l}{\partial T} = \beta_L l \quad \text{and} \quad \frac{\partial V}{\partial T} = \beta_V V$$

where:

l = length
V = volume
T = temperature
β_L = coefficient of linear expansion
β_V = coefficient of volume expansion

Solids

Solids are materials in a state in which the energy of attraction between atoms or molecules is greater than the kinetic energy of the vibrating atoms or molecules. This atomic attraction causes most materials to form into a crystal structure. Noncrystalline solids are called *amorphous*, including glasses, a majority of plastics, and some metals in a semistable state resulting from being cooled rapidly from the liquid state. Amorphous materials lack a long range order.

Crystalline materials will solidify into one of the following geometric patterns:

- Cubic
- Tetragonal
- Orthorhombic
- Monoclinic
- Triclinic
- Hexagonal
- Rhombohedral

Often the properties of a material will be a function of the density and direction of the lattice plane of the crystal.

Some materials will undergo a change of state while still solid. As it is heated, pure iron changes from body centered cubic to face centered cubic at 912°C with a corresponding increase in atomic radius from 0.12 nm to 0.129 nm due to thermal expansion. Materials that can have two or more distinct types of crystals with the same composition are called *polymorphic*.

Liquids

Liquids are materials in a state in which the energies of the atomic or molecular vibrations are approximately equal to the energy of their attraction. Liquids flow under their own mass. The change from solid to liquid is called melting. Materials need a characteristic amount of heat to be melted, called the *heat of fusion*. During melting the atomic crystal experiences a disorder that increases the volume of most materials. A few materials, like water, with **stereospecific** covalent bonds and low packing factors attain a denser structure when they are thermally excited.

Gases

Gases are materials in a state in which the kinetic energies of the atomic and molecular oscillations are much greater than the energy of attraction. For a given pressure, gas expands in proportion to the absolute temperature. For a given volume, the absolute pressure of a gas varies in proportion to the absolute pressure. For a given temperature, the volume of a given weight of gas varies inversely as the absolute pressure. These three facts can be summed up into the Gas Law:

$$PV = RT$$

where:

P = absolute pressure
V = specific volume
T = absolute temperature
R = universal gas constant t

Materials need a characteristic amount of heat to transform from liquid to solid, called the *heat of vaporization*.

4.3 Other Material Properties

Insulators

Insulators are materials with resistivities greater than about 10^7 $\Omega \cdot$ cm. Most ceramics, plastics, various oxides, paper, and air are all insulators. Alumina (Al_2O_3) and beryllia (BeO) are ceramics used as substrates and chip carriers. Some ceramics and plastic films serve as the dielectric for capacitors.

Dielectric Constant

A capacitor consists of two conductive plates separated by a dielectric. Capacitance is directly proportional to the dielectric constant of the insulating material. Ceramic compounds doped with barium titanate have high dielectric constants and are used in capacitors. Plastics, such as mica, polystyrene, polycarbonate, and polyester films also serve as dielectrics for capacitors. Capacitor values are available with both positive and negative changes in value with increased temperature. See the first subsection in Sec. 4.4 for a method to calculate the change in capacitor values at different temperatures.

Resistivity

The resistivity of insulators typically decreases with increasing temperature. Figure 4.2 is a chart of three ceramic compounds indicating the reduced resistivity.

Semiconductors

Semiconductors are materials that range in resistivity from approximately 10^{-4} to 10^{+7} $\Omega \cdot$ cm. Silicon (Si), Germanium (Ge), and Gallium Arsenide (GaAs) are typical semiconductors. The resistivity and its inverse, the conductivity, vary over a wide range, due primarily to doping of other elements. The conductivity of intrinsic Si and Ge follows an exponential function of temperature,

$$\sigma = \sigma_0 e^{\frac{E_g}{2kT}}$$

Thermal Properties

where:

- σ = conductivity
- σ_0 = constant t
- E_g = 1.1 eV for Si
- k = Bolzmann's constant t
- T = temperature °K

Thus, the electrical conductivity of Si increases by a factor of 2400 when the temperature rises from 27 to 200 K.

Conductors

Conductors have resistivity value less than 10^{-4} $\Omega \cdot$ cm and include metals, metal oxides, and conductive nonmetals. The resistivity of conductors typically increases with increasing temperature as shown in Fig. 4.1.

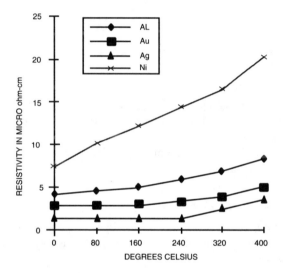

FIGURE 4.1 Resistivity as a function of temperature.

Melting Point

Solder is an important material used in electronic systems. The tin-lead solder system is the most used solder composition. The system's equilibrium diagram shows a typical eutectic at 61.9% Sn. Alloys around the eutectic are useful for general soldering. High Pb content solders have up to 10% Sn and are useful as high-temperature solders. High Sn solders are used in special cases such as in high corrosive environments. Some useful alloys are listed in Table 4.1.

TABLE 4.1 Alloys Useful as Solder

% Sn	% Pb	% Ag	°C
60	40	—	190
60	38	2	192
10	90	—	302
90	10	—	213
95	5	5	230

4.4 Engineering Data

Graphs of resistivity and dielectric constant vs temperature are difficult to translate to values of electronic components. The electronic design engineer is more concerned with how much a resistor changes with temperature and if the change will drive the circuit parameters out of specification. The following defines the commonly used terms for components related to temperature variation.

Temperature Coefficient of Capacitance

Capacitor values vary with temperature due to the change in the dielectric constant with temperature change. The temperature coefficient of capacitance (TCC) is expressed as this change in capacitance with a change in temperature.

$$\text{TCC} = \frac{1}{C}\frac{\partial C}{\partial T}$$

where:

TCC = temperature coefficient of capacitance
C = capacitor value
T = temperature

The TCC is usually expressed in parts per million per degree Celsius (ppm/°C). Values of TCC may be positive, negative, or zero. If the TCC is positive, the capacitor will be marked with a P preceding the numerical value of the TCC. If negative, N will precede the value. Capacitors are marked with NPO if there is no change in value with a change in temperature. For example, a capacitor marked N1500 has a $-1500/1,000,000$ change in value per each degree Celsius change in temperature.

Temperature Coefficient of Resistance

Resistors change in value due to the variation in resistivity with temperature change. The temperature coefficient of resistance (TCR) represents this change. The TCR is usually expressed in parts per million per degree Celsius (ppm/°C).

$$\text{TCR} = \frac{1}{R}\frac{\partial R}{\partial T}$$

where:

TCR = temperature coefficient of resistance
R = resistance value
T = temperature

Values of TCR may be positive, negative, or zero. TCR values for often used resistors are shown in Table 4.2. The last three TCR values refer to resistors imbedded in silicon monolithic integrated circuits.

TABLE 4.2 TCR for Various Resistor Types

Resistor Type	TCR, ppm/°C		
Carbon composition	+500	to	+2000
Wire wound	+200	to	+500
Thick film	+20	to	+200
Thin film	+20	to	+100
Base diffused	+1500	to	+2000
Emitter diffused	+600		
Ion implanted	±100		

Thermal Properties

Temperature Compensation

Temperature compensation refers to the active attempt by the design engineer to improve the performance and stability of an electronic circuit or system by minimizing the effects of temperature change. In addition to utilizing optimum TCC and TCR values of capacitors and resistors, the following components and techniques can also be explored.

- Thermistors
- Circuit design stability analysis
- Thermal analysis

Thermistors

Thermistors are semiconductor resistors that have resistor values that vary over a wide range. They are available with both positive and negative temperature coefficients and are used for temperature measurements and control systems, as well as for temperature compensation. In the latter they are utilized to offset unwanted increases or decreases in resistance due to temperature change.

Circuit Analysis

Analog circuits with semiconductor devices have potential problems with bias stability due to changes in temperature. The current through junction devices is an exponential function as follows:

$$i_D = I_S\left(e^{\frac{q v_D}{nkT}} - 1\right)$$

where:

i_D = junction current
I_S = saturation current
v_D = junction voltage
q = electron charge
n = emission coefficient
k = Boltzmann's constant
T = temperature, in 0 K

Junction diodes and bipolar junction transistor currents have this exponential form. Some biasing circuits have better temperature stability than others. The designer can evaluate a circuit by finding its fractional temperature coefficient,

$$\text{TC}_F = \frac{1}{v(T)} \frac{\partial v(T)}{\partial T}$$

where:

$v(T)$ = circuit variable
TC_F = temperature coefficient
T = temperature

Commercially available circuit simulation programs are useful for evaluating a given circuit for the result of temperature change. SPICE, for example, will run simulations at any temperature with elaborate models included for all circuit components.

Thermal Analysis

Electronic systems that are small or that dissipate high power are subject to increases in internal temperature. Thermal analysis is a technique in which the designer evaluates the heat transfer from active devices that dissipate power to the ambient. Chapter 5, *Heat Management,* discusses thermal analysis of electronic packages.

Defining Terms

Eutectic: Alloy composition with minimum melting temperature at the intersection of two solubility curves.

Stereospecific: Directional covalent bonding between two atoms.

References

Guy, A.G. 1967. *Elements of Physical Metallurgy,* 2nd ed., pp. 255–276. Addison-Wesley, Reading, MA.

Incropera, F.P. and Dewitt, D.P. 1990. *Fundamentals of Heat and Mass Transfer,* 3rd ed., pp. 44–66. Wiley, New York.

Further Information

Additional information on the topic of thermal properties of materials is available from the following sources:

Banzhaf, W. 1990. *Computer-Aided Circuit Analysis Using Psice.* Prentice-Hall, Englewood Cliffs, NJ.

Smith, W.F. 1990. *Principles of Material Science and Engineering.* McGraw-Hill, New York.

Van Vlack, L.H. 1980. *Elements of Material Science and Engineering.* Addison-Wesley, Reading, MA.

5
Heat Management

	5.1	Introduction 5-1
	5.2	Heat Transfer Fundamentals 5-3
		Basic Heat Flow Relations, Data for Heat Transfer Modes
	5.3	Study of Thermal Effects in Packaging Systems 5-12
		Thermal Resistance • Thermal Modeling/Simulation • Experimental Characterization
Zbigniew J. Staszak	5.4	Heat Removal/Cooling Techniques in Design of Packaging Systems 5-18
Technical University of Gdansk	5.5	Concluding Remarks... 5-20

5.1 Introduction

Thermal *cooling/heat* **management** is one of four major functions provided and maintained by a packaging structure or system, with the other three being: (1) mechanical support, (2) electrical interconnection for power and signal distribution, and (3) environmental protection, all aimed at effective transfer of the semiconductor chip performance to the system. Heat management can involve a significant portion of the total packaging design effort and should result in providing efficient heat transfer paths at all packaging levels, that is, level 1 packaging [chip(s), carrier—both single-chip and multichip modules], level 2 (the module substrate—card), level 3 (board, board-to-board interconnect structures), and level 4 (box, rack, or cabinet housing the complete system), to an ultimate **heat sink** outside, while maintaining internal device temperature at acceptable levels in order to control thermal effects on circuits and system performance.

The evolution of chip technology and of packaging technology is strongly interdependent. Very large-scale integrated (VLSI) packaging and interconnect technology are driven primarily by improvements in chip and module technologies, yield, and reliability.

Figures 5.1 (a)–5.1 (c) show the increase in number of transistors per chip and the corresponding increase in chip areas for principal logic circuits, direct random access memories (DRAMs), and microprocessors, as well as power densities at chip and module levels, displayed as a function of years. Although the feature length of these circuits continues to reduce in order to increase the memory capacity of a chip or the number of gates, the greater increase in complexity requires additional area. Currently, the number of transistors per chip is in the range of 100×10^6, whereas chip area can reach 200–300 mm². Despite lowering of supply voltages, one of the results of continual increase in chip packing densities is excess heat generated by the chips that must be successfully dissipated. Moreover, increased chip areas also lead to enhanced stresses and strains due to nonuniformities in temperature distributions and mismatches in mechanical properties of various package materials. This is especially true in multilayer packages at the interfaces of the chip and bonding materials. Power densities for high-performance circuits are currently on the order of 10–50 W/cm² for the chip level, and 0.5–10 W/cm² for the module level.

Trends in circuit performance and their impact on high-performance packaging show that similar to the integration levels, speed, die size, I/O pincount, packing, and dissipated power densities continue to

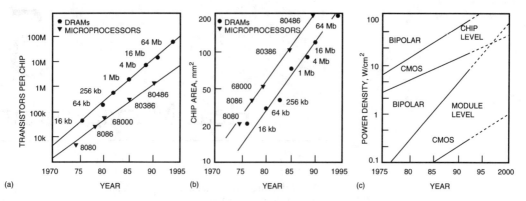

FIGURE 5.1 IC complexity trends: (a) transistors per chip, (b) chip area for DRAMs and microprocessors, (c) power density at chip and module levels for CMOS and bipolar technologies.

TABLE 5.1 Extrapolated VLSI Chips and Packaging Characteristics for Late 1990s

- Chip complexity reaching one billion transistor chips.
- Very large chip areas, 1.5–2.0 cm on a side employing less than 0.35 μm features, and substrate area 10–20 cm on a side for multichip packaging.
- Chip internal gate delays of approximately 50 ps, and chip output driver rise time of 100–200 ps.
- 1000–2000 terminations per chip, 20–100 k I/Os per module, and 10,000–100,000 chip connections per major system.
- Packing densities of 50 and 100 kGates/cm^2, wiring densities of 500 and 1000 cm/cm^2 for single- and multichip modules, respectively.
- Power dissipation at the chip level of 100 W/cm^2 (and more) with chip temperatures required to remain below 100–125°C or less (preferably 65–85°C), and 20–50 W/cm^2 at the module level

increase with each new generation of circuits [Ohsaki 1991, Tummala 1991, Wesseley et al. 1991, Hagge 1992]. Some of the extrapolated, leading-edge VLSI chip characteristics that will have an impact on requirements for single-chip and multichip packaging within the next five years are summarized in Table 5.1.

Packaging performance limits are thus being pushed for high-speed (subnanosecond edge speeds), high-power (~50–100 W/cm^2) chips with reliable operation and low cost. However, there is a tradeoff between power and delay; the highest speed demands that gates be operated at high power. Though packaging solutions are dominated by applications in digital processing for computer and aerospace (avionics) applications, analog and mixed analog/digital, power conversion and microwave applications cannot be overlooked. The net result of current chip and substrate characteristics is that leading-edge electrical and thermal requirements that have to be incorporated in the packaging of the 1990s can be met with some difficulty and only with carefully derived *electrical design rules* [Davidson and Katopis 1989] and complex thermal management techniques [ISHM 1984].

The drive toward high functional density, extremely high speed, and highly reliable operation is constrained by available **thermal design** and *heat removal techniques*. A clear understanding of thermal control technologies, as well as the thermal potential and limits of advanced concepts, is critical. Consequently, the choice of thermal control technology and the particular decisions made in the course of evolving the thermal packaging design often have far-reaching effects on both the reliability and cost of the electronic system or assembly [Nakayama 1986; Antonetti, Oktay, and Simons 1989].

Temperature changes encountered during device fabrication, storage, and normal operation are often large enough to place limits on device characterization and lifetime. Temperature related effects, inherent in device physics, lead to performance limitations, for example, propagation delays, degradation of noise margins, decrease in electrical isolation provided by reverse-biased junctions in an IC, and other effects. **Thermally** enhanced **failures**, such as oxide wearout, fracturing, package delamination, wire bond breakage, deformation of metallization on the chip, and cracks and voids in the chip, substrate, die bond,

Heat Management

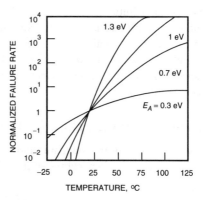

FIGURE 5.2 Normalized failure rate $\lambda(T)/\lambda(T_R)$ as a function of temperature showing an exponential dependence of failure rate with temperature; for example, for $E_A = 0.5$ eV an increase of 10°C in temperature almost doubles, whereas an increase of 20°C more than triples the failure rate at about room temperature.

and solder joints lead to reliability limitations [Jeannotte, Goldmann, and Howard 1989]. The desired system-level reliability, expressed as the **mean time to failure (MTTF)**, is currently aimed at several thousand hours for military equipment and 40,000–60,000 hours for commercial computers [Bar-Cohen 1987, 1993].

The dependence of the **failure rate** on temperature is a complex relationship involving, among other things, material properties and environmental conditions, and in a general form can be described by the Arrhenius equation depicted in Fig. 5.2. This relation is well suited for device-related functional failures; however, for thermally induced structural failures it has to be applied with care since these failures depend both on localized temperature as well as the history of fabrication and assembly, thermal cycling for test purposes (thermal shock testing), and powering up during operation (operational thermal stress), thus being more complicated in nature.

$$\lambda(T) = \lambda(T_R)\exp\left[\frac{E_A}{k}\left(\frac{1}{T_R} - \frac{1}{T}\right)\right]$$

where:

$\lambda(T)$ = failure rate (FIT) at temperature T, K
$\lambda(T_R)$ = failure rate (FIT) at temperature T_R, K
E_A = activation energy, eV, typical values for integrated circuits are 0.3–1.6 eV
k = Boltzmann's constant, 8.616×10^{-5} eV/K

In the definition of $\lambda(T)$, FIT is defined as one failure per billion device hours (10^{-9}/h) or 0.0001%/1000 h. Improvements in thermal parameters and in removable heat densities require enhancements in chip technology, package design, packaging materials compositions, assembly procedures, and cooling techniques (systems). These must be addressed early in the design process to do the following:

- Reduce the rise in temperature and control the variation in the devices' operating temperature across all the devices, components, and packaging levels in the system.
- Reduce deformations and stresses produced by temperature changes during the fabrication process and normal operation.
- Reduce variations in electrical characteristics caused by thermal stresses.

5.2 Heat Transfer Fundamentals

In the analysis of construction of VLSI-based chips and packaging structures, all modes of heat transfer must be taken into consideration, with natural and forced air/liquid convection playing the main role in the cooling process of such systems. The temperature distribution problem may be calculated by applying the (energy) conservation law and equations describing heat conduction, convection, and radiation (and, if required, phase change). Initial conditions comprise the initial temperature or its distribution, whereas boundary conditions include adiabatic (no exchange with the surrounding medium, i.e., surface isolated, no heat flow across it), isothermal (constant temperature), or/and miscellaneous (i.e., exchange with external bodies, adjacent layers or surrounding medium). Material physical parameters, and thermal conductivity, **specific heat**, thermal coefficient of expansion, and **heat transfer coefficients**, can be functions of temperature.

Basic Heat Flow Relations, Data for Heat Transfer Modes

Thermal transport in a solid (or in a stagnant fluid: gas or liquid) occurs by *conduction*, and is described in terms of the *Fourier equation* here expressed in a differential form as

$$q = -k\nabla T(x, y, z)$$

where:

q = heat flux (power density) at any point x, y, z, W/m²
k = thermal conductivity of the material of conducting medium (W/m-degree), here assumed to be independent of x, y, z (although it may be a function of temperature)
T = temperature, °C, K

In the one-dimensional case, and for the transfer area A(m) of heat flow path length L (m) and thermal conductivity k not varying over the heat path, the temperature difference ΔT(°C, K) resulting from the conduction of heat Q(W), normal to the transfer area, can be expressed in terms of a *conduction* **thermal resistance** θ (degree/W). This is done by analogy to electrical current flow in a conductor, where heat flow Q (W) is analogous to electric current I (A), and temperature T (°C, K) to voltage V (V), thus making thermal resistance θ analogous to electrical resistance $R(\Omega)$ and thermal conductivity k (W/m-degree) analogous to electrical conductivity, $\sigma(1/\Omega - m)$:

$$\theta = \frac{\Delta T}{Q} = \frac{L}{kA}$$

Expanding for multilayer (n layer) composite and rectilinear structure,

$$\theta = \sum_{i=1}^{n} \frac{\Delta l_i}{k_i A_i}$$

where:

Δl_i = thickness of the ith layer, m
k_i = thermal conductivity of the material of the ith layer, W/m degree
A_i = cross-sectional area for heat flux of the ith layer, m²

In semiconductor packages, however, the heat flow is not constrained to be one dimensional because it also spreads laterally. A commonly used estimate is to assume a 45° heat spreading area model, treating the flow as one dimensional but using an effective area A_{eff} that is the arithmetic mean of the areas at the top and bottom of each of the individual layers Δl_i of the flow path. Assuming the heat generating source to be square, and noting that with each successive layer A_{eff} is increased with respect to the cross-sectional area A_i for heat flow at the top of each layer, the thermal (spreading) resistance θ_{sp} is expressed as follows:

$$\theta_{sp} = \frac{\Delta l_i}{kA_{eff}} = \frac{\Delta l_i}{kA_i\left(1 + \frac{2\Delta l_i}{\sqrt{A_i}}\right)}$$

On the other hand, if the heat generating region can be considered much smaller than the solid to which heat is spreading, then the semi-infinite heat sink case approach can be employed. If the **heat flux** is applied through a region of radius R, then either $\theta_{sp} = 1/\pi kR$ for uniform heat flux and the maximum temperature occurring at the center of the region, or $\theta_{sp} = 1/4kR$ for uniform temperature over the region of the heat source [Carslaw and Jaeger 1967].

The preceding relations describe only static heat flow. In some applications, however, for example, switching, it is necessary to take into account transient effects. When heat flows into a material volume $V(m^3)$ causing a temperature rise, thermal energy is stored there, and if the heat flow is finite, the time required to effect the temperature change is also finite, which is analogous to an electrical circuit having a capacitance that must be charged in order for a voltage to occur. Thus the required power/heat flow Q to cause the temperature ΔT in time Δt is given as follows:

$$Q = \rho c_p V \frac{\Delta T}{\Delta t} = C_\theta \frac{\Delta T}{\Delta t}$$

where:

C_θ = thermal capacitance, W-s/degree
ρ = density of the medium, g/m³
c_p = specific heat of the medium, W-s/g-degree

Again, we can make use of electrical analogy, noting that thermal capacitance C_θ is analogous to electrical capacitance $C(F)$.

A rigorous treatment of multidimensional heat flow leads to a time-dependent heat flow equation in a conducting medium, which in Cartesian coordinates, and for $Q_V(W/m^3)$ being the internal heat source/generation, is expressed in the form of:

$$k\Delta^2 T(x, y, z, t) = -Q_V(x, y, z, t) + \rho c_p \frac{\partial T(x, y, z, t)}{\partial t}$$

An excellent treatment of analytical solutions of heat transfer problems has been given by Carslaw and Jaeger [1967]. Although analytical methods provide results for relatively simple geometries and idealized boundary/initial conditions, some of them are useful [Newell 1975, Kennedy 1960]. However, thermal analysis of complex geometries requires multidimensional numerical computer modeling limited only by the capabilities of computers and realistic CPU times. In these solutions, the designer is normally interested in the behavior of device/circuit/package over a wide range of operating conditions including temperature dependence of material parameters, finite dimensions and geometric complexity of individual layers, nonuniformity of thermal flux generated within the active regions, and related factors.

Figure 5.3 displays temperature dependence of thermal material parameters of selected packaging materials, whereas Table 5.2 summarizes values of parameters of insulator, conductor, and semiconductor materials, as well as gases and liquids, needed for thermal calculations, all given at room temperature. Note the inclusion of the thermal coefficient of expansion $\beta(°C^{-1}, K^{-1})$, which shows the expansion and contraction ΔL of an unrestrained material of the original length L_o while heated and cooled according to the following equation:

$$\Delta L = \beta L_o(\Delta T)$$

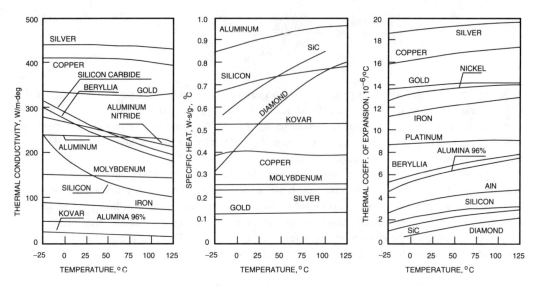

FIGURE 5.3 Temperature dependence of thermal conductivity k, specific heat c_p, and coefficient of thermal expansion (CTE) β of selected packing materials.

Convection heat flow, which involves heat transfer between a moving fluid and a surface, and *radiation heat flow*, where energy is transferred by electromagnetic waves from a surface at a finite temperature, with or without the presence of an intervening medium, can be accounted for in terms of heat transfer coefficients h (W/m²-degree). The values of the heat transfer coefficients depend on the local transport phenomena occurring on or near the package/structure surface. Only for simple geometric configurations can these values be analytically obtained. Little generalized heat transfer data is available for VLSI type conditions, making it imperative to create the ability to translate real-life designs to idealized conditions (e.g., through correlation studies). Extensive use of empirical relations in determining heat transfer correlations is made through the use of dimensional analysis in which useful design correlations relate the transfer coefficients to geometrical/flow conditions [Furkay 1984].

For *convection*, both free-air and forced-air (gas) or -liquid convection have to be considered, and both flow regimes must be treated; **laminar flow** and **turbulent flow**. The detailed nature of convection flow is heavily dependent on the geometry of the thermal duct, or whatever confines the fluid flow, and it is nonlinear. What is sought here are crude estimates, however, just barely acceptable for determining whether a problem exists in a given packaging situation, using as a model the relation of *Newton's law of cooling* for convection heat flow Q_c(W)

$$Q_c = h_c A_S (T_S - T_A)$$

where:

h_c = average convective heat transfer coefficient, W/m²-degree
A_s = cross-sectional area for heat flow through the surface, m²
T_S = temperature of the surface, °C, K
T_A = ambient/fluid temperature, °C, K

For *forced convection* cooling applications, the designer can relate the temperature rise in the coolant temperature $\Delta T_{coolant}$ (°C, K), within an enclosure/heat exchanger containing subsystem(s) that obstruct the fluid flow, to a volumetric flow rate G(m³/s) or fluid velocity v(m/s) as

$$\Delta T_{coolant} = T_{coolant-out} - T_{coolant-in} = \frac{Q_{flow}}{\rho G c_p} = \frac{Q_{flow}}{\rho v A c_p} = \frac{Q_{flow}}{\dot{m} c_p}$$

TABLE 5.2 Selected Physical and Thermal Parameters of Some of the Materials Used in VLSI Packaging Applications (at Room Temperature, $T = 27°C$, 300 K)[a]

Material Type	Density, ρ, g/cm³	Thermal Conductivity, k, W/m-°C	Specific Heat, c_p, W-s/g-°C	Thermal Coeff. of Expansion, β, 10^{-6}/°C
Insulator Materials:				
Aluminum nitride	3.25	100–270	0.8	4
Alumina 96%	3.7	30	0.85	6
Beryllia	2.86	260–300	1.02–1.12	6.5
Diamond (IIa)	3.5	2000	0.52	1
Glass-ceramics	2.5	5	0.75	4–8
Quartz: (fused)	2.2	1.46	0.67–0.74	0.54
Silicon carbide	3.2	90–260	0.69–0.71	2.2
Conductor Materials:				
Aluminum	2.7	230	0.91	23
Beryllium	1.85	180	1.825	12
Copper	8.93	397	0.39	16.5
Gold	19.4	317	0.13	14.2
Iron	7.86	74	0.45	11.8
Kovar	7.7	17.3	0.52	5.2
Molybdenum	10.2	146	0.25	5.2
Nickel	8.9	88	0.45	13.3
Platinum	21.45	71.4	0.134	9
Silver	10.5	428	0.234	18.9
Semiconductor Materials (lightly doped):				
GaAs	5.32	50	0.322	5.9
Silicon	2.33	150	0.714	2.6
Gases:				
Air	0.00122	0.0255	1.004	3.4×10^3
Nitrogen	0.00125	0-025	1.04	10^2
Oxygen	0.00143	0.026	0.912	10^2
Liquids:				
FC-72	1.68	0.058	1.045	1600
Freon	1.53	0.073	0.97	2700
Water	0.996	0.613	4.18	270

[a] Approximate values, depending on exact composition of the material. (*Source:* Compiled based in part on: Touloukian, Y.S. and Ho, C.Y. 1979. *Master Index to Materials and Properties*. Plenum Publishing, New York.)

where:

$T_{\text{coolant-out/in}}$ = the outlet/inlet coolant temperatures, respectively, °C, K
Q_{flow} = total heat flow/dissipation of all components within the enclosure upstream of the component of interest, W
$\dot{m}$ = mass flow rate of the fluid, g/s

Assuming a fixed temperature difference, the **convective heat transfer** may be increased either by obtaining a greater heat transfer coefficient h_c or by increasing the surface area. The heat transfer coefficient may be increased by increasing the fluid velocity, changing the coolant fluid, or utilizing *nucleate boiling*, a form of immersion cooling.

For nucleate boiling, which is a liquid-to-vapor phase change at a heated surface, increased heat transfer rates are the results of the formation and subsequent collapsing of bubbles in the coolant adjacent to the heated surface. The bulk of the coolant is maintained below the boiling temperature of the coolant, while the heated surface remains slightly above the boiling temperature. The **boiling heat transfer** rate Q_b can be approximated by a relation of the following form:

$$Q_b = C_{sf}A_S(T_S - T_{sat})^n = h_b A_S(T_S - T_{sat})$$

where:

C_{sf} = constant, a function of the surface/fluid combination, W/m²-K^n [Rohsenow and Harnett 1973]
T_{sat} = temperature of the boiling point (saturation) of the liquid, °C, K
n = coefficient, usual value of 3
h_b = boiling heat transfer coefficient, $C_{sf}(T_S - T_{sat})^{n-1}$, W/m²-degree

Increased heat transfer of surface area in contact with the coolant is accomplished by the use of *extended surfaces*, plates or pin fins, giving the heat transfer rate Q_f by a fin or fin structure as

$$Q_f = h_c A \eta (T_b - T_f)$$

where:

A = full wetted area of the extended surfaces, m²
η = fin efficiency
T_b = temperature of the fin base, °C, K
T_f = temperature of the fluid coolant, °C, K

Fin efficiency η ranges from 0 to 1; for a straight fin $\eta = \tanh mL/mL$, where $m = \sqrt{2h_c/k\delta}$, L = the fin length (m), and δ = the fin thickness (m) [Kern and Kraus 1972].

Formulas for heat convection coefficients h_c can be found from available empirical correlation and/or theoretical relations and are expressed in terms of dimensional analysis with the dimensionless parameters: Nusselt number Nu, Rayleigh number Ra, Grashof number Gr, Prandtl number Pr and Reynolds number Re, which are defined as follows:

$$Nu = \frac{h_c L_{ch}}{k}, \quad Ra = GrPr, \quad Gr = \frac{g\beta\rho^2}{\mu^2}L_{ch}^3 \Delta T, \quad Pr = \frac{\mu c_p}{k}, \quad Re = vL_{ch}\frac{\rho}{\mu}$$

where:

L_{ch} = characteristic length parameter, m
g = gravitational constant, 9.81 m/s²
μ = fluid dynamic viscosity, g/m-s
$\Delta T = T_S - T_A$, degree

Examples of such expressions for selected cases used in VLSI packaging conditions are presented next. Convection heat transfer coefficients h_c averaged over the plate characteristic length are written in terms of correlations of an average value of Nu vs Ra, Re, and Pr.

1. For natural (air) convection over external flat horizontal and vertical platelike surfaces,

$$Nu = C(Ra)^n$$
$$h_c = (k/L_{ch})Nu = (k/L_{ch})C(Ra)^n = C'(L_{ch})^{3n-1}\Delta T^n$$

where:

C, n = constants depending on the surface orientation (and geometry in general), and the value of the Rayleigh number, see Table 5.3.
$C' = kC[(g\beta\rho^2/\mu^2)Pr]^n$.

TABLE 5.3 Constants for Average Nusselt Numbers for Natural Convection[a] and Simplified Equations for Average Heat Transfer Coefficients h_c (W/m²-degree) for Natural Convection to Air over External Flat Surfaces (at Atmospheric Pressure)[b]

Configuration	L_{ch}	Flow Regime	C	n	h_c for Natural Convection Air Cooling		
					h_c	$C'(27°C)$	$C'(75°C)$
Vertical plate	H						
		$10^4 < Ra < 10^9$ laminar	0.59	0.25	$C'(\Delta T/L_{ch})^{0.25}$	1.51	1.45
		$10^9 < Ra < 10^{12}$ turbulent	0.13	0.33	$C'(\Delta T)^{0.33}$	1.44	1.31
Horizontal plate (heated side up)	$WL/[2(W+L)]$						
		$10^4 < Ra < 10^7$ laminar	0.54	0.25	$C'(\Delta T/L_{ch})^{0.25}$	1.38	1.32
		$10^7 < Ra < 10^{10}$ turbulent	0.15	0.33	$C'(\Delta T)^{0.33}$	1.66	1.52
(heated side down)							
		$10^5 < Ra < 10^{10}$	0.27	0.25	$C'(\Delta T/L_{ch})^{0.25}$	0.69	0.66

Where H, L, and W are height, length, and width of the plate, respectively.

[a] Compilation based on two sources. (*Sources*: McAdams, W.H. 1954. *Heat Transmission*. McGraw-Hill, New York, and Kraus, A.D. and Bar-Cohen, A. 1983. *Thermal Analysis and Control of Electronic Equipment*. Hemisphere, New York.)
[b] Physical properties of air and their temperature dependence, with units converted to metric system, are depicted in Fig. 5.4. (*Source*: Keith, F. 1973. *Heat Transfer*. Harper & Row, New York.)

Most standard applications in electronic equipment, including packaging structures, appear to fall within the laminar flow region. Ellison [1987] has found that the preceding expressions for natural air convection cooling are satisfactory for cabinet surfaces; however, they significantly underpredict the heat transfer from small surfaces. By curve fitting the empirical data under laminar conditions, the following formula for natural convection to air for small devices encountered in the electronics industry was found:

$$h_c = 0.83 f (\Delta T/L_{ch})^n (\text{W/m}^2 - \text{degree})$$

where $f = 1.22$ and $n = 0.35$ for vertical plate; $f = 1.00$ and $n = 0.33$ for horizontal plate facing upward, that is, upper surface $T_S > T_A$, or lower surface $T_S < T_A$; and $f = 0.50$ and $n = 0.33$ for horizontal plate facing downward, that is, lower surface $T_S > T_A$, or upper surface $T_S < T_A$.

2. For forced (air) convection (cooling) over external flat plates:

$$Nu = C(Re)^m (Pr)^n$$
$$h_c = (k/L_{ch}) C(Re)^m (Pr)^n = C''(L_{ch})^{m-1} v^m$$

where:

C, m, n = constants depending on the geometry and the Reynolds number, see Table 5.4
C'' = $kC(\rho/\mu)^m (Pr)^n$
L_{ch} = length of plate in the direction of fluid flow (characteristic length)

Experimental verification of these relations, as applied to the complexity of geometry encountered in electronic equipment and packaging systems, can be found in literature.

Ellison [1987] has determined that for *laminar forced airflow*, better agreement between experimental and calculated results is obtained if a correlation factor f, which depends predominantly on air velocity, is used, in which case the heat convection coefficient h_c becomes

$$h_c = f(k/L_{ch}) Nu$$

TABLE 5.4 Constants for Average Nusselt Numbers for Forced Convections[a], and Simplified Equations for Average Heat Transfer Coefficients h_c (W/m²-degree) for Forced Convection Air Cooling over External Flat Surfaces and at Most Practical Fluid Speeds (at Atmospheric Pressure)[b]

Flow Regime		C	m	n	h_c for Forced Convection Air Cooling		
					h_c	C''(27°C)	C''(75°C)
$Re < 2 \times 10^5$	laminar	0.664	0.5	0.333	$C''(v/L_{ch})^{0.5}$	3.87	3.86
$Re > 3 \times 10^5$	turbulent	0.036	0.8	0.333	$C''v^{0.8}/(L_{ch})^{0.2}$	5.77	5.34

[a] *Compilation based on four sources*: Keith, F. 1973. *Heat Transfer*. Harper & Row, New York. Rohsenow, W.M. and Choi, H. 1961. *Heat, Mass, and Momentum Transfer*. Prentice-Hall, Englewood Cliffs, NJ. Kraus, A.D. and Bar-Cohen, A. 1983. *Thermal Analysis and Control of Electronic Equipment*. Hemisphere, New York. Moffat, R.J. and Ortega, A. 1988. Direct air cooling in electronic systems. In *Advances in Thermal Modeling of Electronic Components and Systems*, ed. A. Bar-Cohen and A.D. Kraus, Vol. 1, pp. 129-282. Hemisphere, New York.

[b] Physical properties of air and their temperature dependence, with units converted to metric system, are depicted in Fig. 5.4. (*Source*: Keith, F. 1973. *Heat Transfer*, Harper & Row, New York.)

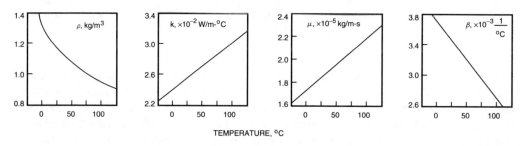

FIGURE 5.4 Temperature dependence of physical properties of air: density ρ, thermal conductivity k, dynamic viscosity μ, and CTE β. Note that specific heat c_p is almost constant for given temperatures, and as such is not displayed here.

Though the correlation factor f was determined experimentally for laminar flow over 2.54 × 2.54 cm ceramic substrates producing the following values: $f = 1.46$ for $v = 1$ m/s, $f = 1.56 - v = 2$ m/s, $f = 1.6 - v = 2.5$ m/s, $f = 1.7 - v = 5$ m/s, $f = 1.78 - v = 6$ m/s, $f = 1.9 - v = 8$ m/s, $f = 2.0 - v = 10$ m/s, and we expect the correlation factor to be somewhat different for other materials and other plate sizes, the quoted values are useful for purposes of estimation.

Buller and Kilburn [1981] performed experiments determining the heat transfer coefficients for laminar flow forced air cooling for integrated circuit packages mounted on printed wiring boards (thus for conditions differing from that of a flat plate), and correlated h_c with the air speed through use of the Colburn J factor, a dimensionless number, in the form of

$$h_c = J\rho c_p v(Pr)^{-2/3} = 0.387(k/L_{ch})Re^{0.54}Pr^{0.333}$$

where:

$J = 0.387*(Re)^{-0.46}$

L_{ch} = redefined characteristic length, allows to account for the three-dimensional nature of the package, $[(A_F/C_F)(A_T/L)]^{0.5}$

W, H = width and height of the frontal area, respectively

A_F = W, H, frontal area normal to air flow

C_F = $2(W + H)$, frontal circumference

A_T = $2H(W + L) + (W + L)$, total wetted surface area exposed to flow

L = length in the direction of flow.

Finally, Hannemann, Fox, and Mahalingham [1991] experimentally determined average heat transfer coefficient h_c for finned heat sink of the fin length $L(m)$, for forced convection in air under laminar conditions at moderate temperatures and presented it in the form of:

$$h_c = 4.37(v/L)^{0.5} (\text{W/m}^2 - \text{degree})$$

Radiation heat flow between two surfaces or between a surface and its surroundings is governed by the *Stefan–Boltzmann equation* providing the (nonlinear) **radiation heat transfer** in the form of

$$Q_r = \sigma A F_T (T_1^4 - T_2^4) = h_r A (T_1 - T_2)$$

where:

Q_r = radiation heat flow, W
σ = Stefan-Boltzmann constant, 5.67×10^{-8} W/m²-K⁴
A = effective area of the emitting surface, m²
F_T = exchange radiation factor describing the effect of geometry and surface properties
T_1 = absolute temperature of the external/emitting surface, K
T_2 = absolute temperature of the ambient/target, K
h_r = average **radiative heat transfer** coefficient, $\sigma F_T (T_1^4 - T_2^4)/(T_1 - T_2)$, W/m²-degree

For two-surface radiation exchange between plates, F_T is given by:

$$F_T = \cfrac{1}{\cfrac{1-\varepsilon_1}{\varepsilon_1} + \cfrac{1-\varepsilon_2}{\varepsilon_2}\cfrac{A_1}{A_2} + \cfrac{1}{F_{1-2}}}$$

where:

ε_1 = emissivity of material 1
ε_2 = emissivity of material 2
A_1 = radiation area of material 1
A_2 = radiation area of material 2
F_{1-2} = geometric view factor [Ellison 1987, Siegal and Howell 1981]

The emissivities of common packaging materials are given in Table 5.5. Several approximations are useful in dealing with radiation:

- For a surface that is smaller compared to a surface by which it is totally enclosed (e.g., by a room or cabinet), $\varepsilon_2 = 1$ (no surface reflection) and $F_{1-2} = 1$, then $F_T = \varepsilon_1$.
- For most packaging applications in a human environment, $T_1 \approx T_2$, then $h_r = 4\sigma F_T T_1^3$.
- For space applications, where the target is space with T_2 approaching 0 K, $Q_r = \sigma F_T T_1^4$.

For purposes of physical reasoning, convection and radiation heat flow can be viewed as being represented by (nonlinear) thermal resistances, *convective* θ_c and *radiational* θ_r, respectively,

$$\theta_{c,r} = \frac{1}{hA_S}$$

where h is either the convective or radiative heat transfer coefficients, h_c or h_r, respectively, the total heat transfer, both convective and radiative, $h = h_c + h_r$. Note that for nucleate boiling ηh_c should be used.

TABLE 5.5 Emissivities of Some Materials Used in Electronic Packaging

Aluminum, polished	0.039–0.057
Copper, polished	0.023–0.052
Stainless steel, polished	0.074
Steel, oxidized	0.80
Iron, polished	0.14–0.38
oxidized	0.31–0.61
Porcelain, glazed	0.92
Paint, flat black lacquer	0.96–0.98
Quartz, rough, fused	0.93–0.075

5.3 Study of Thermal Effects in Packaging Systems

The *thermal evaluation* of solid-state devices and integrated circuits (ICs), and VLSI-based packaging takes two forms: *theoretical analysis* and *experimental characterization*. Theoretical analysis utilizes various approaches: from simple to complex closed-form analytical solutions and numerical analysis techniques, or a combination of both. Experimental characterization of the device/chip junction/surface temperature(s) of packaged/unpackaged structures takes both direct, infrared microradiometry, or liquid crystals and thermographic phosphorous, or, to a lesser extent, thermocouples, and indirect (parametric) electrical measurements.

Thermal Resistance

A *figure-of-merit* of thermal performance, thermal resistance is a measure of the ability of its mechanical structure (material, package, and external connections) to provide heat removal from the active region and is used to calculate the device temperature for a given mounting arrangement and operating conditions to ensure that a maximum safe temperature is not exceeded [Baxter 1977]. It is defined as

$$\theta_{JR} = \frac{T_J - T_R}{P_D}$$

where:

θ_{JR} = thermal resistance between the junction or any other circuit element/active region that generates heat and the reference point, °C/W, K/W
T_J, T_R = temperature of the junction and the reference point, respectively, °C
P_D = power dissipation in the device, W

The conditions under which the device is thermally characterized have to be clearly described. In fact, thermal resistance is made up of constant terms that are related to device and package materials and geometry in series with a number of variable terms. These terms are related to the heat paths from the package boundary to some point in the system that serves as the reference temperature and are determined by the method of mounting, the printed-wiring-board (PWB) if used, other heat generating components on the board or in the vicinity, air flow patterns, and related considerations. For discrete devices, thermal parameters—**junction temperature**, thermal resistance, *thermal time constants*—are well defined since the region of heat dissipation and measurement is well defined. For integrated circuits, however, heat is generated at multiple points (resistors, diodes, or transistors) at or near the chip surface resulting in a nonuniform temperature distribution. Thus θ_{JR} can be misleading unless temperature uniformity is assured or thermal resistance is defined with respect to the hottest point on the surface of the chip. A similar situation exists with a multichip package, where separate thermal resistances should be defined for each die. As a result, the data on spatial temperature distribution are necessary, as well as the need to standardize the definition of a suitable reference temperature and the configuration of the test arrangement, including the means of mounting the device(s) under test.

Heat Management

In thermal evaluation of microelectronic packages/modules, the *overall* (*junction-to-coolant*) thermal resistance θ_{tot}, including forced convection and characterizing the requirements to cool a chip within a single- or multichip module, can be described in terms of the following:

- **Internal thermal resistance** θ_{int}, largely bulk conduction from the circuits to the case/package surface, thus also containing thermal spreading and contact resistances

$$\theta_{int} = \frac{T_{chip} - T_{case/pckg}}{P_{chip}}$$

Thermal contact resistance R_c, occurring at material interface, is dependent on a number of factors including the surface roughness, the properties of the interfacing materials, and the normal stress at the interface. The formula(s) relating the *interfacial normal pressure P* to R_c have been given [Yovanovich and Antonetti 1988] as

$$R_c = 1/C[P/H]^{0.95}$$

where C is a constant related to the surface texture, and the conductivities of the adjoining materials, k_1 and k_2; and H is the microhardness of the softer surface.

- **External thermal resistance** θ_{ext}, primarily convective, from the case/package surface to the coolant fluid (gas or liquid)

$$\theta_{ext} = \frac{T_{case/pckg} - T_{coolant-out}}{P_m}$$

- **Flow (thermal) resistance** θ_{flow}, referring to the transfer of heat from the fluid coolant to the ultimate heat sink, thus associated with the heating of the fluid as it absorbs energy passing through the heat exchanger

$$\theta_{flow} = \frac{T_{coolant-out} - T_{coolant-in}}{Q_{flow}}$$

thus expressing the device junction/component temperature as follows:

$$T_j = \Delta T_{j-chip} + P_{chip}\theta_{int} + P_m\theta_{ext} + Q_{flow}\theta_{flow} + T_{coolant-in}$$

where:

T_{chip} = the chip temperature
$T_{case/pckg}$ = the case/package temperature (at a defined location)
P_{chip}, P_m = the power dissipated within the chip and the module, respectively
ΔT_{j-chip} = the junction to chip temperature rise (can be negligible if power levels are low)

Note that in the literature, the total thermal resistance θ_{tot}, usually under natural convection cooling conditions, is also referred to as θ_{j-a} (junction-ambient), θ_{int} as θ_{j-c} (junction–case), and θ_{ext} as θ_{c-a} (case–ambient).

Thermal Modeling/Simulation

The problem of theoretical thermal description of solid-state devices, circuits, packages, and assemblies requires the solution of heat transfer equations with appropriate initial and boundary conditions and is solved with the level of sophistication ranging from approximate analytical through full numerical of increased sophistication and complexity. To fully theoretically describe the thermal state of a solid-state/VLSI device and, in particular, the technology requirements for large-area packaged chips, thus ensuring optimum design and performance of VLSI-based packaging structures, thermal modeling/simulation has to be coupled with thermal/mechanical modeling/simulation leading to analysis of thermal stress and tensions, material fatigue, reliability studies, **thermal mismatch**, etc., and to designs both low thermal resistance and low thermal stress coexist.

Development of a suitable calculational model is basically a trial-and-error process. In problems involving complex geometries, it is desirable to limit the calculation to the smallest possible region, by eliminating areas weakly connected to the region of interest, or model only part of the structure if geometric symmetry, symmetrical loading, symmetric power/temperature distribution, isotropic material properties, and other factors can be applied. In addition, it is often possible to simplify the geometry considerably by replacing complex shapes with simpler shapes (e.g., of equivalent volume and resistance to heat flow). All of these reduce the physical structure to a numerically (or/and analytically) tractable model. Also, decoupling of the component level problem from the substrate problem, as the chip and the package can be assumed to respond thermally as separate components, might be a viable concept in evaluation of temperature fields. This is true if thermal time constants differ by orders of magnitude. Then the superposition can be used to obtain a temperature response of the total package.

The accuracy with which any given problem can be solved depends on how well the problem can be modeled, and in the case of *finite-difference* (with electrical equivalents) or *finite-element models*, which are the most common approaches to thermal (and thermal–mechanical) analysis of packaging systems, on the fineness of the spatial subdivisions and control parameters used in the computations. To improve computational efficiency and provide accurate results in local areas of complexity some type of *grid continuation strategy* should be used. A straightforward use of uniform discretization meshes seems inappropriate, since a mesh sufficiently fine to achieve the desired accuracy in small regions, where the solution is rough, introduces many unnecessary unknowns in regions where the solution is smooth, thus unnecessarily increasing storage and CPU time requirements. A more sophisticated approach involves starting with coarse grids with relatively few unknowns, inexpensive computational costs, and low accuracy, and using locally finer grids on which accurate solutions may be computed. Coarse grids are used to obtain good starting iterates for the finer meshes and can also be used in the creation of the finer meshes by indicating regions where refinement is necessary.

Electrical network finite-difference models for study of various phenomena occurring in solid-state devices, circuits, and systems have been widely reported in literature [Ellison 1987, Fukuoka and Ishizuka 1984, Riemer 1990]. One of the advantages of such a technique is a simple physical interpretation of the phenomena in question in terms of electrical signals and parameters existing in the network/circuit model (see Fig. 5.5). For all but very simple cases, the equivalent circuits are sufficiently complex that computer solution is required. It is important to note, however, that once the equivalent circuit is established the analysis can readily be accomplished by existing network analysis programs, such as SPICE [SPICE2G User's Manual].

Finite element models (FEMs) have been gaining recognition as tools to carry out analyses of VLSI packages due to the versatility of FEM procedures for the study of a variety of electronic packaging problems, including wafer manufacturing, chip packaging, connectors, board-level simulation, and system-level simulation. Thermal/mechanical design of VLSI packages [Miyake, Suzuki, and Yamamoto 1985] requires an effective method for the study of specific regions of complexity (e.g., sharp corners, material discontinuity, thin material layers, and voids). As fully refined models are too costly, especially in complex two- and three-dimentional configurations, *local analysis procedures* [Simon et al. 1988] can be used for the problems of this kind. The local analysis approach to finite element analysis allows a very refined mesh to be introduced in an overall coarse FEM so that geometric and/or material complexities can be accurately modeled. This method allows efficient connection of two bordering lines with differing nodal densities in order to form an effective transition from coarse to refined meshes. Linear (or) arbitrary constraints are imposed on the bordering lines in order to connect the lines and reduce the number of independent freedoms in the model. Refined mesh regions can be nested one within another as multilevel refined meshes to achieve greater accuracy in a single FEM, as illustrated in Fig. 5.6.

Heat Management

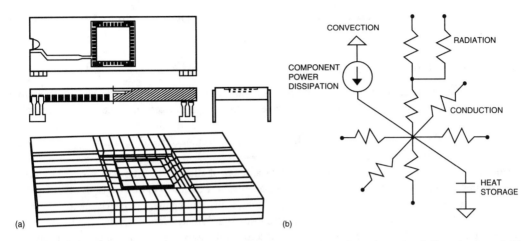

FIGURE 5.5 (a) Views of a ceramic package and its general mesh in three-dimensional discretization of this rectangular type packaging structure. (b) A lumped RC thermal model of a unit cell/volume with (nonlinear) thermal resistances representing conduction, convection, and radiation; capacitances account for time-dependent thermal phenomena, and current sources represent heat generation/dissipation properties (thermal current). Nodal points, with voltages (and voltage sources) corresponding to temperatures, can be located either in the center or at corners of unit cells. Each node may be assigned an initial (given) temperature, heat generation/dissipation rate, and modes of heat transport in relation to other nodes. Voltage controlled current sources could be used to replace resistances representing convection and radiation.

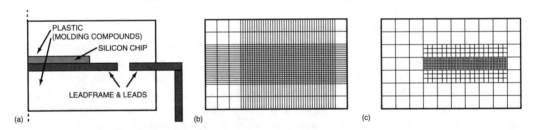

FIGURE 5.6 Example of grid continuation strategy used in discretization of a packaging structure: (a) a two-dimensional cross-section of a plastic package, (b) fine refinement in the area of interest, that is, where gradients of the investigated parameter(s), thermal or thermal-mechanical, are expected to be most significant, (c) a two-dimensional two-level refinement.

Several analytical and numerical computer codes varying widely in complexity and flexibility are available for use in the thermal modeling of VLSI packaging. The programs fall into the two following categories:

1. *Thermal characterization* including, analytical programs such as TXYZ [Albers 1984] and TAMS [Ellison 1987], as well as numerical analysis programs based on a finite-difference (with electrical equivalents) approach such as TRUMP (Edwards 1969], TNETFA [Ellison 1987], SINDA [COSMIC 1982b].
2. *General nonlinear codes*, such as ANSYS [Swanson] ADINAT [Adina], NASTRAN [COSMIC 1982a], NISA [COSMIC 1982c], and GIFTS [Casa] (all based on finite-element approach), to name only a few.

Because of the interdisciplinary nature of electronic packaging engineering, and the variety of backgrounds of packaging designers, efforts have been made to create interface programs optimized to address specific design and research problems of VLSI-packaging structures that would permit inexperienced users to generate and solve complex packaging problems [Shiang, Staszak, and Prince 1987; Nalbandian 1987; Godfrey et al. 1993].

Experimental Characterization

Highly accurate and versatile thermal measurements, in addition to modeling efforts, are indispensable to ensure optimal thermal design and performance of solid-state devices and VLSI-based packaging structures. The need for a systematic analysis of flexibility, sensibility, and reproducibility of current methods and development of appropriate method(s) for thermal characterization of packages and assemblies is widely recognized [Oettinger and Blackburn 1990].

The basic problem of measurement of thermal parameters of all solid-state devices and VLSI-chips is the measurement of temperatures of the active components, for example p–n junctions, or the integrated circuit chip surface temperature. Nonelectrical techniques, which can be used to determine the operating temperature of structures, involve the use of infrared microradiometry, liquid crystals, and other tools, and require that the surface of the operating device chip is directly accessible. The electrical techniques for measuring the temperature of semiconductor chips can be performed on fully packaged devices, and use a *temperature sensitive electrical parameter* (TSEP) of the device, which can be characterized/calibrated with respect to temperature and subsequently used as the temperature indicator.

Methodology for experimental evaluation of transient and steady-state thermal characteristics of solid-state and VLSI-based devices and packaging structures is dependent on the device/structure normal operating conditions. In addition, validity of the data obtained has to be unchallengeable; each thermal measurement is a record of a unique set of parameters, so as to assure repeatability, everything must be the same each time, otherwise the data gathered will have little relevance. Also, the choice of thermal environmental control and mounting arrangement, still air, fluid bath, temperature controlled heat sink, or wind tunnel, must be addressed.

In general, the idea of thermal measurements of all devices is to measure the active region temperature rise $\Delta T(t)$ over the reference (case, or **ambient**) **temperature** T_R, caused by the power P_D being dissipated there, for example, during the bias pulse width. Expanding the time scale, steady-state values can be obtained. Employing electrical methods in thermal measurements, the TSEP that is to be used should be consistent with the applied heating method, whereas the measurements have to be repeatable and meaningful. Most commonly used TSEPs are p–n junction forward voltage or emitter-base voltage (bipolar devices), threshold voltage or intrinsic drain source diode voltage [metal oxide semiconductors (MOSs)]. There are two phases in the experimental evaluation process:

- *Calibration* of the TSEP (no power or negligibly low power dissipated in the device) characterizing the variation of TSEP with temperature.
- *Actual measurement* (device self-heated).

The TSEP is monitored/recorded during or immediately after the removal of a **power** (heating) **dissipation** operating condition, thus determining the increase (change) in the junction temperature in the device under test (DUT). The process of characterization/calibration is carried out in a temperature controlled measurement head (chamber, oven) and should be performed under thermal equilibrium conditions, when the equivalence of the structure active region temperature with the structure case temperature is assured. TSEP can exhibit deviation from its normally linear slope, thus the actual calibration curve should be plotted or fitting techniques applied. Also, location of the temperature sensor and the way it is attached or built-in affects the results.

Figure 5.7 illustrates the idea of the emitter-only-switching method [Oettinger and Blackburn 1990], which is a popular and fairly accurate approach for thermal characterization of transistors. Its principle can also be used in thermal measurement of other device/circuits as long as there is an accessible p–n junction that can be used as a TSEP device. This method is intended both for steady-state and transient measurements. The circuit consists of two current sources of different values, heating current i_H and measurement current i_M, and a diode D, and a switch. When the switch is open, the current flowing

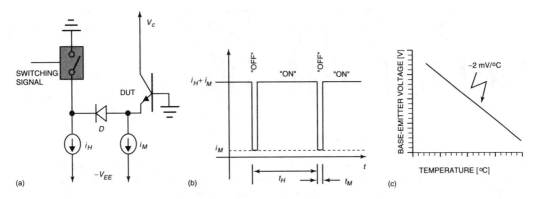

FIGURE 5.7 (a) Schematic of measurement circuit using the emitter-switching-only method for measuring temperature of an *npn* bipolar transistor, (b) heating and measurement currents waveforms; heating time t_H, measurement time t_M, (c) calibration plot of the TSEP: base-emitter voltage vs temperature.

through the device-under-test is equal to $i_H + i_M$ (power dissipation phase). When the switch is closed, the diode is reverse biased, and the current of the DUT is equal to i_M (measurement phase); i_M should be small enough to keep power dissipation in the DUT during this phase negligible.

An approach that has been found effective in experimental evaluation of packaged IC chips employs the use of specially designed test structures that simulate the materials and processing of real chips, but not the logical functions. These look like product chips in fabrication and in assembly and give (electrical) readouts of conditions inside the package [Oettinger 1984]. Any fabrication process is possible that provides high fabrication yield. Such devices may be mounted to a wide variety of test vehicles by many means. In general, they can be used to measure various package parameters—thermal, mechanical, and electrical—and may be thought of as a process monitor since the obtained data reflects the die attach influence or lead frame material influence, or any number of factors and combination of factors.

For thermal measurements, the use of these structures overcomes inherent difficulties in measuring temperatures of active integrated chips. Many IC chips have some type of accessible TSEP element (*p–n* junction), parasitic, input protection, base-emitter junction, output clamp, or isolation diode (substrate diode). Such TSEP devices, as in discrete devices, have to be switched from the operating (heating) mode to measurement (sensing) mode, resulting in unavoidable delays and temperature drops due to cooling processes. Also, the region of direct electrical access to the TSEP does not always agree with the region of highest temperature.

In thermal test chips, heating and sensing elements might be (or are) separated. This eliminates the necessity of electrical switching, and allows measurements to be taken with the DUT under continuous biasing. Also, heating and sensing devices can be placed on the chip in a prescribed fashion, and access to temperature sensing devices can be provided over all of the chip area resulting in variable heat dissipation and sensed temperature distribution. This provides thermal information from all areas of the chip, including the ones of greatest **thermal gradients**. In addition, arrays of test chips can be built up from a standard chip size. It is desirable to use a die of the same or nearly same size as the production or proposed die for thermal testing and to generate a family of test dies containing the same test configuration on different die sizes. The particular design of the (thermal) test chip depends on the projected application, for example, chip die attachment, or void study, or package characterization. Examples of pictorial representations of thermal test chips are given in Fig. 5.8.

More sophisticated, multisensor test chips contain test structures for the evaluation of thermal resistance, electrical performance, corrosion, and thermomechanical stress effects of packaging technologies [Mathuna 1992]. However, multisensor chips have to be driven by computerized test equipment [Alderman, Tustaniwskyi, and Usell 1986; Boudreaux et al. 1991] in order to power the chip in a manner that permits spatially uniform/nonuniform power dissipation and to sense the test chip temperature at multiple space points, all as a function of time, and to record temperatures associated with separate

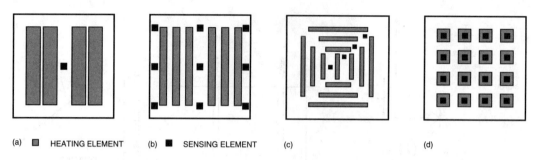

(a) ▫ HEATING ELEMENT   (b) ▪ SENSING ELEMENT   (c)   (d)

FIGURE 5.8 Pictorial representations of examples of thermal test chips: (a) diffused resistors as heating elements and a single *p-n* junction (transistor, diode) as a sensing element [Oettinger 1984], (b) diffused resistors as heating elements and an array of sensings elements—diodes [Alderman, Tustaniwskyi, and Usell 1986], (c) polysilicon or implant resistor network as heating elements and diodes as sensing elements [Boudreaux et al. 1991], (d) an array of heaters and sensors; each heater/sensor arrangement consists of two (bipolar) transistors having common collector and base terminals, but separate emitters for heating and sensing. After [Staszak et al. 1987].

thermal time constants of various parts of the tested structure, ranging from tens and hundreds of nanoseconds on the transistor level, tens and hundreds of microsecond for the chip itself, tens to hundreds of microsecond for the substrate, and seconds and minutes for the package.

5.4 Heat Removal/Cooling Techniques in Design of Packaging Systems

Solutions for adequate cooling schemes and optimum thermal design require an effective combination of expertise in electrical, thermal, mechanical, and materials effect, as well as in industrial and manufacturing engineering. Optimization of packaging is a complex issue: compromises between electrical functions, an optimal heat flow and heat related phenomena, and economical manufacturability have to be made, leading to packaging structures that preserve the performance of semiconductors with minimum package delay to the system. Figure 5.9 shows the operating ranges of a variety of heat removal/cooling techniques. The data is presented in terms of heat transfer coefficients that correlate dissipated power or heat flux with a corresponding increase in temperature.

Current single-chip packages, dual in line package (DIP) (cerdip), pin grid array (PGA) (ceramic, TAB, short pin, pad-grid array), and flat packages [fine chip ceramic, quad flat package (QFP)], surface-mounted onto cards or board and enhanced by heat sinks, are capable of dissipating not more than a few watts per square centimeter, thus are cooled by natural and forced-air cooling. Multichip packaging, however, demands power densities beyond the capabilities provided by air cooling. The state-of-the-art of multichip packaging, multilayer ceramic (cofired multilayer thick-film ceramic substrate, glass-ceramics/copper substrate, polyimide-ceramics substrate), thin-film copper/aluminium-polyimide substrates, chip on board with TAB, and silicon substrate (silicon-on-silicon), require that in order to comply with the preferential increase in temperature not exceeding, for example, 50°C, sophisticated cooling techniques are indispensable. Traditional liquid forced cooling that employs a cold plate or heat exchanger, which physically contacts the device to be cooled, is good for power densities below 20 W/cm^2. For greater power densities, a possible approach is a water jet impingement to generate a fluid interface with a high rate of heat transfer [Mahalingham 1985]. In most severe cases, forced liquid cooling (and boiling) is often the only practical cooling approach: either indirect water cooling (microchannel cooling [Tuckerman and Pease 1981]) or immersion cooling [Yokouchi, Kamehara, and Niwa 1987] (in dielectric liquids like fluorocarbon). Cryogenics operation of CMOS devices in liquid nitrogen is a viable option [Clark et al. 1992].

A major factor in the cooling technology is the quality of the substrate, which also affects the electrical performance. All of the substrate materials (e.g., alumina/polyimide, epoxy/kevlar, epoxy-glass, beryllia,

Heat Management

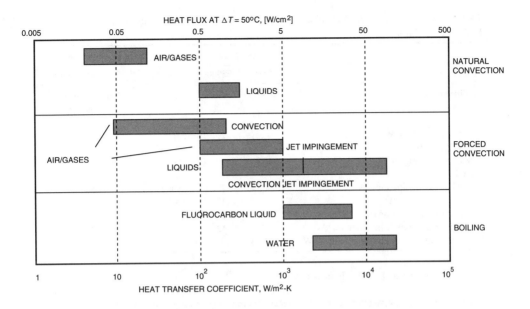

FIGURE 5.9 Heat transfer coefficient for various heat removal/cooling techniques.

alumina, aluminum nitride, glass-ceramics, and, recently, diamond or diamondlike films [Tzeng et al. 1991]) have the clear disadvantage of a relative dielectric constant significantly greater than unity. This results in degraded chip-to-chip propagation times and in enhanced line-to-line capacitive coupling. Moreover, since chip integration will continue to require more power per chip and will use large-area chips, the substrate must provide for a superior thermal path for heat removal from the chip by employing high **thermal conductivity** materials as well as assuring thermal expansion compatibility with the chip. Aluminum nitride and silicon carbide ceramic with good thermal conductivity ($k = 100-200$ W/m-°C and 260 W/m-°C, respectively) and good thermal expansion match [**cofficient of thermal expansion (CTE)** $= 4 \times 10^{-6}/°C$ and $2.6 \times 10^{-6}/°C$, respectively] to the silicon substrate (CTE $= 2.6 \times 10^{-6}/°C$), copper-invar-copper ($k = 188$ W/m-°C and thermal expansion made to approximate that of silicon), and diamond for substrates and heat spreaders ($k = 1000-2000$ W/m-°C) are prime examples of the evolving trends. However, in cases where external cooling is provided not through the substrate base but by means of heat sinks bonded on the back side of the chip or package, high conductivity substrates are not required, and materials such as alumina ($k = 30$ W/m-°C and CTE $= 6 \times 10^{-6}/°C$) or glass-ceramics ($k = 5$ W/m-°C and a tailorable CTE $= 4 - 8 \times 10^{-6}/°C$, matching that required by silicon or GaAs, CTE $= 5.9 \times 10^{-6}/°C$), with low dielectric constants for low propagation delays, are utilized. Designers should also remember thermal enhancement techniques for minimizing the thermal contact resistance (including greases, metallic foils and screens, composite materials, and surface treatment [Fletcher 1990]).

The evolution of packaging and interconnects over the last decade comprises solutions ranging from air-cooled single-chip packages on a printed wiring board, custom or semicustom chips and new materials and technologies, that is, from surface mount packages through silicon-on-silicon (also known as wafer scale integration) and chip-on-board packaging techniques, to perfected multichip module technology with modules operating in liquid-cooled systems. These include such complex designs as IBM's **thermal conduction module** (**TCM**) [Blodgett 1993] and its more advanced version, IBM ES/9000 system [Tummala and Ahmed 1992], the NEC SX **liquid cooling modules** (**LCM**) [Watari and Murano 1995, Murano and Watari 1992], or the liquid-nitrogen-cooled ETA-10 supercomputer system [Carlson et al. 1989]. Table 5.6 shows comparative characteristics of cooling designs for those high-performance mainframe multichip modules.

TABLE 5.6 Characteristics of Cooling Design for High-Performance Mainframe Multichip Modules

System	IBM TCM		NEC SX-3 LCM	ETA-10 Supercomputer
	3081	ES9000		
Cooling system	water	water	liquid	liquid-nitrogen nucleate boiling
Chip:				
Power dissipation, W	4	25	33/38	
Power density, W/cm^2	18.5	60	17/19.5	2[a]
Max. no. of chips per substrate	100	121	100	
Module:				
Total power dissipation, W	300	2000	3800	500[b]
Power density, W/m^2	4.2	17	7.9	
No. of modules per board	9	6		2[c]
Max. power dissipation in a system, kW	2.7	12		4
Thermal resistance junction-coolant, °C/W	11.25		0.6	2

[a] Nucleate boiling peak heat flux limits ~12 W/cm^2.
[b] Per board.
[c] Boards per cryostat; cryogenerator for the total system can support up to four cryostats.

5.5 Concluding Remarks

Active research areas in heat management of electronic packaging include all levels of research from the chip through the complete system. The trend toward increasing packing densities, large-area chips, and increased power densities, imply further serious heat removal and heat-related considerations. These problems can only be partly handled by brute-force techniques, coupled with evolutionary improvements in existing packaging technologies. However, requirements to satisfy the demands of high performance functional density, speed, power/heat dissipation, and levels of customization give rise to the need for exploring new approaches to the problem and for refining and better understanding existing approaches.

Techniques employing forced air and liquid convection cooling through various advanced schemes of cooling systems, including direct cooling of high-power dissipation chips (such as immersion cooling that allow also for uniformity of heat transfer coefficients and coolant temperature throughout the system), very fine coolant channels yielding an extremely low thermal resistance, and nucleate boiling heat transfer as well as cryogenic operation of CMOS devices, demand subsequent innovations in the format of packages and/or associated components/structures in order to accommodate such technologies.

Moreover, the ability to verify packaging designs, and to optimize materials and geometries before hardware is available, requires further developments of models and empirical tools for thermal design of packaging structures and systems. Theoretical models have to be judiciously combined with experimentally determined test structures for thermal (and thermal stress) measurements, creating conditions for equivalence of theoretical and empirical models. This in turn will allow generation of extensive heat transfer data for a variety of packaging structures under various conditions, thus translating real-world designs into modeling conditions.

Nomenclature

A	=	area, m^2
A_{eff}	=	effective cross-sectional area for heat flow, m^2
A_s	=	cross-sectional area for heat flow through the surface, m^2
A_i	=	cross-sectional area of the material of the ith layer, m^2
C	=	capacitance, F; constant

Heat Management

C_θ	=	thermal capacitance, W-s/degree
C_{sf}	=	constant, a function of the surface/fluid combination in the boiling heat transfer, W/m²-K^n
c_P	=	specific heat, W-s/g-degree
E_A	=	activation energy, eV
F_T	=	total radiation factor; exchange factor
F_{1-2}	=	radiation geometric view factor
f	=	constant
G	=	volumetric flow rate, m³/s
Gr	=	Grashof number $(g\beta\rho^2/\mu^2)(L_{ch})^3 \Delta T$, dimensionless
g	=	gravitational constant, m/s²
H	=	height (of plate), m
h	=	heat transfer coefficient, W/m²-degree
h_b	=	boiling heat transfer coefficient, W/m²-degree
h_c	=	average convective heat transfer coefficient, W/m²-degree
h_r	=	average radiative heat transfer coefficient W/m²-degree
I, i	=	current, A
i_M	=	measurement current, A
i_H	=	heating currents, A
J	=	Colburn factor, dimensionless
k	=	thermal conductivity, W/m²-degree
k_i	=	thermal conductivity of the material of the ith layer, W/m²-degree
k	=	Boltzmann's constant, 8.616×10^{-5} eV/K
L	=	length (of plate, fin), m
L_{ch}	=	characteristic length parameter, m
m	=	mass flow, g/s; constant, $\sqrt{hc/k\delta}$
n	=	constant; coefficient
Nu	=	Nusselt number, dimensionless
P_{chip}	=	power dissipated within the chip, W
P_D	=	power dissipation, W
P_m	=	power dissipated within the module, W
Pr	=	Prandtl number, dimensionless, $\mu c_p/k$
Q	=	heat flow, W
Q_b	=	boiling heat transfer rate, W
Q_C	=	convection heat flow, W
Q_f	=	heat transfer rate by a fin/fin structure, W
Q_{flow}	=	total heat flow/dissipation of all components within the enclosure upstream of the component of interest, W
Q_r	=	radiation heat flow, W
Q_V	=	volumetric heat source generation, W/m³
q	=	heat flux, W/m²
R	=	resistance, Ω; radius, m
R_c	=	thermal contact resistance, °C/W, K/W
Ra	=	Raleigh number, $GrPr$, dimensionless
Re	=	Reynolds number, $vL_{ch}(\rho/\mu)$, dimensionless
T	=	temperature, °C, K
T_A	=	ambient temperature, °C, K
T_b	=	temperature of the fin base, °C, K
$T_{case/pckg}$	=	temperature of the case/package at a defined location, °C, K
T_{chip}	=	temperature of the chip, °C, K
$T_{coolant\text{-}in/out}$	=	temperature of the coolant inlet/outlet, °C, K

T_f	=	temperature of the fluid coolant, °C, K
T_J	=	temperature of the junction, °C, K
T_R	=	reference temperature, °C, K
T_S	=	temperature of the surface, °C, K
T_{sat}	=	temperature of the boiling point saturation of the liquid, °C, K
t	=	time, s
t_M	=	measurement time, s
t_H	=	heating time, s
V	=	voltage, V; volume, m³
v	=	fluid velocity, m/s
W	=	width (of plate), m
β	=	coefficient of thermal expansion, 1/°C
Δl_i	=	thickness of the ith layer, m
ΔT	=	temperature difference, °C, K
$\Delta T_{j\text{-chip}}$	=	temperature rise between the junction and the chip, °C, K
Δt	=	time increment, s
δ	=	fin thickness, m
ε	=	emissivity
η	=	fin efficiency
θ	=	thermal resistance, °C/W, K/W
δ	=	fin thickness, m
$\theta_{\text{ext/int}}$	=	external/internal thermal resistance, °C/W, K/W
θ_{JR}	=	thermal resistance between the junction or any other circuit element/active region that generates heat and the reference point, °C/W, K/W
θ_{flow}	=	flow (thermal) resistance, °C/W, K/W
θ_{sp}	=	thermal spreading resistance, °C/W, K/W
λ	=	failure rate
μ	=	dynamic viscosity, g/m-s
ρ	=	density, g/m³
σ	=	electrical conductivity, $1/\Omega$ m; Stefan–Boltzmann constant, 5.673×10^{-8} W/m²-K⁴

Defining Terms[1]

Ambient temperature: Temperature of atmosphere in intimate contact with the electrical part/device.

Boiling heat transfer: A liquid-to-vapor phase change at a heated surface, categorized as either *pool boiling* (occurring in a stagnant liquid) or *flow boiling*.

Coefficient of thermal expansion (CTE): The ratio of the change in length dimensions to the change in temperature per unit starting length.

Conduction heat transfer: Thermal transmission to heat energy from a hotter region to a cooler region in a conducting medium.

Convection heat transfer: Transmission of thermal energy from a hotter to a cooler region through a moving medium (gas or liquid).

External thermal resistance: A term used to represent thermal resistance from a convenient point on the outside surface of an electronic package to an ambient reference point.

Failure, thermal: The temporary or permanent impairment of device or system functions caused by thermal disturbance or damage.

---

[1]Based in part on *Hybrid Microcircuit Design Guide* published by ISHM and IPC, and Tummala, R.R. and Rymaszewski, E.J. 1989. *Microelectronics Packaging Handbook.* Van Nostrand Reinhold, New York.

Failure rate: The rate at which devices from a given population can be expected or were found to fall as a function of time; expressed in FITs, where FIT is defined as one failure per billion device hours (10^{-9}/h or 0.0001%/1000 h of operation).

Flow regime, laminar: A constant and directional flow of fluid across a clean workbench. The flow is usually parallel to the surface of the bench.

Flow regime, turbulent: Flow where fluid particles are disturbed and fluctuate.

Heat flux: The outward flow of heat energy from a heat source across or through a surface.

Heat sink: The supporting member to which electronic components, or their substrate, or their package bottom is attached. This is usually a heat conductive metal with the ability to rapidly transmit heat from the generating source (component).

Heat transfer coefficient: Thermal parameter that encompasses all of the complex effects occurring in the convection heat transfer mode, including the properties of the fluid gas/liquid, the nature of the fluid motion, and the geometry of the structure.

Internal thermal resistance: A term used to represent thermal resistance from the junction or any heat generating element of a device, inside an electronic package, to a convenient point on the outside surface of the package.

Junction temperature: The temperature of the region of transistor between the p and n type semiconductor material in a transistor or diode element.

Mean time to failure (MTTF): A term used to express the reliability level. It is the arithmetic average of the lengths of time to failure registered for parts or devices of the same type, operated as a group under identical conditions. The reciprocal of the failure rate.

Power dissipation: The dispersion of the heat generated from a film circuit when a current flows through it.

Radiation heat transfer: The combined process of emission, transmission, and absorption of thermal energy between bodies separated by empty space.

Specific heat: The quantity of heat required to raise the temperature of 1 g of a substance 1°C.

Temperature cycling: An environmental test where the (film) circuit is subjected to several temperature changes from a low temperature to a high temperature over a period of time.

Thermal conductivity: The rate with which a material is capable of transferring a given amount of heat through itself.

Thermal design: The schematic heat flow path for power dissipation from within a (film) circuit to a heat sink.

Thermal gradient: The plot of temperature variances across the surface or the bulk thickness of a material being heated.

Thermal management or control: The process or processes by which the temperature of a specified component or system is maintained at the desired level.

Thermal mismatch: Differences of coefficients of thermal expansion of materials that are bonded together.

Thermal network: Representation of a thermal space by a collection of conveniently divided smaller parts, each representing the thermal property of its own part and connected to others in a prescribed manner so as not to violate the thermal property of the total system.

Thermal resistance: A thermal characteristic of a heat flow path, establishing the temperature drop required to transport heat across the specified segment or surface; analogous to electrical resistance.

References

Adina. *ADINAT*, User's Manual. Adina Engineering, Inc., Watertown, MA.

Albers, J. 1984. *TXYZ: A Program for Semiconductor IC Thermal Analysis*. NBS Special Pub. 400-76, Washington, DC.

Alderman, J., Tustaniwskyi, J., and Usell, R. 1986. Die attach evaluation using test chips containing localized temperature measurement diodes. *IEEE Trans. Comp., Hybrids, and Manuf. Technol.* 9(4):410–415.

Antonetti, V.W., Oktay, S., and Simons, R.E. 1989. Heat transfer in electronic packages. In *Microelectronics Packaging Handbook*, eds. R.R. Tummala and E.J. Rymaszewski, pp. 167–223. Van Nostrand Reinhold, New York.

Bar-Cohen, A. 1987. Thermal management of air- and liquid-cooled multichip modules. *IEEE Trans. Comp., Hybrids, and Manuf. Technol.* 10(2):159–175.

Bar-Cohen, A. 1993. Thermal management of electronics. In *The Engineering Handbook*, ed. R.C. Dorf, pp. 784-797. CRC Press, Boca Raton, FL.

Baxter, G.K. 1977. A recommendation of thermal measurement techniques for IC chips and packages. In *Proceedings of the 15th IEEE Annual Reliability Physics Symposium*, pp. 204–211, Las Vegas, NV. IEEE Inc.

Blodgett, A.J. 1983. Microelectronic packaging. *Scientific American* 249(1):86–96.

Boudreaux, P.J., Conner, Z., Culhane, A., and Leyendecker, A.J. 1991. Thermal benefits of diamond inserts and diamond-coated substrates to IC packages. In *1991 Gov't Microcircuit Applications Conf. Digest of Papers*, pp. 251–256.

Buller, M.L. and Kilburn, R.F. 1981. Evaluation of surface heat transfer coefficients for electronic modular packages. In *Heat Transfer in Electronic Equipment*, HTD, Vol. 20, pp. 25–28. ASME, New York.

Carlson, D.M., Sullivan, D.C., Bach, R.E., and Resnick, D.R. 1989. The ETA 10 liquid-cooled supercomputer system. *IEEE Trans. Electron Devices* 36(8):1404–1413.

Carslaw, H.S. and Jaeger, J.C. 1967. *Conduction of Heat in Solids*. Oxford Univ. Press, Oxford, UK.

Casa. *GIFTS*, User's Reference Manual. Casa Gifts, Inc., Tucson, AZ.

Clark, W.F., El-Kareh, B., Pires, R.G., Ticomb, S.L., and Anderson, R.L. 1992. Low temperature CMOS—A brief review. *IEEE Trans. Comp., Hybrids, and Manuf. Technol.* 15(3):397–404.

COSMIC. 1982a. *NASTRAN Thermal Analyzer*, COSMIC Program Abstracts No. GSC-12162. Univ. of Georgia, Athens, GA.

COSMIC. 1982b. *SINDA—Systems Improved Numerical Differencing Analyzer*, COSMIC Prog. Abst. No. GSC-12671. Univ. of Georgia, Athens, GA.

COSMIC. 1982c. *NISA: Numerically Integrated Elements for Systems Analysis*, COSMIC Prog. Abst. Univ. of Georgia, Athens, GA.

Davidson, E.E. and Katopis, G.A. 1989. Package electrical design. In *Microelectronics Packaging Handbook*, ed. R.R. Tummala and E.J. Rymaszewski, pp. 111–166. Van Nostrand Reinhold, New York.

Edwards, A.L. 1969. *TRUMP: A Computer Program for Transient and Steady State Temperature Distribution in Multidimensional Systems; User's Manual*, Rev. II. Univ. of California, LRL, Livermore, CA.

Ellison, G.N. 1987. *Thermal Computations for Electronic Equipment*. Van Nostrand Reinhold, New York.

Fletcher, L.S. 1990. A review of thermal enhancement techniques for electronic systems. *IEEE Trans. Comp., Hybrids, and Manuf. Technol.* 13(4):1012–1021.

Fukuoka, Y. and Ishizuka, M. 1984. Transient temperature rise for multichip packages. In *Thermal Management Concepts in Microelectronic Packaging*, pp. 313–334. ISHM Technical Monograph Series 6984-003, Silver Spring, MD.

Furkay, S.S. 1984. Convective heat transfer in electronic equipment: An overview. In *Thermal Management Concepts in Microelectronic Packaging*, pp. 153–179. ISHM Technical Monograph Series 6984-003, Silver Spring, MD.

Godfrey, W.M., Tagavi, K.A., Cremers, C.J., and Menguc, M.P. 1993. Interactive thermal modeling of electronic circuit boards. *IEEE Trans. Comp., Hybrids, Manuf. Technol.* CHMT-16(8)978–985.

Hagge, J.K. 1992. State-of-the-art multichip modules for avionics. *IEEE Trans. Comp., Hybrids, and Manuf. Technol.* 15(1):29–41.

Hannemann, R., Fox, L.R., and Mahalingham, M. 1991. Thermal design for microelectronic components. In *Cooling Techniques for Components*, ed. W. Aung, pp. 245–276. Hemisphere, New York.

ISHM. 1984. *Thermal Management Concepts in Microelectronic Packaging*, ISHM Technical Monograph Series 6984-003. International Society for Hybrid Microelectronics, Silver Spring, MD.

Jeannotte, D.A., Goldmann, L.S., and Howard, R.T. 1989. Package reliability. In *Microelectronics Packaging Handbook*, ed. R.R. Tummala and E.J. Rymaszewski, pp. 225–359. Van Nostrand Reinhold, New York.

Keith, F. 1973. *Heat Transfer*. Harper & Row, New York.

Kennedy, D.P. 1960. Spreading resistance in cylindrical semiconductor devices. *J. App. Phys.* 31(8):1490–1497.

Kern, D.Q. and Kraus, A.D. 1972. *Extended Surface Heat Transfer*. McGraw-Hill, New York.

Krus, A.D. and Bar-Cohen, A. 1983. *Thermal Analysis and Control of Electronic Equipment*. Hemisphere, New York.

Mahalingham, M. 1985. Thermal management in semiconductor device packaging. *Proceedings of the IEEE* 73(9):1396–1404.

Mathuna, S.C.O. 1992. Development of analysis of an automated test system for the thermal characterization of IC packaging technologies. *IEEE Trans. Comp., Hybrids, and Manuf. Technol.* 15(5):615–624.

McAdams, W.H. 1954. *Heat Transmission*. McGraw-Hill, New York.

Miyake, K., Suzuki, H., and Yamamoto, S. 1985. Heat transfer and thermal stress analysis of plastic-encapsulated ICs. *IEEE Trans. Reliability* R-34(5):402–409.

Moffat, R.J. and Ortega, A. 1988. Direct air cooling in electronic systems. In *Advances in Thermal Modeling of Electronic Components and Systems*, ed. A. Bar-Cohen and A.D. Kraus, Vol. 1, pp. 129–282. Hemisphere, New York.

Murano, H. and Watari, T. 1992. Packaging technology for the NEC SX-3 supercomputers. *IEEE Trans. Comp., Hybrids, and Manuf. Technol.* 15(4):401–417.

Nakayama, W. 1986. Thermal management of electronic equipment: A review of technology and research topics. *Appl. Mech. Rev.* 39(12):1847–1868.

Nalbandian, R. 1987. Automatic thermal and dynamic analysis of electronic printed circuit boards by ANSYS finite element program. In *Proceeding of the Conference on Integrating Design and Analysis*, pp. 11.16–11.33, Newport Beach, CA, March 31–April 3.

Newell, W.E. 1975. Transient thermal analysis of solid-state power devices—making a dreaded process easy. *IEEE Trans. Industry Application* IA-12(4):405–420.

Oettinger, F.F. 1984. Thermal evaluation of VLSI packages using test chips—A critical review. *Solid State Tech.* (Feb):169–179.

Oettinger, F.F. and Blackburn, D.L. 1990. *Semiconductor Mesurement Technology: Thermal Resistance Measurements*. NIST Special Pub. 400-86, Washington, DC.

Ohsaki, T. 1991. Electronic packaging in the 1990s: The perspective from Asia. *IEEE Trans. Comp., Hybrids, and Manuf. Technol.* 14(2):254–261.

Riemer, D.E. 1990. Thermal-Stress Analysis with Electrical Equivalents. *IEEE Trans. Comp., Hybrids, Manuf. Technol.* 13(1):194–199.

Rohsenow, W.M. and Choi, H. 1961, Heat, Mass, and Momentum Transfer. Prentice-Hall, Englewood Cliffs, NJ.

Rohsenow, W.M. and Hartnett, J.P. 1973. *Handbook of Heat Transfer*. McGraw-Hill, New York.

Shiang, J.-J., Staszak, Z.J., and Prince, J.L. 1987. APTMC: An interface program for use with ANSYS for thermal and thermally induced stress modeling simulation of VLSI packaging. In *Proceedings of the Conference on Integration Design and Analysis*, pp. 11.55–11.62, Newport Beach, CA, March 31–April 3.

Siegal, R. and Howell, J.R. 1981. *Thermal Radiation Heat Transfer*. Hemisphere, New York.

Simon, B.R., Staszak, Z.J., Prince, J.L., Yuan, Y., and Umaretiya, J.R. 1988. Improved finite element models of thermal/mechanical effects in VLSI packaging. In *Proceedings of the Eighth International Electronic Packaging Conference*, pp. 3–19, Dallas, TX.

SPICE. *SPICE2G User's Manual*. Univ. of California, Berkeley, CA.

Staszak, Z.J., Prince, J.L., Cooke, B.J., and Shope, D.A. 1987. Design and performance of a system for VLSI packaging thermal modeling and characterization. *IEEE Trans. Comp., Hybrids, and Manuf. Technol.* 10(4):628–636.

Swanson. *ANSYS Engineering Analysis System*, User's Manual. Swanson Engineering Analysis System, Houston, PA.

Touloukian, Y.S. and Ho, C.Y. 1979. *Master Index to Materials and Properties*. Plenum Publishing, New York.

Tuckerman, D.P. and Pease, F. 1981. High performance heat sinking for VLSI. *IEEE Electron Device Lett.* EDL-2(5):126–129.

Tummala, R.R. 1991. Electronic packaging in the 1990s: The perspective from America. *IEEE Trans. Comp., Hybrids, and Manuf. Technol.* 14(2):262–271.

Tummala, R.R. and Ahmed, S. 1992. Overview of packaging of the IBM enterprise system/9000 based on the glass-ceramic copper/thin film thermal conduction module. *IEEE Trans. Comp., Hybrids, and Manuf. Technol.* 15(4):426–431.

Tzeng, Y., Yoshikawa, M., Murakawa, M., and Feldman, A. 1991. *Application of Diamond Films and Related Materials*. Elsevier Science, Amsterdam.

Watari T. and Murano, H. 1995. Packaging technology for the NEC SX supercomputers. *IEEE Trans. Comp., Hybrids, and Manuf. Technol.* 8(4):462–467.

Wessely, H., Fritz, O., Hor, M., Klimke, P., Koschnick, W., and Schmidt, K.-K. 1991. Electronic packaging in the 1990s: The perspective from Europe. *IEEE Trans. Comp., Hybrids, and Manuf. Technol.* 14(2):272–284.

Yokouchi, K., Kamehara, N., and Niwa, K. 1987. Immersion Cooling for High-Density Packaging. *IEEE Trans. Comp., Hybrids, and Manuf. Technol.* 10(4):643–646.

Yovanovich, M.M. and Antonetti, V.M. 1988. Application of thermal contact resistance theory to electronic packages. In *Advances in Thermal Modeling of Electronic Components and Systems*, Vol. 1, eds. A. Bar-Cohen and A.D. Kraus, pp. 79–128. Hemisphere, New York.

Further Information

Additional information on the topic of heat management and heat related phenomena in electronic packaging systems is available from the following sources:

Journals:

IEEE Transactions on Components, Packaging, and Manufacturing Technology, published by the Institute of Electrical and Electronics Engineers, Inc., New York (formerly: *IEEE Transactions on Components, Hybrids, and Manufacturing Technology*).

International Journal for Hybrid Microelectronics, published by the International Society for Hybrid Micro-electronics, Silver Spring, MD.

Journal of Electronic Packaging, published by the American Society of Mechanical Engineers, New York.

Conferences:

IEEE Semiconductor Thermal and Temperature Measurement Symposium (SEMI-THERM), organized annually by IEEE Inc.

Intersociety Conference on Thermal Phenomena in Electronic Systems (I-THERM), organized biannually in the U.S. by IEEE Inc, ASME, and other engineering societies.

Electronics Components and Technology Conference (formerly: Electronic Components Conference), organized annually by IEEE Inc.

International Electronic Packaging Conference, organized annually by the International Electronic Packaging Society.

European Hybrid Microelectronics Conference, organized by the International Society for Hybrid Microelectronics.

In addition to the references, the following books are recommended:

Aung, W. ed. 1991. *Cooling Techniques for Computers*. Hemisphere, New York.

Dean, D.J. 1985. *Thermal Design of Electronic Circuit Boards and Packages*. Electrochemical Publications, Ayr, Scotland.

Grigull, U. and Sandner, H. 1984. *Heat Conduction*, International Series in Heat and Mass Transfer. Springer-Verlag, Berlin.

Seely, J.H. and Chu, R.C. 1972. *Heat Transfer in Microelectronic Equipment*. Marcel Dekker, New York.

Sloan, J.L. 1985. *Design and Packaging of Electronic Equipment*. Van Nostrand Reinhold, New York.

Steinberg, D.S. 1991. *Cooling Techniques for Electronic Equipment*. Wiley, New York.

6
Shielding and EMI Considerations

Don White
emf-emi control, Inc.

6.1 Introduction 6-1
6.2 Cables and Connectors 6-1
6.3 Shielding 6-2
  Shielding Effectiveness • Different Levels of Shielding
6.4 Aperture-Leakage Control 6-6
  Controlling Aperture Leakages • Viewing Apertures • Ventilating Holes • Electrical Gaskets • Shielding Composites
6.5 System Design Approach 6-12

6.1 Introduction

Two of many requirements of electronic-system packaging are to ensure that the corresponding equipments are in compliance with applicable national and international **electromagnetic compatibility (EMC)** regulations for emission and immunity control. In the U.S., this refers to FCC, Rules and Regulations, Parts 15.B on emission, and internationally to CISPR 22 and IEC-1000-4 (formerly, IEC 801 family) for emission and immunity control.

Emission and immunity (susceptibility) control each have two parts: conducted (on hard wire) and radiated (radio-wave coupling). This section deals only with the latter on **electromagnetic interference (EMI) radiated emission** and susceptibility control.

In general, reciprocity applies for EMI and its control. When a shield is added to reduce radiated emission, it also reduces radiated immunity. Quantitatively speaking, however, reciprocity does not apply since the bilateral conditions do not exist, viz., the distance from the source to the shield does not equal the distance from the shield to the victim. Thus, generally, expect different shielding effectiveness values for each direction.

6.2 Cables and Connectors

One principal method of dealing with EMI is to shield the source, the victim, or both, so that only intentional radiation will couple between them. The problem, then, is to control unintentional radiation coupling into or out of victim circuits, their PCBs, their housings, and equipment interconnecting cables.

Generalizing, the greatest cause of EMI failure is from interconnecting cables. Cables are often referred to as *the antenna farm* since they serve to unintentionally pick up or radiate emissions to or from the outside world thereby causing EMI failures. However, cables are not the subject of this section. Connectors are loosely referred to as devices to stop outside **conducted emissions** from getting in or vice versa.

6.3 Shielding

If the system is designed correctly in the first place, little shielding may be needed to clean up the remaining EMI. But, to design shielding into many places from the outset may be an inadvisable shortcut in that the design engineer then does not have to do his homework. One classical example is the design and layout of printed circuit boards (PCBs). Multilayer boards radiate at levels 20–35 dB below levels from double-sided, single-layer boards. Since logic interconnect trace radiation is the principal PCB radiator, it follows that multilayer boards are cost effective vs having to shield the outer box housing.

Shielding Effectiveness

Figure 6.1 shows the concept of shielding and its measure of effectiveness, called **shielding effectiveness** (**SE**). The hostile source in the illustration is at the left and the area to be protected is on the right. The metal barrier in the middle is shown in an edge-elevation view.

The oncoming electromagnetic wave (note the orthogonal orientation of the electric and magnetic fields and the direction of propagation: $E \times H$ = Poynting's vector) is approaching the metal barrier. The wave impedance ($Z_w = E/H$) is 377 Ω in the **far field** (distance $> \lambda/2\pi$). Z_w is greater in the **near field** (distance $< \lambda/2\pi$) for high-impedance waves or E fields and less than 377 Ω in the near field for low-impedance waves or H fields. The wave impedance is related to the originating circuit impedance at the frequency in question.

The wavelength λ, in meters, used in the previous paragraph is related to the frequency in megahertz.

$$\lambda_m = 300/f_{\text{MHz}} \qquad (6.1)$$

Since the wave impedance and metal barrier impedance Z_b (usually orders of magnitude lower) are not matched, there is a large reflection loss RL_{dB} established at the air–metal interface

$$RL_{dB} = 20\log_{10}[Z_w/(4 * Z_b)] \qquad (6.2)$$

where Z_w is the wave impedance equal to 377 Ω in the far field.

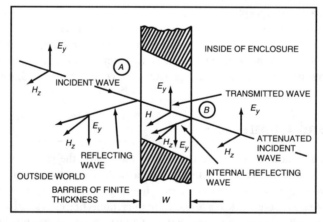

FIGURE 6.1 Shielding effectiveness analysis.

Figure 6.1 shows that what is not reflected at the metal barrier interface continues to propagate inside where the electromagnetic energy is converted into heat. This process is called *absorption loss* AL_{dB} defined as

$$AL_{dB} = 8.6t/\delta \qquad (6.3)$$

$$\delta = 1/\sqrt{\pi f \sigma \mu} \qquad (6.4)$$

where:

- t = metal thickness in any units
- δ = skin depth of metal in same units
- f = frequency, Hz
- σ = absolute conductivity of the metal, mho
- μ = absolute permeability of the metal, H/m

The shielding effectiveness SE_{dB} is the combination of both the reflection and absorption losses,

$$SE_{dB} = RL_{dB} + AL_{dB} \qquad (6.5)$$

Figure 6.2 shows the general characteristics of shielding effectiveness. As mentioned before, the interface between near and far fields occur at the distance $r = \lambda/2\pi$. The electric field is shown decreasing at -20 dB/decade with increasing frequency while the magnetic field increases at $+20$ dB/decade. The two fields meet at the far-field interface, called *plane waves* at higher frequencies.

Figure 6.2 shows the absorption loss (dotted line) appearing when the metal thickness is approaching a fraction of one skin depth (adds 8.6 dB attenuation).

Different Levels of Shielding

Shielding may be performed at the component level, PCB or backplane level, card-cage level, box level or entire system level (such as a shielded room). The latter is excluded from this discussion.

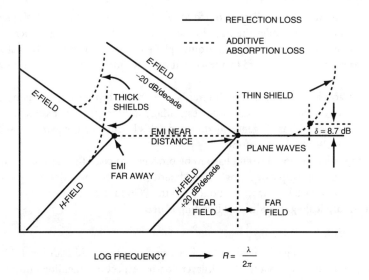

FIGURE 6.2 The general characteristics of shielding effectiveness.

FIGURE 6.3 PCB component shielding. (Courtesy emf-emi control.)

Component Shielding

Figure 6.3 shows a shield at the component level used on a clock on a PCB. The shield is grounded to the signal return, V_0, or board ground traces, as most conveniently located.

The argument for shielding at the component level is that after using a multilayer board only the components can significantly radiate if further radiation reduction is needed.

Tests indicate that shielding effectiveness of component shields at radio frequencies above 10 MHz may vary from 10 to 25 dB over radiation from no shields.

Printed-Circuit Board Shielding

The reason that multilayer PCBs radiate at much lower levels than double-sided, single-layer boards is that the trace return is directly under the trace in the V_{cc} or V_0 planes, forming a microstrip line. This is similar to the technique of using a phantom ground plane on single-layer boards as shown in Fig. 6.4. The trace height is now small compared to a trace that must wander over one or more rows as in single-layer boards.

The remaining traces left to radiate in multilayer boards are the top traces (for example, north–south) and the bottom traces on the solder side (east–west). Many military spec PCBs shield both the top and bottom layers, but industrial boards generally do not do this because of the extra costs. They also make maintenance of the PCBs nearly impossible.

To shield the top side, the board would have to be broached to clear the surface-mount components. The bottom side shield can simply be a single-sided board (foil on the outside) brought up to the solder tabs. The board can be secured in place and bonded to the PCB (at least at the four corners). For greater shielding (see **aperture leakage**), additional bonding is needed.

Card-Cage Shielding

Card-cage shielding (metalizing all six sides of a plastic card cage) can add additional shielding effectiveness if the cards are single-layer boards. If multilayer boards, expect no shielding improvement since the trace-to-ground dimensions are much less than those of the card cage.

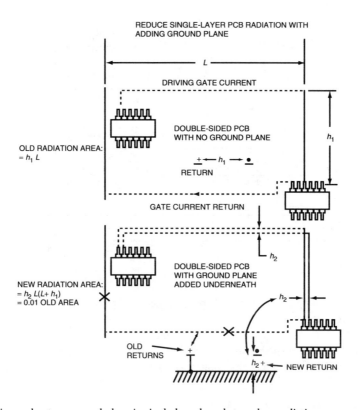

FIGURE 6.4 Using a phantom-ground plane in single-layer boards to reduce radiation.

Shielding at the Box or Housing Level

Classically, the most effective level to shield is at the box-housing level. The attitude often is what radiated EMI you do not stop at the lower levels can be accomplished at the box level by adding conductive coatings and aperture-leakage control.

How much shielding is required at the box level to comply with FCC Parts 15.B or CISPR 22, Class B (residential locations)? Naturally, this depends on the complexity and number of PCBs, the clock frequency, the logic family, etc. At the risk of generalizing, however, if little EMI control is exercised at the PCB level, roughly 40 dB of shielding effectiveness will be required for the entire box over the frequency range of roughly 50–500 MHz.

To achieve a 40-dB overall box shielding, a deposited conductive coating would have to provide at least 50 dB. The coating impedance Z_b in ohms/square is determined from Eq. (6.2)

$$SE_{dB} = 20 \log_{10} [Z_w/(4 * Z_b)] \tag{6.2}$$

where Z_w is the wave impedance equal to 377 Ω in the far field. For SE = 50 dB, solving Eq. (6.2) for Z_b

$$Z_b = Z_w/(4 * 10^{SE/20})$$
$$94/10^{2.5} = 0.3 \Omega/\text{sq.} \tag{6.6}$$

This means that the conductive coating service house needs to provide the equivalent of a conductive coating of about 0.3 Ω/sq.

The electroless process is one of the principal forms of conductive coatings in use today. This process coats both inside and outside the plastic housing. One mil (24 μm) of copper is more than enough to

PROCESS	METALLIZING COSTS COST (1985) INCLUDES APPLICATION COST (FOR PAINTS/SPRAY, MANUAL OPERATION ASSUMED)	
	Dollars/sq. ft.	(Dollars/sq. m)
ELECTROPLATING	$0.25–$2.00	($2.5–$20.00)
ELECTROLESS PLATING (COPPER & NICKEL, INSIDE AND OUTSIDE)	$1.30	($13.00)
METAL SPRAY	$1.50–$4.00	($15.00–$40.00)
VACUUM DEPOSITION	$0.50–$2.25	($5.00–$22.00)
CARBON COATING (0.5 mil = 12.7 μm)	$0.05–$0.50	($0.50–$5.00)
GRAPHITE COATING (2 mil = 51μm)	$0.10–$1.10	($1.00–$11.00)
COPPER COATING (2 mil = 51μm) OVERCOATED WITH CONDUCTIVE GRAPHITE AT 0.2 mil = 5 μm	$0.30–$1.50	($3.00–$15.00)
NICKEL COATING	$0.70	($7.00)
SILVER COATING (0.5 = 12.7 μm)	$1.25–$3.00	($12.5–$30.00)

FIGURE 6.5 Conductive coating processes.

achieve this objective as long as it is uniformly or homogeneously deposited. Figure 6.5 shows a number of other competitive coating processes.

Sometimes, a flashing of nickel is used over an aluminum or copper shield to provide abrasion protection to scratches and normal abuse. The flashing may be of the order of 1 μm. Nickel does not compromise the aluminum or copper shielding. It simply adds to the overall shielding effectiveness.

6.4 Aperture-Leakage Control

As it develops, selecting and achieving a given shielding effectiveness of the conductive coating is the easy part of the box or housing shielding design. Far more difficult is controlling the many different aperture leakages depicted in Fig. 6.6. Therein it is seen that there are large apertures, such as needed for windows for cathode ray tube (CRT) and light emitting diode (LED) displays, holes for convection cooling, and slots for mating panel members.

To predict aperture leakages, the geometry is first defined in Fig. 6.7. The longest aperture dimension is l_{mm}, given in millimeters. The shortest aperture dimension is h_{mm}, given in millimeters. The thickness of the metal or conductive coating is d_{mm}, also given in millimeters.

The aperture leakage SE_{dB} is predicted by using

$$SE_{dB} = 100 - 20 \log 10(l_{mm} f_{MHz}) + 20 \log_{10}[1 + ln_e \ (l_{mm}/h_{mm})] + 30 l_{mm}/d_{mm} \qquad (6.7)$$

with all terms having been previously defined.

To gain a better mental picture of the shielding effectiveness of different apertures, Eq. (6.7) is plotted in Fig. 6.8.

Illustrative Example 1

Even small scratches can destroy the performance of conductive coatings. For example, suppose that a single (inside) conductive coating has been used to achieve 50 dB of shielding effectiveness in the 50 MHz–1 GHz spectrum. Further, suppose that a screwdriver accidentally developed a 4-in (102-mm) $\times$ 10-mil (0.25-mm) scratch. From Eq. (6.7), the resulting shielding effectiveness to the seventh harmonic of a 90-MHz onboard clock is:

$$SE_{dB} = 100 - 20 \log_{10}(102 * 630) + 20 \log_{10}[1 + ln_e \ (102/0.25)]$$
$$100 - 96 + 17 = 21 dB$$

Shielding and EMI Considerations

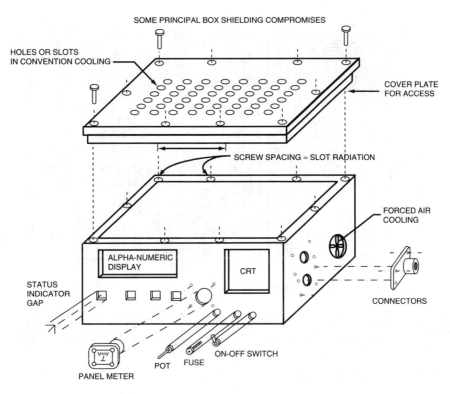

FIGURE 6.6 Principal aperture leakages in boxes.

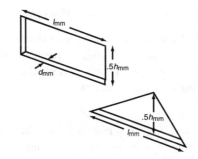

FIGURE 6.7 Aperture dimensions used in mathematic models.

Controlling Aperture Leakages

The success in achieving desired shielding effectiveness, then, is to control all major aperture leakages. As the *SE* requirement increases, it is also necessary to control minor aperture leakages as well.

Most aperture leakage control is divided into four sections:

- Viewing windows
- Ventilating apertures
- Slots and seams
- All other apertures

Viewing Apertures

The first of the four aperture-leakage categories is viewing aperture windows. This applies to apertures such as CRTs, LED displays, and panel meters.

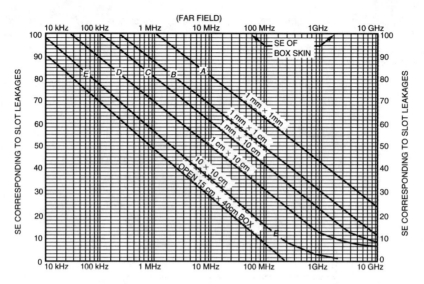

FIGURE 6.8 Aperture leakages using Eq.(6.7).

There are two ways to shore up leaky windows: (1) knitted wire mesh or screens and (2) thin-conductive films. In both cases, the strategy is to block the RF while passing the optical viewing with minimum attenuation.

Knitted-Wire Meshes and Screens

The amount of RF blocking (aperture leakage) from a knitted-wire mesh or screen is

$$SE_{dB} = 20 \log_{10}(\lambda/2/D)$$
$$= 103 - (f_{MHz} D_{mm}) \text{ dB} \qquad (6.8)$$

where D_{mm} is the screen mesh wire separation in millimeter.

Figure 6.9 shows one method for mounting screens and wire mesh. Mesh also comes in the form of metalized textiles with wire-to-wire separations as small as 40 μm, hence the wires are not even discernible, but appear more like an antiglare screen.

Thin Films

Another approach to providing RF shielding is to make a thin-film shield, which provides a surface impedance while passing most of the optical light through the material. Operating on the reflection-loss principle, typical shielding effectiveness of about 30 dB may be obtained [from Eq. (6.2), film impedances correspond to about 3 Ω/square]. When the impedance is lowered to gain greater *SE*, optical blockage begins to exceed 50%. Figure 6.10 shows the performance of typical thin films.

As also applicable to screens and meshes, thin films must be bonded to the aperture opening 360° around the periphery. Otherwise, the film is compromised and *SE* suffers significantly.

Ventilating Holes

Convection cooling requires many small holes or slots at the top cover or upper panels of an electronic enclosure. If the array of holes is tight with relatively little separation (the subtends is less than 10% of a wavelength), the aperture leakage corresponds to that of a single hole defined in Eq. (6.7).

Additional attenuation is achievable by making the depth of the hole comparable to or exceeding the dimension of the hole or slot. This, then, generates the waveguide-beyond-cutoff effect in which the last term of Eq. (6.7) now applies as well.

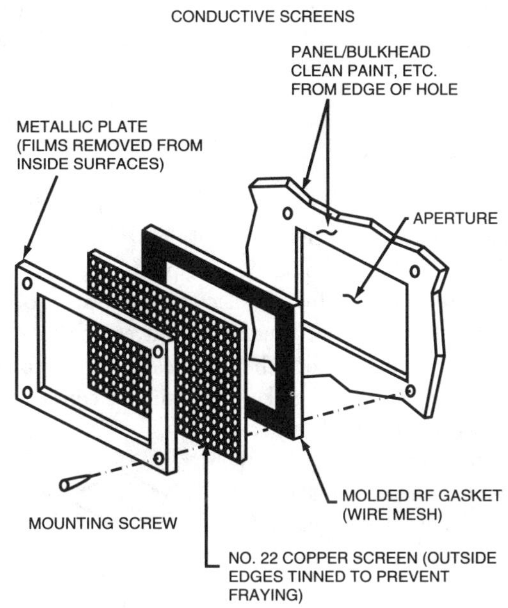

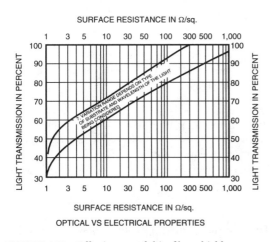

FIGURE 6.9 Method of mounting wire screen over an aperture.

FIGURE 6.10 Effectiveness of thin-films shields.

FIGURE 6.11 Hexagonal honeycomb shields.

Illustrative Example 2

Suppose an egg-crate array of square holes is placed in the top of a monitor for convection cooling. The squares are 4 mm on a side and 5 mm deep. Determine the shielding effectiveness of this configuration at 1.7 GHz.

From Eq. (6.7)

$$SE_{dB} = 100 - 20 \log_{10}(4 * 1700) + 20 \log_{10}[1 + ln_e(4/4)] + 30 * 5/4$$
$$= 100 - 77 + 0 + 38 = 61 \text{ dB}$$

This is only possible with the electroless process of conductive coating of a plastic enclosure since the metal must be plated through the hole as well for the waveguide-beyond-cuttoff concept to work.

Commercial versions of this process are shown in Fig. 6.11 for the hexagonal honeycomb used in all shielded rooms and most shielded electronic enclosures. Air flow is smoothly maintained while SE exceeds 100 dB.

Electrical Gaskets

In some respects electrical gaskets are similar to mechanical gaskets. First, the tolerances of mating metal members may be relaxed considerably to reduce the cost of the fabrication. A one penny rubber washer makes the leakless mating of a garden hose with its bib a snap. The gasket does not work, however, until the joint is tightened enough to prevent blowby. The electrical equivalent of this is the joint must have sufficient pressure to achieve an *electrical seal* (meaning the advertised SE of the gasket is achievable). Second, the gasket should not be over-tightened. Otherwise *coldflow* or distortion will result and the life of the gasket may be greatly reduced. Electrical gaskets have similar considerations.

Figure 6.12 shows the cross-sectional view of an electrical gasket in place between two mating metal members. The gap between the members is to be filled with a metal bonding material (the gasket), which provides a low-impedance path in the gap between the members, thereby avoiding a slot radiator discussed in earlier sections.

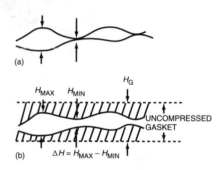

Usually, it is not necessary to use electrical gaskets to achieve SE of 40 dB below 1 GHz. It is almost always necessary to use electrical gaskets to achieve 60-dB SE. This raises the question of what to do to achieve between 40 and 60 dB overall SE?

FIGURE 6.12 Electrical gasket filling the gap between mating metal parts: (a) Joint unevenness = ΔH just touching, (b) compressed gasket in place.

Gaskets can be relatively expensive, increase cost of installation and maintenance and are subject to corrosion and aging effects. Therefore, trying to avoid their use in the 40–60 dB range becomes a motivated objective. This can often be achieved, for example, by replacing a cover plate with an inverted shallow pan, which is less expensive than the gasket. Other mechanical manipulations to force the electromagnetic energy to go around corners is often worth 10–20 dB more in equivalent SE.

Electrical gaskets are available in three different forms: (1) spring-finger stock, (2) wire-mesh gaskets, and (3) conductive impregnated elastomers.

Spring-Finger Stock

The oldest and best known of the electrical gasket family is the spring-finger stock shown in Fig. 6.13. This gasket is used in applications where many openings and closings of the mating metal parts are anticipated. Examples include shielded room doors, shielded electronic enclosures, and cover pans.

Spring-finger gaskets may be applied by using adhesive backing, sweat soldering into place, riveting, or eyeleting.

Wire-Mesh Gaskets

Another popular electrical gasket is the multilayer, knitted-wire mesh. Where significant aperture space needs to be absorbed, the gasket may have a center core of spongy material such as neophrem or rubber. Figure 6.14 shows a number of methods of mounting wire-mesh gaskets.

Conductive-Impregnated Elastomers

A third type of gasket is the conductive-impregnated elastomers. In this configuration, the conductive materials are metal flakes or powder dispersed throughout the carrier. This gasket and some of its applications are shown in Fig. 6.15.

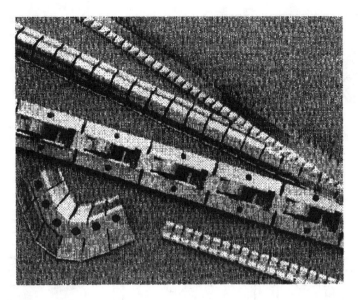

FIGURE 6.13 Spring-finger gaskets.

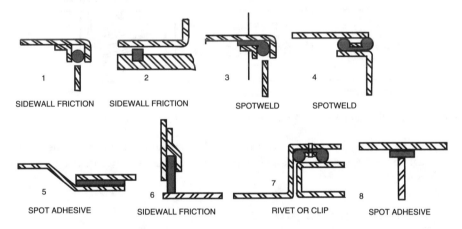

FIGURE 6.14 Knitted-wire mesh mountings.

Shielding Composites

Another material for achieving shielding is called composites. Here the plastic material is impregnated with flakes or powder of a metal such as copper or aluminum. Early versions were not particularly successful since it was difficult to homogeneously deploy the filler without developing leakage spots or clumps of the metal flakes. This led to making the composites into multilayers of sheet material to be formed around a mold.

The problem seems to be corrected today. Injection molding of uniformly deployed copper or aluminum flakes (about 20% by volume) is readily achievable.

A more useful version of these composites is to employ metal filaments with aspect ratios (length-to-diameter) of about 500. For comparable shielding effectiveness, only about 5% loading of the plastic carrier is needed.

Filament composites exhibit the added benefit of providing wall cooling about 50 times that of conductive coatings. One difficulty, however, is making good electrical contact to the composite at the periphery. This may require electroplating or application of an interface metal tape.

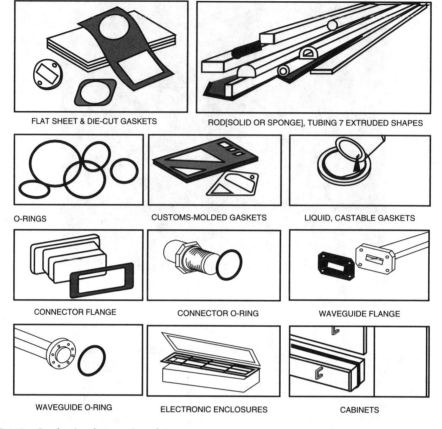

FIGURE 6.15 Conductive elastomeric gaskets.

Shielding composites may become the preferred shielding method in the next decade. As of now, it has not yet caught on in any significant way. Thus, pricewise, it is not as competitive as other conductive coating processes.

6.5 System Design Approach

If the overall shielding effectiveness of a box or equipment housing is to be X dB, then what is the objective for the design of each of the leaky apertures? And, what must be the shielding effectiveness of the conductive coating of the housing?

Leaky apertures combine like resistors in parallel. The overall SE of the enclosure is poorer than any one aperture:

$$SE_{\text{overall}} = 20 \log_{10} \left[\sum \text{antilog } SE_N \right] \tag{6.9}$$

where SE_N is the SE of the Nth aperture.

The required SE_N may be approximated by assuming that each aperture leakage must meet the same shielding amount, hence,

$$SE_N = SE_{\text{overall}} + 20 \log_{10} N \tag{6.10}$$

Shielding and EMI Considerations 6-13

Illustrative Example 3

A box housing has three separate aperture leakages: the CRT, convection cooling holes, and seams at mating conductive coating parts. Determine the aperture leakage requirements of each to obtain an overall *SE* of 40 dB.

From Eq. (6.9)

$$SE_{each} = 40\,dB + 20\log_{10}(3) = 50\,dB$$

To fine tune this value, the aperture leakage can be relaxed in one area if tightened in another. This is often desired where a relaxation can avoid an expensive additive such as an electrical gasket. In this example, it may be desired to relax the seam gasket to produce 45 dB, if the other two leakages are tightened to 60 dB.

Finally, the basic *SE* of the conductive coating must exceed the *SE* of each aperture leakage (or the most stringent of the aperture leakages). A margin of 10 dB is suggested since 6 dB will produce a 4-dB degradation at the aperture, whereas 10 dB will allow a 3-dB degradation. Thus, in example 3, the *SE* of the base coating would have to be 60 dB to permit controlling three apertures to 50 dB to result in an overall *SE* of 40 dB.

Defining Terms

Aperture leakage: The compromise in shielding effectiveness resulting from holes, slits, slots, and the like used for windows, cooling openings, joints, and components.

Broadband EMI: Electrical disturbance covering many octaves or decades in the frequency spectrum or greater than receiver bandwidth.

Common mode: As applied to two or more wires, all currents flowing therein which are in-phase.

Conducted emission (CE): The potential EMI which is generated inside the equipment and is carried out of the equipment over I/O lines, control leads, or power mains.

Coupling path: The conducted or radiated path by which interfering energy gets from a source to a victim.

Crosstalk: The value expressed in dB as a ratio of coupled voltage on a victim cable to the voltage on a nearby culprit cable.

Differential mode: On a wire pair when the voltages or currents are of opposite polarity.

Electromagnetic compatibilty (EMC): Operation of equipments and systems in their installed environments which cause no unfavorable response to or by other equipments and systems in same.

Electromagnetic environmental effects (E3): A broad umbrella term used to cover EMC, EMI, RFI, electromagnetic pulse (EMP), electrostatic discharge (ESD), radiation hazards (RADHAZ), lightning, and the like.

Electromagnetic interference (EMI): When electrical disturbance from a natural phenomenon or an electrical/electronic device of system causes an undesired response to another.

Electrostatic discharge (ESD): Fast risetime, intensive discharges from humans, clothing, furniture, and other charged dielectric sources.

Far field: In radiated-field vernacular, EMI point source distance greater than about 1/6 wavelength.

Ferrites: Powdered magnetic material in the form of beads, rods, and blocks used to absorb EMI on wires and cables.

Field strength: The radiated voltage or current per meter corresponding to electric or magnetic fields.

Narrowband EMI: Interference whose emission bandwidth is less than the bandwidth of EMI measuring receiver or spectrum analyzer.

Near field: In radiated-field vernacular, EMI point source distance less than about 1/6 wavelength.

Noise-immunity level: The voltage threshold in digital logic families above which a logic zero may be sensed as a one and vice versa.

Normal mode: The same as differential mode, emission or susceptibility from coupling to/from wire pairs.

Radiated emission (RE): The potential EMI that radiates from escape coupling paths such as cables, leaky aperatures, or inadequately shielded housings.

Radiated susceptibility: Undesired potential EMI that is radiated into the equipment or system from hostile outside sources.

Radio frequency interference (RFI): Exists when either the transmitter or receiver is carrier operated, causing unintentional responses to or from other equipment or systems.

Shielding effectiveness (*SE*): Ratio of field strengths before and after erecting a shield. *SE* consists of absorption and reflection losses.

Transients: Short surges of conducted or radiated emission during the period of changing load conditions in an electrical or electronic system or from other sources.

References

EEC. Predicting, analyzing, and fixing shielding effectiveness and aperture leakage control. EEC Software Program #3800.

Kakad, A. 1995. ePTFE gaskets: A competitive alternate. *emf-emi control Journal* (Jan.–Feb.):11.

Mardiguian, M. *Electromagnetic Shielding*, Vol. 3, EMC handbook series. EEC Press.

Squires, C. 1994. Component aperture allocation for overall box shielding effectiveness, part 1, strategy. *emf-emi control Journal* (Jan.–Feb.):18.

Squires, C. 1994. Component aperture leakage allocation for overall box shielding effectiveness, part 2, design. *emf- emi control Journal* (March–April): 18.

Squires, C. 1994. EMF containment or avoidance. *emf-emi control Journal* (July–Aug.):27.

Squires, C. 1995. The many facets of shielding. *emf-emi control Journal* (March–April): 11.

Squires, C. 1995. The ABCs of shielding. *emf-emi control Journal* (March–April):28.

Squires, C. 1995. Predicting and controlling PCB radiation. *emf-emi control Journal* (Sep.–Oct.):17.

Stephens, E. 1994. Cable shield outside the telco plant. *emf-emi control Journal* (Jan.–Feb.):18.

Vitale, L. 1995. Magnetic shielding of offices and apartments. *emf-emi control Journal* (March–April):12.

White, D. 1994. Building EM advances from shielding to power conditioning. *emf-emi control Journal* (Jan.–Feb.):17.

White, D. 1994. RF vs. low frequency magnetic shielding. *emf-emi control Journal* (Jan.–Feb.).

White, D. 1994. Shielding to RF/microwave radiation exposure. *emf-emi control Journal*.

White, D. 1994. What to expect from a magnetic shielded room. *emf-emi control Journal*.

White, D. 1995. How much cable shielding is required. *emf-emi control Journal* (March–April):16.

White, W. Fire lookout tower personnel exposed to RE radiation. *emf-emi control Journal*.

Further Information

There exist a few magazines which emphasize EMC, EMI, and its control. The most prolific source is *emf-emi control Journal*, a bimonthly publication containing 15–20 short design and applications-oriented technical articles per issue. For subscriptions contact The EEC Press, phone 540-347-0030 or fax 540-347-5813. Two other sources are ITEM (phone: 610-825-1960) and Compliance Engineering (phone: 508-264-4208).

The Institute of Electrical and Electronics Engineers (IEEE) conducts an annual symposium on EMC. Copies of their Convention Record are available from IEEE, Service Center, 445 Hoes Lane, Piscataway, NJ 08855.

For EMC-related short training courses and seminars, handbooks, CD-ROMs, video tapes and problem-solving software, contact *emf-emi control* at the above phone and fax numbers. IEEE Press also publishes a few books on EMC-related subjects.

Resistors and Resistive Materials

Jerry C. Whitaker
Editor-in-Chief

7.1 Introduction 7-1
7.2 Resistor Types 7-1
    Wire-Wound Resistor • Metal Film Resistor • Carbon Film
    Resistor • Carbon Composition Resistor • Control and Limiting
    Resistors • Resistor Networks • Adjustable Resistors
    • Attenuators

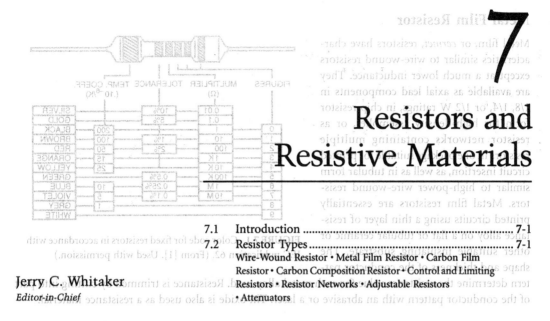

7.1 Introduction

Resistors are components that have a nearly 0° phase shift between voltage and current over a wide range of frequencies with the average value of resistance independent of the instantaneous value of voltage or current. Preferred values of ratings are given ANSI standards or corresponding ISO or MIL standards. Resistors are typically identified by their construction and by the resistance materials used. Fixed resistors have two or more terminals and are not adjustable. Variable resistors permit adjustment of resistance or voltage division by a control handle or with a tool.

7.2 Resistor Types

There are a wide variety of resistor types, each suited to a particular application or group of applications. Low-wattage fixed resistors are usually identified by color-coding on the body of the device, as illustrated in Fig. 7.1. The major types of resistors are identified in the following sections.

Wire-Wound Resistor

The resistance element of most wire-wound resistors is resistance wire or ribbon wound as a single-layer helix over a ceramic or fiberglass core, which causes these resistors to have a residual series inductance that affects phase shift at high frequencies, particularly in large-size devices. Wire-wound resistors have low noise and are stable with temperature, with temperature coefficients normally between ±5 and 200 ppm/°C. Resistance values between 0.1 and 100,000 W with accuracies between 0.001 and 20% are available with power dissipation ratings between 1 and 250 W at 70°C. The resistance element is usually covered with a vitreous enamel, which can be molded in plastic. Special construction includes items such as enclosure in an aluminum casing for heatsink mounting or a special winding to reduce inductance. Resistor connections are made by self-leads or to terminals for other wires or printed circuit boards.

Metal Film Resistor

Metal film, or *cermet*, resistors have characteristics similar to wire-wound resistors except at a much lower inductance. They are available as axial lead components in 1/8, 1/4, or 1/2 W ratings, in chip resistor form for high-density assemblies, or as resistor networks containing multiple resistors in one package suitable for printed circuit insertion, as well as in tubular form similar to high-power wire-wound resistors. Metal film resistors are essentially printed circuits using a thin layer of resistance alloy on a flat or tubular ceramic or other suitable insulating substrate. The shape and thickness of the conductor pattern determine the resistance value for each metal alloy used. Resistance is trimmed by cutting into part of the conductor pattern with an abrasive or a laser. Tin oxide is also used as a resistance material.

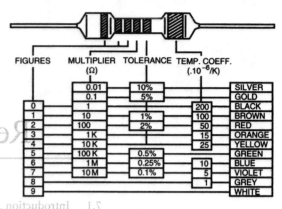

FIGURE 7.1 Color code for fixed resistors in accordance with IEC publication 62. (From [1]. Used with permission.)

Carbon Film Resistor

Carbon film resistors are similar in construction and characteristics to axial lead metal film resistors. Because the carbon film is a granular material, random noise may be developed because of variations in the voltage drop between granules. This noise can be of sufficient level to affect the performance of circuits providing high grain when operating at low signal levels.

Carbon Composition Resistor

Carbon composition resistors contain a cylinder of carbon-based resistive material molded into a cylinder of high-temperature plastic, which also anchors the external leads. These resistors can have noise problems similar to carbon film resistors, but their use in electronic equipment for the last 50 years has demonstrated their outstanding reliability, unmatched by other components. These resistors are commonly available at values from 2.7 Ω with tolerances of 5, 10, and 20% in 1/8-, 1/4-, 1/2-, 1-, and 2-W sizes.

Control and Limiting Resistors

Resistors with a large negative temperature coefficient, *thermistors*, are often used to measure temperature, limit inrush current into motors or power supplies, or to compensate bias circuits. Resistors with a large positive temperature coefficient are used in circuits that have to match the coefficient of copper wire. Special resistors also include those that have a low resistance when cold and become a nearly open circuit when a critical temperature or current is exceeded to protect transformers or other devices.

Resistor Networks

A number of metal film or similar resistors are often packaged in a single module suitable for printed circuit mounting. These devices see applications in digital circuits, as well as in fixed attenuators or padding networks.

Adjustable Resistors

Cylindrical wire-wound power resistors can be made adjustable with a metal clamp in contact with one or more turns not covered with enamel along an axial stripe. Potentiometers are resistors with a movable

Resistors and Resistive Materials

arm that makes contact with a resistance element, which is connected to at least two other terminals at its ends. The resistance element can be circular or linear in shape, and often two or more sections are mechanically coupled or ganged for simultaneous control of two separate circuits. Resistance materials include all those described previously.

Trimmer potentiometers are similar in nature to conventional potentiometers except that adjustment requires a tool.

Most potentiometers have a *linear taper*, which means that resistance changes linearly with control motion when measured between the movable arm and the "low," or counterclockwise, terminal. Gain controls, however, often have a *logarithmic taper* so that attenuation changes linearly in decibels (a logarithmic ratio). The resistance element of a potentiometer may also contain taps that permit the connection of other components as required in a specialized circuit.

Attenuators

Variable attenuators are adjustable resistor networks that show a calibrated increase in attenuation for each switched step. For measurement of audio, video, and RF equipment, these steps may be decades of 0.1, 1, and 10 dB. Circuits for unbalanced and balanced fixed attenuators are shown in Fig. 7.2. Fixed attenuator networks can be cascaded and switched to provide step adjustment of attenuation inserted in a constant-impedance network.

Audio attenuators generally are designed for a circuit impedance of 150 Ω, although other impedances can be used for specific applications. Video attenuators are generally designed to operate with unbalanced 75-Ω grounded-shield coaxial cable. RF attenuators are designed for use with 75- or 50-Ω coaxial cable.

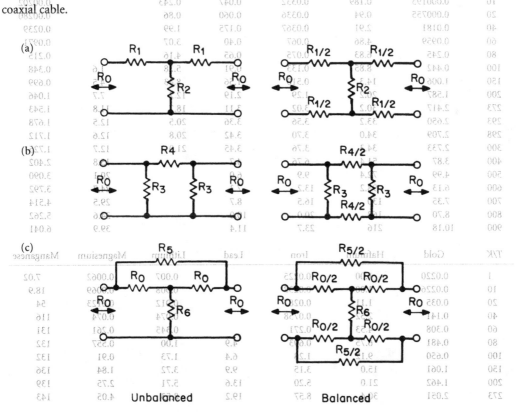

FIGURE 7.2 Unbalanced and balanced fixed attenuator networks for equal source and load resistance: (a) *T* configuration, (b) π configuration, and (c) bridged-*T* configuration.

TABLE 7.1 Resistivity of Selected Ceramics

Ceramic	Resistivity, $\Omega \cdot cm$
Borides	
Chromium diboride (CrB_2)	21×10^{-6}
Hafnium diboride (HfB_2)	$10-12 \times 10^{-6}$ at room temp.
Tantalum diboride (TaB_2)	68×10^{-6}
Titanium diboride (TiB_2) (polycrystalline)	
85% dense	$26.5-28.4 \times 10^{-6}$ at room temp.
85% dense	9.0×10^{-6} at room temp.
100% dense, extrapolated values	$8.7-14.1 \times 10^{-6}$ at room temp.
	3.7×10^{-6} at liquid air temp.
Titanium diboride (TiB_2) (monocrystalline)	
Crystal length 5 cm, 39 deg. and 59 deg. orientation with respect to growth axis	$6.6 \pm 0.2 \times 10^{-6}$ at room temp.
Crystal length 1.5 cm, 16.5 deg. and 90 deg. orientation with respect to growth axis	$6.7 \pm 0.2 \times 10^{-6}$ at room temp.
Zirconium diboride (ZrB_2)	9.2×10^{-6} at 20°C
	1.8×10^{-6} at liquid air temp.
Carbides: boron carbide (B_4C)	$0.3-0.8$

(From [1]. Used with permission.)

TABLE 7.2 Electrical Resistivity of Various Substances in $10^{-8} \, \Omega \cdot m$

T/K	Aluminum	Barium	Beryllium	Calcium	Cesium	Chromium	Copper
1	0.000100	0.081	0.0332	0.045	0.0026		0.00200
10	0.000193	0.189	0.0332	0.047	0.243		0.00202
20	0.000755	0.94	0.0336	0.060	0.86		0.00280
40	0.0181	2.91	0.0367	0.175	1.99		0.0239
60	0.0959	4.86	0.067	0.40	3.07		0.0971
80	0.245	6.83	0.075	0.65	4.16		0.215
100	0.442	8.85	0.133	0.91	5.28	1.6	0.348
150	1.006	14.3	0.510	1.56	8.43	4.5	0.699
200	1.587	20.2	1.29	2.19	12.2	7.7	1.046
273	2.417	30.2	3.02	3.11	18.7	11.8	1.543
293	2.650	33.2	3.56	3.36	20.5	12.5	1.678
298	2.709	34.0	3.70	3.42	20.8	12.6	1.712
300	2.733	34.3	3.76	3.45	21.0	12.7	1.725
400	3.87	51.4	6.76	4.7		15.8	2.402
500	4.99	72.4	9.9	6.0		20.1	3.090
600	6.13	98.2	13.2	7.3		24.7	3.792
700	7.35	130	16.5	8.7		29.5	4.514
800	8.70	168	20.0	10.0		34.6	5.262
900	10.18	216	23.7	11.4		39.9	6.041

T/K	Gold	Hafnium	Iron	Lead	Lithium	Magnesium	Manganese
1	0.0220	1.00	0.0225		0.007	0.0062	7.02
10	0.0226	1.00	0.0238		0.008	0.0069	18.9
20	0.035	1.11	0.0287		0.012	0.0123	54
40	0.141	2.52	0.0758		0.074	0.074	116
60	0.308	4.53	0.271		0.345	0.261	131
80	0.481	6.75	0.693	4.9	1.00	0.557	132
100	0.650	9.12	1.28	6.4	1.73	0.91	132
150	1.061	15.0	3.15	9.9	3.72	1.84	136
200	1.462	21.0	5.20	13.6	5.71	2.75	139
273	2.051	30.4	8.57	19.2	8.53	4.05	143

TABLE 7.2 (continued) Electrical Resistivity of Various Substances in $10^{-8}\ \Omega \cdot m$

T/K	Gold	Hafnium	Iron	Lead	Lithium	Magnesium	Manganese
293	2.214	33.1	9.61	20.8	9.28	4.39	144
298	2.255	33.7	9.87	21.1	9.47	4.48	144
300	2.271	34.0	9.98	21.3	9.55	4.51	144
400	3.107	48.1	16.1	29.6	13.4	6.19	147
500	3.97	63.1	23.7	38.3		7.86	149
600	4.87	78.5	32.9			9.52	151
700	5.82		44.0			11.2	152
800	6.81		57.1			12.8	
900	7.86					14.4	

T/K	Molybdenum	Nickel	Palladium	Platinum	Potassium	Rubidium	Silver
1	0.00070	0.0032	0.0200	0.002	0.0008	0.0131	0.00100
10	0.00089	0.0057	0.0242	0.0154	0.0160	0.109	0.00115
20	0.00261	0.0140	0.0563	0.0484	0.117	0.444	0.0042
40	0.0457	0.068	0.334	0.409	0.480	1.21	0.0539
60	0.206	0.242	0.938	1.107	0.90	1.94	0.162
80	0.482	0.545	1.75	1.922	1.34	2.65	0.289
100	0.858	0.96	2.62	2.755	1.79	3.36	0.418
150	1.99	2.21	4.80	4.76	2.99	5.27	0.726
200	3.13	3.67	6.88	6.77	4.26	7.49	1.029
273	4.85	6.16	9.78	9.6	6.49	11.5	1.467
293	5.34	6.93	10.54	10.5	7.20	12.8	1.587
298	5.47	7.12	10.73	10.7	7.39	13.1	1.617
300	5.52	7.20	10.80	10.8	7.47	13.3	1.629
400	8.02	11.8	14.48	14.6			2.241
500	10.6	17.7	17.94	18.3			2.87
600	13.1	25.5	21.2	21.9			3.53
700	15.8	32.1	24.2	25.4			4.21
800	18.4	35.5	27.1	28.7			4.91
900	21.2	38.6	29.4	32.0			5.64

T/K	Sodium	Strontium	Tantalum	Tungsten	Vanadium	Zinc	Zirconium
1	0.0009	0.80	0.10	0.000016		0.0100	0.250
10	0.0015	0.80	0.102	0.000137	0.0145	0.0112	0.253
20	0.016	0.92	0.146	0.00196	0.039	0.0387	0.357
40	0.172	1.70	0.751	0.0544	0.304	0.306	1.44
60	0.447	2.68	1.65	0.266	1.11	0.715	3.75
80	0.80	3.64	2.62	0.606	2.41	1.15	6.64
100	1.16	4.58	3.64	1.02	4.01	1.60	9.79
150	2.03	6.84	6.19	2.09	8.2	2.71	17.8
200	2.89	9.04	8.66	3.18	12.4	3.83	26.3
273	4.33	12.3	12.2	4.82	18.1	5.46	38.8
293	4.77	13.2	13.1	5.28	19.7	5.90	42.1
298	4.88	13.4	13.4	5.39	20.1	6.01	42.9
300	4.93	13.5	13.5	5.44	20.2	6.06	43.3
400		17.8	18.2	7.83	28.0	8.37	60.3
500		22.2	22.9	10.3	34.8	10.82	76.5
600		26.7	27.4	13.0	41.1	13.49	91.5
700		31.2	31.8	15.7	47.2		104.2
800		35.6	35.9	18.6	53.1		114.9
900			40.1	21.5	58.7		123.1

(From [1]. Used with permission.)

TABLE 7.3 Electrical Resistivity of Various Metallic Elements at (approximately) Room Temperature

Element	T/K	Electrical Resistivity $10^{-8}\ \Omega \cdot m$	Element	T/K	Electrical Resistivity $10^{-8}\ \Omega \cdot m$
Antimony	273	39	Polonium	273	40
Bismuth	273	107.0	Praseodymium	290–300	70.0
Cadmium	273	6.8	Promethium	290–300	75
Cerium	290–300	82.8	Protactinium	273	17.7
Cobalt	273	5.6	Rhenium	273	17.2
Dysprosium	290–300	92.6	Rhodium	273	4.3
Erbium	290–300	86.0	Ruthenium	273	7.1
Europium	290–300	90.0	Samarium	290–300	94.0
Gadolinium	290–300	131	Scandium	290–300	56.2
Gallium	273	13.6	Terbium	290–300	115
Holmium	290–300	81.4	Thallium	273	15
Indium	273	8.0	Thorium	273	14.7
Iridium	273	4.7	Thulium	290–300	67.6
Lanthanum	290–300	61.5	Tin	273	11.5
Lutetium	290–300	58.2	Titanium	273	39
Mercury	273	94.1	Uranium	273	28
Neodymium	290–300	64.3	Ytterbium	290–300	25.0
Niobium	273	15.2	Yttrium	290–300	59.6
Osmium	273	8.1			

(From [1]. Used with permission.)

TABLE 7.4 Electrical Resistivity of Selected Alloys in Units of $10^{-8}\ \Omega \cdot m$

	273K	293K	300K	350K	400K		273K	293K	300K	350K	400K
	Alloy—Aluminum-Copper						Alloy—Copper-Nickel				
Wt % Al						Wt % Cu					
99[a]	2.51	2.74	2.82	3.38	3.95	99[c]	2.71	2.85	2.91	3.27	3.62
95[a]	2.88	3.10	3.18	3.75	4.33	95[c]	7.60	7.71	7.82	8.22	8.62
90[b]	3.36	3.59	3.67	4.25	4.86	90[c]	13.69	13.89	13.96	14.40	14.81
85[b]	3.87	4.10	4.19	4.79	5.42	85[c]	19.63	19.83	19.90	20.32	20.70
80[b]	4.33	4.58	4.67	5.31	5.99	80[c]	25.46	25.66	25.72	26.12[n]	26.44[n]
70[b]	5.03	5.31	5.41	6.16	6.94	70[i]	36.67	36.72	36.76	36.85	36.89
60[b]	5.56	5.88	5.99	6.77	7.63	60[i]	45.38	45.53	45.35	45.20	45.01
50[b]	6.22	6.55	6.67	7.55	8.52	50[i]	50.19	50.05	50.01	49.73	49.50
40[c]	7.57	7.96	8.10	9.12	10.2	40[c]	47.42	47.73	47.82	48.28	48.49
30[c]	11.2	11.8	12.0	13.5	15.2	30[i]	40.19	41.79	42.34	44.51	45.40
25[f]	16.3[n]	17.2	17.6	19.8	22.2	25[c]	33.46	35.11	35.69	39.67[n]	42.81[n]
15[h]	—	12.3	—	—	—	15[c]	22.00	23.35	23.85	27.60	31.38
19[g]	1.8[n]	11.0	11.1	11.7	12.3	10[c]	16.65	17.82	18.26	21.51	25.19
5[e]	9.43	9.61	9.68	10.2	10.7	5[c]	11.49	12.50	12.90	15.69	18.78
1[b]	4.46	4.60	4.65	5.00	5.37	1[c]	7.23	8.08	8.37	10.63[n]	13.18[n]
	Alloy—Aluminum-Magnesium						Alloy—Copper-Palladium				
Wt % Al						Wt % Cu					
99[c]	2.96	3.18	3.26	3.82	4.39	99[c]	2.10	2.23	2.27	2.59	2.92
95[c]	5.05	5.28	5.36	5.93	6.51	95[c]	4.21	4.35	4.40	4.74	5.08
90[c]	7.52	7.76	7.85	8.43	9.02	90[c]	6.89	7.03	7.08	7.41	7.74
85	—	—	—	—	—	85[c]	9.48	9.61	9.66	10.01	10.36
80	—	—	—	—	—	80[c]	11.99	12.12	12.16	12.51[n]	12.87
70	—	—	—	—	—	70[c]	16.87	17.01	17.06	17.41	17.78
60	—	—	—	—	—	60[c]	21.73	21.87	21.92	22.30	22.69
50	—	—	—	—	—	50[c]	27.62	27.79	27.86	28.25	28.64

TABLE 7.4 (continued) Electrical Resistivity of Selected Alloys in Units of 10^{-8} Ω · m

Alloy—Aluminum-Magnesium

Wt % Al					
40	—	—	—	—	—
30	—	—	—	—	—
25	—	—	—	—	—
15	—	—	—	—	—
10[b]	17.1	17.4	17.6	18.4	19.2
5[b]	13.1	13.4	13.5	14.3	15.2
1[a]	5.92	6.25	6.37	7.20	8.03

Alloy—Copper-Palladium

Wt % Cu					
40[c]	35.31	35.51	35.57	36.03	36.47
30[c]	46.50	46.66	46.71	47.11	47.47
25[c]	46.25	46.45	46.52	46.99[n]	47.43[n]
15[c]	36.52	36.99	37.16	38.28	39.35
10[c]	28.90	29.51	29.73	31.19[n]	32.56[n]
5[c]	20.00	20.75	21.02	22.84[n]	24.54[n]
1[c]	11.90	12.67	12.93[n]	14.82[n]	16.68[n]

Alloy—Copper-Gold

Wt % Cu					
99[c]	1.73	1.86[n]	1.91[n]	2.24[n]	2.58[n]
95[c]	2.41	2.54[n]	2.59[n]	2.92[n]	3.26[n]
90[c]	3.29	4.42[n]	3.46[n]	3.79[n]	4.12[n]
85[c]	4.20	4.33	4.38[n]	4.71[n]	5.05[n]
80[c]	5.15	5.28	5.32	5.65	5.99
70[c]	7.12	7.25	7.30	7.64	7.99
60[c]	9.18	9.13	9.36	9.70	10.05
50[c]	11.07	11.20	11.25	11.60	11.94
40[c]	12.70	12.85	12.90[n]	13.27[n]	13.65[n]
30[c]	13.77	13.93	13.99[n]	14.38[n]	14.78[n]
25[c]	13.93	14.09	14.14	14.54	14.94
15[c]	12.75	12.91	12.96[n]	13.36[n]	13.77
10[c]	10.70	10.86	10.91	11.31	11.72
5[c]	7.25	7.41[n]	7.46	7.87	8.28
1[c]	3.40	3.57	3.62	4.03	4.45

Alloy—Copper-Zinc

Wt % Cu					
99[b]	1.84	1.97	2.02	2.36	2.71
95[b]	2.78	2.92	2.97	3.33	3.69
90[b]	3.66	3.81	3.86	4.25	4.63
85[b]	4.37	4.54	4.60	5.02	5.44
80[b]	5.01	5.19	5.26	5.71	6.17
70[b]	5.87	6.08	6.15	6.67	7.19
60	—	—	—	—	—
50	—	—	—	—	—
40	—	—	—	—	—
30	—	—	—	—	—
25	—	—	—	—	—
15	—	—	—	—	—
10	—	—	—	—	—
5	—	—	—	—	—
1	—	—	—	—	—

Alloy—Gold-Palladium

Wt % Au					
99[c]	2.69	2.86	2.91	3.32	3.73
95[c]	5.21	5.35	5.41	5.79	6.17
90[j]	8.01	8.17	8.22	8.56	8.93
85[b]	10.50[n]	10.66	10.72[n]	11.10[n]	11.48[n]
80[b]	12.75	12.93	12.99	13.45	13.93
70[c]	18.23	18.46	18.54	19.10	19.67
60[b]	26.70	26.94	27.02	27.63[n]	28.23[n]
50[a]	27.23	27.63	27.76	28.64[n]	29.42[n]
40[a]	24.65	25.23	25.42	26.74	27.95
30[b]	20.82	21.49	21.72	23.35	24.92
25[b]	18.86	19.53	19.77	21.51	23.19
15[a]	15.08	15.77	16.01	17.80	19.61
10[a]	13.25	13.95	14.20[n]	16.00[n]	17.81[n]
5[a]	11.49[n]	12.21	12.46[n]	14.26[n]	16.07[n]
1[a]	10.07	10.85[n]	11.12[n]	12.99[n]	14.80[n]

Alloy—Gold-Silver

Wt % Au					
99[b]	2.58	2.75	2.80[n]	3.22[n]	3.63[n]
95[a]	4.58	4.74	4.79	5.19	5.59
90[j]	6.57	6.73	6.78	7.19	7.58
85[j]	8.14	8.30	8.36[n]	8.75	9.15
80[j]	9.34	9.50	9.55	9.94	10.33
70[j]	10.70	10.86	10.91	11.29	11.68[n]
60[j]	10.92	11.07	11.12	11.50	11.87
50[j]	10.23	10.37	10.42	10.78	11.14
40[j]	8.92	9.06	9.11	9.46[n]	9.81
30[a]	7.34	7.47	7.52	7.85	8.19
25[a]	6.46	6.59	6.63	6.96	7.30[n]
15[a]	4.55	4.67	4.72	5.03	5.34
10[a]	3.54	3.66	3.71	4.00	4.31
5[j]	2.52	2.64[n]	2.68[n]	2.96[n]	3.25[n]
1[b]	1.69	1.80	1.84[n]	2.12[n]	2.42[n]

Alloy—Iron-Nickel

Wt % Fe					
99[a]	10.9	12.0	12.4	—	18.7
95[c]	18.7	19.9	20.2	—	26.8
90[c]	24.2	25.5	25.9	—	33.2
85[c]	27.8	29.2	29.7	—	37.3
80[c]	30.1	31.6	32.2	—	40.0
70[b]	32.3	33.9	34.4	—	42.4
60[c]	53.8	57.1	58.2	—	73.9
50[d]	28.4	30.6	31.4	—	43.7
40[d]	19.6	21.6	22.5	—	34.0

TABLE 7.4 (continued) Electrical Resistivity of Selected Alloys in Units of $10^{-8} \Omega \cdot m$

Alloy—Iron-Nickel

Wt % Fe					
30[e]	15.3	17.1	17.7	—	27.4
25[b]	14.3	15.9	16.4	—	25.1
15[e]	12.6	13.8	14.2	—	21.1
10[e]	11.4	12.5	12.9	—	18.9
5[e]	9.66	10.6	10.9	—	16.1[n]
1[b]	7.17	7.94	8.12	—	12.8

(From [1]. Used with permission.)

[a] Uncertainty in resistivity is ±2%.
[b] Uncertainty in resistivity is ±3%. [c] Uncertainty in resistivity is ±5%.
[d] Uncertainty in resistivity is ±7% below 300 K and ±5% at 300 and 400 K.
[e] Uncertainty in resistivity is ±7%.
[f] Uncertainty in resistivity is ±8%.
[g] Uncertainty in resistivity is ±10%.
[h] Uncertainty in resistivity is ±12%.
[i] Uncertainty in resistivity is ±4%.
[j] Uncertainty in resistivity is ±1%.
[k] Uncertainty in resistivity is ±3% up to 300 K and ±4% above 300 K.
[m] Crystal usually a mixture of α-hcp and fcc lattice.
[n] In temperature range where no experimental data are available.

TABLE 7.5 Resistivity of Semiconducting Minerals

Mineral	$\rho, \Omega \cdot m$	Mineral	$\rho, \Omega \cdot m$
Diamond (C)	2.7	Gersdorffite, NiAsS	1 to 160 × 10⁻⁶
Sulfides		Glaucodote, (Co, Fe)AsS	5 to 100 × 10⁻⁶
Argentite, Ag_2S	1.5 to 2.0 × 10⁻³	Antimonide	
Bismuthinite, Bi_2S_3	3 to 570	Dyscrasite, Ag_3Sb	0.12 to 1.2 × 10⁻⁶
Bornite, $Fe_2S_3 \cdot nCu_2S$	1.6 to 6000 × 10⁻⁶	Arsenides	
Chalcocite, Cu_2S	80 to 100 × 10⁻⁶	Allemonite, $SbAs_2$	70 to 60,000
Chalcopyrite, $Fe_2S_3 \cdot Cu_2S$	150 to 9000 × 10⁻⁶	Lollingite, $FeAs_2$	2 to 270 × 10⁻⁶
Covellite, CuS	0.30 to 83 × 10⁻⁶	Nicollite, NiAs	0.1 to 2 × 10⁻⁶
Galena, PbS	6.8 × 10⁻⁶ to 9.0 × 10⁻²	Skutterudite, $CoAs_3$	1 to 400 × 10⁻⁶
Haverite, MnS_2	10 to 20	Smaltite, $CoAs_2$	1 to 12 × 10⁻⁶
Marcasite, FeS_2	1 to 150 × 10⁻³	Tellurides	
Metacinnabarite, 4HgS	2 × 10⁻⁶ to 1 × 10⁻³	Altaite, PbTe	20 to 200 × 10⁻⁶
Millerite, NiS	2 to 4 × 10⁻⁷	Calavarite, $AuTe_2$	6 to 12 × 10⁻⁶
Molybdenite, MoS_2	0.12 to 7.5	Coloradoite, HgTe	4 to 100 × 10⁻⁶
Pentlandite, $(Fe, Ni)_9S_8$	1 to 11 × 10⁻⁶	Hessite, Ag_2Te	4 to 100 × 10⁻⁶
Pyrrhotite, Fe_7S_8	2 to 160 × 10⁻⁶	Nagyagite, $Pb_6Au(S, Te)_{14}$	20 to 80 × 10⁻⁶
Pyrite, FeS_2	1.2 to 600 × 10⁻³	Sylvanite, $AgAuTe_4$	4 to 20 × 10⁻⁶
Sphalerite, ZnS	2.7 × 10⁻³ to 1.2 × 10⁴	Oxides	
Antimony-sulfur compounds		Braunite, Mn_2O_3	0.16 to 1.0
Berthierite, $FeSb_2S_4$	0.0083 to 2.0	Cassiterite, SnO_2	4.5 × 10⁻⁴ to 10,000
Boulangerite, $Pb_5Sb_4S_{11}$	2 × 10³ to 4 × 10⁴	Cuprite, Cu_2O	10 to 50
Cylindrite, $Pb_3Sn_4Sb_2S_{14}$	2.5 to 60	Hollandite, $(Ba, Na, K)Mn_8O_{16}$	2 to 100 × 10⁻³
Franckeite, $Pb_5Sn_3Sb_2S_{14}$	1.2 to 4	Ilmenite, $FeTiO_3$	0.001 to 4
Hauchecornite, $Ni_9(Bi, Sb)_2S_8$	1 to 83 × 10⁻⁶	Magnetite, Fe_3O_4	52 × 10⁻⁶
Jamesonite, $Pb_4FeSb_6S_{14}$	0.020 to 0.15	Manganite, $MnO \cdot OH$	0.018 to 0.5
Tetrahedrite, Cu_3SbS_3	0.30 to 30,000	Melaconite, CuO	6000
Arsenic-sulfur compounds		Psilomelane, $KMnO \cdot MnO_2 \cdot nH_2$	0.04 to 6000
Arsenopyrite, FeAsS	20 to 300 × 10⁻⁶	Pyrolusite, MnO_2	0.007 to 30
Cobaltite, CoAsS	6.5 to 130 × 10⁻³	Rutile, TiO_2	29 to 910
Enargite, Cu_3AsS_4	0.2 to 40 × 10⁻³	Uraninite, UO	1.5 to 200

After Carmichael, R.S., Ed., 1982. *Handbook of Physical Properties of Rocks*, Vol. I, CRC Press, Boca Raton, FL.

References

1. Whitaker, J. C. (Ed.), *The Electronics Handbook*, CRC Press, Boca Raton, FL, 1996.

Further Information

Benson, K. B. and J. C. Whitaker, *Television and Audio Handbook for Technicians and Engineers*, McGraw-Hill, New York, 1990.

Benson, K. B., *Audio Engineering Handbook*, McGraw-Hill, New York, 1988.

Whitaker, J. C. and K. B. Benson (Eds.), *Standard Handbook of Video and Television Engineering*, McGraw-Hill, New York, 2000.

Whitaker, J. C., *Television Engineers' Field Manual*, McGraw-Hill, New York, 2000.

References

1. Whitaker, J. C. (Ed.), The Electronics Handbook, CRC Press, Boca Raton, FL, 1996.

Further Information

Benson, K. B. and J. C. Whitaker, Television and Audio Handbook for Technicians and Engineers, McGraw-Hill, New York, 1990.

Benson, K. B., Audio Engineering Handbook, McGraw-Hill, New York, 1988.

Whitaker, J. C. and K. B. Benson (Eds.), Standard Handbook of Video and Television Engineering, McGraw-Hill, New York, 2000.

Whitaker, J. C., Television Engineers' Field Manual, McGraw-Hill, New York, 2000.

Capacitance and Capacitors

Jerry C. Whitaker
Editor-in-Chief

8.1	Introduction 8-1
8.2	Practical Capacitors 8-2
	Polarized/Nonpolarized Capacitors • Operating Losses • Film Capacitors • Foil Capacitors • Electrolytic Capacitors • Ceramic Capacitors • Polarized-Capacitor Construction • Aluminum Electrolytic Capacitors • Tantalum Electrolytic Capacitors • Capacitor Failure Modes • Temperature Cycling • Electrolyte Failures • Capacitor Life Span

8.1 Introduction

A system of two conducting bodies (which are frequently identified as *plates*) located in an electromagnetic field and having equal charges of opposite signs $+Q$ and $-Q$ can be called a *capacitor* [1]. The capacitance C of this system is equal to the ratio of the charge Q (absolute value) to the voltage V (again, absolute value) between the bodies; that is,

$$C = \frac{Q}{V} \qquad (8.1)$$

Capacitance C depends on the size and shape of the bodies and their mutual location. It is proportional to the dielectric *permittivity* ε of the media where the bodies are located. The capacitance is measured in *farads* (F) if the charge is measured in coulombs (C) and the voltage in volts (V). One farad is a very big unit; practical capacitors have capacitances that are measured in micro- (μF, or 10^{-6}F), nano- (nF, or 10^{-9}F), and picofarads (pF, or 10^{-12}F).

The calculation of capacitance requires knowledge of the electrostatic field between the bodies. The following two theorems [2] are important in these calculations.

The integral of the flux density D over a closed surface is equal to the charge Q enclosed by the surface (the Gauss theorem), that is,

$$\oint D \, ds = Q \qquad (8.2)$$

This result is valid for linear and nonlinear dielectrics. For linear and isotropic media $D = \varepsilon E$, where E is the electric field. The magnitude E of the field is measured in volt per meter, the magnitude D of the flux in coulomb per square meter, and the dielectric permittivity has the dimension of farad per meter. The dielectric permittivity is usually represented as $\varepsilon = \varepsilon_0 K_d$ where ε_0 is the permittivity of air ($\varepsilon_0 = 8.86 \times 10^{-12}$ F/m) and K_d is the dielectric constant.

The electric field is defined by an electric potential φ. The directional derivative of the potential taken with the minus sign is equal to the component of the electric field in this direction. The voltage V_{AB} between the points A and B, having the potentials ϕ_A and ϕ_B, respectively (the potential is also measured in volts), is equal to

$$V_{AB} = \int_A^B E\,dl = \phi_A - \phi_B \tag{8.3}$$

This result is the second basic relationship. The left-hand side of Eq. (8.3) is a *line integral*. At each point of the line AB there exist two vectors: *E* defined by the field and *dl* which defines the direction of the line at this point.

8.2 Practical Capacitors

A wide variety of capacitors are in common usage. Capacitors are passive components in which current leads voltage by nearly 90° over a wide range of frequencies. Capacitors are rated by capacitance, voltage, materials, and construction.

A capacitor may have two voltage ratings:

- *Working voltage*—the normal operating voltage that should not be exceeded during operation
- Test or *forming voltage*—which stresses the capacitor and should occur only rarely in equipment operation

Good engineering practice dictates that components be used at only a fraction of their maximum ratings. The primary characteristics of common capacitors are given in Table 8.1. Some common construction practices are illustrated in Fig. 8.1.

Polarized/Nonpolarized Capacitors

Polarized capacitors can be used in only those applications where a positive sum of all DC and peak-AC voltages is applied to the positive capacitor terminal with respect to its negative terminal. These capacitors include all tantalum and most aluminum electrolytic capacitors. These devices are commonly used in power supplies or other electronic equipment where these restrictions can be met.

Nonpolarized capacitors are used in circuits where there is no direct voltage bias across the capacitor. They are also the capacitor of choice for most applications requiring capacity tolerances of 10% or less.

Operating Losses

Losses in capacitors occur because an actual capacitor has various resistances. These losses are usually measured as the *dissipation factor* at a frequency of 120 Hz. Leakage resistance in parallel with the capacitor defines the time constant of discharge of a capacitor. This time constant can vary between a small fraction of a second to many hours depending on capacitor construction, materials, and other electrical leakage paths, including surface contamination.

The *equivalent series resistance* of a capacitor is largely the resistance of the conductors of the capacitor plates and the resistance of the physical and chemical system of the capacitor. When an alternating current is applied to the capacitor, the losses in the equivalent series resistance are the major causes of heat developed in the device. The same resistance also determines the maximum attenuation of a filter or bypass capacitor and the loss in a coupling capacitor connected to a load.

The *dielectric absorption* of a capacitor is the residual fraction of charge remaining in a capacitor after discharge. The residual voltage appearing at the capacitor terminals after discharge is of little concern in

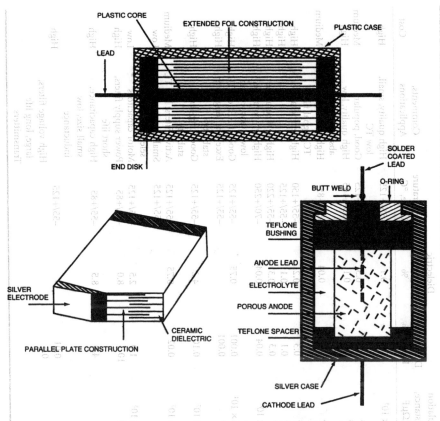

FIGURE 8.1 Construction of discrete capacitors. (From [1]. Used with permission.)

most applications but can seriously affect the performance of *analog-to-digital* (A/D) converters that must perform precision measurements of voltage stored in a sampling capacitor.

The *self-inductance* of a capacitor determines the high-frequency impedance of the device and its ability to bypass high-frequency currents. The self-inductance is determined largely by capacitor construction and tends to be highest in common metal foil devices.

Film Capacitors

Plastic is a preferred dielectrical material for capacitors because it can be manufactured with minimal imperfections in thin films. A metal-foil capacitor is constructed by winding layers of metal and plastic into a cylinder and then making a connection to the two layers of metal. A *metallized foil capacitor* uses two layers, each of which has a very thin layer of metal evaporated on one surface, thereby obtaining a higher capacity per volume in exchange for a higher equivalent series resistance. Metallized foil capacitors are self-repairing in the sense that the energy stored in the capacitor is often sufficient to burn away the metal layer surrounding the void in the plastic film.

Depending on the dielectric material and construction, capacitance tolerances between 1 and 20% are common, as are voltage ratings from 50 to 400 V. Construction types include axial leaded capacitors with a plastic outer wrap, metal-encased units, and capacitors in a plastic box suitable for printed circuit board insertion.

Polystyrene has the lowest dielectric absorption of 0.02%, a temperature coefficient of −20 to −100 ppm/°C, a temperature range to 85°C, and extremely low leakage. Capacitors between 0.001 and 2 µF can be obtained with tolerances from 0.1 to 10%.

TABLE 8.1 Parameters and Characteristics of Discrete Capacitors

Capacitor Type	Rated Voltage, V_R Range	TC ppm/°C	Tolerance, ±%	Insulation Resistance, MΩμF	Dissipation Factor, %	Dielectric Absorption, %	Temperature Range, °C	Comments, Applications	Cost
Polycarbonate	100 pF–30μF	±50	10	5×10^5	0.2	0.1	−55/+125	High quality, small, low TC	High
Polyester/Mylar	1000pF–50μF	+400	10	10^5	0.75	0.3	−55/+125	Good, popular	Medium
Polypropylene	100 pF–50μF	−200	10	10^5	0.2	0.1	−55/+105	High quality, low absorption	High
Polystyrene	10pF–2.7μF	−100	10	10^6	0.05	0.04	−55/+85	High quality, large, low TC, signal filters	Medium
Polysulfone	1000pF–1μF	+80	5	10^5	0.3	0.2	−55/+150	High temperature	High
Parylene	5000pF–1μF	±100	10	10^5	0.1	0.1	−55/+125	High temperature	High
Kapton	1000 pF–1μF	+100	10	10^5	0.3	0.3	−55/+220	High temperature, lowest absorption	High
Teflon	1000pF–2μF	−200	10	5×10^6	0.04	0.04	−70/+250		High
Mica	5 pF–0.01μF	−50	5	2.5×10^4	0.001		−55/+125	Good at RF, low TC	High
Glass	5 pF–1000pF	+140	5	10^6	0.001		−55/+125	Excellent long-term stability	High
Porcelain	100 pF–0.1μF	+120		5×10^5	0.10	4.2	−55/+125	Good long-term stability	High
Ceramic (NPO)	100 pF–1μF	±30		5×10^3	0.02	0.75	−55/+125	Active filters, low TC	Medium
Ceramic	10 pF–1μF	+800, +2500	−10/+100	5×10^3	1.0	2.5	−55/+125	Small, very popular selectable TC	Low
Paper	0.01μF–10μF			100	10	8.0	−40/+85	Motor capacitors	Low
Aluminum	0.1μF–1.6 F							Power supply filters, short life	High
Tantalum (Foil)	0.1μF–1000μF	+800		20	4.0	8.5	−55/+85	High capacitance, small size, low inductance	High
Thin-film	10 pF–200 pF	+100	10	10^6	0.01		−55/+125		High
Oil	0.1μF–20μF				0.5			High voltage filters, large, long life	
Vacuum	1 pF–1000 pF							Transmitters	High

(From [1]. Used with permission.)

Polycarbonate has an upper temperature limit of 100°C, with capacitance changes of about 2% up to this temperature. Polypropylene has an upper temperature limit of 85°C. These capacitors are particularly well suited for applications where high inrush currents occur, such as switching power supplies. Polyester is the lowest-cost material with an upper temperature limit of 125°C. Teflon and other high-temperature materials are used in aerospace and other critical applications.

Foil Capacitors

Mica capacitors are made of multiple layers of silvered mica packaged in epoxy or other plastic. Available in tolerances of 1 to 20% in values from 10 to 10,000 pF, mica capacitors exhibit temperature coefficients as low as 100 ppm. Voltage ratings between 100 and 600 V are common. Mica capacitors are used mostly in high-frequency filter circuits where low loss and high stability are required.

Electrolytic Capacitors

Aluminum foil electrolytic capacitors can be made nonpolar through the use of two cathode foils instead of anode and cathode foils in construction. With care in manufacturing, these capacitors can be produced with tolerance as tight as 10% at voltage ratings of 25 to 100 V peak. Typical values range from 1 to 1000 µF.

Ceramic Capacitors

Barium titanate and other ceramics have a high dielectric constant and a high breakdown voltage. The exact formulation determines capacitor size, temperature range, and variation of capacitance over that range (and consequently capacitor application). An alphanumeric code defines these factors, a few of which are given here.

- Ratings of Y5V capacitors range from 1000 pF to 6.8 µF at 25 to 100 V and typically vary +22 to −82% in capacitance from −30 to +85°C.
- Ratings of Z5U capacitors range to 1.5 µF and vary +22 to −56% in capacitance from +10 to +85°C. These capacitors are quite small in size and are used typically as bypass capacitors.
- X7R capacitors range from 470 pF to 1 µF and vary 15% in capacitance from −55 to +125°C.

Nonpolarized (NPO) rated capacitors range from 10 to 47,000 pF with a temperature coefficient of 0 to +30 ppm over a temperature range of −55 to +125°C.

Ceramic capacitors come in various shapes, the most common being the radial-lead disk. Multilayer monolithic construction results in small size, which exists both in radial-lead styles and as chip capacitors for direct surface mounting on a printed circuit board.

Polarized-Capacitor Construction

Polarized capacitors have a negative terminal—the cathode—and a positive terminal—the anode—and a liquid or gel between the two layers of conductors. The actual dielectric is a thin oxide film on the cathode, which has been chemically roughened for maximum surface area. The oxide is formed with a forming voltage, higher than the normal operating voltage, applied to the capacitor during manufacture. The direct current flowing through the capacitor forms the oxide and also heats the capacitor.

Whenever an electrolytic capacitor is not used for a long period of time, some of the oxide film is degraded. It is reformed when voltage is applied again with a leakage current that decreases with time. Applying an excessive voltage to the capacitor causes a severe increase in leakage current, which can cause the electrolyte to boil. The resulting steam may escape by way of the rubber seal or may otherwise damage the capacitor. Application of a reverse voltage in excess of about 1.5 V will cause forming to begin on the unetched anode electrode. This can happen when pulse voltages superimposed on a DC voltage cause a momentary voltage reversal.

Aluminum Electrolytic Capacitors

Aluminum electrolytic capacitors use very pure aluminum foil as electrodes, which are wound into a cylinder with an interlayer paper or other porous material that contains the electrolyte (see Fig. 8.2). Aluminum ribbon staked to the foil at the minimum inductance location is brought through the insulator to the anode terminal, while the cathode foil is similarly connected to the aluminum case and cathode terminal.

Electrolytic capacitors typically have voltage ratings from 6.3 to 450 V and rated capacitances from 0.47 μF to several hundreds of microfarads at the maximum voltage to several farads at 6.3 V. Capacitance tolerance may range from ±20 to +80/−20%. The operating temperature range is often rated from −25 to +85°C or wider. Leakage current of an electrolytic capacitor may be rated as low as 0.002 times the capacity times the voltage rating to more than 10 times as much.

Tantalum Electrolytic Capacitors

Tantalum electrolytic capacitors are the capacitors of choice for applications requiring small size, 0.33- to 100-μF range at 10 to 20% tolerance, low equivalent series resistance, and low leakage current. These devices are well suited where the less costly aluminum electrolytic capacitors have performance issues. Tantalum capacitors are packaged in hermetically sealed metal tubes or with axial leads in epoxy plastic, as illustrated in Fig. 8.3.

Capacitor Failure Modes

Mechanical failures relate to poor bonding of the leads to the outside world, contamination during manufacture, and shock-induced short-circuiting of the aluminum foil plates. Typical failure modes include short-circuits caused by foil impurities, manufacturing defects (such as burrs on the foil edges or tab connections), breaks or tears in the foil, and breaks or tears in the separator paper.

Short-circuits are the most frequent failure mode during the useful life period of an electrolytic capacitor. Such failures are the result of random breakdown of the dielectric oxide film under normal stress. Proper

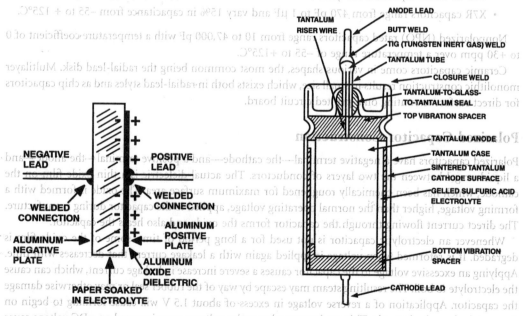

FIGURE 8.2 The basic construction of an aluminum electrolytic capacitor.

FIGURE 8.3 Basic construction of a tantalum capacitor.

capacitor design and processing will minimize such failures. Short-circuits also can be caused by excessive stress, where voltage, temperature, or ripple conditions exceed specified maximum levels.

Open circuits, although infrequent during normal life, can be caused by failure of the internal connections joining the capacitor terminals to the aluminum foil. Mechanical connections can develop an oxide film at the contact interface, increasing contact resistance and eventually producing an open circuit. Defective weld connections also can cause open circuits. Excessive mechanical stress will accelerate weld-related failures.

Temperature Cycling

Like semiconductor components, capacitors are subject to failures induced by thermal cycling. Experience has shown that thermal stress is a major contributor

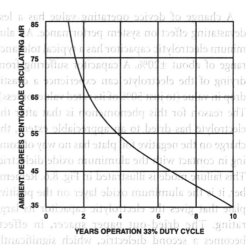

FIGURE 8.4 Life expectancy of an electrolytic capacitor as a function of operating temperature.

to failure in aluminum electrolytic capacitors. Dimensional changes between plastic and metal materials can result in microscopic ruptures at termination joints, possible electrode oxidation, and unstable device termination (changing series resistance). The highest-quality capacitor will fail if its voltage and/or current ratings are exceeded. Appreciable heat rise (20°C during a 2-h period of applied sinusoidal voltage) is considered abnormal and may be a sign of incorrect application of the component or impending failure of the device.

Figure 8.4 illustrates the effects of high ambient temperature on capacitor life. Note that operation at 33% duty cycle is rated at 10 years when the ambient temperature is 35°C, but the life expectancy drops to just 4 years when the same device is operated at 55°C. A common rule of thumb is this: In the range of +75°C through the full-rated temperature, stress and failure rates double for each 10°C increase in operating temperature. Conversely, the failure rate is reduced by half for every 10°C decrease in operating temperature.

Electrolyte Failures

Failure of the electrolyte can be the result of application of a reverse bias to the component, or of a drying of the electrolyte itself. Electrolyte vapor transmission through the end seals occurs on a continuous basis throughout the useful life of the capacitor. This loss has no appreciable effect on reliability during the useful life period of the product cycle. When the electrolyte loss approaches 40% of the initial electrolyte content of the capacitor, however, the electrical parameters deteriorate and the capacitor is considered to be worn out.

As a capacitor dries out, three failure modes may be experienced: leakage, a downward change in value, or *dielectric absorption*. Any one of these can cause a system to operate out of tolerance or fail altogether.

The most severe failure mode for an electrolytic is increased leakage, illustrated in Fig. 8.5. Leakage can cause loading of the power supply, or upset the DC bias of an amplifier. Loading of a supply line often causes additional current to flow through the capacitor, possibly resulting in dangerous overheating and catastrophic failure.

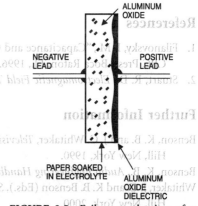

FIGURE 8.5 Failure mechanism of a leaky aluminum electrolytic capacitor. As the device ages, the aluminum oxide dissolves into the electrolyte, causing the capacitor to become leaky at high voltages.

A change of device operating value has a less devastating effect on system performance. An aluminum electrolytic capacitor has a typical tolerance range of about ±20%. A capacitor suffering from drying of the electrolyte can experience a drastic drop in value (to just 50% of its rated value, or less). The reason for this phenomenon is that after the electrolyte has dried to an appreciable extent, the charge on the negative foil plate has no way of coming in contact with the aluminum oxide dielectric. This failure mode is illustrated in Fig. 8.6. Remember, it is the aluminum oxide layer on the positive plate that gives the electrolytic capacitor its large rating. The dried-out paper spacer, in effect, becomes a second dielectric, which significantly reduces the capacitance of the device.

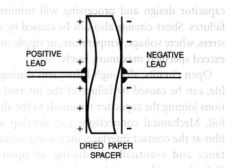

FIGURE 8.6 Failure mechanism of an electrolytic capacitor exhibiting a loss of capacitance. After the electrolyte dries, the plates can no longer come in contact with the aluminum oxide. The result is a decrease in capacitor value.

Capacitor Life Span

The life expectancy of a capacitor—operating in an ideal circuit and environment—will vary greatly, depending upon the grade of device selected. Typical operating life, according to capacitor manufacturer data sheets, ranges from a low of 3 to 5 years for inexpensive electrolytic devices to a high of greater than 10 years for computer-grade products. Catastrophic failures aside, expected life is a function of the rate of electrolyte loss by means of vapor transmission through the end seals, and the operating or storage temperature. Properly matching the capacitor to the application is a key component in extending the life of an electrolytic capacitor. The primary operating parameters include:

Rated voltage—the sum of the DC voltage and peak AC voltage that can be applied continuously to the capacitor. Derating of the applied voltage will decrease the failure rate of the device.

Ripple current—the rms value of the maximum allowable AC current, specified by product type at 120 Hz and +85°C (unless otherwise noted). The ripple current may be increased when the component is operated at higher frequencies or lower ambient temperatures.

Reverse voltage—the maximum voltage that can be applied to an electrolytic without damage. Electrolytic capacitors are polarized and must be used accordingly.

References

1. Filanovsky, I. M., "Capacitance and Capacitors," in *The Electronics Handbook*, J. C. Whitaker (Ed.), CRC Press, Boca Raton, FL, 1996.
2. Stuart, R. D., *Electromagnetic Field Theory*, Addison-Wesley, Reading, MA, 1965.

Further Information

Benson, K. B. and J. C. Whitaker, *Television and Audio Handbook for Technicians and Engineers*, McGraw-Hill, New York, 1990.
Benson, K. B., *Audio Engineering Handbook*, McGraw-Hill, New York, 1988.
Whitaker, J. C. and K. B. Benson (Eds.), *Standard Handbook of Video and Television Engineering*, McGraw-Hill, New York, 2000.
Whitaker, J. C., *Television Engineers' Field Manual*, McGraw-Hill, New York, 2000.

9
Inductors and Magnetic Properties

Jerry C. Whitaker
Editor-in-Chief

9.1 Introduction 9-1
   Electromagnetism • Magnetic Shielding
9.2 Inductors and Transformers.. 9-2
   Losses in Inductors and Transformers • Air-Core Inductors
   • Ferromagnetic Cores • Shielding

9.1 Introduction

The elemental magnetic particle is the spinning electron. In magnetic materials, such as iron, cobalt, and nickel, the electrons in the third shell of the atom are the source of magnetic properties. If the spins are arranged to be parallel, the atom and its associated domains or clusters of the material will exhibit a magnetic field. The magnetic field of a magnetized bar has lines of magnetic force that extend between the ends, one called the north pole and the other the south pole, as shown in Fig. 9.1a. The lines of force of a magnetic field are called *magnetic flux lines*.

Electromagnetism

A current flowing in a conductor produces a magnetic field surrounding the wire as shown in Fig. 9.2a. In a coil or solenoid, the direction of the magnetic field relative to the electron flow (– to +) is shown in Fig. 9.2b. The attraction and repulsion between two iron-core electromagnetic solenoids driven by direct currents is similar to that of two permanent magnets.

The process of magnetizing and demagnetizing an iron-core solenoid using a current being applied to a surrounding coil can be shown graphically as a plot of the magnetizing field strength and the resultant magnetization of the material, called a *hysteresis loop* (Fig. 9.3). It will be found that the point where the field is reduced to zero, a small amount of magnetization, called *remnance*, remains.

Magnetic Shielding

In effect, the shielding of components and circuits from magnetic fields is accomplished by the introduction of a magnetic short circuit in the path between the field source and the area to be protected. The flux from a field can be redirected to flow in a partition or shield of magnetic material, rather than in the normal distribution pattern between north and south poles. The effectiveness of shielding depends primarily upon the thickness of the shield, the material, and the strength of the interfering field.

Some alloys are more effective than iron. However, many are less effective at high flux levels. Two or more layers of shielding, insulated to prevent circulating currents from magnetization of the shielding, are used in low-level audio, video, and data applications.

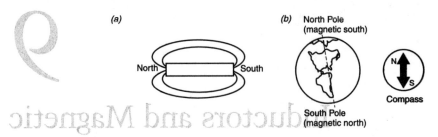

FIGURE 9.1 The properties of magnetism: (a) lines of force surrounding a bar magnet, (b) relation of compass poles to the earth's magnetic field.

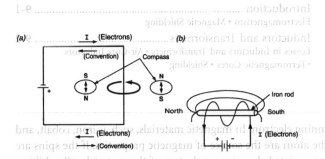

FIGURE 9.2 Magnetic field surrounding a current-carrying conductor: (a) Compass at right indicates the polarity and direction of a magnetic field circling a conductor carrying direct current. *I* Indicates the direction of electron flow. Note: The convention for flow of electricity is from + to −, the reverse of the actual flow. (b) Direction of magnetic field for a coil or solenoid.

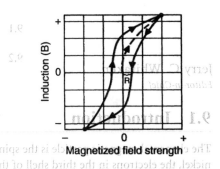

FIGURE 9.3 Graph of the magnetic hysteresis loop resulting from magnetization and demagnetization of iron. The dashed line is a plot of the induction from the initial magnetization. The solid line shows a reversal of the field and a return to the initial magnetization value. *R* is the remaining magnetization (remnance) when the field is reduced to zero.

9.2 Inductors and Transformers

Inductors are passive components in which voltage leads current by nearly 90° over a wide range of frequencies. Inductors are usually coils of wire wound in the form of a cylinder. The current through each turn of wire creates a magnetic field that passes through every turn of wire in the coil. When the current changes, a voltage is induced in the wire and every other wire in the changing magnetic field. The voltage induced in the same wire that carries the changing current is determined by the inductance of the coil, and the voltage induced in the other wire is determined by the *mutual inductance* between the two coils. A transformer has at least two coils of wire closely coupled by the common magnetic core, which contains most of the magnetic field within the transformer.

Inductors and transformers vary widely in size, weighing less than 1 g or more than 1 ton, and have specifications ranging nearly as wide.

Losses in Inductors and Transformers

Inductors have resistive losses because of the resistance of the copper wire used to wind the coil. An additional loss occurs because the changing magnetic field causes *eddy currents* to flow in every conductive material in the magnetic field. Using thin magnetic laminations or powdered magnetic material reduces these currents.

Losses in inductors are measured by the Q, or quality, factor of the coil at a test frequency. Losses in transformers are sometimes given as a specific insertion loss in decibels. Losses in power transformers are given as core loss in watts when there is no load connected and as a regulation in percent, measured as the relative voltage drop for each secondary winding when a rated load is connected.

Transformer loss heats the transformer and raises its temperature. For this reason, transformers are rated in watts or volt-amperes and with a temperature code designating the maximum hotspot temperature allowable for continued safe long-term operation. For example, class A denotes 105°C safe operating temperature. The volt-ampere rating of a power transformer must always be larger than the DC power output from the rectifier circuit connected because volt-amperes, the product of the rms currents and rms voltages in the transformer, are larger by a factor of about 1.6 than the product of the DC voltages and currents.

Inductors also have capacitance between the wires of the coil, which causes the coil to have a self-resonance between the winding capacitance and the self-inductance of the coil. Circuits are normally designed so that this resonance is outside of the frequency range of interest. Transformers are similarly limited. They also have capacitance to the other winding(s), which causes *stray coupling*. An electrostatic shield between windings reduces this problem.

Air-Core Inductors

Air-core inductors are used primarily in radio frequency applications because of the need for values of inductance in the microhenry or lower range. The usual construction is a multilayer coil made self-supporting with adhesive-covered wire. An inner diameter of 2 times coil length and an outer diameter 2 times as large yields maximum Q, which is also proportional to coil weight.

Ferromagnetic Cores

Ferromagnetic materials have a permeability much higher than air or vacuum and cause a proportionally higher inductance of a coil that has all its magnetic flux in this material. Ferromagnetic materials in audio and power transformers or inductors usually are made of silicon steel laminations stamped in the form of the letters E or I (Fig. 9.4). At higher frequencies, powdered ferric oxide is used. The continued magnetization and remagnetization of silicon steel and similar materials in opposite directions does not follow the same path in both directions but encloses an area in the magnetization curve and causes a hysteresis loss at each pass, or twice per AC cycle.

All ferromagnetic materials show the same behavior; only the numbers for permeability, core loss, saturation flux density, and other characteristics are different. The properties of some common magnetic materials and alloys are given in Table 9.1.

Shielding

Transformers and coils radiate magnetic fields that can induce voltages in other nearby circuits. Similarly, coils and transformers can develop voltages in their windings when subjected to magnetic fields from another transformer, motor, or power circuit. Steel mounting frames or chassis conduct these fields, offering less reluctance than air.

The simplest way to reduce the stray magnetic field from a power transformer is to wrap a copper strip as wide as the coil of wire around the transformer enclosing all three legs of the core. Shielding occurs by having a short circuit turn in the stray magnetic field outside of the core.

material in the magnetic field. Using thin magnetic laminations or powdered magnetic material reduces these currents.

Losses in inductors are measured by the quality factor of the coil at a test frequency. Losses in transformers are sometimes given as a specific insertion loss in decibels. Losses in power transformers are given as core loss in watts when there is no load connected and as a regulation in percent, measured as the relative voltage drop from each secondary feeding when a rated load is connected.

Transformer loss heats the transformer and with it the temperature. For this reason, transformers are rated in watts or volt-amperes and with letters which are used designating the maximum hotspot temperature allowable for continuous long-term operation. For example, class A denotes 105°C safe operating temperature. The volt-ampere rating of a power transformer must always be larger than the DC power output from the rectifier circuit connected to it. For rms volt-amperes, the product of the rms currents and rms voltages in the transformer windings are larger by a factor of about 1.6 than the product of the DC voltages and currents.

Inductors also have capacitance between the windings of the coil, which causes the coil to have a self-resonance between the winding capacitance and the self-inductance of the coil. Circuits are normally designed so that this resonance is outside of the frequency range of interest. Transformers are similarly limited. They also have capacitance to the other winding(s), which causes stray coupling. An electrostatic shield between windings reduces this problem.

Air-Core Inductors

Air-core inductors are used primarily in radio frequency applications because of the need for values of inductance in the microhenry range. The simplest construction is a multilayer coil made self-supporting with adhesive-covered wire. An inner diameter of 0.5 times coil length and an outer diameter 2 times as large yields maximum inductance; it is also practical from coil weight.

Ferromagnetic Cores

Ferromagnetic materials have a permeability much higher than air or vacuum and cause a proportionally higher inductance of a coil that has all its magnetic flux in this material. Ferromagnetic materials in power transformers also cause hysteresis loss, principally because of Weiss' elementary magnets, which

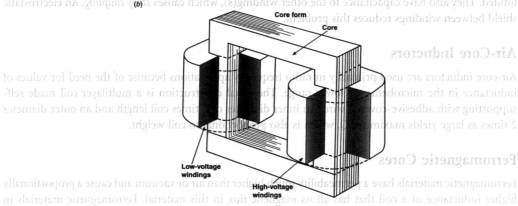

FIGURE 9.4 Physical construction of a power transformer: (*a*) E-shaped device with the low- and high-voltage windings stacked as shown. (*b*) Construction using a box core with physical separation between the low- and high-voltage windings.

follow the same path in each pass, or twice per AC cycle. The difference in the magnetization curve due to magnetization and remagnetization of silicon steel and similar materials in opposite directions encloses an area in the magnetization curve and causes a hysteresis loss at each pass, or twice per AC cycle.

All ferromagnetic materials show the same behavior; only the numbers for permeability, core loss, saturation flux density and other characteristics are different. The properties of some common magnetic materials and alloys are given in Table 9.1.

Shielding

Transformers and coils radiate magnetic fields that can induce voltages in other nearby circuits. Similarly coils and transformers can develop voltages in their windings when subjected to magnetic fields from another transformer, motor, or power circuit. Steel mounting frames or chassis conduct these fields, offering less reluctance than air.

The simplest way to reduce the stray magnetic field from a power transformer is to wrap a copper strip as wide as the coil of wire around the transformer enclosing all three legs of the core. Shielding occurs by having a short circuit turn in the stray magnetic field outside of the core.

Inductors and Magnetic Properties

TABLE 9.1 Properties of Magnetic Materials and Magnetic Alloys

Material (Composition)	Initial Relative Permeability, μ_r/μ_0	Maximum Relative Permeability, μ_{max}/μ_0	Coercive Force H_c, A/m (Oe)	Residual Field B_r, Wb/m²(G)	Saturation Field B_s, Wb/m²(G)	Electrical Resistivity $\rho \times 10^{-8}\ \Omega \cdot m$	Uses
Commercial iron (0.2 imp.)	250	9,000	≈80 (1)	0.77 (7,700)	2.15 (21,500)	10	Relays
Purified iron (0.05 imp.)	10,000	200,000	4 (0.05)	—	2.15 (21,500)	10	
Silicon-iron (4 Si)	1,500	7,000	20 (0.25)	0.5 (5,000)	1.95 (19,500)	60	Transformers
Silicon-iron (3 Si)	7,500	55,000	8 (0.1)	0.95 (9,500)	2 (20,000)	50	Transformers
Silicon-iron (3 Si)	—	116,000	4.8 (0.06)	1.22 (12,200)	2 (20,100)	50	Transformers
Mu metal (5 Cu, 2 Cr, 77 Ni)	20,000	100,000	4 (0.05)	0.23 (2,300)	0.65 (6,500)	62	Transformers
78 Permalloy (78.5 Ni)	8,000	100,000	4 (0.05)	0.6 (6,000)	1.08 (10,800)	16	Sensitive relays
Supermalloy (79 Ni, 5 Mo)	100,000	1,000,000	0.16 (0.002)	0.5 (5,000)	0.79 (7,900)	60	Transformers
Permendur (50 Cs)	800	5,000	160 (2)	1.4 (14,000)	2.45 (24,500)	7	Electromagnets
Mn-Zn ferrite	1,500	2,500	16 (0.2)	—	0.34 (3,400)	20×10^6	Core material for coils
Ni-Zn ferrite	2,500	5,000	8 (0.1)	—	0.32 (3,200)	10^{11}	

After Plonus, M.A., *Applied Electromagnetics*, McGraw-Hill, New York, 1978.

TABLE 9.2 Magnetic Properties of Transformer Steels

	Ordinary Transformer Steel		High Silicon Transformer Steels		
B (Gauss)	H (Oersted)	Permeability = B / H	B	H	Permeability
2,000	0.60	3,340	2,000	0.50	4,000
4,000	0.87	4,600	4,000	0.70	5,720
6,000	1.10	5,450	6,000	0.90	6,670
8,000	1.48	5,400	8,000	1.28	6,250
10,000	2.28	4,380	10,000	1.99	5,020
12,000	3.85	3,120	12,000	3.60	3,340
14,000	10.9	1,280	14,000	9.80	1,430
16,000	43.0	372	16,000	47.4	338
18,000	149	121	18,000	165	109

(From [1]. Used with permission.)

TABLE 9.3 Characteristics of High-Permeability Materials

Material	Form	Approximate % Composition					Typical Heat Treatment, °C	Permeability at $B = 20$, G	Maximum Permeability	Saturation Flux Density B, G	Hysteresis[a] Loss, W_b, ergs/cm³	Coercive[a] Force H_cO	Resistivity $\mu \cdot \Omega$cm	Density, g/cm³
		Fe	Ni	Co	Mo	Other								
Cold rolled steel	Sheet	98.5	—	—	—	—	950 Anneal	180	2,000	21,000	20	1.8	10	7.88
Iron	Sheet	99.91	—	—	—	—	950 Anneal	200	5,000	21,500	35,000	1.0	10	7.88
Purified iron	Sheet	99.95	—	—	—	—	1480 H_2 + 880	5,000	180,000	21,500	300	0.05	10	7.88
4% Silicon-iron	Sheet	96	—	—	—	4 Si	800 Anneal	500	7,000	19,700	3,500	0.5	60	6.65
Grain oriented[b]	Sheet	97	—	—	—	3 Si	800 Anneal	1,500	30,000	20,000	—	0.15	47	7.67
45 Permalloy	Sheet	54.7	45	—	—	0.3 Mn	1050 Anneal	2,500	25,000	15,000	1,200	0.3	45	8.17
45 Permalloy[c]	Sheet	54.7	45	—	—	0.3 Mn	1200 H_2 Anneal	4,000	50,000	15,000	—	0.07	45	8.17
Hipernik	Sheet	50	50	—	—	—	1200 H_2 Anneal	4,500	70,000	15,000	220	0.05	50	8.25
Monimax	Sheet	—	—	—	—	—	1125 H_2 Anneal	2,000	35,000	15,000	—	0.1	80	8.27
Sinimax	Sheet	—	—	—	—	—	1125 H_2 Anneal	3,000	35,000	11,000	—	—	90	—
78 Permalloy	Sheet	21.2	78.5	—	—	0.3 Mn	1050 + 600 Q[d]	8,000	100,000	10,700	200	0.05	16	8.60
4-79 Permalloy	Sheet	16.7	79	—	4	0.3 Mn	1100 + Q	20,000	100,000	8,700	200	0.05	55	8.72
Mu metal	Sheet	18	75	—	—	2 Cr, 5 Cu	1175 H_2	20,000	100,000	6,500	—	0.05	62	8.58
Supermalloy	Sheet	15.7	79	—	5	0.3 Mn	1300 H_2 + Q	100,000	800,000	8,000	—	0.002	60	8.77
Permendur	Sheet	49.7	—	50	—	0.3 Mn	800 Anneal	800	5,000	24,500	12,000	2.0	7	8.3
2 V Permendur	Sheet	49	—	49	—	2 V	800 Anneal	—	4,500	24,000	6,000	2.0	26	8.2
Hiperco	Sheet	64	—	34	—	Cr	850 Anneal	650	10,000	24,000	—	1.0	25	8.0
2-81 Permalloy	Insulated powder	17	81	—	2	—	650 Anneal	125	130	8,000	—	<1.0	10⁶	7.8
Carbonyl iron	Insulated powder	99.9	—	—	—	—	—	55	132	—	—	—	—	7.86
Ferroxcube III	Sintered powder					MnFe₂O₄ + ZnFe₂O₄	—	1,000	1,500	2,500	—	0.1	10⁸	5.0

(From [1]. Used with permission.)

[a] At saturation.
[b] Properties in direction of rolling.
[c] Similar properties for Nicaloi, 4750 alloy, Carpenter 49, Armco 48.
[d] Q, quench or controlled cooling.

TABLE 9.4 Characteristics of Permanent Magnet Alloys

Material	% Composition (remainder Fe)	Heat Treatment[a] (temperature, °C)	Magnetizing Force H_{max} O	Coercive Force H_c O	Residual Induction B_r, G	Energy Product $BH_{max} \times 10^{-6}$	Method of Fabrication[b]	Mechanical Properties[c]	Weight lb/In.^{3}
Carbon steel	1 Mn, 0.9 C	Q 800	300	50	10,000	0.20	HR, M, P	H, S	0.280
Tungsten steel	5 W, 0.3 Mn, 0.7 C	Q 850	300	70	10,300	0.32	HR, M, P	H, S	0.292
Chromium steel	3.5 Cr, 0.9 C, 0.3 Mn	Q 830	300	65	9,700	0.30	HR, M, P	H, S	0.280
17% Cobalt steel	17 Co, 0.75 C, 2.5 Cr, 8 W		1,000	150	9,500	0.65	HR, M, P	H, S	0.296
36% Cobalt steel	36 Co, 0.7 C, 4 Cr, 5 W	Q 950	1,000	240	9,500	0.97	HR, M, P	H, S	0.295
Remalloy or Comol	17 Mo, 12 Co		1,000	250	10,500	1.1	HR, M, P	H	0.249
Alnico I	12 Al, 20 Ni, 5 Co	Q 1200, B 700	2,000	440	7,200	1.4	C, G	H, B	0.256
Alnico II	10 Al, 17 Ni, 2.5 Co, 6 Cu	A 1200, B 700	2,000	550	7,200	1.6	C, G	H, B	0.249
Alnico II (sintered)	10 Al, 17 Ni, 2.5 Co, 6 Cu	A 1300	2,000	520	6,900	1.4	Sn, G	H	0.253
Alnico IV	12 Al, 28 Ni, 5 Co	Q 1200, B 650	3,000	700	5,500	1.3	Sn, C, G	H	0.264
Alnico V	8 Al, 14 Ni, 24 Co, 3 Cu	AF 1300, B 600	2,000	550	12,500	4.5	C, G	H, B	0.268
Alnico VI	8 Al, 15 Ni, 24 Co, 3 Cu, 1 Ti		3,000	750	10,000	3.5	C, G	H, B	0.26
Alnico XII	6 Al, 18 Ni, 35 Co, 8 Ti		3,000	950	5,800	1.5	C, G	H, B	0.295
Vicalloy I	52 Co, 10 V	B 600	1,000	300	8,800	1.0	C, CR, M, P	D	0.292
Vicalloy II (wire)	52 Co, 14 V	CW + B 600	2,000	510	10,000	3.5	C, CR, M, P	D	0.311
Cunife (wire)	60 Cu, 20 Ni	CW + B 600	2,400	550	5,400	1.5	C, CR, M, P	D, M	0.300
Cunico	50 Cu, 21 Ni, 29 Co		3,200	660	3,400	0.80	C, CR, M, P	D, M	0.113
Vectolite	30Fe$_2$O$_3$, 44Fe$_3$O$_4$, 26Co$_2$O$_3$		3,000	1,000	1,600	0.60	Sn, G	W	0.325
Silmanal	86.8Ag, 8.8Mn, 4.4Al		20,000	6,000[d]	550	0.075	C, CR, M	D, M	0.176
Platinum-cobalt	77 Pt, 23 Co	Q 1200, B 650	15,000	3,600	5,900	6.5	C, CR, M	D	
Hyflux	Fine powder		2,000	390	6,600	0.97		D	

(From [1]. Used with permission.)

[a] Q, quenched in oil or water; A, air cooled; B, baked; F, cooled in magnetic field; CW, cold worked.
[b] HR, hot rolled or forged; CR, cold rolled or drawn; M, machined; G, must be ground; P, punched; C, cast; Sn, sintered.
[c] H, hard; B, brittle; S, strong; D, ductile; M, malleable; W, weak.
[d] Value given is intrinsic H_c.

TABLE 9.5 Properties of Antiferromagnetic Compounds

Compound	Crystal Symmetry	θ_N^a K	θ_p^b K	$(P_A)_{eff}^c \mu_B$	$P_A^d \mu_B$
$CoCl_2$	Rhombohedral	25	−38.1	5.18	3.1 ± 0.6
CoF_2	Tetragonal	38	50	5.15	3.0
CoO	Tetragonal	291	330	5.1	3.8
Cr	Cubic	475			
Cr_2O_3	Rhombohedral	307	485	3.73	3.0
CrSb	Hexagonal	723	550	4.92	2.7
$CuBr_2$	Monoclinic	189	246	1.9	
$CuCl_2 \cdot 2H_2O$	Orthorhombic	4.3	4–5	1.9	
$CuCl_2$	Monoclinic	~70	109	2.08	
$FeCl_2$	Hexagonal	24	−48	5.38	4.4 ± 0.7
FeF_2	Tetragonal	79–90	117	5.56	4.64
FeO	Rhombohedral	198	507	7.06	3.32
$\alpha\text{-}Fe_2O_3$	Rhombohedral	953	2940	6.4	5.0
α-Mn	Cubic	95			
$MnBr_2 \cdot 4H_2O$	Monoclinic	2.1	$\begin{Bmatrix} 2.5 \\ 1.3 \end{Bmatrix}$	5.93	
$MnCl_2 \cdot 4H_2O$	Monoclinic	1.66	1.8	5.94	
MnF_2	Tetragonal	72–75	113.2	5.71	5
MnO	Rhombohedral	122	610	5.95	5.0
β-MnS	Cubic	160	982	5.82	5.0
MnSe	Cubic	~173	361	5.67	
MnTe	Hexagonal	310–323	690	6.07	5.0
$NiCl_2$	Hexagonal	50	−68	3.32	
NiF_2	Tetragonal	78.5–83	115.6	3.5	2.0
NiO	Rhombohedral	533–650	~2000	4.6	2.0
$TiCl_3$		100			
V_2O_3		170			

(From [1]. Used with permission.)

[a] θ_N = Néel temperature, determined from susceptibility maxima or from the disappearance of magnetic scattering.
[b] θ_p = a constant in the Curie-Weiss law written in the form $C_A = C_A/(T + \theta_p)$, which is valid for antiferromagnetic material for $T > \theta_N$.
[c] $(P_A)_{eff}$ = effective moment per atom, derived from the atomic Curie constant $C_A = (P_A)^2_{eff}(N^2/3R)$ and expressed in units of the Bohr magneton, $\mu_B = 0.9273 \times 10^{-20}$ erg G^{-1}.
[d] P_A = magnetic moment per atom, obtained from neutron diffraction measurements in the ordered state.

TABLE 9.6 Saturation Constants for Magnetic Substances

Substance	Field Intensity	Induced Magnetization
	(For Saturation)	
Cobalt	9,000	1,300
Iron, wrought	2,000	1,700
cast	4,000	1,200
Manganese steel	7,000	200
Nickel, hard	8,000	400
annealed	7,000	515
Vicker's steel	15,000	1,600

(From [1]. Used with permission.)

TABLE 9.7   Saturation Constants and Curie Points of Ferromagnetic Elements

Element	σ_s^a (20°C)	M_s^b (20°C)	σ_s (0 K)	n_B^c	Curie point, °C
Fe	218.0	1,714	221.9	2.219	770
Co	161	1,422	162.5	1.715	1,131
Ni	54.39	484.1	57.50	0.604	358
Gd	0	0	253.5	7.12	16

(From [1]. Used with permission.)

[a] σ_s = saturation magnetic moment/gram.
[b] M_s = saturation magnetic moment/cm^3, in cgs units.
[c] n_B = magnetic moment per atom in Bohr magnetons.

References

1. Whitaker, J. C. (Ed.), *The Electronics Handbook*, CRC Press, Boca Raton, FL, 1996.

Further Information

Benson, K. Blair, and J. C. Whitaker, *Television and Audio Handbook for Technicians and Engineers*, McGraw-Hill, New York, 1990.
Benson, K. B., *Audio Engineering Handbook*, McGraw-Hill, New York, 1988.
Whitaker, J. C., *Television Engineers' Field Manual*, McGraw-Hill, New York, 2000.

10
Printed Wiring Boards

Ravindranath Kollipara
LSI Logic Corporation

Vijai Tripathi
Oregon State University, Corvallis

10.1 Introduction 10-1
10.2 Board Types, Materials, and Fabrication 10-2
10.3 Design of Printed Wiring Boards 10-6
10.4 PWB Interconnection Models 10-8
10.5 Signal-Integrity and EMC Considerations 10-14

10.1 Introduction

Printed wiring board (PWB) is, in general, a layered dielectric structure with internal and external wiring that allows electronic components to be mechanically supported and electrically connected internally to each other and to the outside circuits and systems. The components can be complex packaged very large-scale integrated (VLSI), RF, and other ICs with multiple I/Os or discrete surface mount active and passive components. PWBs are the most commonly used packaging medium for electronic circuits and systems. Electronic packaging has been defined as the design, fabrication, and testing process that transforms an electronic circuit into a manufactured assembly. The main functions of the packagings include signal distribution to electronic circuits that process and store information, power distribution, heat dissipation, and the protection of the circuits.

The type of assembly required depends on the electronic circuits, which may be discrete, integrated, or hybrid. Integrated circuits are normally packaged in plastic or ceramic packages, which are electrically connected to the outside I/O connections and power supplies with pins that require plated through holes (PTH) or pins and pads meant for surface mounting the package by use of the surface mount technology (SMT). These plastic or ceramic packages are connected to the PWBs by using the pins or the leads associated with the packages.

In addition to providing a framework for interconnecting components and packages, such as IC packages, PWBs can also provide a medium (home) for component design and placement such as inductors and capacitors. PWBs are, generally, a composite of organic and inorganic dielectric material with multiple layers. The interconnects or the wires in these layers are connected by via holes, which can be plated and are filled with metal to provide the electrical connections between the layers. In addition to the ground planes, and power planes used to distribute bias voltages to the ICs and other discrete components, the signal lines are distributed among various layers to provide the interconnections in an optimum manner. The properties of importance that need to be minimized for a good design are the signal delay, distortion, and crosstalk noise induced primarily by the electromagnetic coupling between the signal lines. Coupling through substrate material and ground bounce and switching noise due to imperfect power/ground planes also leads to degradation in signal quality.

In this chapter, the basic properties and applications of PWBs are introduced. These include the physical properties, the wiring board design, and the electrical properties. The reader is referred to other sources for detailed discussion on these topics.

10.2 Board Types, Materials, and Fabrication

The PWBs can be divided into four types of boards: (1) rigid boards, (2) flexible and rigid-flex boards, (3) metal-core boards, and (4) injection molded boards. The board that is most widely used is the rigid board. The boards can be further classified into single-sided, double-sided, or multilayer boards. The ever increasing packaging density and faster propagation speeds, which stem from the demand for high-performance systems, have driven the evolution of the boards from single-sided to double-sided to multilayer boards. On single-sided boards, all of the interconnections are on one side. Double-sided boards have connections on both sides of the board and allow wires to cross over each other without the need for jumpers. This was accomplished at first by Z-wires, then by eyelets, and at present by PTHs. The increased pin count of the ICs has increased the routing requirements, which led to multilayer boards. The necessity of a controlled impedance for the high-speed traces, the need for bypass capacitors, and the need for low inductance values for the power and ground distribution networks have made the requirement of power and ground planes a must in high-performance boards, These planes are possible only in multilayer boards. In multilayer boards, the PTHs can be buried (providing interconnection between inner layers), semiburied (providing interconnection from one of the two outer layers to one of the internal layers), or through vias (providing interconnection between the outer two layers).

The properties that must be considered when choosing PWB substrate materials are their mechanical, electrical, chemical, and thermal properties. Early PWBs consisted of copper-clad phenolic and paper laminate materials. The copper was patterned using resists and etched. Holes were punched in the laminate to plug the component leads, and the leads were soldered to the printed copper pattern. At present, the copper-clad laminates and the prepregs are made with a variety of different matrix resin systems and reinforcements [ASM 1989]. The most commonly used resin systems are fire resistant (FR-4) difunctional and polyfunctional epoxies. Their glass transition temperatures T_g range from 125 to 150°C. They have well-understood processability, good performance, and low price. Other resins include high-temperature, one-component epoxy, polyimide, cyanate esters, and polytetrafluoroethylene (PTFE) (trade name Teflon). Polyimide resins have high T_g, long-term thermal resistance, low coefficient of thermal expansion (CTE), long PTH life and high reliability, and are primarily used for high-performance multilayer boards with a large number of layers. Cyanate esters have low dielectric constants and high T_g, and are used in applications where increased signal speed and improved laminate-dimensional stability are needed. Teflon has the lowest dielectric constant, low dissipation factor, and excellent temperature stability but is difficult to process and has high cost. It is used mainly in high-performance applications where higher densities and transmission velocities are required.

The most commonly used reinforcement is continuous filament electrical (E-) glass. Other high-performance reinforcements include high strength (S-) glass, high modulus (D-) glass, and quartz [Seraphim et al. 1989]. The chemical composition of these glasses determines the key properties of CTE and dielectric constant. As the level of silicon dioxide (SiO_2) in the glass increases, the CTE and the dielectric constant decrease. The substrates of most rigid boards are made from FR-4 epoxy resin-impregnated E-glass cloth. Rolls of glass cloth are coated with liquid resin (A-stage). Then the resin is partially cured to a semistable state (B-stage or prepreg). The rolls are cut into large sheets and several sheets are stacked to form the desired final thickness. If the laminates are to be copper clad, then copper foils form the outside of the stack. The stack is then laminated and cured irreversibly to form the final resin state (C-stage) [ASM 1989].

The single-sided boards typically use phenolic or polyester resins with random mat glass or paper reinforcement. The double-sided boards are usually made of glass-reinforced epoxy. Most multilayer boards are also made of glass-reinforced epoxy. The internal circuits are made on single- or double-sided copper-clad laminates. The inner layers are stacked up with B-stage polymer sheets separating the layers. Rigid pins are used to establish layer-to-layer orientation. The B-stage prepreg melts during lamination and reflows. When it is cured, it glues the entire package into a rigid assembly [ASM 1989]. An alternative approach to pin-parallel composite building is a sequential buildup of the layers, which allows buried vias. Glass reinforced polyimide is the next most used multilayer substrate material due to its excellent

Printed Wiring Boards

handling strength and its higher temperature cycling capability. Other laminate materials include Teflon and various resin combinations of epoxy, polyimide, and cyanate esters with reinforcements. The dielectric thickness of the flat laminated sheets ranges from 0.1 to 3.18 mm (0.004 to 0.125 in), with 1.5 mm (0.059 in) being the most commonly used for single- and double-sided boards. The inner layers of a multilayer board are thinner with a typical range of 0.13–0.75 mm (0.005–0.030 in) [ASM 1989]. The number of layers could be 20 or more. The commonly used substrate board materials and their properties are listed in Table 10.1.

TABLE 10.1 Wiring Board Material Properties

Material	ε_r^*	$\varepsilon_r^*/\varepsilon_r^*$	CTE ($\times 10^{-6}/°C$) x, y	z	$T_g, °C$
FR-4 epoxy-glass	4.0–5.0	0.02–0.03	16–20	60–90	125–135
Polyimide-glass	3.9–4.5	0.005–0.02	12–14	60	>260
Teflon	2.1	0.0004–0.0005	70–120		—
Benzocyclobutane	2.6	0.0004	35–60		>350
High-temperature one-component epoxy-glass	4.45–4.55	0.02–0.022			170–180
Cyanate ester-glass	3.5–3.9	0.003–0.007			240–250
Ceramic	~10.0	0.0005	6–7		
Copper	—	—	17		
Copper/Invar/Copper	—	—	3–6		—

The most common conductive material used is copper. The substrate material can be purchased as copper-clad laminates or unclad laminates. The foil thickness of copper-clad laminates are normally expressed in ounces of copper per square foot. The available range is from 1/8 to 5 oz with 1-oz copper (0.036 mm or 0.0014 in) being the most commonly used cladding. The process of removing copper in the unwanted regions from the copper-clad laminates is called subtractive technology. If copper or other metal is added on to the unclad laminates, the process is called additive technology. Other metals used during electroplating may include Sn, Pb, Ni, Au, and Pd. Selectively screened or stenciled pastes that contain silver or carbon are also used as conducting materials [ASM 1989].

Solder masks are heat- and chemical-resistant organic coatings that are applied to the PWB surfaces to limit solderability to the PTHs and surface-mount pads, provide a thermal and electrical insulation layer isolating adjacent circuitry and components, and protect the circuitry from mechanical and handling damage, dust, moisture, and other contaminants. The coating materials used are thermally cured epoxies or ultraviolet curable acrylates. Pigments or dyes are added for color. Green is the standard color. Fillers and additives are added to modify rheology and improve adhesion [ASM 1989].

Flexible and rigid-flex boards are required in some applications. A flexible printed board has a random arrangement of printed conductors on a flexible insulating base with or without cover layers [ASM 1989]. Like the base material, conductor, adhesive, and cover layer materials also should be flexible. The boards can be single-sided, double-sided or multilayered. However, multilayered boards tend to be too rigid and are prone to conductor damage. Rigid-flex boards are like multilayered boards with bonding and connections between layers confined to restricted areas of the wiring plane. Connections between the rigid laminated areas are provided by multiconductor layers sandwiched between thin base layers and are flexible. The metal claddings are made of copper foil, beryllium copper, aluminium, Inconel, or conductive polymer thick films with copper foil being the most commonly used. Typical adhesives systems used include polyester, epoxy/modified epoxy, acrylic, phenolics, polyimide, and fluorocarbons. Laminates, which eliminate the use of adhesives by placing conductors directly on the insulator, are called adhesiveless laminates. Dielectric base materials include polyimide films, polyester films, aramids, reinforced composites, and fluorocarbons. The manufacturing steps are similar to those of the rigid boards. An insulating film or coating applied over the conductor side acts as a permanent protective cover. It protects the conductors from moisture, contamination, and damage and reduces stress on conductors during flexing.

Pad access holes and registration holes are drilled or punched in an insulating film coated with an adhesive, and the film is aligned over the conductor pattern and laminated under heat and pressure. Often, the same base material is used as the insulating film. When coatings are used instead of films, they are screen printed onto the circuit, leaving pad areas exposed. The materials used are acrylated epoxy, acrylated polyurethane, and thiolenes, which are liquid polymers and are cured using ultraviolet (UV) radiation or infrared (IR) heating to form a permanent, thin, tough coating [ASM 1989].

When selecting a board material, the thermal expansion properties of the material must be a consideration. If the components and the board do not have a closely matched CTE, then the electrical and/or mechanical connections may be broken and reliability of the board will suffer. The widely used epoxy-glass PWB material has a CTE that is larger than that of the encapsulating material (plastic or ceramic) of the components. When leaded devices in dual-in-line (DIP) package format are used, the mechanical forces generated by the mismatches in the CTEs are taken by the PTHs, which accommodate the leads and provide electrical and mechanical connection. When surface mount devices (SMDs) are packaged, the solder joint, which provides both the mechanical and electrical connection, can accommodate little stress without deformation. In such cases, the degree of component and board CTE mismatch and the thermal environment in which the board operates must be considered and if necessary, the board must be CTE tailored for the components to be packaged. The three typical approaches to CTE tailoring are: constraining dielectric structures, constraining metal cores, and constraining low-CTE metal planes [ASM 1989, Seraphim et al. 1989].

Constraining inorganic dielectric material systems include cofired ceramic printed boards and thick-film ceramic wiring boards. Cofired ceramic multilayer board technology uses multiple layers of green ceramic tape into which via holes are made and on which circuit metallization is printed. The multiple layers are aligned and fired at high temperatures to yield a much higher interconnection density capability and much better controlled electrical properties. The ceramic unit contains more than 90% of alumina and has a much higher thermal conductivity than that of epoxy-glass making heat transfer efficient. The thick-film ceramic wiring boards are usually made of cofired alumina and are typically bonded to an aluminium thermal plane with a flexible thermally conducting adhesive when heat removal is required. Both organic and inorganic fiber-reinforcements can be used with the conventional PWB resin for CTE tailoring. The organic fibers include several types of aramid fibers, notably Kevlar 29 and 49 and Technora HM-50. The inorganic fiber most commonly used for CTE tailoring is quartz [ASM 1989].

In the constraining metal core technology, the PWB material can be any one of the standard materials like epoxy-glass, polyimide-glass, or Teflon-based materials. The constraining core materials include metals, composite metals, and low-CTE fiber-resin combinations, which have a low-CTE material with sufficient strength to constrain the module. The two most commonly used constraining core materials are copper-invar-copper (CIC) and copper-molybdenum-copper (CMC). The PWB and the core are bonded with a rigid adhesive. In the constraining low-CTE metal plane technology, the ground and power planes in a standard multilayer board are replaced by an approximately 0.15-mm-thick CIC layer. Both epoxy and polyimide laminates have been used [ASM 1989].

Molded boards are made with resins containing fillers to improve thermal and mechanical properties. The resins are molded into a die or cavity to form the desired shape, including three-dimensional features. The board is metalized using conventional seeding and plating techniques. Alternatively, three-dimensional prefabricated films can also be transfer molded and further processed to form structures with finer dimensions. With proper selection of filler materials and epoxy compounds or by molding metal cores, the molded boards can be CTE tailored with enhanced thermal dissipation properties [Seraphim et al. 1989].

The standard PWB manufacturing primarily involves five technologies [ASM 1989].

1. Machining. This involves drilling, punching and routing. The drill holes are required for PTHs. The smaller diameter holes cost more and the larger aspect ratios (board thickness-to-hole diameter ratios) resulting from small holes make plating difficult and less reliable. They should be limited to thinner boards or buried vias. The locational accuracy is also important, especially

because of the smaller features (pad size or annular ring), which are less tolerant to misregistration. The registration is also complicated by substrate dimensional changes due to temperature and humidity fluctuations, and material relaxation and shifting during manufacturing operations. The newer technologies for drilling are laser and water-jet cutting. CO_2 laser radiation is absorbed by both glass and epoxy and can be used for cutting. The typical range of minimum hole diameters, maximum aspect ratios, and locational accuracies are given in Table 10.2.

2. Imaging. In this step the artwork pattern is transferred to the individual layers. Screen printing technology was used early on for creating patterns for print-and-etch circuits. It is still being used for simpler single- and double-sided boards because of its low capital investment requirements and high volume capability with low material costs. The limiting factors are the minimum line width and spacings that can be achieved with good yields. Multilayer boards and fine line circuits are processed using photoimaging. The photoimageable films are applied by flood screen printing, liquid roller coating, dip coating, spin coating, or roller laminating of dry films. Electrophoresis is also being used recently. Laminate dimensional instability contributes to the misregistration and should be controlled. Emerging photoimaging technologies are direct laser imaging and imaging by liquid crystal light valves. Again, Table 10.2 shows typical minimum trace widths and separations.

3. Laminating. It is used to make the multilayer boards and the base laminates that make up the single- and double-sided boards. Prepregs are sheets of glass cloth impregnated with B-stage epoxy resin and are used to bond multilayer boards. The types of pressing techniques used are hydraulic cold or hot press lamination with or without vacuum assist and vacuum autoclave nomination. With these techniques, especially with autoclave, layer thicknesses and dielectric constants can be closely controlled. These features allow the fabrication of controlled-impedance multilayer boards of eight layers or more. The autoclave technique is also capable of laminating three-dimensional forms and is used to produce rigid-flex boards.

4. Plating. In this step, metal finishing is applied to the board to make the necessary electrical connections. Plating processes could be wet chemical processes (electroless or electrolytic plating) or dry plasma processes (sputtering and chemical vapor deposition). Electroless plating, which does not need an external current source, is the core of additive technology. It is used to metalize the resin and glass portions of a multilayer board's PTHs with high aspect ratios (>15:1) and the three-dimensional circuit paths of molded boards. On the other hand, electrolytic plating, which requires an external current source, is the method of choice for bulk metallization. The advantages of electrolytic plating over electroless plating include greater rate of plating, simpler and less expensive process, and the ability to deposit a broader variety of metals. The newer plasma processes can offer pure, uniform, thin foils of various metals with less than 1-mil line widths and spacings and have less environmental impacts.

5. Etching. It involves removal of metals and dielectrics and may include both wet and dry processes. Copper can be etched with cupric chloride or other isotropic etchants, which limit the practical feature sizes to more than two times the copper thickness. The uniformity of etching is also critical for fine line feature circuits. New anisotropic etching solutions are being developed to extend the fine line capability.

TABLE 10.2 Typical Limits of PWB Parameters

Parameter	Limit
Minimum trace width, mm	0.05–0.15
Minimum trace separation, mm	~0.25
Minimum PTH diameter, mm	0.2–0.85
Location accuracy of PTH, mm	0.015–0.05
Maximum aspect ratios	3.5–15.0
Maximum number of layers	20 or more
Maximum board thickness, mm	7.0 or more

The typical range of minimum trace widths, trace separations, printed through hole diameters, and maximum aspect ratios are shown in Table 10.2. The board cost for minimum widths, spacings, and PTH diameters, and small tolerances for the pad placements is high and reliability may be less.

10.3 Design of Printed Wiring Boards

The design of PWBs has become a challenging task, especially when designing high-performance and high-density boards. The designed board has to meet signal-integrity requirements [crosstalk, simultaneously switching output (SSO) noise, delay, and reflections], **electromagnetic compatibility (EMC)** requirements [meeting **electromagnetic interference (EMI)** specifications and meeting minimum **susceptibility** requirements], thermal requirements (being able to handle the mismatches in the CTEs of various components over the temperature range, power dissipation, and heat flow), mechanical requirements (strength, rigidity/flexibility), material requirements, manufacturing requirements (ease of manufacturability, which may affect cost and reliability), testing requirements (ease of testability, incorporation of **test coupons**), and environmental requirements (humidity, dust). When designing the circuit, the circuit design engineer may not have been concerned with all of these requirements when verifying his circuit design by simulation. But the PWB designer, in addition to wiring the board, must ensure that these requirements are met so that the assembled board functions to specification. Computer-aided design (CAD) tools are indispensable for the design of all but simple PWBs.

The PWB design process is part of an overall system design process. As an example, for a digital system design, the sequence of events may roughly follow the order shown in Fig. 10.1 [ASM 1989, Byers 1991]. The logic design and verification are performed by a circuit designer. The circuit designer and the signal-integrity and EMC experts should help the board designer in choosing proper parts, layout, and electrical rule development. This will reduce the number of times the printed circuit has to be reworked and speeds up the product's time to market. The **netlist** generation is important for both circuit simulation and PWB design programs. The netlist contains the nets (a complete set of physical interconnectivity of the various parts in the circuit) and the circuit devices. Netlist files are usually divided into the component

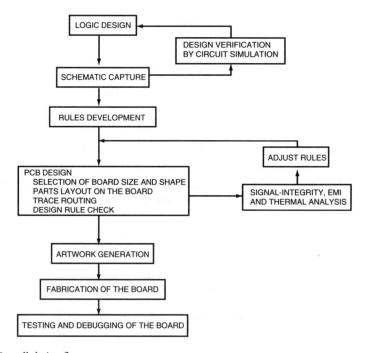

FIGURE 10.1 Overall design flow.

listings and the pin listings. For a list of the **schematic** capture software and PWB software packages, refer to Byers [1991].

Once the circuit is successfully simulated by the circuit designer, the PWB designer starts on the board design. The board designer should consult with the circuit designer to determine critical path circuits, line-length requirements, power and fusing requirements, bypassing schemes, and any required guards or shielding. First the size and shape of the board are chosen from the package count and the number of connectors and switches in the design netlist. Some designs may require a rigidly defined shape, size and defined connector placement, for example, PWB outlines for computer use. This may compromise the optimum layout of the circuit [Byers 1991].

Next the netlist parts are chosen from the component library and placed on the board so that the interconnections form the simplest pattern. If a particular part is not available in the library, then it can be created by the PCB program's library editor or purchased separately. Packages containing multiple gates are given their gate assignments with an eye toward a simpler layout. There are software packages that automize and speed up the component placement process. However, they may not achieve complete component placement. Problems with signal path length and clock skewing may be encountered, and so part of the placement may have to be completed manually. Placement can also be done interactively [Byers 1991].

Once the size and shape of the board is decided on, any areas on the board that cannot be used for parts are defined using fill areas that prevent parts placement within their perimeter. The proper parts placement should result in a cost-effective design and meet manufacturing standards. Critical lines may not allow vias in which case they must be routed only on the surface. Feedback loops must be kept short and equal line length requirements and shielding requirements, if any, should be met. Layout could be on a grid with components oriented in one direction to increase assembly speed, eliminate errors, and improve inspection. Devices should be placed parallel to the edges of the board [ASM 1989, Byers 1991].

The actual placement of a component may rely on the placement of a part before it. Usually, the connectors, user accessible switches and controls, and displays may have rigidly defined locations. Thus, these should be placed first. Next I/O interface chips, which are typically next to the connectors, are placed. Sometimes a particular group of components may be placed as a block in one area, for example, memory chips. Then the automatic placement program may be used to place the remaining components. The algorithms can be based on various criteria. Routability, the relative ease with which all of the traces can be completed, may be one criterion. Overall total connection length may also be used to find an optimal layout. Algorithms based on the requirements for distances, density, and number of crossings can be formulated to help with the layout. After the initial placement, a placement improvement program can be run that fine tunes the board layout by component swapping, logic gate swapping, or when allowed even pin swapping. Parts can also be edited manually. Some programs support block editing, macros and changing of package outlines. The PWB designer may check with the circuit design engineer after a good initial layout is completed to make sure that all of the requirements will be met when the board is routed [Byers 1991].

The last step in the PWB layout process is the placement of the interconnection traces. Most PCB programs have an autorouter software that does most, if not all, of the trace routing, The software uses the wire list netlist and the placement of the parts on the board from the previous step as inputs and decides which trace should go where based on an algorithm. Most routers are point-to-point routers. The most widely used autorouter algorithm is the Lee algorithm. The Lee router works based on the cost functions that change its routing parameters. Cost functions may be associated with direction of traces (to force all of the tracks on a particular layer of the board vertically, the cost of the horizontal tracks can be set high), maximum trace length, maximum number of vias, trace density, or other criteria. In this respect, Lee's algorithm is quite versatile. Lee routers have a high trace completion rate (typically 90% or better) but are slow. There are other autorouters that are faster than the Lee router but their completion rate may not be as high as the Lee router. These include Hightower router, pattern router, channel router, and gridless router [ASM 1989, Byers 1991]. In many cases, fast routers can be run first to get the bulk of the work done in a short time and then the Lee router can be run to finish routing the remaining traces.

The autorouters may not foresee that placing one trace may block the path of another. There are some routers called clean-up routers that can get the offending traces out of the way by rip-up or shove-aside. The rip-up router works by removing the offending trace, completing the blocked trace, and proceeding to find a new path for the removed trace. The rip-up router may require user help in identifying the trace that is to be removed. Shove-aside routers work by moving traces when an uncompleted trace has a path to its destination but does not have enough room for the track. Manual routing and editing may be needed if the automatic trace routing is not 100% complete. Traces may be modified without deleting them (pushing and shoving, placing or deleting vias, or moving segments from one layer to another) and trace widths may be adjusted to accommodate a connection. For example, a short portion of a trace width may be narrowed so that it may pass between the IC pads without shorting them and is called necking down [Byers 1991].

Once the parts layout and trace pattern are found to be optimal, the board's integrity is verified by subjecting the trace pattern to **design rule** check far digital systems. A CAD program checks to see that the tracks, vias, and pads have been placed according to the design rule set. The program also makes sure that all of the nodes in each net are connected and that there are no shorted or broken traces [Byers 1991]. Any extra pins are also listed. The final placement is checked for signal-integrity, EMI compliance, and thermal performance. If any problems arise, the rules are adjusted and the board layout and/or trace routing are modified to correct the problems.

Next a netlist file of the PWB artwork is generated, which contains the trace patterns of each board layer. All of the layers are aligned using placement holes or datum drawn on each layer. The artwork files are sent to a Gerber or other acceptable format photoplotter, and photographic transparencies suitable for PWB manufacturing are produced. From the completed PWB design, solder masks, which apply an epoxy film on the PWB prior to flow soldering to prevent solder bridges from forming between adjacent traces, are also generated. Silkscreen mats could also be produced for board nomenclature. The PWB programs may also support drill file formats that numerically control robotic drilling machines used to locate and drill properly sized holes in the PWB [Byers 1991].

Finally, the board is manufactured and tested. Then the board is ready for assembly and soldering. Finished boards are tested for system functionality and specifications.

10.4 PWB Interconnection Models

The principal electrical properties of the PWB material are relative dielectric constant, loss tangent, and dielectric strength. When designing high-speed and high-performance boards, the wave propagation along the traces should be well controlled. This is done by incorporating power and signal planes in the boards, and utilizing various transmission line structures with well-defined properties. The electrical analog and digital signals are transferred by all of the PWB interconnects and the fidelity of these connections is dependent on the electrical properties of interconnects, as well as the types of circuits and signals. The interconnects can be characterized as electrically short and modeled as lumped elements in circuit simulators if the signal wavelength is large as compared to the interconnect length ($\lambda > 30*l$). For digital signals the corresponding criterion can be expressed in terms of signal rise time being large as compared to the propagation delay (e.g., $T_r > 10\ T_d$). Propagation delay represents the time taken by the signal to travel from one end of the interconnect to the other end and is obviously equal to the interconnect length divided by the signal velocity, which in most cases corresponds to the velocity of light in the dielectric medium. Vias, bond wires, pads, short wires and traces, and bends in PWBs shown in Fig. 10.2 can be modeled as lumped elements, whereas long traces must be modeled as distributed circuits or transmission lines.

The transmission lines are characterized by their characteristic impedances and propagation constants, which can also be expressed in terms of the associated distributed parameters R, L, G, and C per unit

Printed Wiring Boards

length of the lines [Magnuson, Alexander, and Tripathi 1992]. In general, the characteristic parameters are expressed as

$$\gamma \equiv \alpha + j\beta = \sqrt{(R + j\omega L)(G + j\omega C)}$$

$$Z_0 = \sqrt{\frac{R + j\omega L}{G + j\omega C}}$$

Propagation constant $\gamma = \alpha + j\beta$ characterizes the amplitude and phase variation associated with an AC signal at a given frequency or the amplitude variation and signal delay associated with a digital signal. The characteristic impedance is the ratio of voltage to current associated with a wave and is equal to the impedance the lines must be terminated in for zero reflection. The signal amplitude, in general, decreases and it lags behind in phase as it travels along the interconnect with a velocity, in general, equal to the group velocity. For example, the voltage associated with a wave at a given frequency can be expressed as

$$V = V_0 e^{-\alpha z} \cos(\omega t - \beta z)$$

where, V_0 is the amplitude at the input ($z = 0$) and z is the distance.

For low loss and lossless lines the signal velocity and characteristic impedance can be expressed as

$$v = \frac{1}{\sqrt{LC}} = \frac{1}{\sqrt{\mu_0 \varepsilon_0 \varepsilon_{r\,\text{eff}}}}$$

$$Z_0 = \frac{1}{\sqrt{C}} = \frac{\sqrt{\mu_0 \varepsilon_0 \varepsilon_{r\,\text{eff}}}}{C}$$

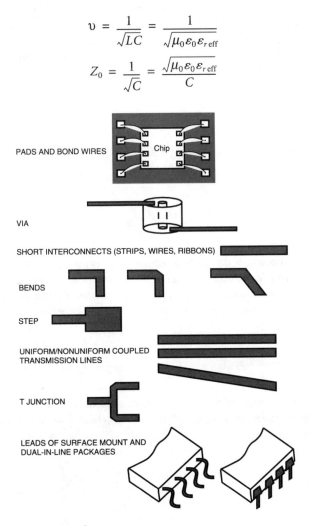

FIGURE 10.2 PWB interconnect examples.

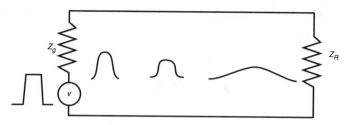

FIGURE 10.3 Signal degradation due to losses and dispersion as it travels along an interconnect.

The effective dielectric constant $\varepsilon_{r\,\text{eff}}$ is an important parameter used to represent the overall effect of the presence of different dielectric mediums surrounding the interconnect traces. For most traces $\varepsilon_{r\,\text{eff}}$ is either equal to or approximately equal to the ε_r, the relative dielectric constant of the board material, except for the traces on the top layer where $\varepsilon_{r\,\text{eff}}$ is a little more than the average of the two media.

The line losses are expressed in terms of the series resistance and shunt conductance per unit length (R and G) due to conductor and dielectric loss, respectively. The signals can be distorted or degraded due to these conductor and dielectric losses, as illustrated in Fig. 10.3 for a typical inter-connect. The resistance is, in general, frequency dependent since the current distribution across the conductors is nonuniform and depends on frequency due to the skin effect. Because of this exclusion of current and flux from the inside of the conductors, resistance increases and inductance decreases with increasing frequency. In the high-frequency limit, the resistance and inductances can be estimated by assuming that the current is confined over the cross section one skin depth from the conductor surface. The skin depth is a measure of how far the fields and currents penetrate into a conductor and is given by

$$\delta = \sqrt{\frac{2}{\omega\mu\sigma}}$$

The conductor losses can be found by evaluating R per unit length and using the expression for γ or by using an incremental inductance rule, which leads to

$$\alpha_c = \frac{\beta \Delta Z_0}{2 Z_0}$$

where α_c is the attenuation constant due to conductor loss, Z_0 is the characteristic impedance, and ΔZ_0 is the change in characteristic impedance when all of the conductor walls are receded by an amount $\delta/2$. This expression can be readily implemented with the expression for Z_0. The substrate loss is accounted for by assigning the medium a conductivity σ, which is equal to $\omega \varepsilon_0 \varepsilon_r''$. For many conductors buried in near homogeneous medium

$$G = C \frac{\sigma}{\varepsilon}$$

If the lines are not terminated in their characteristic impedances, there are reflections from the terminations. These can be expressed in terms of the ratio of reflected voltage or current to the incident voltage or current and are given as

$$\frac{V_{\text{reflected}}}{V_{\text{incident}}} = -\frac{I_{\text{reflected}}}{I_{\text{incident}}} = \frac{Z_R - Z_0}{Z_R + Z_0}$$

where Z_R is the termination impedance and Z_0 is the characteristic impedance of the line. For a perfect match, the lines must be terminated in their characteristic impedances. If the signal is reflected, the signal received by the receiver is different from that sent by the driver. That is, the effect of mismatch includes signal distortion, as illustrated in Fig. 10.4, as well as ringing and increase in crosstalk due to multiple reflections resulting in an increase in coupling of the signal to the passive lines.

The electromagnetic coupling between the interconnects is the factor that sets the upper limit to the number of tracks per channel or, in general, the interconnect density. The time-varying voltages and currents result in capacitive and inductive coupling between the interconnects. For longer interconnects, this coupling is distributed and modeled in terms of distributed self- and mutual-line constants of the multiconductor transmission line systems. In general, this coupling results in both the near- and far-end crosstalk as illustrated in Fig. 10.5 for two coupled microstrips. Crosstalk increases noise margins and degrades signal quality. Crosstalk increases with longer trace coupling distances, smaller separation between traces, shorter pulse rise and fall times, larger magnitude currents or voltages being switched, and decreases with the use of adjacent power and ground planes or with power and ground traces interlaced between signal traces on the same layer.

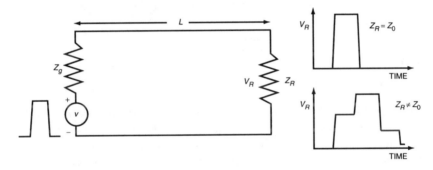

FIGURE 10.4 Voltage waveform is different when lines are not terminated in their characteristic impedance.

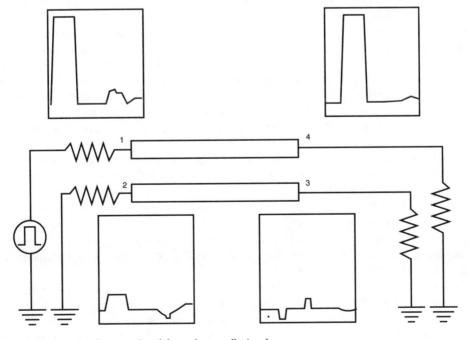

FIGURE 10.5 Example of near-end and far-end crosstalk signal.

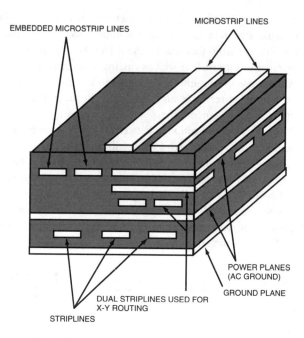

FIGURE 10.6 Example transmission line structures in PWBs.

The commonly used PWB transmission line structures are microstrip, embedded microstrip, stripline, and dual stripline, whose cross sections are shown in Fig. 10.6. The empirical CAD oriented expressions for transmission line parameters, and the models for wires, ribbons, and vias are given in Table 10.3.

The traces on the outside layers (microstrips) offer faster clock and logic signal speeds than the stripline traces. Hooking up components is also easier in microstrip structure than in stripline structure. Stripline offers better noise immunity for RF emissions than microstrip. The minimum spacing between the signal traces may be dictated by maximum crosstalk allowed rather than by process constraints. Stripline allows closer spacing of traces than microstrip for the same layer thickness. Lower characteristic impedance structures have smaller spacing between signal and ground planes. This makes the boards thinner allowing drilling of smaller diameter holes, which in turn allow higher circuit densities. Trace width and individual layer thickness tolerances of $\pm 10\%$ are common. Tight tolerances ($\pm 2\%$) can be specified, which would result in higher board cost. Typical impedance tolerances are $\pm 100\%$. New statistical techniques have been developed for designing the line structures of high-speed wiring boards [Mikazuki and Matsui 1994].

The low dielectric constant materials improve the density of circuit interconnections in cases where the density is limited by crosstalk considerations rather than by process constraints. Benzocyclobutene is one of the low dielectric constant polymers with excellent electrical, thermal, and adhesion properties. In addition, water absorption is lower by a factor of 15 compared to conventional polyimides. Teflon also has low dielectric constant and loss, which are stable over wide ranges of temperature, humidity, and frequency [ASM 1989, Evans 1994].

In surface mount technology, the components are soldered directly to the surface of a PWB, as opposed to through-hole mounting. This allows efficient use of board real estate, resulting in smaller boards and simpler assembly. Significant improvement in electrical performance is possible with the reduced package parasitics and the short interconnections. However, the reliability of the solder joint is a concern if there are CTE mismatches [Seraphim et al. 1989].

In addition to establishing an impedance-reference system for signal lines, the power and ground planes establish stable voltage levels for the circuits [Montrose 1995]. When large currents are switched, large voltage drops can be developed between the power supply and the components. The planes minimize the voltage drops by providing a very small resistance path and by supplying a larger capacitance and lower inductance contribution when two planes are closely spaced. Large decoupling capacitors are also

Printed Wiring Boards

TABLE 10.3 Interconnect Models

Interconnect type	Model
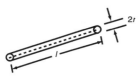 Round wire	$L \cong 0.002l\left[\ell n\left(\dfrac{2l}{r}-0.75\right)\right], \mu H; l > r$ l, r in cm.
 Straight rectangular bar or ribbon	$L \cong 0.002l\left[\ell n \dfrac{2l}{b+c}+0.5+0.2235\dfrac{b+c}{l}\right], \mu H$ b, c, l in cm.
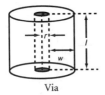 Via	$L = 0.002l\left[\ell n \dfrac{2l}{r+W}-1+\xi\right], \mu H$ where $\xi = 0.25\left[\cos\left(\dfrac{r}{r+W}\dfrac{\pi}{2}\right)-0.07\sin\left(\dfrac{r}{r+W}\pi\right)\right]$ l, r, W in cm.
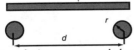 Round wire over a ground plane and parallel round wires	$Z_0 = \dfrac{120}{\sqrt{\varepsilon_r}}\cosh^{-1}\dfrac{d}{2r}$, ohm $= \dfrac{120}{\sqrt{\varepsilon_r}}\ell n \dfrac{d}{r}$, ohm; $d \gg r$
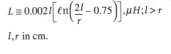 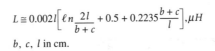 Microstrip	$\varepsilon_{\text{reff}} = \dfrac{\varepsilon_r+1}{2}+\dfrac{\varepsilon_r-1}{2}\left[1+\dfrac{10h}{W_e}\right]^{-a.b}$ where $\dfrac{W_e}{h} = \dfrac{W}{h}+\dfrac{1.25}{\pi}\dfrac{t}{h}\left[1+\ell n\dfrac{2[h^4+(2\pi W)^4]^{0.25}}{t}\right]$ $a = 1+\dfrac{1}{49}\ell n\left\{\dfrac{\left[\left(\dfrac{W_e}{h}\right)^4+\left(\dfrac{W_e}{52h}\right)^2\right]}{\left[\left(\dfrac{W_e}{h}\right)^4+0.432\right]}\right\}$ $+\dfrac{1}{18.7}\ell n\left[1+\left(\dfrac{W_e}{18.1h}\right)^3\right]$ $b = 0.564\left[\dfrac{\varepsilon_r-0.9}{\varepsilon_r+3}\right]^{0.053}$ $Z_0 = \dfrac{60}{\sqrt{\varepsilon_{\text{reff}}}}\ell n\left[\dfrac{F_1 h}{W_e}+\sqrt{1+\left(\dfrac{2h}{W_e}\right)^2}\right]$ with $F_1 = 6+(2\pi-6)e^{-\left(30.666\dfrac{h}{W_e}\right)^{0.7528}}$

TABLE 10.3 Interconnect Models (continued)

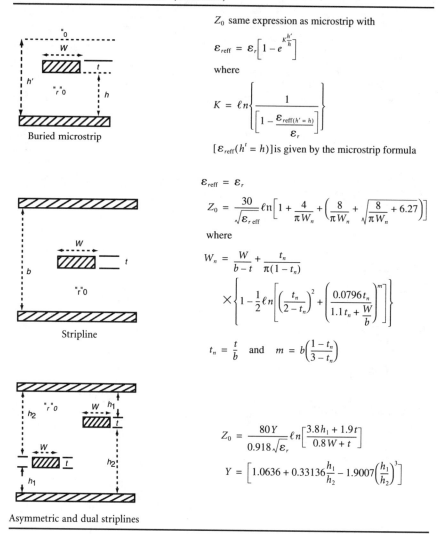

added between the power and ground planes for increased voltage stability. High-performance and high-density boards require accurate computer simulations to determine the total electrical response of the components and complex PWB structures involving various transmission lines, vias, bends, and planes with vias.

10.5 Signal-Integrity and EMC Considerations

The term signal-integrity refers to the issues of timing and quality of the signal. The timing analysis is performed to ensure that the signal arrives at the destination within the specified time window. In addition, the signal shape should be such that it causes correct switching and avoids false switching or multicrossing of the threshold. The board features that influence the signal-integrity issues of timing and quality of signal are stub length, parallelism, vias, bends, planes, and termination impedances. Increasing clock speeds and signal edge rates make the signal-integrity analysis a must for the design of high speed PWBs. The signal-integrity tools detect and help correct problems associated with timing, crosstalk, ground bounce, dispersion, resonance, ringing, clock skew, and susceptibility [Maliniak 1995]. Some

CAD tool vendors offer stand-alone tools that interface with simulation and layout tools. Other vendors integrate signal-integrity capabilities into the PWB design flow. Both prelayout and postlayout tools are available. Prelayout tools incorporate signal-integrity design rules up front and are used before physical design to investigate tradeoffs. Postlayout tools are used after board layout and provide a more in-depth analysis of the board. A list of commercially available signal-integrity tools is given in Maliniak [1995] and Beckert [1993]. A new automated approach for generating wiring rules of high-speed PWBs based on a priori simulation-based characterization of the interconnect circuit configurations has also been developed [Simovich et al. 1994].

The selection of the signal-integrity tools depends on the accuracy required and on the run times, in addition to the cost. Tools employing full-wave approach solve the Maxwell's field equations and provide accurate results. However, the run times are longer because they are computation intensive. Accurate modeling of radiation losses and complex three-dimensional structures including vias, bends, and tees can only be performed by full-wave analysis. Tools employing a hybrid approach are less accurate but run times are short. If the trace dimensions are smaller (roughly by a factor of 10) compared to the free space wavelength of the fields, then quasistatic methods can be used to compute the distributed circuit quantities like self- and mutual inductances and capacitances. These are then used in a circuit simulator like SPICE (simulated program for integrated circuit emphasis) to compute the time-domain response, from which signal delay and distortion are extracted [Maliniak 1995].

The inputs for the signal-integrity tools are the device SPICE or behavioral models and the layout information including the thickness of the wires, the metal used, and the dielectric constant of the board. The input/output buffer information specification (IBIS) is a standard used to describe the analog behavior of a digital IC's I/O buffers. The IBIS models provide a nonproprietary description of board drivers and receivers that are accurate and faster to simulate. The models are based on DC I-V curves, AC transition data, and package parasitic and connection information. A SPICE-to-IBIS translation program is also available, which uses existing SPICE models to perform simulation tests and generate accurate IBIS models. An IBIS on-line repository, which maintains the IBIS models from some IC vendors, is also available. Other IC vendors provide the IBIS models under nondisclosure agreements. Additional information on IBIS and existing IBIS models can be obtained by sending an e-mail message to ibis-info@vhdl.org [Malinak 1995].

The PWBs, when installed in their systems, should meet the North American and international EMC compliance requirements. EMC addresses both emissions, causing EMI and susceptibility, vulnerability to EMI. The designer needs to know the manner in which electromagnetic fields transfer from the circuit boards to the chassis and/or case structure. Many different layout design methodologies exist for EMC [Montrose 1995]. To model EMC, the full electromagnetic behavior of components and interconnects, not simple circuit models, must be considered. A simple EMC model has three elements: (1) a source of energy, (2) a receptor, and (3) a coupling path between the source and receptor. All three elements must be present for interference to exist. The emissions could be coupled by either radiation or by conduction. Emissions are **suppressed** by reducing noise source level and/or reducing propagation efficiency. Immunity is enhanced by increasing susceptor noise margin and/or reducing propagation efficiency. In general, the higher the frequency, the greater is the likelihood of a radiated coupling path and the lower the frequency, the greater is the likelihood of a conducted coupling path. The five major considerations in EMI analysis are frequency, amplitude, time, impedance, and dimensions [Montrose 1995].

The use of power and ground planes embedded in the PWB with proper use of stripline and microstrip topology is one of the important methods of suppression of the RF energy internal to the board. Radiation may still occur from bond wires, lead frames, sockets, and interconnect cables. Minimizing lead inductance of PWB components reduces radiated emissions. Minimizing trace lengths of high-speed traces reduces radiation. On single- or double-sided boards, the power and ground traces should be adjacent to each other to minimize loop currents, and signal paths should parallel the power and ground traces. In multilayer boards, signal routing planes should be adjacent to solid image planes. If logic devices in multilayer boards have asymmetrical pull-up and pull-down current ratios, then an imbalance is created in the power and ground plane structure. The flux cancellation between RF currents that exist within the board traces, components, and circuits, in relation to a plane is not optimum when pull-up and pull-down current ratios are different and traces are routed adjacent to a power plane. In such cases, high-speed signal and clock traces should be routed adjacent to a ground plane for optimal EMI suppression [Montrose 1995].

To minimize fringing between the power and ground planes and the resulting RF emissions, power planes should be physically smaller than the closest ground planes. Typical reduction size is 20 times the thickness between the power and the ground plane on each side. Any signal traces on the adjacent routing plane located over the absence of copper area must be rerouted. Selecting proper type of grounding is also critical for EMI suppression. The distance between any two neighboring ground point locations should not exceed 1/20 of the highest frequency wavelength generated on the board in order to minimize RF ground loops. Moats are acceptable in an image plane only if the adjacent signal routing layer does not have traces that cross the moated area. Partition the board into high-bandwidth, medium-bandwidth, and low-bandwidth areas, and isolate each section using partitions or moats. The peak power currents generated when logic gates switch states may inject high-frequency switching noise into the power planes. Fast edge rates generate greater amount of spectral bandwidth of RF energy. The slowest possible logic family that meets adequate timing margins should be selected to minimize EMI effects [Montrose 1995].

Surface mount devices have lower lead inductance values and smaller loop areas due to smaller package size as compared to through hole mounted ones and, hence, have lower RF emissions. Decoupling capacitors remove RF energy generated on the power planes by high-frequency components and provide a localized source of DC power when simultaneous switching of output pins occur. Bypassing capacitors remove unwanted RF noise that couples component or cable common-mode EMI into susceptible areas. Proper values and proper placement on the board of decoupling and bypassing capacitors prevent RF energy transfer from one circuit to another and enhance EMC. The capacitance between the power and ground planes also acts as a decoupling capacitor [Montrose 1995].

Clock circuits may generate significant EMI and should be designed with special attention. A localized ground plane and a shield may be required over the clock generation circuits. Clock traces should be considered critical and routed manually with characteristic terminations. Clock traces should be treated as transmission lines with good impedance control and with no vias or minimum number of vias in the traces to reduce reflections and ringing, and creation of RF common-mode currents. Ground traces should be placed around each clock trace if the board is single or double sided. This minimizes crosstalk and provides a return path for RF current. Shunt ground traces may also be provided for additional RF suppression. Bends should be avoided, if possible, in high-speed signal and clock traces and, if present, bend angles should not be smaller than 90° and they must be chamfered. Analog and input/output sections should be isolated from the digital section. PWBs must be protected from electrostatic discharge (ESD) that might enter at I/O signal and electrical connection points. For additional design techniques and guidelines for EMC compliance and a listing of North American and international EMI requirements and specifications, refer to Montrose [1995].

Signal-integrity and EMC considerations should be an integral part of printed circuit board (PCB) design. Signal integrity, EMI problems, and thermal performance should be addressed early in the development cycle using proper CAD tools before physical prototypes are produced. When boards are designed taking these considerations into account, the signal quality and signal-to-noise ratio are improved. Whenever possible, EMC problems should be solved at the PCB level than at the system level. Proper layout of PWB ensures EMC compliance at the level of cables and interconnects.

Defining Terms

Design rules: A set of electrical or mechanical rules that must be followed to ensure the successful manufacturing and functioning of the board. These may include minimum track widths and track spacings, track width required to carry a given current, maximum length of clock lines, and maximum allowable distance of coupling between a pair of signal lines.

Electromagnetic compatibility (EMC): The ability of a product to coexist in its intended electromagnetic environment without causing or suffering functional degradation or damage.

Electromagnetic interference (EMI): A process by which disruptive electromagnetic energy is transmitted from one electronic device to another via radiated or conducted paths or both.

Netlist: A file of component connections generated from a schematic. The file lists net names and the pins, which are a part of each net in the design.

Schematic: A drawing or set of drawings that shows an electrical circuit design.

Suppression: Designing a product to reduce or eliminate RE energy at the source without relying on a secondary method such as a metal housing.

Susceptibility: A relative measure of a device or system's propensity to be disrupted or damaged by EMI exposure.

Test coupon: Small pieces of board carrying a special pattern, made alongside a required board, which can be used for destructive testing.

Trace: A node-to-node connection, which consists of one or more tracks. A track is a metal line on the PWB. It has a start point, an end point, a width, and a layer.

Via: A hole through one or more layers on a PWB that does not have a component lead through it. It is used to make a connection from a track on one layer to a track on another layer.

References

ASM. 1989. *Electronic Materials Handbook*, Vol. 1, Packaging, Sec. 5, Printed Wiring Boards, pp. 505–629. ASM International, Materials Park, OH.

Beckert, B.A. 1993. Hot analysis tools for PCB design. *Comp. Aided Eng.* 12(1):44–49.

Byers, T.J. 1991. *Printed Circuit Board Design With Microcomputers.* Intertext, New York.

Evans, R. 1994. Effects of losses on signals in PWBs. *IEEE Trans. Comp. Pac., Man. Tech. Pt. B:Adv. Pac.* 17(2):217–222.

Magnuson, P.C., Alexander, G.C., and Tripathi, V.K. 1992. *Transmission Lines and Wave Propagation*, 3rd ed. CRC Press, Boca Raton, FL.

Maliniak, L. 1995. Signal analysis: A must for PCB design success. *Elec. Design* 43(19):69–82.

Mikazuki, T. and Matsui, N. 1994. Statistical design techniques for high speed circuit boards with correlated structure distributions. *IEEE Trans. Comp. Pac., Man. Tech.* 17(1):159–165.

Montrose, M.I. 1995. *Printed Circuit Board Design Techniques for EMC Compliance.* IEEE Press, New York.

Seraphim, D.P., Barr, D.E., Chen, W.T., Schmitt, G.P., and Tummala, R.R. 1989. Printed-circuit board packaging. In *Microelectronics Packaging Handbook*, eds. Rao R. Tummala and Eugene 1. Rymaszewski, pp. 853–921. Van Nostrand–Reinhold, New York.

Simovich, S., Mehrotra, S., Franzon, P., and Steer, M. 1994. Delay and reflection noise macromodeling for signal integrity management of PCBs and MCMs. *IEEE Trans. Comp. Pac., Man. Tech. Pt. B:Adv. Pac.* 17(1):15–20.

Further Information

Electronic Packaging and Production journal.

IEEE Transactions on Components, Packaging, and Manufacturing Technology—Part A.

IEEE Transaction on Components, Packaging, and Manufacturing Technology—Part B: Advanced Packaging.

Harper, C.A. 1991. *Electronic Packaging and Interconnection Handbook.*

PCB Design Conference, Miller Freeman Inc., 600 Harrison Street, San Francisco, CA 94107, (415)905-4994. *Printed Circuit Design* journal.

Proceedings of the International Symposium on Electromagnetic Compatibility, sponsored by IEEE Transactions on EMC.

11
Hybrid Microelectronics Technology

11.1	Introduction	11-1
11.2	Substrates for Hybrid Applications	11-2
11.3	Thick-Film Materials and Processes	11-3
	Thick-Film Materials • Thick-Film Conductor Materials • Thick-Film Resistor Material • Properties of Thick-Film Resistors • Properties of Thick-Film Dielectric Materials • Processing Thick-Film Circuits	
11.4	Thin-Film Technology	11-11
	Deposition Technology • Photolithographic Processes • Thin-Film Resistors	
11.5	Resistor Trimming	11-15
11.6	Comparison of Thick-Film and Thin-Film Technologies	11-16
11.7	The Hybrid Assembly Process	11-16
	The Chip-and-Wire Process • Tape Automated Bonding and Flip-Chip Bonding	

Jerry E. Sergent
BBS PowerMod, Incorporated

11.1 Introduction

Hybrid microelectronics technology is one branch of the electronics packaging discipline. As the name implies, the hybrid technology is an integration of two or more technologies, utilizing a film-deposition process to fabricate conductor, resistor, and dielectric patterns on a ceramic substrate for the purpose of mounting and interconnecting semiconductors and other devices as necessary to form an electronics circuit.

The method of deposition is what differentiates the hybrid circuit from other packaging technologies and may be one of two types: *thick film* or *thin film*. Other methods of metallizing a ceramic substrate, such as direct bond copper, active metal brazing, and plated copper, may also be considered to be in the hybrid family, but do not have a means for directly fabricating resistors and are not considered here. Semiconductor technology provides the **active components**, such as integrated circuits, transistors, and diodes. The **passive components**, such as resistors, capacitors, and inductors, may also be fabricated by thick- or thin-film methods or may be added as separate components.

The most common definition of a hybrid circuit is a circuit that contains two or more components, one of which must be active, which are mounted on a substrate that has been metallized by one of the film technologies. This definition is intended to eliminate single-chip packages, which contain only one active component, and also to eliminate such structures as resistor networks. A hybrid circuit may, therefore, be as simple as a diode-resistor network or as complicated as a multilayer circuit containing in excess of 100 integrated circuits.

The first so-called hybrid circuits were manufactured from polymer materials for use in radar sets in World War II and were little more than carbon-impregnated epoxies printed across printed circuit traces. The modern hybrid era began in the 1960s when the first commercial cermet thick-film materials became available. Originally developed as a means of reducing the size and weight of rocket payloads, hybrid circuits are now used in a variety of applications, including automotive, medical, telecommunication, and commercial, in addition to continued use in space and military applications. Today, the hybrid microelectronics industry is a multibillion dollar business and has spawned new technologies, such as the multichip module technology, which have extended the application of microelectronics into other areas.

11.2 Substrates for Hybrid Applications

The foundation for the hybrid microcircuit is the substrate, which provides the base on which the circuit is formed. The substrate is typically ceramic, a mixture of metal oxides and/or nitrides and glasses that are formulated to melt at a high temperature. The constituents are formed into powders, mixed with an organic binder, patterned into the desired shape, and fired at an elevated temperature. The result is a hard, brittle composition, usually in the configuration of a flat plate, suitable for metallization by one of the film processes. The ceramics used for hybrid applications must have certain mechanical, chemical, and electrical characteristics as described in Table 11.1.

TABLE 11.1 Desirable Properties of Substrates for Hybrid Applications

- Electrical insulator
- Chemically and mechanically stable at processing temperatures
- Good thermal dissipation characteristics
- Low dielectric constant, especially at high frequencies
- Capable of being manufactured in large sizes and maintain good dimensional stability throughout processing
- Low cost
- Chemically inert to postfiring fabrication steps
- Chemically and physically reproducible to a high degree

The most common material is 96% alumina (Al_2O_3), with a typical composition of 96% Al_2O_3, 0.8% MgO, and 3.2% SiO_2. The magnesia and silica form a magnesium-alumino-silicate glass with the alumina when the substrate is manufactured at a firing temperature of 1600°C. The characteristics of alumina are presented in Table 11.2.

Other materials used in hybrid applications are described as follows:

- *Beryllium oxide, BeO, (Beryllia)*. Beryllia has excellent thermal properties (almost 10 times better than alumina at 250°C), and also has a lower **dielectric constant** (6.5).
- *Steatite*. Steatite has a lower concentration of alumina and is used primarily for low-cost thick-film applications.
- *Glass*. Borosilicate or quartz glass can be used where light transmission is required (e.g., displays). Because of the low melting point of these glasses, their use is restricted to the thin-film process and some low-temperature thick-film materials.
- *Enamel steel*. This is a glass or porcelain coating on a steel core. It has been in commercial development for about 15 years and has found some limited acceptance in thick-film applications.
- *Silicon carbide, SiC*. SiC has a higher **thermal conductivity** than beryllia, but also has a high dielectric constant (40), which renders it unsuitable for high-speed applications.
- *Silicon*. Silicon has a very high thermal conductivity and matches the thermal expansion of semiconductors perfectly. Silicon is commonly used in thin-film applications because of the smooth surface.

- *Aluminum nitride (AlN)*. AlN substrates have a thermal conductivity less than that of beryllia, but greater than alumina. At higher temperatures, however, the gap between the thermal conductivities of AlN and BeO narrows somewhat. In addition, the **temperature coefficient of expansion (TCE)** of aluminum nitride is close to that of silicon, which lowers the stress on the bond under conditions of temperature extremes. Certain of the copper technologies, such as plated copper or **direct-bonded copper (DBC)**, may be used as a conductor system if the AlN surface is properly prepared.

TABLE 11.2 Typical Properties of Alumina

Property	96% Alumina	99% Alumina
Electrical Properties		
Resistivity	10^{14} Ω-cm	10^{16} Ω-cm
Dielectric constant	9.0 at 1 MHz, 25°C	9.2 at 1 MHz, 25°C
	8.9 at 10 GHz, 25°C	9.0 at 10 GHz, 25°C
Dissipation factor	0.0003 at 1 MHz, 25°C	0.0001 at 1 MHz, 25°C
	0.0006 at 10 GHz, 25°C	0.0002 at 10 GHz, 25°C
Dielectric strength	16 V/mm at 60 Hz, 25°C	16 V/mm at 60 Hz, 25°C
Mechanical Properties		
Tensile strength	1760 kg/cm^2	1760 kg/cm^2
Density	3.92 g/cm^3	3.92 g/cm^3
Thermal conductivity	0.89 W/°C-in at 25°C	0.93 W/°C-in at 25°C
Coefficient of thermal expansion	6.4 ppm/°C	6.4 ppm/°C
Surface Properties		
Surface roughness, avg	0.50 mm CLA*	0.23 mm CLA*
Surface flatness, avg (Camber)	0.002 cm/cm	0.002 cm/cm

*Center line average.

11.3 Thick-Film Materials and Processes

Thick-film technology is an additive process that utilizes screen printing methods to apply conductive, resistive, and insulating films, initially in the form of a viscous paste, onto a ceramic substrate in the desired pattern. The films are subsequently dried and fired at an elevated temperature to activate the adhesion mechanism to the substrate.

There are three basic types of thick-film materials:

- Cermet thick-film materials, a combination of glass ceramic and metal (hence the name cermet). Cermet films are designed to be fired in the range 850–1000°C.
- **Refractory** thick-film materials, typically tungsten, molybdenum, and titanium, which also may be alloyed with each other in various combinations. These materials are designed to be cofired with ceramic substrates at temperatures ranging up to 1600°C and are postplated with nickel and gold to allow component mounting and wire bonding.
- Polymer thick films, a mixture of polymer materials with conductor, resistor, or insulating particles. These materials cure at temperatures ranging from 85–300°C.

This chapter deals primarily with cermet thick-film materials, as these are most directly associated with the hybrid technology.

Thick-Film Materials

A cermet thick-film paste has four major ingredients: an active element, an adhesion element, an organic binder, and a solvent or thinner. The combination of the organic binder and thinner are often referred to as the vehicle, since it acts as the transport mechanism of the active and adhesion elements to the substrate. When these constituents are mixed together and milled for a period of time, the result is a thick, viscous mixture suitable for screen printing.

Active Element

The nature of the material that makes up the active element determines the electrical properties of the fired film. If the active material is a metal, the fired film will be a conductor; if it is a conductive metal oxide, it will be a resistor; and, if it is an insulator, it will be a dielectric. The active metal is in powder form ranging from 1 to 10 μm, with a mean diameter of about 5 μm.

Adhesion Element

The adhesion element provides the adhesion mechanism of the film to the substrate. There are two primary constituents used to bond the film to the substrate. One adhesion element is a glass, or frit, with a relatively low melting point. The glass melts during firing, reacts with the glass in the substrate, and flows into the irregularities on the substrate surface to provide the adhesion. In addition, the glass flows around the active material particles, holding them in contact with each other to promote sintering and to provide a series of three-dimensional continuous paths from one end of the film to the other.

A second class of conductor materials utilizes metal oxides to provide the adhesion. In this case, a pure metal is placed in the paste and reacts with oxygen atoms on the surface of the substrate to form an oxide. The primary adhesion mechanism of the active particles is sintering, both to the oxide and to each other, which takes place during firing. Conductors of this type offer improved adhesion and have a pure metal surface for added bondability, solderability, and conductivity. Conductors of this type are referred to as fritless materials, oxide-bonded materials, or molecular-bonded materials.

A third class of conductor materials utilizes both reactive oxides and glasses. These materials, referred to as mixed-bonded systems, incorporate the advantages of both technologies and are the most frequently used conductor materials.

Organic Binder

The organic binder is a **thixotropic** or **pseudoplastic fluid** and serves two purposes: it acts as a vehicle to hold the active and adhesion elements in suspension until the film is fired, and it gives the paste the proper fluid characteristics for screen printing. The organic binder is usually referred to as the nonvolatile organic since it does not readily evaporate, but begins to burn off at about 350°C. The binder must oxidize cleanly during firing, with no residual carbon, which could contaminate the film. Typical materials used in this application are ethyl cellulose and various acrylics.

For nitrogen-fireable films, where the firing atmosphere can contain only a few ppm of oxygen, the organic vehicle must decompose and thermally depolymerize, departing as a highly volatile organic vapor in the nitrogen blanket provided as the firing atmosphere.

Solvent or Thinner

The organic binder in its usual form is too thick to permit screen printing, which necessitates the use of a solvent or thinner. The thinner is somewhat more volatile than the binder, evaporating rapidly above 70–100°C. Typical materials used for this application are terpineol, butyl carbitol, and certain of the complex alcohols into which the nonvolatile phase can dissolve.

The ingredients of the thick-film paste are mixed together in proper proportions and milled on a three-roll mill for a sufficient period of time to ensure that they are thoroughly mixed and that no agglomeration exists.

Thick-Film Conductor Materials

Most thick-film conductor materials are noble metals or combinations of noble metals, which are capable of withstanding the high processing temperatures in an oxidizing atmosphere. The exception is copper, which must be fired in a pure nitrogen atmosphere.

Thick-film conductors must perform a variety of functions:

- Provide a path for electrical conduction
- Provide a means for component attach
- Provide a means for terminating thick-film resistors

The selection of a conductor material depends on the technical, reliability, and cost requirements of a particular application. Figure 11.1 illustrates the process compatibility of the various types of conductors.

Gold

Gold (Au) is used in applications where a high degree of reliability is required, such as military and space applications, and where eutectic die bonding is necessary. Gold is also used where gold ball bonding is required or desired. Gold has a strong tendency to form intermetallic compounds with other metals used in the electronic assembly process, especially tin (Sn) and aluminum (Al), and the characteristics of these alloys may be detrimental to reliability. Therefore, when used in conjunction with tin or aluminum, gold must frequently be alloyed with other noble metals, such as platinum (Pt) or palladium (Pd) to prevent AuSn or AuAl alloys from forming. It is common to use a PtAu alloy when PbSn solder is to be used, and to use a PdAu alloy when Al wire bonding is required to minimize intermetallic compound formation.

Silver

Silver (Ag) shares many of the properties of gold in that it alloys readily with tin and leaches rapidly into molten Sn/Pb solders in the pure state. Pure silver conductors must have special bonding materials in the adhesion mechanism or must be alloyed with Pd and/or Pt in order to be used with Sn/Pb solder. Silver is also susceptible to **migration** when moisture is present. Although most metals share this characteristic to a greater or lesser degree, Ag is highly susceptible due to its low ionization potential.

Alloys of Silver with Platinum and/or Palladium

Alloying Ag with Pd and/or Pt slows down both the leaching rate and the migration rate, making it practical to use these alloys for soldering. These are used in the vast majority of commercial applications, making them by far the most commonly used conductor materials. Until recently, silver-bearing materials have been rarely used in multilayer applications due to their tendency to react with and diffuse into dielectric materials, causing short circuits and weakened voltage handling capability. Advancements in these materials have improved this property, which will lead to the increased use of silver-based materials in multilayer structures, resulting in a significant cost reduction. A disadvantage of Pd/Pt/Ag alloys is that the electrical resistance is significantly increased. In some cases, pure Ag is nickel (Ni) plated to increase leach resistance.

		AU WIRE BONDING	AL WIRE BONDING	EUTECTIC BONDING	SN/PB SOLDER	EPOXY BONDING
MATERIAL	AU	Y	N	Y	N	Y
	PD/AU	N	Y	N	Y	Y
	PT/AU	N	Y	N	Y	Y
	AG	Y	N	N	Y	Y
	PD/AG	N	Y	N	Y	Y
	PT/AG	N	Y	N	Y	Y
	PT/PD/AG	N	Y	N	Y	Y
	CU	N	Y	N	Y	N

FIGURE 11.1 Process compatibility of thick-film conductors.

Copper

Originally developed as a low-cost substitute for gold, copper is now being selected when solderability, leach resistance, and low resistivity are required. These properties are particularly attractive for power hybrid circuits. The low resistivity allows the copper conductor traces to handle higher currents with a lower voltage drop, and the solderability allows power devices to be soldered directly to the metallization for better thermal transfer.

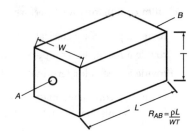

FIGURE 11.2 Resistance of a rectangular solid.

Thick-Film Resistor Material

Thick-film resistors are formed by adding metal oxide particles to glass particles and firing the mixture at a temperature/time combination sufficient to melt the glass and to sinter the oxide particles together. The resulting structure consists of a series of three-dimensional chains of metal oxide particles embedded in a glass matrix. The higher the metal oxide-to-glass ratio, the lower will be the resistivity and vice versa.

Referring to Fig. 11.2, the electrical resistance of a material in the shape of a rectangular solid is given by the classic formula

$$R = \frac{\rho_B L}{WT} \tag{11.1}$$

where:

R = electrical resistance, Ω
ρ_B = bulk resistivity of the material, ohms-length
L = length of the sample in the appropriate units
W = width of the sample in the appropriate units
T = thickness of the sample in the appropriate units

A bulk property of a material is one that is independent of the dimensions of the sample. When the length and width of the sample are much greater than the thickness, a more convenient unit to use is the sheet resistance, which is equal to the bulk resistivity divided by the thickness,

$$\rho_s = \frac{\rho_B}{T} \tag{11.2}$$

where ρ_s is the sheet resistance in ohms/square/unit thickness.

The sheet resistance, unlike the bulk resistivity, is a function of the dimensions of the sample. Thus, a sample of a material with twice the thickness as another sample will have half the sheet resistivity, although the bulk resistivity is the same. In terms of the sheet resistivity, the electrical resistivity is given by

$$R = \frac{\rho_s L}{W} \tag{11.3}$$

If the length is equal to the width (the sample is a square), the electrical resistance is the same as the sheet resistivity independent of the actual dimensions of the sample. This is the basis of the units of sheet resistivity, ohms/square/unit thickness. For thick-film resistors, the standard adopted for unit thickness is 0.001 in or 25 mm of *dried* thickness. The specific units for thick-film resistors are $\Omega/\square/0.001$ (read ohms per square per mil) of dried thickness. For convenience, the units are generally referred to as, simply, $\Omega/\square$.

A group of thick-film materials with identical chemistries that are blendable is referred to as a *family* and will generally have a range of values from 10 $\Omega/\square$ to 1 M$\Omega/\square$ in decade values, although intermediate values are available as well. There are both high and low limits to the amount of material that may be added. As more and more material is added, a point is reached where there is not enough glass to maintain the structural integrity of the film. A practical lower limit of sheet resistivity of resistors formed in this manner is about 10 $\Omega/\square$. Resistors with a sheet resistivity below this value must have a different chemistry and often are not blendable with the regular family of materials. At the other extreme, as less and less material is added, a point is reached where there are not enough particles to form continuous chains, and the sheet resistance rises very abruptly. Within most resistor families, the practical upper limit is about 2 M$\Omega/\square$. Resistor materials are available to about 20 M$\Omega/\square$, but the chemical composition of these materials is not amenable to blending with lower value resistors.

The active phase for resistor formulation is the most complex of all thick-film formulations due to the large number of electrical and performance characteristics required. The most common active material used in air-fireable resistor systems is ruthenium, which can appear as RuO_2 (ruthenium dioxide) or as $BiRu_2O_7$ (bismuth ruthenate).

Properties of Thick-Film Resistors

The conduction mechanisms in thick-film resistors are very complex and have not yet been well defined. Mechanisms with metallic, semiconductor, and insulator properties have all been identified. High ohmic value resistors tend to have more of the properties associated with semiconductors, whereas low ohmic values tend to have more of the properties associated with conductors:

- High ohmic value resistors tend to have a more negative *temperature coefficient of resistance* (*TCR*) than low ohmic value resistors. This is not always the case in commercially available systems due to the presence of TCR modifiers, but always holds true in pure metal oxide–glass systems.
- High ohmic value resistors exhibit substantially more current noise than low ohmic value resistors.
- High ohmic value resistors are more susceptible to high voltage pulses and static discharge than low ohmic value resistors. Some resistor families can be reduced in value by more than an order of magnitude when exposed to an *electrostatic discharge* (*ESD*) charge of only moderate value.

Temperature Coefficient of Resistance

All materials exhibit a change in resistance with material, either positive or negative, and many are nonlinear to a high degree. By definition, the TCR at a given temperature is the slope of the resistance–temperature curve at that temperature. The TCR of thick-film resistors is generally linearized over one or more intervals of temperature, typically in the range from –55 to +125°C, as shown in Eq. (11.4) as

$$\text{TCR} = \frac{R(T_2) - R(T_1)}{R(T_1)(T_2 - T_1)} \times 10^6 \qquad (11.4)$$

where:

TCR = temperature coefficient of resistance, ppm/°C
$R(T_2)$ = resistance at temperature T_2
$R(T_1)$ = resistance at temperature T_1
T_1 = temperature at which $R(T_1)$ is measured, reference temperature
T_2 = temperature at which $R(T_2)$ is measured

A commercial resistor paste carefully balances the metallic, nonmetallic, and semiconducting fractions to obtain a TCR as close to zero as possible. This is not a simple task, and the hot TCR may be quite different from the cold TCR. Paste manufacturers give both the hot and cold values when describing a resistor paste. Although this does not fully define the curve, it is adequate for most design work.

It is important to note that the TCR of most materials is not linear, and the single point measurement is at best an approximation. The only completely accurate method of describing the temperature characteristics of a material is to examine the actual graph of temperature vs resistance. The TCR for a material may be positive or negative. By convention, if the resistance increases with increasing temperature, the TCR is positive. Likewise, if the resistance decreases with increasing temperature, the TCR is negative.

Voltage Coefficient of Resistance (VCR)

The expression for the VCR is similar in form to the TCR and may be represented by Eq. (11.5) as

$$\text{VCR} = \frac{R(V_2) - R(V_1)}{R(V_1)(V_2 - V_1)} \tag{11.5}$$

where:

$R(V_1)$ = resistance at V_1
$R(V_2)$ = resistance at V_2
V_1 = voltage at which $R(V_1)$ is measured
V_2 = voltage at which $R(V_2)$ is measured

Because of the semiconducting nature of resistor pastes, the VCR is always negative. As V_2 is increased, the resistance decreases. Also, because higher resistor decade values contain more glass and oxide constituents, and are more semiconducting, higher paste values tend to have more negative VCRs than lower values.

The VCR is also dependent on resistor length. The voltage effect on a resistor is a gradient; it is the volt/mil rather than the absolute voltage that causes resistor shift. Therefore, long resistors show less voltage shift than short resistors, for similar compositions and voltage stress.

Resistor Noise

Resistor noise is an effective method of measuring the quality of the resistor and its termination. On a practical level, noise is measured according to MIL-STD-202 on a Quantech Noise Meter. The resistor current noise is compared to the noise of a standard low noise resistor and reported as a **noise index** in decibel.

The noise index is expressed as microvolt/volt/frequency decade. The noise measured by MIL-STD-202 is 1/f noise, and the measurement assumes this is the only noise present. However, there is thermal or white noise present in all materials that is not frequency dependent and adds to the 1/f noise. Measurements are taken at low frequencies to minimize the effects of thermal noise. The noise index of low-value resistors is lower than high-value resistors because the low-value resistors have more metal and more free electrons. Noise also decreases with increasing resistor area (actually resistor volume).

High-Temperature Drift

The high-temperature drift characteristic of a resistor is an important material attribute, as it affects the long-term performance of the circuit. A standard test condition is 125°C for 1000 h at normal room humidity. A more aggressive test would be 150°C for 1000 h or 175°C for 40 h. A trimmed thick-film resistor is expected to remain within 1% of the original value after testing under these conditions.

Power Handling Capability

Drift due to high power is probably due to internal resistor heating. It is different from thermal aging in that the heat is generated at the point-to-point metal contacts within the resistor film. When a resistor is subjected to heat from an external source, the whole body is heated to the test temperature. Under power, local heating can result in a much higher temperature. Because lower value resistors have more metal and, therefore, many more contacts, low-value resistors tend to drift less than higher value resistors under similar loads.

Resistor pastes are generally rated at 50 W/in^2 of active resistor area. This is a conservative figure, however, and the resistor can be rated at 100 or even 200 W/in^2. A burn-in process will substantially reduce subsequent drift since most of the drift occurs in the first hours of load.

Process Considerations

The process windows for printing and firing thick-film resistors are extremely critical in terms of both temperature control and atmosphere control. Small variations in temperature or time at temperature can cause significant changes in the mean value and distribution of values. In general, the higher the ohmic value of the resistor, the more dramatic will be the change. As a rule, high ohmic values tend to decrease as temperature and/or time is increased, whereas very low values (<100 $\Omega/\square$) may tend to increase.

Thick-film resistors are very sensitive to the firing atmosphere. For resistor systems used with air fireable conductors, it is critical to have a strong oxidizing atmosphere in the firing zone of the furnace. In a neutral or reducing atmosphere, the metallic oxides that comprise the active material will reduce to pure metal at the temperatures used to fire the resistors, dropping resistor values by more than an order of magnitude. Again, high ohmic value resistors are more sensitive than low ohmic value. Atmospheric contaminants, such as vapors from hydrocarbons or halogenated hydrocarbons, will break down at firing temperatures, creating a strong reducing atmosphere. For example, one of the breakdown components of hydrocarbons is carbon monoxide, one of the strongest reducing agents known. Concentrations of fluorinated hydrocarbons in a firing furnace of only a few ppm can drop the value of a 100 kΩ resistor to below 10 kΩ.

Properties of Thick-Film Dielectric Materials

Multilayer Dielectric Materials

Thick-film dielectric materials are used primarily as insulators between conductors, either as simple crossovers or in complex multilayer structures. Small openings, or vias, may be left in the dielectric layers so that adjacent conductor layers may interconnect. In complex structures, as many as several hundred vias per layer may be required. In this manner, complex interconnection structures may be created. Although the majority of thick-film circuits can be fabricated with only three layers of metallization, others may require several more. If more than three layers are required the yield begins dropping dramatically with a corresponding increase in cost.

Dielectric materials used in this application must be of the devitrifying or recrystallizable type. These materials in paste form are a mixture of glasses that melt at a relatively low temperature. During firing, when they are in the liquid state, they blend together to form a uniform composition with a higher melting point than the firing temperature. Consequently, on subsequent firings they remain in the solid state, which maintains a stable foundation for firing sequential layers. By contrast, vitreous glasses always melt at the same temperature and would be unacceptable for layers either to sink and short to conductor layers underneath, or to swim and form an open circuit. Additionally, secondary loading of ceramic particles is used to enhance devitrification and to modify the TCE.

Dielectric materials have two conflicting requirements in that they must form a continuous film to eliminate short circuits between layers while, at the same time, they must maintain openings as small as 0.010-in. In general, dielectric materials must be printed and fired twice per layer to eliminate pinholes and prevent short circuits between layers.

The TCE of thick-film dielectric materials must be as close as possible to that of the substrate to avoid excessive bowing, or warpage, of the substrate after several layers. Excessive bowing can cause severe problems with subsequent processing, especially where the substrate must be held down with a vacuum or where it must be mounted on a heated stage. In addition, the stresses created by the bowing can cause the dielectric material to crack, especially when it is sealed within a package. Thick-film material manufacturers have addressed this problem by developing dielectric materials that have an almost exact TCE match with alumina substrates. Where a serious mismatch exists, matching layers of dielectric must be printed on the bottom of the substrate to minimize bowing, which obviously increases the cost.

Dielectric materials with higher dielectric constants are also available for manufacturing thick-film capacitors. These generally have a higher loss tangent than chip capacitors and utilize a great deal of space. Although the initial tolerance is not good, thick-film capacitors can be trimmed to a high degree of accuracy.

Overglaze Materials

Dielectric overglaze materials are vitreous glasses designed to fire at a relatively low temperature, usually around 550°C. They are designed to provide mechanical protection to the circuit, to prevent contaminants and water from spanning the area between conductors, to create solder dams, and to improve the stability of thick-film resistors after trimming.

When soldering a device with many leads, it is imperative that the volume of solder under each lead be the same. A well-designed overglaze pattern can prevent the solder from wetting other circuit areas and flowing away from the pad, keeping the solder volume constant. In addition, overglaze can help to prevent solder bridging between conductors.

Overglaze material has long been used to stabilize thick-film resistors after laser trim. In this application, a green or brown pigment is added to enhance the passage of a yttrium–aluminum–garnet (YAG) laser beam. Colors toward the shorter wavelength end of the spectrum, such as blue, tend to reflect a portion of the YAG laser beam and reduce the average power level at the resistor. There is some debate as to the effectiveness of overglaze in enhancing the resistor stability, particularly with high ohmic values. Several studies have shown that, although overglaze is undoubtedly helpful at lower values, it can actually increase the resistor drift of high-value resistors by a significant amount.

Processing Thick-Film Circuits

Screen Printing

The thick-film fabrication process begins with the generation of artwork from the drawings of the individual layers created during the design stage. The artwork represents the exact pattern to be deposited on a 1:1 scale. A stainless-steel wire mesh screen ranging from 80 to 400 mesh count is coated with a photosensitive material and exposed using the artwork for the layer to be printed. The selection of the mesh count depends on the linewidth to be printed. For 0.010 in lines and spaces, a 325 mesh is adequate. For lines down to 0.005 in, a 400 mesh screen is required, and for coarse materials such as solder paste, an 80 mesh screen is needed. The unexposed photoresistive material is washed away leaving openings in the screen where paste is to be deposited. The screen is mounted into a screen printer that has, at a minimum, provisions for controlling squeegee pressure, squeegee speed, and the snapoff distance. Most thick-film materials are printed by the off-contact process in which the screen is separated from the substrate by a distance called the snapoff. As the squeegee is moved, it stretches the screen by a small amount, creating tension in the wire mesh. As the squeegee passes, the tension in the screen causes the mesh to snap back, leaving the paste on the screen. The snapoff distance is probably the most critical parameter in the entire screen printing process. If it is too small, the paste will tend to be retained in the screen mesh, resulting in an incomplete print; if it is too large, the screen will be stretched too far, resulting in a loss of tension. A typical snapoff distance for an 8×10 in screen is 0.025 in.

Drying

Thick-film pastes are dried prior to firing to evaporate the volatile solvents from the printed films. If the volatile solvents are allowed to enter the firing furnace, flash evaporation may occur, leaving pits or craters in the film. In addition, the by-products of these materials may result in reduction of the oxides that comprise the fired film. Most solvents in thick-film pastes have boiling points in the range of 180–250°C. Because of the high surface area/volume of deposited films, drying at 80–160°C for a period of 10–30 min is adequate to remove most of the solvents from wet films.

Hybrid Microelectronics Technology

FIGURE 11.3 Thick-film circuits: Hermetic (left) and nonhermetic.

Firing

Belt furnaces having independently controlled heated zones through which a belt travels at a constant speed are commonly used for firing thick films. By adjusting the zone temperature and the belt speed, a variety of time vs temperature profiles can be achieved. The furnace must also have provisions for atmosphere control during the firing process to prevent reduction (or, in the case of copper, oxidation) of the constituents.

Figure 11.3 illustrates two typical thick-film circuits.

11.4 Thin-Film Technology

Thin-film technology, in contrast to thick-film technology, is a subtractive technology in that the entire substrate is coated with several layers of material and the unwanted material is etched away in a succession of photoetching processes. The use of photolithographic processes to form the patterns enables much narrower and more well-defined lines than can be formed by the thick-film process. This feature promotes the use of thin-film technology for high-density and high-frequency applications.

Thin-film circuits typically consist of three layers of material deposited on a substrate. The bottom layer serves two purposes: it is the resistor material and also provides the adhesion to the substrate. The adhesion mechanism of the film to the substrate is an oxide layer, which forms at the interface between the film and the substrate. The bottom layer must, therefore, be a material that oxidizes readily. The middle layer acts as an interface between the resistor layer and the conductor layer, either by improving the adhesion of the conductor or by preventing diffusion of the resistor material into the conductor. The top layer acts as the conductor layer.

Deposition Technology

The term thin film refers more to the manner in which the film is deposited onto the substrate, as opposed to the actual thickness of the film. Thin films are typically deposited by a vacuum deposition technique or by electroplating.

Sputtering

Sputtering is the principal method by which thin films are applied to substrates. In the most basic form of sputtering, a current is established in a conducting plasma formed by striking an arc in a partial vacuum of approximately 10-μ pressure with a potential applied. The gas used to establish the plasma is typically an inert gas, such as argon, that does not react with the target material. The substrate and a target material are situated in the plasma with the substrate at ground potential and the target at a high potential, which may be AC or DC. The high potential attracts the gas ions in the plasma to the point where they collide with the target with sufficient kinetic energy to dislodge microscopically sized particles with enough residual kinetic energy to travel the distance to the substrate and adhere. This process is referred to as *triode sputtering*.

Ordinary triode sputtering is a very slow process, requiring hours to produce useable films. By utilizing magnets at strategic points, the plasma can be concentrated in the vicinity of the target, greatly speeding up the deposition process. The potential that is applied to the target is typically RF energy at a frequency of approximately 13 MHz, which may be generated by a conventional electronic oscillator or by a magnetron. The magnetron is capable of generating considerably more power with a correspondingly higher deposition rate.

By adding small amounts of other gases, such as oxygen and nitrogen, to the argon, it is possible to form oxides and nitrides of certain target materials on the substrate. It is this technique, called *reactive sputtering*, which is used to form tantalum nitride, a common resistor material.

Evaporation

The evaporation of a material into the surrounding area occurs when the vapor pressure of the material exceeds the ambient pressure and can take place from either the solid state or the liquid state. In the thin film process, the material to be evaporated is placed in the vicinity of the substrate and heated until the vapor pressure of the material is considerably above the ambient pressure. The evaporation rate is directly proportional to the difference between the vapor pressure of the material and the ambient pressure and is highly dependent on the temperature of the material.

Evaporation must take place in a relatively high vacuum ($<10^{-6}$ torr) for three reasons:

1. To lower the vapor pressure required to produce an acceptable evaporation rate, thereby lowering the temperature required to evaporate the material.
2. To increase the mean free path of the evaporated particles by reducing the scattering due to gas molecules in the chamber. As a further result, the particles tend to travel in more of a straight line, improving the uniformity of the deposition.
3. To remove atmospheric contaminants and components, such as oxygen and nitrogen, which tend to react with the evaporated film.

At 10^{-7} torr, a vapor pressure of 10^{-2} torr is required to produce an acceptable evaporation rate. For most metals, this temperature is in excess of 1000°C. The refractory metals, or metals with a high melting point such as tungsten, titanium, or molybdenum, are frequently used as carriers, or *boats*, to hold other metals during the evaporation process. To prevent reactions with the metals being evaporated, the boats may be coated with alumina or other ceramic materials.

In general, the kinetic energy of the evaporated particles is substantially less than that of sputtered particles. This requires that the substrate be heated to about 300°C to promote the growth of the oxide adhesion interface. This may be accomplished by direct heating of the substrate mounting platform or by radiant infrared heating.

There are several techniques by which evaporation can be accomplished. The two most common of these are resistance heating and electron-beam (E-beam) heating.

Evaporation by resistance heating usually takes place from a boat made with a refractory metal, a ceramic crucible wrapped with a wire heater, or a wire filament coated with the evaporant. A current is passed through the element, and the generated heat heats the evaporant. It is somewhat difficult to monitor the temperature of the melt by optical means due to the propensity of the evaporant to coat the inside of the chamber, and control must be done by empirical means.

The E-beam evaporation method takes advantage of the fact that a stream of electrons accelerated by an electric field tend to travel in a circle when entering a magnetic field. This phenomenon is utilized to direct a high-energy stream of electrons onto an evaporant source. The kinetic energy of the electrons is converted into heat when they strike the evaporant. E-beam evaporation is somewhat more controllable since the resistance of the boat is not a factor, and the variables controlling the energy of the electrons are easier to measure and control. In addition, the heat is more localized and intense, making it possible to evaporate metals with higher 10^{-2} torr temperatures and lessening the reaction between the evaporant and the boat.

Comparison Between Sputtering and Evaporation

The adhesion of a sputtered film is superior to that of an evaporated film, and is enhanced by presputtering the substrate surface by random bombardment of argon ions prior to applying the potential to the target. This process removes several atomic layers of the substrate surface, creating a large number of broken oxygen bonds and promoting the formation of the oxide interface layer. The oxide formation is further enhanced by the residual heating of the substrate that results from the transfer of the kinetic energy of the sputtered particles to the substrate when they collide.

It is difficult to evaporate alloys such as NiCr due to the difference between the 10^{-2} torr temperatures. The element with the lower temperature tends to evaporate somewhat faster, causing the composition of the evaporated film to be different from the composition of the alloy. To achieve a particular film composition, the composition of the melt must contain a higher portion of the material with the higher 10^{-2} torr temperature and the temperature of the melt must be tightly controlled. By contrast, the composition of a sputtered film is identical to that of the target.

Evaporation is limited to the metals with lower melting points. Refractory metals and ceramics are virtually impossible to deposit by evaporation.

Reactive deposition of nitrides and oxides is very difficult to control.

Electroplating

Electroplating is accomplished by applying a potential between the substrate and the anode, which are suspended in a conductive solution of the material to be plated. The plating rate is a function of the potential and the concentration of the solution. In this manner, most metals can be plated to a metal surface.

This is considerably more economical and results in much less target usage. For added savings, some companies apply photoresist to the substrate and electroplate gold only where actually required by the pattern.

Photolithographic Processes

In the photolithographic process, the substrate is coated with a photosensitive material that is exposed through a pattern formed on a glass plate. Ultraviolet light, X rays, and electron beams may all be used to expose the film. The photoresist may be of the positive or negative type, with the positive type being prevalent due to its inherently higher resistance to the etchant materials. The unwanted material that is not protected by the photoresist may be removed by wet, or chemical etching, or by dry, or sputter etching, in which the unwanted material is removed by ion bombardment. In essence, the exposed material acts as a sputtering target. The photoresist, being more compliant, simply absorbs the ions and protects the material underneath.

Thin-Film Materials

Virtually any inorganic material may be deposited by the sputtering process, although RE sputtering is necessary to deposit dielectric materials or metals, such as aluminum, that oxidize readily. Organic materials are difficult to sputter because they tend to release absorbed gases in a high vacuum, which interferes with the sputtering process. A wide variety of substrate materials are also available, but these in general must contain or be coated with an oxygen compound to permit adhesion of the film.

Thin-Film Resistors

Materials used for thin-film resistors must perform a dual role in that they must also provide the adhesion to the substrate, which narrows the choice to those materials that form oxides. The resistor film begins forming as single points on the substrate in the vicinity of substrate faults or other irregularities, which might have an excess of broken oxygen bonds. The points expand into islands, which in turn join to form continuous films. The regions where the islands meet are called *grain boundaries* and are a source of collisions for the electrons. The more grain boundaries that are present, the more negative will be the TCR. Unlike thick-film resistors, however, the boundaries do not contribute to the noise level. Further, laser trimming does not create microcracks in the glass-free structure, and the inherent mechanisms for resistor drift are not present in thin films. As a result, thin-film resistors have better stability, noise, and TCR characteristics than thick-film resistors.

The most common types of resistor material are nichrome (NiCr) and tantalum nitride (TaN). Although NiCr has excellent stability and TCR characteristics, it is susceptible to corrosion by moisture if not passivated by sputtered quartz or by evaporated silicon monoxide (SiO). TaN, on the other hand, may be passivated by simply baking in air for a few minutes. This feature has resulted in the increased use of TaN at the expense of NiCr, especially in military programs. The stability of passivated TaN is comparable to that of passivated NiCr, but the TCR is not as good unless annealed for several hours in a vacuum to minimize the effect of the grain boundaries. Both NiCr and TaN have a relatively low maximum sheet resistivity on alumina, about 400 $\Omega/\square$ for NiCr and 200 $\Omega/\square$ for TaN. This requires lengthy and complex patterns to achieve a high value of resistance, resulting in a large area and the potential for low yield.

The TaN process is the most commonly used due to its inherently high stability. In this process N_2 is introduced into the argon gas during the sputtering process, forming TaN by reacting with pure Ta atoms on the surface of the substrate. By heating the film in air at about 425°C for 10 min, a film of TaO is formed over the TaN, which is virtually impervious to further O_2 diffusion at moderately high temperatures, which helps to maintain the composition of the TaN film and to stabilize the value of the resistor. TaO is essentially a dielectric and, during the stabilization of the film, the resistor value is increased. The amount of increase for a given time and temperature is dependent on the thickness and sheet resistivity of the film. Films with a lower sheet resistivity increase proportionally less than those with a higher sheet resistivity. The resistance increases as the film is heated longer, making it possible to control the sheet resistivity to a reasonable accuracy on a substrate-by-substrate basis.

Other materials, based on chromium and rhenium alloys, are also available. These materials have a higher sheet resistivity than NiCr and TaN and offer the same degree of stability.

Barrier Materials

When Au is used as the conductor material, a barrier material between the Au and the resistor is required. When gold is deposited directly on NiCr, the Cr has a tendency to diffuse through the Au to the surface, which interferes with both wire bonding and eutectic die bonding. To alleviate this problem, a thin layer of pure Ni is deposited over the NiCr, which also improves the solderability of the surface considerably. The adhesion of Au to TaN is very poor. To provide the necessary adhesion, a thin layer of 90Ti/10W may be used between the Au and the TaN.

FIGURE 11.4 A thin-film circuit.

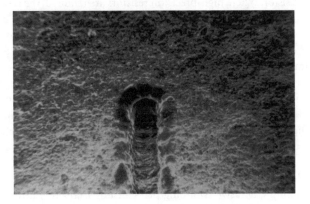

FIGURE 11.5 A laser-trimmed thick-film resistor.

Conductor Materials

Gold is the most common conductor material used in thin-film hybrid circuits because of the ease of wire and die bonding and its high resistance to tarnish and corrosion. Aluminum and copper are also used in some applications. A thin-film circuit is illustrated in Fig. 11.4.

11.5 Resistor Trimming

One of the advantages of the hybrid technology over other **packaging** technologies is the ability to adjust the value of resistors to a more precise value by a process called *trimming*. By removing a portion of the resistor with a laser or by abrasion, as shown in Fig. 11.5, the value can be increased from the as-processed state to a predetermined value. The laser trimming process can be highly automated and can trim a resistor to a tolerance of better than 1% in less than a second.

More than any other development, laser trimming has contributed to the rapid growth of the hybrid microelectronics industry over the past few years. Once it became possible to trim resistors to precision values economically, hybrid circuits became cost competitive with other forms of packaging and the technology expanded rapidly.

Lasers used to trim thick-film resistors are generally a YAG crystal, which has been doped with neodymimum. YAG lasers are of lower power and are capable of trimming film resistors without causing serious damage if the parameters are adjusted properly. Even with the YAG laser, it is necessary to lower the maximum power and spread out the beam to avoid overtrimming and to increase the trimming speed. This is accomplished by a technique called *Q-switching*, which decreases the peak power and widens the pulse width such that the overall average power is the same. One of the effects of this technique is that the first and last pulses of a given trim sequence have substantially greater energy than the intervening ones. The first pulse does not create a problem, but the last pulse occurs within the body of the resistor, penetrating deeper into the substrate and creating the potential for resistor drift by creating minute cracks (microcracks) in the resistor, which emanate from the point of trim termination.

The microcracks occur as a result of the large thermal gradient, which exists between the area of the last pulse and the remainder of the resistor. Over the life of the resistor, the cracks will propagate through the body of the resistor until they reach a termination point, such as an active particle or even another microcrack. The propagation distance can be up to several mils in a high-value resistor with a high concentration of glass. The microcracks cause an increase in the value of the resistor and also increase the noise by increasing the current density in the vicinity of the trim. With proper design of the resistor, and with proper selection of the cut mode, the amount of drift can be held to less than 1%.

The propagation of the microcracks can be accelerated by the application of heat, which also has the effect of stabilizing the resistor. Where high-precision resistors are required, the resistors may be trimmed slightly low, exposed to heat for a period of time, and retrimmed to a value in a noncritical area.

The amount of drift depends on the amount of propagation distance relative to the distance from the termination point of the trim to the far side of the resistor. This can be minimized by either making the resistor larger or by limiting the distance that the laser penetrates into the resistor. As a rule of thumb, the value of the resistor should not be increased by more than a factor of two to minimize drift.

In the abrasive trimming process, a fine stream of sand propelled by compressed air is directed at the resistor, abrading away a portion of the resistor and increasing the value. Trimming resistors to a high precision is more difficult with abrasive trimming due to the size of the particle stream, although tolerances of 1% can be achieved with proper setup and slow trim speeds. It has also proven difficult to automate the abrasive trimming process to a high degree, and it remains a relatively slow process. In addition, substantially larger resistors are required.

Despite these shortcomings, abrasive trimming plays an important role in the hybrid industry. The cost of an abrasive trimmer is substantially less than that of a laser trimmer, and the setup time is much less. In a developmental mode where only a few prototype units are required, abrasive trimming is generally more economical and faster than laser trimming.

In terms of performance, abrasively trimmed resistors are more stable than laser-trimmed resistors and generate less noise because no microcracks are generated during the abrasive trimming process. In the power hybrid industry, resistors that carry high currents or that dissipate high power are frequently abrasively trimmed by trimming a groove in the middle of the resistor. This technique minimizes current crowding and further enhances the stability.

11.6 Comparison of Thick-Film and Thin-Film Technologies

Most circuits can be fabricated with either the thick- or the thin-film technology. The ultimate result of both processes is an interconnection pattern on a ceramic substrate with integrated resistors, and, to a degree, they can be considered to be competing technologies. Given a particular set of requirements, however, the choice between technologies is usually quite distinct. Table 11.3 summarizes the particular advantages of each technology.

11.7 The Hybrid Assembly Process

The assembly of a hybrid circuit involves mechanically mounting the components on the substrate and electrically connecting the components to the proper substrate traces.

TABLE 11.3 Comparison of Thick-Film and Thin-Film Circuits

Thick Film	Thin Film
Multilayer structures are much simpler and more economical to fabricate than with the thin-film technology.	The lines and spaces are much smaller than can be attained by the thick-film process. In a production environment, the thin-film process can produce lines and spaces 0.001 in width, whereas the thick film process is limited to 0.01 in.
The range of resistor values is wider. Thin-film designers are usually limited to a single value of sheet resistivity from which to design all of the resistors in the circuit.	The line definition available by the thin-film process is considerably better than that of the thick-film process. Consequently, the thin-film process operates much better at high frequencies.
For a given application, the thick-film process is usually less expensive	The electrical properties of thin-film resistors are substantially better than thick-film resistors in terms of noise, stability, and precision.

In the most fundamental configuration, the **semiconductor die** is taken directly from the wafer without further processing (generally referred to as a bare die). In this process, the die is mechanically mounted to the substrate with epoxy, solder, or by direct eutectic bonding of the silicon to the substrate metallization, and the electrical connections are made by bonding small wires from the bonding pads to the appropriate conductor. This is referred to as chip-and-wire technology, illustrated in Fig. 11.4.

Other approaches require subsequent processing of the die at the wafer level. Two of the most common configurations are *tape automated bonding* (*TAB*), shown in Fig. 11.7, and the so-called *flip chip* or bumped chip approach, depicted in Fig. 11.8. These are discussed later.

The Chip-and-Wire Process

Bare semiconductor die may be mounted with epoxy, solder, or by direct eutectic bonding. The most common method of both active and passive component attachment is with epoxy, both conductive and nonconductive. This method offers a number of advantages over other techniques, including ease of repair, reliability, productivity, and low process temperature. Most epoxies have a filler added or electrical and/or thermal conductivity. The most common material used for conductive epoxies is silver, which provides both. Other conductive filler materials include gold, palladium–silver for reduced silver migration, and copper plated with a surface coating of tin. Nonconductive filler materials include aluminum oxide and magnesium oxide for improved thermal conductivity. For epoxy attach, it is preferred that the bottom surface of the die be gold. Solder attachment of a die to a substrate is commonly used in the power hybrid industry. For this application, a coating of Ti/Ni/Ag alloy on the bottom is favored to gold. Since many power die are, by necessity, physically large, a compliant solder, such as one of the PbIn alloys is preferred. Eutectic bonding refers to the direct bonding of a silicon device to gold metallization, which takes place at 370°C with a combination of 94% gold and 6% silicon by weight. Gold is the only metal to which the eutectic temperature in combination with silicon is sufficiently low to be practical.

Wire bonding is used to make the electrical connections from the aluminum contacts to the substrate metallization or to a lead frame, from other components, such as chip resistors or chip capacitors, to substrate metallization, from package terminals to the substrate metallization, or from one point on the substrate metallization to another.

There are two basic methods of *wire bonding, thermocompression wire bonding* and *ultrasonic wire bonding*. Thermocompression wire bonding, as the name implies, utilizes a combination of heat and pressure to form an intermetallic bond between the wire and a metal surface. In pure thermocompression bonding, a gold wire is passed through a hollow capillary and a ball formed on the end by means of an electrical arc. The substrate is heated to about 300°C, and the ball is forced into contact with the bonding pad on the device with sufficient force to cause the two metals to bond. The capillary is then moved to the bond site on the substrate, feeding the wire as it goes, and as the wire is bonded to the substrate by the same process, except that the bond is in the form of a stitch, as opposed to the ball on the device. The wire is then clamped and broken at the stitch by pulling and another ball formed as described. Thermocompression bonding is rarely used for the following reasons:

- The high substrate temperature precludes the use of epoxy for device mounting.
- The temperature required for the bond is above the threshold temperature for gold–aluminum intermetallic compound formation. The diffusion rate for aluminum into gold is much greater than for gold into aluminum. The aluminum contact on a silicon device is very thin, and when it diffuses into the gold, voids, called Kirkendall voids, are created in the bond area, increasing the electrical resistance of the bond and decreasing the mechanical strength.
- The thermocompression bonding action does not effectively remove trace surface contaminants that interfere with the bonding process.

The ultrasonic bonding process uses ultrasonic energy to vibrate the wire (primarily aluminum wire) against the surface to blend the lattices together. Localized heating at the bond interface caused by the scrubbing action, aided by the oxide on the aluminum wire, assists in forming the bond. The substrate itself is not heated. Intermetallic compound formation is not as critical, as with the thermocompression bonding process, as both the wire and the device metallization are aluminum. Kirkendall voiding on an aluminum wire bonded to gold substrate metallization is not as critical, since there is substantially more aluminum available to diffuse than on device metallization. Ultrasonic bonding makes a stitch on both the first and second bonds because of the fact that it is very difficult to form a ball on the end of an aluminum wire because of the tendency of aluminum to oxidize. For this reason, ultrasonic bonding is somewhat slower than thermocompression, since the capillary must be aligned with the second bond site when the first bond is made. Ultrasonic bonding to package leads may be difficult if the leads are not tightly clamped, since the ultrasonic energy may be propagated down the leads instead of being coupled to the bond site.

The use of thermosonic bonding of gold wire overcomes the difficulties noted with thermocompression bonding. In this process, the substrate is heated to 150°C and ultrasonic energy is coupled to the wire through the transducer action of the capillary, scrubbing the wire into the metal surface and forming a ball–stitch bond from the device to the substrate, as in thermocompression bonding.

Thermosonic gold bonding is the most widely used bonding technique, primarily because it is faster than ultrasonic aluminum bonding. Once the ball bond is made on the device, the wire may be moved in any direction without stress on the wire, which greatly facilitates automatic wire bonding, as the movement need only be in the x and y directions.

By contrast, before the first ultrasonic stitch bond is made on the device, the circuit must be oriented so that the wire will move toward the second bond site only in the direction of the stitch. This necessitates rotational movement, which not only complicates the design of the bonder, but increases the bonding time as well.

The wire size is dependent on the amount of current that the wire is to carry, the size of the bonding pads, and the throughput requirements. For applications where current and bonding size is not critical, wire 0.001 diameter is the most commonly used size. Although 0.0007-in wire is less expensive, it is difficult to thread through a capillary without frequent breaking and consequent line stoppages.

For high-volume applications, gold is the preferred method due to bonding speed. Once a ball bond is made, the wire can be dressed in any direction. Stitch bonding, on the other hand, requires that the second bond be lined up with the first prior to bonding, which necessitates rotation of the holding platform with a corresponding loss in bonding rate.

Hybrid Microelectronics Technology

FIGURE 11.6 Ultrasonic bonding of heavy aluminum wire.

Figure 11.6 illustrates aluminum ultrasonic bonding of heavy wire, whereas thermosonic bonding is illustrated in Figs. 11.3 and 11.4.

Devices in die form have certain advantages over packaged devices in hybrid circuit applications, including:

- Availability: Devices in the chip form are more readily available than other types and require no further processing.
- **Thermal management**: Devices in the chip form are mounted in intimate contact with the substrate, which maximizes the contact area and improves the heat flow out of the circuit.
- Size: Except for the flip-chip approach, the chip-and-wire approach utilizes the least substrate area.
- Cost: Since they require no special processing, devices in the chip form are less expensive than packaged devices.

Coupled with these advantages, however, are at least two disadvantages:

- Fragility: Devices in the chip form are susceptible to mechanical handling, static discharge, and corrosion as a result of contamination. In addition, the wire bonds require only a few grams of force in the normal direction to fail.
- Testability: This is perhaps the most serious problem with devices in the die form. It is difficult to perform functional testing at high speeds and at temperature extremes as a result of the difficulties in probing. Testing at cold temperatures is virtually prohibitive as a result of moisture condensation. The net result when several devices must be utilized in a single hybrid is lower yields at first electrical test. If the yield of a single chip is 0.98, then the initial yield that can be expected in a hybrid circuit with 10 devices is $0.98^{10} = 0.817$. This necessitates added troubleshooting and repair time with a corresponding detrimental effect on both cost and reliability.

Tape Automated Bonding and Flip-Chip Bonding

The tape automated bonding process was developed in the early 1970s, and has since been adopted by several companies as a method of chip packaging. There are two basic approaches to the TAB process, the *bumped-chip process* and the *bumped-tape process*.

In one form of the bumped-chip process, the wafer is passivated with silicon nitride and windows are opened up over the bonding pads by photoetching. The wafer is then coated with thin layers of titanium, palladium, and gold applied in succession by sputtering. The titanium and palladium act as barrier layers to prevent the formation of intermetallic compounds between the gold and the aluminum. Using a dry film photoresist with a thickness of about 0.0015 in, windows are again opened over the bonding pads

FIGURE 11.7 Tape automated bonding technique.

and gold is electroplated to a thickness of about 0.001 in at these points. The photoresist is removed, and the wafer is successively dipped in gold, palladium, and titanium etches, leaving only the gold bumps over the bonding pads. To prepare the tape, a copper foil is laminated to a 35-mm polyimide film with holes punched at intervals. A lead frame, which is designed to align with the bumps on the chip and to give stress relief after bonding to the substrate, is etched in the copper foil, and the ends of the leads are gold plated. The remainder of the leads may be either tin plated or gold plated, depending on whether the leads are intended to be attached to the substrate by soldering or by thermocompression bonding. The wafer is mounted to a ceramic block with a low-temperature wax, and the die is separated by sawing through completely. The leads are attached to the bumps by thermocompression bonding (usually referred to as the inner-lead bonding process) with a heated beryllia **thermode** whose tip size is the same as that of the die. During the bonding process, the wax melts allowing the chip to be removed from the ceramic block. The inner-lead bonding process may be highly automated by using pattern recognition systems to locate the chip with respect to the lead frame. A TAB assembly mounted to the substrate is shown in Fig. 11.7.

The bumped-tape process is somewhat less expensive than the bumped-chip process, and the technology is not as complex. Early efforts at bumped-tape bonding utilized the same gold-plated leads as in the bumped-chip process and proved somewhat unreliable due to the direct gold–aluminum contact. More recently, the use of sputtered aluminum on the lead frame has eliminated this problem, and has further reduced the processing temperature by utilizing ultrasonic bonding to attach the leads. Low-cost devices may use only a strip of copper or aluminum, whereas devices intended for pretesting use a multilayer copper-polyimide structure. The so-called area array TAB, developed by Honeywell, utilizes a multilayer structure, which can form connections within the periphery of the chip, greatly increasing the lead density without increasing the size of the chip and simplifying the interconnection structure on the chip itself.

TAB technology has other advantages over chip-and-wire technology in the areas of high frequency and high-lead density. TAB may be used on pads about one-third the size required by gold thermosonic bonding, and the larger size of the TAB leads lowers the series inductance, increasing the overall operating frequency. These factors contribute to the use of TAB in the very high-speed integrated circuit (VHSIC) program, which utilizes chips with in excess of 400 bonds operating at high speeds. In addition, the devices are in direct contact with the substrate, easing the problem of removing heat from the device. This is a critical feature, since high-speed operation is usually accompanied by increased power dissipation.

The technology and capital equipment required to mount TAB devices to substrates with the outer-lead bonding process are not expensive. Several companies have recognized this and have begun offering chips in the TAB form for use by other companies. If these efforts are successful, TAB technology should

Hybrid Microelectronics Technology

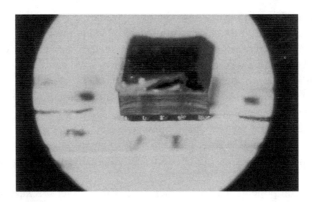

11.8 Flip-chip bonding technique.

grow steadily. If not, TAB will be limited to those companies willing to make the investment and will grow at a less rapid pace.

Flip-chip technology, as shown in Fig. 11.8, is similar to TAB technology in that successive metal layers are deposited on the wafer, ending up with solder-plated bumps over the device contacts. One possible configuration utilizes an alloy of nickel and aluminum as an interface to the aluminum bonding pads. A thin film of pure nickel is plated over the Ni/Al, followed by copper and solder. The copper is plated to a thickness of about 0.0005 in, and the solder is plated to about 0.003 in. The solder is then reflowed to form a hemispherical bump. The devices are then mounted to the substrate face down by reflow solder methods. During reflow, the face of the device is prevented from contacting the substrate metallization by the copper bump. This process is sometimes referred to as the *controlled collapse* process.

Testing of flip-chip devices is not as convenient compared to TAB devices, since the solder bumps are not as amenable to making ohmic contact as the TAB leads. Testing is generally accomplished by lining the devices over a **bed of nails**, which is interfaced to a testing system. At high speeds, this does not realistically model the configuration that the device actually sees on a substrate, and erroneous data may result.

Flip-chip technology is the most space efficient of all of the packaging technologies, since no room outside the boundaries of the chip is required. Further, the contacts may be placed at any point on the chip, reducing the net area required and simplifying the interconnection pattern. IBM has succeeded in fabricating chips with an array in excess of 300 contacts.

There are several problems that have hampered the widespread use of flip-chip-technology. The chips have not been widely available, since only a few companies have developed the technology, primarily for in-house use. New technologies in which the user applies the solder paste, or conductive epoxy, are emerging, which may open up the market for these devices to a wider range of applications.

Other limitations include difficulty in inspecting the solder joints and thermal management. While the joints can be inspected by X-ray methods, military specifications require that each interconnection be visually inspected. The paths for removal of heat are limited to the solder joints themselves, which limits the amount of power that the chips can dissipate.

Surface mount techniques have also been successfully used to assembly hybrid circuits. The metallized ceramic substrate is simply substituted for the conventional printed circuit board and the process of screen printing solder, component placement, and reflow solder is identical.

Defining Terms

Active component: An electronic component that can alter the waveshape of an electronic signal, such as a diode, a transistor, or an integrated circuit. By contrast, ideal resistors, capacitors, and inductors leave the waveform intact and are referred to as passive components.

Bed of nails: A method of interfacing a component or electronic circuit to another assembly, consisting of inverted, spring-loaded probes located at predetermined points.

Dielectric constant: A measurement of the ability of a material to hold electric charge.

Direct bond copper (DBC): A method of bonding copper to a substrate, which blends a portion of the copper with a portion of the substrate at the interface.

Migration: Transport of metal ions from a positive surface to a negative surface in the presence of moisture and electric potential.

Noise index: A measurement of the amount of noise generated in a component as a result of random movement of carriers from one energy state to another.

Packaging: The technology of converting an electronic circuit from a schematic or prototype to a finished form.

Passive components: Components that, in the ideal form, cannot alter the waveshape of an electronic signal.

Plasma: A gas that has been ionized to produce a conducting medium.

Pseudoplastic fluid: A fluid in which the viscosity is nonlinear, with the rate of change lessening as the pressure is increased.

Refractory metal: In this context, a metal with a high melting point.

Semiconductor die: A semiconductor device in the unpackaged state.

Temperature coefficient of expansion (TCE): A measurement of the dimensional change of a material as a result of temperature change.

Thermal conductivity: A measurement of the ability of a material to conduct heat.

Thermal management: The technology of removing heat from the point of generation in an electronics circuit.

Thermode: A device used to transmit heat from one surface to another to facilitate lead or wire bonding.

Thixotropic fluid: A fluid in which the viscosity is nonlinear, with the rate of change increasing as the pressure is increased.

Vapor pressure: A measurement of the tendency of a material to evaporate. A material with a low vapor pressure will evaporate more readily.

References

Elshabini-Riad, A. 1996. *Handbook of Thin Film Technology*. McGraw-Hill, New York.
Hablanian, M. 1990. *High-Vacuum Technology*. Marcel Dekker, New York.
Harper, C. 1996. *Handbook of Electronic Packaging*. McGraw-Hill, New York.
Sergent, J. E. and Harper, C. 1995. *Handbook of Hybrid Microelectronics*, 2nd ed. McGraw-Hill, New York.

Further Information

For further information on hybrid circuits see the following:

ISHM Modular Series on Microelectronics.
Licari, J.J. 1995. *Multichip Module Design, Fabrication, and Testing*. McGraw-Hill, New York.
Proceedings of ISHM Symposia, 1967–1995. *Journal of Hybrid Microelectronics*, published by ISHM.
Tummala, R.R. and Rymaszewski, E. 1989. *Microelectronics Packaging Handbook*. Van Nostrand Reinhold, New York.

12
Surface Mount Technology

12.1	Introduction	12-1
12.2	Definition	12-1
	Considerations in the Implementation of SMT	
12.3	Surface Mount Device (SMD) Definitions	12-4
12.4	Substrate Design Guidelines	12-5
12.5	Thermal Design Considerations	12-7
12.6	Adhesives	12-10
12.7	Solder Paste and Joint Formation	12-12
12.8	Parts Inspection and Placement	12-13
	Parts Placement	
12.9	Reflow Soldering	12-14
	Postreflow Inspection	
12.10	Prototype Systems	12-17

Glenn R. Blackwell
Purdue University

12.1 Introduction

This chapter is intended for the practicing engineer who is familiar with standard **through-hole** (insertion mount) circuit board design, manufacturing, and test, but who now needs to learn more about the realm of surface mount technology (SMT). Numerous references will be given, which will also be of help to the engineer who may be somewhat familiar with SMT, but needs more in-depth information in specific areas. The reader with little knowledge about SMT is referred to a basic introductory article [Mims 1987] and to a journal series which covered design of an SMT memory board [Leibson 1987].

12.2 Definition

Surface mount technology is a collection of scientific and engineering methods needed to design, build, and test products made with electronic components, which mount to the surface of the printed circuit board without holes for leads [Higgins 1991]. This definition notes both the breadth of topics necessary to understand SMT, as well as that the successful implementation of SMT will require the use of *concurrent engineering* [Classon 1993, Shina 1991]. Concurrent engineering means that a team of design, manufacturing, test, and marketing people will concern themselves with board layout, parts and parts placement issues, soldering, cleaning, test, rework, and packaging before any product is made. The careful control of all of these issues improves both yield and reliability of the final product. In fact, SMT cannot be reasonably implemented without the use of concurrent engineering and/or the principles contained in *design for manufacturability (DFM)* and *design for testability (DFT)* and, therefore, any facility that has not embraced these principles should do so if SMT is to be successfully implemented.

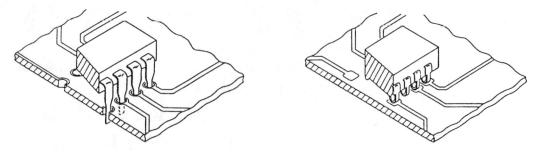

FIGURE 12.1 Comparison of DIP and SMT. (After Intel. 1994. *Packaging Handbook.* Intel Corp., Santa Clara, CA.)

Considerations in the Implementation of SMT

The main reasons to consider implementation of SMT include the following:

- Reduction in circuit board size
- Reduction in circuit board weight
- Reduction in number of layers in the circuit board
- Reduction in trace lengths on the circuit board

Note that not all of these reductions may occur in any given product redesign from **through-hole technology** (THT) to SMT. Reduction in circuit board size can vary from 40 to 60% [TI 1984]. By itself, this reduction presents many advantages in packaging possibilities. Camcorders and digital watches are only possible through the use of SMT. The reduction in weight means that circuit boards and components are less susceptible to vibration problems. Reduction in the number of layers in the circuit board means the bare board will be significantly less expensive to build. Reduction in trace lengths means that high-frequency signals will have fewer problems or that the board will be able to operate at higher frequencies than a through-hole board with longer trace/lead lengths.

Of course there are some disadvantages using SMT. During the assembly of a through-hole board, either the component leads go through the holes or they do not, and the component placement machines can typically detect the difference in force involved and signal for help. During SMT board assembly, the placement machine does not have such direct feedback, and accuracy of final soldered placement becomes a stochastic (probability-based) process, dependent on such items as the following:

- Component pad design
- Accuracy of the PCB artwork and fabrication, which affects the accuracy of trace location
- Accuracy of solder paste deposition location and deposition volume
- Accuracy of adhesive deposition location and volume if adhesive is used
- Accuracy of placement machine vision system(s)
- Variations in component sizes from the assumed sizes
- Thermal issues in the solder reflow process.

In THT test there is a through-hole at every potential test point, making it easy to align a bed-of-nails tester. In SMT designs there are not holes corresponding to every device lead. The design team must consider form, fit and function, time-to-market, existing capabilities, and the cost and time to characterize a new process when deciding on a change of technologies.

Once circuit design is complete, substrate design and fabrication, most commonly of a printed circuit board (PCB), enters the process. Generally, PCB assembly configurations using SMDs are classified as shown in Fig. 12.2.

Surface Mount Technology

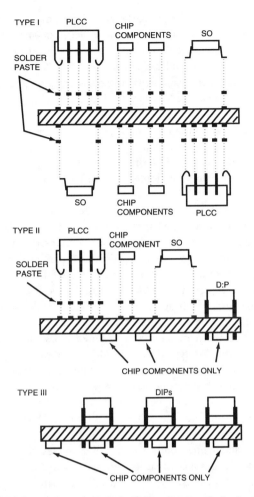

FIGURE 12.2 Type I, II, and III SMT circuit boards. (After Intel. 1994. *Packaging Handbook*. Intel Corp., Santa Clara, CA.)

- **Type I:** Only SMDs are used, typically on both sides of the board. No through-hole components are used. Top and bottom may contain both large and small active and passive SMDs. This type board uses reflow soldering only.
- **Type II:** A double-sided board, with SMDs on both sides. The top side may have all sizes of active and passive SMDs, as well as through-hole components, whereas the bottom side carries passive SMDs and small active components such as transistors. This type of board requires both reflow and wave soldering, and will require placement of bottom-side SMDs in adhesive.
- **Type III:** Top side has only through-hole components, which may be active and/or passive, whereas the bottom side has passive and small active SMDs. This type of board uses wave soldering only, and also requires placement of the bottom-side SMDs in adhesive.

A type I bare board will first have solder paste applied to the component pads on the board. Once solder paste has been deposited, active and passive parts are placed in the paste. For prototype and low-volume lines this can be done with manually guided X–Y tables using vacuum needles to hold the components, whereas in medium and high-volume lines automated placement equipment is used. This equipment will pick parts from reels, sticks, or trays, then place the components at the appropriate pad locations on the board, hence the term *pick and place* equipment.

After all parts are placed in the solder paste, the entire assembly enters a reflow oven to raise the temperature of the assembly high enough to reflow the solder paste and create acceptable solder joints at the component lead/pad transitions. Reflow ovens most commonly use convection and IR heat sources to heat the assembly above the point of solder liquidus, which for 63/37 tin–lead eutectic solder is 183°C. Because of the much higher thermal conductivity of the solder paste compared to the IC body, reflow soldering temperatures are reached at the leads/pads before the IC chip itself reaches damaging temperatures. The board is inverted and the process repeated.

If mixed-technology type II is being produced, the board will then be inverted, an adhesive will be dispensed at the centroid of each SMD, parts will be placed, the adhesive will be cured, the assembly will be rerighted, through-hole components will be mounted, and the circuit assembly will then be wave soldered, which will create acceptable solder joints for both the through-hole components and bottom-side SMDs.

A type III board will first be inverted, adhesive dispensed, SMDs placed on the bottom-side of the board, the adhesive cured, the board rerighted, through-hole components placed, and the entire assembly wave-soldered. It is imperative to note that only passive components and small active SMDs can be successfully bottom-side wave-soldered without considerable experience on the part of the design team and the board assembly facility. It must also be noted that successful wave soldering of SMDs requires a dual-wave machine with one turbulent wave and one laminar wave.

It is common for a manufacturer of through-hole boards to convert first to a type II or type III substrate design before going to an all-SMD type I design. This is especially true if amortization of through-hole insertion and wave-soldering equipment is necessary. Many factors contribute to the reality that most boards are mixed-technology type II or type III boards. Although most components are available in SMT packages, through-hole connectors are still commonly used for the additional strength the through-hole soldering process provides, and high-power devices such as three-terminal regulators are still commonly through-hole due to off-board heat-sinking demands. Both of these issues are actively being addressed by manufacturers and solutions are foreseeable that will allow type I boards with connectors and power devices [Holmes 1993].

Again, it is imperative that all members of the design, build, and test teams be involved from the design stage. Today's complex board designs mean that it is entirely possible to exceed the ability to adequately test a board if test is not designed-in, or to robustly manufacture a board if in-line inspections and handling are not adequately considered. Robustness of both test and manufacturing are only assured with full involvement of all parties to overall board design and production.

12.3 Surface Mount Device (SMD) Definitions

The new user of surface mount designs (SMDs) must rapidly learn the packaging sizes and types for SMDs. Resistors, capacitors, and most other passive devices come in two-terminal packages, as shown in Fig. 12.3, which have end-terminations designed to rest on substrate pads/lands. SMD ICs come in a wide variety of packages, from 8-pin *small outline packages* (SOLs) to 200+ pin packages in a variety of sizes and lead configurations, as shown in Fig. 12.4. The most common commercial packages currently include **plastic leaded chip carriers** (**PLCCs**), *small outline* packages (SOs), **quad flat packs** (**QFPs**), and *plastic quad flat packs* (PQFPs) also known as *bumpered quad flat packs* (BQFPs). Add in *tape automated bonding* (TAB), *ball grid array* (BGA), and other newer technologies, and the IC possibilities become overwhelming. Examples of all of these technologies are not possible in this handbook The reader is referred to the standards of the Institute for Interconnecting and Packaging Electronic Circuits (IPC) to find the latest package standards.

Each IC manufacturer's data books will have packaging information for their products. The engineer should be familiar with the term *lead pitch*, which means the center-to-center distance between IC leads. Pitch may be in thousandths of an inch, also known as mils, or may be in millimeters. Common pitches are 0.050 in (50-mil pitch); 0.025 in (25-mil pitch), frequently called fine pitch; and 0.020 in and smaller,

Surface Mount Technology

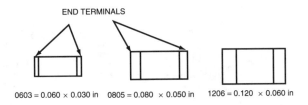

FIGURE 12.3 Example of passive component sizes (not to scale).

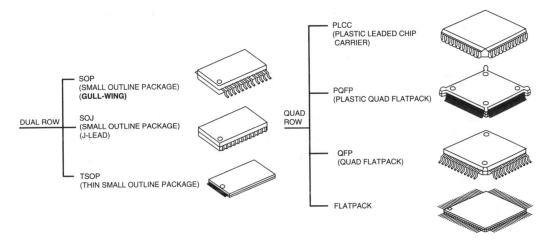

FIGURE 12.4 Examples of SMT plastic packages. (After Intel. 1994. *Packaging Handbook.* Intel Corp., Santa Clara, CA.)

frequently called ultrafine pitch. Metric equivalents are 1.27mm, 0.635 mm, and 0.508 mm, and smaller. Conversions from metric to inches are easily approximated if one remembers that 1 mm is approximately 40 mils.

12.4 Substrate Design Guidelines

As noted previously, substrate (typically PCB) design has an effect not only on board/component layout, but also on the actual manufacturing process. Incorrect land design or layout can negatively affect the placement process, the solder process, the test process or any combination of the three. Substrate design must take into account the mix of surface mount devices that are available for use in manufacturing.

The considerations which will be noted here as part of the design process are neither all encompassing, nor in sufficient detail for a true SMT novice to adequately deal with all of the issues involved in the process. They are intended to guide an engineer through the process, allowing access to more detailed information, as necessary. General references are noted at the end of this chapter, and specific references will be noted as applicable. In addition, conferences such as the National Electronics Production and Productivity Conference (NEPCON) and Surface Mount International (SMI) are invaluable sources of information for both the beginner and the experienced SMT engineer. It should be noted that although these guidelines are noted as steps, they are not necessarily in an absolute order, and several back-and-forth iterations among the steps may be required to result in a final satisfactory process and product.

After the circuit design (schematic capture) and analysis, step one in the process is to determine whether all SMDs will be used in the final design making a type I board, or whether a mix of SMDs and

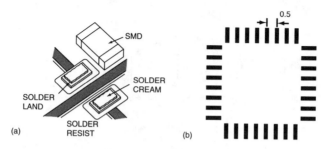

FIGURE 12.5 SMT footprint considerations: (a) Land and resist. (b) QFP footprint (After Fig. 12.5(a) Philips. 1991. *Surface Mount Process and Application Notes*. Philips Semiconductor Corporation. With permission. Figure 12.5(b) Intel. 1994. *Packaging Handbook*. Intel Corp., Santa Clara, CA.)

through hole parts will be used, leading to a type II or type III board. This is a decision that will be governed by some or all of the following considerations:

- Current parts stock
- Existence of current through hole (TH) placement and/or wave solder equipment
- Amortization of current TH placement and solder equipment
- Existence of reflow soldering equipment
- Cost of new reflow soldering equipment
- Desired size of the final product
- Panelization of smaller type I boards
- Thermal issues related to high-power circuit sections on the board

It may be desirable to segment the board into areas based on function: RF, low power, high power, etc., using all SMDs where appropriate, and mixed-technology components as needed. Typically, power portions of the circuit will point to the use of through-hole components, although circuit board materials are available that will sink substantial amounts of heat. Using one solder technique (reflow or wave) can be desirable from a processing standpoint, and may outweigh other considerations.

Step two in the SMT process is to define all of the footprints of the SMDs under consideration for use in the design. The footprint is the copper pattern, commonly called the *land*, on the circuit board upon which the SMD will be placed. Footprint examples are shown in Figs. 12.5(a) and 12.5(b), and footprint recommendations are available from IC manufacturers and in the appropriate data books. They are also available in various ECAD packages used for the design process, as well as in several references that include an overview of the SMT process [Hollomon 1995, Capillo 1990]. However, the reader is cautioned about using the general references for anything other than the most common passive and active packages. Even the position of pin 1 may be different among IC manufacturers of the same package. The footprint definition may also include the position of the solder resist pattern surrounding the copper pattern. Footprint definition sizing will vary depending on whether reflow or wave solder process is used. Wave solder footprints will require recognition of the direction of travel of the board through the wave, to minimize solder shadowing in the final fillet, as well as to meet requirements for solder thieves. The copper footprint must allow for the formation of an appropriate, inspectable solder fillet.

If done as part of the electronic design automation (EDA) process, using appropriate electronic computer-aided design (CAD) software, the software will automatically assign copper directions to each component footprint, as well as appropriate coordinates and dimensions. These may need adjustment based on considerations related to wave soldering, test points, RF and/or power issues, and board production limitations. Allowing the software to select 10-mil traces when the board production facility to be used can only reliably do 15-mil traces would be inappropriate. Likewise, the solder resist patterns must be governed by the production capabilities.

Surface Mount Technology

Final footprint and trace decisions will:

- Allow for optimal solder fillet formation
- Minimize necessary trace and footprint area
- Allow for adequate test points
- Minimize board area, if appropriate
- Set minimum interpart clearances for placement and test equipment to safely access the board
- Allow adequate distance between components for postreflow operator inspections
- Allow room for adhesive dots on wave-soldered boards
- Minimize solder bridging

Design teams should restrict wave-solder-side SMDs to passive components and transistors. Although small SMT ICs can be successfully wave soldered, this is inappropriate for an initial SMT design and is not recommended by some IC manufacturers.

Decisions that will provide optimal footprints include a number of mathematical issues:

- Component dimension tolerances
- Board production capabilities, both artwork and physical tolerances across the board relative to a 0-0 fiducial
- Solder deposition volume consistencies refillet sizes
- Placement machine accuracies
- Test probe location control

These decisions may require a statistical computer program, which should be used if available to the design team. The stochastic nature of the overall process suggests a statistical programmer will be of value.

12.5 Thermal Design Considerations

Thermal management issues remain major concerns in the successful design of an SMT board and product. Consideration must be taken of the varibles affecting both board temperature and junction temperature of the IC.

The design team must understand the basic heat transfer characteristics of most SMT IC packages [ASME 1993]. Since the silicon chip of an SMD is equivalent to the chip in an identical-function DIP package, the smaller SMD package means the internal lead frame metal has a smaller mass than the lead frame in a DIP package, as shown in Fig. 12.7. This lesser ability to conduct heat away from the chip is somewhat offset by the leadframe of many SMDs being constructed of copper, which has a lower thermal resistance than the Kovar and Alloy 42 materials commonly used for DIP packages. However, with less metal and shorter lead lengths to transfer heat to ambient air, more heat is typically transferred to the circuit board itself. Several board thermal analysis software packages are available, and are highly recommended for boards that are expected to develop high thermal gradients [Flotherm].

Since all electronics components generate heat in use, and elevated temperatures negatively affect the reliability and failure rate of semiconductors, it is important that heat generated by

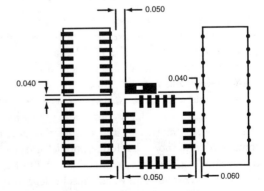

FIGURE 12.6 Minimum land-to-land clearance examples. (After Intel. 1994. *Packaging Handbook*. Intel Corp., Santa Clara, CA.)

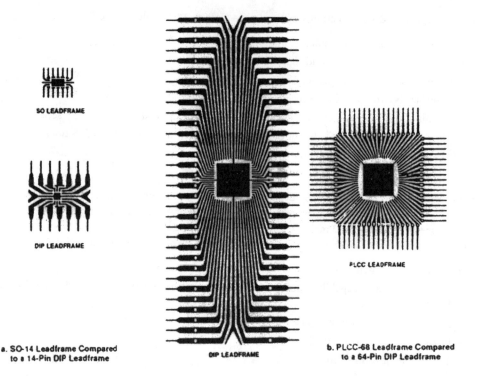

FIGURE 12.7 Lead frame comparison. (After Philips. 1991. *Surface Mount Process and Application Notes.* Philips Semiconductor Corporation.)

SMDs be removed as efficiently as possible. In most electronic devices and/or assemblies, heat is removed primarily by some combination of conduction and convection, although some radiation effects are present.

The design team needs to have expertise with the variables related to thermal transfer:

- Junction temperature: T_j
- Thermal resistances: $\Theta_{jc}, \Theta_{ca}, \Theta_{cs}, \Theta_{sa}$
- *Temperature sensitive parameter* (TSP) method of determining Θ
- Power dissipation: P_D
- Thermal characteristics of substrate material

SMT packages have been developed to maximize heat transfer to the substrate. These include PLCCs SOICs with integral heat spreaders, the SOT-89 power transistor package, and various power transistor packages. Analog ICs are also available in power packages. Note that all of these devices are designed primarily for processing with the solder paste process, and some specifically recommend against their use with wave-solder applications.

In the conduction process, heat is transferred from one element to another by direct physical contact between the elements. Ideally the material to which heat is being transferred should not be adversely affected by the transfer. As an example, the **glass transition temperature** T_g of FR-4 is 125°C. Heat transferred to the board has little or no detrimental effect as long as the board temperature stays at least 50°C below T_g. Good heat sink material exhibits high thermal conductivity, which is not a characteristic of fiberglass. Therefore, the traces must be depended on to provide the thermal transfer path [Choi, Kim, and Ortega 1994]. Conductive heat transfer is also used in the transfer of heat from THT IC packages to heat sinks, which also requires use of thermal grease to fill all air gaps between the package and the flat surface of the sink.

The discussion of lead properties of course does not apply to leadless devices such as *leadless ceramic chip carriers* (LCCCs). Design teams using these and similar packages must understand the better heat transfer properties of the alumina used in ceramic packages and must match **coefficients of thermal expansion** (CTEs or TCEs) between the LCCC and the substrate since there are no leads to bend and absorb mismatches of expansion.

Since the heat transfer properties of the system depend on substrate material properties, it is necessary to understand several of the characteristics of the most common substrate material, FR-4 fiberglass. The glass transition temperature has already been noted, and board designers must also understand that multilayer FR-4 boards do not expand identically in the X, Y, and Z directions as temperature increases. Plate-through-holes will constrain z-axis expansion in their immediate board areas, whereas nonthrough-hole areas will expand further in the z axis, particularly as the temperature approaches and exceeds T_g [Lee et al. 1984]. This unequal expansion can cause delamination of layers and plating fracture.

If the design team knows that there will be a need for higher abilities to dissipate heat and/or needs for higher glass transition temperatures and lower coefficients of thermal expansion than FR-4 possesses, many other materials are available, examples of which are shown in Table 12.1.

Note in the table that copper-clad Invar has both variable T_g, and variable thermal conductivity depending on the volume mix of copper and Invar in the substrate. Copper has a high TCE, and Invar has a low TCE, and so the TCE increases with the thickness of the copper layers. In addition to heat transfer considerations, board material decisions must also be based on the expected stress and humidity in the application.

TABLE 12.1 Thermal Properties of Common PCB Materials

Substrate Material	Glass Transition, Temperature, T_g, °C	TCE—Thermal Coefficient of X–Y Expansion, PPM/°C	Thermal Conductivity, W/M°C	Moisture Absorption, %
FR-4 epoxy glass	125	13–18	0.16	0.10
Polyimide glass	250	12–16	0.35	0.35
Copper-clad Invar	Depends on resin	5–7	160XY–15-20Z	NA
Poly aramid fiber	250	3–8	0.15	1.65
Alumina/ceramic	NA	5–7	20–45	NA

Convective heat transfer involves transfer due to the motion of molecules, typically airflow over a heat sink. Convective heat transfer, like conductive, depends on the relative temperatures of the two media involved. It also depends on the velocity of air flow over the boundary layer of the heat sink. Convective heat transfer is primarily effected when forced air flow is provided across a substrate and when convection effects are maximized through the use of heat sinks. The rules that designers are familiar with when designing THT heat-sink device designs also apply to SMT design.

The design team must consider whether passive conduction and convection will be adequate to cool a populated substrate or whether forced-air cooling or liquid cooling will be needed. Passive conductive cooling is enhanced with thermal layers in the substrate, such as the previously mentioned copper/Invar. There will also be designs that will rely on the traditional through-hole device with heat sink to maximize heat transfer. An example of this would be the typical three-terminal voltage regulator mounted on a heat sink or directly to a metal chassis for heat conduction, for which standard calculations apply [Motorola 1993].

Many specific examples of heat transfer may need to be considered in board design and, of course, most examples involve both conductive and convective transfer. For example, the air gap between the bottom of a standard SMD and the board affects the thermal resistance from the case to ambient, Θ_{ca}. A wider gap will result in a higher resistance, due to poorer convective transfer, whereas filling the gap with a thermal-conductive epoxy will lower the resistance by increasing conductive heat transfer. Thermal-modeling software is the best way to deal with these types of issues, due to the rigorous application of computational fluid dynamics (CFD) [Lee 1994].

12.6 Adhesives

In the surface mount assembly process, type II and type III boards will always require adhesive to mount the SMDs for passage through the solder wave. This is apparent when one envisions components on the bottom side of the substrate with no through hole leads to hold them in place. Adhesives will stay in place after the soldering process and throughout the life of the substrate and the product, since there is no convenient means for adhesive removal once the solder process is complete. This means the adhesive used must meet a number of both physical and chemical characteristics:

- Electrically nonconductive
- Thermal cofficient of expansion similar to the substrate and the components
- Stable in both storage and after application, prior to curing
- Stable physical drop shape, retains drop height and fills z-axis distance between the board and the bottom of the component. Thixotropic with no adhesive migration.
- Noncorrosive to substrate and component materials
- Nonconductive electrically
- Chemically inert to flux, solder, and cleaning materials used in the process
- Curable as appropriate to the process: UV, oven, or air cure
- Removable for rework and repair
- Once cured, unaffected by temperatures in the solder process

Adhesive can be applied by screening techniques similar to solder paste screen application, by pin-transfer techniques, and by syringe deposition. Screen and pin-transfer techniques are suitable for high-volume production lines with few product changes over time. Syringe deposition, an X–Y table riding over the board with a volumetric pump and syringe tip, is more suitable for lines with a varying product mix, prototype lines, and low-volume lines where the open containers of adhesive necessary in pin-transfer and screen techniques are avoided. Newer syringe systems are capable of handling high-volume lines. See Fig. 12.8 for methods of adhesive deposition.

If type II or type III assemblies are used and thermal transfer between components and the substrate is a concern, the design team must consider thermally conductive adhesives.

Regardless of the type of assembly, the type of adhesive used, or the curing technique used, adhesive volume and height must be carefully controlled. Slump of adhesive after application is undesirable, since the adhesive must stay high enough to solidly contact the bottom of the component, and must not spread and contaminate any pad associated with the component:

If:

X = adhesive dot height
Y = substrate metal height
Z = SMD termination thickness

Then, $X > Y + Z$, allowing for all combinations of potential errors, for example:

- End termination minimum and maximum thickness
- Adhesive dot minimum and maximum height
- Substrate metal minimum and maximum height

A common variation on the design shown in Fig. 12.9 is to place dummy copper pads under the center of the part. Since these pads are etched and plated at the same time as the actual solder pads, the variation in metal height Y is eliminated as an issue. Adhesive dots are placed on the dummy pads and $X > Z$ is the primary concern.

Surface Mount Technology

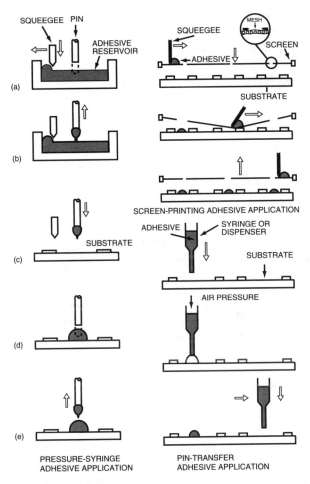

FIGURE 12.8 Methods of adhesive deposition. (After Philips. 1991. *Surface Mount Process and Application Notes.* Philips Semiconductor Corporation.)

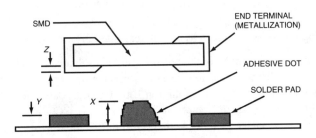

FIGURE 12.9 Relation of adhesive dot, substrate, and component.

One-part adhesives are easier to work with than two-part adhesives, since an additional process step is not required. The user must verify that the adhesive has sufficient shelf life and pot life for the user's perceived process requirements. Both epoxy and acrylic adhesives are available as one-part or two-part systems, and must be cured thermally. Generally, epoxy adhesives are cured by oven-heating, whereas acrylics may be formulated to be cured by long-wave UV light or heat.

Typically, end termination thickness variations are available from the part manufacturer. Solder pad thickness variations are a result of the board manufacturing process, and will vary not only on the type of board metallization (standard etch vs plated-through-hole) but also on the variations within each type. For adequate dot height, which will allow for some dot compression by the part, X should be between 1.5 and 2.5 times the total $Y + Z$, or just Z when dummy tracks are used.

12.7 Solder Paste and Joint Formation

Solder joint formation is the culmination of the entire process. Regardless of the quality of the design, or any other single portion of the process, if high-quality reliable solder joints are not formed, the final product is not reliable. It is at this point that PPM levels take on their finest meaning. For a medium-size substrate (nominal 6×8 in), with a medium density of components, a typical mix of active and passive parts on the topside and only passive and 3- or 4-terminal active parts on bottomside, there may be in excess of 1000 solder joints/board. If solder joints are manufactured at the 3 sigma level (99.73% good joints, or 0.27% defect rate, or 2700 defects/1 million joints) there will be 2.7 defects per board! At the 6 sigma level, of 3.4 PPM, there will be a defect on 1 board out of every 294 boards produced. If your anticipated production level is 1000 units/day, you will have 3.4 rejects based solely on solder joint problems, not counting other sources of defects.

Solder paste may be deposited by syringe, or by screen or stencil printing techniques. Stencil techniques are best for high-volume/speed production although they do require a specific stencil for each board design. Syringe and screen techniques may be used for high-volume lines and are also suited to mixed-product lines where only small volumes of a given board design are to have solder paste deposited. Syringe deposition is the only solder paste technique that can be used on boards that already have some components mounted. It is also well suited for prototype lines and for any use that requires only software changes to develop a different deposition pattern.

Solder joint defects have many possible origins:

- Poor or inconsistent solder paste quality
- Inappropriate solder pad design/shape/size/trace connections
- Substrate artwork or production problems: for example, mismatch of copper and mask, warped substrate
- Solder paste deposition problems: for example, wrong volume or location
- Component lead problems: for example, poor coplanarity or poor tinning of leads
- Placement errors: for example, part rotation or X–Y offsets
- Reflow profile: for example, preheat ramp too fast or too slow; wrong temperatures created on substrate
- Board handling problems: for example, boards get jostled prior to reflow

Once again, a complete discussion of all of the potential problems that can affect solder joint formation is beyond the scope of this chapter. Many references are available which address the issues. An excellent overview of solder joint formation theory is found in Lau [1991]. Update information on this and all SMT topics is available each year at conferences, such as SMI and NEPCON.

Although commonly used solder paste for both THT and SMT production contains 63–37 eutectic tin-lead solder, other metal formulations are available, including 96-4 tin–silver (*silver solder*). The fluxes available are also similar, with typical choices being made between RMA, water-soluble, and no-clean fluxes. The correct decision rests as much on the choice of flux as it does on the proper metal mixture. A solder paste supplier can best advise on solder pastes for specific needs. Many studies are in process to determine a no-lead replacement for lead-based solder in commercial electronic assemblies. The design should investigate the current status of these studies as well as the status of no-lead legislation as part of the decision making process.

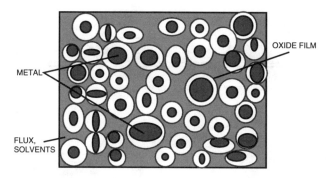

FIGURE 12.10 The make-up of SMT solder paste.

To better understand solder joint formation, one must understand the make-up of solder paste used for SMT soldering. The solder paste consists of microscopic balls of solder, most commonly tin–lead with the accompanying oxide film, flux, and activator and thickener solvents as shown in Fig. 12.10.

Regardless of the supplier, frequent solder paste tests are advisable, especially if the solder is stored for prolonged periods before use. At a minimum, viscosity, percent metal, and solder sphere formation should be tested [Capillo 1990]. Solder sphere formation is particularly important since acceptable particle sizes will vary depending on the pitch of the smallest pitch part to be used, and the consistency of solder sphere formation will affect the quality of the final solder joint. Round solder spheres have the smallest surface area for a given volume, and therefore will have the least amount of oxide formation. Uneven distribution of sphere sizes within a given paste can lead to uneven heating during the reflow process, with the result that the unwanted solder balls will be expelled from the overall paste mass at a given pad/lead site. Fine-pitch paste has smaller ball sizes and, consequently, more surface area on which oxides can form.

It should be noted at this point that there are three distinctly different solder balls referred to in this chapter and in publications discussing SMT. The solder sphere test refers to the ability of a volume of solder to form a ball shape due to its inherent surface tension when reflowed (melted). This ball formation is dependent on minimum oxides on the microscopic metal balls that make up the paste, the second type of solder ball. It is also dependent on the ability of the flux to reduce the oxides that are present, as well the ramp up of temperature during the preheat and drying phases of the reflow oven profile. Too steep a time/temperature slope can cause rapid escape of entrapped volatile solvents, resulting in expulsion of small amounts of metal, which will form undesirable solder balls of the third type, that is, small metal balls scattered around the solder joint(s) on the substrate itself rather than on the tinned metal of the joint.

12.8 Parts Inspection and Placement

Briefly, all parts must be inspected prior to use. Functional parts testing should be performed on the same basis as for through-hole devices. Each manufacturer of electronic assemblies is familiar with the various processes used on through-hole parts, and similar processes must be in place on SMDs. Problems with solderability of leads and lead **planarity** are two items which can lead to the largest number of defects in the finished product. Solderability is even more important with SMDs than with through-hole parts, since all electrical and mechanical strength rests within the solder joint, there being no hole-with-lead to add mechanical strength.

Lead coplanarity is defined as follows. If a multilead part, for example, an IC, is placed on a planar surface, lack of coplanarity exists if the solderable part of any lead does not touch that surface. Coplanarity requirements vary depending on the pitch of the component leads and their shape, but generally out-of-plane measurements should not exceed 4 mils (0.004 in) for 50-mil pitch devices, and 2 mils for 25-mil pitch devices.

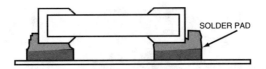

FIGURE 12.11 Part placed *into* solder paste with a passive part.

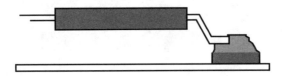

FIGURE 12.12 Part placed *into* solder paste with an active part.

All SMDs undergo thermal shocking during the soldering process, and particularly if the SMDs are to be wave soldered (type I or type II boards), which means they will be immersed in the molten solder wave for 2–4 s. Therefore, all plastic-packaged parts must be controlled for moisture content. If the parts have not been stored in a low-humidity environment (<25% relative humidity) the absorbed moisture will expand during the solder process and crack the package, a phenomenon known as popcorning since the crack is accompanied by a loud pop, and the package expands due to the expansion of moisture, just like popcorn expands.

Parts Placement

Proper parts placement not only places the parts within an acceptable window relative to the solder pad pattern on the substrate, the placement machine will apply enough downward pressure on the part to force it halfway into the solder paste (see Figs. 12.11 and 12.12).

This assures both that the part will sit still when the board is moved and that acceptable planarity offsets among the leads will still result in an acceptable solder joint. Parts placement may be done manually for prototype or low-volume operations, although the author suggests the use of guided X–Y tables with vacuum part pickup for even the smallest operation. Manual placement of SMDs does not lend itself to repeatable work. For medium- and high-volume work, a multitude of machines are available. See Fig. 12.13 for the four general categories of automated placement equipment.

One good source for manufacturer's information on placement machines and most other equipment used in the various SMT production and testing phases is the annual *Directory of Suppliers to the Electronics Manufacturing Industry* published by *Electronic Packaging and Production* (see Further Information section). Among the elements to consider in the selection of placement equipment, whether fully automated or X–Y-vacuum assist tables, are:

- Volume of parts to be placed/hour
- Conveyorized links to existing equipment
- Packaging of components to be handled; tubes, reels, trays, bulk, etc.
- Ability to download placement information from CAD/CAM systems
- Ability to modify placement patterns by the operator
- Vision capability needed, for board fiducials and/or fine-pitch parts

12.9 Reflow Soldering

Once SMDs have been placed in solder paste, the assembly will be reflow soldered. This can be done in either batch-type ovens, or conveyorized continuous-process ovens. The choice depends primarily on

Surface Mount Technology

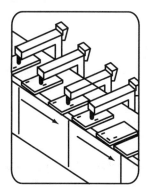

(a)
- MOVING BOARD/FIXED HEAD
- EACH HEAD PLACES ONE COMPONENT
- 1.8 TO 4.5 SECONDS/BOARD

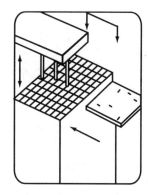

(b)
- FIXED TABLE/HEAD
- ALL COMPONENTS PLACED SIMULTANEOUSLY
- SEVEN TO 10 SECONDS/BOARD

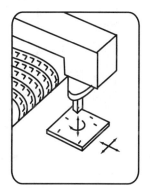

(c)
- X–Y MOVEMENT OF TABLE/HEAD
- COMPONENTS PLACED IN SUCCESSION INDIVIDUALLY
- 0.3 TO 1.8 SECONDS/COMPONENT

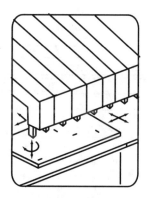

(d)
- X–Y TABLE/FIXED HEAD
- SEQUENTIAL/SIMULTANEOUS FIRING OF HEADS
- 0.2 SECONDS/COMPONENT

FIGURE 12.13 Four major categories of placement equipment: (a) In-line placement equipment, (b) simultaneous placement equipment, (c) sequential placement equipment, (d) sequential simultaneous placement equipment. (After Intel. 1994. *Packaging Handbook.* Intel Corp., Santa Clara, CA.)

the board throughput/hour required. Whereas many early ovens were of the vapor phase type, most ovens today use IR heating, convection heating, or a combination of the two. The ovens are zoned to provide a thermal profile necessary for successful SMD soldering. An example of an oven thermal profile is shown in Fig. 12.14.

The phases of reflow soldering, which are reflected in the example profile in Fig. 12.14 are:

- *Preheat.* The substrate, components and solder paste preheat.
- *Dry*: Solvents evaporate from the solder paste. Flux activates, reduces oxides, and evaporates. Both low- and high-mass components have enough soak time to reach temperature equilibrium.
- *Reflow*: The solder paste temperature exceeds the liquidus point and reflows, wetting both the component leads and the board pads. Surface tension effects occur, minimizing wetted volume.
- *Cooling.* The solder paste cools below the liquidus point, forming acceptable (shiny and appropriate volume) solder joints.

The setting of the reflow profile is not trivial. It will vary depending on whether the flux is RMA, water soluble, or no clean, and it will vary depending on both the mix of low- and high-thermal mass components, and on how those components are laid out on the board. The profile should exceed the

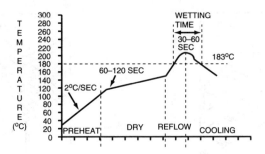

FIGURE 12.14 Typical thermal profile for SMT reflow soldering (types I or II assemblies). (After Cox, R.N. 1992. *Reflow Technology Handbook*. Research Inc., Minneapolis, MN.)

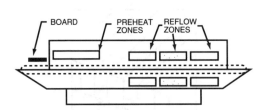

FIGURE 12.15 Conveyorized reflow oven showing the zones that create the profile. (After Intel. 1994. *Packaging Handbook*. Intel Corp., Santa Clara, CA.)

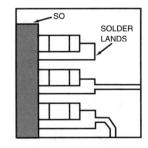

FIGURE 12.16 Solder bridge risk due to misalignment. (After Philips. 1991. *Surface Mount Process and Application Notes*. Philips Semiconductor Corporation.)

liquidus temperature of the solder paste by 20–25°C. Although final setting of the profile will depend on actual quality of the solder joints formed in the oven, initial profile setting should rely heavily on information from the solder paste vendor, as well as the oven manufacturer. Remember that the profile shown in Fig. 12.15 is the profile to be developed on the substrate, and the actual control settings in various stages of the oven itself may be considerably different, depending on the thermal inertia of the product in the oven and the heating characteristics of the particular oven being used.

Defects as a result of poor profiling include:

- Component thermal shock
- Solder splatter
- Solder balls formation
- Dewetted solder
- Cold or dull solder joints

It should be noted that many other problems may contribute to defective solder joint formation. One example would be placement misalignment, which contributes to the formation of solder bridges, as shown in Fig. 12.16.

Other problems that may contribute to defective solder joints include poor solder mask adhesion and unequal solder land areas at opposite ends of passive parts, which creates unequal moments as the paste liquifies and develops surface tension. Wrong solder paste volumes, whether too much or too little, will create defects, as will board shake in placement machines and coplanarity problems in IC components. Many of these problems should be covered and compensated for during the design process and the qualification of SMT production equipment.

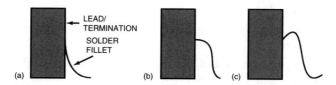

FIGURE 12.17 Solder joint inspection criteria: (a) Good; open angle, >90°, (b) adequate, = 90°, (c) unacceptable, <90°.

Postreflow Inspection

Final analysis of the process is performed based on the quality of the solder joints formed in the reflow process. Whatever criteria may have been followed during the overall process, solder joint quality is the final determining factor of the correctness of the various process steps. As noted earlier, the quality level of solder joint production is a major factor in successful board assembly. A primary criterion is the indication of wetting at the junction of the reflowed solder and the part termination. This same criterion shown in Fig. 12.17 applies to both through hole and SMDs, with only the inspection location being different.

Note that criteria shown in Fig. 12.17 are for any solderable surface, whether component or board, SMT or THT. Some lead surfaces are defined as not solderable, that is, the cut and not-tinned end of an SO or QFP lead is not considered solderable. Parts manufacturers will define whether a given surface is designed to be solderable.

Presentation of criteria for all the various SMD package types and all of the possible solder joint problems is beyond the scope of this chapter. The reader is directed to Hollomon [1995], Hwang [1989], Lau [1991], Klein-Wassink [1989], and Prasad [1989] for an in-depth discussion of these issues.

12.10 Prototype Systems

Systems for all aspects of SMT assembly are available to support low-volume/prototype needs. These systems will typically have manual solder-paste deposition and parts placement systems, with these functions being assisted for the user. Syringe solder paste deposition may be as simple as a manual medical-type syringe dispenser, which must be guided and squeezed freehand. More sophisticated systems will have the syringe mounted on an X–Y arm to carry the weight of the syringe, and will apply air pressure to the top of the syringe with a foot-pedal control, freeing the operator's arm to guide the syringe to the proper location on the substrate and perform the negative z-axis manuever, which will bring the syringe tip into the proper location and height above the substrate. Dispensing is then accomplished by a timed air-pressure burst applied to the top of the syringe under foot-pedal control. Paste volume is likewise determined by trial and error with the time/pressure relation, and depends on the type and manufacturer of paste being dispensed.

Parts placement may be as simple as tweezers and progress to hand-held vacuum probes to allow easier handling of the components. As mentioned previously, X–Y arm/tables are available that have vacuum-pick nozzles to allow the operator to pick a part from a tray, reel, or stick, and move the part over the correct location on the substrate. The part is then moved down into the solder paste, the vacuum is turned off manually or automatically, and the nozzle is raised away from the substrate.

Soldering of prototype/low-volume boards may be done by contact soldering of each component, by a manually guided hot-air tool, or in a small batch or conveyorized oven. Each step up in soldering sophistication is, of course, accompanied by an increase in the investment required.

For manufacturers with large prototype requirements, it is possible to set up an entire line that would involve virtually no hardware changes from one board to another using the following steps:

- ECAD design and analysis produce Gerber files.
- CNC circuit board mill takes Gerber files and mills out two-sided boards.

- Software translation package generates solder-pad centroid information.
- Syringe solder paste deposition system takes translated Gerber file and dispenses appropriate amount at each pad centroid.
- Software translation package generates part centroid information.
- Parts placement equipment places parts based on translated part centroid information.
- Assembly is reflow soldered.

The only manual process in this system is adjustment of the reflow profile based on the results of soldering an assembly. The last step in the process would be to test the finished prototype board. This system could also be used for very small volume production runs, and all components as described are available. With a change from milled boards to etched boards, the system can be used as a flexible assembly system.

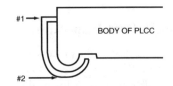

FIGURE 12.18 Planarity of an IC body.

Defining Terms

Coefficient of thermal expansion (CTE or TCE): A measure of the ratio between the measure of a material and its expansion as temperature increases. May be different in X, Y, and Z axes. Expressed in parts per million per degree Celsius, it is a measure that allows comparison of materials that are to be joined.

Coplanarity: A simplified definition of *planarity*, which is difficult to measure. Noncoplanarity is the distance between the highest and lowest leads and is easily measured by placing the IC on a flat surface, such as a glass plate. The lowest leads will then rest on the plate, and the measured difference to the lead highest above the plate is the measurement of noncoplanarity.

Glass transition temperature (T_g): Below T_g a polymer substance, such as fiberglass, is relatively linear in its expansion/contraction due to temperature changes. Above T_g, the expansion rate increases dramatically and becomes nonlinear. The polymer will also lose its stability, that is, an FR-4 board droops above T_g.

Gull wing: An SMD lead shape as shown in Fig. 12.4 for SOPs and QFPs, and in Fig. 12.12. So called because it looks like a gull's wing in flight.

Plastic leaded chip carrier (PLCC): Shown in Fig. 12.4, it is a common SMT IC package and is the only package that has the leads bent back under the IC itself.

Planarity: Lying in the same plane. A plane is defined by the exit of the leads from the body of the IC (arrow 1 in Fig. 12.18). A second plane is defined at the average of the lowest point all leads are below the first plane (arrow 2 in Fig. 12.18). Nonplanarity is the maximum variation in mils or millimeters of any lead of an SMD from the lowest point plane.

Quad flat pack: Any flat pack IC package that has leads on all four sides.

Through hole: Also a *plate through hole* (PTH). A via that extends completely through a substrate and is solder plated.

Through-hole technology: The technology of using leaded components which require holes through the substrate for their mounting (insertion) and soldering.

References

ASME. 1993. Conduction heat transfer measurements for an array of surface mounted heated components. In *Proceedings of the 1993 ASME Annual Meeting*, Vol. 263, pp. 69–78. Am. Soc. of Mech. Engr., Heat Transfer Div.

Capillo, C. 1990. *Surface Mount Technology, Materials, Processes and Equipment*, Chaps. 3, 7, 8. McGraw-Hill, New York.

Choi, C.Y., Kim, S.J., and Ortega, A. 1994. Effects of substrate conductivity on convective cooling of electronic components. *Journal of Electronic Packaging* 116(3):198–205.
Classon, F. 1993. *Surface Mount Technology for Concurrent Engineering and Manufacturing.* McGraw-Hill, New York.
Cox, N.R. 1992. *Reflow Technology Handbook.* Research, Inc., Minneapolis, MN.
Flotherm. Advanced thermal analysis of packaged electronic systems. Flomerics, Inc., Westborough, MA.
Higgins, C. 1991. Signetics Corp. presentation, Nov.
Hollomon, J.K., Jr. 1995. *Surface Mount Technology for PC Board Design.* Prompt, Indianapolis, IN.
Holmes, J.G. 1993. Surface mount solution for power devices. *Surface Mount Technology* 7(9):18–20.
Hwang, J.S. 1989. *Solder Paste in Electronics Packaging.* Van Nostrand Reinhold, New York.
Intel. 1994. *Packaging Handbook.* Intel Corp., Santa Clara, CA.
Klein-Wassink, R.J. 1989. *Soldering in Electronics.* Electrochemical Publishing Co., Ltd.
Lau, J.H., ed. 1991. *Solder Joint Reliability.* Van Nostrand Reinhold, New York.
Lea, C. 1988. *A Scientific Guide to SMT.* Electrochemical Publishing Co., Ltd.
Lee, L.C. et al. 1984. Micromechanics of multilayer printed circuit board. *IBM Journal of Research and Development* 28(6).
Lee, T.Y. 1994. Application of a CFD tool for system-level thermal simulation. *IEEE Trans. Components, Packaging, and Manufacturing Tech.* Part A 17(4):564–571.
Leibson, S.H. 1987. The promise of surface mount technology. *EDN Magazine* 5-28-87, pp. 165–174, five parts through 7-3-87.
Marcoux, P.P. 1992. *Fine Pitch Surface Mount Technology.* Van Nostrand Reinhold, New York.
Mims, F.M., III. 1987. Surface mount technology: An introduction to the packaging revolution. *Radio-Electronics* (11):58–90.
Motorola. 1993. *Linear/Interface IC Device Databook*, Vol. 1, Sec. 3 Addendum. Motorola, Inc.
Philips. 1991. *Signetics Surface Mount Process and Application Notes.* Philips Semiconductor, Sunnyvale CA.
Prasad, R.P. 1989. *Surface Mount Technology Principles and Practice.* Van Nostrand Reinhold, New York.
Rowland, R. 1993. *Applied Surface Mount Assembly.* Van Nostrand Reinhold, New York.
Shina, S.G. 1991. *Concurrent Engineering and Design for Manufacture of Electronic Products.* Van Nostrand Reinhold, New York.
Texas Instruments. 1984. *How to Use Surface Mount Technology.* Texas Instruments, Dallas, TX.

Further Information

Specific journal references are available on any aspect of SMT. A search of the COMPENDEX Engineering Index 1987–present will show over 1500 references specifically to SMT topics.

Education/Training: a partial list of organizations which specialize in education and training directly related to issues in surface mount technology:

Electronic Manufacturing Productivity Facility. 714 North Senate Ave., Indianapolis, IN 46202-3112. 317-226-5607.

SMT Plus, Inc. 5403-F Scotts Valley Drive, Scotts Valley, CA 95066; 408-438-6116

Surface Mount Technology Association (SMTA), 5200 Wilson Rd, Ste 100, Edina, MN 55424-1338. 612-920-7682.

Institute for Interconnecting and Packaging Electronic Circuits (IPC), 7380 N, Lincoln Ave., Lincolnwood, Il 60646-1705, 708-677-2850.

Conferences directly related to SMT:

Surface Mount International (SMI). Sponsored by SMTA.

National Electronics Packaging and Production Conference (NEPCON). Coordinated by Reed Exhibition Co., P.O. Box 5060, Des Plaines, IL 60017-5060. 708-299-9311.

Journals covering various aspects of SMT:

Advanced Packaging, published bimonthly, P.O. Box 159, Libertyville, IL 60048-0159.

Advances in Electronic Packaging, Transactions of the ASME, American Society of Mechanical Engineers.

Circuits Assembly, published monthly, 2000 Powers Ferry Center, Suite 450, Marietta, GA 30067.

Electronic Packaging & Production, published monthly, 1350 E. Touhy Ave., Des Plaines, IL 60018-3358. Also annual.

Directory of Suppliers to the Electronics Manufacturing Industry Cahners, 8773 South Ridgeline Blvd., Highlands Ranch, CO 80126-2329.

IEEE Transactions on Components, Packaging and Manufacturing Technology, Institute for Electrical and Electronic Engineers, 345 East 47th St., New York, NY 10017.

Journal of Electronic Packaging, Transactions of the ASME, American Society of Mechanical Engineers.

Journal of Surface Mount Technology, published quarterly by the Surface Mount Technology Assn., 5200 Willson Rd., Suite 200, Edina, MN 55424-1338.

Printed Circuit Design, published monthly, 2000 Powers Ferry Center, Suite 450, Marietta, GA 30067.

Surface Mount Technology, published monthly, P.O. Box 159, Libertyville, IL 60048-0159.

13
Semiconductor Failure Modes

Jerry C. Whitaker
Editor-in-Chief

13.1 Introduction 13-1
13.2 Terminology... 13-2
13.3 Semiconductor Device Assembly................................. 13-3
   Packaging the Die • Hybrid Devices • VLSI and VHSIC Devices
13.4 Semiconductor Reliability 13-6
   Reliability Analysis • Semiconductor Burn-In • Bond Pad Structure • Cratering • Reliability of VLSI Devices • Case Histories
13.5 Device Ruggedness... 13-12
   Forward Bias Safe Operating Area • Reverse Bias Safe Operating Area • Power Handling Capability • Semiconductor Derating • MOSFET Devices • Safe Operating Area
13.6 Failure Modes... 13-16
   Discrete Transistor Failure Modes • Breakdown Effects • Thermal Stress-Induced Failures • Package Failure Mechanisms • Package Improvements • MOSFET Device Failure Modes • Surface-Mounted Device Failure Modes
13.7 Heat Sink Considerations.. 13-21
   Mounting Power Semiconductors • Fastening Techniques • Air Handing Systems

13.1 Introduction

Engineers have been working to improve the performance and reliability of semiconductor devices for decades. The first formal meeting of the IEEE-sponsored Reliability Physics Symposium in 1962 marked the beginning of a long process that has taken the semiconductor industry from single-device packages to packages containing 100,000 or even 1,000,000 devices on a single chip. Within the foreseeable future, device density is expected to increase by a factor of 10 to 100. A single chip will constitute an entire system. Such advancements place exacting requirements on reliability engineering.

The reliability of semiconductors today is determined by the materials, processes, and quality control of chip manufacturers. Performance in the field is determined by the care taken in hardware design and the operating environment.

13.2 Terminology

Before beginning to examine the mechanics of semiconductor failure, it is appropriate to define key terms.

Acceptance Quality Level (AQL). Sampling based upon total lot size, as opposed to lot tolerance percent defective (LTPD) sampling, which is independent of lot size.

Active device. A device that converts input signal energy into output signal energy through interaction with energy from an auxiliary source.

Bi-metallic contamination. A corrosion that results from the interaction of gold with aluminum that contains more than 2% silicon. At temperatures greater than 167°C the silicon will act as a catalyst to create an aluminum–gold alloy. The resulting metallic migration results in gaps in the gold–aluminum interface. Bi-metallic contamination (also referred to as *purple-plague*) can decrease bond wire adhesion and lower current-carrying capacity at the interface.

Coefficient of thermal expansion. The measurement of the rate at which a given material will expand or contract with a change in temperature. Where two materials with different rates of thermal expansion are joined, expansion or contraction will strain the bond interface.

Contact step. The drop from the surface of the passivation of the semiconductor die to the surface of the contact itself.

Contact window. The opening in the passivation through which the metallization makes contact with the circuit elements.

Crazing. The propagation of small cracks in the surface of the glassivation.

Current-carrying edge. That portion of the metal over a contact window that is closest to the incoming metallization.

Die. The active silicon-based element that is the heart of a semiconductor product. (The plural form of die is *dice*.)

Diffused area. Portions of the die that have had impurities diffused into the surface of the silicon to change the electrical characteristics of the die.

Dual in-line package (DIP). A rectangular integrated circuit package with leads protruding from two opposite sides. The DIP package is by far the most popular in use today. Pinouts range from 8 to 40 or more.

Eutectic. An alloy of two or more metals providing the lowest possible melting point, usually lower than the melting points of any of the constituent metals.

FIT. Failures in time at 10^9 hours, a standard measure of reliability in semiconductor devices.

Glassivation. The protective coating that is placed over the entire surface of the completed die.

Hybrid microcircuit. A collection of components from one or more technologies combined into a single functional device.

Junction. The boundary between a P region and an N region in a silicon substrate.

Land. A portion of the package lead that extends into the package cavity.

MOSFET. Metal oxide field effect transistor.

Pad. An expanded metal area on the die that is not covered with glassivation.

Parasitic elements. Interactions between elements of a semiconductor to produce additional, usually unwanted, effects. The nature of semiconductor design makes it impossible to eliminate all parasitics.

Passivation. The surface coat of silicon dioxide that is deposited over the surface of the die at various diffusion steps.

Post. The portion of the lead inside the package cavity. Although only header packages have true posts, the term is often used to describe all internal lead surfaces.

ULSI. Ultra high scale integration, the next step beyond VLSI (very large scale integration).

Semiconductor Failure Modes

Undercutting. The inward sloping of the metal or silicon on a die such that the surface is wider than at the base.

VHSI. Very high speed integrated circuit, a relatively new class of devices designed for high frequency operation.

13.3 Semiconductor Device Assembly

The intrinsic reliability of a semiconductor begins with the die itself. As illustrated in Fig. 13.1, semiconductors are manufactured by building layer upon layer to produce the desired product. Semiconductor production is an exacting science. A reliable product cannot be assembled from an unreliable die. Reliability is governed by:

- The purity of the raw materials.
- The cleanliness of the fabrication facility (fab).
- The accuracy of process control steps.
- The design of the device itself.

Impurities, whether introduced through raw materials or the fab environment, are the enemies of efficient semiconductor production. Even a non-conductive particle can cause a minute defect during the masking process, resulting in immediate failure or failure during a subsequent operation. Fig. 13.2 illustrates some of the defects that contamination can cause in a device. Any defects in the metallization can lead to eventual device failures. There are several methods for evaluating the metallization, including:

Scanning electron microscope (SEM) inspection. A sample of the die, still in wafer form, is checked for flaws. The SEM allows magnification of 15,000 times and greater, enabling engineers to see defects that would not be detected under an optical microscope.

Internal visual inspection. An SEM can be used only to evaluate the metallization quality within a wafer run on a sample basis. A visual inspection with a 100 to 200 power microscope permits each individual die to be checked for scratches in the metallization surface. Scratches that decrease the total metallization cross-section can result in higher than acceptable current density. The metallization is also examined for evidence of corrosion during this quality control step.

Glassivation. This is typically used to designate any inert surface layer added to a completed wafer. Glassivation protects the surface of the die from chemical contamination, handling damage, and the possibility of short circuits caused by loose particles within the package. Because of its importance, glassivation must be carefully evaluated for thickness, coverage, and integrity. Glassivation is usually checked by visual inspection. Excessive glassivation coverage can be as much of a problem as inadequate coverage because bonding surfaces may be partially obscured.

After a die has passed the preceding quality checks, the electrical performance of the device must be confirmed. Electrical testing begins at the wafer stage and continues throughout the assembly process.

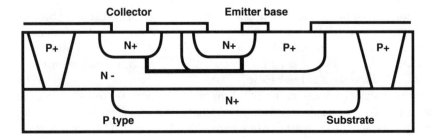

FIGURE 13.1 The construction of a semiconductor device. (After [14].)

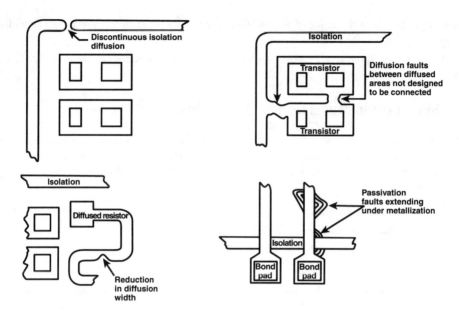

FIGURE 13.2 Types of manufacturing defects that may result from contamination of a semiconductor die. (After [14].)

Each die, prior to scribe and beak, is electrically checked. Dice that meet the required performance characteristics are assembled into packages. Dice that do not are discarded. After final assembly, samples are subjected to burn-in qualification tests.

Packaging the Die

A semiconductor package serves three important functions:

1. It protects the die from mechanical and environmental damage.
2. It dissipates heat generated during operation of the device.
3. It provides connection points for electrically accessing the die itself.

The mechanical assembly process is critical to the ultimate reliability of the device. The die is bonded to the package of lead frame using eutectic, an epoxy compound, or some other suitable medium. The strength (or adhesion) of the die to the package is an important concern because a poorly attached die could lift later as a result of vibration. After attachment, the die is electrically connected to the package leads by bonding aluminum or gold wires to each bonding pad on the die and then to the respective package lead posts. Attachment of the lid completes the mechanical assembly. The device may be hermetically sealed (for military-grade products) using glass, solder, or welding techniques, or encapsulated (for commercial-grade devices) in plastic or epoxy.

Loose particles within the package cavity can present a serious reliability hazard because they could short exposed metallization. Following attachment of the bonding wires, the die attach medium and the package itself are examined for particles that could break loose. Particles of foreign material on the surface of the die are also checked to determine if they are embedded in the die surface or are attached in a manner that would allow them to break loose.

For hermetically sealed packages, the quality of the seal is an important consideration. Damage to the seal can allow chemicals, moisture, or other contaminants from the external environment to enter the package cavity and cause corrosive damage or electrical shorting. Figure 13.3 shows a cutaway view of a DIP IC package.

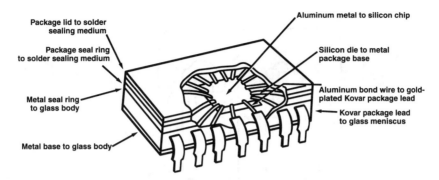

FIGURE 13.3 Cutaway view of a DIP integrated circuit package showing the internal-to-external interface. (After [14].)

Hybrid Devices

A hybrid microcircuit typically utilizes a number of components from more than one technology to perform a function that could not be achieved in monolithic form with the same performance, efficiency, and/or cost. Hybrids can be subdivided into the following general categories:

Multichip device. A product that contains several chips (active or passive), and may contain a substrate. A multichip device will not include thick film elements on the substrate. Figure 13.4 illustrates a multichip hybrid.

Simple hybrid. A product that contains one or more dice, with the dice mounted on a substrate (normally ceramic or alumina) that also has deposited metallization traces and other thin or thick film components, such as resistors and capacitors. Figure 13.5 illustrates a simple hybrid device.

Complex hybrids. A product that contains multiple dice mounted on a substrate with deposited metallization traces and other thin or thick film components. A complex hybrid typically has a seal periphery that exceeds 2 in.

Completed hybrids are usually subjected to 100% screening. Because of the number of dice utilized and the tight electrical tolerances imposed on many components, hybrid yields tend to go down as complexity goes up. Even if electrical yields of 100% for each component used in a hybrid could be guaranteed, the hybrids must still contend with burn-in dropouts. For example, a 10-chip hybrid where each chip yielded 95% reliability through burn-in would have a cumulative yield of only 60%. The use of an intermediate package, referred to as a *leadless chip carrier* (LCC), for each active device in the hybrid permits the individual elements to pass burn-in qualification before being placed into the hybrid. The LCC functions like a conventional package, but is small enough so that it can be installed on a hybrid substrate, which is later placed in a larger package.

When complex dice such as microprocessors are to be used as part of a hybrid, the LCC can provide the hybrid manufacturer the ability to extensively test the die before further assembly. This avoids the necessity for expensive rework if the die fails during qualification testing. With proper screening before assembly, the yields for even complex hybrids can be quite good. Given a reliable hybrid, the overall reliability of the system in which the hybrid is used should improve because of the reduction in implicit components (connectors and solder joints) that circuit concentration provides.

VLSI and VHSIC Devices

Integrated circuit density has been increasing at an exponential rate. The semiconductor industry is moving from the age of VLSI into the age of ULSI and VHSIC. This transition requires still greater attention to semiconductor manufacturing technology. For the sake of comparison, a VLSI device may contain the equivalent of 10,000 to 100,000 logic gates. A ULSI device may contain the equivalent of 100,000 to 1,000,000 gates.

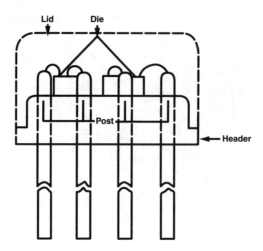

FIGURE 13.4 Basic construction of a multichip hybrid device. (After [14].)

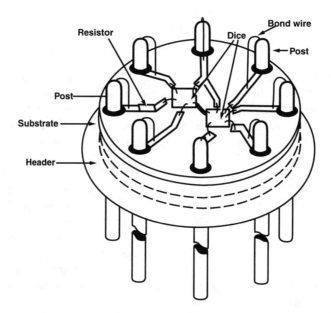

FIGURE 13.5 Basic construction of simple hybrid device. (After [14].)

As advanced devices come into production, the previous clear-cut distinctions between devices and systems begin to disappear. Many of the reliability measurements and considerations previously applied at the system level will be appropriate at the device level as well. Failure rates and failure modes for devices will be categorized much as they currently are for systems, with attention to distinctions such as catastrophic failure vs. recoverable failure.

13.4 Semiconductor Reliability

Active components are the heart of any electronic product, and most—with the exception of high power transmitter stages—employ semiconductors. Although very reliable, discrete semiconductors and integrated circuits are vulnerable to damage from a variety of environmental and electrical forces. The circuit

Semiconductor Failure Modes

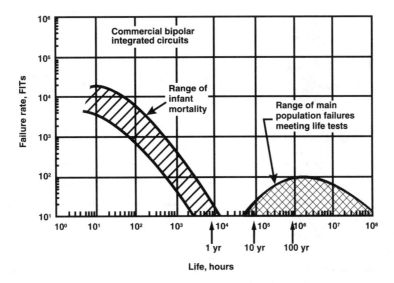

FIGURE 13.6 Operating failure rate for commercial bipolar ICs as a function of time. (After [13].)

density of an IC can have a direct relationship on its survivability under adverse conditions in the field. As chip density goes up, trace widths generally go down. Smaller elements are usually susceptible to damage at lower stress levels than older designs.

Reliability Analysis

Failure rate is the measure of semiconductor reliability. Techniques are available for predicting device failure rates, and for controlling those rates in future designs. Environmental stress screening is used extensively during product development and manufacture to identify latent defects. Infant mortality includes two major elements:

- Dead-on-arrival (DOA). Devices that fail when initially tested after shipment or incorporation in the next assembly level.
- Device operating failures (DOFs). Devices that fail after some period of operating time.

Experience has shown that while DOA levels may range from 0.3 to 1.0% of shipped product, DOF levels observed in normal factory testing and equipment burn-in are typically smaller, ranging from 0.01 to 0.5%. While the length of the infant mortality period cannot be precisely defined, it clearly has the greatest impact in the first year of device use. Figure 13.6 plots failure rates vs. time for commercial bipolar ICs. Infant mortality failures result in added costs for both original equipment manufacturers (OEMs) and users in the field. Costs include those resulting from failures at incoming test, circuit-package test, system test and burn-in, field installation, and in-service failures. Equipment repair costs increase as the device moves from one stage to the next higher in the production chain.

Infant mortality defects result from a variety of device design, manufacturing, handling, and application-related causes. Failure mechanisms differ in type and degree from one product type to another, and from one lot to another. Failures may result from one or more of the following:

- Manufacturing defects, including oxide pinholes, photoresist or etching defects, scratches, weak bonds, conductive debris, or partially cracked chips or ceramics.
- Operator-errors during production.
- Wafer contamination (as described previously).
- Assembly operations, including physical stress on the die and incomplete sealing or encapsulation.

TABLE 13.1 Infant Mortality Failures for a Sample of TTL, CMOS, and Memory Devices

Failure Mode	Product Type		
	TTL	CMOS	Memory
Overstress	4	60	17
Oxide defects	2	1	51
Surface defects	18	0	24
Bonds, beams	37	5	7
Meallization	30	34	0
Other failure modes	9	0	1

(After [13].)

Small variations in the manufacturing process can result in device lots with infant mortality characteristics significantly different from the norm. For example, Table 13.1 lists device failures resulting from various mechanisms for a lot of TTL, CMOS, and memory devices. The table demonstrates the variability of infant mortality failures. There is no fundamental reason that TTL circuits should have a greater percentage of bond failures than CMOS devices. Samples taken from a different lot at a different time might show a significantly different failure mix.

Semiconductor Burn-In

Burn-in is an effective means of screening out defects in semiconductor devices. There are two primary burn-in classifications:

1. *Static burn-in*. A DC bias is applied to the device under test at an elevated temperature. The voltage is used to reverse-bias as many junctions as possible within the device. Static burn-in is particularly effective in identifying defects resulting from corrosion or contamination.
2. *Dynamic burn-in*. The device under test is exercised to simulate actual system operation. Dynamic burn-in provides more complete access to the internal device elements.

Bond Pad Structure

Wire bonding is the process by which connections are made between the die and the lead frame using bond wires. During this process, some amount of stress is imparted onto the bond pad. Cracks on the bond pads or in the silicon substrate may result if the bonding stress is excessive. Figure 13.7 shows a cutaway view of a bonding pad.

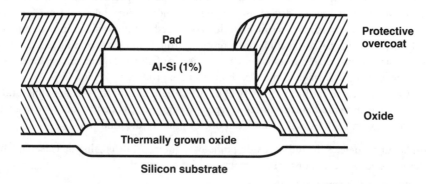

FIGURE 13.7 Cutaway structure of the bonding pad of a semiconductor device. (After [17].)

Semiconductor Failure Modes

Stress during the bonding process is the result of applied force and ultrasonic vibration. This delicate procedure requires a controlled process to enhance reliability. A crack in the bond is a serious failure mechanism. Such micro-cracks on the underlying layers of the bond pad, sometimes extending to the silicon substrate, give rise to functional failures. This class of defect is usually not detected visually unless the crack is continuous, causing the bond to be detached.

Plastic-encapsulated chips are susceptible to mechanical stress-induced failures as a result of plastic cracking, die cracking, thin film cracking, and/or wire problems. These failure modes result from *thermal coefficient of expansion* (TCE) mismatch among the materials of the package. Stress-related problems become more severe as the size of the die increases. Failures can occur during any of the thermal excursions to which the device is subjected during manufacture and use.

Cratering

The failure mechanism known as cratering is caused by a combination of water contamination of the bonding surface, and heat conducted to the bonding surface during soldering of the component. Fig. 13.8 shows the mechanisms involved. The problems begin with a small damage point at the bonding surface. If water contamination subsequently penetrates to the die, it can collect between the bonding ball and the metallization. Given this condition, the application of heat during soldering can cause the water to vaporize, placing significant pressure on the die. Cratering of the silicon substrate may then result.

Reliability of VLSI Devices

The circuit density found in VLSI devices today has pushed semiconductor manufacturing technology close to its fundamental reliability limits. VLSI devices are difficult to manufacture, and are sensitive to more subtle defects. Still, VLSI reliability has steadily improved over the past two decades. Insofar as VLSI technology is concerned, reliability is a moving target. In the early 1970s, it was considered acceptable to have VLSI devices with failure rates of 1000 to 2000 FIT. As system chip counts grew, customers began demanding lower failure rates. Chip manufacturers, meanwhile, gained a better understanding of reliability physics, and lower failure rates became common. Figure 13.9 plots the actual and projected long-term failure rates of one manufacturer (Intel) through the end of this century. Other device manufacturers have their own targets; however, it is clear that customer demands will drive all producers to deliver more reliable devices.

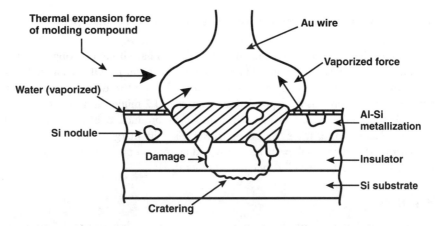

FIGURE 13.8 The physical mechanisms of cratering. (After [18].)

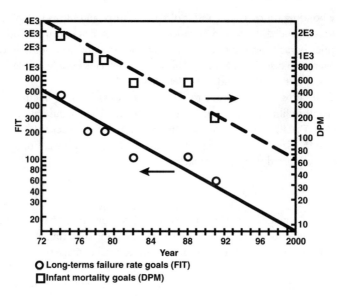

FIGURE 13.9 Long-term failure rate goals for VLSI devices. (After [16].)

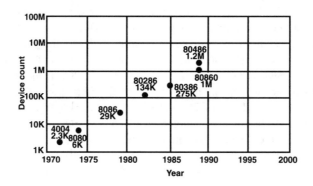

FIGURE 13.10 Microcomputer transistor count per chip as a function of time. (After [16].)

Integrated circuits intended for microcomputer applications have been a driving force in the semiconductor industry. Reliability has become an important selling point for OEMs. Figure 13.10 plots the dramatic increase in device counts that has occurred during the past two decades. The 80286 microprocessor chip, for example, contains the equivalent of more than 1.2 million transistors.

From a reliability standpoint, achieving such high density products does not come free. Each new generation of product demands more from the intrinsic reliability margin present in a given design. As device geometries scale-down, small particles that would present no problems for earlier chips can now cause interlayer short circuits, metal line opens, and other defects.

Case Histories

Although much effort has gone into producing semiconductors with no latent defects, problems still occur in the field. The following examples will illustrate the types of failures that technicians can encounter.

(a)   (b)

FIGURE 13.11 Three views of a hybrid voltage regulator that failed because of a damaged pass transistor: (a) the overall circuit geometry; (b) a closeup of the damaged pass transistor area. (Courtesy of Intertec Publishing. After [8].)

A microprocessor used in the Doppler radar of a strategic bomber was found to be failing in the field at a greater than 50% rate. At least half of the time this radar was returned for repair, the microprocessor was found faulty and replaced. The microprocessor is mounted on a circuit card that runs hot and probably saw at least 3000 temperature cycles prior to failure. Of the 19 chips analyzed, 6 had cracks; 3 of them internal to the device. The cracks, near the output pins, caused the microprocessors to fail, bringing down the radar system. The solution: replace the microprocessor chip with a *hardened* device (a device designed to operate over a wider temperature range).[23]

An operational amplifier that drives the heads-up display in the F-15 fighter aircraft was identified for failure analysis because of its high removal rate. The op amp is a 3-resistor, 3-transistor, plastic-potted 8-pin cylindrical component. The failure mode was described as intermittent. Electrical testing before and after the plastic potting was etched away by a maintenance technician demonstrated that the failure was the result of a cracked solder joint. The root cause of the joint failure was identified as thermal cycling. At 50,000 feet, the ambient temperature can easily be −50°C, and the F-15 probably makes that climb in less than 5 min. An aircraft sitting idle on the runway may experience 50°C or greater. During any one mission, the op amp might experience ten 100°C thermal cycles. The solution: replace the op amp with a hardened device.[23]

Power supply failures were experienced in the field by a manufacturer. An investigation of failed units revealed the cause to be high voltage breakdown in a hybrid semiconductor chip. The microphotographs of Fig. 13.11 (a-c) show the damage points. Failure analysis of several units demonstrated that the regulator pass transistor was over-stressed because of excessive input/output voltage differential. The manufacturer determined that during testing of the system, the regulator load would be removed, causing the unregulated supply voltage to rise, exceeding the input/output capability of the pass transistor. Most of the failures occurred only after the supply was placed into service by the end-user. The solution: change the testing procedures.[7]

Excessive transistor failures were experienced in the pre-driver stage of an RF power amplifier. A series of tests on the failed devices were conducted. Analysis showed the root cause of the problem to be overvoltage stress caused by parasitic oscillations in the stage. The solution: place additional shielding around the transmitter assembly.[24]

13.5 Device Ruggedness

The best constructed device will fail if exposed to stress exceeding its design limits. The safe operating area (SOA) of a power transistor is the single most important parameter in the design of a solid-state amplifier or controller. Fortunately, advances in diffusion technology, masking, and device geometry have enhanced the power-handling capabilities of semiconductor devices.

A bipolar transistor exhibits two regions of operation that must be avoided:

Dissipation region. Where the voltage-current product remains unchanged over any combination of voltage (V) and current (I). Gradually, as the collector-to-emitter voltage increases, the electric field through the base region causes hot spots to form. The carriers can actually punch a hole in the junction by melting silicon. The result is a dead (shorted) transistor.

Second breakdown (I_s/b) *region.* Where power transistor dissipation varies in a non-linear inverse relationship with the applied collector-to-emitter voltage when the transistor is forward-biased.

To get SOA data into some type of useful format, a family of curves at various operating temperatures must be developed and plotted. This exercise gives a clear picture of what the data sheet indicates, compared to what happens in actual practice.

Forward Bias Safe Operating Area

The forward bias safe operating area (FBSOA) describes the ability of a transistor to handle stress when the base is forward biased. Manufacturer FBSOA curves detail maximum limits for both steady state dissipation and turn-on load lines. Because it is possible to have a positive base-emitter voltage and negative base current during the device storage time, forward bias is defined in terms of base current.

Bipolar transistors are particularly sensitive to voltage stress; more so than with stress induced by high currents. This situation is particularly true of switching transistors, and it shows up on the FBSOA curve.

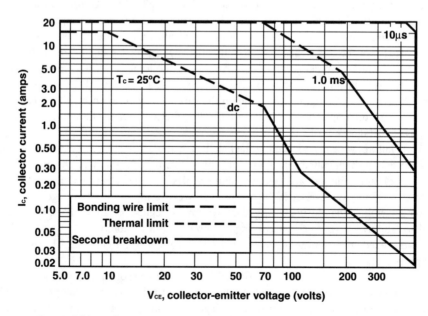

FIGURE 13.12 Forward bias safe operating area curve for a bipolar transistor (MJH16010A). (Courtesy of Motorola.)

Semiconductor Failure Modes

Figure 13.12 shows a typical curve for a common power transistor. In the case of the DC trace, the following observations can be made:

- The power limit established by the *bonding wire limit* portion of the curve permits 135 W maximum dissipation (15 A × 9 V).
- The power limit established by the *thermal limit* portion of the curve permits (at the maximum voltage point) 135 W maximum dissipation (2A × 67.5 V). There is no change in maximum power dissipation.
- The power limit established by the *second breakdown* portion of the curve decreases dramatically from the previous two conditions. At 100 V, the maximum current is 0.42 A, for a maximum power dissipation of 42 W.

Reverse Bias Safe Operating Area

The reverse bias safe operating area (RBSOA) describes the ability of a transistor to handle stress with its base reverse biased. As with FBSOA, RBSOA is defined in terms of current. In many respects, RBSOA and FBSOA are analogous. First among these is voltage sensitivity. Bipolar transistors exhibit the same sensitivity to voltage stress in the reverse bias mode as in the forward bias mode. A typical RBSOA curve is shown in Fig. 13.13. Note that maximum allowable peak instantaneous power decreases significantly as voltage is increased.

Power Handling Capability

The primary factor in determining the amount of power a given device can handle is the size of the active junction(s) on the chip. The same power output from a device may be achieved through the use of several smaller chips in parallel. This approach, however, can result in unequal currents, and uneven distribution of heat. At high power levels, heat management becomes a significant factor in chip design.

Specialized layout geometries have been developed to ensure even current distribution throughout the device. One approach involves the use of a matrix of emitter resistances constructed so that the overall distribution of power among the parallel emitter elements results in even thermal dissipation. Figure 13.14 illustrates this interdigited geometry technique.

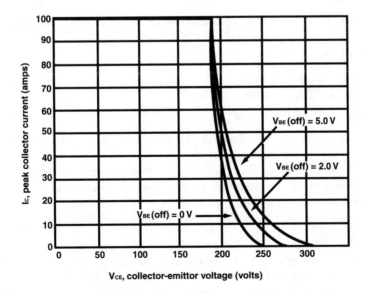

FIGURE 13.13 Reverse bias safe operating area curve for a bipolar transistor (MJH10610A). (Courtesy of Motorola.)

With improvements in semiconductor fabrication processes, output device SOA is primarily a function of the size of the silicon slab inside the package. Package type, of course, determines the ultimate dissipation because of thermal saturation with temperature rise. A good TO-3 or a 2-screw-mounted plastic package will dissipate approximately 350 to 375 W if properly mounted. Figure 13.15 demonstrates the relationships between case size and power dissipation for a TO-3 package.

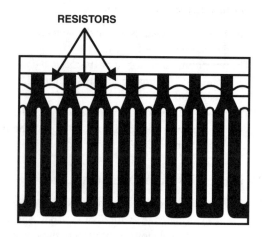

FIGURE 13.14 Interdigited geometry of emitter resistors used to balance currents throughout a power device chip.

Semiconductor Derating

Good design calls for a measure of caution in the selection and application of active devices. Unexpected operating conditions, or variations in the manufacturing process may result in field failures unless a margin of safety is allowed. Derating is a common method of achieving such a margin. The primary derating considerations are:

- *Power derating.* Designed to hold the worst-case junction temperature to a value below the normal permissible rating.
- *Junction-temperature derating.* An allowance for the worst-case ambient temperature or case temperature that the device will likely experience in service.
- *Voltage derating.* An allowance intended to compensate for temperature-dependent voltage sensitivity and other threats to device reliability as a result of instantaneous peak-voltage excursions caused by transient disturbances.

MOSFET Devices

Power MOSFETs have found numerous applications because of their unique performance attributes. A variety of specifications can be used to indicate the maximum operating voltages a specific device can withstand. The most common specifications include:

- Gate-to-source breakdown voltage.
- Drain-to-gate breakdown voltage.
- Drain-to-source breakdown voltage.

These limits mark the maximum voltage excursions possible with a given device before failure. Excessive voltages cause carriers within the depletion region of the reverse biased P-N junction to acquire sufficient kinetic energy to result in ionization. Voltage breakdown can also occur when a *critical electric field* is reached. The magnitude of this voltage is determined primarily by the characteristics of the die itself.

Safe Operating Area

The safe DC operating area of a MOSFET is determined by the rated power dissipation of the device over the entire drain-to-source voltage range (up to the rated maximum voltage). The maximum drain-source voltage is a critical parameter. If exceeded even momentarily, the device can be damaged permanently.

Semiconductor Failure Modes

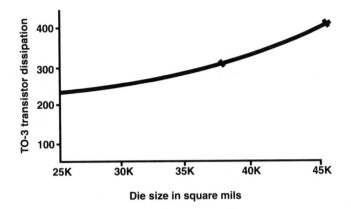

FIGURE 13.15 Relationship between case (die) size and transistor dissipation.

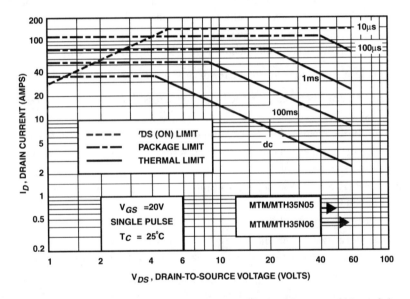

FIGURE 13.16 Safe operating area (SOA) curve for a power FET device. (Courtesy of Motorola.)

Figure 13.16 shows a representative SOA curve for a MOSFET. Notice that limits are plotted for several parameters, including drain-source voltage, thermal dissipation (a time-dependent function), package capability, and drain-source on-resistance. The capability of the package to withstand high voltages is determined by the construction of the die itself, including bonding wire diameter, size of the bonding pad and internal thermal resistances. The drain-source on-resistance limit is simply a manifestation of Ohm's Law; with a given on-resistance, current is limited by the applied voltage.

To a large extent, the thermal limitations described in the SOA chart determine the boundaries for MOSFET use in linear applications. The maximum permissible junction temperature also affects the pulsed current rating when the device is used as a switch. MOSFETs are, in fact, more like rectifiers than bipolar transistors with respect to current ratings; their peak current ratings are not gain limited, but are thermally limited.

In switching applications, total power dissipation is comprised of both switching losses and on-state losses. At low frequencies, switching losses are small. As the operating frequency increases, however, switching losses become a significant factor in circuit design.

13.6 Failure Modes

A semiconductor device can fail in a catastrophic, intermittent, or degraded mode. Such failures are usually opens, shorts, or parameters out of specifications. For integrated circuits and discrete semiconductors, the three most destructive stresses are excessive temperature, voltage, and vibration.

Semiconductor failure modes can be broken down into two basic categories: mechanical (including temperature and vibration) and electrical (including electrostatic discharge and transient over-voltage). Semiconductor manufacturers are able to increase device reliability by analyzing why good parts go bad.

There are, in fact, a frightening number of mechanical construction anomalies that can result in degraded or catastrophic failure of a semiconductor device. Some of the more significant threats include:

- Encapsulation failures caused by humidity and impurity penetration, imperfections in termination materials, stress cracks in the encapsulation material, and differential thermal expansion coefficients of the encapsulant, device leads, or chip.
- Wire bond failures caused by misplaced bonds, crossed wires and oversize bonds.
- Imperfect chip attachment to the device substrate resulting in incomplete thermal contact, stress cracks in the chip or substrate, and solder or epoxy material short circuits.
- Aluminum conductor faults caused by metalization failures at contact windows, electromigration, corrosion, and geometrical misalignment of leads and/or the chip itself.

The principal failure modes for semiconductor devices include the following:

- Internal short circuit between metalized leads or across a junction, usually resulting in system failure.
- Open circuit in the metalization or wire bond, usually resulting in system failure.
- Variation in gain or other electrical parameters, resulting in marginal performance of the system or temperature sensitivity.
- Leakage currents across P-N junctions, causing effects ranging from system malfunctions to out-of-tolerance conditions.
- Shift in turn-on voltage, resulting in random logic malfunctions in digital systems.
- Loss of seal integrity through the ingress of ambient air, moisture, and/or contaminants. The effects range from system performance degradation to complete failure.

Aluminum interconnects in a semiconductor device are the nerves that make integration of complex circuits onto blocks of silicon possible. The integrity of these interconnects is of critical importance to reliable operation of a device.

Discrete Transistor Failure Modes

It is estimated that as much as 95% of all transistor failures in the field are directly or indirectly the result of excessive dissipation or applied voltages in excess of the maximum design limits of the device. There are at least four types of voltage breakdown that must be considered in a reliability analysis of discrete power transistors. Although each type is not strictly independent, they can be treated separately, keeping in mind that each is related to the others.

Avalanche Breakdown

Avalanche is a voltage breakdown that occurs in the collector-base junction, similar to the *Townsend effect* in gas tubes. This effect is caused by the high dielectric field strength that occurs across the collector-base junction as the collector voltage is increased. This high intensity field accelerates the free charge carriers so they collide with other atoms, knocking loose additional free charge carriers that, in turn, are accelerated and have further collisions.

Semiconductor Failure Modes

This multiplication process occurs at an increasing rate as the collector voltage increases until at some voltage, V_a (avalanche voltage), the current suddenly tries to go to infinity. If enough heat is generated in this process, the junction will be damaged or destroyed. A damaged junction will result in higher than normal leakage currents, increasing the steady-state heat generation of the device, which may ultimately destroy the semiconductor junction.

Alpha Multiplication

Alpha multiplication is produced by the same physical phenomenon that produces avalanche breakdown, but differs in circuit configuration. This effect occurs at a lower potential than the avalanche voltage and generally is responsible for collector-emitter breakdown when the base current is equal to zero.

Punch-Through

Punch-through is a voltage breakdown occurring between the collector-base junction because of high collector voltage. As collector voltage is increased, the *space charge region* (collector junction width) gradually increases until it penetrates completely through the base region, touching the emitter. At this point the emitter and collector are effectively shorted together.

This type of breakdown occurs in some PNP junction transistors but generally alpha multiplication breakdown occurs at a lower voltage than punch-through. Because this breakdown occurs between collector and emitter, punch-through is more serious in the common-emitter or common-collector configuration.

Thermal Runaway

Thermal runaway is a regenerative process where an increase in temperature causes an increase in the leakage current which results in increased collector current, which, in turn, causes increased power dissipation. This action raises the junction temperature, causing a further increase in leakage current.

If the leakage current is sufficiently high (resulting from high temperature or high voltage), and the current is not adequately stabilized to counteract increased collector current because of increased leakage current, this process can regenerate to a point that the temperature of the transistor rapidly increases, destroying the device. This type of effect is more prominent in power transistors where the junction is normally operated at high temperatures and where high leakage currents are present because of the large junction area. Thermal runaway is related to the avalanche effect, and is dependent upon circuit stability, ambient temperature, and transistor power dissipation.

Breakdown Effects

The effects of the breakdown modes outlined manifest themselves in various ways on the transistor.

Avalanche breakdown usually results in destruction of the collector-base junction because of excessive currents, which, in turn, results in an open between the collector and base.

Breakdown due to alpha multiplication and thermal runaway most often results in destruction of the transistor because of excessive heat dissipation that shows up electrically as a short between the collector and emitter. This condition, which is most common in transistors that have suffered catastrophic failure, is not always easily detected. In many cases an ohmmeter check may indicate a good transistor. Only after operating voltages are applied will the failure mode be exhibited.

Punch-through breakdown generally does not permanently damage the transistor; it can be a self-healing type of breakdown. After the over-voltage is removed, the transistor will usually operate satisfactorily.

Thermal Stress-Induced Failures

Thermal fatigue represents a threat to any semiconductor component, especially power devices. This phenomenon results from the thermal mismatch between a silicon chip and the device header under temperature-related stresses. Typical failure modes include voids and cracks in solder material within the device, which results in increased thermal resistance to the outside world and the formation of *hot spots* inside the device. Catastrophic failure will occur if sufficient stress is put on the die. Although generally considered a problem just for power devices, thermal fatigue also affects VLSI ICs because of the increased size of the die. Figure 13.17 illustrates the deformation process that results from excessive heat on a semiconductor device.

As a case in point, consider the 2N3055 power transistor. The component is an NPN device rated for 115 W dissipation at 25°C ambient temperature. The component can handle up to 100 V on the collector at 15 A. Although the 2N3055 is designed for demanding applications, the effects of thermal cycling take their toll. Figure 13.18 charts the number of predicted thermal cycles for the 2N3055 vs. temperature change. Note that the lifetime of the device increases from 9000 cycles at 120°C to 30,000 cycles at 50°C.

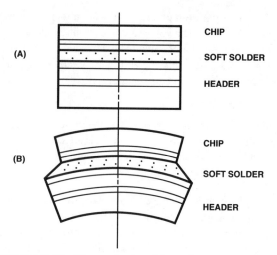

FIGURE 13.17 The mechanics of thermal stress on semiconductor devices: (A) a normal chip/solder/header composite structure; (B) the assembly subjected to a change in temperature. (After [5].)

Package Failure Mechanisms

The majority of failures in electronic packages are caused by mechanical failure mechanisms such as fracture, fatigue, and corrosion. These failures are primarily wear-out phenomenon rather than statistically random phenomenon, and hence cannot be characterized as a constant failure-rate process during any significant segment of device life.

Fatigue failures in microelectronic packages are usually the result of thermomechanical loading from thermal and power excitation, and mechanical loading from vibration. Failures often occur not in the bulk of any material, but at the interface of different materials. Such problems are commonly observed at the interfaces of the die and die attach, substrate or case and the attach, wire and bond-pad, or substrate and bond-pad.

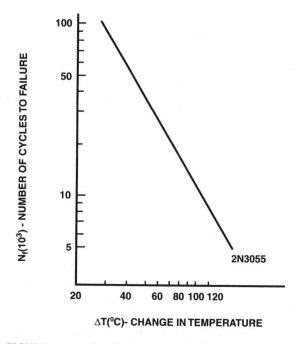

FIGURE 13.18 The effect of thermal cycling on the lifetime of a 2N3055 power transistor. (After [5].)

Package Improvements

As chip design and manufacturing processes improve, the resulting performance benefits tend to be well publicized. At the same time, however, package technology is also evolving. Figure 13.19 illustrates improvements made to the TO-220 package by one manufacturer. Changes include:

Semiconductor Failure Modes

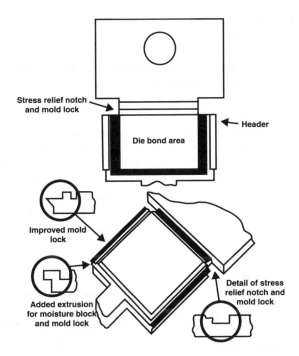

FIGURE 13.19 Improved mechanical packaging for a TO-220. (Courtesy of Motorola.)

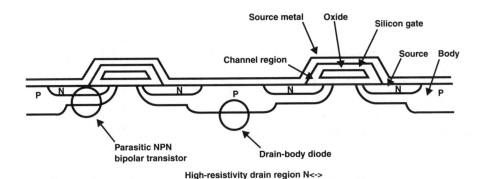

FIGURE 13.20 Cross-section of a power MOSFET device showing the parasitic bipolar transistor and diode inherent in the structure. (After [21].)

Stress relief. In contrast to previous designs that used planar heat-sink surfaces, the new package has a stress relief groove between the die bond area and the tab. Notches on either side of this groove provide additional stress relief.

Mold locks. Improved mechanical design gives the molding compound an optimal metal surface to grip as it curves. This design also extends the distance that any contaminants must travel before penetrating into the die bond area of a plastic package device.

Heat sink. Advanced materials provide higher thermal conductivity, which helps to maximize thermal efficiency.

MOSFET Device Failure Modes

Power MOSFETs have found application in a wide variety of power control and conversion systems. Most of these applications require that the device be switched on and off at a high frequency. The thermal and electrical stresses that a MOSFET experiences during switching can be severe, particularly during turn-off when an inductive load is present.

When power MOSFETs were first introduced, it was usually stated that, because the MOSFET was a majority carrier device, it was immune to second breakdown as observed in bipolar transistors. It must be understood, however, that a parasitic bipolar transistor is inherent in the structure of a MOSFET. This phenomenon is illustrated in Fig. 13.20. The parasitic bipolar transistor can allow a failure mechanism similar to second breakdown. Research has shown[21] that if the parasitic transistor becomes active, the MOSFET may fail. This situation is particularly troublesome if the MOSFET drain-source breakdown voltage is approximately twice the collector-emitter sustaining voltage of the parasitic bipolar transistor. This failure mechanism results, apparently, when the drain voltage snaps-back to the sustaining voltage of the parasitic device. This so-called *negative resistance characteristic* can cause the total device current to constrict to a small number of cells in the MOSFET structure, leading to device failure.

The precipitous voltage drop synonymous with second breakdown is a result of avalanche injection and any mechanism, electrical or thermal, that can cause the current density to become large enough for avalanche injection to occur.

Surface-Mounted Device Failure Modes

Surface-mounted devices (SMDs) are widely used today in products ranging from consumer television sets to airborne radar. SMD components are mounted directly onto a printed wiring board (PWB) and soldered to its surface. This process differs from thru-hole mounting where the device lead is inserted through a plated hole in the PWB and soldered into place. Some SMDs have leads, while others are considered leadless, with the edge of the package attached directly to the board by solder. Three popular SMD designs are shown in Fig. 13.21.

When an SMD component is exposed to thermal cycling, the package and the board materials expand at different rates, causing strain to occur in the solder connection. Cracking of the solder joint is the major failure mode for an SMD. The main design parameters that affect the integrity of the connections include the lead design, package size, board material, and the ability to dissipate heat.

The main failure mechanisms for the solder connection are thermal fatigue and *creep*. Creep is defined as the slow and progressive deformation of a material with time under a constant stress. These combined phenomena accelerate the failure of the solder.

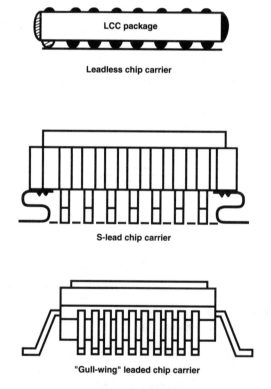

FIGURE 13.21 Three common surface-mounted chip designs. (After [1].)

Finite element analysis (FEA) has been applied to study the reliability of SMD components.[1] FEA is a computer simulation technique that may be used to predict the thermal and mechanical response of structures exposed to various environmental conditions. To assess reliability, a comparison is made between the stresses in the materials and the corresponding material strengths. FEA was performed on three different surface-mount designs: the leadless chip carrier (LCC), the S-lead chip carrier, and the gull-wing chip carrier. For each design, two computer models were created. The first represented the

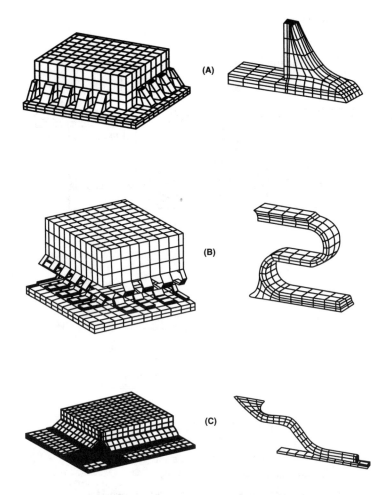

FIGURE 13.22 Finite element analysis plots for three types of SMD chips: (A) leadless chip carrier; (B) S-lead chip carrier; (C) gull-wing leaded chip carrier. (After [1].)

package mounted onto a board, and the second represented a single lead (see Fig. 13.22). The package/board model was thermally cycled and the results inputted into the lead model in order to obtain a detailed stress distribution plot for the solder connection points. The results indicated that the gull-wing and S-lead chip carriers would be reliable when placed in a given temperature environment, but that the leadless chip carrier would have reliability problems after a short period of time.

13.7 Heat Sink Considerations

Heat generated in a power transistor (primarily at the collector junction) must be removed at a sufficient rate to keep the junction temperature within a specific upper limit. This is accomplished primarily by conduction from the junction through the transistor material to a metal mounting base that is designed to provide good thermal contact to an external heat dissipator or heat sink.

Because heat transfer is associated with a temperature difference, a differential will exist between the collector junction and the transistor mounting surface. A temperature differential will also exist between the device mounting surface and the heat sink. Ideally these differentials will be small. They will, however, exist to one extent or another. It follows, therefore, that an increase in dissipated power at the collector junction will result in a corresponding increase in junction temperature. In general, assessing the heat sink requirements of a device or system (and the potential for problems) is a difficult proposition.

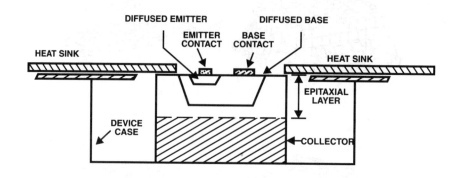

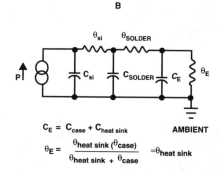

FIGURE 13.23 Simplified model of thermal transmission from the junction of a power transistor (TO-3 case) to a heat sink: (A) the structure of a double-diffused epitaxial planar device; (B) simplified electrical equivalent of the heat transfer mechanism. (After [9].)

Figure 13.23 shows some of the primary elements involved in thermal transmission of energy from the silicon junction to the external heat sink. An electrical analog of the process is helpful for illustration. The model shown in Fig. 13.23 (B) includes two primary elements: *thermal capacitance* and *thermal resistance*. The energy storage property of a given mass, expressed as C, is the basis for the transient thermal properties of transistors. The thermal transmission loss from one surface or material to another, expressed as θ, causes a temperature differential between the various components of the semiconductor model shown in the figure.

Although this model may be used to predict the rise of junction temperature that results from a given increase in power dissipation, it is an extreme over-simplification of the mechanics involved. The elements considered in our example include the silicon transistor die (Si); the solder used inside the transistor to bond the emitter, base, and collector to the outside-world terminals; and the combined effects of the heat sink and transistor case. This model assumes the transistor is directly mounted onto a heat sink, not through a mica (or other style) insulator.

The primary purpose of a heat sink is to increase the effective heat-dissipation area of the transistor. If the full power-handling capability of a transistor is to be achieved, there must be zero temperature differential between the case and the ambient air. This condition exists only when the thermal resistance of the heat sink is zero, requiring an infinitely large heat sink. Although such a device can never be

Semiconductor Failure Modes

realized, the closer the approximation of actual conditions to ideal conditions, the greater the maximum possible operating power.

In typical power transistor applications, the case must be electrically insulated from the heat sink (except for circuits using a grounded-collector configuration). The thermal resistance from case to heat sink, therefore, includes two components: surface irregularities of the insulating material, transistor case, and heat sink and the insulator itself. Thermal resistance resulting from surface irregularities can be minimized through the use of silicon grease compounds. The thermal resistance of the insulator itself, however, can represent a significant problem. Unfortunately, materials that are good electrical insulators are also usually good thermal insulators. The best materials for such applications are mica, beryllium oxide, and anodized aluminum.

Mounting Power Semiconductors

Semiconductor current and power ratings are inseparably linked to their thermal environment. Except for lead-mounted parts used at low currents, a heat sink (or heat exchanger) is required to prevent the junction temperature from exceeding its rated limits. Experience has shown that most early life field failures of power semiconductors can be traced to faulty mounting procedures. High junction temperatures can result in reduced component life. Failures can also result from improperly mounting a device to a warped surface; mechanical damage can cause cracks in the case, resulting in moisture contamination, or a crack in the semiconductor die itself.

Figure 13.24 shows an extreme case of improper mounting for a TO-220 package. Excessive force on the fastening screw has warped the device heat sink, and possibly resulted in a crack in the package. The die will experience excessive heating because of poor heat transfer. Note that only a small portion of the device heat sink is in direct contact (through the mica washer) with the equipment heat sink.

Proper mounting of a power device to a heat sink requires attention to the following points:

Prepare the mounting surface. The surface must be reasonably flat to facilitate efficient heat transfer.

Apply thermal grease. When a significant amount of power must be radiated, something must be done to fill the air voids between the mating surfaces in the thermal path. Thermal joint compounds (thermal grease) are commonly used for this purpose. To prevent accumulation of airborne particulate matter, excess compound should be wiped away using a cloth moistened with acetone or alcohol.

Install the insulator, if used. Devices using the case as a terminal must usually be insulated from the equipment heat sink. Insulating materials include mica, a variety of proprietary films, and beryllium oxide. From the standpoint of heat transfer, beryllium is the best. It is, however, expensive and must be handled with care; beryllium dust is highly toxic.

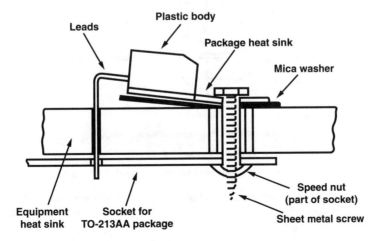

FIGURE 13.24 Extreme case of improperly mounting a TO-220 package semiconductor device. (Courtesy of Motorola.)

Secure the device to the heat sink. Mounting holes should only be large enough to allow clearance of the fastener. Punched holes should be avoided if possible because of the possibility of creating a depression around the hole. Such a crater in the heat sink can distort the semiconductor package and/or result in an inefficient transfer of heat.

Connect the device terminals to the circuit.

Fastening Techniques

Each type of power semiconductor package requires the use of unique fastening hardware. Consider the following guidelines:

Tab-mount device. Characterized by the popular TO-220 package, improper hardware is often used to mount this class of device. The correct method of mounting a TO-220 is shown in Fig. 13.25. The rectangular washer shown in Fig. 13.25 (A) is used to minimize distortion of the mounting flange. Be certain that the tool used to drive the mounting screw does not apply pressure to the body of the device, possibly resulting in a crack in the package.

Flange-mount device. A large variety of devices fall into the flange-mount category, including the popular TO-3 (also known as the TO-204AA). The rugged base and distance between the die and the mounting holes combine to make mounting practices less critical for most flange-mounted

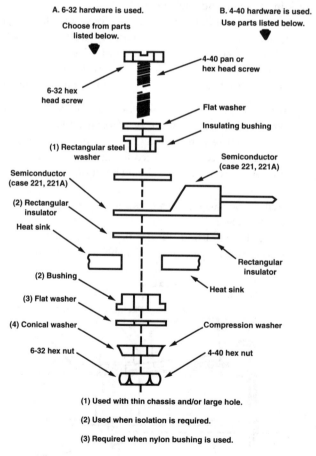

FIGURE 13.25 Mounting hardware for a TO-220 package: (A) preferred arrangement for isolated or non-isolated mounting (screw is at semiconductor case potential); (B) alternate arrangement for isolated mounting (screw is at heat sink potential). (Courtesy of Motorola.)

Semiconductor Failure Modes

power semiconductors. Figure 13.26 shows the mounting hardware used for a TO-3 package. RF power devices are more sensitive to heat sink surface flatness than other packages. Specific mounting requirements for RF devices must be followed carefully to ensure reliable operation.

Stud-mount device. Errors with non-insulated stud-mounted devices are generally confined to the application of excessive torque, or tapping the stud into a threaded heat sink hole. Either practice may result in warping of the hex base, which may crack the semiconductor die. The proper hardware components for mounting a non-insulated device to a heat sink or chassis are shown in Fig. 13.27.

Press-fit device. Mounting a press-fit device into a heat sink is straightforward. Figure 13.28 illustrates the approach for mounting on a heat sink, and for mounting on a chassis. Apply force evenly on the shoulder ring to avoid tilting the case in the hole during assembly. Use thermal joint compound. Pressing force will vary from 250 to 1000 pounds, depending on the heat sink material.

Air Handing Systems

There are two basic methods of cooling electronic equipment: active cooling, where the coolant uses machinery (fans, blowers, or pumps) to remove heat, and passive cooling, where no machinery is used to effect the heat removal process.

Passive cooling can involve conduction, radiation, and natural convection. One major disadvantage of active cooling is that once incorporated, the equipment may be dependent upon the cooling system for survival, thus introducing an additional source of potential failure. Passive cooling, therefore, is generally preferred unless large amounts of heat must be removed.

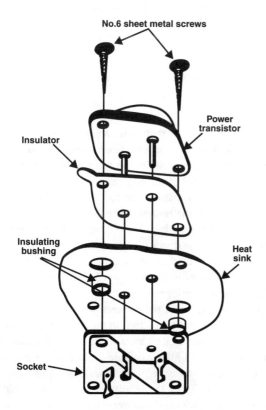

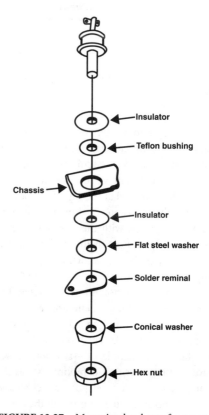

FIGURE 13.26 Mounting hardware for a TO-3 (TO-204A) power semiconductor. (Courtesy of Motorola.)

FIGURE 13.27 Mounting hardware for a non-isolated stud-mounted device. (Courtesy of Motorola.)

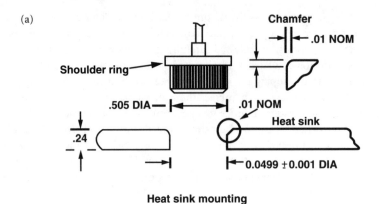

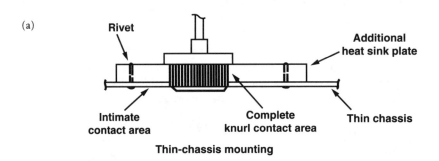

FIGURE 13.28 Mounting a press-fit device: (a) heat sink mounting; (b) thin-chassis mounting. (Courtesy of Motorola.)

Research into passive cooling methods for electronic systems shows that—on average—circuit boards should be mounted vertically with the long dimension horizontal. Unrestricted free air passageways must be provided between circuit boards, including adequate ventilation holes at the top and bottom of the card cage. High wattage components should be mounted along the bottom edge, and low wattage devices should be mounted at the top of the board. Assuming uniform dissipation over the assembly, the hottest point is the top-center region, as shown in Fig. 13.29. Note that the hottest components exceed a 60°C temperature rise, with the power to the board at 25 W. If the same board is redesigned for optimum power dissipation, with the hottest components at the bottom and sides of the board, more efficient cooling is possible. As shown in Fig. 13.30, optimizing for power dissipation permits greater total power input to the board, with cooler operation overall. All components in the redesigned board operate at a 50°C temperature rise. Additional reliability is achieved because uniform temperatures across the assembly reduce thermal gradient stresses, thus improving the fatigue life of the connections.

The construction of an equipment housing may also have a significant effect on the internal operating temperature of an electronic system. As documented previously, increased temperature translates to reduced reliability. For example, excessive failures were noted on a ship-based missile defense control system. Excessive heating of critical components was suspected. Temperature measurements at various points in the equipment rack showed poor air circulation, as illustrated in Fig. 13.31. Because of the chassis layout, virtually no air circulated through the card cage area. Further, heat from the power supply mounted below the assembly added to the heat build-up. A redesign of the ventilation system solved the reliability problem.

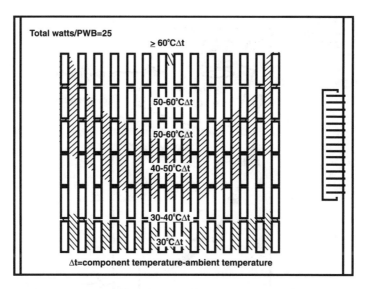

FIGURE 13.29 Operating temperatures of board-mounted components, passively cooled, with the power dissipation set uniformly at 0.5 W/in.^{2}. (After [28].)

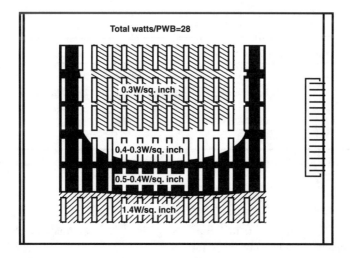

FIGURE 13.30 Printed wiring board component arrangement for constant temperature operation as a passively cooled assembly. Overall power dissipation is adjusted by changing component placement. (After [28].)

The fans and filters used in an electronic system are important to long-term reliability in the field. Depending on the pores per inch rating (PPI) of the material used, fan filters can cause a significant difference in ambient temperature vs internal equipment air temperature, even when clean. Table 13.2 documents typical measurements taken on a piece of electronic equipment. Clearly, the *porosity* (and thickness of the filter material selected) influences the best-case internal temperature of an electronic system. The cubic feet per minute (CFM) rating of the fan or blower used must provide for operation under a wide variety of conditions and temperatures.

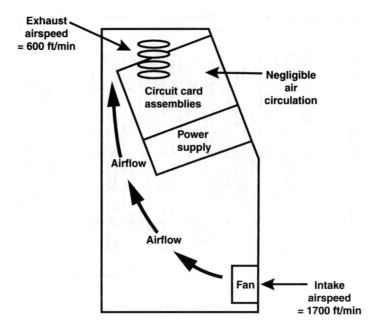

FIGURE 13.31 Control panel airflow for a ship-based missile control system. (After [29].)

TABLE 13.2 Temperature Measurements (in degrees F) on an Electronic System Showing the Differences in Cooling Performance without Fan Filters and with Filters of Various PPI (pores-per-inch) Ratings

		Foam Filters		
	Without Filter	40PPI-$\frac{1}{4}$THK	20PPI-$\frac{1}{4}$THK	10PPI-$\frac{1}{4}$THK
Room Ambient	71°	73°	75°	77°
Equipment air temperature	81°	87°	94°	93°
Temperature rise	10°	24°	19°	16°

Note: The more pores per inch, the greater the pressure drop. Density of material is not related to pore size. (After [8].)

References

1. Bivens, G. A., "Predicting Time-to-Failure Using Finite Element Analysis," *Proc. IEEE Reliability and Maintainability Symp.*, IEEE, New York, 1990.
2. Pantic, D., "Benefits of Integrated Circuit Burn-In to Obtain High Reliability Parts," *IEEE Trans. Reliability*, R-35(1), IEEE, New York, 1986.
3. King, R. and J. Hiatt, "A Practical VLSI Characterization and Failure Analysis System for the IC User," *Proc. IEEE Reliability Phys. Conf.*, IEEE, New York, April 1986.
4. Koch, T., W. Richling, J. Witlock, and D. Hall, "A Bond Failure Mechanism," *Proc. IEEE Reliability Phys. Conf.*, IEEE, New York, April 1986.
5. Guang-bo, G., C. An, and G. Xiang, "A Layer Damage Model for Calculating Thermal Fatigue Lifetime of Power Devices," *Proc. IEEE Reliability Phys. Conf.*, IEEE, New York, April 1986.
6. Koyama, H., Y. Mashiko, and T. Nishioka, "Suppression of Stress Induced Aluminum Void Formation," *Proc. IEEE Reliability Phys. Conf.*, IEEE, New York, April 1986.
7. Sydnor, A., "Voltage Breakdown in Transistors," *Electronic Servicing & Technology*, Intertec Publishing, Overland Park, KS, July 1986.

8. Meeldijk, V., "Why Do Components Fail?," *Electronic Servicing and Technology*, Intertec Publishing, Overland Park, KS, November 1986.
9. Technical staff, *RCA Power Circuits: DC to Microwave*, RCA Electronic Components, Harrison, NJ, 1969.
10. Dasgupta, A., D. Barker, and M. Pecht, "Reliability Prediction of Electronic Packages," *Proc. IEEE Reliability and Maintainability Symp.*, IEEE, New York, 1990.
11. Zins, E., and G. Smith, "R&M Attributes of VHSIC/VLSI Technology," *Proc. IEEE Reliability and Maintainability Symp.*, IEEE, New York, 1987.
12. Smith, W. B., "Integrated Product and Process Design to Achieve High Reliability in Both Early and Useful Life of the Product," *Proc. IEEE Reliability and Maintainability Symp.*, IEEE, New York, 1987.
13. Fink, D. and D. Christiansen (Eds.), *Electronics Engineers' Handbook*, 3rd ed., McGraw-Hill, New York, 1989.
14. Technical staff, Military/Aerospace Products Division, *The Reliability Handbook*, National Semiconductor, Santa Clara, CA, 1987.
15. Technical staff, *Bipolar Power Transistor Reliability Report*, Motorola Semiconductor, Phoenix, AZ, 1988.
16. Crook, D. L., "Evolution of VLSI Reliability Engineering," *Proc. IEEE Reliability Phys. Symp.*, IEEE, New York, 1990.
17. Ching, T. B. and W. H. Schroen, "Bond Pad Structure Reliability," *Proc. IEEE Reliability Phys. Symp.*, IEEE, New York, 1988.
18. Koyama, H., H. Shiozaki, I. Okumura, S. Mizugashira, H. Higuchi, and T. Ajiki, "A New Bond Failure: Wire Crater in Surface Mount Device," *Proc. IEEE Reliability Phys. Symp.*, IEEE, New York, 1988.
19. Lau, J., G. Harkins, D. Rick, and J. Kral, "Thermal Fatigue Reliability of SMT Packages and Interconnections," *Proc. IEEE Reliability Phys. Symp.*, IEEE, New York, 1987.
20. Shirley, C. G. and R. C. Blish, "Thin-Film Cracking and Wire Ball Shear in Plastic DIPs Due to Temperature Cycle and Thermal Shock, *Proc. IEEE Reliability Phys. Symp.*, IEEE, New York, 1987.
21. Blackburn, D. L., "Turn-Off Failure of Power MOSFETs," *IEEE Trans. on Power Electronics*, PE-2(2), IEEE, New York, April 1987.
22. Roehr, B., "Mounting Considerations for Power Semiconductors," *TMOS Power MOSFET Data Handbook*, Motorola Semiconductor, Phoenix, AZ, 1989.
23. Green, T. J., "A Review of IC Fabrication Design and Assembly Defects Manifested as Field Failures in Air Force Avionic Equipment," *Proc. IEEE Reliability Phys. Symp.*, IEEE, New York, 1988.
24. Dick, P., "System Effectiveness: A Key to Assurance," *Proc. IEEE Reliability and Maintainability Sympo.*, IEEE, New York, 1988.
25. Benson, K. B., and J. Whitaker, *Television and Audio Handbook for Engineers and Technicians*, McGraw-Hill, New York, 1989.
26. Whitaker, J., *Radio Frequency Transmission Systems: Design and Operation*, McGraw-Hill, New York, 1990.
27. Jordan, E. C. (Ed.), *Reference Data for Engineers: Radio, Electronics, Computer, and Communications*, 7th ed., Howard W. Sams, Indianapolis, IN, 1985.
28. Leonard, C. T. and M. G. Pecht, "Improved Techniques for Cost Effective Electronics," *Proc. Reliability and Maintainability Soc.*, IEEE, New York, 1991.
29. Fortna, H., R. Zavada, and T. Warren, "An Integrated Analytic Approach for Reliability Improvement," *Proc. IEEE Reliability and Maintainability Symp.*, IEEE, New York, 1990.

14
Power System Protection Alternatives

Jerry C. Whitaker
Editor-in-Chief

14.1 Introduction 14-1
   The Key Tolerance Envelope • Assessing the Lightning Hazard • FIPS Publication 94 • Transient Protection Alternatives
14.2 Motor-Generator Set 14-10
   System Configuration • Motor-Design Considerations • Maintenance Considerations • Motor-Generator UPS
14.3 Uninterruptible Power Systems 14-19
   UPS Configuration • Power-Conversion Methods • Redundant Operation • Output Transfer Switch • Battery Supply
14.4 Dedicated Protection Systems 14-31
   Ferroresonant Transformer • Isolation Transformer • Tap-Changing Regulator • Variable Voltage Transformer • Line Conditioner • Active Power Line Conditioner

14.1 Introduction

Utility companies make a good-faith attempt to deliver clean, well-regulated power to their customers. Most disturbances on the AC line are beyond the control of the utility company. Large load changes imposed by customers on a random basis, PF correction switching, lightning, and accident-related system faults all combine to produce an environment in which tight control over AC power quality is difficult to maintain. Therefore, the responsibility for ensuring AC power quality must rest with the users of sensitive equipment.

The selection of a protection method for a given facility is as much an economic question as it is a technical one. A wide range of power-line conditioning and isolation equipment is available. A logical decision about how to proceed can be made only with accurate, documented data on the types of disturbances typically found on the AC power service to the facility. The protection equipment chosen must be matched to the problems that exist on the line. Using inexpensive basic protectors may not be much better than operating directly from the AC line. Conversely, the use of a sophisticated protector designed to shield the plant from every conceivable power disturbance may not be economically justifiable.

Purchasing transient-suppression equipment is only one element in the selection equation. Consider the costs associated with site preparation, installation, and maintenance. Also consider the operating efficiency of the system. Protection units that are placed in series with the load consume a certain amount of power and, therefore, generate heat. These considerations may not be significant, but they should be taken into account. Prepare a complete life-cycle cost analysis of the protection methods proposed. The study may reveal that the long-term operating expense of one system outweighs the lower purchase price of another.

The amount of money a facility manager is willing to spend on protection from utility company disturbances generally depends on the engineering budget and how much the plant has to lose. Spending $225,000 on system-wide protection for a highly computerized manufacturing center is easily justified. At smaller operations, justification may not be so easy.

The Key Tolerance Envelope

The susceptibility of electronic equipment to failure because of disturbances on the AC power line has been studied by many organizations. The benchmark study was conducted by the Naval Facilities Engineering Command (Washington, D.C.). The far-reaching program, directed from 1968 to 1978 by Lt. Thomas Key, identified three distinct categories of recurring disturbances on utility company power systems. As shown in Table 14.1, it is not the magnitude of the voltage, but the duration of the disturbance that determines the classification.

TABLE 14.1 Types of Voltage Disturbances Identified in the Key Report

Parameter	Type 1	Type 2	Type 3
Definition	Transient and oscillatory overvoltage	Momentary undervoltage or overvoltage	Power outage
Causes	Lightning, power network switching, operation of other loads	Power system faults, large load changes, utility company equipment malfunctions	Power system faults, unacceptable load changes, utility equipment malfunctions
Threshold[a]	200–400% of rated rms voltage or higher (peak instantaneous above or below rated rms)	Below 80–85% and above 110% of rated rms voltage	Below 80–85% of rated rms voltage
Duration	Transients 0.5–200 μs wide and oscillatory up to 16.7 ms at frequencies of 200 Hz to 5 kHz and higher	From 4–6 cycles, depending on the type of power system distribution equipment	From 2–60 s if correction is automatic; from 15 min to 4 h if manual

(After [1].)

[a] The approximate limits beyond which the disturbance is considered to be harmful to the load equipment.

In the study, Key found that most data processing (DP) equipment failure caused by AC line disturbances occurred during bad weather, as shown in Table 14.2. According to a report on the findings, the incidence of thunderstorms in an area can be used to predict the number of failures. The type of power-transmission system used by the utility company was also found to affect the number of disturbances observed on power company lines (Table 14.3). For example, an analysis of utility system problems in Washington, D.C., Norfolk, VA, and Charleston, SC, demonstrated that underground power-distribution systems experienced one-third fewer failures than overhead lines in the same areas. Based on his research, Key developed the "recommended voltage tolerance envelope" shown in Fig. 14.1. The design goals illustrated are recommendations to computer manufacturers for implementation in new equipment.

Assessing the Lightning Hazard

As identified by Key in his Naval Facilities study, the extent of lightning activity in an area significantly affects the probability of equipment failure caused by transient activity. The threat of a lightning flash to a facility is determined, in large part, by the type of installation and its geographic location. The type and character of the lightning flash are also important factors.

The *Keraunic number* of a geographic location describes the likelihood of lightning activity in that area. Figure 14.2 shows the *Isokeraunic map* of the United States, which estimates the number of lightning days per year across the country. On average, 30 storm days occur per year across the continental United States. This number does not fully describe the lightning threat because many individual lightning flashes occur during a single storm.

Power System Protection Alternatives

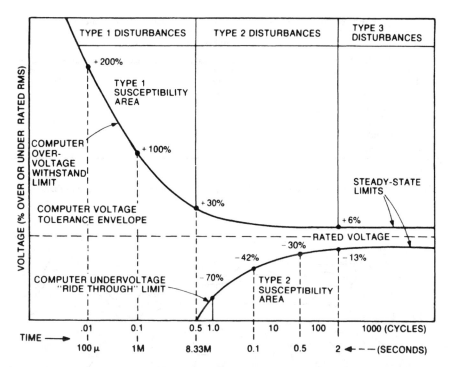

FIGURE 14.1 The recommended voltage tolerance envelope for computer equipment. This chart is based on pioneering work done by the Naval Facilities Engineering Command. The study identified how the magnitude *and* duration of a transient pulse must be considered in determining the damaging potential of a spike. The design goals illustrated in the chart are recommendations to computer manufacturers for implementation in new equipment. (After [1].)

TABLE 14.2 Causes of Power-Related Computer Failures, Northern Virginia, 1976

Recorded Cause	Disturbance		Number of Computer Failures
	Undervoltage	Outage	
Wind and lightning	37	14	51
Utility equipment failure	8	0	8
Construction or traffic accident	8	2	10
Animals	5	1	6
Tree limbs	1	1	2
Unknown	21	2	23
Totals	80	20	100

(After [1].)

The structure of a facility has a significant effect on the exposure of equipment to potential lightning damage. Higher structures tend to collect and even trigger localized lightning flashes. Because storm clouds tend to travel at specific heights above the earth, conductive structures in mountainous areas more readily attract lightning activity. The *plant exposure factor* is a function of the size of the facility and the Isokeraunic rating of the area. The larger the physical size of an installation, the more likely it is to be hit by lightning during a storm. The longer a transmission line (AC or RF), the more lightning flashes it is likely to receive.

The relative frequency of power problems is seasonal in nature. As shown in Fig. 14.3, most problems are noted during June, July, and August. These high problem rates primarily can be traced to increased thunderstorm activity.

TABLE 14.3 Effects of Power-System Configuration on Incidence of Computer Failures

Configuration	Number of Disturbances		Recorded Failures
	Undervoltage	Outage	
Overhead radial	12	6	18
Overhead spot network	22	1	23
Combined overhead (weighted[a])	16	4	20
Underground radial	6	4	10
Underground network	5	0	5
Combined underground (weighted[a])	5	2	7

(After [1].)

[a] The combined averages weighted based on the length of time monitored (30 to 53 months).

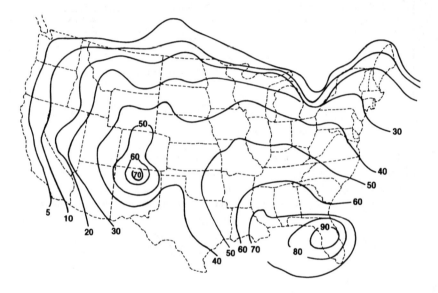

FIGURE 14.2 The Isokeraunic map of the United States, showing the approximate number of lightning days per year.

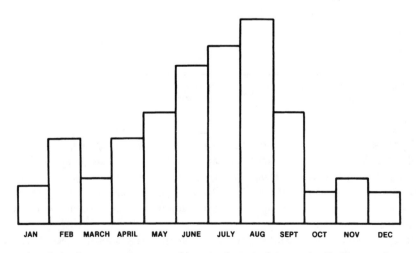

FIGURE 14.3 The relative frequency of power problems in the United States, classified by month.

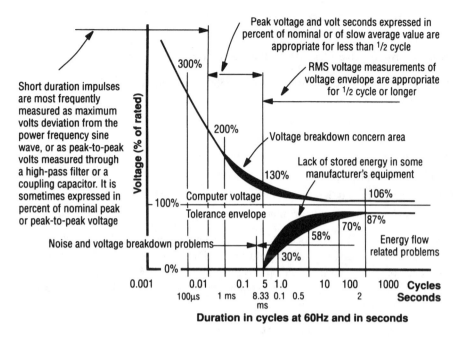

FIGURE 14.4 The CBEMA curve from FIPS Pub. 94. (From [2]. Used with permission.)

FIPS Publication 94

In 1983, the U.S. Department of Commerce published a guideline summarizing the fundamentals of powering, grounding, and protecting sensitive electronic devices [2]. The document, known as Federal Information Processing Standards (FIPS) Publication 94, was first reviewed by governmental agencies and then was sent to the Computer Business Equipment Manufacturers Association (CBEMA) for review. When the CBEMA group put its stamp of approval on the document, the data processing industry finally had an overarching guideline for power quality.

FIPS Pub. 94 was written to cover *automatic data processing equipment* (ADP), which at that time constituted the principal equipment that was experiencing difficulty running on normal utility-supplied power. Since then, IEEE standard P1100 was issued, which applies to all sensitive electronic equipment. FIPS Pub. 94 is a guideline intended to provide a cost/benefit course of action. As a result, it can be relied upon to give the best solution to typical problems that will be encountered, for the least amount of money.

In addition to approving the FIPS Pub. 94 document, the CBEMA group provided a curve that had been used as a guideline for their members in designing power supplies for modern electronic equipment. The CBEMA curve from the FIPS document is shown in Fig. 14.4. Note the similarity to the Key tolerance envelope shown in Fig. 14.1.

The curve is a susceptibility profile. In order to better explain its meaning, the curve has been simplified and redrawn in Fig. 14.14. The vertical axis of the graph is the percent of voltage that is applied to the power circuit, and the horizontal axis is the time factor involved (in µs to s). In the center of the chart is the acceptable operating area, and on the outside of that area is a danger area on top and bottom. The danger zone at the top is a function of the tolerance of equipment to excessive voltage levels. The danger zone on the bottom sets the tolerance of equipment to a loss or reduction in applied power. The CBEMA guideline states that if the voltage supply stays within the acceptable area given by the curve, the sensitive load equipment will operate as intended.

Transient Protection Alternatives

A facility can be protected from transient disturbances in two basic ways: the *systems* approach or the *discrete device* approach. Table 14.4 outlines the major alternatives available:

- UPS (uninterruptible power system) and standby generator
- UPS stand-alone system
- Secondary AC spot network
- Secondary selective AC network
- Motor-generator set
- Shielded isolation transformer
- Suppressors, filters, and lightning arrestors
- Solid-state line-voltage regulator/filter

Table 14.5 lists the relative benefits of each protection method. Because each installation is unique, a thorough investigation of facility needs should be conducted before purchasing any equipment. The systems approach offers the advantages of protection engineered to a particular application and need, and (usually) high-level factory support during equipment design and installation. The systems approach also means higher costs for the end-user.

Specifying System-Protection Hardware

Developing specifications for systemwide power-conditioning/backup hardware requires careful analysis of various factors before a particular technology or a specific vendor is selected. Key factors in this process relate to the load hardware and load application. The electrical power required by a sensitive load may vary widely, depending on the configuration of the system. The principal factors that apply to system specification include the following:

- Power requirements, including voltage, current, power factor, harmonic content, and transformer configuration.
- Voltage-regulation requirements of the load.
- Frequency stability required by the load, and the maximum permissible *slew rate* (the rate of change of frequency per second).
- Effects of unbalanced loading.
- Overload and inrush current capacity.
- Bypass capability.
- Primary/standby path transfer time.
- Maximum standby power reserve time.
- System reliability and maintainability.
- Operating efficiency.

An accurate definition of *critical applications* will aid in the specification process for a given site. The potential for future expansion also must be considered in all plans.

Power requirements can be determined either by measuring the actual installed hardware or by checking the nameplate ratings. Most nameplate ratings include significant safety margins. Moreover, the load will normally include a *diversity factor*; all individual elements of the load will not necessarily be operating at the same time.

TABLE 14.4 Types of Systemwide Protection Equipment Available to Facility Managers and the AC Line Abnormalities That Each Approach Can Handle

System	Type 1	Type 2	Type 3
UPS system and standby generator	All source transients; no load transients	All	All
UPS system	All source transients; no load transients	All	All outages shorter than the battery supply discharge time
Secondary spot network[a]	None	None	Most, depending on the type of outage
Secondary selective network[b]	None	Most	Most, depending on the type of outage
Motor-generator set	All source transients; no load transients	Most	Only brown-out conditions
Shielded isolation transformer	Most source transients; no load transients	None	None
Suppressors, filters, lightning arrestors	Most transients	None	None
Solid-state line voltage regulator/filter	Most source transients; no load transients	Some, depending on the response time of the system	Only brown-out conditions

(After [1] and [2].)

[a] Dual power feeder network.
[b] Dual power feeder network using a static (solid-state) transfer switch.

Every load has a limited tolerance to noise and harmonic distortion. Total harmonic distortion (THD) is a measure of the quality of the waveform applied to the load. It is calculated by taking the geometric sum of the harmonic voltages present in the waveform, and expressing that value as a percentage of the fundamental voltage. Critical DP loads typically can withstand 5% THD, where no single harmonic exceeds 3%. The power-conditioning system must provide this high-quality output waveform to the load, regardless of the level of noise and/or distortion present at the AC input terminals.

If a power-conditioning/standby system does not operate with high reliability, the results can often be disastrous. In addition to threats to health and safety, there is a danger of lost revenue or inventory, and hardware damage. Reliability must be considered from three different viewpoints:

- Reliability of utility AC power in the area.
- Impact of line-voltage disturbances on DP loads.
- Ability of the protection system to maintain reliable operation when subjected to expected and unexpected external disturbances.

The environment in which the power-conditioning system operates will have a significant effect on reliability. Extremes of temperature, altitude, humidity, and vibration can be encountered in various applications. Extreme conditions can precipitate premature component failure and unexpected system shutdown. Most power-protection equipment is rated for operation from 0°C to 40°C. During a commercial power failure, however, the ambient temperature of the equipment room can easily exceed either value, depending on the exterior temperature. Operating temperature derating typically is required for altitudes in excess of 1000 ft.

Table 14.6 lists key power-quality attributes that should be considered when assessing the need for power-conditioning hardware.

TABLE 14.5 Relative Merits of Systemwide Protection Equipment

System	Strong Points	Weak Points	Technical Profile
UPS system and standby generator	Full protection from power outage failures and transient disturbances; ideal for critical DP and life-safety loads	Hardware is expensive and may require special construction; electrically and mechanically complex; noise may be a problem; high annual maintenance costs	Efficiency 80–90%; typical high impedance presented to the load may be a consideration; frequency stability good; harmonic distortion determined by UPS system design
UPS system	Completely eliminates transient disturbances; eliminates surge and sag conditions; provides power outage protection up to the limits of the battery supply; ideal for critical load applications	Hardware is expensive; depending on battery supply requirements, special construction may be required; noise may be a problem; periodic maintenance required	Efficiency 80–90%; typical high impedance presented to the load may be a consideration; frequency stability good; harmonic content determined by inverter type
Secondary spot network[a]	Simple; inexpensive when available in a given area; protects against local power interruptions; no maintenance required by user	Not available in all locations; provides no protection from area-wide utility failures; provides no protection against transient disturbances or surge/sag conditions	Virtually no loss, 100% efficient; presents low impedance to the load; no effect on frequency or harmonic content
Secondary selective network[b]	Same as above; provides faster transfer from one utility line to the other	Same as above	Same as above
Motor-generator set	Electrically simple; reliable power source; provides up to 0.5 s power-fail ride-through in basic form; completely eliminates transient and surge/sag conditions	Mechanical system requires regular maintenance; noise may be a consideration; hardware is expensive; depending upon m-g set design, power-fail ride-through may be less than typically quoted by manufacturer	Efficiency 80–90%; typical high impedance presented to the load may be a consideration; frequency stability may be a consideration, especially during momentary power-fail conditions; low harmonic content
Shielded isolation transformer	Electrically simple; provides protection against most types of transients and noise; moderate hardware cost; no maintenance required	Provides no protection from brown-out or outage conditions	No significant loss, essentially 100% efficient; presents low impedance to the load; no effect on frequency stability; usually low harmonic content
Suppressors, filters, lightning arrestors	Components inexpensive; units can be staged to provide transient protection exactly where needed in a plant; no periodic maintenance required	No protection from Type 2 or 3 disturbances; transient protection only as good as the installation job	No loss, 100% efficient; some units subject to power-follow conditions; no effect on impedance presented to the load; no effect on frequency or harmonic content
Solid-state line voltage regulator/filter	Moderate hardware cost; uses a combination of technologies to provide transient suppression and voltage regulation; no periodic maintenance required	No protection against power outage conditions; slow response time may be experienced with some designs	Efficiency 92–98%; most units present low impedance to the load; usually no effect on frequency; harmonic distortion content may be a consideration

(After [1] and [2].)

[a] Dual power feeder network.
[b] Dual power feeder network using a static (solid state) transfer switch.

Power System Protection Alternatives

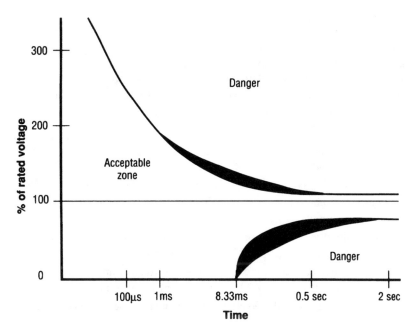

FIGURE 14.5 A simplified version of the CBEMA curve. Voltage levels outside the *acceptable zone* results in potential system shutdown and hardware and software loss. (From [2]. Used with permission.)

TABLE 14.6 Power-Quality Attributes for Data Processing Hardware

Environmental Attribute	Typical Environment	Acceptable Limits for DP Systems	
		Normal	Critical
Line frequency	± 0.1 to ± 3%	± 1%	± 0.3%
Rate of frequency change	0.5 to 20 Hz/s	1.5 Hz/s	0.3 Hz/s
Over- and under-voltage	± 5% to +6, −13.3%	+ 5% to −10%	± 3%
Phase imbalance	2 to 10%	5% max	3% max
Tolerance to low power factor	0.85 to 0.6 lagging	0.8 lagging	less than 0.6 lagging, or 0.9 leading
Tolerance to high steady state peak current	1.3 to 1.6 peak, rms	1.0 to 2.5 peak, rms	Greater than 2.5 peak, rms
Harmonic voltages	0 to 20% total rms	10 to 20% total, 5 to 10% largest	5% max total, 3% largest
Voltage deviation from sine wave	5 to 50%	5 to 10%	3 to 5%
Voltage modulation	Negligible to 10%	3% max	1% max
Surge/sag conditions	+10%, −15%	+20%, −30%	+5%, −5%
Transient impulses	2 to 3 times nominal peak value (0 to 130% Vs)	Varies; 1.0 to 1.5 kV typical	Varies; 200 to 500 V typical
RFI/EMI normal and common modes	10 V up to 20 kHz, less at high freq.	Varies widely, 3 V is typical	Varies widely, 0.3 V typical
Ground currents	0 to 10 A plus impulse noise current	0.001 to 0.5 A or more	0.0035 A or less

(After [2].)

14.2 Motor-Generator Set

As the name implies, a motor-generator (m-g) set consists of a motor powered by the AC utility supply that is mechanically tied to a generator, which feeds the load. (See Fig. 14.6.) Transients on the utility line will have no effect on the load when this arrangement is used. Adding a flywheel to the motor-to-generator shaft will protect against brief power dips (up to 0.5 s on many models). Figure 14.7 shows the construction of a typical m-g set. The attributes of an m-g include the following:

- An independently generated source of voltage can be regulated without interaction with line-voltage changes on the power source. Utility line changes of ± 20% commonly can be held to within ± 1% at the load.
- The rotational speed and inertial momentum of the rotating mass represents a substantial amount of stored rotational energy, preventing sudden changes in voltage output when the input is momentarily interrupted.
- The input and output windings are separated electrically, preventing transient disturbances from propagating from the utility company AC line to the load.
- Stable electrical characteristics for the load: (1) output voltage and frequency regulation, (2) ideal sine wave output, and (3) true 120° phase shift for three-phase models.
- Reduced problems relating to the power factor presented to the utility company power source.

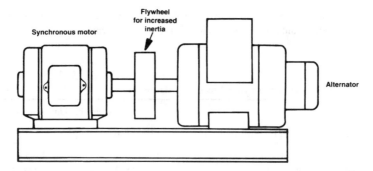

FIGURE 14.6 Two-machine motor-generator set with an optional flywheel to increase inertia and carry-through capability.

FIGURE 14.7 Construction of a typical m-g set. (Courtesy of Computer Power Protection.)

Power System Protection Alternatives

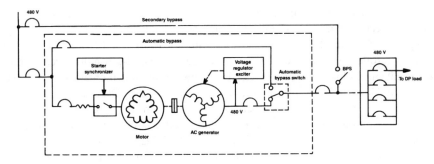

FIGURE 14.8 Simplified schematic diagram of an m-g set with automatic and secondary bypass capability. (After [2].)

The efficiency of a typical m-g ranges from 65 to 89%, depending on the size of the unit and the load. Motor-generator sets have been used widely to supply 415 Hz power to mainframe computers that require this frequency.

System Configuration

There are a number of types of m-g sets, each having its own characteristics, advantages, and disadvantages. A simplified schematic diagram of an m-g is shown in Fig. 14.8. The type of motor that drives the set is an important design element. Direct-current motor drives can be controlled in speed independently of the frequency of the AC power source from which the DC is derived. Use of a DC motor, thereby, gives the m-g set the capability to produce power at the desired output frequency, regardless of variations in input frequency. The requirement for rectifier conversion hardware, control equipment, and commutator maintenance are drawbacks to this approach that must be considered.

The simplest and least expensive approach to rotary power conditioning involves the use of an induction motor as the mechanical source. Unfortunately, the rotor of an induction motor turns slightly slower than the rotating field produced by the power source. This results in the generator being unable to produce 60 Hz output power if the motor is operated at 60 Hz and the machines are directly coupled end-to-end at their shafts, or are belted in a 1:1 ratio. Furthermore, the shaft speed and output frequency of the generator decreases as the load on the generator is increased.

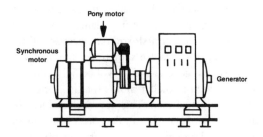

Figure 14.9 Use of a pony motor for an m-g set to aid in system starting and restarting. (After [2].)

This potential for varying output frequency may be acceptable where the m-g set is used solely as the input to a power supply in which the AC is rectified and converted to DC. However, certain loads cannot tolerate frequency changes greater than 1 Hz/s and frequency deviations of more than 0.5 Hz from the nominal 60 Hz value.

Low-slip induction motor-driven generators are available that can produce 59.7 Hz at full load, assuming 60 Hz input. During power interruptions, the output frequency will drop further, depending upon the length of the interruption. The capability of the induction motor to restart after a momentary power interruption is valuable. Various systems of variable-speed belts have been tried successfully. Magnetically controlled slipping clutches have been found to be largely unsatisfactory. Other approaches to make the induction motor drive the load at constant speed have produced mixed results.

Using a synchronous motor with direct coupling or a cogged 1:1 ratio belt drive guarantees that the output frequency will be equal to the motor input frequency. Although the synchronous motor is more expensive, it is more efficient and can be adjusted to provide a unity PF load to the AC source. The starting characteristics and the mechanical disturbance following a short line-voltage interruption depends, to a large extent, on motor design. Many synchronous motors that are not required to start under load have weak starting torque, and may use a *pony motor* to aid in starting. This approach is shown in Fig. 14.9. Those motors designed to start with a load have starting pole face windings that provide starting torque comparable to that of an induction motor. Such motors can be brought into synchronism while under load with proper selection of the motor and automatic starter system. Typical utility company AC interruptions are a minimum of six cycles (0.1 s). Depending upon the design and size of the flywheel used, the ride-through period can be as much as 0.5 s or more. The generator will continue to produce output power for a longer duration, but the frequency and rate of frequency change will most likely fall outside of the acceptable range of most DP loads after 0.5 s.

If input power is interrupted and does not return before the output voltage and frequency begin to fall outside acceptable limits, the generator output controller can be programmed to disconnect the load. Before this event, a warning signal is sent to the DP control circuitry to warn of impending shutdown and to initiate an orderly interruption of active computer programs. This facilitates easy restart of the computer after the power interruption has passed.

It is important for users to accurately estimate the length of time that the m-g set will continue to deliver acceptable power without input to the motor from the utility company. This data facilitates accurate power-fail shutdown routines. It is also important to ensure that the m-g system can handle the return of power without operating overcurrent protection devices because of high inrush currents that may be required to accelerate and synchronize the motor with the line frequency. Protection against the latter problem requires proper programming of the synchronous motor controller to correctly disconnect and then reconnect the field current supply. It may be worthwhile to delay an impending shutdown for 100 ms or so. This would give the computer time to prepare for the event through an orderly interruption. It also would be useful if the computer were able to resume operation without shutdown, in case utility power returns within the ride-through period. Control signals from the m-g controller should be configured to identify these conditions and events to the DP system.

Generators typically used in m-g sets have substantially higher internal impedance than equivalent kVA-rated transformers. Because of this situation, m-g sets are sometimes supplied with an oversized generator that will be lightly loaded, coupled with a smaller motor that is adequate to drive the actual load. This approach reduces the initial cost of the system, decreases losses in the motor, and provides a lower operating impedance for the load.

Motor-generator sets can be configured for either horizontal installation, as shown previously, or for vertical installation, as illustrated in Fig. 14.10.

The most common utility supply voltage used to drive the input of an m-g set is 480 V. The generator output for systems rated at about 75 kVA or less is typically 208 Y/120 V. For larger DP systems, the most economical generator output is typically 480 V. A 480-208 Y/120 V three-phase isolating transformer usually is included to provide 208 Y/120 V power to the computer equipment.

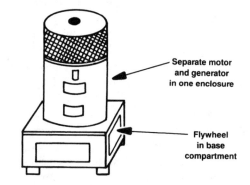

FIGURE 14.10 Vertical m-g set with an enclosed flywheel. (After [2].)

Motor-Design Considerations

Both synchronous and induction motors have been used successfully to drive m-g sets, and each has advantages and disadvantages. The major advantage of the synchronous motor is that while running normally, it is synchronized with the supply frequency. An 1800 rpm motor rotates at exactly 1800 rpm for a supply frequency of exactly 60 Hz. The generator output, therefore, will be exactly 60 Hz. Utility frequencies *average* 60 Hz; utilities vary the frequency slowly to maintain this average value under changing load conditions. Research has shown that utility operating frequencies typically vary from 58.7 to 60.7 Hz. Although frequency tolerances permitted by most computer manufacturers are usually given as ±0.5 Hz on a nominal 60 Hz system, these utility variations are spread over a 24-hour period or longer, and generally do not result in problems for the load.

The major disadvantage of a synchronous motor is that the device is difficult to start. A synchronous motor must be started and brought up to *pull-in* speed by an auxiliary winding on the armature, known as the *armortisseur* winding. The pull-in speed is the minimum speed (close to synchronous speed) at which the motor will pull into synchronization if excitation is applied to the field. The armortisseur winding is usually a squirrel-cage design, although it may be of the wound-rotor type in some cases. This winding allows the synchronous motor to start and come up to speed as an induction motor. When pull-in speed is achieved, automatic sensing equipment applies field excitation, and the motor locks in and runs as a synchronous machine. As discussed previously, some large synchronous motors are brought up to speed by an auxiliary pony motor.

The armortisseur winding can produce only limited torque, so synchronous motors usually are brought up to speed without a load. This requirement presents no problem for DP systems upon initial startup. However, in the event of a momentary power outage, problems can develop. When the utility AC fails, the synchronous motor must be disconnected from the input immediately, or it will act as a generator and feed power back into the line, thus rapidly depleting its stored (kinetic) rotational energy. During a power failure, the speed of the motor rapidly drops below the pull-in speed, and when the AC supply returns, the armortisseur winding must reaccelerate the motor under load until the field can be applied again. This requires a large winding and a sophisticated control system. When the speed of the m-g set is below synchronous operation, the generator output frequency may be too low for proper computer operation.

The induction motor has no startup problems, but it does have *slip*. To produce torque, the rotor must rotate at a slightly lower speed than the stator field. For a nominal 1800 rpm motor, the actual speed will be about 1750 rpm, varying slightly with the load and the applied input voltage. This represents a slip of about 2.8%. The generator, if driven directly or on a common shaft, will have an output frequency of about 58.3 Hz. This is below the minimum permissible operating frequency for most computer hardware. Special precision-built low-slip induction motors are available with a slip of approximately 0.5% at a nominal motor voltage of 480 V. With 0.5% slip, speed at full load will be about 1791 rpm, and the directly driven or common-shaft generator will have an output frequency of 59.7 Hz. This frequency is within tolerance, but close to the minimum permissible frequency.

A belt-and-pulley system adjustable-speed drive is a common solution to this problem. By making the pulley on the motor slightly larger in diameter than the pulley on the generator (with the actual diameters adjustable) the generator can be driven at synchronous speed.

Voltage sags have no effect on the output frequency of a synchronous motor-driven m-g set until the voltage gets so low that the torque is reduced to a point at which the machine pulls out of synchronization. Resynchronization then becomes a problem. On an induction motor, when the voltage sags, slip increases and the machine slows down. The result is a drop in generator output frequency. The adjustable-speed drive between an induction motor and the generator solves the problem for separate machines. If severe voltage sags are anticipated at a site, the system can be set so that nominal input voltage from the utility company produces a frequency of 60.5 Hz, 0.5 Hz on the high side of nominal frequency. Figure 14.11 charts frequency vs. motor voltage for three operating conditions:

- Slip compensation set high (curve *A*)
- Slip compensation set for 60 Hz (curve *B*)
- No slip compensation (curve *C*)

Through proper adjustment of slip compensation, considerable input-voltage margins can be achieved.

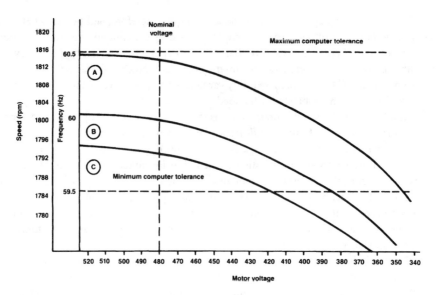

FIGURE 14.11 Generator output frequency vs. motor input voltage for an induction-motor-based m-g set.

Single-Shaft Systems

There are two basic m-g set machine mechanical designs used for DP applications: (1) separate motor-generator systems, and (2) single-shaft, single-housing units. Both designs can use either a synchronous or induction motor. In each case, there are advantages and disadvantages. The separate machine design (discussed previously) uses a motor driving a physically separate generator by means of a coupling shaft or pulley. In an effort to improve efficiency and reduce costs, manufacturers also have produced various types of single-shaft systems.

The basic concept of a single-shaft system is to combine the motor and generator elements into a single unit. A common stator eliminates a number of individual components, making the machine less expensive to produce and mechanically more efficient. The common-stator set substantially reduces mechanical energy losses associated with traditional m-g designs, and it improves system reliability as well. In one design, the stator is constructed so that alternate slots are wound with input and output windings. When it is fed with a three-phase supply, a rotating magnetic field is created, causing the DC-excited rotor to spin at a synchronous speed. By controlling the electrical characteristics of the rotor, control of the output at the secondary stator windings is accomplished.

Common-stator machines offer lower working impedance for the load than a comparable two-machine system. For example, a typical 400-kVA machine has approximately an 800-kVA frame size. The larger frame size yields a relatively low-impedance power source capable of clearing subcircuit fuses under fault conditions. The output of the unit typically can supply up to seven times the full-load current under fault conditions. Despite the increase in frame size, the set is smaller and lighter than comparable systems because of the reduced number of mechanical parts.

Flywheel Considerations

In an effort to achieve higher energy and power densities, m-g set designers have devoted considerable attention to the flywheel element itself. New composite materials and power electronics technologies have resulted in compact flywheel "batteries" capable of high linear velocity at the outside radius of the flywheel (*tip speed*) [3]. The rotational speed of the flywheel is important because the stored energy in a flywheel is proportional to the square of its rotational speed. Therefore, an obvious method for maximizing stored energy is to increase the speed of the flywheel. All practical designs, however, have a limiting speed, which is determined by the stresses developed within the wheel resulting from inertial loads. These loads are also proportional to the square of rotational speed. Flywheels built of composite materials weigh less, and hence develop lower inertial loads at a given speed. In addition, composites are often stronger than conventional engineering metals, such as steel. This combination of high strength and low weight enables extremely high tip speeds, relative to conventional wheels.

For a given geometry, the limiting *energy density* (energy per unit mass) of a flywheel is proportional to the ratio of material strength to weight density, otherwise known as the *specific strength*. Table 14.7 illustrates the advantage that composite materials offer in this respect.

TABLE 14.7 Specific Strength of Selected Materials

Material	Specific Strength (in.^{3})
Graphite/epoxy	3,509,000
Boron/epoxy	2,740,000
Titanium and alloys	1,043,000
Wrought stainless steel	982,000
Wrought high-strength steel	931,000
7000 series aluminum alloys	892,000

(After [4].)

Recent advances in composite materials technology may allow nearly an order of magnitude advantage in the specific strength of composites when compared to even the best common engineering metals. The result of this continuous research in composites has been flywheels capable of operation at rotational speeds in excess of 100,000 rpm, with tip speeds in excess of 1,000 m/s.

These high speeds bring with them new challenges. The ultrahigh rotational speeds that are required to store significant kinetic energy in these systems virtually rule out the use of conventional mechanical bearings. Instead, most systems run on magnetic bearings. This relatively recent innovation uses magnetic forces to "levitate" a rotor, eliminating the frictional losses inherent in rolling element and fluid film bearings. Unfortunately, aerodynamic drag losses force most high-speed flywheels to operate in a partial vacuum, which complicates the task of dissipating the heat generated by ohmic losses in the bearing electromagnets and rotor. In addition, active magnetic bearings are inherently unstable, and require sophisticated control systems to maintain proper levitation.

The integrated generator of these systems is usually a rotating-field design, with the magnetic field supplied by rare-earth permanent magnets. Because the specific strength of these magnets is typically just fractions of that of the composite flywheel, they must spin at much lower tip speeds; in other words, they must be placed very near the hub of the flywheel. This compromises the power density of the generator. An alternative is to mount them closer to the outer radius of the wheel, but contain their inertial loads with the composite wheel itself. Obviously, this forces the designer to either derate the machine speed, or operate closer to the stress limits of the system.

Maintenance Considerations

Because m-g sets require some maintenance that necessitates shutdown, most systems provide bypass capability so the maintenance work can be performed without having to take the computer out of service. If the automatic bypass contactor, solid-state switch, and control hardware are in the same cabinet as other devices that also need to be de-energized for maintenance, a secondary bypass is recommended. After the automatic bypass path has been established, transfer switching to the secondary bypass can be enabled, taking the m-g set and its automatic bypass system out of the circuit completely. Some automatic bypass control arrangements are designed to transfer the load of the generator to the bypass route with minimum disturbance. This requires the generator output to be synchronized with the bypass power before closing the switch and opening the generator output breaker. However, with the load taken off the generator, bypass power no longer will be synchronized with it. Consequently, retransfer of the load back to the generator may occur with some disturbance. Adjustment for minimum disturbance in either direction requires a compromise in phase settings, or a means to shift the phase before and after the transfer.

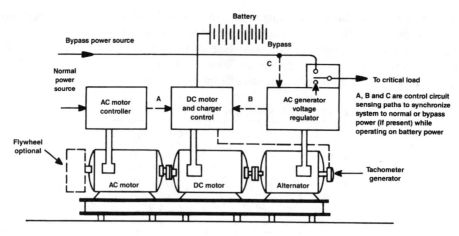

FIGURE 14.12 Uninterruptible m-g set with AC and DC motor drives. (After [2].)

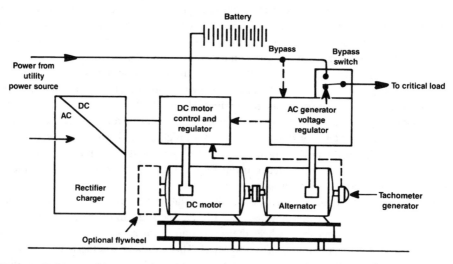

FIGURE 14.13 Uninterruptible m-g set using a single DC drive motor. (After [2].)

The use of rotating field exciters has eliminated the need for slip rings in most m-g designs. Brush inspection and replacement, therefore, are no longer needed. However, as with any rotating machinery, bearings must be inspected and periodically replaced.

Motor-Generator UPS

Critical DP applications that cannot tolerate even brief AC power interruptions can use the m-g set as the basis for an uninterruptible source of power through the addition of a battery-backed DC motor to the line-driven AC motor shaft. This concept is illustrated in Fig. 14.12. The AC motor normally supplies power to drive the system from the utility company line. The shafts of the three devices are all interconnected, as shown in the figure. When AC power is present, the DC motor serves as a generator to charge the battery bank. When line voltage is interrupted, the DC motor is powered by the batteries. Figure 14.13 shows a modified version of this basic m-g UPS using only a DC motor as the mechanical power source. This configuration eliminates the inefficiency involved in having two motors in the system. Power from the utility source is rectified to provide energy for the DC motor, plus power for charging the batteries. A complex control system to switch the AC motor off and the DC motor on in the event of a utility power failure is not needed in this design.

Power System Protection Alternatives

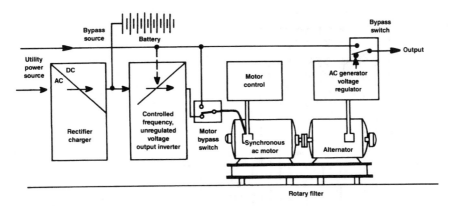

FIGURE 14.14 Uninterruptible m-g set using a synchronous AC motor. (After [2].)

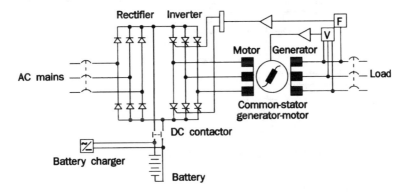

FIGURE 14.15 Common-stator UPS m-g set. The firing angle of the SCR inverters is determined by a feedback voltage from the generator output.

The m-g UPS also can be built around a synchronous AC motor, as illustrated in Fig. 14.14. Utility AC energy is rectified and used to drive an inverter, which provides a regulated frequency source to power the synchronous motor. The output from the DC-to-AC inverter need not be a well-formed sine wave, nor a well-regulated source. The output from the generator will provide a well-regulated sine wave for the load. The m-g set also can be operated in a bypass mode that eliminates the rectifier, batteries, and inverter from the current path, operating the synchronous motor directly from the AC line.

An m-g UPS set using a common stator machine is illustrated in Fig. 14.15. A feedback control circuit adjusts the firing angle of the inverter to compensate for changes in input power. This concept is taken a step further in the system shown in Fig. 14.16. A solid-state inverter bypass switch is added to improve efficiency. During normal operation, the bypass route is enabled, eliminating losses across the rectifier diodes. When the control circuitry senses a drop in utility voltage, the inverter is switched on, and the bypass switch is deactivated. A simplified installation diagram of the inverter/bypass system is shown in Fig. 14.17. Magnetic circuit breakers and chokes are included as shown. An isolation transformer is inserted between the utility input and the rectifier bank. The static inverter is an inherently simple design; commutation is achieved via the windings. Six thyristors are used. Under normal operating conditions, 95% of the AC power passes through the static switch; 5% passes through the inverter. This arrangement affords maximum efficiency, while keeping the battery bank charged and the rectifiers and inverter thyristors preheated. Preheating extends the life of the components by reducing the extent of thermal cycling that occurs when the load suddenly switches to the battery backup supply. The static switch allows for fast disconnect of the input AC when the utility power fails.

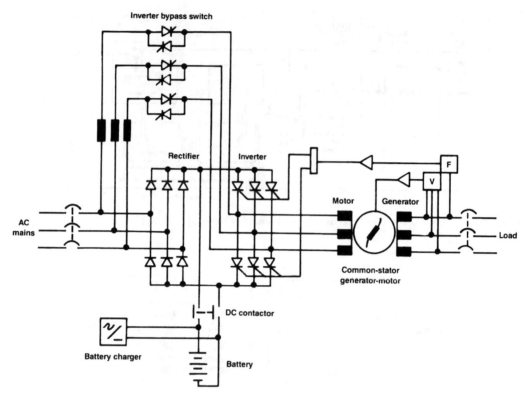

FIGURE 14.16 Common-stator UPS m-g set with a solid-state inverter bypass switch.

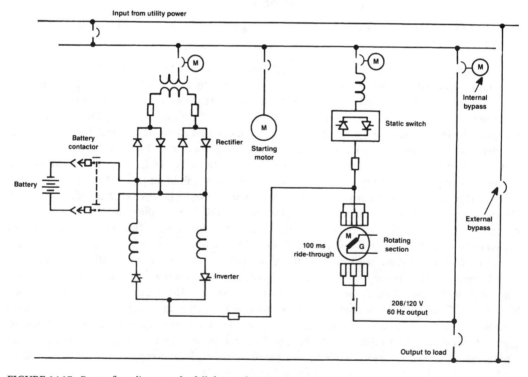

FIGURE 14.17 Power flow diagram of a full-featured UPS m-g set.

Power System Protection Alternatives

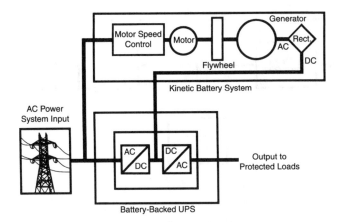

FIGURE 14.18 Functional block diagram of a kinetic battery m-g system for powering a UPS. (After [5].)

Kinetic Battery Storage System

As outlined previously, one of the parameters that limits the ride-through period of an m-g set is the speed decay of the flywheel/generator combination. As the flywheel slows down, the output frequency drops. This limits the useful ride-through period to 0.5 s or so. Figure 14.18 shows an ingenious modification of the classic power conditioning template that extends the potential ride-through considerably. As shown in the figure, an m-g set is used in a UPS-based system as an element of the DC supply grid. The major components of the system include:

- Steel flywheel for energy storage
- Small drive motor, sized at 15 to 20% of the rated system output, to start the flywheel and maintain its normal operating speed
- Variable speed drive (VSD) to slowly ramp the flywheel up to speed and maintain it at the desired rpm
- Generator to convert the kinetic energy stored in the flywheel into electrical energy
- Diode bridge rectifier to convert the AC generator output to DC for use by the UPS bus, which continues to draw usable energy essentially independent of flywheel rpm.

Because the AC output of the generator is converted to DC, frequency is no longer a limiting factor, which allows the DC voltage output to be maintained over a much greater range of the flywheel's rpm envelope.

In operation, the small drive motor spins the flywheel while the variable speed drive maintains the proper motor speed [5]. Because the amount of stored kinetic energy increases by the square of the flywheel rpm, it is possible to greatly increase stored energy, and thus ride-through, by widening the flywheel's usable energy output range. These factors permit typical ride-through times from 10 s to several minutes, depending on loading and other operating conditions. Among the benefits of this approach include the reduced cycling of battery supplies and engine-generators systems.

14.3 Uninterruptible Power Systems

Uninterruptible power systems have become a virtual necessity for powering large or small computer systems where the application serves a critical need. Computers and data communications systems are no more reliable than the power from which they operate. The difference between UPS and *emergency standby power* is that the UPS is always in operation. It reverts to an alternative power source, such as the utility company, only if the UPS fails or needs to be deactivated for maintenance. Even then, the transfer of power occurs so quickly (within milliseconds) that it does not interrupt proper operation of the load.

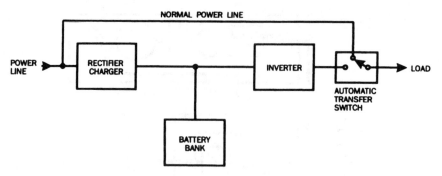

FIGURE 14.19 Forward-transfer UPS system.

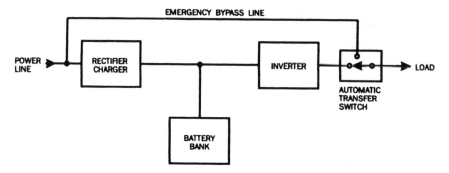

FIGURE 14.20 Reverse-transfer UPS system.

Emergency standby power is normally off and does not start (manually or automatically) until the utility AC feed fails. A diesel generator can be started within 10 to 30 s if the system has been maintained properly. Such an interruption, however, is far too long for DP hardware. Most DP systems cannot ride through more than 8 to 22 ms of power interruption. Systems that can successfully ride through short-duration power breaks, as far as energy continuity is concerned, still may enter a fault condition because of electrical noise created by the disturbance.

UPS hardware is available in a number of different configurations. All systems, however, are variations of two basic designs:

- *Forward-transfer mode:* The load normally is powered by the utility power line, and the inverter is idle. If a commercial power failure occurs, the inverter is started and the load is switched. This configuration is illustrated in Fig. 14.19. The primary drawback of this approach is the lack of load protection from power-line disturbances during normal (utility-powered) operation.

- *Reverse-transfer mode:* The load normally is powered by the inverter. In the event of an inverter failure, the load is switched directly to the utility line. This configuration is illustrated in Fig. 14.20. The reverse-transfer mode is, by far, the most popular type of UPS system in use for large-scale systems.

The type of load-transfer switch used in the UPS system is another critical design parameter. The continuity of AC power service to the load is determined by the type of switching circuit used. An electromechanical transfer switch, shown in Fig. 14.21, is limited to switch times of 20 to 50 ms. This time delay can cause sensitive load equipment to malfunction, and perhaps shut down. A control circuit actuates the relay when the sensed output voltage falls below a preset value, such as 94% of nominal. A static transfer switch, shown in Fig. 14.22, can sense a failure and switch the load in about 4 ms. Most loads will ride through this short delay without any malfunction. To accomplish a smooth transition, the inverter output must be synchronized with the power line.

Power System Protection Alternatives

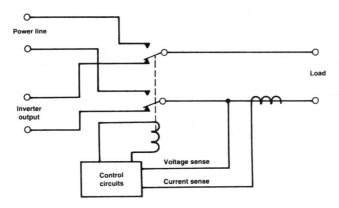

FIGURE 14.21 Electromechanical load-transfer switch.

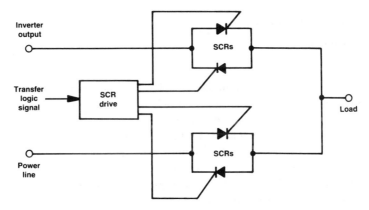

FIGURE 14.22 Static load-transfer switch.

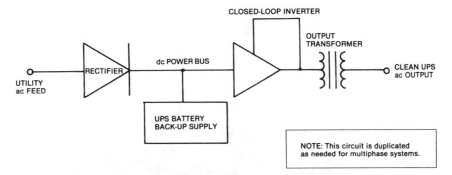

FIGURE 14.23 Block diagram of an uninterruptible power system using AC rectification to float the battery supply. A closed-loop inverter draws on this supply and delivers clean AC power to the protected load.

UPS Configuration

The basic uninterruptible power system is built around a battery-driven inverter, with the batteries recharged by the utility AC line. As shown in Fig. 14.23, AC from the utility feed is rectified and applied to recharge or *float* a bank of batteries. This DC power drives a single- or multiphase closed-loop inverter, which regulates output voltage and frequency. The output of the inverter is generally a sine wave, or pseudo sine wave (a stepped square wave). If the utility voltage drops or disappears, current is drawn from the batteries. When AC power is restored, the batteries are recharged. Many UPS systems incorporate a standby diesel generator that starts as soon as the utility company feed is interrupted. With this arrangement, the batteries are called upon to supply operating current for only 30 s or so, until the generator gets up to speed. A UPS system intended to power a computer center is illustrated in Fig. 14.24.

Power-Conversion Methods

Solid-state UPS systems that do not employ rotating machinery utilize one of several basic concepts. The design of an inverter is determined primarily by the operating power level. The most common circuit configurations include:

- Ferroresonant inverter
- Delta magnetic inverter
- Inverter-fed L/C tank
- Quasi-square wave inverter
- Step wave inverter
- Pulse-width modulation inverter
- Phase modulation inverter

Ferroresonant Inverter

Illustrated in Fig. 14.25, the ferroresonant inverter is popular for low- to medium-power applications. A ferroresonant transformer can be driven with a distorted, unregulated input voltage and deliver a regulated, sinusoidal output when filtered properly. The ferroresonant transformer core is designed so that the secondary section is magnetically saturated at the desired output voltage. As a result, the output level remains relatively constant over a wide range of input voltages and loads. Capacitors, connected across the secondary, help drive the core into saturation and, in conjunction with inductive coupling, provide harmonic filtering. Regulated, sinusoidal three-phase output voltages are derived from two inverters operating into ferroresonant transformers in a *Scott-T* configuration. The basic ferroresonant inverter circuit, shown in Fig. 14.26, consists of an oscillator that controls SCR switches, which feed a ferroresonant transformer and harmonic filter. The saturated operating mode produces a regulated output voltage and inherent current limiting. Efficiency varies from 50 to 83%, depending on the load. Response time of the ferroresonant inverter is about 20 ms.

Although ferroresonant inverters are rugged, simple, and reliable, they do have disadvantages. First, such systems tend to be larger and heavier than electronically controlled inverters. Second, there is a phase shift between the inverter square wave and the output sine wave. This phase shift varies with load

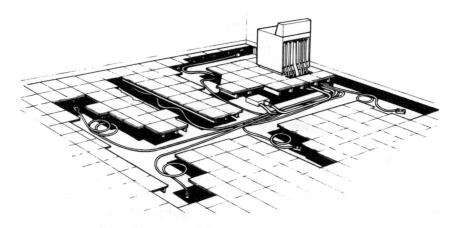

FIGURE 14.24 Installation details for a computer-room UPS power conditioner.

Power System Protection Alternatives

magnitude and power factor. When an unbalanced load is applied to a three-phase ferroresonant inverter, the output phases can shift from their normal 120° relationships. This results in a change in line-to-line voltages, even if individual line-to-neutral voltages are regulated perfectly. This voltage imbalance cannot be tolerated by some loads.

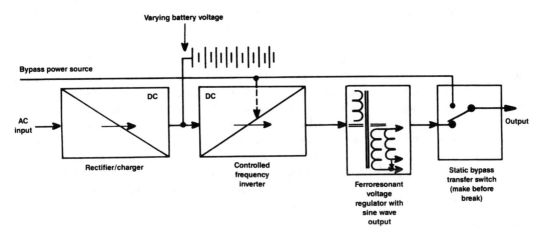

FIGURE 14.25 Simplified diagram of a static UPS system based on a ferroresonant transformer. (After [2].)

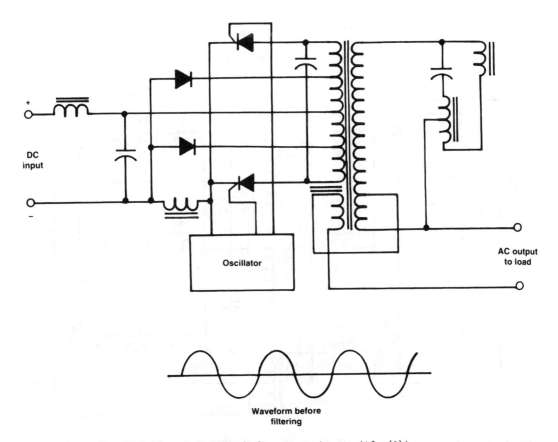

FIGURE 14.26 Simplified schematic diagram of a ferroresonant inverter. (After [2].)

Delta Magnetic Inverter

While most multiphase inverters are single-phase systems adapted to three-phase operation, the delta magnetic inverter is inherently a three-phase system. A simplified circuit diagram of a delta magnetic inverter is shown in Fig. 14.27. Inverter modules A1, B1, and C1 produce square wave outputs that are phase-shifted relative to each other by 120°. The waveforms are coupled to the primaries of transformer T1 through linear inductors. T1 is a conventional three-phase isolation transformer. The primaries of the device are connected in a delta configuration, reducing the third harmonic and all other harmonics that are odd-order multiples of the third. The secondaries of T1 are connected in a wye configuration to provide a four-wire three-phase output. Inductors L4–L9 form a network connected in a delta configuration to high-voltage taps on the secondary windings of T1. Inductors L4–L6 are single-winding saturating reactors, and L7–L9 are double-winding saturating reactors. Current drawn by this saturating reactor network is nearly sinusoidal and varies in magnitude, in a nonlinear manner, with voltage variations. For example, if an increase in load tended to pull the inverter output voltages down, the reduced voltage applied to the reactor network would result in a relatively large decrease in current drawn by the network. This, in turn, would decrease the voltage drop across inductors L1–L3 to keep the inverter output voltage at the proper level. The delta magnetic regulation technique is essentially a three-phase shunt regulator. Capacitors C1–C3 help drive the reactor network into saturation, as well as provide harmonic filtering in conjunction with the primary inductors.

Inverter-Fed L/C Tank

An inductor/capacitor tank is driven by a DC-to-AC inverter, as illustrated in Fig. 14.28. The tank circuit acts to reconstruct the sine wave at the output of the system. Regulation is accomplished by varying the capacitance or the inductance to control partial resonance and/or power factor. Some systems of this type use a saturable reactor in which electronic voltage-regulator circuits control the DC current in the reactor. Other systems, shown in Fig. 14.29, use a DC-to-variable-DC inverter/converter to control the UPS output through adjustment of the boost voltage. This feature permits compensation for changes in battery level.

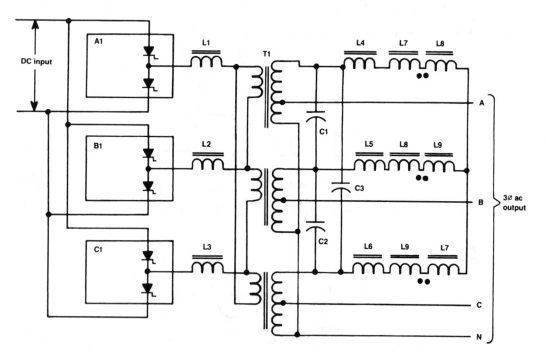

FIGURE 14.27 Simplified schematic diagram of the delta magnetic regulation system.

Quasi-Square Wave Inverter

Shown in Fig. 14.30, the quasi-square wave inverter produces a variable-duty waveshape that must be filtered by tuned series and parallel inductive-capacitive networks to reduce harmonics and form a sinusoidal output. Because of the filter networks present in this design, the inverter responds slowly to load changes; response time in the range of 150 to 200 ms is common. Efficiency is about 80%. This type of inverter requires voltage-regulating and current-limiting networks, which increase circuit complexity and make it relatively expensive.

Step Wave Inverter

A multistep inverter drives a combining transformer, which feeds the load. The general concept is illustrated in Fig. 14.31. The purity of the output sine wave is a function of the number of discrete steps produced by the inverter. Voltage regulation is achieved by a boost DC-to-DC power supply in series with the battery. Figures 14.32 and 14.33 show two different implementations of single-phase units. Each system uses a number of individual inverter circuits, typically three or multiples of three. The inverters are controlled by a master oscillator; their outputs are added in a manner that reduces harmonics, producing a near-sinusoidal output. These types of systems require either a separate voltage regulator on the DC bus or a phase shifter. Response time is about 20 ms. Little waveform filtering is required, and efficiency can be as high as 85%. The step wave inverter, however, is complex and expensive. It is usually found in large three-phase UPS systems only.

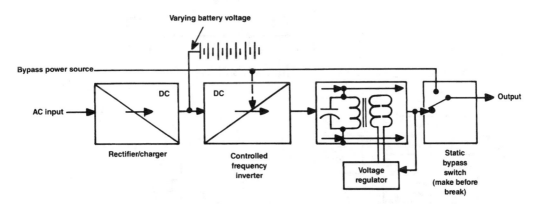

FIGURE 14.28 Static UPS system using saturable reactor voltage control. (After [2].)

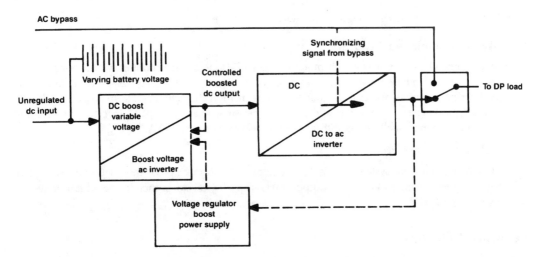

FIGURE 14.29 Static UPS system with DC boost voltage control. (After [2].)

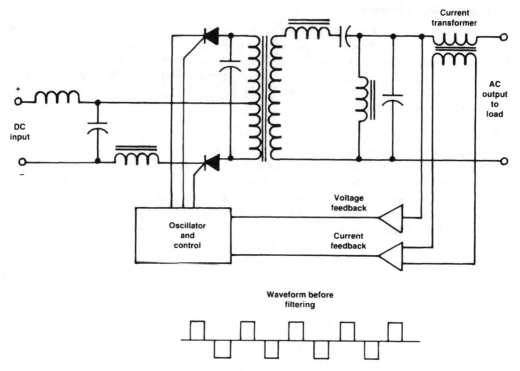

FIGURE 14.30 Simplified schematic diagram of a quasi-square wave inverter.

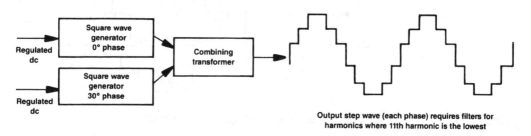

FIGURE 14.31 Static UPS system using a stepped wave output. (After [2].)

Pulse-Width Modulation Inverter

Illustrated in Fig. 14.34, the pulse-width modulation (PWM) circuit incorporates two inverters, which regulate the output voltage by varying the pulse width. The output closely resembles a sine wave. Reduced filtering requirements result in good voltage-regulation characteristics. Response times close to 100 ms are typical. The extra inverter and feedback networks make the PWM inverter complex and expensive. Such systems usually are found at power levels greater than 50 kVA.

Phase Modulation Inverter

Illustrated in Fig. 14.35, this system uses DC-to-AC conversion through phase modulation of two square wave high-frequency signals to create an output waveform. The waveform then is filtered to remove the carrier signal and feed the load.

Redundant Operation

UPS systems can be configured as either a single, large power-conditioning/backup unit, or as several smaller systems arranged in a *parallel redundant* or *isolated redundant* mode. In the parallel redundant

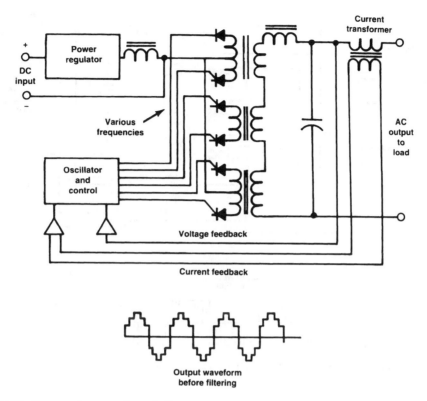

FIGURE 14.32 Step wave inverter schematic diagram.

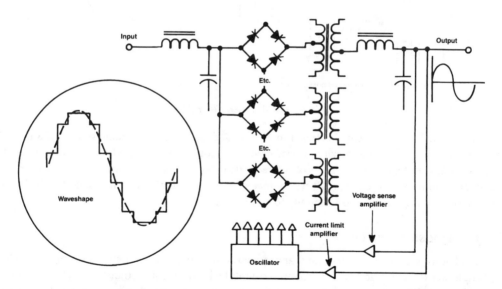

FIGURE 14.33 Step wave inverter and output waveform.

mode, the UPS outputs are connected together and share the total load. See Fig. 14.36. The power-output ratings of the individual UPS systems are selected to provide for operation of the entire load with any one UPS unit out of commission. In the event of expansion of the DP facility, additional UPS units can be added to carry the load. The parallel system provides the ability to cope with the failure of any single unit.

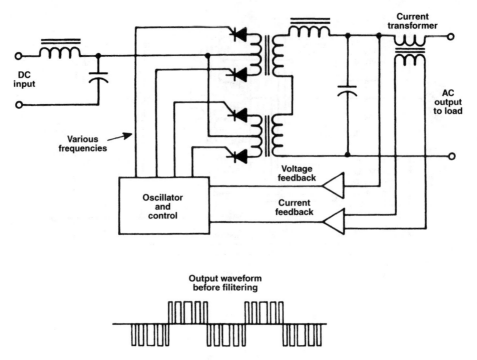

FIGURE 14.34 Simplified diagram of a pulse-width modulation inverter.

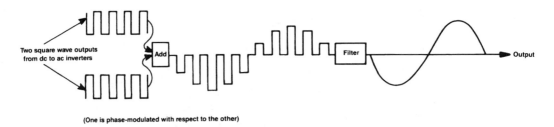

FIGURE 14.35 Static UPS using a phase-demodulated carrier. (After [2].)

An isolated redundant system, illustrated in Fig. 14.37, divides the load among several UPS units. If one of the active systems fails, a static bypass switch will connect the affected load to a standby UPS system dedicated to that purpose. The isolated redundant system does not permit the unused capacity of one UPS unit to be utilized on a DP system that is loading another UPS to full capacity. The benefit of the isolated configuration is its immunity to systemwide failures.

Output Transfer Switch

Fault conditions, maintenance operations, and system reconfiguration require the load to be switched from one power source to another. This work is accomplished with an output transfer switch. As discussed previously, most UPS systems use electronic (static) switching. Electromechanical or motor-driven relays operate too slowly for most DP loads. Static transfer switches can be configured as either of the following:

- *Break-before-make:* Power output is interrupted before transfer is made to the new source.
- *Make-before-break:* The two power sources are overlapped briefly so as to prevent any interruption in AC power to the load.

Figure 14.38 illustrates each approach to load switching.

Power System Protection Alternatives

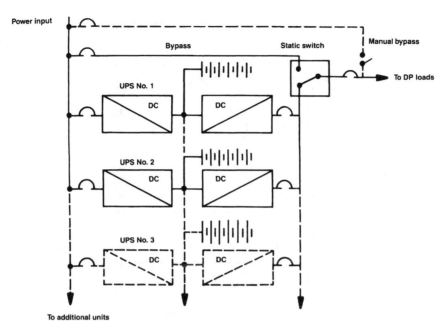

FIGURE 14.36 Configuration of a parallel redundant UPS system. (After [2].)

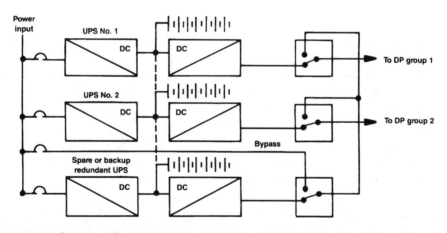

FIGURE 14.37 Configuration of an isolated redundant UPS system. (After [2].)

For critical-load applications, a make-before-break transfer is necessary. For the switchover to be accomplished with minimum disturbance to the load, both power sources must be synchronized. The UPS system must, therefore, be capable of synchronizing to the utility AC power line (or other appropriate power source).

Battery Supply

UPS systems typically are supplied with sufficient battery capacity to carry a DP load for periods ranging from 5 min to 1 hour. Long backup time periods usually are handled by a standby diesel generator. Batteries require special precautions. For large installations, they almost always are placed in a room dedicated to that purpose. Proper temperature control is important for long life and maximum discharge capacity.

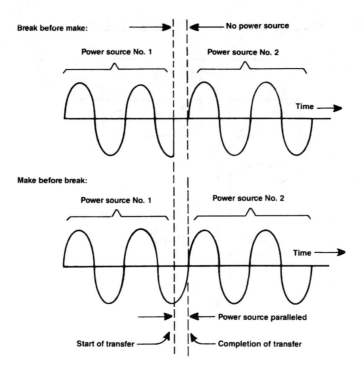

FIGURE 14.38 Static transfer switch modes.

Most rectifier/charger circuits operate in a *constant-current* mode during the initial charge period, and automatically switch to a *constant-voltage* mode near the end of the charge cycle. This provides maximum battery life consistent with rapid recharge. It also prevents excessive battery outgassing and water consumption. The charger provides a *float voltage level* for maintaining the normal battery charge, and sometimes a higher voltage to equalize certain devices.

Four battery types typically are found in UPS systems:

- *Semisealed lead calcium.* A gel-type electrolyte is used that does not require the addition of water. There is no outgassing or corrosion. This type of battery is used when the devices are integral to small UPS units, or when the batteries must be placed in occupied areas. The life span of a semisealed lead calcium battery, under ideal conditions, is about 5 years.
- *Conventional lead calcium.* The most common battery type for UPS installations, these units require watering and terminal cleaning about every 6 months. Expected lifetime ranges up to 20 years. Conventional lead-calcium batteries outgas hydrogen under charge conditions and must be located in a secure, ventilated area.
- *Lead-antimony.* The traditional lead-acid batteries, these devices are equivalent in performance to lead-calcium batteries. Maintenance is required every 3 months. Expected lifetime is about 10 years. To retain their capacity, lead-antimony batteries require a monthly equalizing charge.
- *Nickel-cadmium.* Advantages of the nickel-cadmium battery include small size and low weight for a given capacity. These devices offer excellent high- and low-temperature properties. Life expectancy is nearly that of a conventional lead-calcium battery. Nickel-cadmium batteries require a monthly equalizing charge, as well as periodic discharge cycles to retain their capacity. Nickel-cadmium batteries are the most expensive of the devices typically used for UPS applications.

14.4 Dedicated Protection Systems

A wide variety of power-protection technologies are available to solve specific problems at a facility. The method chosen depends upon a number of factors, not the least of which is cost. Although UPS units and m-g sets provide a high level of protection against AC line disturbances, the costs of such systems are high. Many times, adequate protection can be provided using other technologies at a fraction of the cost of a full-featured, facility-wide system. The applicable technologies include:

- Ferroresonant transformer
- Isolation transformer
- Tap-changing regulator
- Line conditioner

Ferroresonant Transformer

Ferroresonant transformers exhibit unique voltage-regulation characteristics that have proved valuable in a wide variety of applications. Voltage output is fixed by the size of the core, which saturates each half cycle, and by the turns ratio of the windings. Voltage output is determined at the time of manufacture and cannot be adjusted. The secondary circuit resonance depends upon capacitors, which work with the saturating inductance of the device to keep the resonance active. A single-phase ferroresonant transformer is shown in Fig. 14.39.

Load currents tend to demagnetize the core, so output current is automatically limited. A ferroresonant transformer typically cannot deliver more than 125 to 150% of its full-load rated output without going into a current-limiting mode. Such transformers cannot support the normal starting loads of motors without a significant dip in output voltage.

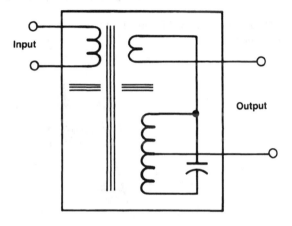

FIGURE 14.39 Basic design of a ferroresonant transformer. (After [2].)

Three-phase versions consisting of single-phase units connected in delta-delta, delta-wye, or wye-wye can be unstable when working into unbalanced loads. Increased stability for three-phase operation can be achieved with zigzag and other special winding configurations.

Shortcomings of the basic ferroresonant transformer have been overcome in advanced designs intended for low-power (2 kVA and below) voltage-regulator applications for DP equipment. Figure 14.40 illustrates one of the more common regulator designs. Two additional windings are included:

- *Compensating winding* W_c, which corrects for minor flux changes that occur after saturation has been reached.
- *Neutralizing winding* (W_n), which cancels out most of the harmonic content of the output voltage. Without some form of harmonic reduction, the basic ferroresonant transformer would be unsuitable for sensitive DP loads.

A unique characteristic of the ferroresonant regulator is its ability to reduce normal-mode impulses. Because the regulating capability of the device is based on driving the secondary winding into saturation, transients and noise bursts are clipped, as illustrated in Fig. 14.41.

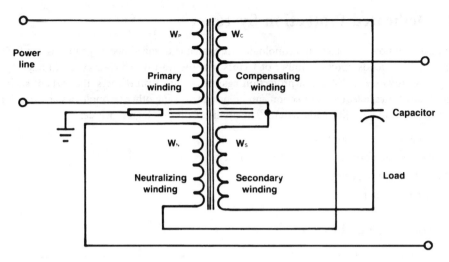

FIGURE 14.40 Improved ferroresonant transformer design incorporating a compensating winding and a neutralizing winding.

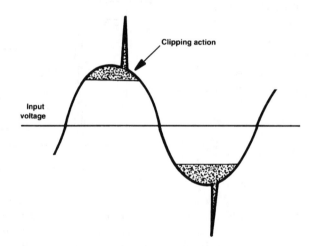

FIGURE 14.41 Clipping characteristics of a ferroresonant transformer.

The typical ferroresonant regulator has a response time of about 25 ms. Because of the tuned circuit at the output, the ferroresonant regulator is sensitive to frequency changes. A 1% frequency change will result (typically) in a 1.5% change in output voltage. Efficiency of the device is about 90% at full rated load, and efficiency declines as the load is reduced. The ferroresonant regulator is sensitive to leading and lagging power factors, and exhibits a relatively high output impedance.

Magnetic-Coupling-Controlled Voltage Regulator

An electronic magnetic regulator uses DC to adjust the output voltage through magnetic saturation of the core. The direct-current fed winding changes the saturation point of the transformer. This action, in turn, controls the AC flux paths through boost or buck coils to raise or lower the output voltage in response to an electronic circuit that monitors the output of the device. A block diagram of the magnetic-coupling-controlled voltage regulator is shown in Fig. 14.42. Changes in output voltage are smooth, although the typical response time of 5 to 10 cycles is too slow to prevent surge and sag conditions from reaching the load.

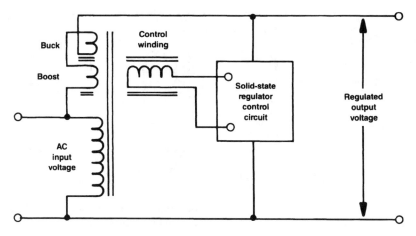

FIGURE 14.42 Block diagram of an electronic magnetic regulator. (After [2].)

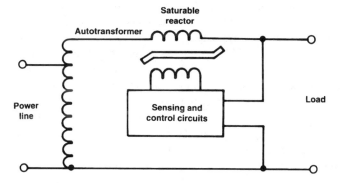

FIGURE 14.43 Autotransformer/saturable reactor voltage regulator.

Figure 14.43 illustrates an electromechanical version of a saturable regulating transformer. Output voltage is controlled by varying the impedance of the saturable reactor winding in series with a step-up autotransformer. Response to sag and surge conditions is 5 to 10 cycles. Such devices usually exhibit high output impedance and are sensitive to lagging load power factor.

Isolation Transformer

Transients, as well as noise (RF and low-level spikes) can pass through transformers, not only by way of the magnetic lines of flux between the primary and the secondary, but through resistive and capacitive paths between the windings as well. There are two basic types of noise signals with which transformer designers must cope:

- *Common-mode* noise. Unwanted signals in the form of voltages appearing between the local ground reference and each of the power conductors, including neutral and the equipment ground.
- *Normal-mode* noise. Unwanted signals in the form of voltages appearing in line-to-line and line-to-neutral signals.

Increasing the physical separation of the primary and secondary windings will reduce the resistive and capacitive coupling. However, it will also reduce the inductive coupling and decrease power transfer.

A better solution involves shielding the primary and secondary windings from each other to divert most of the primary noise current to ground. This leaves the inductive coupling basically unchanged.

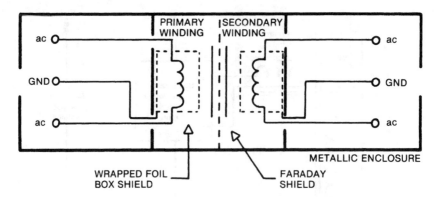

FIGURE 14.44 The shielding arrangement used in a high-performance isolation transformer. The design goal of this mechanical design is high common-mode and normal-mode noise attenuation.

The concept can be carried a step further by placing the primary winding in a shielding box that shunts noise currents to ground and reduces the capacitive coupling between the windings.

One application of this technology is shown in Fig. 14.44, in which transformer noise decoupling is taken a step further by placing the primary and secondary windings in their own wrapped foil box shields. The windings are separated physically as much as possible for the particular power rating and are placed between Faraday shields. This gives the transformer high noise attenuation from the primary to the secondary, and from the secondary to the primary. Figure 14.45 illustrates the mechanisms involved. Capacitances between the windings, and between the windings and the frame, are broken into smaller capacitances and shunted to ground, thus minimizing the overall coupling. The interwinding capacitance of a typical transformer using this technique ranges from 0.001 pF to 0.0005 pF. Common-mode noise attenuation is generally in excess of 100 dB. This high level of attenuation prevents common-mode impulses on the power line from reaching the load. Figure 14.46 illustrates how an isolation transformer combines with the AC-to-DC power supply to prevent normal-mode noise impulses from affecting the load.

A 7.5-kVA noise-suppression isolation transformer is shown in Fig. 14.47. High-quality isolation transformers are available in sizes ranging from 125 VA single-phase to 125 kVA (or more) three-phase. Usually, the input winding is tapped at 2.5% intervals to provide the rated output voltage, despite high or low average input voltages. The total tap adjustment range is typically from 5% above nominal to 10% below nominal. For three-phase devices, typical input-voltage nominal ratings are 600, 480, 240, and 208 V line-to-line for 15 kVA and larger.

Tap-Changing Regulator

The concept behind a tap-changing regulator is simple: adjust the transformer input voltage to compensate for AC line-voltage variations. A tap-changing regulator is shown in Fig. 14.48. Although simple in concept, the actual implementation of the system can become complex because of the timing waveforms and pulses necessary to control the SCR banks. Because of the rapid response time of SCRs, voltage adjustment can be made on a cycle-by-cycle basis in response to changes in both the utility input and the load. Tap steps are typically 2 to 3%. Such systems create no objectionable switching transients in unity power factor loads. For low-PF loads, however, small but observable transients can be generated in the output voltage at the moment the current is switched. This noise usually has no significant effect on DP loads. Other operating characteristics include:

- Low-internal impedance (similar to an equivalent transformer)
- High efficiency from full load to 25% or less

Power System Protection Alternatives

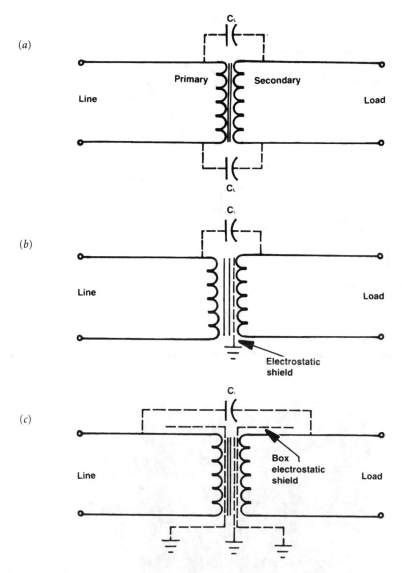

FIGURE 14.45 The elements involved in a noise-suppression isolation transformer: (*a*) conventional transformer with capacitive coupling as shown; (*b*) the addition of an electrostatic shield between the primary and the secondary; (*c*) transformer with electrostatic box shields surrounding the primary and secondary windings.

- Rapid response time (typically one to three cycles) to changes in the input AC voltage or load current
- Low acoustic noise level

An autotransformer version of the tap-changing regulator is shown in Fig. 14.49.

Variable Ratio Regulator

Functionally, the variable ratio regulator is a modified version of the tap-changer. Rather than adjusting output voltage in steps, the motor-driven regulator provides a continuously variable range of voltages. The basic concept is shown in Fig. 14.50. The system is slow, but generally reliable. It is excellent for keeping DP hardware input voltages within the optimum operating range. Motor-driven regulators usually

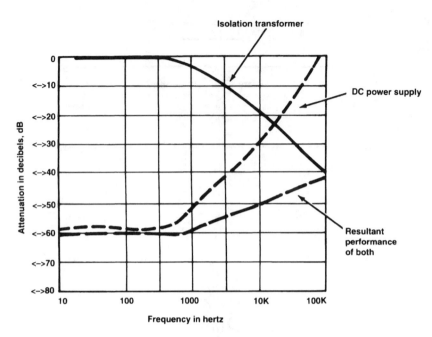

FIGURE 14.46 How the normal-mode noise attenuation of an isolation transformer combines with the filtering characteristics of the AC-to-DC power supply to prevent noise propagation to the load.

FIGURE 14.47 A 7.5-kVA noise-suppression isolation transformer. (Courtesy of Topaz.)

are able to follow the steady rise and fall of line voltages that typically are experienced on utility company lines. Efficiency is normally good, approaching that of a good transformer. The internal impedance is low, making it possible to handle a sudden increase or decrease in load current without excessive undervoltages and/or overvoltages. Primary disadvantages of the variable ratio regulator are limited current ratings, determined by the moving brush assembly, and the need for periodic maintenance.

Variations on the basic design have been marketed, including the system illustrated in Fig. 14.51. A motor-driven brush moves across the exposed windings of an autotransformer, causing the series transformer to buck or boost voltage to the load. The correction motor is controlled by a voltage-sensing circuit at the output of the device.

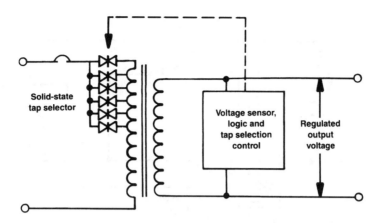

FIGURE 14.48 Simplified schematic diagram of a tap-changing voltage regulator. (After [2].)

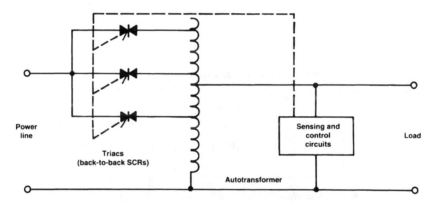

FIGURE 14.49 Tap-changing voltage regulator using an autotransformer as the power-control element.

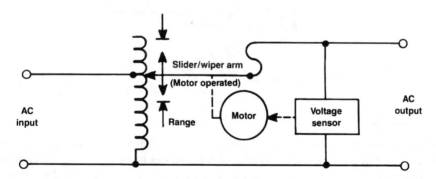

FIGURE 14.50 Variable ratio voltage regulator. (After [2].)

The induction regulator, shown in Figure 14.52, is still another variation on the variable ratio transformer. Rotation of the rotor in one direction or the other varies the magnetic coupling and raises or lowers the output voltage. Like the variable ratio transformer, the induction regulator is slow, but it has no brushes and requires little maintenance. The induction regulator has higher inductive reactance and is slightly less efficient than the variable ratio transformer.

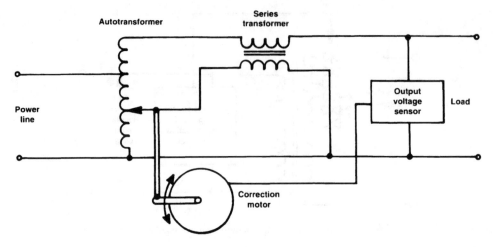

FIGURE 14.51 Motor-driven line-voltage regulator using an autotransformer with a buck/boost series transformer.

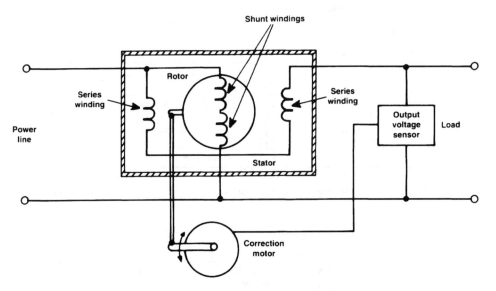

FIGURE 14.52 Rotary induction voltage regulator.

Variable Voltage Transformer

Because of their application in voltage control systems, it is worthwhile to discuss the operation of variable voltage transformers in greater detail. A number of conventional transformers can have—to a limited degree—their primary/secondary ratio changed through the use of taps located (usually) on the primary windings [6]. To obtain a greater degree of flexibility in changing the ratio between the primary and secondary coils, and thus allow greater secondary voltage changes, a variable voltage transformer is used. The amount of voltage change depends upon the basic construction of the device, which typically divides into one of two categories:

- Brush type
- Induction type

Brush Type

To achieve a variable output voltage, one tap on the transformer secondary is fixed and the other tap is connected to a brush that slides along an uninsulated section of the transformer coil [6]. One way of accomplishing this objective is to have the coil wrapped around a toroidal shaped core. The voltage ratio relates to the location of the brush as it rides against the coil and depends upon where on the coil the brush is allowed to make contact. *Limited ratio* variable transformers are available, as well as full range units that can adjust the output voltage from 0 to about 120% of the incoming line voltage. When the output voltage exceeds the input voltage, it means that there are extra turns on the coil that extend beyond the windings that lie between the incoming power terminals (in effect, the unit becomes a step-up transformer).

Ratings start at less than 1 kVA for a 120 V, single-phase variable transformer. Basic units are wired in parallel and/or in series to obtain higher power capacity. Two units in parallel have double the current and kVA rating. The individual units are stacked on top of each other, bolted together, and operated from a common shaft that rotates the brushes. Operating in a configuration that combines parallel and series connections, with several units stacked together, this type of variable voltage transformer, at 480 V, can have a rating exceeding 200 kVA. Sometimes, a control mechanism is attached to the rotor that turns the brushes, allowing automatic adjustment of the output voltage.

An important characteristic of this type of transformer relates to brush contact and the amount of current flowing through the carbon brush. A rating based solely on output kVA can cause serious operational problems because, for a given kVA load, the current drawn depends on the output voltage. Because the output voltage is variable, a load of a given kVA value can draw a safe current at 100% voltage, but at 25% voltage, the current required to serve a load of the same kVA value would require four times as much current, which could cause the brush to overheat.

Induction Type

The induction type variable transformer does not use brushes [6]. The usual voltage change for these units is ±10%, but it can be greater. The device is essentially a variable-ratio autotransformer that uses two separate windings, a primary and a secondary. There is a laminated steel stator on which is wound a winding that serves as a secondary coil. This winding is connected in series with the load. The primary coil is connected across the supply line. The shunt winding is wound around a rotor. Construction is similar to that of a motor except that in this case, the rotor can only turn 180 mechanical and electrical degrees.

As the primary core is rotated, the amount of primary flux passing through the secondary winding is decreased until the core reaches a position at right angles to the secondary winding. In this position, no primary flux passes through the secondary windings and the induced voltage in the coil is zero. The continued rotation of the core in the same direction again increases the amount of flux passing through the secondary, but it is now in the opposite direction and so reverses the direction of the induced voltage. Thus, the output voltage can be varied by adding or subtracting from it the voltage induced in the secondary winding.

Both single and 3-phase transformers are available. Ratings of these types of transformers vary from 8 kVA, 120 V single-phase to 1500 kVA, 480V 3-phase.

Line Conditioner

A line conditioner combines the functions of an isolation transformer and a voltage regulator in one unit. The three basic types of line conditioners for DP applications are:

- *Linear amplifier correction system.* As illustrated in Fig. 14.53, correction circuitry compares the AC power output to a reference source, derived from a 60-Hz sine wave generator. A correction voltage is developed and applied to the secondary power winding to cancel noise and voltage fluctuations. A box shield around the primary winding provides common-mode impulse rejection

(80 to 100 dB typical). The linear amplifier correction system is effective and fast, but the overall regulating range is limited.

- *Hybrid ferroresonant transformer.* As shown in Fig. 14.54, this system consists of a ferroresonant transformer constructed using the isolation techniques discussed in the section titled "Isolation Transformer." The box and Faraday shields around the primary and compensating windings give the transformer essentially the characteristics of a noise-attenuating device, while preserving the voltage-regulating characteristics of a ferroresonant transformer.
- *Electronic tap-changing, high-isolation transformer.* This system is built around a high-attenuation isolation transformer with a number of primary winding taps. SCR pairs control voltage input to each tap, as in a normal tap-changing regulator. Tap changing can also be applied to the secondary, as shown in Fig. 14.55. The electronic tap-changing, high-isolation transformer is an efficient design that effectively regulates voltage output and prevents noise propagation to the DP load.

Hybrid Transient Suppressor

A wide variety of AC power-conditioning systems are available, based on a combination of solid-state technologies. Most incorporate a combination of series and parallel elements to shunt transient energy. One such system is pictured in Fig. 14.56.

Active Power Line Conditioner

The *active power line conditioner* (APLC) provides an adaptive solution to many power quality problems [7]. The major features of APLC include:

- Near-instantaneous voltage regulation
- Source voltage harmonic compensation
- Load current harmonic cancellation
- Distortion power factor correction

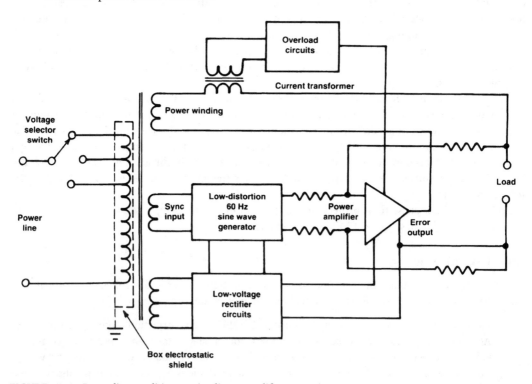

FIGURE 14.53 Power-line conditioner using linear amplifier correction.

The basic concept of the APLC can be expressed in mathematical terms. Figure 14.57 shows a block diagram of the APLC in a simplified power system. As shown, the system has two primary components:

- A component in series with the harmonic sensitive load, which controls the voltage sine wave being supplied by the power system.
- A component in parallel with the harmonic sensitive load, which controls the harmonic current going to the load.

The voltage that is supplied by the power distribution system (V_{in}) contains two components: (1) the fundamental 60 Hz frequency component represented by V_f and (2) the harmonic component of the source voltage represented by V_h. By definition, V_h is any portion of the source voltage that is not of the 60 Hz fundamental frequency. The APLC modifies this incoming voltage by adding two more voltage components:

- The V_h component, which is always equal to the V_h component of V_{in} but of the opposite polarity
- The V_r component, which is the amount of buck or boost required to provide regulation of the output voltage.

Because the two V_h components cancel, the load receives a regulated sinusoidal voltage at the fundamental frequency with no harmonic components.

The load uses current (I_L) in a nonlinear fashion. The harmonic portion of the current I_h is—by definition—everything that is not the fundamental frequency component. The APLC becomes the source of I_h for the load circuit, leaving only the fundamental frequency portion (I_f) to be supplied by the power system. Because the harmonic portion of the load current is supplied locally by the APLC, the distortion power factor of the load is near unity.

A digital signal processor (DSP) is used to monitor the input and output parameters of the APLC. The DSP then makes any needed adjustment in the amount of voltage compensation and current injection by modifying the switching characteristics of the insulated-gate bipolar transistors (IGBTs) in the APLC.

A simplified schematic of the APLC power circuitry is given in Fig. 14.58. The DC bus (V_{dc}) is the energy storage device that supplies the power required for voltage compensation and current injection. The parallel "filter" serves two purposes. First, the IGBTs in this portion of the APLC keep the DC link capacitor charged. Second, it is used as the source for the harmonic currents needed by the load. By using power from the DC link, the series "filter" compensates for input voltage fluctuations through the buck/boost transformer.

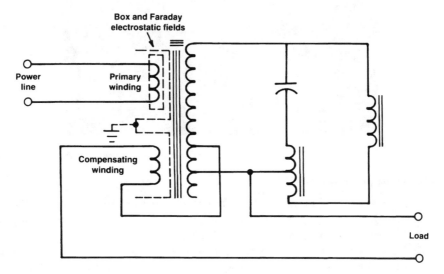

FIGURE 14.54 Line conditioner built around a shielded-winding ferroresonant transformer.

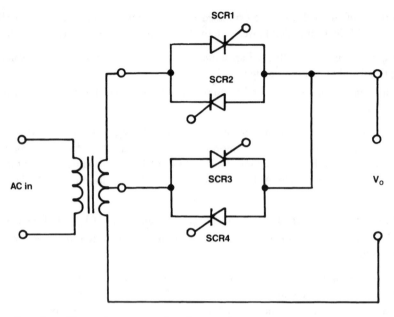

FIGURE 14.55 Secondary-side synchronous tap-changing transformer.

FIGURE 14.56 Interior view of a high-capacity hybrid power conditioner. (Courtesy of Control Concepts.)

Application Considerations

In a conventional power control system system, voltage regulation is performed by using either ferroresonant transformers or tap switching transformers. The fastest response times for either of these solutions is about 1/2 cycle (8 ms). This response time is the result of the tap switching transformer's need for a zero crossing of the waveform to make a change. The use of IGBTs in the APLC allows full power to be switched at any point on the sine wave. Combining these high-speed power transistors with a fast digital signal processor, the voltage can be regulated in less than 1 ms [7]. Furthermore, the input voltage can

Power System Protection Alternatives

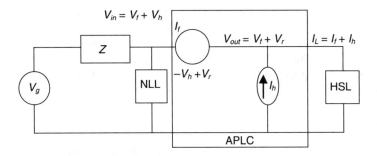

Z = source and line impedance
NLL= non-linear load(s)
V_h = harmonic voltage across Z I_h = harmonic load current
V_g = generator voltage I_f = fundamental load current
V_r = voltage required to regulate V_{out} I_l = load current
V_f = fundamental input voltage   HSL = harmonic sensitive load(s)

FIGURE 14.57 Basic conceptual diagram of the active power line conditioner. (After [7].)

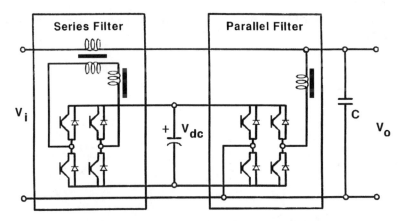

FIGURE 14.58 Series/parallel active power line conditioner power control circuit. (After [7].)

exhibit up to 10% total harmonic distortion (THD), but output of the APLC to the load will typically see a voltage waveform with less than 1.0% THD [7].

As discussed previously, a switching power supply requires current in a nonlinear fashion. These load currents may have significant harmonic components. The APLC can cancel up to 100% load current THD at full load. The source current being supplied by the power distribution system will see a purely linear load with no more than 3% current THD [7].

References

1. Key, T. "The Effects of Power Disturbances on Computer Operation," IEEE Industrial and Commercial Power Systems Conference, Cincinnati, June 7, 1978.
2. Federal Information Processing Standards Publication No. 94, *Guideline on Electrical Power for ADP Installations*, U.S. Department of Commerce, National Bureau of Standards, Washington, D.C., 1983.
3. Plater, B. B., and J. A. Andrews, "Advances in Flywheel Energy Storage Systems," *Power Quality '97*, Intertec International, Ventura, CA, pp. 460–469, 1997.
4. Materials Selector 1987, *Materials Engineering*, Penton Publishing.

5. Weaver, E., J., "Dynamic Energy Storage System," *Power Quality Solutions/Alternative Energy*, Intertec International, Ventura, CA, pp. 373–380, 1996.
6. Berutti, A., *Practical Guide to Applying, Installing and Maintaining Transformers*, Robert B. Morgan (Ed.), Intertec Publishing, Overland Park, KS, p. 13, 1994.
7. Petrecca, R. M., "Mitigating the Effects of Harmonics and Other Distrubances with an Active Power Line Conditioner," *Power Quality Solutions*, Intertec International, Ventura, CA, 1997.

Further Information

"How to Correct Power Line Disturbances," Dranetz Technologies, Edison, NJ, 1985.

Lawrie, R., *Electrical Systems for Computer Installations*, McGraw-Hill, New York, 1988.

Martzloff, F. D., "The Development of a Guide on Surge Voltages in Low-Voltage AC Power Circuits," *14th Electrical/Electronics Insulation Conference*, IEEE, Boston, October 1979.

Newman, P., "UPS Monitoring: Key to an Orderly Shutdown," *Microservice Management*, Intertec Publishing, Overland Park, KS, March 1990.

Noise Suppression Reference Manual, Topaz Electronics, San Diego, CA.

Nowak, S., "Selecting a UPS," *Broadcast Engineering*, Intertec Publishing, Overland Park, KS, April 1990.

Pettinger, W., "The Procedure of Power Conditioning," *Microservice Management*, Intertec Publishing, Overland Park, KS, March 1990.

Smeltzer, D., "Getting Organized About Power," *Microservice Management*, Intertec Publishing, Overland Park, KS, March 1990.

15
Facility Grounding

15.1 Introduction 15-1
   Terms and Codes
15.2 Establishing an Earth Ground 15-6
   Grounding Interface • Chemical Ground Rods • Ufer Ground System • Bonding Ground-System Elements • Exothermic Bonding • Ground-System Inductance • Grounding Tower Elements • Ground-Wire Dressing • Facility Ground Interconnection • Grounding on Bare Rock
15.3 Transmission-System Grounding 15-24
   Transmission Line • Satellite Antenna Grounding
15.4 Designing a Building Ground System 15-26
   Bulkhead Panel • Lightning Protectors • Checklist for Proper Grounding
15.5 AC System Grounding Practices 15-38
   Building Codes • Single-Point Ground • Isolated Grounding • Separately Derived Systems • Grounding Terminology • Facility Ground System • Power-Center Grounding • Grounding Equipment Racks
15.6 Grounding Signal-Carrying Cables 15-50
   Analyzing Noise Currents • Types of Noise • Noise Control • Patch-Bay Grounding • Input/Output Circuits • Cable Routing • Overcoming Ground-System Problems

Jerry C. Whitaker
Editor-in-Chief

15.1 Introduction

The attention given to the design and installation of a facility ground system is a key element in the day-to-day reliability of the plant. A well-designed and -installed ground network is invisible to the engineering staff. A marginal ground system, however, will cause problems on a regular basis. Grounding schemes can range from simple to complex, but any system serves three primary purposes:

- Provides for operator safety.
- Protects electronic equipment from damage caused by transient disturbances.
- Diverts stray radio frequency energy from sensitive audio, video, control, and computer equipment.

Most engineers view grounding mainly as a method to protect equipment from damage or malfunction. However, the most important element is operator safety. The 120 V or 208 V AC line current that powers most equipment can be dangerous—even deadly—if handled improperly. Grounding of equipment and structures provides protection against wiring errors or faults that could endanger human life.

Proper grounding is basic to protection against AC line disturbances. This applies if the source of the disturbance is lightning, power-system switching activities, or faults in the distribution network. Proper grounding is also a key element in preventing radio frequency interference in transmission or computer equipment. A facility with a poor ground system can experience RFI problems on a regular basis. Implementing an effective ground network is not an easy task. It requires planning, quality components, and skilled installers. It is not inexpensive. However, proper grounding is an investment that will pay dividends for the life of the facility.

Any ground system consists of two key elements: (1) the earth-to-grounding electrode interface outside the facility, and (2) the AC power and signal-wiring systems inside the facility.

Terms and Codes

A *facility* can be defined as something that is built, installed, or established to serve a particular purpose [1]. A facility is usually thought of as a single building or group of buildings. The National Electrical Code (NEC) uses the term *premises* to refer to a facility when it defines premises wiring as the interior and exterior (facility) wiring, such as power, lighting, control, and signal systems. Premises wiring includes the service and all permanent and temporary wiring between the service and the load equipment. Premises wiring does not include wiring internal to any load equipment.

The Need for Grounding

The Institute of Electrical and Electronic Engineers (IEEE) defines grounding as a conducting connection, whether intentional or accidental, by which an electric circuit or equipment is connected to the earth, or to some conducting body of relatively large extent that serves in place of the earth. It is used for establishing and maintaining the potential of the earth (or of the conducting body) or approximately that potential, on conductors connected to it, and for conducting ground current to and from the earth (or the conducting body) [2]. Based on this definition, the reasons for grounding can be identified as:

- Personnel safety by limiting potentials between all non-current-carrying metal parts of an electrical distribution system.
- Personnel safety and control of electrostatic discharge (ESD) by limiting potentials between all non-current-carrying metal parts of an electrical distribution system and the earth.
- Fault isolation and equipment safety by providing a low-impedance fault return path to the power source to facilitate the operation of overcurrent devices during a ground fault.

The IEEE definition makes an important distinction between *ground* and *earth*. *Earth* refers to mother earth, and *ground* refers to the equipment grounding system, which includes equipment grounding conductors, metallic raceways, cable armor, enclosures, cabinets, frames, building steel, and all other non-current-carrying metal parts of the electrical distribution system.

There are other reasons for grounding not implicit in the IEEE definition. Overvoltage control has long been a benefit of proper power system grounding, and is described in IEEE Standard 142, also known as the *Green Book* [3]. With the increasing use of electronic computer systems, noise control has become associated with the subject of grounding, and is described in IEEE Standard 1100, the *Emerald Book* [4].

Equipment Grounding

Personnel safety is achieved by interconnecting all non-current-carrying metal parts of an electrical distribution system, and then connecting the interconnected metal parts to the earth [5]. This process of interconnecting metal parts is called equipment grounding and is illustrated in Fig. 15.1, where the equipment grounding conductor is used to interconnect the metal enclosures. Equipment grounding ensures that there is no difference of potential, and thus no shock hazard between non-current-carrying metal parts anywhere in the electrical distribution system. Connecting the equipment grounding system to earth ensures that there is no difference of potential between the earth and the equipment grounding system. It also prevents static charge buildup.

Facility Grounding

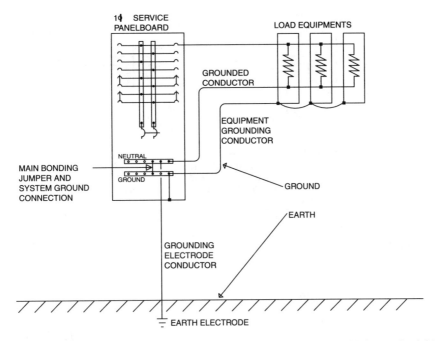

FIGURE 15.1 Equipment grounding and system grounding. (Courtesy of Intertec Publishing. After [5].)

System Grounding

System grounding, which is also illustrated in Fig. 15.1, is the process of intentionally connecting one of the current-carrying conductors of the electrical distribution system to ground [5]. The figure shows the neutral conductor intentionally connected to ground and the earth. This conductor is called the *grounded* conductor because it is intentionally grounded. The purpose of system grounding is overvoltage control and equipment safety through fault isolation. An ungrounded system is subject to serious overvoltages under conditions such as intermittent ground faults, resonant conditions, and contact with higher voltage systems. Fault isolation is achieved by providing a low-impedance return path from the load back to the source, which will ensure operation of overcurrent devices in the event of a ground fault. The system ground connection makes this possible by connecting the equipment grounding system to the low side of the voltage source. Methods of system grounding include solidly grounded, ungrounded, and impedance-grounded.

Solidly grounded means that an intentional zero-impedance connection is made between a current-carrying conductor and ground. The single-phase (1ϕ) system shown in Fig. 15.1 is solidly grounded. A solidly grounded, three-phase, four-wire, wye system is illustrated in Fig. 15.2. The neutral is connected directly to ground with no impedance installed in the neutral circuit. The NEC permits this connection to be made at the service entrance only [6]. The advantages of a solidly grounded wye system include reduced magnitude of transient overvoltages, improved fault protection, and faster location of ground faults. There is one disadvantage of the solidly grounded wye system. For low-level arcing ground faults, the application of sensitive, properly coordinated, ground fault protection (GFP) devices is necessary to prevent equipment damage from arcing ground faults. The NEC requires arcing ground fault protection at 480Y/277V services, and a maximum sensitivity limit of 1200 A is permitted. Severe damage is less frequent at the lower voltage 208V systems, where the arc may be self-extinguishing.

Ungrounded means that there is no intentional connection between a current-carrying conductor and ground. However, charging capacitance will create unintentional capacitive coupling from each phase to ground, making the system essentially a capacitance-grounded system. A three-phase, three-wire system

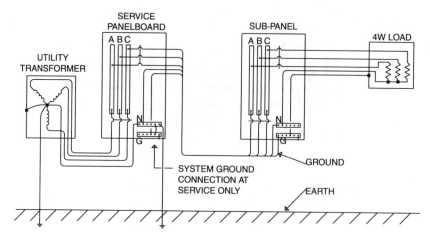

FIGURE 15.2 Solidly grounded wye power system. (Courtesy of Intertec Publishing. After [5].)

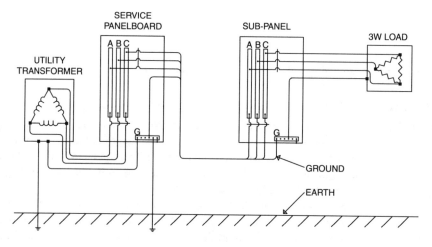

FIGURE 15.3 Ungrounded delta power system. (Courtesy of Intertec Publishing. After [5].)

from an ungrounded delta source is illustrated in Fig. 15.3. The most important advantage of an ungrounded system is that an accidental ground fault in one phase does not require immediate removal of power. This allows for continuity of service, which made the ungrounded delta system popular in the past. However, ungrounded systems have serious disadvantages. Because there is no fixed system ground point, it is difficult to locate the first ground fault and to sense the magnitude of fault current. As a result, the fault is often permitted to remain on the system for an extended period of time. If a second fault should occur before the first one is removed, and the second fault is on a different phase, the result will be a double line-to-ground fault causing serious arcing damage. Another problem with the ungrounded delta system is the occurrence of high transient overvoltages from phase-to-ground. Transient overvoltages can be caused by intermittent ground faults, with overvoltages capable of reaching a phase-to-ground voltage of from six to eight times the phase-to-neutral voltage. Sustained overvoltages can ultimately result in insulation failure and thus more ground faults.

Impedance-grounded means that an intentional impedance connection is made between a current-carrying conductor and ground. The high-resistance grounded wye system, illustrated in Fig. 15.4, is an alternative to solidly grounded and ungrounded systems. High-resistance grounding will limit ground fault current to a few amperes, thus removing the potential for arcing damage inherent in solidly grounded

Facility Grounding

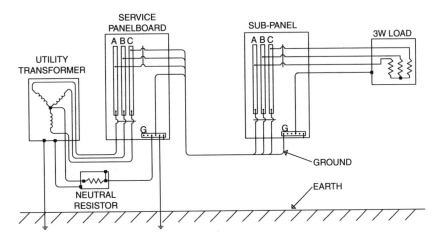

FIGURE 15.4 High-resistance grounded wye power system. (Courtesy of Intertec Publishing. After [5].)

TABLE 15.1 Comparison of System Grounding Methods

	System Grounding Method		
Characteristic Assuming No Fault Escalation	Solidly Grounded	Ungrounded	High Resistance
Operation of overcurrent device on first ground fault	Yes	No	No
Control of internally generated transient overvoltages	Yes	No	Yes
Control of steady-state overvoltages	Yes	No	Yes
Flash hazard	Yes	No	No
Equipment damage from arcing ground-faults	Yes	No	No
Overvoltage (on unfaulted phases) from ground-fault[a]	L-N Voltage	>> L-L Voltage	L-L Voltage
Can serve line-to-neutral loads	Yes	No	No

(After [3].)

[a] L = line, N = neutral

systems. The ground reference point is fixed, and relaying methods can locate first faults before damages from second faults occur. Internally generated transient overvoltages are reduced because the neutral to ground resistor dissipates any charge that may build up on the system charging capacitances.

Table 15.1 compares the three most common methods of system grounding. There is no one *best* system grounding method for all applications. In choosing among the various options, the designer must consider the requirements for safety, continuity of service, and cost. Generally, low-voltage systems should be operated solidly grounded. For applications involving continuous processes in industrial plants or where shutdown might create a hazard, a high-resistance grounded wye system, or a solidly grounded wye system with an alternate power supply, may be used. The high-resistance grounded wye system combines many of the advantages of the ungrounded delta system and the solidly grounded wye system. IEEE Standard 142 recommends that medium-voltage systems less than 15 kV be low-resistance grounded to limit ground fault damage yet permit sufficient current for detection and isolation of ground faults. Standard 142 also recommends that medium-voltage systems over 15 kV be solidly grounded. Solid grounding should include sensitive ground fault relaying in accordance with the NEC.

The Grounding Electrode

The process of connecting the grounding system to earth is called *earthing*, and consists of immersing a metal electrode or system of electrodes into the earth [5]. The conductor that connects the grounding system to earth is called the *grounding electrode conductor*. The function of the grounding electrode conductor is to keep the entire grounding system at earth potential (i.e., voltage equalization during lightning and other transients) rather than for conducting ground-fault current. Therefore, the NEC allows reduced sizing requirements for the grounding electrode conductor when connected to *made* electrodes.

The basic measure of effectiveness of an earth electrode system is called *earth electrode resistance*. Earth electrode resistance is the resistance, in ohms, between the point of connection and a distant point on the earth called *remote earth*. Remote earth, about 25 ft from the driven electrode, is the point where earth electrode resistance does not increase appreciably when this distance is increased. Earth electrode resistance consists of the sum of the resistance of the metal electrode (negligible) plus the contact resistance between the electrode and the soil (negligible) plus the soil resistance itself. Thus, for all practical purposes, earth electrode resistance equals the soil resistance. The soil resistance is nonlinear, with most of the earth resistance contained within several feet of the electrode. Furthermore, current flows only through the electrolyte portion of the soil, not the soil itself. Thus, soil resistance varies as the electrolyte content (moisture and salts) of the soil varies. Without electrolyte, soil resistance would be infinite.

Soil resistance is a function of soil resistivity. A 1 m³ sample of soil with a resistivity ρ of 1 Ωm will present a resistance R of 1 Ω between opposite faces. A broad variation of soil resistivity occurs as a function of soil types, and soil resistivity can be estimated or measured directly. Soil resistivity is usually measured by injecting a known current into a given volume of soil and measuring the resulting voltage drop. When soil resistivity is known, the earth electrode resistance of any given configuration (single rod, multiple rods, or ground ring) can be determined by using standard equations developed by Sunde [6], Schwarz [7], and others.

Earth resistance values should be as low as practicable, but are a function of the application. The NEC approves the use of a single *made* electrode if the earth resistance does not exceed 25 Ω. IEEE Standard 1100 reports that the very low earth resistance values specified for computer systems in the past are not necessary. Methods of reducing earth resistance values include the use of multiple electrodes in parallel, the use of ground rings, increased ground rod lengths, installation of ground rods to the permanent water level, increased area of coverage of ground rings, and the use of concrete-encased electrodes, ground wells, and electrolytic electrodes.

Earth Electrode

Earth electrodes may be *made* electrodes, *natural* electrodes, or *special-purpose* electrodes [5]. Made electrodes include driven rods, buried conductors, ground mats, buried plates, and ground rings. The electrode selected is a function of the type of soil and the available depth. Driven electrodes are used where bedrock is 10 ft or more below the surface. Mats or buried conductors are used for lesser depths. Buried plates are not widely used because of the higher cost when compared to rods. Ground rings employ equally spaced driven electrodes interconnected with buried conductors. Ground rings are used around large buildings, around small unit substations, and in areas having high soil resistivity.

Natural electrodes include buried water pipe electrodes and concrete-encased electrodes. The NEC lists underground metal water piping, available on the premises and not less than 10 ft in length, as part of a preferred grounding electrode system. Because the use of plastic pipe in new water systems will impair the effectiveness of water pipe electrodes, the NEC requires that metal underground water piping be supplemented by an additional approved electrode. Concrete below ground level is a good electrical conductor. Thus, metal electrodes encased in such concrete will function as excellent grounding electrodes. The application of concrete-encased electrodes is covered in IEEE Standard 142.

15.2 Establishing an Earth Ground

The grounding electrode is the primary element of any ground system. The electrode can take many forms. In all cases, its purpose is to interface the electrode (a conductor) with the earth (a semiconductor). Grounding principles have been refined to a science. Still, however, many misconceptions exist regarding grounding. An understanding of proper grounding procedures begins with the basic earth-interface mechanism.

Grounding Interface

The grounding electrode (or ground rod) interacts with the earth to create a hemisphere-shaped volume, as illustrated in Fig. 15.5. The size of this volume is related to the size of the grounding electrode. The length of the electrode has a much greater effect than the diameter. Studies have demonstrated that the earth-to-electrode resistance from a driven ground rod increases exponentially with the distance from that rod. At a given point, the change becomes insignificant. It has been found that for maximum effectiveness of the earth-to-electrode interface, each ground rod requires a hemisphere-shaped volume with a diameter that is approximately 2.2 times the rod length [9].

The constraints of economics and available real estate place practical limitations on the installation of a ground system. It is important, however, to keep the 2.2 rule in mind because it allows the facility design engineer to use the available resources to the best advantage. Figure 15.6 illustrates the effects of locating ground rods too close (less than 2.2 times the rod length). An overlap area is created that effectively wastes some of the earth-to-electrode capabilities of the two ground rods. Research has shown, for example, that two 10-ft ground rods driven only 1 ft apart provide about the same resistivity as a single 10-ft rod.

There are three schools of thought with regard to ground-rod length. The first approach states that extending ground-rod length beyond about 10 ft is of little value for most types of soil. The reasoning behind this conclusion is presented in Fig. 15.7a, where ground resistance is plotted as a function of ground-rod length. Beyond 10 ft in length, a point of diminishing returns is reached.

The second school of thought concludes that optimum earth-to-electrode interface is achieved with long (40 ft or greater) rods, driven to penetrate the local water table. When planning this type of installation, consider the difficulty that may be encountered when attempting to drive long ground rods. The foregoing discussion assumes that the soil around the grounding electrode is reasonably uniform in composition. Depending upon the location, however, this assumption may not hold true.

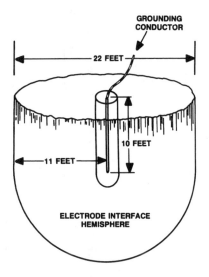

FIGURE 15.5 The effective earth-interface hemisphere resulting from a single driven ground rod. The 90% effective area of the rod extends to a radius of approximately 1.1 times the length of the rod. (After [10].)

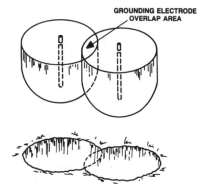

FIGURE 15.6 The effect of overlapping earth interface hemispheres by placing two ground rods at a spacing less than 2.2 times the length of either rod. The overlap area represents wasted earth-to-grounding electrode interface capability. (After [10].)

The third school of thought concludes that the optimum ground rod installation is achieved by using the longest possible rod depth (length). Data to support this conclusion is given in Fig. 15.7b, which plots ground rod performance as a function of depth in soil of uniform resistivity. This curve does not take into account seasonal moisture content, changing chemical composition at different soil layers, and frozen soil conditions.

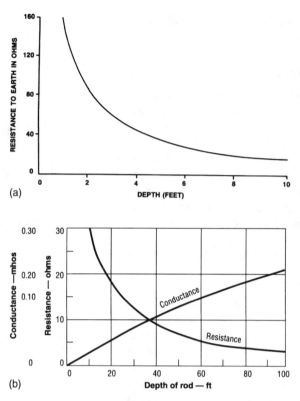

FIGURE 15.7 Charted grounding resistance as a function of ground-rod length: (a) data demonstrating that ground-rod length in excess of 10 ft produces diminishing returns (1-in.-diameter rod) [10], (b) data demonstrating that ground system performance continues to improve as depth increases. (Chart b courtesy of Intertec Publishing. After [12].)

Given these three conflicting approaches, the benefits of hiring an experienced, licensed professional engineer to design the ground system can be readily appreciated.

Horizontal grounding electrodes provide essentially the same resistivity as an equivalent-length vertical electrode, given uniform soil conditions. As Fig. 15.8 demonstrates, the difference between a 10-ft vertical and a 10-ft horizontal ground rod is negligible (275 Ω vs. 250 Ω). This comparison includes the effects of the vertical connection element from the surface of the ground to the horizontal rod. Taken by itself, the horizontal ground rod provides an earth-interface resistivity of approximately 308 Ω when buried at a depth of 36 in.

Ground rods come in many sizes and lengths. The more popular sizes are $1/2$, $5/8$, $3/4$, and 1 in. The $1/2$-in. size is available in steel with stainless-clad, galvanized, or copper-clad rods. All-stainless-steel rods also are available. Ground rods can be purchased in unthreaded or threaded (sectional) lengths. The sectional sizes are typically $9/16$-in. or $1/2$-in. rolled threads. Couplers are made from the same materials as the rods. Couplers can be used to join 8- or 10-ft length rods together. A 40-ft ground rod, for example, is driven one 10-ft section at a time.

The type and size of ground rod used is determined by how many sections are to be connected and how hard or rocky the soil is. Copper-clad $5/8$-in. × 10-ft rods are probably the most popular. Copper cladding is designed to prevent rust. The copper is not primarily to provide better conductivity. Although the copper certainly provides a better conductor interface to earth, the steel that it covers is also an excellent conductor when compared with ground conductivity. The thickness of the cladding is important only insofar as rust protection is concerned.

Facility Grounding

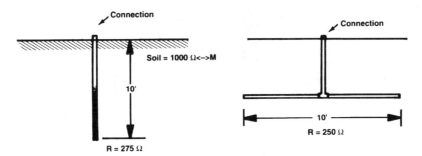

FIGURE 15.8 The effectiveness of vertical ground rods compared with horizontal ground rods. (After [10].)

TABLE 15.2 Typical Resistivity of Common Soil Types

Type of Soil Resistivity in Ω/cm	Average	Minimum	Maximum
Filled land, ashes, salt marsh	2,400	600	7,000
Top soils, loam	4,100	340	16,000
Hybrid soils	6,000	1,000	135,000
Sand and gravel	90,000	60,000	460,000

Wide variations in soil resistivity can be found within a given geographic area, as documented in Table 15.2. The wide range of values shown results from differences in moisture content, mineral content, and temperature.

Temperature is a major concern in shallow grounding systems because it has a significant effect on soil resistivity [11]. During winter months, the ground system resistance can rise to unacceptable levels because of freezing of liquid water in the soil. The same shallow grounding system can also suffer from high resistance in the summer as moisture is evaporated from soil. It is advisable to determine the natural frost line and moisture profile for an area before attempting design of a ground system.

Figure 15.9 describes a four-point method for in-place measurement of soil resistivity. Four uniformly spaced probes are placed in a linear arrangement and connected to a ground resistance test meter. An alternating current (at a frequency other than 60 Hz) is passed between the two most distant probes, resulting in a potential difference between the center potential probes. The meter display in ohms of resistance can then be applied to determine the average soil resistivity in ohm-centimeters for the hemispherical area between the C1 and P2 probes.

Soil resistivity measurements should be repeated at a number of locations to establish a resistivity profile for the site. The depth of measurement can be controlled by varying the spacing between the probes. In no case should the probe length exceed 20% of the spacing between probes.

After the soil resistivity for a site is known, calculations can be made to determine the effectiveness of a variety of ground system configurations. Equations for several driven rod and radial cable configurations are given in [11], which—after the solid resistivity is known—can be used for the purpose of estimating total system resistance. Generally, driven rod systems are appropriate where soil resistivity continues to improve with depth or where temperature extremes indicate seasonal frozen or dry soil conditions. Figure 15.10 shows a typical soil resistivity map for the U.S.

Ground Electrode Testing

Testing of all ground electrodes before they are connected to form a complex network is a fairly simple process that is well described in the documentation included with all ground electrode meters. This instructional process, therefore, will not be described here. Also, the system as a whole should be tested after all interconnections are made, providing a benchmark for future tests.

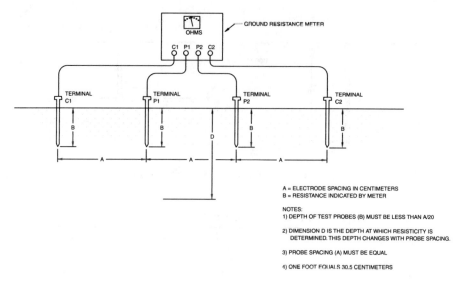

FIGURE 15.9 The four-point method for soil resistivity measurement. (After [11].)

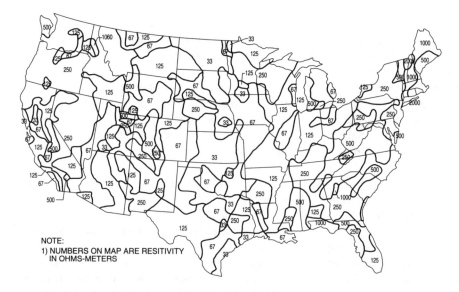

FIGURE 15.10 Typical soil resistivity map for the U.S. (After [11].)

At a new site, it is often advisable to perform ground system tests before the power company ground/neutral conductor is attached to the system. Conduct a before-and-after test with probes in the same position to determine the influence of the power company attachment [11]. It is also worthwhile to install permanent electrodes and marker monuments at the original P2 and C2 probe positions to ensure the repeatability of future tests.

Chemical Ground Rods

A chemically activated ground system is an alternative to the conventional ground rod. The idea behind the chemical ground rod is to increase the earth-to-electrode interface by conditioning the soil surrounding the rod. Experts have known for many years that the addition of ordinary table salt (NaCl) to soil will reduce the resistivity of the earth-to-ground electrode interface. With the proper soil moisture level (4 to 12%), *salting* can reduce soil resistivity from perhaps 10,000 Ω/m to less than 100 Ω/m. Salting the

Facility Grounding

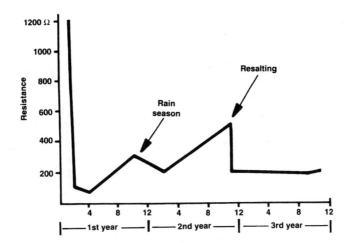

FIGURE 15.11 The effect of soil salting on ground-rod resistance with time. The expected resalting period, shown here as 2 years, varies depending on the local soil conditions and the amount of moisture present. (After [10].)

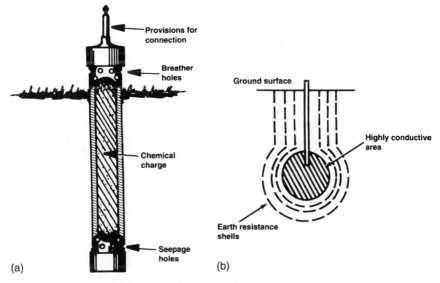

FIGURE 15.12 An air-breathing chemically activated ground rod: (a) breather holes at the top of the device permit moisture penetration into the chemical charge section of the rod; (b) a salt solution seeps out of the bottom of the unit to form a conductive shell. (After [10].)

area surrounding a ground rod (or group of rods) follows a predictable life-cycle pattern, as illustrated in Fig. 15.11. Subsequent salt applications are rarely as effective as the initial salting.

Various approaches have been tried over the years to solve this problem. One such product is shown in Fig. 15.12. This chemically activated grounding electrode consists of a 2-1/2-in.-diameter copper pipe filled with rock salt. Breathing holes are provided on the top of the assembly, and seepage holes are located at the bottom. The theory of operation is simple. Moisture is absorbed from the air (when available) and is then absorbed by the salt. This creates a solution that seeps out of the base of the device and conditions the soil in the immediate vicinity of the rod.

Another approach is shown in Fig. 15.13. This device incorporates a number of ports (holes) in the assembly. Moisture from the soil (and rain) is absorbed through the ports. The metallic salts subsequently absorb the moisture, forming a saturated solution that seeps out of the ports and into the earth-to-electrode hemisphere. Tests have shown that if the moisture content is within the required range, earth

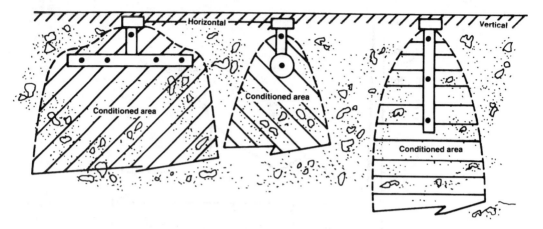

FIGURE 15.13 An alternative approach to the chemically activated ground rod. Multiple holes are provided on the ground-rod assembly to increase the effective earth-to-electrode interface. Note that chemical rods can be produced in a variety of configurations. (After [10].)

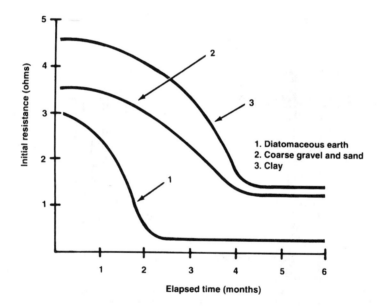

FIGURE 15.14 Measured performance of a chemical ground rod. (After [10].)

resistivity can be reduced by as much as 100:1. Figure 15.14 shows the measured performance of a typical chemical ground rod in three types of soil.

Implementations of chemical ground-rod systems vary depending on the application. Figure 15.15 illustrates a counterpoise ground consisting of multiple leaching apertures connected in a spoke fashion to a central hub. The system is serviceable in that additional salt compound can be added to the hub at required intervals to maintain the effectiveness of the ground. Figure 15.16 shows a counterpoise system made up of individual chemical ground rods interconnected with radial wires buried below the surface.

Ufer Ground System

Driving ground rods is not the only method of achieving a good earth-to-electrode interface [9]. The concept of the *Ufer* ground has gained interest because of its simplicity and effectiveness. The Ufer approach

Facility Grounding

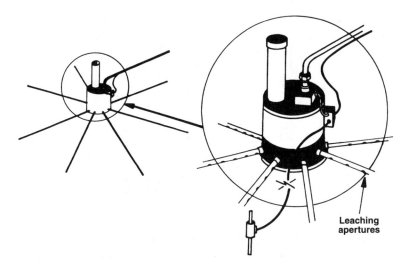

FIGURE 15.15 Hub and spoke counterpoise ground system. (After [10].)

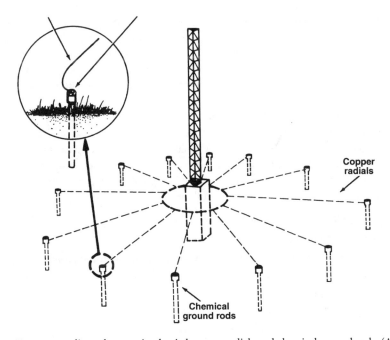

FIGURE 15.16 Tower grounding scheme using buried copper radials and chemical ground rods. (After [10].)

(named for its developer), however, must be designed into a new structure. It cannot be added on later. The Ufer ground takes advantage of the natural chemical- and water-retention properties of concrete to provide an earth ground. Concrete typically retains moisture for 15 to 30 days after a rain. The material has a ready supply of ions to conduct current because of its moisture-retention properties, mineral content, and inherent pH. The large mass of any concrete foundation provides a good interface to ground.

A Ufer system, in its simplest form, is made by routing a solid-copper wire (no. 4 gauge or larger) within the foundation footing forms before concrete is poured. Figure 15.17 shows one such installation. The length of the conductor run within the concrete is important. Typically a 20-ft run (10 ft in each direction) provides a 5 Ω ground in 1000 Ω/m soil.

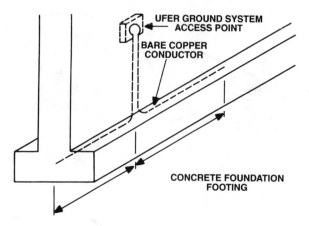

FIGURE 15.17 The basic concept of a Ufer ground system, which relies on the moisture-retentive properties of concrete to provide a large earth-to-electrode interface. Design of such a system is critical. Do not attempt to build a Ufer ground without the assistance of an experienced contractor. (After [9].)

As an alternative, steel reinforcement bars (rebar) can be welded together to provide a rigid, conductive structure. A ground lug is provided to tie equipment to the ground system in the foundation. The rebar must be welded, not tied, together. If it is only tied, the resulting poor connections between rods can result in arcing during a current surge. This can lead to deterioration of the concrete in the affected areas.

The design of a Ufer ground is not to be taken lightly. Improper installation can result in a ground system that is subject to problems. The grounding electrodes must be kept a minimum of 3 in. from the bottom and sides of the concrete to avoid the possibility of foundation damage during a large lightning surge. If an electrode is placed too near the edge of the concrete, a surge could turn the water inside the concrete to steam and break the foundation apart.

The Ufer approach also can be applied to guy-anchor points or a tower base, as illustrated in Fig. 15.18. Welded rebar or ground rods sledged in place after the rebar cage is in position can be used. By protruding below the bottom concrete surface, the ground rods add to the overall electrode length to help avoid thermal effects that can crack the concrete. The maximum length necessary to avoid breaking the concrete under a lightning discharge is determined by the following:

- Type of concrete (density, resistivity, and other factors)
- Water content of the concrete
- How much of the buried concrete surface area is in contact with the ground
- Ground resistivity
- Ground water content
- Size and length of the ground rod
- Size of lightning flash

The last variable is a bit of a gamble. The 50% mean occurrence of lightning strikes is 18 A, but superstrikes can occur that approach 100 to 200 kA.

Before implementing a Ufer ground system, consult a qualified contractor. Because the Ufer ground system will be the primary grounding element for the facility, it must be done correctly.

Bonding Ground-System Elements

A ground system is only as good as the methods used to interconnect the component parts [9]. Do not use soldered-only connections outside the equipment building. Crimped/brazed and *exothermic* (*Cadwelded*) connections are preferred. (Cadweld is a registered trademark of Erico Corp.) To make a

Facility Grounding

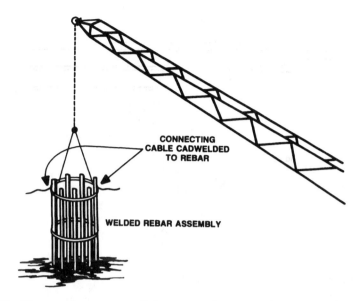

FIGURE 15.18 The Ufer ground system as applied to a transmission-tower base or guy-wire anchor point. When using this type of ground system, bond all rebar securely to prevent arcing in the presence of large surge currents. (After [9].)

proper bond, all metal surfaces must be cleaned, any finish removed to bare metal, and surface preparation compound applied. Protect all connections from moisture by appropriate means, usually sealing compound and heat-shrink tubing.

It is not uncommon for an untrained installer to use soft solder to connect the elements of a ground system. Such a system is doomed from the start. Soft-soldered connections cannot stand up to the acid and mechanical stress imposed by the soil. The most common method of connecting the components of a ground system is silver soldering. The process requires the use of brazing equipment, which may be unfamiliar to many facility engineers. The process uses a high-temperature/high-conductivity solder to complete the bonding process. For most grounding systems, however, the best approach to bonding is the exothermic process.

Exothermic Bonding

Exothermic bonding is the preferred method of connecting the elements of a ground system [9]. Molten copper is used to melt connections together, forming a permanent bond. This process is particularly useful in joining dissimilar metals. In fact, if copper and galvanized cable must be joined, exothermic bonding is the only acceptable means. The completed connection will not loosen or corrode and will carry as much current as the cable connected to it. Figure 15.19 illustrates the bonding that results from the exothermic process.

The bond is accomplished by dumping powdered metals (copper oxide and aluminum) from a container into a graphite crucible and igniting the material by means of a flint lighter. Reduction of the copper oxide by the aluminum produces molten copper and aluminum oxide slag. The molten copper flows over the conductors, bonding them together. The process is illustrated in Fig. 15.20. Figure 15.21 shows a typical mold. A variety of special-purpose molds are available to join different-size cables and copper strap. Figure 15.22 shows the bonding process for a copper-strap-to-ground-rod interface.

Ground-System Inductance

Conductors interconnecting sections or components of an earth ground system must be kept as short as possible to be effective [9]. The inductance of a conductor is a major factor in its characteristic

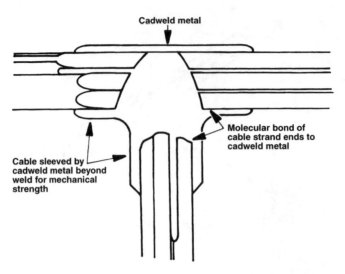

FIGURE 15.19 The exothermic bonding process. (After [9].)

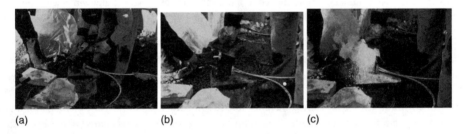

FIGURE 15.20 Exothermic bonding is the preferred method of joining the elements of a ground system. This photo sequence illustrates the procedure: (a) the powdered copper oxide and aluminum compound are added to the mold after the conductors have been mechanically joined; (b) final preparation of the bond before igniting; (c) the chemical reaction that bonds the materials together.

impedance to surge energy. For example, consider a no. 6 AWG copper wire 10 m in length. The wire has a DC resistance of 0.013 Ω and an inductance of 10 μH. For a 1000-A lightning surge with a 1 μs rise time, the resistive voltage drop will be 13 V, but the reactive voltage drop will be 10 kV. Furthermore, any bends in the conductor will increase its inductance and further decrease the effectiveness of the wire. Bends in ground conductors should be gradual. A 90°-bend is electrically equivalent to a 1/4-turn coil. The sharper the bend, the greater the inductance.

Because of the fast rise time of most lightning discharges and power-line transients, the *skin effect* plays an important role in ground-conductor selection. When planning a facility ground system, view the project from an RF standpoint.

The effective resistance offered by a conductor to radio frequencies is considerably higher than the ohmic resistance measured with direct currents. This is because of an action known as the skin effect, which causes the currents to be concentrated in certain parts of the conductor and leaves the remainder of the cross-section to contribute little or nothing toward carrying the applied current.

When a conductor carries an alternating current, a magnetic field is produced that surrounds the wire. This field continually is expanding and contracting as the AC current wave increases from zero to its maximum positive value and back to zero, then through its negative half-cycle. The changing magnetic lines of force cutting the conductor induce a voltage in the conductor in a direction that tends to retard the normal flow of current in the wire. This effect is more pronounced at the center of the conductor. Thus, current within the conductor tends to flow more easily toward the surface of the wire. The higher

Facility Grounding

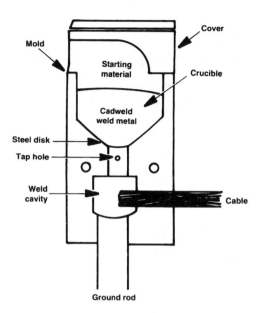

FIGURE 15.21 Typical exothermic bonding mold for connecting a cable to a ground rod. (After [9].)

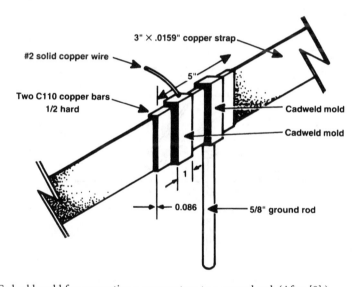

FIGURE 15.22 Cadweld mold for connecting a copper strap to a ground rod. (After [9].)

the frequency, or the faster the rise time of the applied waveform, the greater the tendency for current to flow at the surface. Figure 15.23 shows the distribution of current in a radial conductor.

When a circuit is operating at high frequencies, the skin effect causes the current to be redistributed over the conductor cross-section in such a way as to make most of the current flow where it is encircled by the smallest number of flux lines. This general principle controls the distribution of current regardless of the shape of the conductor involved. With a flat-strip conductor, the current flows primarily along the edges, where it is surrounded by the smallest amount of flux.

Grounding Tower Elements

Guyed towers are better than self-supporting towers at dissipating lightning surge currents [9]. This is true, however, only if the guy anchors are grounded properly. Use of the Ufer technique is one way of effectively

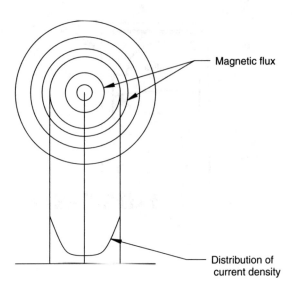

FIGURE 15.23 Skin effect on an isolated round conductor carrying a moderately high-frequency signal.

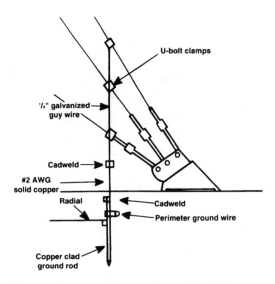

FIGURE 15.24 Recommended guy-anchor grounding procedure. (After [9].)

grounding the anchors. For anchors not provided with a Ufer ground during construction, other, more conventional, techniques can be used. For guyed towers whose guy lines are not electrically continuous between the tower and the base, such as sectionalized towers (certain FM and TV broadcast towers and AM broadcast towers), surge current dissipation is essentially equivalent to a self-supporting tower.

Never rely on the turnbuckles of a guy anchor as a path for lightning energy. The current resulting from a large flash can weld the turnbuckles in position. If the turnbuckles are provided with a safety loop of guy cable (as they should be), the loop can be damaged where it contacts the guys and turnbuckle. Figure 15.24 shows the preferred method of grounding guy wires: tie them together above the loop and turnbuckles. Do not make these connections with copper wire, even if they are Cadwelded. During periods of precipitation, water shed from the top copper wire will carry ions that may react with the lower galvanized (zinc) guy wires. This reaction washes off the zinc coating, allowing rust to develop.

Facility Grounding

The best way to make the connection is with all-galvanized materials. This includes the grounding wire, clamps, and ground rods. It may not be possible to use all galvanized materials because, at some point, a connection to copper conductors will be required. *Battery action* caused by the dissimilar metal junction may allow the zinc to act as a sacrificial anode. The zinc eventually will disappear into the soil, leaving a bare steel conductor that can fall victim to rust.

Ruling out an all-galvanized system, the next best scheme uses galvanized wire (guy-wire material) to tie the guy wires together. Just above the soil, Cadweld the galvanized wire to a copper conductor that penetrates below grade to the perimeter ground system. The height above grade for the connection is determined by the local snowfall or flood level. The electric conductivity of snow, although low, can cause battery action from the copper through the snow to the zinc. The Cadwelded joint should be positioned above the usual snow or flood level.

Ground-Wire Dressing

Figure 15.25 illustrates the proper way to bond the tower base ground leads to the buried ground system [9]. Dress the leads close to the tower from the lowest practical structural element at the base. Keep the conductors as straight and short as possible. Avoid any sharp bends. Attach the ground wires to the tower only at one or more existing bolts (or holes). Do not drill any holes into the tower. Do not loosen any bolts to make the ground-wire attachment. Use at least two 3- to 4-in copper straps between the base of the tower and the buried ground system. Position the straps next to the concrete pier of the tower base. For towers more than 200 ft in height, use four copper straps, one on each side of the pier.

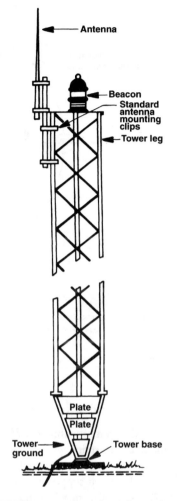

FIGURE 15.25 Ground-conductor dressing for the base of a guyed tower. (After [9].)

Figure 15.26 illustrates the proper way to bond guy wires to the buried ground system. The lead is dressed straight down from the topmost to the lowest guy. It should conform as close to vertical as possible, and be dressed downward from the lower side of each guy wire after connecting to each wire (Fig. 15.24). To ensure that no arcing will occur through the turnbuckle, a connection from the anchor plate to the perimeter ground circle is recommended. No. 2 gauge copper wire is recommended. This helps minimize the unavoidable inductance created by the conductor being in the air. Interconnect leads that are suspended in air must be dressed so that no bending radius is less than 8 in.

A *perimeter* ground—a circle of wire connected at several points to ground rods driven into the earth—should be installed around each guy-anchor point. The perimeter system provides a good ground for the anchor, and when tied together with the tower base radials, acts to rapidly dissipate lightning energy in the event of a flash. Tower base radials are buried wires interconnected with the tower base ground that extend away from the center point of the structure.

The required depth of the perimeter ground and the radials depends upon soil conductivity. Generally speaking, however, about 8 in. below grade is sufficient. In soil with good conductivity, the perimeter wire may be as small as no. 10 gauge. Because no. 2 gauge is required for the segment of conductor

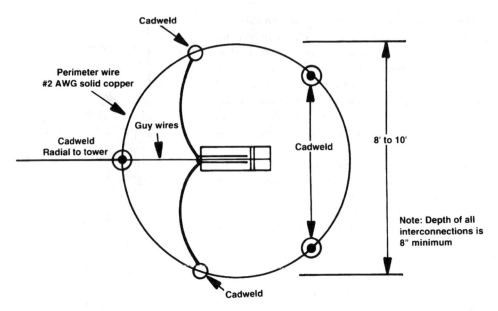

FIGURE 15.26 Top view of proper guy-anchor grounding techniques. A properly dressed and installed ground wire prevents surge currents from welding turnbuckles and damaging safety loops. The perimeter ground connects to the tower base by way of a radial wire. (After [9].)

suspended in air, it may be easier to use no. 2 throughout. An added advantage is that the same size Cadweld molds may be used for all bonds.

Facility Ground Interconnection

Any radial that comes within 2 ft of a conductive structure must be tied into the ground system [9]. Bury the interconnecting wire, if possible, and approach the radial at a 45° angle, pointing toward the expected surge origin (usually the tower). Cadweld the conductor to the radial and to the structure.

For large-base, self-supporting towers, the radials should be split among each leg pad, as shown in Fig. 15.27. The radials may be brought up out of the soil (in air) and each attached at spaced locations around the foot pad. Some radials may have to be tied together first and then joined to the foot pad. Remember, if space between the radial lines can be maintained, less mutual inductance (coupling) will exist, and the system surge impedance will be lower.

It is desirable to have a continuous one-piece ring with rods around the equipment building. Connect this ring to one, and only one, of the tower radials, thus forming no radial loops. See Fig. 15.28. Bury the ring to the same depth as the radials to which it interconnects. Connect power-line neutral to the ring. *Warning*: Substantial current may flow when the power-line neutral is connected to the ring. Follow safety procedures when making this connection.

Install a ground rod (if the utility company has not installed one) immediately outside the generator or utility company vault, and connect this rod to the equipment building perimeter ground ring. Route a no. 1/0 insulated copper cable from the main power panel inside the generator vault to the ground rod outside the vault. Cut the cable to length, strip both ends, and tie one end to the power-line neutral at the main power panel in the generator vault. Connect the other end to the ground rod. *Warning*: Use care when making this connection. Hazardous voltage may exist between the power-line neutral and any point at earth potential.

Do not remove any existing earth ground connections to power-line neutral, particularly if they are installed by the power company. To do so may violate the local electrical code. The goal of this interconnection is to minimize noise that may be present on the neutral, and to conduct this noise as directly as possible outside to earth ground.

Facility Grounding

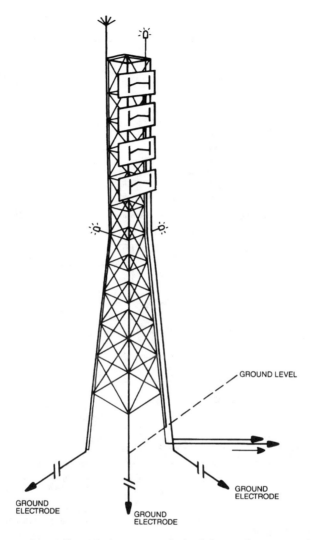

FIGURE 15.27 Interconnecting a self-supporting tower to the buried ground system.

Personnel Protection

The threat to personnel during a lightning strike ranges from the obvious danger of direct contact with a lightning strike to the more obscure effects of *step* and *touch voltages* [11]. Protection from a direct strike when near or within structures is accomplished with traditional rolling sphere concept methods discussed in Chapter 2. Step and touch potentials, however, are created as a lightning current passes through resistive soil and other available paths as it dissipates into the earth. A person in contact with only one point of the gradient will simply rise and fall in potential with the gradient without injury. A person in contact with multiple points on the earth or objects at different potentials along the gradient, however, will become part of the current path and may sustain injury or death.

Figure 15.29 [13, 14] indicates a number of methods for protecting personnel from the direct and secondary effects of lightning. A typical tower/transmitter site is used as an example. A technician responding to a service problem during a thunderstorm would likely exit his or her vehicle outside the gate, unlock and open the gate, and move the vehicle into the inside yard. The technician would then leave the vehicle, and enter the building.

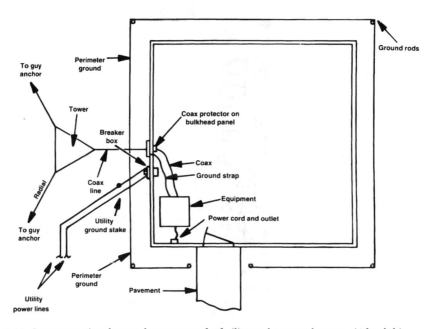

FIGURE 15.28 Interconnecting the metal structures of a facility to the ground system. (After [9].)

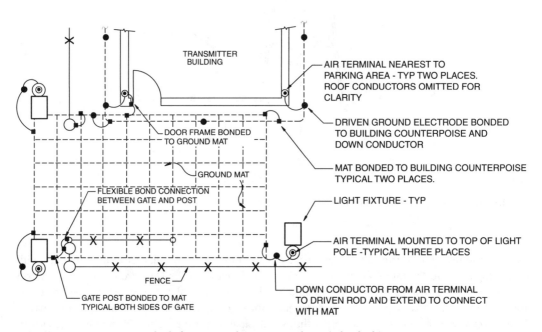

FIGURE 15.29 Protection methods for personnel at an exposed site. (After [11].)

The threat of a direct lightning strike to the technician has been minimized by establishing a protective zone over the areas to be traversed. This zone is created by the tower and air terminals mounted atop light poles.

Step potentials are minimized through the use of a ground mat buried just below the surface of the area where the technician is expected to be outside the vehicle. Ground mats are commercially available, fabricated in a 6-in. × 6-in. square pattern using no. 8 AWG bare copper wire. Each intersection is welded, creating—for all practical purposes—an *equipotential area* that short-circuits the step potential gradient in the area above the mat. The mat, as a whole, will rise and fall in potential as a result of the lightning

Facility Grounding

current discharges; however, there will be very little difference in potential between the technician's feet. Mats should be covered with six in. of crushed stone or pavement.

The threat of dangerous touch potentials is minimized by bonding the following elements to the ground system:

- Personnel ground mat
- Fence at each side of the gate opening
- Door frame of the transmitter building
- Flexible bonding connection between the swing gate and its terminal post

Such bonding will ensure that the object being touched by the technician is at or near the same potential as his or her feet.

Bonding both sides of the gate opening to the mat helps to ensure that the technician and both sides of the gate are at approximately the same potential while the gate is being handled. The flexible bond between the gate and its support post can be accomplished using a commercially available kit or by Cadwelding a short length of flexible 2/0 AWG welding cable between the two elements.

Grounding on Bare Rock

A bare rock mountaintop location provides special challenges to the facility design engineer [9]. There is no soil, thus there are no ground rods. Radials are the only means to develop a ground system. Install a large number of radials, laid straight, but not too taut. The portions not in contact with the rock are in air and form an inductance that will choke the surge current. Because rock is not conductive when it is dry, keep the radials short. Only a test measurement will determine how short the radials should be. A conventional earth-resistance tester will tell only half the story (besides, ground rods cannot be placed in rock for such a measurement). A dynamic ground tester offers the only way to obtain the true surge impedance of the system.

Rock-Based Radial Elements

On bare rock, a radial counterpoise will conduct and spread the surge charge over a large area. In essence, it forms a leaky capacitor with the more conductive earth on or under the mountain [9]. The conductivity of the rock will be poor when dry, but quite good when wet. If the site experiences significant rainfall before a lightning flash, protection will be enhanced. The worst case, however, must be assumed: an early strike under dry conditions.

The surge impedance, measured by a dynamic ground tester, should be 25 Ω or less. This upper-limit number is chosen so that less stress will be placed on the equipment and its surge protectors. With an 18-kA strike to a 25 Ω ground system, the entire system will rise 450 kV above the rest of the world at peak current. This voltage has the potential to jump almost 15.750 in. (0.35 in./10 kV at standard atmospheric conditions of 25°C, 30 in. of mercury and 50% relative humidity).

For nonsoil conditions, tower anchor points should have their own radial systems or be encapsulated in concrete. Configure the encapsulation to provide at least 3 in. of concrete on all sides around the embedded conductor. The length will depend on the size of the embedded conductor. Rebar should extend as far as possible into the concrete. The dynamic ground impedance measurements of the anchor grounds should each be less than 25 Ω.

The size of the bare conductor for each tower radial (or for an interconnecting wire) will vary, depending on soil conditions. On rock, a bare no. 1/0 or larger wire is recommended. Flat, solid-copper strap would be better, but may be blown or ripped if not covered with soil. If some amount of soil is available, no. 6 cable should be sufficient. Make the interconnecting radial wires continuous, and bury them as deep as possible; however, the first 6 to 10 in. will have the most benefit. Going below 18 in. will not be cost-effective, unless in a dry, sandy soil where the water table can be reached and ground-rod penetration is shallow. If only a small amount of soil exists, use it to cover the radials to the extent possible. It is more important to cover radials in the area near the tower than at greater distances. If, however, soil exists only at the outer distances and cannot be transported to the inner locations, use the soil to cover the outer portions of the radials.

If soil is present, install ground rods along the radial lengths. Spacing between ground rods is affected by the depth that each rod is driven; the shallower the rod, the closer the allowed spacing. Because the ultimate depth a rod can be driven cannot always be predicted by the first rod driven, use a maximum spacing of 15 ft when selecting a location for each additional rod. Drive rods at building corners first (within 24 in. but not closer than 6 in. to a concrete footer unless that footer has an encapsulated Ufer ground), then fill in the space between the corners with additional rods.

Drive the ground rods in place. Do not auger; set in place, then back-fill. The soil compactness is never as great on augured-hole rods when compared with driven rods. The only exception is when a hole is augured or blasted for a ground rod or rebar and then back-filled in concrete. Because concrete contains lime (alkali base) and is porous, it absorbs moisture readily, giving it up slowly. Electron carriers are almost always present, making the substance a good conductor.

If a Ufer ground is not being implemented, the radials may be Cadwelded to a subterranean ring, with the ring interconnected to the tower foot pad via a minimum of three no. 1/0 wires spaced at 120° angles and Cadwelded to the radial ring.

15.3 Transmission-System Grounding

Nature can devastate a communications site. Lightning can seriously damage an unprotected facility with little or no warning, leaving an expensive and time-consuming repair job. The first line of defense is proper grounding of the communications system.

Transmission Line

All coax and waveguide lines must include grounding kits for bonding the transmission line at the antenna [9]. On conductive structures, this can be accomplished by bonding the tail of the grounding kit to the structure itself. Remove all nonconductive paint and corrosion before attachment. Do not drill holes, and do not loosen any existing tower member bolts. Antenna clamping hardware can be used, or an all-stainless-steel hose clamp of the appropriate size can be substituted. The location of the tower-top ground is not as critical as the bottom grounding kit.

On nonconductive structures, a no. 1/0 or larger wire must be run down the tower [9]. Bond the transmission-line grounding kit to this down-run. Keep the wire as far away from all other conductive runs (aviation lights, coax, and waveguide) as possible. Separation of 2 ft is preferred; 18 in. is the minimum. If any other ground lines, conduit, or grounded metallic-structure members must be traversed that are closer than 18 in., they too must be grounded to the down-lead ground line to prevent flashover.

At the point where the coax or waveguide separates from the conductive structure (metal tower), a coax or waveguide grounding kit must be installed. Secure the connection to a large vertical structure member with a small number of joints. Attach to a structural member as low as possible on the tower.

Dress the grounding kit tails in a nearly straight 45° downward angle to the tower member. On nonconductive structures, a metal busbar must be used. Ground the bar to one or more of the vertical downconductors, and as close to ground level as possible.

Coaxial cables, lighting conduit, and other lines on the tower must be secured properly to the structure. Figure 15.30 illustrates several common attachment methods.

Cable Considerations

Ground-strap connections must withstand weathering and maintain low electrical resistance between the grounded component and earth [9]. Corrosion impairs ground-strap performance. Braided-wire ground straps should not be used in outside installations. Through capillary action resembling that of a wick, the braid conveys water, which accelerates corrosion. Eventually, advancing corrosion erodes the

Facility Grounding

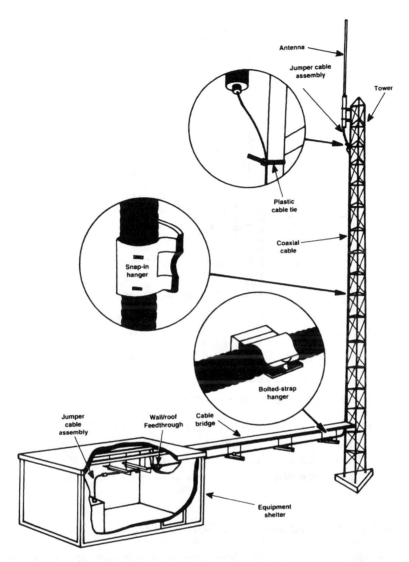

FIGURE 15.30 Transmission-line mounting and grounding procedures for a communications site.

ground-strap cable. Braid also can act as a duct that concentrates water at the bond point. This water speeds corrosion, which increases the electrical resistance of the bond. A jacketed, seven-strand copper wire strap (no. 6 to no. 2) is recommended for transmission-line grounding at the tower.

Satellite Antenna Grounding

Most satellite receiving/transmitting antenna piers are encapsulated in concrete [9]. Consideration should be given, therefore, to implementing a Ufer ground for the satellite dish. A 4-in.-diameter pipe, submerged 4 to 5 ft in an 18-in.-diameter (augered) hole will provide a good start for a Ufer-based ground system. It should be noted that an augered hole is preferred because digging and repacking the soil around the pier will create higher ground resistance. In areas of good soil conductivity (100 Ω/m or less), this basic Ufer system may be adequate for the antenna ground.

Figure 15.31 shows the preferred method: a hybrid Ufer/ground-rod and radial system. A cable connects the mounting pipe (Ufer ground) to a separate driven ground rod. The cable then is connected to

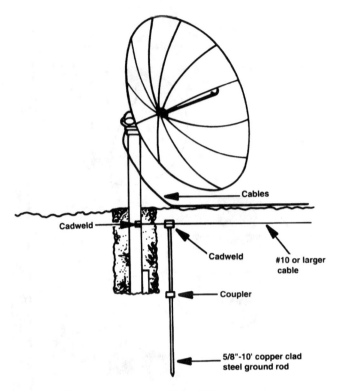

FIGURE 15.31 Grounding a satellite receiving antenna. (After [9].)

the facility ground system. In areas of poor soil conductivity, additional ground rods are driven at increments (2.2 times the rod length) between the satellite dish and the facility ground system. Run all cables underground for best performance. Make the interconnecting copper wire no. 10 size or larger; bury the wire at least 8 in. below finished grade. Figure 15.32 shows the addition of a lightning rod to the satellite dish.

15.4 Designing a Building Ground System

After the required grounding elements have been determined, they must be connected into a unified system [9]. Many different approaches can be taken, but the goal is the same: establish a low-resistance, low-inductance path to surge energy. Figure 15.33 shows a building ground system using a combination of ground rods and buried bare-copper radial wires. This design is appropriate when the building is large or located in an urban area. This approach also can be used when the facility is located in a highrise building that requires a separate ground system. Most newer office buildings have ground systems designed into them. If a comprehensive building ground system is provided, use it. For older structures (constructed of wood or brick), a separate ground system will be needed.

Figure 15.34 shows another approach in which a perimeter ground strap is buried around the building and ground rods are driven into the earth at regular intervals (2.2 times the rod length). The ground ring consists of a one-piece copper conductor that is bonded to each ground rod.

If a transmission or microwave tower is located at the site, connect the tower ground system to the main ground point via a copper strap. The width of the strap must be at least 1% of the length and, in any event, not less than 3 in. wide. The building ground system is not a substitute for a tower ground system, no matter what the size of the tower. The two systems are treated as independent elements, except for the point at which they interconnect.

Facility Grounding

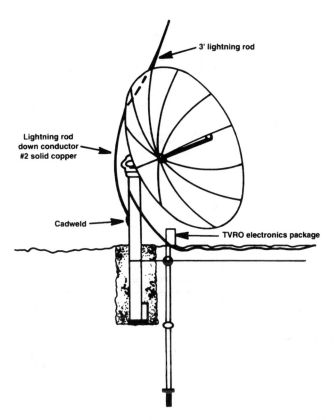

FIGURE 15.32 Addition of a lightning rod to a satellite antenna ground system. (After [9].)

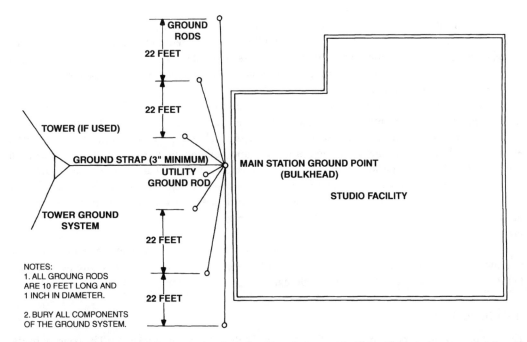

FIGURE 15.33 A facility ground system using the hub-and-spoke approach. The available real estate at the site will dictate the exact configuration of the ground system. If a tower is located at the site, the tower ground system is connected to the building ground as shown.

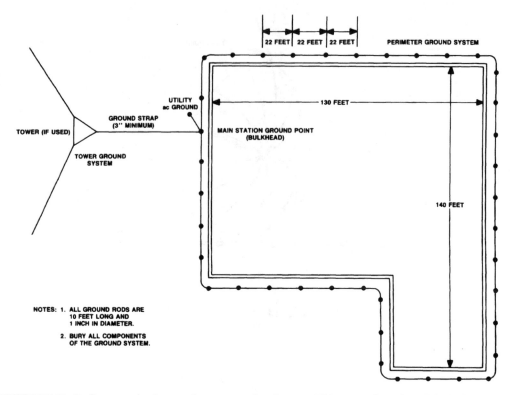

FIGURE 15.34 Facility ground using a perimeter ground-rod system. This approach works well for buildings with limited available real estate.

Connect the utility company power-system ground rod to the main facility ground point as required by the local electrical code. Do not consider the building ground system to be a substitute for the utility company ground rod. The utility rod is important for safety reasons and must not be disconnected or moved. Do not remove any existing earth ground connections to the power-line neutral connection. To do so may violate local electrical code.

Bury all elements of the ground system to reduce the inductance of the overall network. Do not make sharp turns or bends in the interconnecting wires. Straight, direct wiring practices will reduce the overall inductance of the system and increase its effectiveness in shunting fast-rise-time surges to earth. Figure 15.35 illustrates the interconnection of a tower and building ground system. In most areas, soil conductivity is high enough to permit rods to be connected with no. 6 bare-copper wire or larger. In areas of sandy soil, use copper strap. A wire buried in low-conductivity, sandy soil tends to be inductive and less effective in dealing with fast-rise-time current surges. As stated previously, make the width of the ground strap at least 1% of its overall length. Connect buried elements of the system as shown in Fig. 15.36.

For small installations with a low physical profile, a simplified grounding system can be implemented, as shown in Fig. 15.37. A grounding plate is buried below grade level, and a ground wire ties the plate to the microwave tower mounted on the building.

Bulkhead Panel

The bulkhead panel is the cornerstone of an effective facility grounding system [9]. The concept of the bulkhead is simple: establish one reference point to which all cables entering and leaving the equipment building are grounded and to which all transient-suppression devices are mounted. Figure 15.38 shows

Facility Grounding

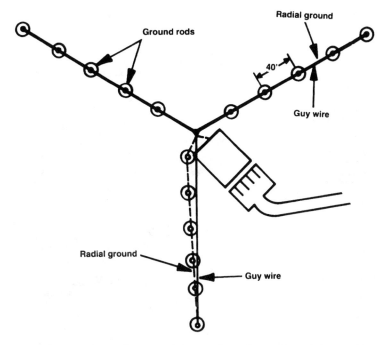

FIGURE 15.35 A typical guy-anchor and tower-radial grounding scheme. The radial ground is no. 6 copper wire. The ground rods are 5/8 in. × 10 ft. (After [9].)

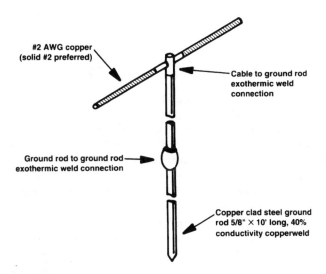

FIGURE 15.36 Preferred bonding method for below-grade elements of the ground system. (After [9].)

a typical bulkhead installation for a broadcast or communications facility. The panel size depends on the spacing, number, and dimensions of the coaxial lines, power cables, and other conduit entering or leaving the building.

To provide a weatherproof point for mounting transient-suppression devices, the bulkhead can be modified to accept a subpanel, as shown in Fig. 15.39. The subpanel is attached so that it protrudes through an opening in the wall and creates a secondary plate on which transient suppressors are mounted and grounded. A typical cable/suppressor-mounting arrangement for a communications site is shown in Fig. 15.40. To handle the currents that may be experienced during a lightning strike or large transient

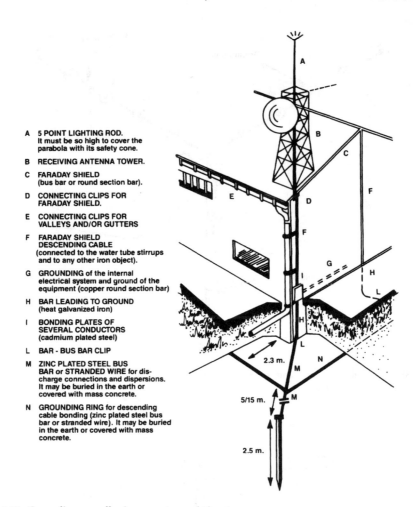

FIGURE 15.37 Grounding a small microwave transmission tower.

on the utility company AC line, the bottommost subpanel flange (which joins the subpanel to the main bulkhead) must have a total surface-contact area of at least 0.75 in.² per transient suppressor.

Because the bulkhead panel will carry significant current during a lightning strike or AC line disturbance, it must be constructed of heavy material. The recommended material is 1/8-in. C110 (solid copper) 1/2-hard. This type of copper stock weighs nearly 5-1/2 lb/ft² and is rather expensive. Installing a bulkhead, while sometimes difficult and costly, will pay dividends for the life of the facility. Use 18-8 stainless-steel mounting hardware to secure the subpanel to the bulkhead.

Because the bulkhead panel establishes the central grounding point for all equipment within the building, it must be tied to a low-resistance (and low-inductance) perimeter ground system. The bulkhead establishes the *main facility ground point*, from which all grounds inside the building are referenced. A typical bulkhead installation for a small communications site is shown in Fig. 15.41.

Bulkhead Grounding

A properly installed bulkhead panel will exhibit lower impedance and resistance to ground than any other equipment or cable grounding point at the facility [9]. Waveguide and coax line grounding kits should be installed at the bulkhead panel as well as at the tower. Dress the kit tails downward at a straight 45° angle using 3/8-in. stainless-steel hardware to the panel. Position the stainless-steel lug at the tail end, flat against a cleaned spot on the panel. Joint compound will be needed for aluminum and is recommended for copper panels.

Facility Grounding

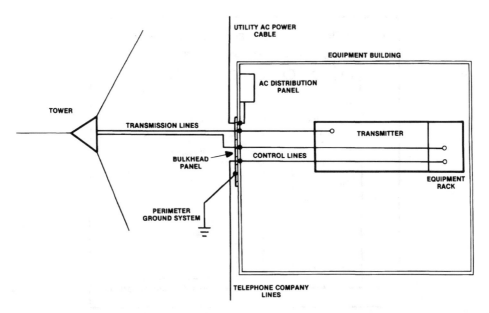

FIGURE 15.38 The basic design of a bulkhead panel for a facility. The bulkhead establishes the grounding reference point for the plant.

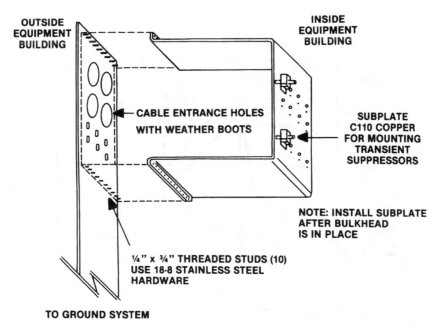

FIGURE 15.39 The addition of a subpanel to a bulkhead as a means of providing a mounting surface for transient-suppression components. To ensure that the bulkhead is capable of handling high surge currents, use the hardware shown. (After [9].)

Because the bulkhead panel will be used as the central grounding point for all the equipment inside the building, the lower the inductance to the perimeter ground system, the better. The best arrangement is to simply extend the bulkhead panel down the outside of the building, below grade, to the perimeter ground system. This will give the lowest resistance and the smallest inductive voltage drop. This approach is illustrated in Fig. 15.42.

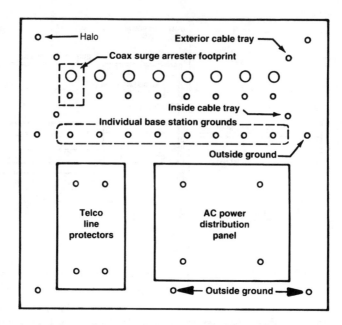

FIGURE 15.40 Mounting-hole layout for a communications site bulkhead subpanel.

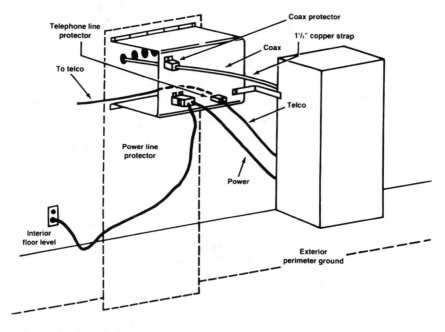

FIGURE 15.41 Bulkhead installation at a small communications site. (After [9].)

If cables are used to ground the bulkhead panel, secure the interconnection to the outside ground system along the bottom section of the panel. Use multiple no. 1/0 or larger copper wire or several solid-copper straps. If using strap, attach with stainless-steel hardware, and apply joint compound for aluminum bulkhead panels. Clamp and Cadweld, or silver-solder for copper/brass panels. If no. 1/0 or larger wire is used, employ crimp lug and stainless-steel hardware. Measure the DC resistance. It should be less than 0.01 Ω between the ground system and the panel. Repeat this measurement on an annual basis.

Facility Grounding

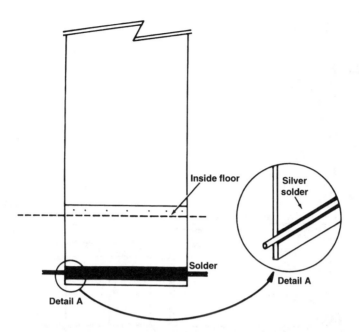

FIGURE 15.42 The proper way to ground a bulkhead panel and provide a low-inductance path for surge currents stripped from cables entering and leaving the facility. The panel extends along the building exterior to below grade. It is silver-soldered to a no. 2/0 copper wire that interconnects with the outside ground system. (After [9].)

If the antenna feed lines do not enter the equipment building via a bulkhead panel, treat them in the following manner:

- Mount a feed-line ground bar on the wall of the building approximately 4 in. below the feed-line entry point.
- Connect the outer conductor of each feed line to the feed-line ground bar using an appropriate grounding kit.
- Connect a no. 1/0 cable or 3- to 6-in.-wide copper strap between the feed-line ground bar and the external ground system. Make the joint a Cadweld or silver-solder connection.
- Mount coaxial arrestors on the edge of the bar.
- Weatherproof all connections.

Lightning Protectors

A variety of lightning arrestors are available for use on coaxial transmission lines, utility AC feeds, and telephone cables. The protector chosen must be carefully matched to the requirements of the application. Do not use air gap protectors because these types are susceptible to air pollution, corrosion, temperature, humidity, and manufacturing tolerances. The turn-on speed of an air gap device is a function of all of the foregoing elements. A simple gas-tube-type arrestor is an improvement, but neither of these devices will operate reliably to protect shunt-fed cavities, isolators, or receivers that include static drain inductors to ground (which most have). Such voltage-sensitive crowbar devices are short-circuited by the DC path to ground found in these circuits. The inductive change in current per unit time ($\Delta di/dt$) voltage drop is usually not enough to fire the protector, but it can be sufficient to destroy the inductor and then the receiver front end. Instead, select a protector that does not have DC continuity on the coaxial center pin from input to output. Such units have a series capacitor between the gas tube and the equipment center pin that will allow the voltage to build up so that the arrestor can fire properly.

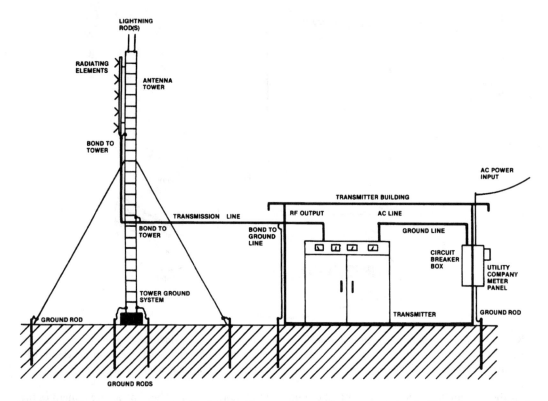

FIGURE 15.43 A common, but not ideal, grounding arrangement for a transmission facility using a grounded tower. A better configuration involves the use of a bulkhead panel through which all cables pass into and out of the equipment building.

Typical Installation

Figure 15.43 illustrates a common grounding arrangement for a remotely located grounded-tower (FM, TV, or microwave radio) transmitter plant. The tower and guy wires are grounded using 10-ft-long copper-clad ground rods. The antenna is bonded to the tower, and the transmission line is bonded to the tower at the point where it leaves the structure and begins the horizontal run into the transmitter building. Before entering the structure, the line is bonded to a ground rod through a connecting cable. The transmitter itself is grounded to the transmission line and to the AC power-distribution system ground. This, in turn, is bonded to a ground rod where the utility feed enters the building. The goal of this arrangement is to strip all incoming lines of damaging overvoltages before they enter the facility. One or more lightning rods are mounted at the top of the tower structure. The rods extend at least 10 ft above the highest part of the antenna assembly.

Such a grounding configuration, however, has built-in problems that can make it impossible to provide adequate transient protection to equipment at the site. Look again at the example. To equipment inside the transmitter building, two grounds actually exist: the utility company ground and the antenna ground. One ground will have a lower resistance to earth, and one will have a lower inductance in the connecting cable or copper strap from the equipment to the ground system.

Using the Fig. 15.43 example, assume that a transient overvoltage enters the utility company meter panel from the AC service line. The overvoltage is clamped by a protection device at the meter panel, and the current surge is directed to ground. But *which ground*, the utility ground or the antenna ground?

The utility ground surely will have a lower inductance to the current surge than the antenna ground, but the antenna probably will exhibit a lower resistance to ground than the utility side of the circuit.

Facility Grounding

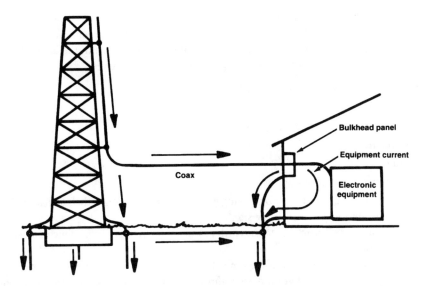

FIGURE 15.44 The equivalent circuit of the facility shown in Fig. 15.43. Note the discharge current path through the electronic equipment.

Therefore, the surge current will be divided between the two grounds, placing the transmission equipment in series with the surge suppressor and the antenna ground system. A transient of sufficient potential will damage the transmission equipment.

Transients generated on the antenna side because of a lightning discharge are no less troublesome. The tower is a conductor, and any conductor is also an inductor. A typical 150-ft self-supporting tower may exhibit as much as 40 µH inductance. During a fast-rise-time lightning strike, an instantaneous voltage drop of 360 kV between the top of the tower and the base is not unlikely. If the coax shield is bonded to the tower 15 ft above the earth (as shown in the previous figure), 10% of the tower voltage drop (36 kV) will exist at that point during a flash. Figure 15.44 illustrates the mechanisms involved.

The only way to ensure that damaging voltages are stripped off all incoming cables (coax, AC power, and telephone lines) is to install a bulkhead entrance panel and tie all transient-suppression hardware to it. Configuring the system as shown in Fig. 15.45 strips away all transient voltages through the use of a single-point ground. The bulkhead panel is the ground reference for the facility. With such a design, secondary surge current paths do not exist, as illustrated in Fig. 15.46.

Protecting the building itself is another important element of lightning surge protection. Figure 15.47 shows a common system using multiple lightning rods. As specified by NFPA 78, the dimensions given in the figure are as follows:

- A = 50-ft maximum spacing between air terminals
- B = 150-ft maximum length of a coursing conductor permitted without connection to a main perimeter or downlead conductor
- C = 20- or 25-ft maximum spacing between air terminals along an edge

Checklist for Proper Grounding

A methodical approach is necessary in the design of a facility ground system. Consider the following points:

1. Install a bulkhead panel to provide mechanical support, electric grounding, and lightning protection for coaxial cables, power feeds, and telephone lines entering the equipment building.

2. Install an internal ground bus using no. 2 or larger solid-copper wire. (At transmission facilities, use copper strap that is at least 3 in. wide.) Form a *star* grounding system. At larger installations, form a *star-of-stars* configuration. Do not allow ground loops to exist in the internal ground bus. Connect the following items to the building internal ground system:

   - Chassis racks and cabinets of all hardware
   - All auxiliary equipment
   - Battery charger
   - Switchboard
   - Conduit
   - Metal raceway and cable tray

3. Install a tower earth ground array by driving ground rods and laying radials as required to achieve a low earth ground impedance at the site.
4. Connect outside metal structures to the earth ground array (towers, metal fences, metal buildings, and guy-anchor points).
5. Connect the power-line ground to the array. Follow local electrical code to the letter.
6. Connect the bulkhead to the ground array through a low-inductance, low-resistance bond.
7. Do not use soldered-only connections outside the equipment building. Crimped, brazed, and exothermic (Cadwelded) connections are preferable. For a proper bond, all metal surfaces must be cleaned, any finish removed to bare metal, and surface preparation compound applied (where necessary). Protect all connections from moisture by appropriate means (sealing compound and heat sink tubing).

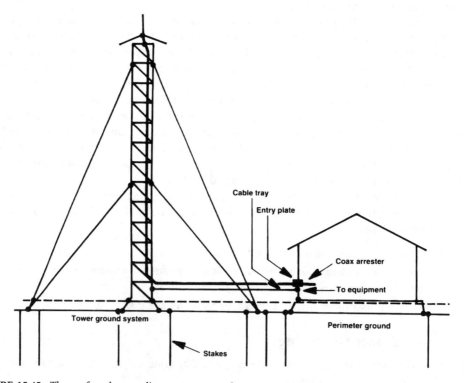

FIGURE 15.45 The preferred grounding arrangement for a transmission facility using a bulkhead panel. With this configuration, all damaging transient overvoltages are stripped off the coax, power, and telephone lines before they can enter the equipment building.

Facility Grounding

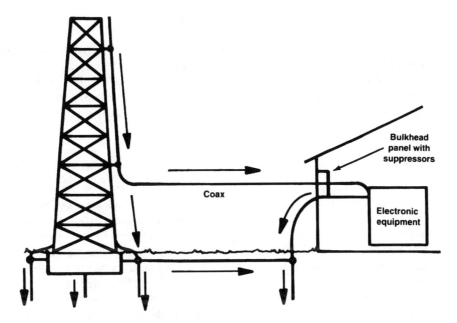

FIGURE 15.46 The equivalent circuit of the facility shown in Fig. 15.45. Discharge currents are prevented from entering the equipment building.

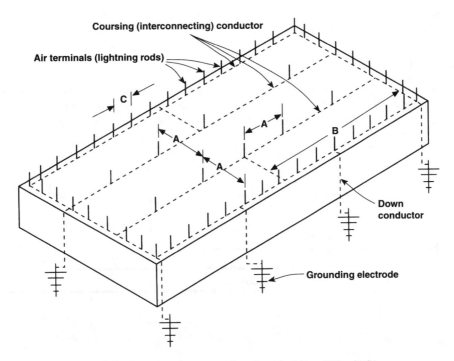

FIGURE 15.47 Conduction lightning-protection system for a large building. (After [12].)

15.5 AC System Grounding Practices

Installing an effective ground system to achieve a good earth-to-grounding-electrode interface is only half the battle for a facility designer. The second, and equally important, element of any ground system is the configuration of grounding conductors inside the building. Many different methods can be used to implement a ground system, but some conventions always should be followed to ensure a low-resistance (and low-inductance) layout that will perform as required. Proper grounding is important whether or not the facility is located in a high-RF field.

Building Codes

As outlined previously in this chapter, the primary purpose of grounding electronic hardware is to prevent electric shock hazard. The National Electrical Code (NEC) and local building codes are designed to provide for the safety of the workplace. Local codes should always be followed. Occasionally, code sections are open to some interpretation. When in doubt, consult a field inspector. Codes constantly are being changed or expanded because new situations arise that were not anticipated when the codes were written. Sometimes, an interpretation will depend upon whether the governing safety standard applies to building wiring or to a factory-assembled product to be installed in a building. Underwriters Laboratories (UL) and other qualified testing organizations examine products at the request and expense of manufacturers or purchasers, and list products if the examination reveals that the device or system presents no significant safety hazard when installed and used properly.

Municipal and county safety inspectors generally accept UL and other qualified testing laboratory certification listings as evidence that a product is safe to install. Without a listing, the end-user might not be able to obtain the necessary wiring permits and inspection sign-off. On-site wiring must conform with local wiring codes. Most codes are based on the NEC. Electrical codes specify wiring materials, wiring devices, circuit protection, and wiring methods.

Single-Point Ground

Single-point grounding is the basis of any properly designed facility ground network. Fault currents and noise should have only one path to the facility ground. Single-point grounds can be described as *star* systems in which radial elements circle out from a central hub. A star system is illustrated in Fig. 15.48.

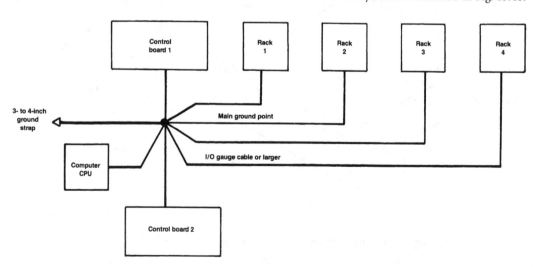

FIGURE 15.48 Typical facility grounding system. The *main facility ground point* is the reference from which all grounding is done at the plant. If a bulkhead entrance panel is used, it will function as the main ground point.

Note that all equipment grounds are connected to a *main ground point*, which is then tied to the facility ground system. Multiple ground systems of this type can be cascaded as needed to form a *star-of-stars* facility ground system. The key element in a single-point ground is that each piece of equipment has one ground reference. Fault energy and noise are then efficiently drained to the outside earth ground system. The single-point ground is basically an extension of the bulkhead panel discussed previously.

Isolated Grounding

Isolated grounding schemes, where the signal reference plane is isolated from equipment ground but connected to an *isolated* electrode in the earth, do not work, are unsafe, and violate the NEC [5]. It is thought by some people that the isolated earth connection is *clean* because there is no connection between it and the *dirty* system ground connection at the service entrance. The clean, isolated earth connection is also viewed (incorrectly) as a point where noise currents can flow into the earth and be dissipated. Kirchoff's current law teaches that any current flowing into the isolated ground must return to the source through another earth connection. Current cannot be dissipated. It must always return to its source. Even lightning current is not dissipated into the earth. It must have a return path (i.e., the electrostatic and electromagnetic fields that created the charge build-up and the lightning strike in the first place).

Consider what might happen if such a system were subjected to a lightning strike. Assume that a transient current of 2000 A flows into the earth and through an earth resistance of 5 Ω between the system ground electrode and the isolated electrode. A more realistic resistance might be even higher, perhaps 25 Ω; 2000 A flowing through 5 Ω results in a voltage drop or transient potential of 10,000 V between the two electrodes. Because this potential is impressed between the equipment frame (system ground electrode) and the signal reference plane (isolated electrode), it could result in equipment damage and personnel hazard. Dangerous potential differences between grounding subsystems can be reduced by bonding together all earth electrodes at a facility. Fine print note (FPN) No. 2, in NEC 250-86, states that the bonding together of separate grounding electrodes will limit potential differences between them and their associated wiring systems [6].

A facility ground system, then, can be defined as an electrically interconnected system of multiple conducting paths to the earth electrode or system of electrodes. The facility grounding system includes all electrically interconnected grounding subsystems such as:

- The equipment grounding subsystem
- Signal reference subsystem
- Fault protection subsystem
- Lightning protection subsystem

Isolated ground (IG) receptacles, which are a version of single-point grounding, are permitted by the NEC. Proper application of IG receptacles is very important. They must be used with an insulated equipment grounding conductor, not a bare conductor. Also, only metallic conduit should be used.

Separately Derived Systems

A separately derived system is a premises wiring system that derives its power from generator, transformer, or converter windings and that has no direct electrical connection, including a solidly connected grounded circuit conductor, to supply conductors originating in another system [6]. Solidly grounded, wye-connected, isolation transformers used to supply power to computer room equipment are examples of separately derived systems [5]. Figure 15.49 illustrates the bonding and grounding requirements of separately derived systems. NEC 250-26 permits the bonding and grounding connections to be made at the source of the separately derived system or at the first disconnecting means. Other examples of separately derived systems include generators and UPS systems. Note that all earth electrodes are bonded together via the equipment grounding conductor system. This is consistent with the recommendations listed in NEC 250-86.

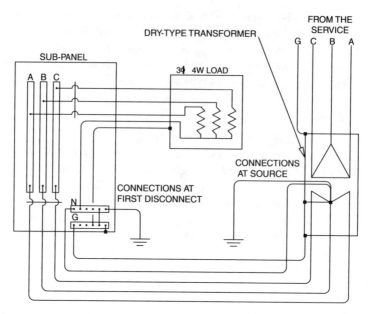

FIGURE 15.49 Basic scheme of the separately derived power system. (Courtesy of Intertec Publishing. After [5].)

Grounding Terminology

- *Grounded conductor.* A system or circuit conductor that is intentionally grounded [6].
- *Equipment grounding conductor.* The conductor used to connect the noncurrent-carrying metal parts of equipment, raceways, and other enclosures to the system grounded conductor, the grounding electrode conductor, or both, at the service equipment or at the source of a separately derived system [6].
- *Main bonding jumper.* The connection between the grounded circuit conductor and the equipment grounding conductor at the service [6].
- *Grounding electrode conductor.* The conductor used to connect the grounding electrode to the equipment grounding conductor, to the grounded conductor, or to both, of the circuit at the service equipment or at the source of a separately derived system [6].
- *Service.* The conductors and equipment for delivering energy from the electricity supply system to the wiring system of the premises served [6].
- *Service conductors.* The supply conductors that extend from the street main or from transformers to the service equipment of the premises supplied [6].
- *Service equipment.* The necessary equipment, usually consisting of a circuit breaker or switch and fuses, and their accessories, located near the point of entrance of supply conductors to a building or other structure, or an otherwise defined area, and intended to constitute the main control and means of cutoff of the supply [6].
- *Equipotential plane.* A mass of conducting material that offers a negligible impedance to current flow, thus producing zero volts (equipotential) between points on the plane.
- *Floating signal grounding.* A non-grounding system in which all electronic signal references are isolated from ground.
- *Single-point grounding.* A grounding system in which all electronic signal references are bonded together and grounded at a single point.
- *Multipoint grounding.* A grounding system in which all electronic signal references are grounded at multiple points.

Facility Ground System

Figure 15.50 illustrates a star grounding system as applied to an AC power-distribution transformer and circuit-breaker panel. Note that a central ground point is established for each section of the system: one in the transformer vault and one in the circuit-breaker box. The breaker ground ties to the transformer vault ground, which is connected to the building ground system. Figure 15.51 shows single-point grounding applied to a data processing center. Note how individual equipment groups are formed into a star grounding system, and how different groups are formed into a star-of-stars configuration. A similar approach can be taken for a data processing center using multiple modular power center (MPC) units, as shown in Fig. 15.52. The terminal mounting wall is the reference ground point for the room.

Grounding extends to all conductive elements of the data processing center. As illustrated in Fig. 15.53, the raised-floor supports are integrated into the ground system, in addition to the metal work of the building. The flooring is bonded to form a mesh grounding plane that is tied to the central ground reference point of the room. This is another natural extension of the bulkhead concept discussed previously. Figure 15.54 shows connection detail for the raised-floor supports. Make sure to use a flooring system that can be bolted together to form secure electrical—as well as mechanical—connections.

Figure 15.55 shows the recommended grounding arrangement for a typical broadcast or audio/video production facility. The building ground system is constructed using heavy-gauge copper wire (no. 4 or larger) if the studio is not located in an RF field, or a wide copper strap (3-in. minimum) if the facility is located near an RF energy source. Figure 15.56 gives a functional view of the plan shown in Fig. 15.55. Note the bulkhead approach.

Run the strap or cable from the perimeter ground to the main facility ground point. Branch out from the main ground point to each major piece of equipment and to the various equipment rooms. Establish a *local ground point* in each room or group of racks. Use a separate ground cable for each piece of equipment (no. 12 gauge or larger). Figure 15.57 shows the grounding plan for a communications facility. Equipment grounding is handled by separate conductors tied to the bulkhead panel (entry plate). A "halo" ground is constructed around the perimeter of the room. Cable trays are tied into the halo. All electronic equipment is grounded to the bulkhead to prevent ground-loop paths. In this application, the halo serves essentially the same function as the raised-floor ground plane in the data processing center.

The AC line ground connection for individual pieces of equipment often presents a built-in problem for the system designer. If the equipment is grounded through the chassis to the equipment-room ground point, a ground loop can be created through the green-wire ground connection when the equipment is plugged in. The solution to this problem involves careful design and installation of the AC power-distribution system to minimize ground-loop currents, while at the same time providing the required protection against ground faults. Some equipment manufacturers provide a convenient solution to the ground-loop problem by isolating the signal ground from the AC and chassis ground. This feature offers

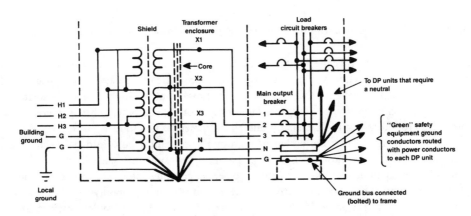

FIGURE 15.50 Single-point grounding applied to a power-distribution system. (*After* [15].)

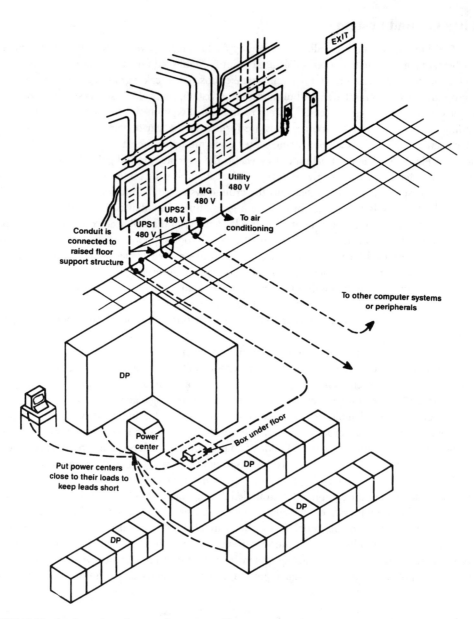

FIGURE 15.51 Configuration of a star-of-stars grounding system at a data processing facility. (After [15].)

the user the best of both worlds: the ability to create a signal ground system and AC ground system that are essentially free of interaction and ground loops. Do not confuse this isolated signal ground with the isolated ground system described in the section entitled "Noise Control." In this context, "isolated" refers only to the equipment input/output (signal) connections, not the equipment chassis; there is still only one integrated ground system for the facility, and all equipment is tied to it.

It should be emphasized that the design of a ground system must be considered as an integrated package. Proper procedures must be used at all points in the system. It takes only one improperly connected piece of equipment to upset an otherwise perfect ground system. The problems generated by a single grounding error can vary from trivial to significant, depending upon where in the system the error exists. This consideration leads, naturally, to the concept of ground-system maintenance for a facility. Check the ground network from time to time to ensure that no faults or errors have occurred.

Facility Grounding

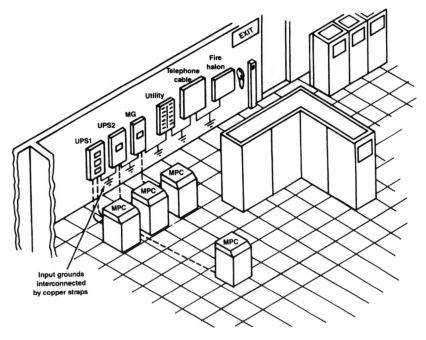

FIGURE 15.52 Establishing a star-based, single-point ground system using multiple modular power centers. (After [15].)

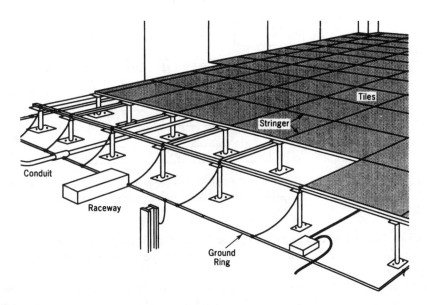

FIGURE 15.53 Grounding plan for a raised-floor support system.

Any time new equipment is installed, or old equipment is removed from service, careful attention must be given to the possible effects that such work will have on the ground system.

Grounding Conductor Size

The NEC and local electrical codes specify the minimum wire size for grounding conductors. The size varies, depending upon the rating of the current-carrying conductors. Code typically permits a smaller ground conductor than hot conductors. It is recommended, however, that the same size wire be used for

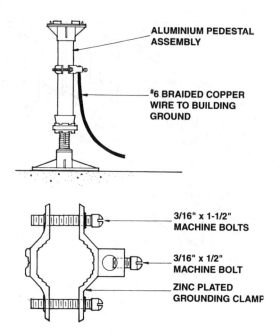

FIGURE 15.54 Bonding detail for the raised-floor support system.

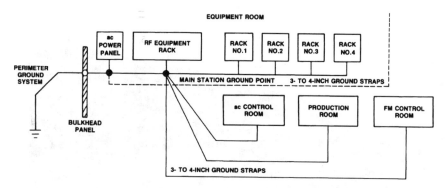

FIGURE 15.55 Typical grounding arrangement for individual equipment rooms at a communications facility. The ground strap from the main ground point establishes a *local ground point* in each room, to which all electronic equipment is bonded.

both ground lines and hot lines. The additional cost involved in the larger ground wire often is offset by the use of a single size of cable. Furthermore, better control over noise and fault currents is achieved with a larger ground wire.

It is recommended that separate insulated ground wires be used throughout the AC distribution system. Do not rely on conduit or raceways to carry the ground connection. A raceway interface that appears to be mechanically sound may not provide the necessary current-carrying capability in the event of a phase-to-ground fault. Significant damage can result if a fault occurs in the system. When the electrical integrity of a breaker panel, conduit run, or raceway junction is in doubt, fix it. Back up the mechanical connection with a separate ground conductor of the same size as the current-carrying conductors. Loose joints have been known to shower sprays of sparks during phase-to-ground faults, creating a fire hazard. Secure the ground cable using appropriate hardware. Clean attachment points of any paint or dirt accumulation. Properly label all cables.

Facility Grounding

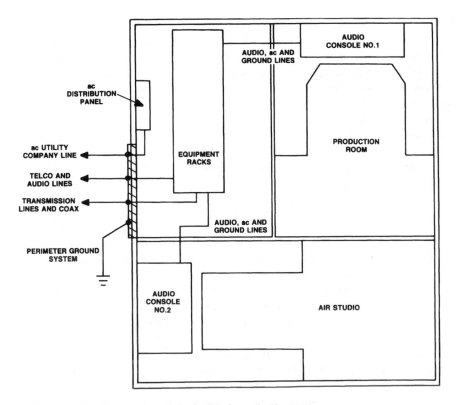

FIGURE 15.56 Physical implementation of the facility shown in Fig. 15.55.

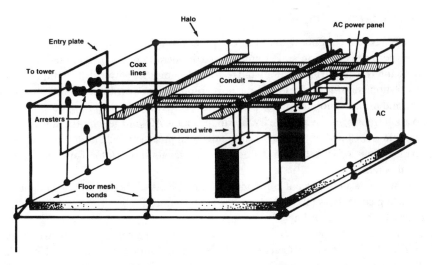

FIGURE 15.57 Bulkhead-based ground system including a grounding halo.

Structural steel, compared with copper, is a poor conductor at any frequency. At DC, steel has a resistivity 10 times that of copper. As frequency rises, the skin effect is more pronounced because of the magnetic effects involved. A no. 6 copper wire can have less RF impedance than a 12-in steel "I" beam. Furthermore, because of their bolted, piecemeal construction, steel racks and building members should not be depended upon alone for circuit returns.

High-Frequency Effects

A significant amount of research has been conducted into why the 60-Hz power grounding system is incapable of conducting RF signals to the common reference point, thereby equalizing existing differences in potential [16]. A properly designed power grounding system is of sufficiently low impedance at 60 Hz to equalize any potential differences so that enclosures, raceways, and all grounded metal elements are at the same reference (ground) potential. However, at higher frequencies, equalization is prevented because of increased impedance.

The impedance of a conductor consists of three basic components: resistance, capacitive reactance, and inductive reactance. Although the inductance L (in Henrys) will be constant for a conductor of a given length and cross-sectional area, the inductive reactance X_L will vary according to the frequency f of the applied voltage as follows:

$$X_L = 2\pi f L \qquad (15.1)$$

Therefore, at 60 Hz, the inductive reactance will be $377 \times L$; and at 30 MHz, it will be $(188.5 \times 10^6) \times L$. It is evident, then, that at 60 Hz the equipment grounding conductor is a short-circuit, but at 30 MHz, it is effectively an open circuit.

In addition to increased inductive reactance at higher frequencies, there is also stray capacitance and stray inductance between adjacent conductors and/or between conductors and adjacent grounded metal, as well as resonance effects. These factors combine to yield an increase in apparent conductor impedance at higher frequencies.

If conductors are connected in a mesh or grid to form a multitude of low-impedance loops in parallel, there will be little voltage difference between any two points on the grid at all frequencies from 60 Hz up to a frequency where the length of one side of the square represents about $1/10$ wavelength. A grid made up of 2-ft squares (such as a raised computer floor) will at any point provide an effective equipotential ground reference point for signals up to perhaps 30 MHz.

Power-Center Grounding

A modular power center, commonly found in computer-room installations, provides a comprehensive solution to AC power-distribution and ground-noise considerations. Such equipment is available from several manufacturers, with various options and features. A computer power-distribution center generally includes an isolation transformer designed for noise suppression, distribution circuit breakers, power-supply cables, and a status-monitoring unit. The system concept is shown in Fig. 15.58. Input power is fed to an isolation transformer with primary taps to match the AC voltage required at the facility. A bank of circuit breakers is included in the chassis, and individual preassembled and terminated cables supply AC power to the various loads. A status-monitoring circuit signals the operator if any condition is detected outside normal parameters.

The ground system is an important component of the MPC. A unified approach, designed to prevent noise or circulating currents, is taken to grounding for the entire facility. This results in a clean ground connection for all equipment on-line.

The use of a modular power center can eliminate the inconvenience associated with rigid conduit installations. Distribution systems also are expandable to meet future facility growth. If the plant is ever relocated, the power center can move with it. MPC units usually are expensive. However, considering the costs of installing circuit-breaker boxes, conduit, outlets, and other hardware on-site by a licensed electrician, the power-center approach may be economically viable. The use of a power center also will make it easier to design a standby power system for the facility. Many computer-based operations do not have a standby generator on site. Depending on the location of the facility, it might be difficult or even impossible to install a generator to provide standby power in the event of a utility company outage. However, by using the power-center approach to AC distribution for computer and other critical load equipment, an uninterruptible power system can be installed easily to power only the loads that are

Facility Grounding

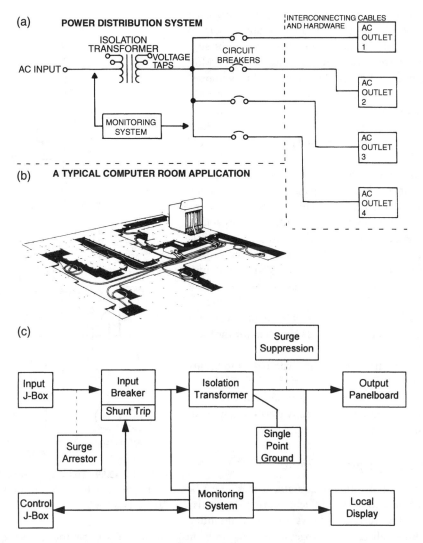

FIGURE 15.58 The basic concept of a computer-room modular power center: (*a*) basic line drawing of system, (*b*) typical physical implementation, (*c*) functional block diagram. Both single- and multiphase configurations are available. When ordering an MPC, the customer can specify cable lengths and terminations, making installation quick and easy. (After [17].)

required to keep the facility operating. With a conventional power-distribution system—where all AC power to the building, or a floor of the building, is provided by a single large circuit-breaker panel—separating the critical loads from other nonessential loads (such as office equipment, lights, and heating equipment) can be an expensive detail.

Isolation Transformers

One important aspect of an MPC is the isolation transformer. The transformer serves to:

- Attenuate transient disturbances on the AC supply lines.
- Provide voltage correction through primary-side taps.
- Permit the establishment of a local ground reference for the facility served.

Whether or not an MPC is installed at a facility, consideration should be given to the appropriate use of an isolation transformer near a sensitive load.

The AC power supply for many buildings originates from a transformer located in a basement utility room. In large buildings, the AC power for each floor can be supplied by transformers closer to the loads they serve. Most transformers are 208 Y/120 V three-phase. Many fluorescent lighting circuits operate at 277 V, supplied by a 408 Y/277 V transformer. Long feeder lines to DP systems and other sensitive loads raise the possibility of voltage fluctuations based on load demand and ground-loop-induced noise.

Figure 15.59 illustrates the preferred method of power distribution in a building. A separate dedicated isolation transformer is located near the DP equipment, providing good voltage regulation and permitting the establishment of an effective single-point star ground in the data processing center. Note that the power-distribution system voltage shown in the figure (480 V) is maintained until the DP step-down isolation transformer. Use of this higher voltage provides more efficient transfer of electricity throughout the plant. At 480 V, the line current is about 43% of the current in a 208 V system for the same conducted power.

There are a number of tradeoffs with regard to facility AC power system design. An experienced, licensed electrical contractor and/or registered professional engineer should be consulted during the early stages of any renovation or new construction project.

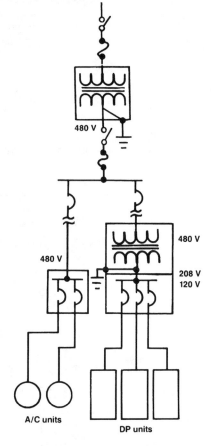

FIGURE 15.59 Preferred power-distribution configuration for a data processing site. (After [15].)

Grounding Equipment Racks

The installation and wiring of equipment racks must be planned carefully to avoid problems during day-to-day operations. Figure 15.60 shows the recommended approach. Bond adjacent racks together with $3/8$- to $1/2$-in.-diameter bolts. Clean the contacting surfaces by sanding down to bare metal. Use lockwashers on both ends of the bolts. Bond racks together using at least six bolts per side (three bolts for each vertical rail). After securing the racks, repaint the connection points to prevent corrosion.

Run a ground strap from the *main facility ground point*, and bond the strap to the base of each rack. Spot-weld the strap to a convenient spot on one side of the rear portion of each rack. Secure the strap at the same location for each rack used. A mechanical connection between the rack and the ground strap can be made using bolts and lockwashers, if necessary. Be certain, however, to sand down to bare metal before making the ground connection. Because of the importance of the ground connection, it is recommended that each attachment be made with a combination of crimping and silver-solder. Keep the length of strap between adjacent bolted racks as short as possible by routing the strap directly under the racks.

Install a vertical ground bus in each rack (as illustrated in Fig. 15.60). Use about 1-$1/2$-in.-wide, $1/4$-in.-thick copper busbar. Size the busbar to reach from the bottom of the rack to about 1 ft short of the top. The exact size of the busbar is not critical, but it must be sufficiently wide and rigid to permit the drilling of $1/8$-in. holes without deforming.

Mount the ground busbar to the rack using insulated standoffs. Porcelain standoffs commonly found in high-voltage equipment are useful for this purpose. Porcelain standoffs are readily available and

Facility Grounding

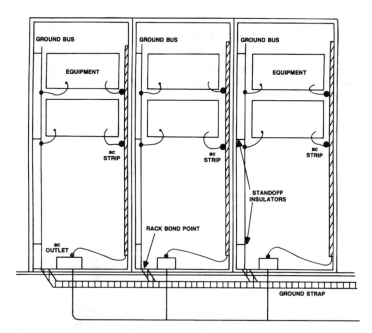

FIGURE 15.60 Recommended grounding method for equipment racks. To make assembly of multiple racks easier, position the ground connections and AC receptacles at the same location in all racks.

reasonably priced. Attach the ground busbar to the rack at the point where the facility ground strap attaches to the rack. Silver-solder the busbar to the rack and strap at the same location in each rack used.

Install an orange-type isolated AC receptacle box at the bottom of each rack. The orange-type outlet isolates the green-wire power ground from the receptacle box. Use insulated standoffs to mount the AC outlet box to the rack. The goal of this arrangement is to keep the green-wire AC and facility system grounds separate from the AC distribution conduit and metal portions of the building structure. Try to route the power conduit and facility ground cable or strap via the same physical path. Keep metallic conduit and building structures insulated from the facility ground line, except at the bulkhead panel (main grounding point). Carefully check the local electrical code before proceeding.

Although the foregoing procedure is optimum from a signal-grounding standpoint, it should be pointed out that under a ground fault condition, performance of the system can be unpredictable if high currents are being drawn in the current-carrying conductors supplying the load. Vibration of AC circuit elements resulting from the magnetic field effects of high-current-carrying conductors is insignificant as long as all conductors are within the confines of a given raceway or conduit. A ground fault will place return current outside of the normal path. If sufficiently high currents are being conducted, the consequences can be devastating. Sneak currents from ground faults have been known to destroy wiring systems that were installed exactly to code. As always, consult an experienced electrical contractor.

Mount a vertical AC strip inside each rack to power the equipment. Insulate the power strip from the rack using porcelain standoffs or use a strip with an insulated (plastic) housing. Such a device is shown in Fig. 15.61. Power equipment from the strip using standard three-prong grounding AC plugs. Do not defeat the safety ground connection. Equipment manufacturers use this ground to drain transient energy. Furthermore, defeating the safety ground will violate local electrical codes.

Mount equipment in the rack using normal metal mounting screws. If the location is in a high-RF field, clean the rack rails and equipment-panel connection points to ensure a good electrical bond. This is important because in a high-RF field, detection of RF energy can occur at the junctions between equipment chassis and the rack.

Connect a separate ground wire from each piece of equipment in the rack to the vertical ground busbar. Use no. 12 stranded copper wire (insulated) or larger. Connect the ground wire to the busbar by drilling a hole in the busbar at a convenient elevation near the equipment. Fit one end of the ground wire with an enclosed-hole solderless terminal connector (no. 10-sized hole or larger). Attach the ground wire to the busbar using no. 10 (or larger) hardware. Use an internal-tooth lockwasher between the busbar and the nut. Fit the other end of the ground wire with a terminal that will be accepted by the ground terminal at the equipment. If the equipment has an isolated *signal* ground terminal, tie it to the ground busbar as well.

Whenever servicing equipment in the rack, make certain to disconnect the AC power cord before removing the unit from the rack and/or disconnecting the rack ground wire. During any service work, make the first step removal of the AC power plug; when reassembling the unit, make the last step reinsertion of the plug. Do not take chances.

Figure 15.62 shows each of the grounding elements discussed in this section integrated into one diagram. This approach fulfills the requirements of personnel safety and equipment performance.

Follow similar grounding rules for simple one-rack equipment installations. Figure 15.63 illustrates the grounding method for a single open-frame equipment rack. The vertical ground bus is supported by insulators, and individual jumpers are connected from the ground rail to each chassis.

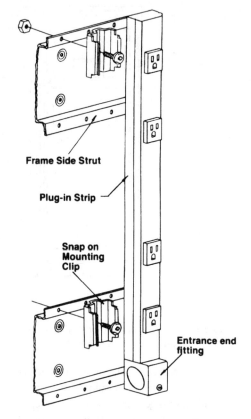

FIGURE 15.61 Rack mounting detail for an insulated-shell AC power strip.

15.6 Grounding Signal-Carrying Cables

Proper ground-system installation is the key to minimizing noise currents on signal-carrying cables. Audio, video, and data lines are often subject to AC power noise currents and RFI. The longer the cable run, the more susceptible it is to disturbances. Unless care is taken in the layout and installation of such cables, unacceptable performance of the overall system can result.

Analyzing Noise Currents

Figure 15.64 shows a basic source and load connection. No grounds are present, and both the source and the load float. This is the optimum condition for equipment interconnection. Either the source or the load can be tied to ground with no problems, provided only one ground connection exists. *Unbalanced* systems are created when each piece of equipment has one of its connections tied to ground, as shown in Fig. 15.65. This condition occurs if the source and load equipment have unbalanced (single-ended) inputs and outputs. This type of equipment utilizes chassis ground (or common) for one of the conductors. Problems are compounded when the equipment is separated by a significant distance.

As shown in Fig. 15.65, a difference in ground potential causes current flow in the ground wire. This current develops a voltage across the wire resistance. The ground-noise voltage adds directly to the signal itself. Because the ground current is usually the result of leakage in power transformers and line filters, the 60-Hz signal gives rise to hum of one form or another. Reducing the wire resistance through a heavier ground conductor helps the situation, but cannot eliminate the problem.

Facility Grounding

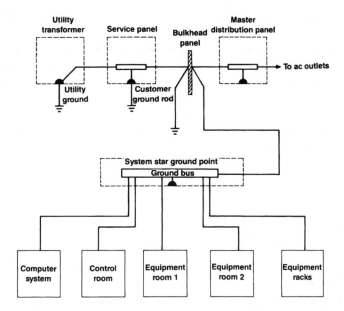

FIGURE 15.62 Equivalent ground circuit diagram for a medium-sized commercial/industrial facility.

By amplifying both the high side and the ground side of the source and subtracting the two to obtain a *difference signal*, it is possible to cancel the ground-loop noise. This is the basis of the *differential input* circuit, illustrated in Fig. 15.66. Unfortunately, problems still can exist with the unbalanced-source-to-balanced-load system. The reason centers on the impedance of the unbalanced source. One side of the line will have a slightly lower amplitude because of impedance differences in the output lines. By creating an output signal that is out of phase with the original, a balanced source can be created to eliminate this error. See Fig. 15.67. As an added benefit, for a given maximum output voltage from the source, the signal voltage is doubled over the unbalanced case.

Types of Noise

Two basic types of noise can appear on AC power, audio, video, and computer data lines within a facility: *normal mode* and *common mode*. Each type has a particular effect on sensitive load equipment. The normal-mode voltage is the potential difference that exists between pairs of power (or signal) conductors. This voltage also is referred to as the *transverse-mode* voltage. The common-mode voltage is a potential difference (usually noise) that appears between the power or signal conductors and the local ground reference. The differences between normal-mode and common-mode noise are illustrated in Fig. 15.68.

The common-mode noise voltage will change, depending upon what is used as the ground reference point. It is often possible to select a ground reference that has a minimum common-mode voltage with respect to the circuit of interest, particularly if the reference point and the load equipment

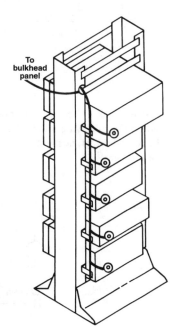

FIGURE 15.63 Ground bus for an open-frame equipment rack.

are connected by a short conductor. Common-mode noise can be caused by electrostatic or electromagnetic induction.

In practice, a single common-mode or normal-mode noise voltage is rarely found. More often than not, load equipment will see both common-mode and normal-mode noise signals. In fact, unless the facility wiring system is unusually well-balanced, the noise signal of one mode will convert some of its energy to the other mode.

Common-mode and normal-mode noise disturbances typically are caused by momentary impulse voltage differences among parts of a distribution system that have differing ground potential references. If the sections of a system are interconnected by a signal path in which one or more of the conductors are grounded at each end, the ground offset voltage can create a current in the grounded signal conductor.

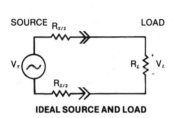

FIGURE 15.64 A basic source and load connection. No grounds are indicated, and both the source and the load float.

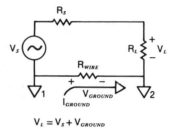

$$V_L = V_S + V_{GROUND}$$

FIGURE 15.65 An unbalanced system in which each piece of equipment has one of its connections tied to ground.

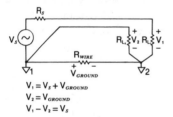

$$V_1 = V_S + V_{GROUND}$$
$$V_2 = V_{GROUND}$$
$$V_1 - V_2 = V_S$$

FIGURE 15.66 Ground-loop noise can be canceled by amplifying both the high side and the ground side of the source and subtracting the two signals.

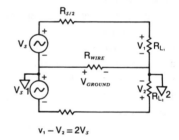

$$V_1 - V_2 = 2V_S$$

FIGURE 15.67 A balanced source configuration where the inherent amplitude error of the system shown in Fig. 15.66 is eliminated.

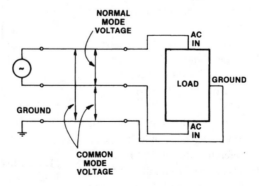

FIGURE 15.68 The principles of normal-mode and common-mode noise voltages as they apply to AC power circuits.

Facility Grounding

If noise voltages of sufficient potential occur on signal-carrying lines, normal equipment operation can be disrupted. See Fig. 15.69.

Noise Control

Noise control is an important aspect of computer and electronic system design and maintenance [5]. The process of noise control through proper grounding techniques is more correctly called *referencing*. For this discussion, electronic systems can be viewed as a multiplicity of signal sources transmitting signals to a multiplicity of loads. Practically speaking, the impedance of the signal return path is never zero, and dedicated return paths for each source-load pair are not always practical. Packaged electronics systems typically incorporate a common signal reference plane that serves as a common return path for numerous source-load pairs. (See Fig. 15.70a.) The signal reference plane may be a large dedicated area

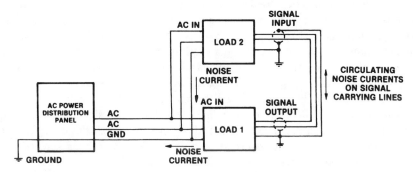

FIGURE 15.69 An illustration of how noise currents can circulate within a system because of the interconnection of various pieces of hardware.

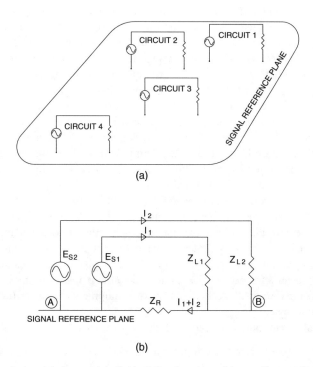

FIGURE 15.70 The equipotential plane: (a) individual signal sources, (b) net effect on the signal reference plane. (Courtesy of Intertec Publishing. After [5].)

on a circuit board, the metal chassis or enclosure of the electronic equipment, or the metal frame or mounting rack that houses several different units. Ideally, the signal reference plane offers zero impedance to the signal current. Practically, however, the signal reference plane has a finite impedance. The practical result is illustrated in Fig. 15.70b, and is called *common-impedance* or *conductive coupling*.

Because a practical signal reference plane has a finite impedance, current flow in the plane will produce potential differences between various points on the plane. Source-load pairs referenced to the plane will, therefore, experience interference as a result. Z_R is common to both circuits referenced to the plane in Fig. 15.70b. Thus, I_1 and I_2 returning to their respective sources will produce interference voltages by flowing through Z_R. The total interference voltage drop seen across Z_R causes the source reference A to be at a different potential than the load reference B. This difference in potential is often called *ground voltage shift* (even though ground may not even be involved), and is a major source of noise and interference in electronic circuits.

Ground voltage shifts can also be caused by electromagnetic or electrostatic fields in close proximity to the source-load pairs. The interference source induces interference voltages into any closed loop by antenna action. This loop is called a *ground loop* (even though ground may not be involved). Interference voltages can be minimized by reducing the loop area as much as possible. This can be very difficult if the loop includes an entire room. The interference voltage can be eliminated entirely by breaking the loop.

Within individual electronic equipments, the signal reference plane consists of a metal plate or the metal enclosure or a rack assembly as previously discussed. Between units of equipment that are located in different rooms, on different floors, or even in different buildings, the signal reference planes of each unit must be connected together via interconnected wiring such as coax shields or separate conductors. This action, of course, increases the impedance between signal reference planes and makes noise control more difficult. Reducing noise caused by common-impedance coupling is a matter of reducing the impedance of the interconnected signal reference planes.

Regardless of the configuration encountered (i.e., circuit board, individual electronic equipment, or equipments remotely located within a facility or in separate buildings), the next question to be answered is: should the signal reference be connected to ground? Floating signal grounding, single-point grounding, and multipoint grounding are methods of accomplishing this signal-reference-to-ground connection.

IEEE Standard 1100 recommends multipoint grounding for most computers. For effective multipoint grounding, conductors between separate points desired to be at the same reference potential should be less than 0.1 wavelength ($\lambda/10$) of the interference or noise-producing signal. Longer conductor lengths will exhibit significant impedance because of the antenna effect and become ineffective as equalizing conductors between signal references. On printed circuit cards or within individual electronics equipments, short conductor lengths are usually easy to achieve. For remotely located equipments, however, short equalizing conductor lengths may be impossible. Equipments in relatively close proximity can be equalized by cutting bonding conductors to an odd multiple of $\lambda/2$ at the noise frequency. The distribution of standing waves on bonding conductors is illustrated in Fig. 15.71. The impedance, Z, is minimum for odd multiples of $\lambda/2$ but maximum for odd multiples of $\lambda/4$.

For remotely located electronics equipment within the same room, equalizing potential between respective signal reference systems may have to be accomplished with an equipotential plane. An ideal equipotential signal reference plane is one that has zero volts (thus zero impedance) between any two points on the plane. Because an *ideal* equipotential plane is not attainable, a *nominal* equipotential plane is accepted. Multipoint grounding connections are made to the plane, which ensures minimum ground voltage shift between signal reference systems connected to the plane. Collectively, the current flow in an equipotential plane can be quite large. Between any two equipments, however, the current flow should be low because of the many current paths available.

Practical versions of an equipotential plane include the following:

- Bolted stringer system of a raised computer floor
- Flat copper strips bonded together at 2 ft centers
- Copper conductors bonded together at 2 ft centers
- Single or multiple, flat copper strips connected between equipment

Facility Grounding

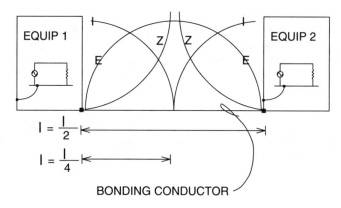

FIGURE 15.71 Standing waves on a bonding conductor resulting from the antenna effect. (After [5].)

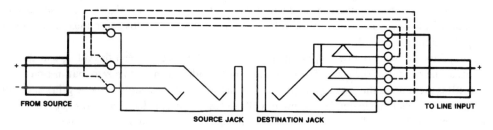

FIGURE 15.72 Patch-panel wiring for seven-terminal normalling jack fields. Use patch cords that connect ground (sleeve) at both ends.

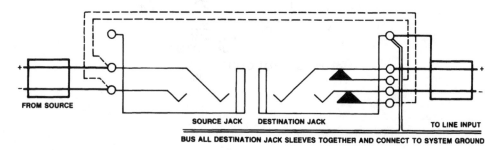

FIGURE 15.73 Patch-panel wiring for conventional normalling jack fields. Use patch cords that connect ground (sleeve) at both ends.

Patch-Bay Grounding

Patch panels for audio, video, and data circuits require careful attention to planning to avoid built-in grounding problems. Because patch panels are designed to tie together separate pieces of equipment, often from remote areas of a facility, the opportunity exists for ground loops. The first rule of patch-bay design is to never use a patch bay to switch low-level (microphone) signals. If mic sources must be patched from one location to another, install a bank of mic-to-line amplifiers to raise the signal levels to 0 dBm before connection to the patch bay. Most video output levels are 1 V P-P, giving them a measure of noise immunity. Data levels are typically 5 V. Although these line-level signals are significantly above the noise floor, capacitive loading and series resistance in long cables can reduce voltage levels to a point that noise becomes a problem.

Newer-design patch panels permit switching of ground connections along with signal lines. Figure 15.72 illustrates the preferred method of connecting an audio patch panel into a system. Note that the source and destination jacks are *normalled* to establish ground signal continuity. When another signal is plugged into the destination jack, the ground from the new source is carried to the line input of the destination jack. With such an approach, jack cords that provide continuity between sleeve (ground) points are required.

If only older-style, conventional jacks are available, use the approach shown in Fig. 15.73. This configuration will prevent ground loops, but because destination shields are not carried back to the source when normalling, noise will be higher. Bus all destination jack sleeves together, and connect to the local (rack) ground. The wiring methods shown in Figs. 15.72 and 15.73 assume balanced input and output lines with all shields terminated at the load (input) end of the equipment.

Input/Output Circuits

Common-mode rejection ratio (CMRR) is the measure of how well an input circuit rejects ground noise. The concept is illustrated in Fig. 15.74. The input signal to a differential amplifier is applied between the plus and minus amplifier inputs. The stage will have a certain gain for this signal condition, called the *differential gain*. Because the ground-noise voltage appears on both the plus and minus inputs simultaneously, it is common to both inputs.

The amplifier subtracts the two inputs, giving only the difference between the voltages at the input terminals at the output of the stage. The gain under this condition should be zero, but in practice it is not. CMRR is the ratio of these two gains (the differential gain and the common-mode gain) in decibels. The larger the number, the better. For example, a 60-dB CMRR means that a ground signal common to the two inputs will have 60 dB less gain than the desired differential signal. If the ground noise is already

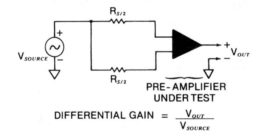

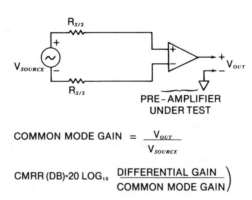

FIGURE 15.74 The concept of common-mode rejection ratio (CMRR) for an active-balanced input circuit: (a) differential gain measurement; (b) calculating CMRR.

40 dB below the desired signal level, the output noise level will be 100 dB below the desired signal level. If, however, the noise is already part of the differential signal, the CMRR will do nothing to improve it.

Active-balanced I/O circuits are the basis of nearly all professional audio interconnections (except for speaker connections), many video signals, and increasing numbers of data lines. A wide variety of circuit designs have been devised for active-balanced inputs. All have the common goal of providing high CMRR and adequate gain for subsequent stages. All also are built around a few basic principles.

Figure 15.75 shows the simplest and least expensive approach, using a single operational amplifier (op-amp). For a unity gain stage, all the resistors are the same value. This circuit presents an input impedance to the line that is different for the two sides. The positive input impedance will be twice that of the negative input. The CMRR is dependent on the matching of the four resistors and the balance of the source impedance. The noise performance of this circuit, which usually is limited by the resistors, is a tradeoff between low loading of the line and low noise.

Another approach, shown in Fig. 15.76, uses a buffering op-amp stage for the positive input. The positive signal is inverted by the op-amp, then added to the negative input of the second inverting amplifier stage. Any common-mode signal on the positive input (which has been inverted) will cancel when it is added to the negative input signal. Both inputs have the same impedance. Practical resistor-matching limits the CMRR to about 50 dB. With the addition of an adjustment potentiometer, it is possible to achieve 80 dB CMRR, but component aging will usually degrade this over time.

Adding a pair of buffer amplifiers before the summing stage results in an *instrumentation-grade* circuit, as shown in Fig. 15.77. The input impedance is increased substantially, and any source impedance effects are eliminated. More noise is introduced by the added op-amp, but the resistor noise usually can be decreased by reducing impedances, causing a net improvement in system noise.

Early active-balanced output circuits used the approach shown in Fig. 15.78. The signal is buffered to provide one phase of the balanced output. This signal then is inverted with another op-amp to provide the other phase of the output signal. The outputs are taken through two resistors, each of which is half of the desired source impedance. Because the load is driven from the two outputs, the maximum output voltage is double that of an unbalanced stage.

The circuit in Fig. 15.78 works reasonably well if the load is always balanced, but it suffers from two problems when the load is not balanced. If the negative output is shorted to ground by an unbalanced load connection, the first op-amp is likely to distort. This produces a distorted signal at the input to the other op-amp. Even if the circuit is arranged so that the second op-amp is grounded by an unbalanced

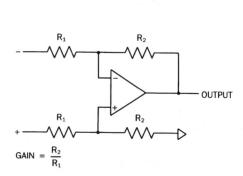

FIGURE 15.75 The simplest and least expensive active-balanced input op-amp circuit. Performance depends on resistor-matching and the balance of the source impedance.

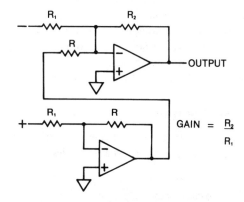

FIGURE 15.76 An active-balanced input circuit using two op-amps; one to invert the positive input terminal and the other to buffer the difference signal. Without adjustments, this circuit will provide about 50 dB CMRR.

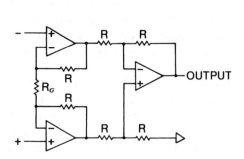

FIGURE 15.77 An active-balanced input circuit using three op-amps to form an instrumentation-grade circuit. The input signals are buffered, then applied to a differential amplifier.

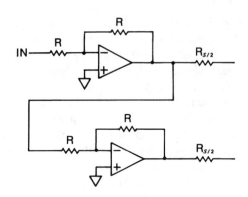

FIGURE 15.78 A basic active-balanced output circuit. This configuration works well when driving a single balanced load.

load, the distorted output current probably will show up in the output from coupling through grounds or circuit-board traces. Equipment that uses this type of balanced stage often provides a second set of output connections that are wired to only one amplifier for unbalanced applications.

The second problem with the circuit in Fig. 15.78 is that the output does not *float*. If any voltage difference (such as power-line hum) exists between the local ground and the ground at the device receiving the signal, it will appear as an addition to the signal. The only ground-noise rejection will be from the CMRR of the input stage at the receive end.

The preferred basic output stage is the electronically balanced and floating design shown in Fig. 15.79. The circuit consists of two op-amps that are cross-coupled with positive and negative feedback. The output of each amplifier is dependent on the input signal and the signal present at the output of the other amplifier. These designs may have gain or loss, depending on the selection of resistor values. The output impedance is set via appropriate selection of resistor values. Some resistance is needed from the output terminal to ground to keep the output voltage from floating to one of the power-supply rails. Care must be taken to properly compensate the devices. Otherwise, stability problems can result.

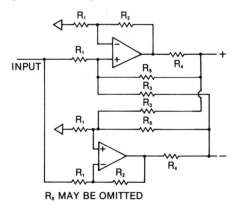

FIGURE 15.79 An electronically balanced and floating output circuit. A stage such as this will perform well even when driving unbalanced loads.

Cable Routing

Good engineering practice dictates that different signal levels be grouped and separated from each other. It is common practice to separate cables into the following groups:

- AC power
- Speaker lines
- Line-level audio
- Microphone-level audio
- Video lines
- Control and data lines

Always use two-conductor shielded cable for all audio signal cables. This includes both balanced and unbalanced circuits, and microphone-level and line-level cables. On any audio cable connecting two pieces of equipment, tie the shield at one end only. Connect at the receiving end of signal transmission. On video coaxial cables running to outlet jacks mounted on plates, isolate the connector from the plate. The shield should connect to ground only at the equipment input/output or patch panel. For data cables, carefully follow the recommendations of the equipment manufacturer. The preferred interconnection method for long data cables is fiber optics, which eliminates ground-loop problems altogether.

Overcoming Ground-System Problems

Although the concept of equipment grounding seems rather basic, it can become a major headache if not done correctly. Even if all of the foregoing guidelines are followed to the letter, there is the possibility of ground loops and objectionable noise on audio, video, or data lines. The larger the physical size of the facility, the greater the potential for problems. An otherwise perfect ground system can be degraded by a single wiring error. An otherwise clean signal ground can be contaminated by a single piece of equipment experiencing a marginal ground fault condition.

If problems are experienced with a system, carefully examine all elements to track down the wiring error or offending load. Do not add AC line filters and/or signal line filters to correct a problem system. In a properly designed system, even one in a high-RF field, proper grounding and shielding techniques will permit reliable operation. Adding filters merely hides the problem. Instead, correct the problem at its source. In a highly complex system such as a data processing facility, the necessity to interconnect a large number of systems may require the use of fiber-optic transmitters and receivers.

References

1. Webster's New Collegiate Dictionary.
2. IEEE Standard 100, *Definitions of Electrical and Electronic Terms*, IEEE, New York.
3. IEEE Standard 142, "Recommended Practice for Grounding Industrial and Commercial Power Systems," IEEE, New York, 1982.
4. IEEE Standard 1100, "Recommended Practice for Powering and Grounding Sensitive Electronics Equipment," IEEE, New York, 1992.
5. DeWitt, William E., "Facility Grounding Practices," in *The Electronics Handbook*, Jerry C. Whitaker (Ed.), CRC Press, Boca Raton, FL, pp. 2218–2228, 1996.
6. NFPA Standard 70, "The National Electrical Code," National Fire Protection Association, Quincy, MA, 1993.
7. Sunde, E. D., *Earth Conduction Effects in Transmission Systems*, Van Nostrand Co., New York, 1949.
8. Schwarz, S. J., "Analytical Expression for Resistance of Grounding Systems," *AIEE Transactions*, 73, Part III-B, 1011–1016, 1954.
9. Block, Roger, "The Grounds for Lightning and EMP Protection," PolyPhaser Corporation, Gardnerville, NV, 1987.
10. Carpenter, Roy, B., "Improved Grounding Methods for Broadcasters," *Proceedings, SBE National Convention*, Society of Broadcast Engineers, Indianapolis, 1987.
11. Lobnitz, Edward A., "Lightning Protection for Tower Structures," in *NAB Engineering Handbook*, 9th ed., Jerry C. Whitaker (Ed.), National Association of Broadcasters, Washington, D.C., 1998.
12. DeDad, John A., (Ed.), "Basic Facility Requirements," in *Practical Guide to Power Distribution for Information Technology Equipment*, PRIMEDIA Intertec, Overland Park, KS, p. 24, 1997.
13. Military Handbook 419A, "Grounding, Bonding, and Shielding for Electronic Systems," U.S. Government Printing Office, Philadelphia, PA, December 1987.
14. IEEE 142 (Green Book), "Grounding Practices for Electrical Systems," IEEE, New York.

15. Federal Information Processing Standards Publication No. 94, *Guideline on Electrical Power for ADP Installations*, U.S. Department of Commerce, National Bureau of Standards, Washington, D.C., 1983.
16. DeDad, John A., (Ed.), "ITE Room Grounding and Transient Protection," in *Practical Guide to Power Distribution for Information Technology Equipment*, PRIMEDIA Intertec, Overland Park, KS, pp. 83–84, 1997.
17. Gruzs, Thomas M., "High Availability, Fault-Tolerant AC Power Distribution Systems for Critical Loads, *Proceedings*, Power Quality Solutions/Alternative Energy, Intertec International, Ventura, CA, pp. 22, September 1996.

Further Information

Benson, K. B., and Jerry C. Whitaker, *Television and Audio Handbook for Engineers and Technicians*, McGraw-Hill, New York, 1989.

Block, Roger, "How to Ground Guy Anchors and Install Bulkhead Panels," *Mobile Radio Technology*, PRIMEDIA Intertec, Overland Park, KS, February 1986.

Davis, Gary, and Ralph Jones, *Sound Reinforcement Handbook*, Yamaha Music Corporation, Hal Leonard Publishing, Milwaukee, WI, 1987.

Defense Civil Preparedness Agency, "EMP Protection for AM Radio Stations," Washington, D.C., TR-61-C, May 1972.

Fardo, S., and D. Patrick, *Electrical Power Systems Technology*, Prentice-Hall, Englewood Cliffs, NJ, 1985.

Hill, Mark, "Computer Power Protection," *Broadcast Engineering*, PRIMEDIA Intertec, Overland Park, KS, April 1987.

Lanphere, John, "Establishing a Clean Ground," *Sound & Video Contractor*, PRIMEDIA Intertec, Overland Park, KS, August 1987.

Lawrie, Robert, *Electrical Systems for Computer Installations*, McGraw-Hill, New York, 1988.

Little, Richard, "Surge Tolerance: How Does Your Site Rate?" *Mobile Radio Technology*, Intertec Publishing, Overland Park, KS, June 1988.

Midkiff, John, "Choosing the Right Coaxial Cable Hanger," *Mobile Radio Technology*, Intertec Publishing, Overland Park, KS, April 1988.

Mullinack, Howard G., "Grounding for Safety and Performance," *Broadcast Engineering*, PRIMEDIA Intertec, Overland Park, KS, October 1986.

Schneider, John, "Surge Protection and Grounding Methods for AM Broadcast Transmitter Sites," *Proceedings*, SBE National Convention, Society of Broadcast Engineers, Indianapolis, 1987.

Sullivan, Thomas, "How to Ground Coaxial Cable Feedlines," *Mobile Radio Technology*, PRIMEDIA Intertec, Overland Park, KS, April 1988.

Technical Reports LEA-9-1, LEA-0-10, and LEA-1-8, Lightning Elimination Associates, Santa Fe Springs, CA.

16
Network Switching Concepts

Tsong-Ho Wu
Bellcore

16.1 Introduction 16-1
16.2 Broadband Switching Layer Model and Formats 16-3
   Multilayer Transport Model • Physical Path vs Virtual Path
   • Channel Formats for STM and ATM
16.3 STM System Configuration and Operations 16-8
16.4 ATM Cell Switching System Configuration
   and Operations 16-9
16.5 Comparison Between STM and ATM 16-10
16.6 Protection Switching Concept 16-10
16.7 Summary and Remarks 16-12

16.1 Introduction

Switching is the control or routing of signals in either physical or virtual circuit form to transmit data between specific points in a network. A switched network is a communication network that has switching facilities or centers to perform the function of message switching, packet switching, or circuit switching. The primary purpose of the network switching function is to maximize the transmission system efficiency by sharing the capacity of the transmission system among communications circuits.

In general, there are two major types of network switching techniques: packet switching and circuit switching. The message switching system may be viewed as a special case of the packet switching system. The traditional circuit switching concept is designed to handle and transport stream-type traffic, such as voice and video. A circuit-switched connection is set up with a fixed bandwidth for the duration of a connection, which provides fixed throughput and constant delay. In other words, network resources are dedicated in circuit-switched systems. In contrast, packet switching is primarily designed to efficiently carry data communications traffic. Its characteristics are access with buffering, statistical multiplexing, and variable throughput and delay, which is a consequence of the dynamic sharing of communication resources to improve utilization. Today, it is used exclusively for data applications. Examples of circuit-switched networks include telephony networks and private line networks. Examples of packet-switched networks include X.25 (with variable packet sizes) and **asynchronous transfer mode (ATM)** networks (with the fixed-cell size).

Today's telecommunications services are primarily supported by three separate switched networks using two major switching concepts. As depicted in Fig. 16.1, the circuit-switched network supports voice transport, which includes copper, radio, and/or fiber transmission systems using asynchronous or synchronous optical network (SONET) transmission technology. The signaling system needed to support switched voice services is supported by a specialized (delay-sensitive) packet-switched SS7 network. The

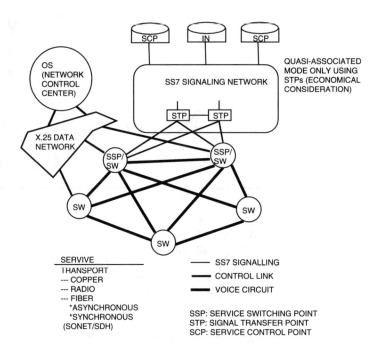

FIGURE 16.1 Today's telecommunications network infrastructure.

network supporting the access to the operations systems from the service transport system is a packet-switched X.25 network.

In the future telecommunication network infrastructure, both the narrowband services of today and emerging broadband services will be supported by an integrated B-integrated services digital network (B-ISDN). ATM has been adopted by the International Telecommunications Union (ITU-T) as the core technology to support B-ISDN transport. ATM is a high-speed, integrated multiplexing and switching technology that transmits information using fixed-length cells in a connection-oriented manner. Physical interfaces for the user-network interface (UNI) of 155.52 Mb/s and 622.08 Mb/s provide integrated support for high-speed information transfer and various communication modes, such as circuit and packet modes, and constant/variable/burst bit-rate communications.

In the ATM network, services, control, signaling, and operations and maintenance (OAM) messages are carried by the same physical network, but they are logically separated from each other by using different virtual channels (VCs) or virtual paths (VPs) with different quality of service (QoS) requirements.

ATM is growing and maturing rapidly. It has been implemented today in products by many equipment suppliers and deployed by customers who anticipate such business advantages as:

- Enabling high-bandwidth applications including desktop video, digital libraries, and real-time image transfer
- Coexistence of different types of traffic on a single network platform to reduce both the transport and operations costs
- Long-term network scalability and architectural stability

In addition, ATM has been used in both local and wide area networks. It can support a variety of high layer protocols and will cope with future network speeds of gigabits per second.

In North American standards, as specified by the American National Standards Institute (ANSI) T1S1 committee, ATM cells are transported over the SONET system as the physical layer. [Note that in ITU-T standards, there are two options for the transmission standards: synchronous digital hierarchy (SDH) and cell-based transmission. SDH is functionally equivalent to SONET, although some of the rates are different and the use of some overhead and multiplexing schemes may also be different. For the cell-based

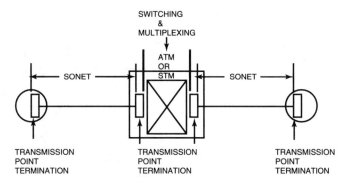

FIGURE 16.2 Relationship between SONET transmission, STM, and ATM switching.

transmission, ATM cells will not be accommodated by the SDH frame within the transmission link.] SONET is the North American standard synchronous transmission network technology for optical signal transmission [Ballart and Ching 1989]. In this multilayer network environment, the switching of the service connection may be implemented at the SONET layer using synchronous transfer mode (STM) or at the ATM layer using ATM. STM is a time division multiplexing and switching technology that uses a similar switching concept as traditional circuit switching systems.

Figure 16.2 depicts the relationship between SONET transmission, STM and ATM switching. The end-to-end ATM connection is established and transported through a set of transmission links that are terminated at nodes for processing and switching (called transfer mode). The nodal transfer mode may be existing STM or emerging ATM, depending on types of services and QoS being supported. STM switching technology has been widely deployed in today's telecommunications networks, whereas ATM is expected to be widely deployed in the near future. Without loss of generality, we will assume in this section that STM is operated on the SONET system.

To help understand how the broadband SONET/ATM network is used to support high-speed data and video services (discussed subsequently), this section focuses on reviewing switching concepts used in the SONET layer (STM switching) and the ATM layer (ATM switching). The switching concept is based on the characteristics of the signal switching format (e.g., time slot or virtual connection). Thus, this section will first review signal switching formats and their bandwidth relationship in a multilayer SONET/ATM network environment. The operations of STM and ATM switching concepts will then be explained. The protection switching concept for SONET/ATM networks will also be discussed.

16.2 Broadband Switching Layer Model and Formats

Multilayer Transport Model

Figure 16.3 shows the B-ISDN signal transport protocol reference model defined in CCITT Rec. I.321. [The International Telegraph and Telephone Consultative Committee (CCITT) has been renamed the International Telecommunication Union (ITU) since 1994.] Three layers are defined in the model: physical, ATM, and ATM adaptation layer (AAL), where the physical and ATM layers form the ATM transport platform.

In North America, the physical layer uses SONET standards, whereas in ITU-T, it uses the SDH or cell-based transmission system. The AAL layer is a service-dependent layer. Details of this B-ISDN transport reference model can be found in Boudec [1992]. SONET defines a progressive hierarchy of optical signal and line rates. The basic SONET building block is the synchronous transport signal at level 1 (STS-1) signal operating at 51.840 Mb/s. Higher rate signals (STS-N) are multiples (N) of this basic rate. Values of N currently used in standards are 1, 3, 12, 24, and 48, with 192 likely in the near future. Broadband services requiring more than one STS-1 payload capacity are transported by concatenated

STS-1s. For example, high-definition television (HDTV) signals requiring 135 Mb/s can be carried by three concatenated STS-1s, denoted by STS-3c, whose transport overheads and payload envelopes are aligned. The function of the SONET layer is to carry ATM cells or non-ATM frames (e.g., STS-1) in a high-speed and transparent manner, as well as to provide very fast protection switching capability to ATM cells or non-ATM frames whenever needed. SONET offers opportunities to implement new network architectures in a cost-effective manner [Wu 1992] and has been widely deployed by local exchange carriers (LECs) since 1991.

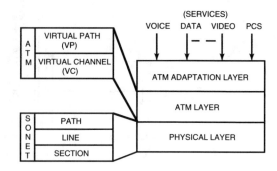

FIGURE 16.3 B-ISDN reference model.

The major function of the ATM layer is to provide fast multiplexing and routing for data transfer based on the header information. The ATM layer is divided into two sublayers: virtual path and virtual channel. A VC is a generic term used to describe a unidirectional communication capability for the transport of ATM cells. The VC identifies an end-to-end connection that can be set via either provisioning (permanent VC) or near real-time call setup (switched VC). Either way the bandwidth capacity of that connection is determined in real-time depending on traffic to be transported over that connection.

The VP accommodates a set of different VCs, all with the same terminations (source and destination). The VPs are managed by network systems, whereas VCs can be managed, end-to-end, by users with ATM terminals. For example, a business user may be provisioned with a VP to another user location to provide the equivalent of leased circuits, and another VP to the serving central office for switched services. Each VP may include several VCs for wide area network (WAN), MAN, private branch exchange (PBX), and video conference traffic.

The VC and VP are identified through virtual channel identifier (VCI) and virtual path identifier (VPI), respectively. For large networks, VCIs and VPIs are assigned on a per link basis (i.e., local significance). Thus, translation of VPI (or VCI) is necessary at intermediate ATM switches for an end-to-end VP (or VC). The switching operation for ATM cells will be discussed in Sec. 16.4.

Figure 16.4 shows the relationship among the virtual channel, virtual path, SONET's STS path, and SONET link. The SONET physical link [e.g., an optical carrier level-48 (OC-48) optical system] provides a bit stream to carry digital signals. This bit stream may include one of multiple digital paths, such as SONET STS-3c, STS-12c, or STS-48c. Each digital path may carry ATM cells via multiple VPs, each having multiple VCs. The switching method used for SONET's STS paths is STM using a hierarchical **time slot interchange** (TSI) concept, whereas the switching method for VPs/VCs uses a nonhierarchical ATM switching concept. In addition, STM performs network rerouting through physical network reconfiguration, whereas ATM performs network rerouting using logical network reconfiguration through the update of the routing table.

Physical Path vs Virtual Path

The switching principles used for SONET's STS Paths (STM) and ATM VPs/VCs (ATM) are completely different due to different characteristics of corresponding path structures. The STS path uses a physical path structure, whereas the VP/VC uses a logical path structure, as depicted in Fig. 16.5. The physical path concept used in the SONET STM system has a hierarchical structure with a fixed capacity for each physical path. For example, the VT1.5 and STS-1 have the capacity of 1.728 Mb/s and 51.84 Mb/s, respectively. To transport the optical signals over fiber, VT1.5s are multiplexed to an STS-1 and then to STS-12, STS-48 with other multiplexed streams for optical transport. Thus, a SONET transport node may equip a variety of switching equipment needed for each hierarchy of the signals. In contrast, the VP transport system is physically nonhierarchical [see Fig. 16.5(b)] and its capacity can be varied, which

Network Switching Concepts

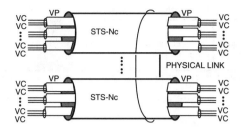

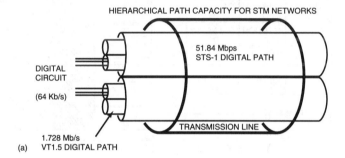

FIGURE 16.4 Relationship among the VC, VP, digital path, and physical link.

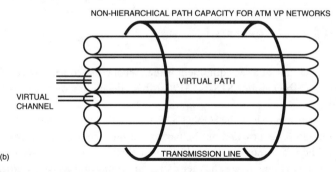

FIGURE 16.5 Physical path vs. virtual path: (a) Physical path structure, (b) virtual path structure.

ranges from 0 (for protection) up to the line rate or STS-Nc, depending on applications. This nonhierarchical multiplexing structure provides a natural grooming characteristic, which may simplify the nodal system design. Detailed discussions between STM and ATM VP path structures and their implications on network designs can be found in Sato, Ueda, and Yoshikai [1991].

Channel Formats for STM and ATM

In SONET/STM, data channels are represented by time slots and identified through the relative time slot location within the frame, as depicted in Fig. 16.6(a). Thus, STM switching is performed through time

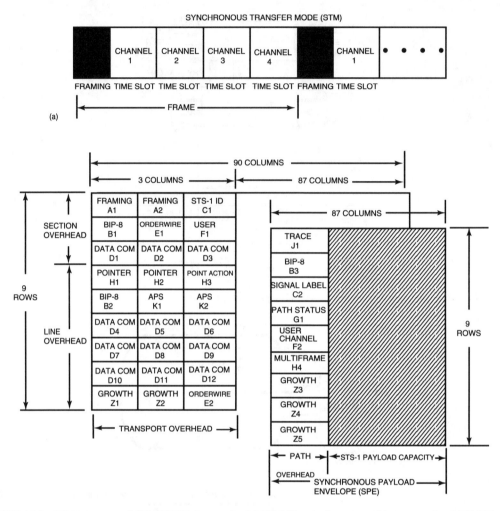

FIGURE 16.6 STM concept and SONET STS-1 example: (a) STM (framing) concept, (b) an example of SONET.

slot interchange. One example of STM equipment is the digital cross-connect system (DCS), which cross connects VT1.5s (1.728 Mb/s) and/or STS-1s (51.84 Mb/s), depending on the type of DCSs (wideband DCS or broadband DCS).

Figure 16.6(b) shows an example of the SONET STS-1 frame that corresponds to a frame in Fig. 16.6(a). In this example, each VT1.5 path carried on the payload of this STS-1 frame represents a channel (time slot) in Fig. 16.6(a). The STS-1 frame is divided into two portions: transport overhead and information payload, where the transport overhead is further divided into line and section overheads. The STS frame is transmitted every 125 μs. The first three columns of the STS-1 frame contain transport overhead and the remaining 87 columns and 9 rows (a total of 783 bytes) carry the STS-1 synchronous payload envelope (SPE). The SPE contains a 9-byte path overhead that is used for end-to-end service performance monitoring. Optical carrier level-1 (OC-1) is the lowest level optical signal used at equipment and network interfaces. An OC-1 is obtained from a STS-1 after scrambling and electrical-to-optical conversion.

In contrast, ATM channels are represented by a set of fixed-size cells [see Fig. 16.7(a)] and are identified through the channel indicator in the cell header [see Fig. 16.7(b)]. Thus, ATM switching is performed on a cell-by-cell basis based on routing information in the cell header.

Network Switching Concepts

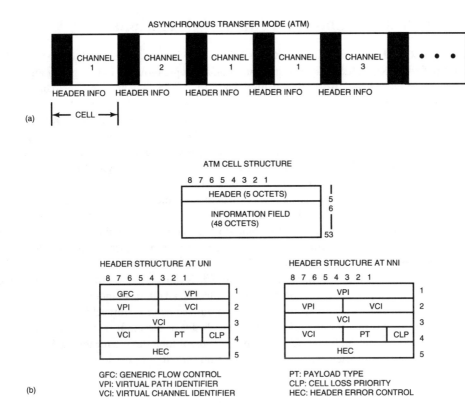

FIGURE 16.7 ATM concept and cell formats: (a) ATM (fixed cell) concept, (b) ATM cell format.

The ATM cell has two parts: the header and the payload. The header has 5 bytes and the payload has 48 bytes. As mentioned earlier, the major function of the ATM layer is to provide fast multiplexing and routing for data transfer based on information included in the 5-byte header. This 5-byte header includes information not only for routing (i.e., VPI and VCI fields), but also fields used to (1) indicate the type of information contained in the cell payload (e.g., user data or network operations information), (2) assist in controlling the flow of traffic at the UNI, (3) establish priority for the cell, and (4) facilitate header error control and cell delineation functions. One key feature of this ATM layer is that ATM cells can be independently labeled and transmitted on demand. This allows facility bandwidth to be allocated as needed, without fixed hierarchical channel rates required for STM networks. Since ATM cells may be sent either periodically or randomly (i.e., in bursts), both constant and variable bit-rate services are supported at a broad range of bit rates. The connections supported at the VP layer are either permanent or semipermanent connections that do not require call control, real-time bandwidth management, and processing capabilities. The connections at the VC layer may be permanent, semipermanent, or switched virtual connections (SVCs). SVCs require a signaling system to support its establishment, tear-down, and capacity management.

Figure 16.8 shows an example of ATM cell mapping in SONET payloads for high-speed transport. In ANSI T1S1, ATM cells are transported through the SONET STS-3c or STS-12c path. Note that ATM Forum has specified other ATM cell mapping formats such as DS1, DS3, and TAXI interfaces.

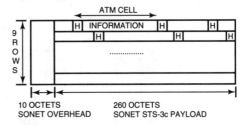

FIGURE 16.8 ATM cell mapping on SONET payloads.

16.3 STM System Configuration and Operations

One example of SONET equipment that performs STS path switching is a SONET DCS. Figure 16.9 depicts a simplified SONET DCS system configuration and an operation for facility grooming. In this figure, an incoming OC-48 optical signal is demultiplexed to 48 STS-1s, which are cross connected to the appropriate output ports destined for appropriate destinations. The STS-1 path cross connection is performed through the TSI switching matrix. The TSI switching matrix within the DCS interfaces STS-1s/STS-Ncs and cross connects STS-1s/STS-Ncs. The SONET DCS is primarily deployed at a hub for facility grooming and test access. For example, in the figure, STS 1#7 (carried by an OC-48 fiber system between CO1 and the hub) and STS1#52 (carried by an OC-12 fiber system between CO2 and the hub) are designated for CO3. STS1#7 and STS1#52 terminate at different input ports of the DCS and are cross connected to two output ports that connect to the same fiber system for CO3.

Conceptually, a TSI can be viewed as a buffer, which reads from a single input and writes onto a single output. The input is framed into m fixed-length time slots. The number contained in each input time slot is the output time slot for information delivery. The information in each input time slot is read sequentially into consecutive slots (cyclically) of a buffer of m slots. The output is framed into n time slots, and information from the appropriate slot in the buffer is transmitted on the output slot. Hence, over the duration of an output frame, the content of the buffer is read out in a random manner according to a read-out sequence as shown in Fig. 16.10. This read-out sequence is uniquely determined by the connection pattern. In this manner, the information in each slot of the input frame is rearranged into the appropriate slot in the output frame, achieving the function of time slot interchanging. Since the TSI switching matrix performs physical STS path switching from one input port to one output port, TSI rerouting involves physical switch reconfiguration.

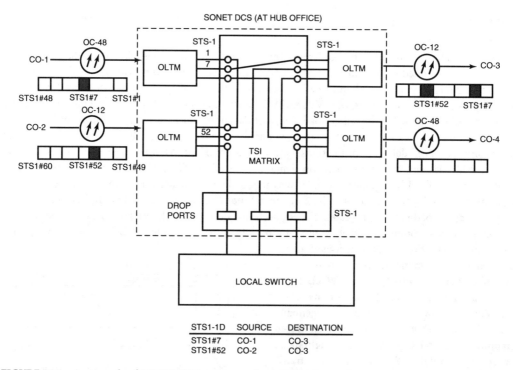

FIGURE 16.9 An example of SONET/STM system configuration and operation.

16.4 ATM Cell Switching System Configuration and Operations

Figure 16.11 illustrates a simplified ATM cell switching system configuration and a switching principle. The ATM cell entering input port 3 of the switch arrives with VPI = 9 and VCI = 4 (for example, the first ATM cell in the figure). The call processor has been alerted through the routing table that the cell must leave the ATM switch with VPI = 4 and VCI = 5 on output port 4. The call processor directs the virtual channel identifier converter to remove the 9 VPI and 4 VCI and replaces them with 4 and 5, respectively. If the switch is a large multistage switch, the call processor further directs the virtual channel identifier converter to create a tag that travels with the cell, identifying the internal multistage routing within the ATM switch matrix. Note that ATM cells belonging to the same VP must have the same VPIs in the input port (e.g., VPI = 9 in input port 3) or the output port (e.g., VPI = 4 in output port 4).

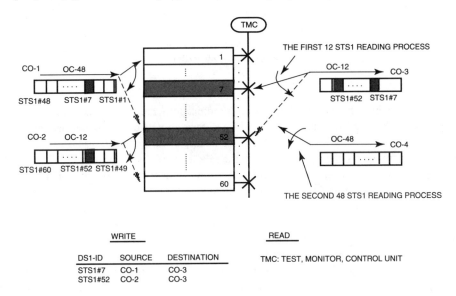

FIGURE 16.10 TSI switching concept.

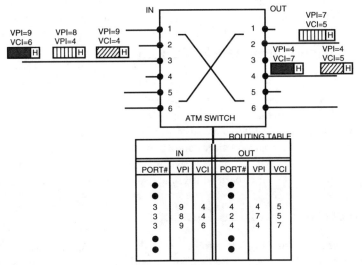

FIGURE 16.11 An example of ATM cell switching principle.

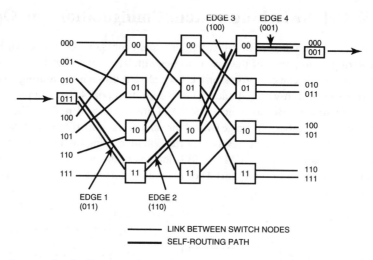

FIGURE 16.12 An example of self-routing concept.

The ATM switch implemented for large telecommunications networks usually uses a **self-routing** switching matrix. (Note: ATM switches designed for enterprise networks may not use the self-routing switching matrix.) An ATM cell is self-routing if the header of the cell contains all of the information needed to route through a switching network. In the following, we use a simple example to explain the self-routing concept, as depicted in Fig. 16.12. The example switching network is a shuffle exchange network, which has $N = 2^n$ inputs and 2^n outputs, interconnected by n stages of 2^{n-1} binary switch nodes. In this example, $n = 3$ and $N = 8$. We assume that the ATM cell is to be routed from input port 4 (i.e., code 011) to output port 2 (i.e., code 001). When the cell arrives at input port 4 (011), the call processor will check with VPI/VCI routing table, identify the destination output 2 (001), and generate a routing label which combines the source and destination addresses: 011001. Through this code, the system identifies edges in each stage where each edge is represented by three consecutive digits from the left side to the right of the routing label. In this case, 4 edges are identified: edge 1: 011, edge 2: 110, edge 3: 100, and edge 4: 001. Each edge (ijk) connects switch node (ij) in the $M - 1$ stage and switch node (jk) in stage M, if this edge is edge M. For example, edge 2 connects switch node (11) in stage 1 to switch node (10) in stage 2. Thus, a self-routing path is established as follows: Input port (011)-switch node (11) of stage 1-switch node (10) of stage 2-switch node (00) of stage 3-output port (001), as shown in the figure.

16.5 Comparison Between STM and ATM

Table 16.1 summarizes differences between STM and ATM switching concepts. In general, the STM system switches signals on a physical and hierarchical basis, whereas ATM switches signals on a logical and nonhierarchical basis, due to their corresponding switching path structures. Thus, ATM may be more flexible than its STM counterpart in terms of bandwidth on demand, bandwidth allocation, and transmission system efficiency, but a relatively more complex control system is needed. Since STM physically switches the signals, its rerouting requires physical switch reconfiguration that may make network rerouting much slower than its ATM counterpart, which only requires an update of the routing table.

16.6 Protection Switching Concept

The function of **protection switching** is to reroute new and existing connections around the failure area if a network failure occurs. The basic requirement for network protection switching is to restore services as fast as possible in a cost-effective manner. ATM technology is more flexible and efficient than STM technology in terms of bandwidth allocation and transmission efficiency. However, these advantages

TABLE 16.1 Comparison Between STM and ATM

System Parameters	STM	ATM
Switching unit	Time slot (STS paths)	Fixed-size cell (VPs/VCs)
Path structure	Physical and hierarchical	Logical and nonhierarchical
Number of allowable path (STS, VP) capacities	Limited	More
Switching system complexity	Simpler	Complex
Network nodal complexity	Complex (hierarchical)	Simpler (nonhierarchical)
Transmission system efficiency (on average)	Lower	Higher
Method of rerouting	Physical network reconfiguration	Logical network reconfiguration
Bandwidth on demand	More difficult	Easier

come at an expense of requiring much more complex control systems, especially when network restoration is an important consideration in system requirements. Unlike STM transport, the service quality in ATM networks is affected by both network component failure and network congestion (because resources are not dedicated).

In addition to fast restoration times, other possible network restoration objectives include the maximal restoration ratio, and the minimal spare capacity and overhead associated with network restoration. A network restoration system may be designed to optimize tradeoffs of these objectives or may be designed for some specific objective (such as minimum protection switching time), depending on the network planning strategy and service requirements being supported.

For SONET/ATM networks, network restoration can be performed at the physical (SONET) or ATM layer. For physical layer protection, service restoration can be performed at the STS-3Nc level (where $N = 1, 4,$ or 16) or at the physical transmission link level. For ATM layer protection, network restoration can be performed at the VC or VP (or group of VPs) level. Service restoration at the VC level is typically slower and more expensive than that performed at the VP level due to greater VC-based ATM network complexity. The restoration system complexity at the VP level is similar to that at the STS-1/STS-3c level. Determining which layer (SONET, VP, or VC) is an appropriate restoration layer depends on the SONET/ATM network architecture and other factors such as costs and QoS.

The major differences between STM and ATM network restoration technologies are in the restoration unit (e.g., STS-1 or VP) and the network reconfiguration technology. SONET networks restore traffic on physical paths (e.g., STS-3c) and/or links (e.g., SONET links) through the physical network reconfiguration using time slot interchange technology. In contrast, ATM networks restore traffic carried on VPs/VCs through the logical network reconfiguration by modifying the VPI/VCI routing table.

SONET protection switching systems, such as automatic protection switching systems (APSs) and SONET self-healing rings, have been widely deployed in LECs. The impacts of emerging ATM technology on network restoration (compared to SONET network restoration) may be significant due to following factors:

- Separation of capacity allocation and physical route assignment for VPs/VCs that would reduce required spare capacity (e.g., the capacity of protection VPs/VCs can be zero in normal conditions)
- More OAM bandwidth with allocation on demand that would reduce delays for restoration message exchanges, and the detection of system degradation (i.e., soft failure) in the ATM network much more quickly than its SONET counterpart
- Nonhierarchical path multiplexing (for VPs) that would simplify the survivable network design and help reduce intranode processing delays and required spare capacity.

Table 16.2 summarizes the relative comparison between STM and ATM network protection technologies. In general, ATM layer protection may require less spare capacity than SONET layer protection at the expense of a slower restoration time and a more complex control system. How best to use a combination

TABLE 16.2 Comparison of SONET and ATM Layer Protection Technologies

Protection Layer	ATM Layer	SONET Layer (STM)
Protection protocol complexity	Moderate	Simple (APS/rings) moderate (DCS mesh)
Restoration time	Moderate	Fast (APS/rings) Slow (DCS mesh)
Spare capacity needed	Least	Moderate
Equipment needed	Less	More
Network management systems/OSs	Under development	Already developed
Equipment availability	Available soon	Available now
Protection targets	Node and link	Node and link

of these two layer protection schemes in the same network is a challenging task for network planners and engineers who are seeking cost-effective solutions for SONET/ATM network protection. More details of network protection schemes and operations for SONET/ATM networks can be found in Wu [1992, 1995]; Sato, Ueda, and Yoshikai [1991]; and Wu et al. [1994].

16.7 Summary and Remarks

We have reviewed the basic switching concepts that are used in emerging SONET/ATM broadband transport networks. Implementations of such switching concepts may be different between enterprise networks and public telecommunications networks due to traditionally different QoS perspectives and operations experiences from computer and telecommunications industries. Such differences may result in two sets of ATM switching network infrastructures that could coexist for a period of time. An open question is how and when they may be merged as an integrated ATM network infrastructure. This challenge may require the change of traditional QoS perspectives, definitions, and requirements, and requirements from both the computer and telecommunications industries.

Defining Terms

Asynchronous transfer mode (ATM):   An international standard high-speed, integrated multiplexing and switching technology that transmits information using fixed length cells in a connection-oriented manner.

Protection switching: A function to reroute new and existing connections around the failure area if a network failure occurs.

Self-routing: An ATM cell is self-routing if the header of the cell contains all of the information needed to route through a switching network.

Switching: Control or routing of signals in either physical or virtual circuit form to transmit data between specific points in a network.

Synchronous optical network (SONET):   A North American standard synchronous transmission network technology for optical signal transmission.

Synchronous transfer mode (STM):   A time division multiplexing and switching technology that uses a similar switching concept as traditional circuit switching systems.

Time slot interchange (TSI):   A technique to exchange the location of time slots within a frame.

References

Ballart, R. and Ching, Y.-C. 1989. SONET: Now it's the standard optical network. *IEEE Comm. Mag.* (March):8–15.

Boudec, J.-Y.L. 1992. The asynchronous transfer mode: A tutorial. *Computer Networks and ISDN Systems* 24:279–309.

Sato, K., Ueda, H., and Yoshikai, N. 1991. The role of virtual path crossconnection. *IEEE Mag. Lightwave Telecomm. Sys.* 2(3):44–54.

Wu, T.-H. 1992. *Fiber Network Service Survivability.* Artech House, May.

Wu, T.-H. 1995. Emerging technologies for fiber network survivability. *IEEE Comm. Mag.* (Feb.).

Wu, T.-H., McDonald, J.C., Sato, K., and Flanagan, T.P., eds. 1994. Integrity of public telecommunications networks. *IEEE J. Selected Areas in Comm.* (Jan.).

Further Information

Additional information on the topic of network switching concept is available from the following sources:

IEEE Communications Magazine, a monthly periodical dealing with broadband communication networking technology. The magazine is published by IEEE.

IEEE Journal on Selected Areas in Communications, a monthly periodical dealing with current research and development on broadband communication networking technology. The magazine is published by IEEE.

17

Network Concepts

17.1	Introduction 17-1
17.2	OSI Model 17-1
	Physcial Layer • Data Link Layer • Network Layer • Transport Layer • Session Layer • Presentation Layer • Application Layer
17.3	Network Classificaitons 17-5

James E. Goldman
Purdue University

17.1 Introduction

The open system interconnections (OSI) model is the most broadly accepted explanation of LAN transmissions in an open system. The reference model was developed by the International Organization for Standardization (ISO) to define a framework for computer communication. The OSI model divides the process of data transmission into the following steps:

- Physical layer
- Data link layer
- Network layer
- Transport layer
- Session layer
- Presentation layer
- Application layer

17.2 OSI Model

The OSI model allows data communications technology developers, as well as standards developers, to talk about the interconnection of two networks or computers in common terms without dealing in proprietary vendor jargon [1]. These common terms are the result of the layered architecture of the seven-layer OSI model. The architecture breaks the task of two computers communicating with each other into separate but interrelated tasks, each represented by its own layer. The top layer (layer 7) represents network services provided to the application program running on each computer and is therefore aptly named the *application layer*. The bottom layer (layer 1) is concerned with the actual physical connection of the two computers or networks and is therefore named the *physical layer*. The remaining layers (2–6) may not be as obvious but, nonetheless, represent a sufficiently distinct logical group of functions required to connect two computers so as to justify a separate layer.

To use the OSI model, a network analyst lists the known protocols for each computing device or network node in the proper layer of its own seven-layer OSI model. The collection of these known protocols in their proper layers in known as the *protocol stack* of the network node. For example, the physical media employed, such as unshielded twisted pair, coaxial cable, or fiber optic cable, would be entered as a layer 1 protocol, whereas Ethernet or token ring network architectures might be entered as a layer 2 protocol.

The OSI model allows network analysts to produce an accurate inventory of the protocols present on any given network node. This protocol profile represents a unique personality of each network node and gives the network analyst some insight into what *protocol conversion*, if any, may be necessary in order to allow any two network nodes to communicate successfully. Ultimately, the OSI model provides a structured methodology for determining what hardware and software technologies will be required in the physical network design in order to meet the requirements of the logical network design.

The basic elements and parameters of each layer are detailed in the following sections.

Physical Layer

Layer 1 of the OSI model is responsible for carrying an electric current through the computer hardware to perform an exchange of information [2]. The physical layer is defined by the following parameters:

- Bit transmission rate
- Type of transmission medium (twisted pair, coaxial cable, or fiber optic cable), sometimes referred to as Layer 0
- Electrical specifications, including voltage- or current-based, and balanced or unbalanced
- Type of connectors used (for example, RJ-45 or DB-9)

Many different implementations exist at the physical layer

Layer 1 can exhibit error messages as a result of over-usage. For example, if a file server is being burdened with requests from workstations, the results may show up in error statistic that reflect the server's inability to handle all incoming requests. An overabundance of *response timeouts* may also be noted in this situation. A response timeout (in this context) is a message sent back to the workstation stating that the waiting period allotted for a response from the file server has passed without action from the server.

Error messages of this sort, which can be gathered by any number of commercially available software diagnostic utilities, can indicate an overburdened file server or a hardware flaw within the system. Intermittent response timeout errors can also be caused by a corrupted *network interface card* (NIC) in the server. A steady flow of timeout errors throughout all nodes on the network may indicate the need for another server or bridge.

Hardware problems are among the easiest to locate in a networked system. In a simple configuration where something has suddenly gone wrong, the physical layer and the data link layer are usually the first suspects.

Data Link Layer

Layer 2 of the OSI model, the data-link layer, describes hardware that enables data transmission (NICs and cabling systems) [2]. This layer integrates data packets into messages for transmission and checks them for integrity. Sometimes layer 2 will also send an "arrived safely" or "did not arrive correctly" message back to the transport layer (layer 4), which monitors this communications layer. The data link layer must define the frame (or package) of bits that is transmitted down the network cable. Incorporated within the frame are several important fields:

- Addresses of source and destination workstations
- Data to be transmitted between workstations
- Error control information, such as a *cyclic redundancy check* (CRC), which assures the integrity of the data

Network Concepts

The data link layer must also define the method by which the network cable is accessed because only one workstation can transmit at a time on a baseband LAN. The two predominant schemes are:

- *Token passing*, used with token ring and related networks
- *Carrier sense multiple access with collision detection* (CSMA/CD), used with Ethernet and and related networks

At the data link layer, the true identity of the LAN begins to emerge.

Because most functions of the data link layer (in a PC-based system[1]) take place in integrated circuits on NICs, software analysis is generally not required in the event of a failure. As mentioned previously, when something happens on the network, the data link layer is among the first to suspect. Because of the complexities of linking multiple topologies, cabling systems, and operating systems, the following failure modes may be experienced:

- RF disturbance. Transmitters, AC power controllers, and other computers can all generate energy that may interfere with data transmitted on the cable. RF interference (RFI) is usually the single biggest problem in a broadband network. This problem can manifest itself through excessive checksum errors and/or garbled data.
- Excessive cable runs. Problems related to the data link layer can result from long cable runs. Ethernet runs can stretch 1000 ft or more, depending on the cable and the Ethernet implementation. A basic token ring system can stretch 600 ft or so with the same qualification. The need for additional distance can be accommodated by placing a *bridge, gateway, active hub*, equalizer, or amplifier on the line.

The data link layer usually includes some type of routing hardware, including one or more of the following:

- Active hub
- Passive hub
- Multiple access units (for token ring-type networks
- Layer 2 switch

Ethernet (IEEE 802.3) is the most common data link layer protocol.

Network Layer

Layer 3 of the OSI model guarantees the delivery of transmissions as requested by the upper layers of the OSI [2]. The network layer establishes the physical path between the two communicating endpoints through the *communications subnet*, the common name for the physical, data link, and network layers taken collectively. As such, layer 3 functions (routing, switching, and network congestion control) are critical. From the viewpoint of a single LAN, the network layer is not required. Only one route—the cable—exists. Inter-network connections are a different story, however, because multiple routes are possible. The Internet Protocol (IP) and Internet Packet Exchange (IPX) are two examples of layer 3 protocols.

The network layer confirms that signals get to their designated targets, and then translates logical addresses into physical addresses. The physical address determines where the incoming transmission is stored. Lost data or similar errors can usually be traced back to the network layer, in most cases incriminating the network operating system. The network layer is also responsible for statistical tracking, and communications with other environments, including gateways. Layer 3 decides which route is the best to take, given the needs of the transmission. If router tables are being corrupted or excessive time

---

[1] In this context, the term "PC" is used to describe any computer, workstation, or laptop device.

is required to route from one network to another, an operating system error on the network layer may be involved. Layer 3 switchers and routers are examples of network layer hardware.

Transport Layer

Layer 4, the transport layer, acts as an interface between the bottom three and the upper three layers, ensuring that the proper connections are established and maintained [2]. It does the same work as the network layer, only on a local level. The network operating system driver performs transport layer tasks.

Connection difficulties between computers on a network can sometimes be attributed to the shell driver. The transport layer may have the ability to save transmissions that were en route in the case of a system crash, or re-route a transmission to its destination in case of primary route failure. The transport layer also monitors transmissions, checking to make sure that packets arriving at the destination node are consistent with the *build specifications* given to the sending node in layer 2. The data link layer in the sending node builds a series of packets according to specifications sent down from higher levels, then transmits the packets to a *destination node*. The transport layer monitors these packets to ensure they arrive according to specifications indicated in the original build order. If they do not, the transport layer calls for a retransmission. Some operating systems refer to this technique as a *sequenced packet exchange* (SPX) transmission, meaning that the operating system guarantees delivery of the packet. Transmission control protocol (TCP) is the transport layer protocol associated with the IP network layer protocol. Port addresses unique to different services are processed on layer 4 by layer 4 switches, also known as *traffic shapers* or *load-balancing switches*.

Session Layer

Layer 5 is responsible for turning communications on and off between communicating parties [2]. Unlike other levels, the session layer can receive instructions from the application layer through the network basic input/output operation system (netBIOS), skipping the layer directly above it. The netBIOS protocol allows applications to "talk" across the network. The session layer establishes the session, or logical connection, between communicating host processors. Name-to-address translation is another important function; most communicating processors are known by a common name, rather than a numerical address.

Multi-vendor problems often crop up in the session layer. Failures relating to gateway access usually fall into layer 5 for the OSI model, and are typically related to compatibility issues. Remote Procedure Call (RPC) is another example of a layer 5 protocol.

Presentation Layer

Layer 6 translates application layer commands into syntax understood throughout the network [2]. It also translates incoming transmissions for layer 7. The presentation layer masks other devices and software functions. Layer 6 software controls printers and other peripherals, and may handle encryption and special file formatting. Data compression, encryption, and translations are examples of presentation layer functions.

Failures in the presentation layer are often the result of products that are not compatible with the operating system, an interface card, a resident protocol, or another application.

Application Layer

At the top of the seven-layer stack is the application layer. It is responsible for providing protocols that facilitate user applications [2]. Print spooling, file sharing, and e-mail are components of the application layer, which translates local application requests into network application requests. Layer 7 provides the first layer of communications into other open systems on the network.

Failures at the application layer usually center around software quality and compatibility issues. The program for a complex network may include latent faults that will manifest only when a given set of conditions are present. The compatibility of the network software with other programs is another source of potential problems. Network File System (NFS) is an example of a layer 7 protocol.

17.3 Network Classifications

Although there are no hard and fast rules for network categorization, some general parameters are usually accepted for most applications. The following are a few of the more common categories of networking [1]:

- **Remote connectivity:** A single remote user wishes to access local network resources. This type of networking is particularly important to mobile professionals such as sales representatives, service technicians, field auditors, etc.
- **Local area networking:** Multiple users' computers are interconnecting for the purpose of sharing applications, data, or networked technology such as printers or mass storage. Local area networks (LANs) can have anywhere from two or three users to several hundred (or more). LANs are often limited to a single department or floor in a building, although technically any single-location corporation could be networked via a LAN.
- **Internetworking:** Also known as LAN-to-LAN networking or connectivity, internetworking involves the connection of multiple LANs and is common in corporations in which users on individual departmental LANs need to share data or otherwise communicate. The challenge of internetworking is in getting departmental LANs of different protocol stacks (as determined by use of the OSI model) to talk to each other, while only allowing authorized users access to the internetwork and other LANs. Variations of internetworking also deal with connecting LANs to mainframes or minicomputers rather than to other LANs.
- **Wide area networking:** Also known as *enterprise networking*, this involves the connection of computers, network nodes, or LANs over a sufficient distance as to require the purchase of wide area network (WAN) service from the local phone company or an alternative carrier. In some cases, the wide area portion of the network may be owned and operated by the corporation itself. Nonetheless, the geographic distance between nodes is the determining factor in categorizing a wide area network. A subset of WANs, known as *metropolitan area networks* (MANs), is confined to a campus or metropolitan area of usually not more than a few miles in diameter. MAMs often employ optical networking hardware and media.

The important thing to remember is that categorization of networking is somewhat arbitrary and that what really matters is that the proper networking technology (hardware and software) is specified in any given system in order to meet stated business objectives.

References

1. Goldman, J.E.: "Network Communication," in *The Electronics Handbook*, J.C. Whitaker (Ed.), CRC Press, Boca Raton, FL, 1996.
2. Dahlgren, M.W.: "Servicing Local Area Networks," *Broadcast Engineering*, Intertec Publishing, Overland Park, KS, November 1989.

Further Information

Goldman, J.: *Applied Data Communications: A Business Oriented Approach*, 3rd ed., Wiley, New York, 2001.
Goldman, J.: *Local Area Networks: A Business Oriented Approach*, 2nd ed., Wiley, New York, 2000.
Held, G.: *Ethernet Networks: Design Implementation, Operation and Management*, Wiley, New York, 1994.
Held, G.: *Internetworking LANs and WANs*, Wiley, New York, 1993.

Held, G.: *Local Area Network Performance Issues and Answers*, Wiley, New York, 1994.

Held, G.: *The Complete Modem Reference*, Wiley, New York, 1994.

International Organization for Standardization: "Information Processing Systems—Open Systems Interconnection—Basic Reference Model," ISO 7498, 1984.

Miller, M.A.: *LAN Troubleshooting Handbook*, M&T Books, Redwood City, CA, 1990.

Miller, M.A.: "Servicing Local Area Networks," *Microservice Management*, Intertec Publishing, Overland Park, KS, February 1990.

18
Data Acquisition

18.1	Fundamentals of Data Acquisition 18-1	
	Signals • Plug-In DAQ Boards • Types of ADCs • Analog Input Architecture • Basic Analog Specifications	
18.2	Data Acquisition Software 18-6	
	Board Register-Level Programming • Driver Software	
18.3	Digital Sampling... 18-6	
	Real-Time Sampling Techniques • Preventing Aliasing • Software Polling • External Sampling • Continuous Scanning • Multirate Scanning • Simultaneous Sampling • Interval Scanning • Seamless Changing of the Sampling Rate • Considerations When Seamlessly Changing the Sampling Rate	
18.4	Equivalent-Time Sampling.. 18-14	
	ETS Counter/Timer Operations • Hardware Analog Triggering • Retriggerable Pulse Generation • Autoincrementing • Flexible Data Acquisition Signal Routing • Considerations When Using ETS • ETS Application Examples	
18.5	Factors Influencing the Accuracy of Your Measurements.. 18-16	
	Linearity • Differential Nonlinearity • Relative Accuracy • Integral Nonlinearity • Settling Time • Noise	
18.6	Common Issues with PC-based Data Acquisition 18-22	
	The Low-Noise Voltmeter Standard • Where Does Noise Come from? • Proper Wiring and Signal Connections • Source Impedance Considerations • Pinpointing Your Noise Problems • Noise Removal Strategies • Signal Amplification • Averaging • Filtering • Grounding • Grounded and Floating Signal Sources • Types of Measurement Inputs • Nonreferenced Single-Ended Inputs • Measuring Grounded Signal Sources, Avoiding Loops • Measuring Floating Signals • Basic Signal Conditioning Functions • Transducer Excitation and Interfacing • Linearization • Variety of Signal Conditioning Architectures • Signal Conditioners Offer I/O Expansion	

Edward McConnell
National Instruments

Dave Jernigan
National Instruments

18.1 Fundamentals of Data Acquisition

The fundamental task of a data acquisition system is the measurement or generation of real-world physical signals. Before a physical signal can be measured by a computer-based system, you must use a sensor or transducer to convert the physical signal into an electrical signal, such as voltage or current. Often only the plug-in **data acquisition** (**DAQ**) board is considered the data acquisition system; however, it is only one of the components in the system. Unlike stand-alone instruments, signals often cannot be directly

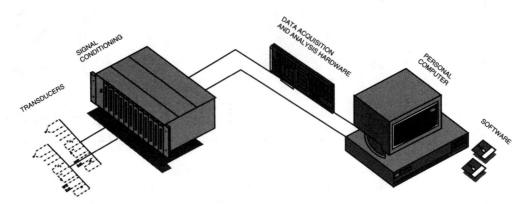

FIGURE 18.1 Components of data acquisition system.

connected to the DAQ board. The signals may need to be conditioned by some signal conditioning accessory before they are converted to digital information by the plug-in DAQ board. Finally, software controls the data acquisition system—acquiring the raw data, analyzing the data, and presenting the results. The components are shown in Fig. 18.1.

Signals

Signals are defined as any physical variable whose magnitude or variation with time contains information. Signals are measured because they contain some types of useful information. Therefore, the first question you should ask about your signal is: What information does the signal contain, and how is it conveyed? The functionality of the system is determined by the physical characteristics of the signals and the type of information conveyed by the signals. Generally, information is conveyed by a signal through one or more of the following signal parameters: state, rate, level, shape, or frequency content.

All signals are, fundamentally, analog, time-varying signals. For the purpose of discussing the methods of signal measurement using a plug-in DAQ board, a given signal should be classified as one of five signal types.

Because the method of signal measurement is determined by the way the signal conveys the needed information, a classification based on these criteria is useful in understanding the fundamental building blocks of a data acquisition system.

As shown in the Fig. 18.2, any signal can generally be classified as analog or digital. A digital, or binary, signal has only two possible discrete levels of interest, a high (on) level and a low (off) level. An analog signal, on the other hand, contains information in the continuous variation of the signal with time. The two digital signal types are the on–off signals and the pulse train signal. The three analog signal types are the DC signal, the time-domain signal, and the frequency-domain signal. The two digital types and three analog types of signals are unique in the information conveyed by each. The category to which a signal belongs depends on the characteristic of the signal to be measured. You can closely parallel the five types of signals with the five basic types of signal information: state, rate, level, shape, and frequency content.

Plug-In DAQ Boards

The fundamental component of a data acquisition system is the plug-in DAQ board. These boards plug directly into a slot in a PC and are available with analog, digital, and timing inputs and outputs. The most versatile of the plug-in DAQ boards is the multifunction input/output (I/O) board. As the name implies, this board typically contains various combinations of analog-to-digital convertors (ADCs), digital-to-analog convertors (DACs), digital I/O lines, and counters/timers. ADCs and DACs measure and generate analog voltage signals, respectively. The digital I/O lines sense and control digital signals. Counters/timers measure pulse rates, widths, delays, and generate timing signals. These many features make the multifunction DAQ board useful for a wide range of applications.

Data Acquisition

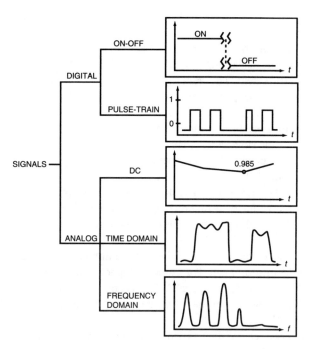

FIGURE 18.2 Classes of signals.

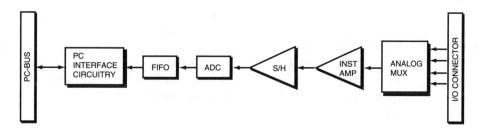

FIGURE 18.3 Analog input section of a plug-in DAQ board.

Multifunction boards are commonly used to measure analog signals. It is done by the ADC, which converts the analog voltage level into a digital number that the computer can interpret. The analog multiplexer (MUX), the instrumentation amplifier, the sample-and-hold (S/H) circuitry, and the ADC comprise the analog input section of a multifunction board (see Fig. 18.3).

Typically, multifunction DAQ boards have one ADC. Multiplexing is a common technique for measuring multiple channels (generally 16 single ended or 8 differential) with a single ADC. The analog mux switches between channels and passes the signal to the instrumentation amplifier and the sample-and-hold circuitry. The multiplexer architecture is the most common approach taken with plug-in DAQ boards. Although plug-in boards typically include up to only 16 single-ended or 8 differential inputs, you can further expand the number of analog input channels with external multiplexer accessories.

Instrumentation amplifiers typically provide a differential input and selectable gain by jumper or software. The differential input rejects small common-mode voltages. The gain is often software programmable. In addition, many DAQ boards also include the capability to change the amplifier gain while scanning channels at high rates. Therefore, you can easily monitor signals with different ranges of amplitudes. The output of the amplifier is sampled, or held at a constant voltage, by the sample-and-hold device at measurement time so that voltage does not change during digitization.

The ADC digitizes the analog signal into a digital value, which is ultimately sent to computer memory. There are several important parameters of A/D conversion. The fundamental parameter of an ADC is the number of bits. The number of bits of an ADC determines the range of values for the binary output of the ADC conversion. For example, many ADCs are 12 b, so a voltage within the input range of the ADC will produce a binary value that has one of $2^{12} = 4096$ different values. The more bits that an ADC has, the higher the resolution of the measurement. The resolution determines the smallest amount of change that can be detected by the ADC. Depending on your background, you may be more familiar with resolution expressed as number of digits of a voltmeter or dynamic range in decibels, than with bits. Table 18.1 shows the relation between bits, number of digits, and dynamic range in decibels.

TABLE 18.1 Relation Between Bits, Number of Digits, and Dynamic Range

Bits	Digits	dB
20	6.0	120
16	4.5	96
12	3.5	72
8	2.5	48

The resolution of the A/D conversion is also determined by the input range of the ADC and the gain. DAQ boards usually include an instrumentation amplifier that amplifies the analog signal by a gain factor prior to the conversion. You use this gain to amplify low-level signals so that you can make more accurate measurements.

Together, the input range of the ADC, the gain, and the number of bits of the board determine the maximum accuracy of the measurement. For example, suppose you are measuring a low-level ± 30 mV signal with a 12-b A/D convertor that has a ± 5 V input range. If the system includes an amplifier with a gain of 100, the resulting resolution of the measurement will be range/(gain $\times 2^{bits}$) = resolution, or 10 V/(100 $\times 2^{12}$) = 0.0244 mV.

Finally, an important parameter of digitization is the rate at which A/D conversions are made, referred to as the *sampling rate*. The A/D system must be able to sample the input signal fast enough to accurately measure the important waveform attributes. To meet this criterion, the ADC must be able to convert the analog signal to digital form quickly enough.

When scanning multiple channels with a multiplexing data acquisition system, other factors can affect the throughput of the system. Specifically, the instrumentation amplifier must be able to settle to the needed accuracy before the A/D conversion occurs. With the multiplexed signals, multiple signals are being switched into one instrumentation amplifier. Most amplifiers, especially when amplifying the signals with larger gains, will not be able to settle to the full accuracy of the ADC when scanning channels at high rates. To avoid this situation, consult the specified setting times of the DAQ board for the gains and sampling rates required by your application.

Types of ADCs

Different DAQ boards use different types of ADCs to digitize the signal. The most popular type of ADC on plug-in DAQ boards is the successive approximation ADC, because it offers high speed and high resolution at a modest cost. *Subranging* (also called *half-flash*) ADCs offer very high-speed conversion with sampling speeds up to several million samples per second. The state-of-the-art technology in ADCs is *delta–sigma modulating* ADCs. These ADCs sample at high rates, are able to achieve high resolution, and offer the best linearity of all ADCs. *Integrating and flash* ADCs are mature technologies still used on DAQ boards today. Integrating ADCs are able to digitize with high resolution but must sacrifice sampling speed to obtain it. Flash ADCs are able to achieve the highest sampling rate (GHz) but are available only with low resolution. The different types of ADCs are summarized in Table 18.2.

Analog Input Architecture

With the typical DAQ board, the multiplexer switches among analog input channels. The analog signal on the channel selected by the multiplexer then passes to the programmable gain instrumentation amplifier (PGIA), which amplifies the signal. After the signal is amplified, the sample and hold keeps the analog signal constant so that the A/D converter can determine the digital representation of the analog signal. A good DAQ board will then place the digital signal in a first-in first-out (FIFO) buffer, so that no data

TABLE 18.2 Types of ADCs

Type of ADC	Advantages	Features
Successive approximation	High resolution High speed Easily multiplexed	200-kHz sampling rate 12-b resolution
Subranging	Higher speed	55-kHz sampling rate 16-b resolution 1-MHz sampling rate 12-b resolution
Delta–Sigma	High resolution Excellent linearity Built in antialiasing State-of-the-art technology	48-kHz sampling rate 16-b resolution
Integrated	High resolution Good noise rejection Mature technology	15-kHz sampling rate
Flash	Higher speed Mature technology	125-kHz sampling rate

will be lost if the sample cannot transfer immediately over the PC I/O channel to computer memory. Having a FIFO becomes especially important when the board is run under operating systems that have large interrupt latencies, such as Microsoft Windows.

Basic Analog Specifications

Almost every DAQ board data sheet specifies the number of channels, the maximum sampling rate, the resolution, and the input range and gain.

The number of channels, which is determined by the multiplexer, is usually specified in two forms, differential and single ended. Differential inputs are inputs that have different reference points for each channel, none of which is grounded by the board. Differential inputs are the best way to connect signals to the DAQ board because they provide the best noise immunity.

Single-ended inputs are inputs that are referenced to a common ground point. Because single-ended inputs are referenced to a common ground, they are not as good as differential inputs for rejecting noise. They do have a larger number of channels, however. You can use the single-ended inputs when the input signals are high level (greater than 1 V), the leads from the signal source to the analog input hardware are short (less than 15 ft), and all input signals share a common reference.

Some boards have pseudodifferential inputs, which have all inputs referenced to the same common—like single-ended inputs—but the common is not referenced to ground. Using these boards, you have the benefit of a large number of input channels, like single-ended inputs, and the ability to remove some common mode noise, especially if the common mode noise is consistent across all channels. Differential inputs are still preferable to pseudodifferential, however, because differential is more immune to magnetic noise.

Sampling rate determines how fast the analog signal is converted to a digital signal. If you are measuring AC signals, you will want to sample at least two times faster than the highest frequency of your input signal. Even if you are measuring DC signals, you can sample faster than you need to and then average the samples to increase the accuracy of the signal by reducing the effects of noise.

If you have multiple DC-class signals, you will want to select a board with *interval scanning*. With interval scanning, all channels are scanned at one sample interval (usually the fastest rate of the board), with a second interval (usually slow) determining the time before repeating the scan. Interval scanning gives the effects of simultaneously sampling for slowly varying signals without requiring the additional cost of input circuitry for true simultaneous sampling.

Resolution is the number of bits that are used to represent the analog signal. The higher the resolution, the higher the number of divisions the input range is broken into and, therefore, the smaller the possible detectable voltage. Unfortunately, some data acquisition specifications are misleading when they specify the resolution associated with the DAQ board. Many DAQ board specifications state the resolution of the ADC without stating the linearity and noise and, therefore, do not give you the information you need to determine the resolution of the entire board. Resolution of the ADC, combined with the settling time, integral nonlinearity, differential nonlinearity, and noise will give you an understanding of the accuracy of the board.

Input range and gain tell you what level of signal you can connect to the board. Usually, the range and gain are specified separately, so you must combine the two to determine the actual signal input range as

$$\text{signal input range} = \text{range/gain}$$

For example, a board using an input range of ± 10 V with a gain of 2 will have a signal input range of ± 5V. The closer the signal input range is to the range of your signal, the more accurate your readings from the DAQ board will be. If your signals have different input ranges, you will want to look for a DAQ board that has the capability of different gains per channel.

18.2 Data Acquisition Software

The software is often the most critical component of the data acquisition system. Properly chosen software can save you a great deal of time and money. Likewise, poorly chosen software can cost you time and money. A whole spectrum of software options exists, with important tradeoffs and advantages.

Board Register-Level Programming

The first option is not to use vendor-supplied software and program the DAQ board yourself at the hardware level. DAQ boards are typically register based, that is, they include a number of digital registers that control the operation of the board. The developer may use any standard programming language, such as C or BASIC, to write series of binary codes to the DAQ board to control its operation. Although this method affords the highest level of flexibility, it is also the most difficult and time consuming, especially for the inexperienced programmer. The programmer must know the details of programming all hardware, including the board, the PC interrupt controller, the direct memory access (DMA) controller, and PC memory.

Driver Software

If you do not have the time or interest to learn the details of your computer and plug-in board hardware and you want to take full advantage of the hardware capabilities of your board, you can use data acquisition driver software. Driver software typically consists of a library of function calls usable from a standard programming language. These function calls provide a high-level interface to control the standard functions of the plug-in board. For example, a function called SCAN_OP may configure, initiate, and complete a multiple channel scanning data acquisition operation of a predetermined number of points. The function call would include parameters to indicate the channels to be scanned, the amplifier gains to be used, the sampling rate, and the total number of data points to collect. The driver responds to this one function call by programming the plug-in board, the DMA controller, the interrupt controller, and CPU to scan the channels as requested.

18.3 Digital Sampling

Every DAQ system has the task of gathering information about analog signals. To do this, the system captures a series of instantaneous snapshots or samples of the signal at definite time intervals. Each

Data Acquisition

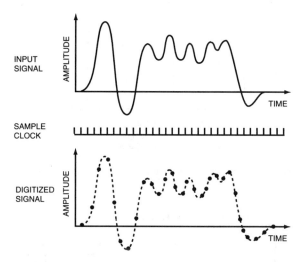

FIGURE 18.4 Consecutive discrete samples recreate the input signal.

sample contains information about the signal at a specific instant. Knowing the exact **conversion time** and the value of the sample, you can reconstruct, analyze, and display the digitized waveform.

Two classifications of sample timing techniques are used to control the ADC conversions, real-time sampling and equivalent-time sampling (ETS). Depending on the type of signal you acquire and the rate of acquisition, one sampling technique may be better than the other.

Real-Time Sampling Techniques

In real-time sampling, you immediately see the changes, as the signal changes (Fig. 18.4). According to the **Nyquist theorem**, you must sample at least twice the rate of the maximum frequency component in that signal to prevent **aliasing**. The frequency at one-half the sampling frequency is referred to as the Nyquist frequency. Theoretically, it is possible to recover information about those signals with frequencies at or below the Nyquist frequency. Frequencies above the Nyquist frequency will alias to appear between DC and the Nyquist frequency.

For example, assume the sampling frequency fs, is 100 Hz. Also assume the input signal to be sampled contains the following frequencies: 25, 70,160, and 510 Hz. Figure 18.5 shows a spectral representation of the input signal. The mathematics of sampling theory show us that a sampled signal is shifted in the frequency domain by an amount equal to integer multiples of the sampling frequency, fs. Figure 18.6 shows the spectral content of the input signal after sampling. Frequencies below 50 Hz, the Nyquist frequency ($fs/2$), appear correctly. However, frequencies above the Nyquist appear as aliases below the Nyquist frequency. For example, F1 appears correctly; however, F2, F3, and F4 have aliases at 30, 40, and 10 Hz, respectively.

The resulting frequency of aliased signals can be calculated with the following formula:

Apparent (Alias) Freq. = ABS (Closest Integer Multiple of Sampling Freq.-Input Freq.)

For the example of Figs. 18.5 and 18.6

$$\text{alias } F2 = |100 - 70| = 30 \text{ Hz}$$
$$\text{alias } F3 = |(2)100 - 160| = 40 \text{ Hz}$$
$$\text{alias } F4 = |(5)100 - 510| = 10 \text{ Hz}$$

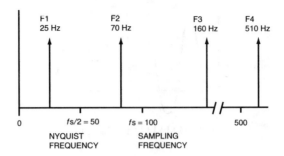

FIGURE 18.5 Spectral of signal with multiple frequencies.

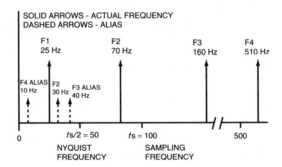

FIGURE 18.6 Spectral of signal with multiple frequencies after sampled at $fs = 100$ Hz.

Preventing Aliasing

You can prevent aliasing by using filters on the front end of your DAQ system. These antialiasing filters are set to cut off any frequencies above the Nyquist frequency (half the sampling rate). The perfect filter would reject all frequencies above the Nyquist; however, because perfect filters exist only in textbooks, you must compromise between sampling rate and selecting filters. In many applications, one- or two-pole passive filters are satisfactory. By oversampling (5–10 times) and using these filters, you can sample adequately in most cases.

Alternatively, you can use active antialiasing filers with programmable cutoff frequencies and very sharp attenuation of frequencies above the cutoff. Because these filters exhibit a very steep rolloff, you can sample at two to three times the filter cutoff frequency. Figure 18.7 shows a transfer function of a high-quality antialiasing filter.

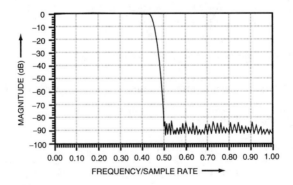

FIGURE 18.7 Transfer function of an antialiasing filter.

Data Acquisition

The computer uses digital values to recreate or to analyze the waveform. Because the signal could be anything between each sample, the DAQ board may be unaware of any changes in the signal between samples. There are several sampling methods optimized for the different classes of data; they include software polling, external sampling, continuous scanning, multirate scanning, simultaneous sampling, interval scanning, and seamless changing of the sample rate.

Software Polling

A software loop polls a timing signal and starts the ADC conversion via a software command when the edge of the timing signal is detected. The timing signal may originate from the computer's internal clock or from a clock on the DAQ board. Software polling is useful in simple, low-speed applications, such as temperature measurements.

The software loop must be fast enough to detect the timing signal and trigger a conversion. Otherwise, a window of uncertainty, also known as *jitter*, will exist between two successive samples. Within the window of uncertainty, the input waveform could change enough to drastically reduce the accuracy of the ADC.

Suppose a 100-Hz, 10-V full-scale sine wave is digitized (Fig. 18.8). If the poking loop takes 5 ms to detect the timing signal and to trigger a conversion, then the voltage of the input sine wave changes as much as 31 mV $[\Delta V - 10\sin(2\pi \times 100 \times 5 \times 10^{-6})]$. For a 12-b ADC operating over an input range of 10 V and a gain of 1, 1 least significant bit (LSB) of error represents 2.44 mV,

$$\left(\frac{\text{input range}}{\text{gain} \times 2^n}\right) = \left(\frac{10 \text{ V}}{1 \times 2^{12}}\right) = 2.44 \text{ mV}$$

But because the voltage error due to jitter is 31 mV, the accuracy error is 13 LSB,

$$\left(\frac{31 \text{ mV}}{2.44 \text{ mV}}\right)$$

This represents uncertainty in the last 4b of a 12-b ADC. Thus, the effective accuracy of the system is no longer 12 b, but rather 8 b.

External Sampling

Some DAQ applications must perform a conversion based on another physical event that triggers the data conversion. The event could be a pulse from an optical encoder measuring the rotation of a cylinder. A sample would be taken every time the encoder generates a pulse corresponding to *n* degrees of rotation. External triggering is advantageous when trying to measure signals whose occurrence is relative to another physical phenomenon.

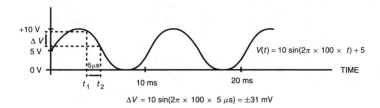

FIGURE 18.8 Jitter reduces the effective accuracy of the DAQ board.

Continuous Scanning

When a DAQ board acquires data, several components on the board convert the analog signal to a digital value. These components include the analog multiplexer (mux), the instrumentation amplifier, the sample-and-hold circuitry, and the ADC. When acquiring data from several input channels, the analog mux connects each signal to the ADC at a constant rate. This method, known as *continuous scanning*, is significantly less expensive than having a separate amplifier and ADC for each input channel.

Continuous scanning is advantageous because it eliminates jitter and is easy to implement. However, it is not possible to simultaneously sample multiple channels. Because the mux switches between channels, a time skew occurs between any two successive channel samples. Continuous scanning is appropriate for applications where the time relationship between each sampled point is unimportant or where the skew is relatively negligible compared to the speed of the channel scan.

If you are using samples from two signals to generate a third value, then continuous scanning can lead to significant errors if the time skew is large. In Fig. 18.9, two channels are continuously sampled and added together to produce a third value. Because the two sine waves are 90° out of phase, the sum of the signals should always be zero. But because of the skew time between the samples, an erroneous sawtooth signal results.

Multirate Scanning

Multirate scanning, a method that scans multiple channels at different scan rates, is a special case of continuous scanning. Applications that digitize multiple signals with a variety of frequencies use multirate scanning to minimize the amount of buffer space needed to store the sampled signals. You can use channel independent ADCs to implement hardware multirate scanning; however, this method is extremely expensive. Instead of multiple ADCs, only one ADC is used. A channel/gain configuration register stores the scan rate per channel and software divides down the scan clock based on the per-channel scan rate. Software-controlled multirate scanning works by sampling each input channel at a rate that is a fraction of the specified scan rate.

Suppose you want to scan channels 0–3 at 10 kilosamples/second, channel 4 at 5 kilosamples/second, and channels 5–7 at 1 kilosamples/second. You should choose a base scan rate of 10 kilosamples/second. Channels 0–3 are acquired at the base scan rate. Software and hardware divide the base scan rate by 2 to sample channel 4 at 5 kilosamples/second, and by 10 to sample channels 5–7 at 1 kilosamples/second.

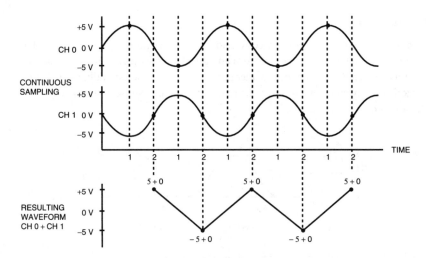

FIGURE 18.9 If the channel skew is large compared to the signal, then erroneous conclusions may result.

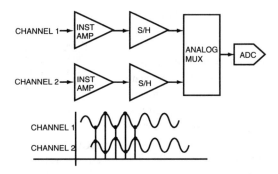

FIGURE 18.10 Block diagram of DAQ components used to simultaneously sample multiple channels.

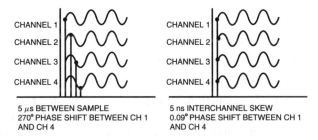

FIGURE 18.11 Comparison of continuous scanning and simultaneous sampling.

Simultaneous Sampling

For applications where the time relationship between the input signals is important, such as phase analysis of AC signals, you must use simultaneous sampling. DAQ boards capable of simultaneous sampling typically use independent instrumentation amplifiers and sample-and-hold circuitry for each input channel, along with an analog mux, which routes the input signals to the ADC for conversion (as shown in Fig. 18.10).

To demonstrate the need for a simultaneous-sampling DAQ board, consider a system consisting of four 50-kHz input signals sampled at 200 kilosamples/second. If the DAQ board uses continuous scanning, the skew between each channel is $5\,\mu s$ (1 S/200 kilosamples/second) which represents a 270° (15 $\mu s/20\,\mu s \times 360°$) shift in phase between the first channel and fourth channel. Alternately, with a simultameous-sampling board with a maximum 5 ns interchannel time offset, the phase shift is only 0.09° (5 ns/20 $\mu s \times 360°$). This phenomenon is illustrated in Fig. 18.11.

Interval Scanning

For low-frequency signals, interval scanning creates the effect of simultaneous sampling, yet maintains the cost benefits of a continuous scanning system. This method scans the input channels at one rate and uses a second rate to control when the next scan begins. You can scan the input channels at the fastest rate of the ADC, creating the effect of simultaneously sampling. Interval scanning is appropriate for slow moving signals, such as temperature and pressure. Interval scanning results in a jitter-free sample rate and minimal skew time between channel samples. For example, consider a DAQ system with 10 temperature signals. Using interval scanning, you can set up the DAQ board to scan all channels with an interchannel delay of 5 μs, then repeat the scan every second. This method creates the effect of simultaneously sampling 10 channels at 1 S/s, as shown in Fig. 18.12.

To illustrate the difference between continuous and interval scanning, consider an application that monitors the torque and revolutions per minute (RPMs) of an automobile engine and computes the engine horsepower, Two signals, proportional to torque and RPM, are easily sampled by a DAQ board at a rate of 1000 S/s. The values are multiplied together to determine the horsepower as a function of time. A continuously scanning DAQ board must sample at an aggregate rate of 2000 S/s. The time between which the torque signal is sampled and the RPM signal is sampled will always be 0.5 ms (1/2000). If either signal changes within 0.5 ms, then the calculated horsepower is incorrect. But using interval scanning at a rate of 1000 S/s, the DAQ board samples the torque signal every 1 ms, and the RPM signal is sampled as quickly as possible after the torque is sampled. If a 5-μs interchannel delay exists between the torque and RPM samples, then the time skew is reduced by 99% [(0.5 ms − 5μs)/0.5 ms], and the chance of an incorrect calculation is reduced.

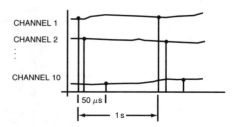

FIGURE 18.12 Interval scanning: all 10 channels are scanned within 45 μs; this is insignificant relative to the overall acquisition rate of 1 S/s.

Seamless Changing of the Sampling Rate

This technique, a variation of real-time sampling, is used to vary the sampling rate of the ADC without having to stop and reprogram the counter/timer for different conversion rates. For example, you may want to start sampling slowly, and then, following a trigger, begin sampling quickly; this is particularly useful when performing transient analysis. The ADC samples slowly until the input crosses a voltage level, and then the ADC samples quickly to capture the transient.

Traditional DAQ boards, ones which use the 8253 or 8254 counter/timer chips, are not capable of seamlessly changing the sampling rate. Changing the sampling rate with these boards requires that you stop the onboard counter and load it with a new count value. Stopping and reloading the counters causes nondeterministic reprogramming delays and yields missed samples. However, the DAQ-STC™ [an application specific integrated circuit (ASIC) designed by National Instruments specifically for DAQ system timing and control along with analog triggering circuitry] accommodates the complex timing and triggering signals necessary to seamlessly change the sampling rate on DAQ boards.

Four complex timing and triggering mechanisms are necessary to seamlessly change the sampling rate. They include hardware analog triggering, frequency ship keying (FSK), flexible data acquisition signal routing, and buffered relative-time stamping.

Hardware Analog Triggering. A trigger level serves as the reference point at which the sampling rate changes. Analog trigger circuitry monitors the input voltage of the waveform and generates a transistor-transmitter logic (TTL) high level whenever the voltage input is greater than the trigger voltage V set; and a TTL low level when the voltage input is less than the trigger voltage. Therefore, as the waveform crosses the trigger level, the analog trigger circuitry signals the counter to count at a different frequency.

Frequency Shift Keying. Frequency shift keying (FSK) occurs when the frequency of a digital pulse varies over time. Frequency modulation (FM) in the analog domain is analogous to FSK in the digital domain. FSK determines the frequency of a generated pulse train relative to the level present at the gate of a counter. For example, when the gate signal is a TTL high level, the pulse train frequency is three times greater than the pulse train frequency when the gate signal is a TTL low level.

Flexible Data Acquisition Signal Routing. For continuous or interval scanning, a simple, dedicated sample timer directly controls ADC conversions. But for more complex sampling techniques, such as seamlessly changing the sampling rate, signals from other parts of the DAQ board control ADC conversions. The DAQ-STC ASIC provides 20 possible signal sources to time each ADC conversion. One of the sources is the output of the general-purpose counter 0, which is more flexible than a dedicated timer. In particular, the general-purpose counter generates the FSK signal that is routed to the ADC to control the sampling rate.

Data Acquisition

Buffered Relative-Time Stamping. Because different sample rates are used to acquire the signal, keeping track of the various acquisition rates is a challenge for the board and software. The sampled values must have an acquisition time associated with them in order for the signal to be correctly displayed. While values are sampled by the ADC, the DAQ-STC counter/timers measure the relative time between each sample using a technique called *buffered relative-time stamping*. The measured time is then transferred from the counter/timer registers to PC memory via direct memory access.

The counter continuously measures the time interval between successive, same-polarity transitions of the FSK pulse with a measurement resolution of 50 ns. Countings begin at 0. The counter contents are stored in a buffer after an edge of the appropriate polarity is detected. Then, the counting begins again at 0. Software routines use DMA to transfer data from the counter to a buffer until the buffer is filled.

For example, in Fig. 18.13, the period of an FSK signal is measured. The first period is 150 ns (3 clock cycles × 50-ns resolution); the second, third, and fourth periods are 100 ns (2 clock cycles × 50 ns); the fifth period is 150 ns. For a 10-MHz board that can change its sampling rate seamlessly, the FSK signal determines the ADC conversions; the effective sampling rate is 6.7 MHz for the first part of the signal and 10 MHz for the second part of the signal.

Figure 18.14 details the timing signals necessary to change the sampling rate of the ADC without missing data. As the input waveform crosses the trigger voltage 1 the analog trigger circuitry generates a low on its output. The output signal is routed to a general-purpose counter, which generates pulses at a predetermined frequency 2. Each high-to-low transition of the FSK signal causes the ADC to sample the waveform 3. When the input waveform crosses the trigger level again, the analog trigger circuit generates a high, which causes the general purpose counter to generate pulses at a second frequency. This timing process continues until a predetermined number of samples has been acquired.

Considerations When Seamlessly Changing the Sampling Rate

The intention of seamlessly changing the sampling rate is to switch between rates, yet not miss significant changes in the signal. Selecting the various rates and the instance at which the rate changes requires some thought. For instance, when switching between a sampling rate of 10 S/s and 20 kilosamples/second, the analog trigger circuitry checks for a trigger condition every 1/10 of a second. This means that, at most, 0.1 s will pass before the DAQ board is aware of the trigger and increases the sampling rate to 20 kilosamples/second. If the ADC is switching between a rate of 20 kilosamples/second and 10 S/s, the trigger condition is checked every 50 ms, and the board will take, at most, 50 ms to switch to the 10 S/s rate. Thus, you should set the trigger condition so you can detect the trigger and start the faster rate before the signal changes significantly.

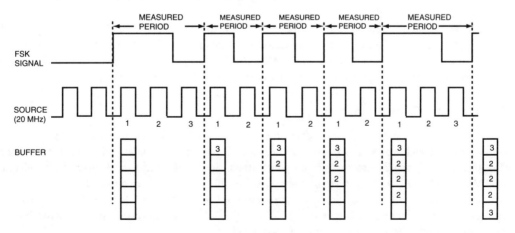

FIGURE 18.13 The number of counts is stored in a buffer. These counts are used to determine the time between conversions.

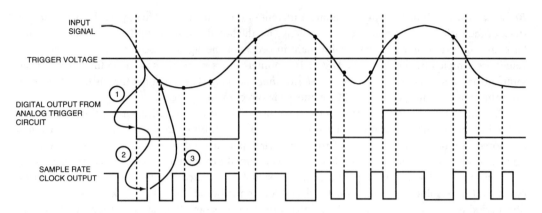

FIGURE 18.14 Complex triggering, counting, and timing are combined to seamlessly change the sampling rate.

Suppose automatic braking systems (ABS) are tested by monitoring signal transients. The test requires a few samples to be recorded before and after the transient, as well as the transient signal. The test specifies a sample rate of 400 kilosamples/second. If the DAQ board continuously samples the system, you must sample the entire signal at 400 kilosamples/second. For a signal that is stable for 1 min before the transient, then 24×10^6 samples are acquired before the transient even occurs. If the data is stored on disk, a large hard disk is needed. But if the stable portion of the signal is sampled more slowly, such as 40 S/s, the amount of unnecessary data acquired is reduced. If the sampling rate is changed on the fly, then the board can sample at 40 S/s before and after the transient and at 400 kilosamples/second during the transient. In the 1 min before the transient, only 2400 samples are logged. Once the transient occurs, the board samples at 400 kilosamples/second. Once the transient passes, the ADC begins to sample at 40 S/s again. Using this method, exactly the number of samples needed relative to the signal are logged to disk. Extra equipment, such as a large hard disk, is not necessary to store the digitized information.

18.4 Equivalent-Time Sampling

If real-time sampling techniques are not fast enough to digitize the signal, then consider ETS as an approach. Unlike continuous, interval, or multirate scanning, complex counter/timers control ETS conversions, and ETS strictly requires that the input waveform be repetitive throughout the entire sampling period.

In ETS mode, an analog trigger arms a counter, which triggers an ADC conversion at progressively increasing time intervals beyond the occurrence of the analog trigger (as shown in Fig. 18.15). Instead of acquiring samples in rapid succession, the ADC digitizes only one sample per cycle. Samples from several cycles of the input waveform are then used to recreate the shape of the signal.

ETS Counter/Timer Operations

Four complex timing and triggering mechanisms are necessary for ETS. They include hardware analog triggering, retriggerable pulse generation, autoincrementing, and flexible data acquisition signal routing. Most plug-in DAQ boards lack the necessary circuitry to implement these three mechanisms. Neither the AMD 9513A nor the 8253/54 are by themselves capable of such triggering and timing operations. Typically, you must add external circuitry to the multifunction DAQ board to supply the A/D counter with appropriate trigger signals. However, the DAQ-STC and analog triggering circuitry accommodate the complex timing and triggering signals necessary to implement ETS.

Data Acquisition

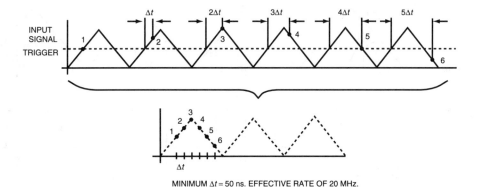

FIGURE 18.15 In ETS mode, a waveform is recreated from samples taken from several cycles of the input waveform.

Hardware Analog Triggering

Analog trigger circuitry monitors the input voltage of the waveform and generates a pulse whenever the trigger conditions are met. Each time the repetitive waveform crosses the trigger level, the analog trigger circuitry arms the board for another acquisition.

Retriggerable Pulse Generation

The ADC samples when it receives a pulse for the conversion counter. In real-time sampling, the conversion counter generates a series of continuous pulses that cause the ADC to successively digitize multiple values. In ETS, however, the conversion counter generates only one pulse, which corresponds to one sample from one cycle of the waveform. This pulse is regenerated in response to a signal from the analog trigger circuitry, which occurs after each new cycle of the waveform.

Autoincrementing

If the ADC sampled every occurrence of the waveform using only retriggerable pulse generation, the same point along the repetitive signal would be digitized, namely, the point corresponding to the analog trigger. A method is needed to make the ADC sample different points along different cycles of the waveform. This method, known as *autoincrementing*, is the most important counter/timer functional for controlling ETS. An autoincrementing counter produces a series of delayed pulses.

Flexible Data Acquisition Signal Routing

As with seamlessly changing the sample rate, ETS uses signals from other parts of the board to control ADC conversions. The DAQ-STC generates autoincrementing retriggerable pulses.

Figure 18.16 details the timing signals used for ETS. As the input waveform crosses the trigger voltage 1, the analog trigger circuitry generates a pulse at the gate input of the autoincrementing counter. This counter generates a conversion pulse 2, which is used to trigger the ADC to take a sample 3. The timing process continues until a predetermined number of samples has been acquired. The delay time Dt is directly related to the relative sampling rate set. For example, for an effective sampling rate of 20 MS/s, Dt is equal to 50 ns. Because the waveform is repetitive over the complete acquisition, the points in the recreated waveform appear as if they were taken every 50 ns.

Considerations When Using ETS

Although ETS is useful in a number of applications, ETS users need to be aware of a few issues. First, the input waveform must be repetitive. ETS will not correctly reproduce a waveform that is nonrepetitive

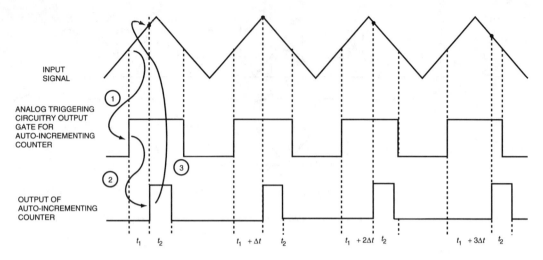

FIGURE 18.16 Complex triggering, counting, and timing are combined to control the ADC in ETS.

because the analog trigger will never occur at the same place along the waveform. One-shot acquisitions are not possible, because ETS digitizes a sample from cycles of a repetitive waveform.

The accuracy of the conversion is limited to the accuracy of the counter/timer. Jitter in the DAQ-STC counter/timer causes the sample to be taken within a multiple of 50 ns from the trigger. The sample may occur along a slightly different portion of the input waveform than expected. For example, using a 12-b board to sample the 100-Hz, 10-V peak-to-peak sine wave in Fig. 18.8, the accuracy error is 0.13 LSB.

$$\left(\frac{\Delta V}{2.44 \text{ mV}} = \frac{10\sin(72\pi \times 100 \times 50 \times 10^{-9})}{244 \text{ mV}}\right)$$

The frequency range of the signals that you can acquire is determined by the bandwidth of the analog input circuitry. On boards implementing only real-time sampling techniques, the speed of the ADC limits the frequency range of waveforms, and the input bandwidth of the analog circuitry is never an issue. But now that ETS has increased the performance of DAQ boards, knowing the input bandwidth of the board is essential. For example, if a board has an input bandwidth of 1 MHz and ETS is used to digitize a 1-MHz square wave, then the digitized waveform appears as only a 1-MHz sine wave. All frequent components higher than 1 MHz are attenuated and cannot be digitized.

ETS Application Examples

ETS techniques are useful in measuring the rise and fall times of TTL signals. ETS is also used to measure the impulse response of a 1-MHz, low-pass filter subject to a 2-ms repetitive impulse. The impulse response is a repetitive signal of pulses that also have a duration of 2 ms. Using a 20-MHz ETS conversion rate results in 40 digitized samples per input impulse. This is enough samples to display the impulse response in the time domain. Other applications for ETS include disk drive testing, nondestructive testing, ultrasonic testing, vibration analysis, laser diode characterization, and impact testing.

18.5 Factors Influencing the Accuracy of Your Measurements

How do you tell if the plug-in data acquisition board that you already have or the board that you are considering integrating into your system will give you the results you want? With a sophisticated measuring device like a plug-in DAQ board, you can obtain significantly different accuracies depending on which board you are using. For example, you can purchase DAQ products on the market today with 16-b analog to digital converters and get less than 12 b of useful data, or you can purchase a product with a 16-b ADC and actually get 16 b of useful data. This difference in accuracies causes confusion in the PC industry where everyone is used to switching out PCs, video cards, printers, and so on, and experiencing similar results between equipment.

The most important thing to do is to scrutinize more specifications than the resolution of the A/D converter that is used on the DAQ board. For DC-class measurements, you should at least consider the settling time of the instrumentation amplifier, **differential nonlinerarity (DNL), relative accuracy, integral nonlinearity (INL)**, and noise. If the manufacturer of the board you are considering does not supply you with each of these specifications in the data sheets, you can ask the vendor to provide them or you can run tests yourself to determine these specifications of your DAQ board.

Linearity

Ideally, as you increase the level of voltage applied to a DAQ board, the digital codes from the ADC should also increase linearly. If you were to plot the voltage vs the output code from an ideal ADC, the plot would be a straight line. Deviations from this ideal straight line are specified as the nonlinearity. Three specifications indicate how linear a DAQ board's transfer function is: differential nonlinearity, relative accuracy, and integral nonlinearity.

Differential Nonlinearity

For each digital output code, there is a continuous range of analog input values that produce it. This range is bounded on either side by transitions. The size of this range is known as the *code width*. Ideally, the width of all binary code values is identical and is equal to the smallest detectable voltage change,

$$\text{ideal code width} = \frac{\text{input voltage range}}{2^n}$$

where n is the resolution at ADC. For example, a board that has a 12-b ADC, input range of 0–10, and a gain of 100 will have an ideal code width of

$$\frac{(10 \text{ V})/100}{2^{12}} = 24 \ \mu\text{V}$$

This ideal analog code width defines an analog unit called the least significant bit. DNL is a measure in LSB of the worst-case deviation of code widths from their ideal value of 1 LSB. An ideal DAQ board has a DNL of 0 LSB. Practically, a good DAQ board will have a DNL within ±0.5 LSB.

There is no upper limit on how wide a code can be. Codes do not have widths of less than 0 LSB, so the DNL is never worse than −1 LSB. A poorly performing DAQ board may have a code width equal to or very near zero, which indicates a *missing code*. No matter what voltage you input to the DAQ board with a missing code, the board will never quantize the voltage to the value represented by this code. Sometimes DNL is specified by stating that a DAQ board has no missing codes, which means that the DNL is bounded below by −1 LSB but does not make any specifications about the upper bound.

If the DAQ board in the previous example had a missing code at 120 μV, then increasing the voltage from 96 to 120 μV would not be detectable. Only when the voltage is increased another LSB, or in this example, 24 μV, will the voltage change be detectable (Fig. 18.17). As you can see, poor DNL reduces the resolution of the board.

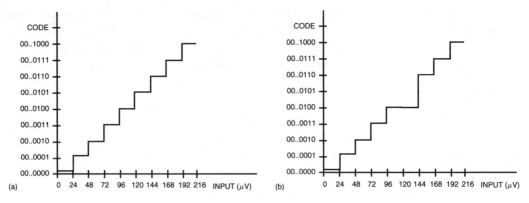

Figure 18.17 Demonstration of a missing code: (a) an ideal transfer curve with DNL of 0 and no missing codes, (b) a poor transfer curve with a missing code at 120 μV.

To run your own DNL test:

1. Input a high-resolution, highly linear triangle wave into one channel of the DAQ board. The frequency of the triangle should be low and the amplitude should swing from minus full scale to plus full scale of the input to the DAQ board.
2. Start an acquisition on the plug-in board so that you acquire a large number of points. Recommended values are 1×10^6 samples for a 12-b board and 20×10^6 million samples for a 16-b board.
3. Make a histogram of the acquired binary codes. This will give you the relative frequency of each code occurrence.
4. Normalize the histogram by dividing by the averaged value of the histogram.
5. The DNL is the greatest deviation from the value of 1 LSB. Because the input was a triangle wave, it had a uniform distribution over the DAQ board codes. The probability of a code occurring is directly proportional to the code width, and therefore shows the DNL.

Figure 18.18 shows a section of the DNL plot of two products with the same 16-b ADC yet significantly different DNL. Ideally, the codewidth plot will be a straight line at 1 LSB. Figure 18.18(a) shows the codewidth plot of a product that has DNL of less than ±0.5 LSB. Because the codewidth is never 0, the product also has no missing codes. Figure 18.18(b) shows the codewidth plot of a product that has poor DNL. This product has many missing codes, a code width that is as much as 4.2 LSB at the code value 0, and is clearly not 16-b linear.

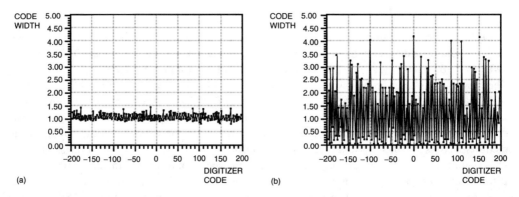

FIGURE 18.18 Section of DNL plots at the critical point (zero crossing) of two products that have the same ADC: (a) the product is the National Instruments AT-MIO-16X, which has a DNL of less than ±0.5 and has no missing codes, (b) the product has very poor DNL and many missing codes.

Data Acquisition

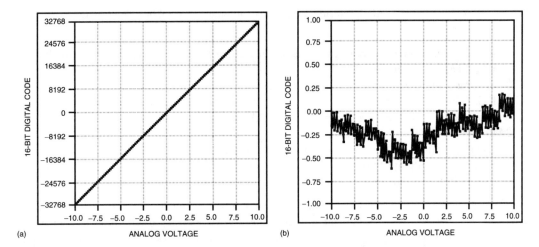

FIGURE 18.19 Determining the relative accuracy of a DAQ board: (a) the plot generated by sweeping the input and how it appears as a straight line, (b) how the straight line of Fig. 18.19(a) is not actually straight by subtracting an actual straight line.

Relative Accuracy

Relative accuracy is a measure in LSBs of the worst-case deviation from the ideal DAQ board transfer function, a straight line. To run your own relative accuracy test:

1. Connect a high-accuracy analog voltage-generation source to one channel of the DAQ board. The source should be very linear and have a higher solution than the DAQ board that you wish to test.
2. Generate a voltage that is near minus full scale.
3. Acquire at least 100 samples from the DAQ board and average them. The reason for averaging is to reduce the effects of any noise that is present.
4. Increase the voltage slightly and repeat step 3. Continue steps 3 and 4 until you have swept through the input range of the DAQ board.
5. Plot the averaged points on the computer. You will have a straight line, as shown in Fig. 18.19(a), unless the relative accuracy of your DAQ board is astonishingly bad.
6. To see the deviation from a straight line, you must first generate an actual straight line in software that starts at the minus full-scale reading and ends at the plus full-scale reading using a straight-line endpoint fit analysis routine.
7. Subtract the actual straight line from the waveform that you acquired. If you plot the resulting array, you should see a plot similar to Fig. 18.19(b).
8. The maximum deviation from zero is the relative accuracy of the DAQ board.

The driver software for a DAQ board will translate the binary code value of the ADC to voltage by multiplying by a constant. Good relative accuracy is important for a DAQ board because it ensures that the translation from the binary code of the ADC to the voltage value is accurate. Obtaining good relative accuracy requires that both the ADC and the surrounding analog support circuitry be designed properly.

Integral Nonlinearity

INL is a measure in LSB of the straightness of a DAQ board's transfer function. Specifically, it indicates how far a plot of the DAQ board's *transitions* deviates from a straight line. Factors that contribute to poor

INL on a DAQ board are the ADC, multiplexer, and instrumentation amplifier. The INL test is the most difficult to make:

1. Input a signal from a digital to analog converter that has higher resolution and linearity than the DAQ board that you are testing.
2. Starting at minus full scale, increase the codes to the DAC until the binary reading from the ADC flickers between two consecutive values. When the binary reading flickers evenly between codes, you have found the LSB transition for the DAQ board.
3. Record the DAC codes of all transitions.
4. Using an analysis routine, make an endpoint fit on the recorded DAC codes.
5. Subtract the endpoint fit line from the recorded DAC values.
6. The farthest deviation from zero is the INL.

Even though highly specified, the INL specification does not have as much value as the relative accuracy specification and is much more difficult to make. The relative accuracy specification is showing the deviation from the ideal straight line whereas the INL is showing the deviation of the transitions from an ideal straight line. Therefore, relative accuracy takes the quantization error, INL, and DNL into consideration.

Settling Time

On a typical plug-in DAQ board, an analog signal is first selected by a multiplexer, and then amplified by an instrumentation amplifier before it is converted to a digital signal by the ADC. This instrumentation amplifier must be able to track the output of the multiplexer as the multiplexer switches channels, and the instrumentation amplifier must be able to settle to the accuracy of the ADC. Otherwise, the ADC will convert an analog signal that has not settled to the value that you are trying to measure with your DAQ board. The time required for the instrumentation amplifier to settle to a specified accuracy is called the *settling time*. Poor settling time is a major problem because the amount of inaccuracy usually varies with gain and sampling rate. Because the errors occur in the analog stages of the DAQ board, the board cannot return an error message to the computer when the instrumentation amplifier does not settle.

The instrumentation amplifier is most likely not to settle when you are sampling multiple channels at high gains and high rates. When the application is sampling multiple channels, the multiplexer is switching among different channels that can have significant differences in voltage levels (Fig. 18.20). Instrumentation amplifiers can have difficulty tracking this significant difference in voltage. Typically, the higher the gain and the faster the channel switching time, the less likely it is that the instrumentation

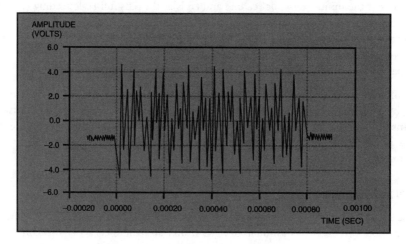

FIGURE 18.20 The input to an instrumentation amplifier that is multiplexing 40 DC signals appears to be a high-frequency AC signal.

Data Acquisition

amplifier will settle. In fact, no off-the-shelf programmable-gain instrumentation amplifier can settle to 12-b accuracy in 5 μs when amplifying at a gain of 100.

It is important to be aware of settling time problems so you know at what multichannel rates and gains you can run your DAQ board and maintain accurate readings. Unfortunately, only a few DAQ board suppliers specify the maximum multichannel acquisition rates. Running your settling time test is relatively easy:

1. Apply a signal to one channel of your DAQ board that is nearly full scale. Be sure to take the range and gain into account when you select the levels of the signal you will apply. For example, if you are using a gain of 100 on a ± 10V input range on the DAQ board, apply a signal slightly less than 10 V/100 or +0.1-V signal to the channel.
2. Acquire at least 1000 samples from one channel only and average them. The averaged value will be the expected value of the DAQ board when the instrumentation amplifier settles properly.
3. Apply a minus full-scale signal to a second channel. Be sure to take range and gain into account, as described earlier.
4. Acquire at least 1000 samples from the second channel only and average them. This will give you the expected value for the second channel when the instrumentation amplifier settles properly.
5. Have the DAQ board sample both channels at the highest possible sampling rate so that the multiplexer is switching between the first and the second channels.
6. Average at least 100 samples from each channel.
7. The deviation between the values returned from the board when you sampled the channels using a single channel acquisition and the values returned when you sampled the channels using a multichannel acquisition will be the settling time error.

To plot the settling time error, repeat steps 5 and 6, but reduce the sampling rate each time. Then plot the greatest deviation at each sampling rate. A DAQ board settles to the accuracy of the ADC only when the settling time error is less than ± 0.5 LSB. By plotting the settling time error, you can determine at what rate you can sample with your DAQ board and settle to the accuracy that your application requires. The settling-time plot will usually vary with gain and so you will want to repeat the test at different gains.

Figure 18.21 shows an example of two DAQ boards that have a 12-b resolution ADC, yet have significantly different settling times. For the tests, both boards were sampling two channels with a gain of 100. The settling time plot for the first board, shown in Fig. 18.21(a), uses an off-the-shelf instrumentation amplifier, which has 34 LSBs of settling-time error when sampling at 100 kHz. Because the board was using an input range of 20 V at a gain of 100, 1 LSB = 48.8 μV. Therefore, data is inaccurate by as much as 1.7 mV due to settling time when the board is sampling at 100 kHz at a gain of 100. If you use this board you must either increase the sampling period to greater than 20 μs so that the board can settle, or be able to accept the settling-time inaccuracy. The settling-time plot for the second board,

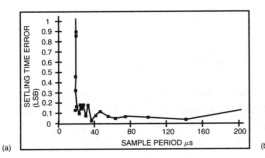
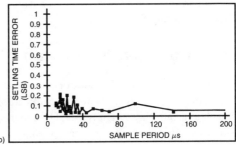

FIGURE 18.21 Settling plots for two DAQ boards that have the same resolution ADC yet significantly different settling times. Both boards were using a gain of 100. The plot in (b) is of the National Instruments AT-MIO-16E-2 which settles properly, whereas the plot in (a) does not. (a) Settling time of off-the-shelf instrumentation amplifier, (b) settling time of NI-PGIA instrumentation amplifier.

shown in Fig. 18.21(b), settles properly because it uses the Nl-PGIA™, a custom instrumentation amplifier designed specifically to settle in DAQ board applications. Because the board settles to within ±0.5 LSB at the maximum sampling rate, you can run the board at all sampling rates and not have any detectable settling-time error.

Applications that are sampling only one channel, applications that are sampling multiple channels at very slow rates, or applications that are sampling multiple channels at low gains usually do not have a problem with settling time. For the other applications, your best solution is to purchase a DAQ board that will settle at all rates and all gains. Otherwise, you will have to either reduce your sampling rate or realize that your readings are not accurate to the specified value of the ADC.

Noise

Noise is any unwanted signal that appears in the digitized signal of the DAQ board. Because the PC is a noisy digital environment, acquiring data on a plug-in board takes a very careful layout on multilayer DAQ boards by skilled analog designers. Simply slapping an A/D converter, instrumentation amplifier, and bus interface circuitry on a one- or two-layer board will most likely result in a very noisy DAQ board. Designers can use metal shielding on a DAQ board to help reduce noise. Proper shielding should not only be added around sensitive analog sections on a DAQ board, but must also be built into the DAQ board with ground planes.

Of all of the tests to run, a DC-class signal noise test is the easiest to run:

1. Connect the + and − inputs of the DAQ board directly to ground. If possible, make the connection directly on the I/O connector of the board. By doing this, you can measure only the noise introduced by the DAQ board instead of noise introduced by your cabling.
2. Acquire a large number of points (1×10^6) with the DAQ board at the gain you choose in your application.
3. Make a histogram of the data and normalize the array of samples by dividing by the total number of points acquired.
4. Plot the normalized array.
5. The deviation, in LSB, from the highest probability of occurrence is the noise.

Figure 18.22 shows the DC-noise plot of two DAQ products, both of which use the same ADC. You can determine two qualities of the DAQ board from the noise plots—range of noise and the distribution. The plot in Fig. 18.22(a) has a high distribution of samples at 0 and a very small number of points occurring at other codes. The distribution is Gaussian, which is what is expected from random noise. From the plot, the noise level is within ±3 LSB. The plot in Fig. 18.22(b) is a very noisy DAQ product, which does not have the expected distribution and has a noise greater than ±20 LSB, with many samples occurring at points other than the expected value. For the DAQ product in Fig. 18.22(b), the tests were run with an input range of ±10 V and a gain of 10. Therefore, 1 LSB = 31 μV, thus a noise level of 20 LSB is equivalent to 620 μV of noise.

18.6 Common Issues with PC-based Data Acquisition

Aside from software and programming, the most common problem users run into when putting together DAQ systems is a noisy measurement. Unfortunately, noise in DAQ systems is a complicated issue and difficult to avoid. However, it is useful to understand where noise typically comes from, how much noise is to be expected, and some general techniques to reduce or avoid noise corruption.

The Low-Noise Voltmeter Standard

One of the most common questions asked of technical support personnel of DAQ vendors concerns the voltmeter. Users connect their signal leads to a handy voltmeter or digital multimeter (DMM), without

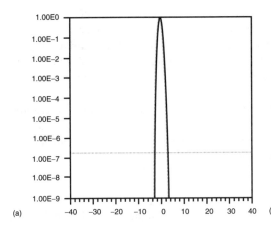 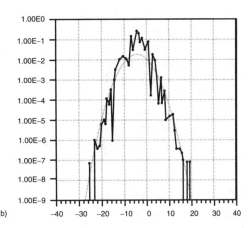

FIGURE 18.22 Noise plots of two DAQ products, which have significantly different noise performance even though they use the same 16-b ADC:(a) The National Instruments AT-MIO-16X, which has codes ranging from −3 LSB to +3 LSB. The codes at ±3 LSB have less than a 10^4 and a 10^{-7} probability of occurrence. (b) The DAQ product has noise as high as ±20 LSB and has a high probability (10^{-4}) of codes occurring as much as 15 LSB from the expected value.

worrying too much about cabling and grounding, and obtain a rock solid reading with little jitter. The user then duplicates the experiment with a DAQ board and is disappointed to find that the readings returned by the DAQ board look very noisy and very unstable.

The user decides there is a problem with the DAQ board and calls the technical support line for help. In fact, the user has just demonstrated the effects of two different measurement techniques, each with advantages and disadvantages. DAQ boards are designed as flexible, general-purpose measurement devices. The DAQ board front end typically consists of a gain amplifier and a *sampling* analog-to-digital converter. A sampling ADC takes an instantaneous measurement of the input signal. If the signal is noisy, the sampling ADC will digitize the signal as well as a noise. The digital voltmeter, on the other hand, will use an *integrating* ADC that integrates the signal over a given time period. This integration effectively filters out any high-frequency noise that is present in the signal.

Although the integrating input of the voltmeter is useful for measuring static, or DC signals, it is not very useful for measuring changing signals or digitizing waveforms, or for capturing transient signals. The plug-in DAQ board, with its sampling ADC, has the flexibility to perform all of these types of measurements. With a little software, the DAQ board can also emulate the operation of the integrating voltmeter by digitizing the static signal at a higher rate and performing the integration, or averaging, of the signal in software.

Where Does Noise Come from?

There are basically four possible sources of noise in a DAQ system:

- Signal source, or transducer
- Environment (noise induced onto signal leads)
- PC environment
- DAQ board

Although the signal source is commonly a significant source of noise, that topic is beyond the scope of this chapter. Most measurement noise problems are the result of noise that is radiated, conducted, or coupled onto signal wires attaching the sensor or transducer to the DAQ equipment. Signal wires basically act as antennas for noise.

Placing a sensitive analog measurement device, like a plug-in DAQ board, inside a PC chassis might seem like asking for trouble. The high-speed digital traffic and power supplies inside a PC are prime candidates for noise radiation. For example, it is a good idea to not install your DAQ board directly next to your video card. Probably the most dangerous area for your analog signals is not inside the PC, but on top of it. Keep your signal wires clear of your video monitor, which can radiate large high-frequency noise levels onto your signal.

The DAQ board itself can be a source of measurement noise. Poorly designed boards, for example, may not properly shield the analog sections from the digital logic sections that radiate high-frequency switching noise. Properly designed boards, with well-designed shielding and grounding, can provide very low-noise measurements in the relatively noisy environment of the PC.

Proper Wiring and Signal Connections

In most cases, the major source of noise is the environment through which the signal wires must travel. If your signal leads are relatively long, you will definitely want to pay careful attention to your cabling scheme. A variety of cable types are available for connecting sensors to DAQ systems. Unshielded wires or ribbon cables are inexpensive and work fine for high-level signals and short to moderate cable lengths. For low-level signals or longer signal paths, you will want to consider shielded or twisted-pair wiring. Tie the shield for each signal pair to ground reference at the source. Practically speaking, consider shielded, twisted-pair wiring if the signal is less than 1 V or must travel farther than approximately 1 m. If the signal has a bandwidth greater than 100 kHz, however, you will want to use coaxial cables.

Another useful tip for reducing noise corruption is to use a *differential* measurement. Differential inputs are available on most signal conditioning modules and DAQ boards. Because both the (+) and (–) signal lines travel from signal source to the measurement system, they pick up the same noise. A differential input will reject the voltages that are common to both signal lines. Differential inputs are also best when measuring signals that are referenced to ground. Differential inputs will avoid ground loops and reject any difference in ground potentials. On the other hand, single-ended measurements reference the input to ground, causing ground loops and measurement errors.
Other wiring tips:

- If possible, route your analog signals separately from any digital I/O lines. Separate cables for analog and digital signals are preferred.
- Keep signal cables as far as possible from AC and other power lines.
- Take caution when shielding analog and digital signals together. With a single-ended (not differential) DAQ board, noise coupled from the digital signals to the analog signals via the shield will appear as noise. If using a differential input DAQ board, the coupled noise will be rejected as common-mode noise (assuming the shield is tied to ground at one end only).

Source Impedance Considerations

When time-varying electric fields, such as AC power lines, are in the vicinity of your signal leads, noise is introduced onto the signal leads via capacitive coupling. The capacitive coupling increases in direct proportion to the frequency and amplitude of the noise source and to the impedance of the measurement circuit. Therefore, the source impedance of your sensor or transducer has a direct effect on the susceptibility of your measurement circuit to noise pickup. The higher the source impedance of your sensor or signal source, the larger amount of capacitive coupling. The best defense against capacitive noise coupling is the shield that is grounded at the source end. Table 18.3 lists some common transducers and their impedance characteristics.

Data Acquisition

TABLE 18.3 Impedance Characteristics of Transducers

Transducer	Impedance Characteristic
Thermocouples	Low (<20 Ω)
RTDs	Low (<1 kΩ)
Strain gauges	Low (<1 kΩ)
Thermistors	Moderate to high (>1 kΩ)
Solid-state pressure transducer	High (<1 kΩ)
Potentiometer	High (500 to 100 kΩ)
Glass pH electrode	Very high (10^9 Ω)

Pinpointing Your Noise Problems

If your system is resulting in noisy measurements, follow these steps to determine the source of the noise and how best to reduce it. The first step can also give you an idea of the noise performance of your DAQ board itself. The steps are:

1. Short one of the analog input channels of the DAQ board to ground directly at the I/O connector of the board. Then, take a number of readings and plot the results. The amount of noise present is the amount of noise introduced by the PC and the DAQ board itself with a very low-impedance input. Typical results are shown in Fig. 18.23. This plot shows a reading that jumps between 0.00 and 2.44 mV. Because this particular board uses a 12-b ADC and the amplifier was set to a gain of 1, this deviation corresponds to only 1 b, or LSB. In other words, the 12-b ADC toggled between binary values 0 and 1. If this test yields large amount of noise, your DAQ board is not operating properly, or another plug-in board in the PC may be radiating noise onto the DAQ board. Try removing other PC boards to see if the noise level decreases.
2. Attach your signal wires to the DAQ board. At your signal source or signal conditioning unit, ground or short the input leads. Acquire and plot a number of readings as in step 1. If the observed noise levels are roughly the same as those with the actual signal source instead of the short in place, the cabling and/or the environment in which the cabling is run is the culprit. You may need to try relocating your cabling farther from potential noise sources. If the noise source is not known, spectral analysis can help identify the source of the noise.
3. If the noise level in step 2 is less than with the actual signal source, replace the short with a resistor approximately equal to the output impedance of the signal source. This setup will show whether capacitive coupling in the cable due to high impedance is the problem. If the observed noise level is still less than with the actual signal source, then cabling and the environment can be dismissed as the problem. In this case, the culprit is either the signal source itself, or improper grounding configuration.

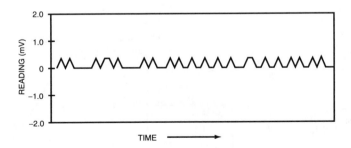

FIGURE 18.23 Measured noise with shorted inputs to DAQ board.

Noise Removal Strategies

After you have optimized your cabling and hardware setup, you may still need additional techniques to reduce noise that is unavoidable with proper cabling and grounding.

Signal Amplification

If you must pass very low-level signals through long signal leads, you will want to consider amplifying the signals near the source. An amplifying signal conditioner could boost the signal level before it is subject to the noise corruption of the environment. The same amount of noise will be radiated onto the signal, but will have a much smaller effect on the high-level signal.

Averaging

A very powerful technique for making low-noise measurements of static, or DC, signals is data averaging. For example, suppose you were monitoring the output of a thermocouple in an environment known to contain high amounts of 60-Hz power line noise. For each required temperature reading, therefore, you collect 100 readings over a time period of $i/60$ s, where i is some integer, and average the 100 data readings. Because the data was collected over an integer number of 60-Hz power cycles, the averaging of the data will average out any 60 Hz noise to zero. For 50-Hz power noise, collect the readings over a time period equal to $i/50$ s. This averaging has the same filtering effect as the integrating voltmeter.

Filtering

Of course, one method of removing noise from a electrical signal is with a hardware filter. There are a couple of options. First, you can use commercial signal conditioners that implement low-pass filters. Or, for simple filtering needs (moderate amounts of noise), you might consider building a simple RC filter on the input of your DAQ board. Figure 18.24 shows a single-pole RC filter that you could easily build and would attenuate signals with a frequency higher than the cutoff frequency F_c. F_c will be equal to

$$\frac{1}{2\pi RC}$$

For $R = 5\ \text{k}\Omega$ and $C = 4.7\ \mu\text{F}$, the cutoff frequency can be calculated as

$$F_c = \frac{1}{2\pi RC} = \frac{1}{2\pi(5000\Omega)(4.7 \cdot 10^{-6})} = 6.8\ \text{Hz}$$

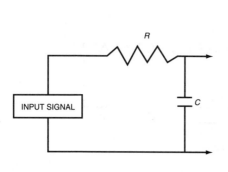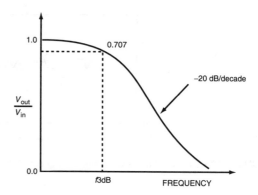

FIGURE 18.24 Simple low-pass RC filter.

Data Acquisition

Figure 18.24 also shows the transfer function, which shows the attenuation of the signal (defined as V_{out}/V_{in}) as a function of frequency. If a signal of frequency F_C, for example, is applied to the input, the signal output of the RC filter will be equal to $0.707(V_{in})$. The equation that tells us the amount of attenuation as a function of input frequency F_{in}, for the RC filter is

$$\left|\frac{V_{out}}{V_{in}}\right| = \frac{1}{\sqrt{1 + (F_{in}/F_c)^2}}$$

For the single-pole RC filter with $F_c = 6.8$ Hz subject to 60-Hz noise input, the equation above is calculated to be 0.11. Therefore, 60-Hz noise introduced into the signal will be attenuated by 89%.

Grounding

Another very common source of problems in DAQ systems is grounding. In fact, noisy measurement problems are often due to improper grounding of the measurement circuit. You can avoid most grounding problems with the following steps before you configure your DAQ system:

1. Determine your signal source type—grounded or floating.
2. Identify and use the proper input measurement mode—nonreferenced (differential or nonreferenced single ended) or ground referenced (single ended).
3. If using a differential measurement system with a floating signal source, provide a ground reference for the signal.

Grounded and Floating Signal Sources

Signal sources can be grouped into two types, *grounded* or *floating*. A grounded source is one in which the voltage signal is referenced to the building system ground. Because they are connected to the building ground, they share a common ground with the DAQ board. The most common examples of a grounded source are devices that plug into the building ground, such as signal generators and power supplies.

A floating source is a source in which the voltage signal is not referred to an absolute reference, such as Earth or building ground. Some common examples of floating signal sources are batteries, battery-powered signal sources, thermocouples, transformers, isolation amplifiers, and any instrument that explicitly floats its output signal. Notice that neither terminal of the source is referred to the electrical outlet ground. Thus, each terminal is independent of Earth.

Types of Measurement Inputs

In general, with regard to ground-referencing of the inputs, there are three types of measurement systems. Following is a description of each type.

Differential Inputs

A differential, or nonreferenced, measurement system has neither of its inputs tied to a fixed reference such as Earth or building ground. DAQ boards with instrumentation amplifiers can be configured as differential measurement systems.

An ideal differential measurement system responds only to the potential *difference* between its two terminals, the (+) and (−) inputs. Any voltage measured with respect to the instrumentation amplifier ground present at both amplifier inputs is referred to as a common-mode voltage. The term *common-mode* voltage range measures the ability of a DAQ board in differential mode to reject the common-mode voltage signal.

Ground-Referenced Inputs

A grounded or ground-referenced measurement system is similar to a grounded source, in that the measurement is made with respect to ground. This is also referred to as a *ground-referenced single-ended* (GRSE) measurement system.

Nonreferenced Single-Ended Inputs

A variant of the single-ended measurement technique, known as *nonreferenced single-ended* (NRSE) measurement system, is often found in DAQ boards. In an NRSE measurement system, all measurements are still made with respect to a single-node analog sense, but the potential at this node can vary with respect to the measurement system ground.

Now that we have identified the different types of signal sources and measurement systems, we can discuss the proper measurement system for each type of signal source.

Measuring Grounded Signal Sources, Avoiding Loops

A grounded signal source is best measured with a differential or NRSE measurement system. With this configuration, the measured voltage V_m is the sum of the signal voltage V_s and the potential difference ΔV_g that exists between the signal source ground and the measurement system ground. This potential difference is generally not a DC level; thus, the result is a noisy measurement system often showing power-line frequency (50 or 60 Hz) components in the readings. Ground-loop introduced noise may have both AC and DC components, thus introducing offset errors as well as noise in the measurements. The potential difference between the two grounds causes a current to flow in the interconnection. This current is called *ground-loop* current.

The preferable input mode for a grounded signal is differential or NRSE mode. With either of these configurations, any potential difference between references of the source and the measuring device appears as common-mode voltage to the measurement system and is subtracted from the measured signal.

Measuring Floating Signals

You can use differential or single-ended inputs to measure a floating signal source. With a ground-referenced input, the DAQ board provides the one ground reference. When using differential inputs to measure signals that are not ground referenced, however, you must explicitly provide a ground reference to make accurate measurements. The differential input can be referenced by simple grounding of the (−) lead of the signal input. Alternatively, resistors can be connected from each signal lead to ground. This configuration maintains a balanced input and may be desirable for high-impedance signal sources. Many signal conditioning accessories include provisions for installing these resistors or direct connections to ground. Figure 18.25 summarizes the analog input connections.

Basic Signal Conditioning Functions

In general, signal conditioners exist to interface raw signals from transducers to a general-purpose measurement device, such as a plug-in DAQ board, while simultaneously boosting the quality and reliability of the measurement. To accomplish this goal, signal conditioners perform a number of functions, including the following.

Signal Amplification

Many transducers output very small voltages that can be difficult to measure accurately. For example, a J-type thermocouple signal varies only about 50 μV/°C over most of its range. Most signal conditioners, therefore, include amplifiers to boost the signal level to better match the input range of the analog-to-digital converter and improve resolution and sensitivity. Although many DAQ boards and I/O devices include onboard amplifiers for this reason, it may be necessary to locate an additional signal conditioner

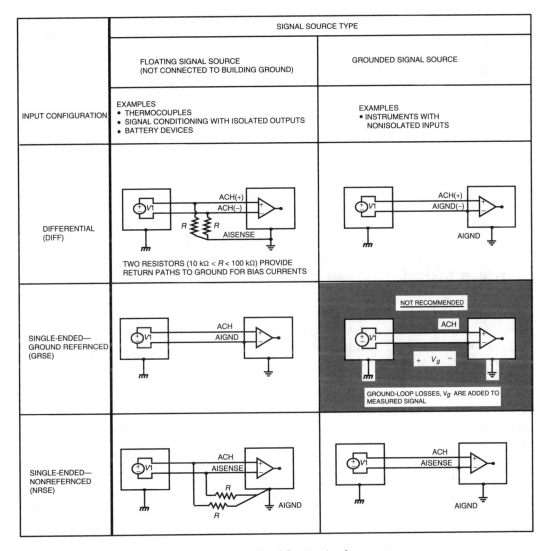

FIGURE 18.25 Summary of connections of grounded and floating signal sources.

with amplification near the source of low-level signals, such as thermocouples, to increase their immunity to electrical noise from the environment. Otherwise, any small amount of noise picked up on lead wires can corrupt your data.

Filtering

Additionally, signal conditioners can include filters to reject unwanted noise within a certain frequency range. For example, most conditioners include low-pass filters to reduce high-frequency noise, such as the very common 60- or 50-Hz periodic noise from power systems or machinery. Some signal conditioners that are used for more dynamic measurements, such as vibration monitoring, include special antialiasing that feature programmable bandwidth (variable according to the sampling rate) and very sharp filter rolloff.

Isolation

One of the most common causes of measurement problems, noise, and damaged I/O equipment is improper grounding of the system. These nagging problems tend to disappear when isolated signal conditioners are introduced into the measurement system. Isolated conditioners pass the signal from its source to the measurement device without a galvanic or physical connection. Besides breaking ground loops, isolation blocks high-voltage surges and rejects high common-mode voltage, protecting expensive DAQ instrumentation.

For example, suppose you are to monitor the temperature of an extrusion process. Although you are using thermocouples with output signals of 0 and 50 mV, the thermocouples are soldered to the extruder. The extruder machines are powered by a dedicated power system and your thermocouple leads are actually sitting at 50 V. Connecting the thermocouple leads directly to nonisolated DAQ board would probably damage the board. However, you can connect the thermocouple leads to an isolated signal, which rejects the common-mode voltage (50 V), safely passing the differential 50-mV differential signal on the DAQ board for measurement.

A common method for circuit isolation is using optical, capacitive, or transformer isolators. Capacitive and transformer isolators modulate the signal to convert it from a voltage to a frequency value. The frequency signal is then coupled across capacitors or a transformer, where it is then converted back to the proper voltage value. Optical isolators, commonly used for digital signals, use LEDs to convert the voltage on/off information into light signals to couple the signal across the isolation barrier.

Transducer Excitation and Interfacing

Many types of sensors and transducers have particular signal conditioning requirements. For example, thermocouples require cold-junction compensation for the thermoelectric voltages created where the thermocouple wires are connected to the data acquisition equipment. Resistive temperature devices (RTDs) require an accurate current excitation source to convert their small changes in electrical resistance into measurable changes in voltage. To avoid errors caused by the resistance in the lead wires, RTDs are often used in a 4-wire configuration. The 4-wire RTD measurement avoids lead resistance errors because two additional leads carry current to the RTD device, so that current does not flow in the sense, or measurement. Strain gauge transducers, on the other hand, are used in a Wheatstone bridge configuration with a constant voltage or current power source. The signal conditioning requirements for these and other common transducers are listed in Table 18.1.

Linearization

Most sensors exhibit an output that is nonlinear with respect to the measured. Therefore, many signal conditioners include circuitry or onboard intelligence to linearize the transfer function of the sensor. This onboard linearization is designed to offload some of the processing requirements of the DAQ system. With the increased use of PCs, however, this need is diminished and you can easily perform this linearization function in software. Unlike hardware linearization, software linearization is a very flexible solution, making it possible for a single signal conditioning module to be easily adapted via software for a wide variety of sensors. In fact, you can even implement your own customized transducer linearization routines if necessary.

Variety of Signal Conditioning Architectures

Signal conditioning systems come in all different forms, ranging from single-channel I/O modules to multichannel chassis-based systems with sophisticated signal routing capabilities. In addition, several products commonly classified as signal conditioners include an onboard ADC with digital communications interface. Here, we will concentrate on nondigitizing conditioning systems used as a front-end for data acquisition and control systems, such as plug-in DAQ boards.

The typical single-channel I/O module is a fixed-function conditioner designed for a particular type of transducer and signal range. You cable the conditioned output signal, usually a voltage signal, directly to a DAQ board input channel. Some modules are DIN rail-mountable, whereas others install into a backplane that holds 8–16 modules. Newer versions of the single-channel modules feature programmability and added intelligence for scaling and diagnostics. Because the modules do not incorporate signal multiplexing, they are best suited for applications with fewer I/O channels.

Many DAQ board vendors supply specialized signal conditioning boards for use with their DAQ boards. These signal conditioning boards, usually designed for a particular transducer type, tend to provide a less flexible system. Meanwhile, other DAQ vendors incorporate signal conditioning directly on the PC plug-in DAQ board. Although this approach can provide a low-cost system for simpler applications, the benefits of locating your high-voltage isolation barrier inside the PC are questionable. In addition, you do not have the option of amplifying your low-level sensor signals before they enter the potentially noisy PC.

Signal Conditioners Offer I/O Expansion

A final class of signal conditioners incorporate signal multiplexing to significantly expand the I/O capabilities of the DAQ system. These systems consist of a chassis that houses a variety of signal conditioning modules. Instead of simply passing the conditioned signals to an outgoing connector, each module multiplexes the conditioned signals onto a single analog output channel. You can cable the multiplexed output directly to a DAQ board or pass it to the chassis backplane bus. This backplane bus routes the conditioned analog signals, as well as digital communications and timing control signals, among the modules. Such a system is expandable; adding channels is accomplished by plugging a new multiplexing module into the backplane bus. Because the bus also incorporates a digital communications path, you can also incorporate digital I/O and analog output modules into the same chassis.

The multiplexing architecture of these signal conditioning systems is especially well suited for applications involving larger numbers of channels. For example, some systems can multiplex 3072 channels into a single PC plug-in DAQ board. More importantly, these multiplexing signal conditioners offer significant advantages in cost and physical space requirements. By switching multiple inputs into a single processing block, including amplification, filtering, isolation, and ADC, you can achieve a very low cost per channel not attainable with single-channel modules. Even though single-channel modules are being developed that are smaller and slimmer, systems with a multiplexing architecture will always be able to pack many more I/O channels into a given physical space.

Defining Terms

Alias: A false lower frequency component that appears in sampled data acquired at too low a sampling rate.

Conversion time: The time required, in an analog input or output system, from the moment a channel is interrogated (such as with a read instruction) to the moment that accurate data is available.

Data acquisition (DAQ): (1) Collecting and measuring electrical signals from sensors, transducers, and test probes or fixtures and inputting them to a computer for processing. (2) Collecting and measuring the same kinds of electrical signals with A/D and/or DIO boards plugged into a PC, and possibly generating control signals with D/A and/or DIO boards in the same PC.

Differential nonlinearity (DNL): A measure in LSB of the worst-case deviation of code widths from their ideal value of 1 LSB.

Integral nonlinearity (INL): A measure in LSB of the worst-case deviation from the ideal A/D or D/A transfer characteristic of the analog I/O circuitry.

Nyquist sampling theorem: A law of sampling theory stating that if a continuous bandwidth-limited signal contains no frequency components higher than half the frequency at which it is sampled, then the original signal can be recovered without distortion.

Relative accuracy: A measure in LSB of the accuracy of an ADC. It includes all nonlinearity and quantization errors. It does not include offset and gain errors of the circuitry feeding the ADC.

References

House, R. 1993. Understanding important DA specifications. *Sensors* 10(6):11–16.
House, R. 1994. Understanding inaccuracies due to settling time, linearity, and noise. *National Instruments European User Symposium Proceedings*, Nov. 10–11:17–26.
McConnell, E. 1994. PC-based data acquisition users face numerous challenges. *ECN* 38(8):11–12.
McConnell, E. 1995. Choosing a data-acquisition method. *Electronic Design* 43(13):147–156.
Potter, D. and Razdan, A. 1994. Fundamentals of PC-based data acquisition. *Sensors* 11(2).
Potter, D. 1994. Sensor to PC—Avoiding some common pitfalls. *Sensors Expo Proceedings*, Sept. 20:12–20.
Potter, D. 1995. Signal conditioners expand DAQ system capabilities. *I&CS* 68(8).

Further Information

Johnson, G.W. 1994. *LabVIEW Graphical Programming*. McGraw-Hill, NewYork.
McConnell, E. 1994. New achievements in counter/timer data acquisition technology. *MessComp 1994 Proceedings*, Sept. 13–15.
McConnell, E. 1995. Equivalent time sampling extends DA performance. *Sensors* 12(6).

19
Computer-Based Circuit Simulation

19.1 Introduction 19-1
19.2 PSPICE 19-1
 Circuit Entry (Netlist and Schematic Capture) • Passive Components Description • Semiconductor Component Description • Independent Sources • Graphics Outputs of Simulation Results: Probe • Running PSpice: Control Shell • Simulation Example 1 • Op-Amp Description • Simulation Example 2 • Time-Domain Analysis • Simulation Example 3 • Fourier Analysis • Simulation Example 4 • Subcircuits • Simulation Example 5 • Simulation vs Practical • PSpice Error Messages
19.3 Other Simulators 19-24
 Digital Simulators • Simulation Example 6 • Mixed Analog–Digital Simulators • Simulation Example 7 • Block-Diagram Simulators • Simulation Example 8

Bashir Al-Hashimi
Staffordshire University

19.1 Introduction

Simulation allows designs to be analyzed and modified without having to go to the effort, expense, and time of building a prototype. Simulation can also be used to perform analyses that are not always possible or desirable on a physical design such as worst-case component tolerances sensitivity analyses and simulating faulty circuits. Simulation packages are usually classified into four types: **analog, digital, mixed analog–digital** (A/D), and **block-diagram simulators**. The aim of this chapter is to provide an introduction to the various simulation types and demonstrate their use through a number of worked examples. Section 19.2 deals with analog simulation and the *PSpice* simulator. Different types of simulation are discussed, including frequency response, and transient and Fourier analyses. The correlation between simulation and measured results depend on the accuracy of the component models used in the simulation. To give the reader insight into how good or bad a simulation is, a comparison between simulated and measured results will be presented where appropriate. Section 19.3 covers digital, mixed analog–digital and block-diagram simulators. To illustrate the use of these simulators, a number of simulation examples are included. For digital simulation, the simulator *Quicksim*, is chosen, whereas the simulators *PSpice A/D* and *Tesla* have been selected for the mixed analog–digital and block-diagram simulations, respectively.

19.2 PSPICE

The **simulation program with integrated circuit emphasis (SPICE)** is a powerful computer package used to analyze electrical and electronic circuits [Nagel 1975]. SPICE was developed in the early 1970s in the

Department of Electrical Engineering of the University of California at Berkeley, progressing through various versions, culminating with SPICE 2G.6 in 1981. Although SPICE is a powerful and robust program, it did not gain popularity in nonacademic circles for a number of reasons: the program needs to run on mainframe computers, has user-unfriendly interfaces, and has no component models libraries. A number of commercial organizations could see the potential of the core software provided these shortcomings were rectified and versions are now available for use on a variety of computers ranging from personal computers (PCs) to mainframes. Table 19.1 shows some PC-based Spices. PSpice is often considered to be the standard for PC-based analog simulation and, therefore, is chosen here.

TABLE 19.1 Some Commercially Available PC-Based Spices

Vendor	Simulator Name
Contec Microelectronics	ContecSpice
Intusoft	IsSpice
Meta-Software	HSpice
MicroSim	PSpice
Spectrum Software	Micro-Cap
Tanner Research	TSpice

Circuit Entry (Netlist and Schematic Capture)

To carry out a simulation, the circuit must first be described to the simulator. This is usually achieved using two methods: **netlist** and **schematic capture**. A netlist allows the user to describe a circuit using description statements, where each statement expresses a component or a signal source in terms of a name, node connections, and a value. The description statements are typed manually using a standard text editor, with the resulting file being called the circuit input file or the netlist. Because netlists are entered manually, it is easy for a user to make data entry mistakes; this problem increases over time as designs get larger. Schematic capture programs were introduced with the intention of simplifying the process of generating netlists. A schematic capture program allows a designer to draw a circuit using a mouse and symbols picked from the simulator library to represent the circuit. Once the circuit is drawn, the schematic capture program generates the circuit netlist automatically minimizing the input errors that occur when typing a netlist.

The choice of circuit entry for simulation depends on the user preference, with both methods requiring some time to learn. However, it is accepted that understanding the basic rules of creating circuit netlists often allow a better appreciation of the simulation process and principles, and it is, therefore, recommended. The simulation examples given in this section have been tested using the DOS evaluation copy of PSpice, Version 6.1. The examples should also run with the earlier versions of PSpice and other SPICE-like simulators with minor modifications to the circuit netlists.

Passive Components Description

A passive component is described using the following description statement:

$$<component\ name>\ <node\ 1>\ <node2>\ <value>$$

The parameter *<component name>* is the name of the component, and this name must start with a letter recognized by PSpice. Table 19.2 shows some of the letters recognized by PSpice as components, for example, a capacitor is identified by the letter C, whereas a resistor is identified by the letter R. Component names can be up to eight characters long, but must start with a letter recognized by PSpice. For example, the resistors in the circuit of Fig. 19.1 could be R1 and R2 or Rin and Rout, respectively. Note that each component must have a unique name. The parameters *<node 1>* and *<node 2>* show

Computer-Based Circuit Simulation

TABLE 19.2 Some Components Recognized by PSpice

Letter	Description
B	Gallium arsenide field effect transistor (GaAsFET)
C	Capacitor
D	Diode
J	junction field effect transistor (JEFT)
K	Coefficient of coupling for mutual inductance
L	Inductor
M	Metal oxide semiconductor field effect transistor (MOSFET)
Q	Bipolar junction transistor
R	Resistor
S	Voltage controlled switch
T	Transmission line
W	Current controlled switch

how a component is connected within a circuit. A node is defined as a point of connection between two or more circuit components. For example, R1 of Fig. 19.1 is connected between nodes 1 and 2, whereas R2 is connected between nodes 4 and 0. PSpice always assumes the circuit ground is at node 0. The order in which nodes are numbered is arbitrary, but each node number must be unique. Using the passive component description statement, the circuit in Fig. 19.1 is represented as

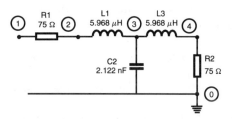

FIGURE 19.1 RLC circuit.

R1 1 2 75
L1 2 3 5.968uH
C2 3 0 2.122nF
L3 3 4 5.968uH
R2 4 0 75

This shows that the resistor R1 is connected between nodes 1 and 2 and its value is 75 Ω, inductor L1 is connected between nodes 2 and 3 and its value is 5.968 µH. Similarly, the capacitor C2 is connected between nodes 3 and 0 and its value is 2.122 nF. It is not essential to describe components using upper-case letters; lower-case letters would have been just as valid. PSpice allows the user to express component values using a number of notations. One such notation is the scale suffix used in the preceding circuit description. For example, the scale suffix "u" used in expressing the value of L1, will multiply the number 5.968 by 10^{-6}. Note that the program represents the symbol μ by the letter u. PSpice recognizes all of the scale suffixes usually used in practice, examples are k = 10^3, m = 10^{-3}, p = 10^{-12}, and so on. Another notation often used in expressing component values is the exponential (E) notation, where E denotes the power of 10. For example, the values of L1 and C2 could have been specified as 5.968E-6 and 2.122E-9, respectively.

The R1 and R2 description statements show that the resistors have no units (Ω) after their values, this is because the program assumes the basic electrical units for component values as shown in Table 19.3. This means the values of L1 and C2, for example, could have described as 5.968u and 2.122n, respectively, without altering the meaning of the statements.

TABLE 19.3 PSpice Default Units

Type	Units
Resistor	Ohm
Inductor	Henry
Capacitor	Farad

Semiconductor Component Description

Table 19.2 shows that PSpice is capable of simulating semiconductor components such as diodes and various types of transistors. To describe semiconductor components, two statements are required, unlike passive components, which are described using one statement. The two statements are: a component description statement, which defines the type and input and output nodes of the component; and a model statement, which describes the component characteristics in terms of a set of parameters. To demonstrate how semiconductor components are described, bipolar transistors are considered. For information on the description of the other semiconductor components, see Microsim [1994], Al-Hashimi [1995], Tuinenga [1992], and Rashid [1990]. The basic form of the bipolar transistor component description statement is

$$Q<name> <NC> <NB> <NE> <model\ name>$$

where Q is the PSpice symbol for a bipolar transistor and <name> is the transistor name, which is chosen arbitrarily by the user. The parameters <NC>, <NB>, and <NE> are the collector, base, and emitter nodes of the transistor, respectively. It is essential that the transistor nodes are entered in the correct order. The parameter <model name> is a descriptive name of the transistor chosen arbitrarily by the user.

The basic form of the bipolar transistor model statement is

$$.model <model\ name> <type> [model\ parameters]$$

where the parameter <model name> is the name given to the bipolar transistor in the description statement. The parameter <type> determines the transistor type and could be either NPN or PNP. The simulator has a detailed and accurate model of the bipolar transistor and is used to model many effects ranging from current gain, capacitance junctions to noise and temperature characteristics. The transistor model has over 40 parameters [Microsim 1994], each of which has a specific default value. For example, the default value of the transistor current gain is 100. If no model parameters are specified by the user using the [model parameters] part of the .model statement, the program will set the parameters to their default values giving typical transistor performance. Note that the .symbol in the .model statement is a necessary part of the statement line, and the [ ] symbol in a statement indicates the provision of parameters by the user is optional. If no parameters are supplied, PSpice will use the default values. To illustrate how transistors are represented, consider the circuit in Fig. 19.2. The description is

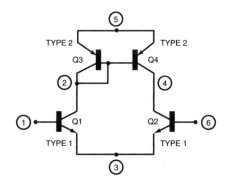

FIGURE 19.2 Bipolar transistors circuit.

    Q1 2 1 3 type1
    Q2 4 6 3 type1
    .model type1 NPN
    Q3 2 2 5 type2
    Q4 4 2 5 type2
    .model type2 PNP

The bipolar transistor Q1 has its collector, base, and emitter at nodes 2, 1, and 3, respectively, and has the model name of type1, which has been chosen arbitrarily. The transistor is an NPN with all its model parameters set to default values, since no model parameters were specified in the .model statement. Similarly, Q2 is an NPN with the collector, base, and emitter at nodes 4, 6, and 3, respectively. Since Q1 and Q2 use the same default transistor model type1, it is only necessary to write one .model statement. Transistors Q3 and Q4 are PNP type and both use the same default transistor model, called type2, which is defined using the second .model statement. This example has shown how user-defined bipolar transistors are described, with the .model statement allowing the user to specify an optional value of the transistor model. If no values are specified, as is the case in the example given, the program will use the default parameter values yielding typical transistor performance. For simple circuits simulation, it is sufficient to allow the transistor parameters to have their default values. There are situations, however, that would require accurate transistor models and in this case, the user must have a good understanding of the model and its parameters. For a detailed discussion of SPICE semiconductor models, see Antognetti and Massobrio [1988]. PSpice, however, offers the user a wide choice of commercially available transistor models. These models are in the bipolar.lib library and are written in plain text that can be easily viewed or modified. To use a library transistor model, first, the transistor must be described using the description statement discussed earlier, with the <model name> parameter, which must now be the name of the transistor to be selected from the library. Second, the transistor model library must be specified in the circuit input file using the PSpice. LIB statement so that the specified model is referenced. Note that the .model statement is not required when a library transistor model is used. To illustrate how a library transistor model is used, assume the transistor Q1 of the circuit in Fig. 19.2 is the commercially available 2N2222A. Using the PSpice evaluation copy, the transistor description is

Q1 2 1 3 Q2N2222A
.LIB C:\miseval61\eval.lib

The PSpice evaluation copy has a limited library of component models including diodes, transistors, and op-amps. The library is called eval.lib, which must be referenced in the netlist using the .LIB statement. The .LIB statement directs the program to where the model of the transistor 2N2222A is, and in this case it is C:\miseval6l\eval.lib. Note that in the transistor description statement, the model name is Q2N2222A rather than 2N2222A. This is because of a PSpice requirement that the model name must begin with a letter. As a result all library transistor models are prefixed with the letter Q.

Independent Sources

Thus far, the descriptions of some simple passive and active circuits have been discussed. Test signals and power supplies are specified in PSpice using independent sources. There are three types of independent sources available: DC, AC, and transient sources.

DC Independent Source

This source is used to specify a fixed DC value of voltage or current. The description statement of a DC source is:

$$<source\ name> <+node> <-node>\ DC\ <magnitude>$$

where <source name> is the independent voltage or current source. These sources are identified as components starting with the letters V and I, respectively. The parameters <+node> and <−node> are the positive and negative nodes of the independent source. The parameter, DC, specifies a DC type source, and <magnitude> represents a fixed voltage or current value, which can be positive or negative. DC independent sources are often used to specify power supplies, for example,

$$Vcc\ 10\ 0\ DC\ +\ 12\ V$$

$$Vee\ 20\ 0\ DC\ -\ 12\ V$$

The first statement specifies a positive power supply named Vcc connected between nodes 10 and 0 (ground) with a value of +12 V. The second statement describes a negative power supply connected between nodes 20 and 0 with a value of −12 V. Incidentally, it is just as valid to define the negative power supply using the following statement:

$$\text{Vee 0 20 DC} + 12 \text{ V}$$

AC Independent Source

This source is used to define an AC signal of specific amplitude and phase. The description statement of an AC source is

$$<source\ name> <+node> <-node>\ AC\ <magnitude> <phase>$$

where $<source\ name>$, $<+node>$, and $<-node>$ are as defined in the case of the DC independent source. The parameter, *AC*, specifies an AC type source, $<magnitude>$ describes the amplitude of the signal in volt or ampere, and $<phase>$ represents the phase angle of the AC source in degrees. As an example, the statement

$$\text{Vin 10 AC 1V 0}$$

describes an AC independent voltage source called Vin, connected between nodes 1 and 0 and has a magnitude of 1 V and phase angle of 0°. It is just as valid to rewrite the preceding statement as

$$\text{Vin 1 0 AC}$$

This is because PSpice assumes default values of 1 V (magnitude) and 0° (phase) if no values are specified by the user.

AC independent sources are used when the response of a circuit over a range of frequencies is required. This type of **analysis** is called a **frequency response simulation** and the results of such simulation are graphs of amplitude and/or phase against frequency. These graphs are commonly known as Bode plots [Bogart 1990].

The frequency range of an AC independent source is specified using the following statement:

$$.AC\ <sweep\ type> <N> <F\ start> <F\ stop>$$

where the parameter $<sweep\ type>$ defines how the frequency points are determined in the range of $<F\ start>$ and $<F\ stop>$. The program has two types of frequency sweeps, linear and logarithmic. To specify a linear frequency scale, the .AC statement becomes

$$.AC\ LIN\ <N> <F\ start> <F\ stop>$$

where $<LIN>$ specifies a linear frequency scale. The parameter $<N>$ defines the number of frequency points in the linear sweep starting at $<F\ start>$ and finishing at $<F\ stop>$. As an example, the statement

$$.AC\ LIN\ 500\ 1\ Hz\ 1\ kHz$$

instructs PSpice (together with an AC independent source description statement) to perform an AC analysis over the frequency range from 1 Hz to 1 kHz with 500 frequency points in the range. Note that the program does not allow the AC analysis to start at DC (i.e., 0 Hz), therefore, $<F\ start>$ must be a some number >0.

Computer-Based Circuit Simulation

There are two options available in which a logarithmic frequency scale is specified. The first is a decade scale (a tenfold increase in frequency), and the second is an octave scale (a twofold increase in frequency). To specify a decade frequency scale, the .AC statement becomes

$$.AC\ DEC\ <N>\ <F\ start>\ <F\ stop>$$

where $<DEC>$ specifies a decade logarithmic frequency scale and $<N>$ is the number of frequency points per decade. As an example, the statement

$$.AC\ DEC\ 10\ 10\ kHz\ 10\ MHz$$

causes PSpice to divide the frequency range (10 kHz–10 MHz) into decades, and in this case there are 3 decades, each of which has 10 frequency points.

To specify an octave logarithmic frequency sweep, the following statement is required:

$$.AC\ OCT\ <N>\ <F\ start>\ <F\ stop>$$

where $<OCT>$ specifies an octave logarithmic frequency scale, and $<N>$ is the number of points per octave. As an example, the statement:

$$.AC\ OCT\ 5\ 1\ kHz\ 16\ kHz$$

instructs the program to divide the frequency range (1–16 kHz) into octaves, and in this case there are 4 octaves, each of which has 5 frequency points.

Transient Independent Source

This source is used to describe time-dependent signals such as sine and square waveforms. Transient sources are discussed later in the section when dealing with **time-domain analysis**.

Graphics Outputs of Simulation Results: Probe

Once a circuit is entered and a simulation is run, PSpice allows the user to view the simulation results graphically using the program Probe. This program is part of the PSpice package. To plot the simulation results, the user must specify the statement

$$.PROBE\ [output\ variables]$$

in the circuit input file. The optional parameter *[output variables]* can be node voltages and/or component currents. For example, the statement

$$.PROBE\ V(3,0)\ V(R1)\ I(C1)$$

will allow the user to plot the voltage between nodes 3 and 0, the voltage across Rl, and the current through Cl. Note that if no *[output variables]* are specified in the .PROBE statement, the user will be able to plot all of the node voltages and component currents of the circuit at the expense of higher simulation time.

Running PSpice: Control Shell

The circuit description in terms of components, input signal(s), analysis type(s), and .PROBE statement is commonly referred to as the circuit input file or the circuit netlist. A circuit input file is usually created using a text editor. This represents the first step in the simulation process. Having created the input file, the next step is to run PSpice and obtain the simulation results. The program automates the simulation

FIGURE 19.3 Control shell start-up screen.

steps using the control shell software [Microsim 1994]. The control shell consists of an editor used for creating input files, and shell program from which PSpice and Probe can be run. Figure 19.3 shows the start-up screen for the control shell. For example, File allows a circuit input file to be created, and Run is used to run PSpice and Probe.

Simulation Example 1

Consider the common-emitter transistor amplifier [Bogart 1990] shown in Fig. 19.4. Use PSpice to obtain the frequency response (gain and phase) of the amplifier over the frequency range from 1 Hz to 10 MHz.

To create the circuit input file, first load control shell and Fig. 19.3 will appear. The circuit is ready for entry using the various description statements, as shown in Fig. 19.5. The first line of the circuit input file must be a title and the last line must be an .END statement. These statements mark the beginning and the finish of the netlist, and in between the circuit can be described. For example, the AC independent source description statement (i.e., the Vin statement), together with the .AC statement, instructs PSpice to perform a frequency response simulation. The simulation is to be carried over the specified frequency range from 1 Hz to 10 meg using a logarithmic decade scale. The transistor has been described using two statements. The first is the Q1 description statement, which defines the nodes and the model name of the transistor, and the second is the .LIB statement, which directs the program to where the model of the transistor 2N2222A is located. In this case, it is C:\miseval6l\eval.lib. The .PROBE statement allows the amplifier output voltage, V(5), to be plotted.

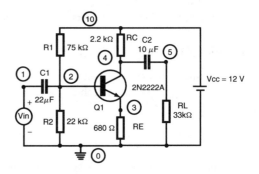

FIGURE 19.4 Circuit of simulation example 1.

Once the circuit description is finished, a name must be given to the circuit input file. Let us assume, in this case, the input file is called example1. To run PSpice and Probe, select Run from the Tool Bar, and then the command Simulator from the menu. If errors are encountered in the input file, PSpice will terminate and a list of errors will be produced in the PSpice circuit output file. If there are no errors, PSpice will then perform the simulation and move on to Probe for plotting the simulation results. Probe

Computer-Based Circuit Simulation

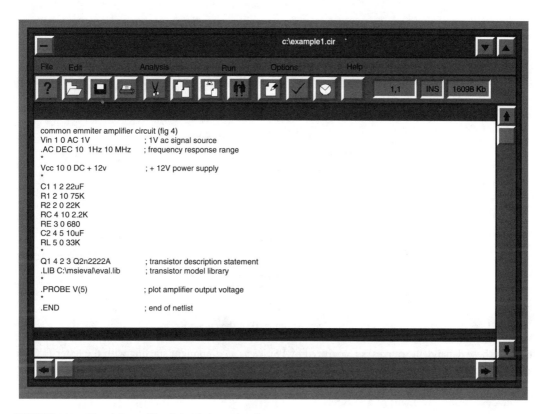

FIGURE 19.5 Circuit input file of simulation example 1.

usually plots the sweep variable of the simulation as the x axis (frequency for AC analysis and time for transient analysis), and in this example, the x axis is frequency. The *y* axis can be any node voltages and current of the circuit, which depends on the output variables specified in the .PROBE statement. In this simulation example, the amplifier output voltage, V(5), is specified. To obtain a graph of the amplifier output voltage against frequency, first select Add_trace from the Probe menu, then type VdB(5). The variable VdB(5) means the voltage at node 5 in decibel. Figure 19.6 shows the simulated frequency response (gain and phase) of the amplifier. The phase response of the amplifier is obtained by typing VP(5). To divide the screen for the various plots, as in Fig. 19.6, select the option Plot_control from the Probe main menu, and then Add_plot from the submenu, followed by Exit to go back to the main menu to start plotting.

PSpice calculates the amplifier DC bias node voltages by default, and they are included in the circuit output file.

Op-Amp Description

PSpice is capable of simulating amplifiers using dependent sources. Unlike the independent sources discussed previously, the output of a dependent source is controlled by a number of inputs. The program has four dependent or controlled sources:

- Voltage controlled voltage source (VCVS)
- Voltage controlled current source (VCCS)
- Current controlled current source (CCCS)
- Current controlled voltage source (CCVS)

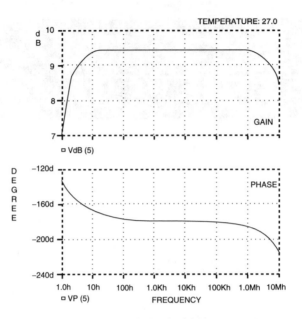

FIGURE 19.6 Simulated frequency response (gain and phase) of the circuit in Fig. 19.4.

Each of these dependent sources is recognized by PSpice as a component starting with the letter E, G, F, and H, respectively. Only VCVS is discussed here; for information on the other sources, see Microsim [1994], Al-Hashimi [1995], Tuinenga [1992], and Rashid [1990].

The description statement of a VCVS is

$$E{<}name{>} \ {<}{+}node{>} \ {<}{-}node{>} \ {<}{+}controlling\ node{>} \ {<}{-}controlling\ node{>} \ {<}gain{>}$$

where E is the symbol for a VCVS and $<name>$ is the VCVS name chosen arbitrarily by the user. The parameters $<+node>$ and $<-node>$ are the output nodes of the VCVS, whereas $<+controlling\ node>$ and $<-controlling\ node>$ are the input nodes of the VCVS. VCVS are usually used to describe the operation of ideal op-amps. To demonstrate this, consider the amplifier circuit shown in Fig. 19.7. The PSpice description of the op-amp is

$$E\ 3\ 0\ 0\ 2\ 200\ k$$

The op-amp has been called E with the inverting input at node 2 and the noninverting input at node 0. The VCVS statement requires positive and negative output nodes to be specified. Most op-amps are single ended and in this example the positive output of the op-amp is at node 3 and the negative node is at node 0. It has been assumed that the op-amp has a gain of 200 k. Using simple equivalent circuit models, the E component describes amplifiers with ideal characteristics (infinite input impedance, zero output impedance, infinite gain vs frequency, etc.). To define amplifiers with nonideal characteristics, a more complex op-amp model is needed. Such a model can be developed, for example, using the internal transistor circuitry of the op-amp. The resulting model, however, would require a large amount of simulation time and memory space (typical op-amp consists of approximately 30 transistors). To reduce simulation time and memory usage, a compressed version of the transistor op-amp has been developed. The compressed op-amp model is commonly known as

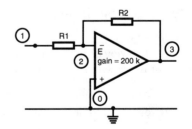

FIGURE 19.7 Amplifier circuit.

Computer-Based Circuit Simulation

a macromodel and was first described by Boyle et al. [1974]. Figure 19.8 represents the Boyle macromodel, which shows that the operation of the op-amp is now defined using only two transistors and various independent controlled sources. PSpice has a wide choice of commercially available **op-amp macromodels**. These models are part of the full package and are classified as individual libraries according to their semiconductor manufacturers; examples are Analogue Devices, Burr–Brown, National Semiconductor and others.

To use an op-amp macromodel in a simulation, an X call statement is required. This statement has the following format:

$$X<name> \ <+node> \ <-node> \ <Vcc> \ <Vee> \ <output> \ <model\ name>$$

where $<+node>$ and $<-node>$ are the noninverting and inverting inputs of the op-amp, respectively. The parameters $<Vcc>$ and $<Vee>$ define the positive and negative power supplies of the op-amp. The parameter $<output>$ is the op-amp output node, and $<model\ name>$ is the library model name of the op-amp required. The appropriate op-amp library must therefore be specified in the circuit input file using the .LIB statement.

Simulation Example 2

This example shows how op-amp-based circuits are simulated; consider the circuit shown in Fig. 19.9. It is a second-order low-pass active filter [Van Valkenburg 1982] with -3-dB frequency point at 1 kHz and DC gain of 1. Use PSpice to simulate the frequency response of the filter using:

1. Ideal op-amps.
2. Macromodel op-amps, assume the filter will be realized using 741 op-amps. PSpice allows the user to combine a number of circuit input files so that multiple analyses of the same circuit can be performed. Listing A shows the input files of the filter based on an ideal op-amp and on a macromodel op-amp. Note that each circuit netlist must start with a comment line and finishes with an .END statement. The first netlist, which describes the filter based on ideal op-amp (Fig. 19.9), shows that the op-amp has been described using a VCVS (E component). The op-amp output is at node 4 with respect to ground (node 0), whereas the noninverting and inverting inputs are at nodes 3 and 4, respectively. It has been assumed that the op-amp has a gain of 200,000.

The second netlist, which describes the filter based on macromodel op-amp (Fig. 19.10), shows that the 741 op-amp has been described using an X statement with the noninverting input at node 3,

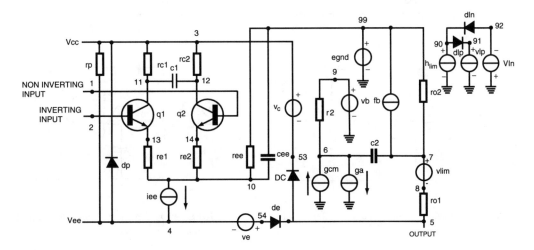

FIGURE 19.8 Boyle op-amp macromodel.

Listing A Input Circuit Files of Simulation Example 2

1st Netlist, Low-pass Filter	
+ (Fig. 19.9) Simulated	
+ Using Ideal op-amps	Comment Line
Vin 1 0 AC 1	;1 V AC signal
.AC LIN 500 1 50 kHz	Linear frequency range
+	with 500 points
*	
R1 1 2 10 k	
R2 2 3 10 k	
C1 2 4 22.5 nF	
C2 3 0 11.25 nF	
*	
E 4 0 3 4 200 k	; Ideal op-amp description
*	
.PROBE V(4)	; Plot filter output, V(4)
.END	; End of first netlist
2nd netlist, lowpass filter	
+ (Fig.19.10) simulated	
+ Using macromodel op-amp	; Comment line
*	
Vin 1 0 AC 1	;1V AC signal
.AC LIN 500 1 50 kHz	;Linear frequency range
+	with 500 points
*	
Vcc 10 0 DC + 12 V	; +12-V power supply
Vee 20 0 DC − 12 V	; −12-V negative power
+	supply
*	
R1 1 2 10 k	
R2 2 3 10 k	
C1 2 4 22.5 nF	
C2 3 0 11.25 nF	
*	
X 3 4 10 20 4 uA741	; Macromodel op-amp
+	(741) description
.LIB C:\miseval61\eval.lib	; Library of op-amp model
*	
.PROBE V(4)	; Plot filter output, V(4)
.END	; End of second netlist

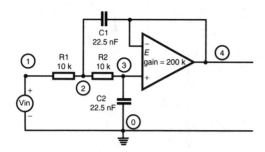

FIGURE 19.9 Circuit of simulation example 2 based on ideal op-amp.

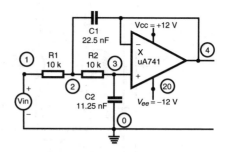

FIGURE 19.10 Circuit of simulation example 2 based on op-amp macromodel.

Computer-Based Circuit Simulation

whereas the inverting input is at node 4. The positive and negative power supply are at nodes 10 and 20, respectively. Both of these power supplies have been described using DC independent sources as shown in the listing. The op-amp output is at node 4. The model name of the 741 op-amp in the "eval.lib" is uA741, and the library is referred to using the .LIB statement. Figure 19.11 shows the simulated magnitude frequency response of the filter using ideal and macromodel op-amps. As can be seen, there is a very small difference between the two simulations, both predict a filter with −3-dB frequency point at 1 kHz, as designed. It should be noted, however, that the simulation time of the ideal filter was 7 s, whereas the macromodel filter was 13.9 s. Both simulations were performed on a 486DX, 33 MHz IBM PC. Clearly, the difference in simulation time will be much greater for more complex circuits. This example shows the importance of not overspecifying the op-amp models, and the user must decide when to use ideal or macromodel depending on the application. This point will be considered further later in the section when dealing with simulation vs practical.

Op-Amp Macromodels Limitations and Capabilities

A Boyle op-amp macromodel is capable of simulating the following parameters:

- Open loop gain-frequency characteristics.
- Slew rate.
- Maximum output voltage swing.
- Input and output impedance.
- Bias current and common-mode gain.
- DC quiescent power.

It does not simulate the following parameters:

- Input offset voltage and offset bias current.
- Noise and distortion.
- Variations in performance with temperature.

Some semiconductor manufacturers have improved the Boyle macromodel to the extent that they are capable of modeling the second list of parameters and other characteristics, see Biagi and Stiff [1990]; Jung [1990]; Buxton [1992]; Peic [1991]; and Robinson, Sanchez-Sinencio, and Kao [1993].

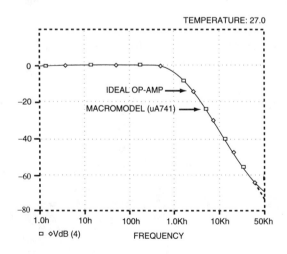

FIGURE 19.11 Simulated magnitude frequency response of the circuits in Figs. 19.9 and 19.10.

Time-Domain Analysis

Thus far we have seen how PSpice is used to perform frequency response simulation. The program is also capable of performing transient analysis or time-domain simulation. This type of simulation allows the user to predict the response of a circuit over a specified time range. The result is a graph of amplitude against time; therefore, this simulation is the same as using an oscilloscope in practice. Two statements are required to perform transient analysis; the first is a transient independent source description statement and the second is a .TRAN statement. The basic form of a transient independent voltage source description statement is

$$<V> <+node> <-node> <type> <value>$$

where $<V>$ is the independent voltage source and $<+node>$ and $<-node>$ show how the source is connected within the circuit. There are five types of time dependent sources available:

- Exponential (EXP)
- Pulse (PULSE)
- Sinusoidal (SIN)
- Single frequency frequency-modulated (SFFM)
- Piecewise linear (PWL)

Only pulse and sinusoidal transient sources are discussed here; for information on the other sources, see Microsim [1994], Al-Hashimi [1995], Tuinenga [1992], and Rashid [1990].

Pulse Independent Source (PULSE)

This source, as the name indicates, defines pulse waveforms. The description statement of a pulse independent voltage source is

$$<V> <+node> <-node> \text{ PULSE } (v1\ v2\ td\ tr\ tf\ pw\ per)$$

where PULSE specifies a pulse independent source. The pulse signal parameters ($v1\ v2\ \ldots$) are explained in Table 19.4. As an example, the following statement:

$$\text{Vin 1 0 PULSE } (-1\text{ V } 1\text{ V } 0\text{ s } 10\text{ ns } 10\text{ ns } 0.5\text{ ms } 1\text{ ms})$$

describes a voltage independent source called Vin connected between nodes 1 and 0. This source produces a square waveform of −1-V to +1-V amplitude and 1-kHz frequency. It has assumed that the waveform has 0-s delay and rise and fall time of 10 ns.

TABLE 19.4 Pulse Signal Parameters

Parameter	Meaning	Default Value	Units
v1	Initial value	None	Volt
v2	Final value	None	Volt
td	Time delay	0.0	Second
tr	Rise time	\<print interval\> (see text)	Second
tf	Fall time	\<print interval\> (see text)	Second
pw	Pulse width	\<final time\> (see text)	Second
per	Period	\<final time\> (see text)	Second

To generate a 1-MHz triangular waveform with amplitude of −2 V to +2 V, and rise time and fall time of 10 ns, the following statement is required:

Vin 1 0 PULSE (−2 V 2 V 0 s 10 ns 10 ns 1 ps 1 us)

The pulse width (pw) has been set to a very small value, 1 ps, and the waveform has been assumed to have 0 s delay.

Sinusoidal Independent Source (SIN)

This source defines sine wavefoms; the description statement of a sinusoidal independent voltage source is

<V> <+node> <−node> SIN (vo va freq td df phase)

where SIN specifies a sinusoidal independent source. The signal parameters are explained in Table 19.5. As an example the statement

Vin 1 0 SIN (0 V 2 V 10 kHz)

produces a sine waveform of 2-V peak amplitude and 10-kHz frequency. It has been assumed that the waveform has 0-V offset voltage.

It should be noted that the SIN waveform is for transient analysis only and has no effect during AC analysis. To perfom AC analysis (i.e., frequency response simulation), see simulation examples 1 and 2.

The .TRAN Statement

This statement is used to specify the period of time over which the circuit transient simulation is to be performed. The basic form of the .TRAN statement is

.TRAN <print interval> <final time> [no-print] [time step]

As mentioned earlier, the symbol [ ] in a PSpice statement indicates an optional provision of parameters by the user. If no parameters are supplied, the program will use the default values. The .TRAN statement is one of the few statements that is not immediately clear in its application and is best illustrated by the use of an example. Consider the waveform shown in the top of Fig. 19.12; this waveform has been generated by the statements

Vin 1 0 SIN (0 V 1 V 2 kHz)

.TRAN 5 us 1 m 0 ms 5 us

The first statement describes a sine wave of 1-V peak amplitude and 2-kHz frequency. The .TRAN statement is explained as follows: <print interval> determines the number of print points across the span. In this example, 5 us means 200 points across the printout of 1 ms. Note, if the print interval is chosen to be too coarse, erroneous curves may result. Figure 19.12 (middle) shows the same waveform with a <print interval> and < time step> of 50 us.

TABLE 19.5 Sinusoidal Signal Parameters

Parameter	Meaning	Default Value	Units
vo	Offset voltage	None	Volt
va	Peak amplitude	None	Volt
freq	Frequency	1/<final time>	Hertz
td	Delay time	0.0	Second
df	Damping factor	0.0	1/second
phase	Phase	0.0	Degree

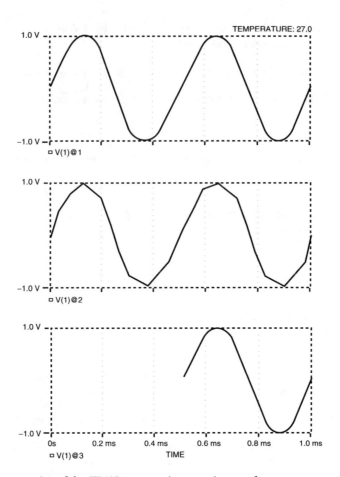

FIGURE 19.12 Demonstration of the .TRAN statement in generating waveforms.

Here, <*final time*> is the span of the analysis and in this example is 1 ms. In addition to the compulsory parameters of <*print interval*> and <*final time*>, the following optional parameters can be specified:

First, [*No print interval*], the program always starts the transient analysis at time = 0 s; it is possible, however, to suppress an early part of the simulation when printed or plotted. In this example, the [*no print interval*] has been set to 0 ms, the default value. Figure 19.12 (bottom) shows the original waveform with [*no print interval*] set to 0.5 ms.

Next, [*time step*] is the maximum period between individual calculations the program uses when performing transient simulation. If not specified, PSpice uses the default value of <*final time/50*>, which would have been 20 us for the example given if it had not been specified as 5 us. There is a tradeoff when deciding on the value of the [*time step*] parameter for a simulation; a small step time will ensure accuracy at the expense of longer simulation time, whereas a large step time has the opposite effects.

Simulation Example 3

To illustrate the use of transient analysis, consider the circuit of the low-pass active filter shown in Fig. 19.9. Obtain the filter transient response when a square waveform of 2-V (peak to peak) amplitude and 200-Hz frequency is applied to the input. Assume the wavefom has a 10-ns rise and fall time, and 0-s delay. The circuit input file is given in Listing B. The PULSE and .TRAN statements instructs PSpice to perform time-domain simulation. The simulation is carried out over 10 ms (2 cycles of the input signal) with a time step of 0.01 ms. Figure 19.13 gives the simulated input and output waveforms of the filter. As expected, the output waveform has some overshoot [Van Valkenburg 1982].

Listing B Input File of Example 3

Low-pass filter (Fig. 19.9) Simulated Using +Ideal op-amps	; Comment Line
*	
Vin 1 0 PULSE (–1 V 1 V 0 s 10 ns 10 ns +2.5 ms 5 ms)	; 2 Vpk-pk, 200-Hz input square waveform
.TRAN 0.01 ms 10 ms 0 s 0.01 ms	; Transient analysis range
*	
R1 1 2 10 k	
R2 2 3 10 k	
C1 2 4 22.5 nF	
C2 3 0 11.25 nF	
*	
E 4 0 3 4 200 k	; Ideal op-amp description
*	
.PROBE V(1) V(4)	;Plot filter input and
+	output waveforms
*	
.END	; End of netlist

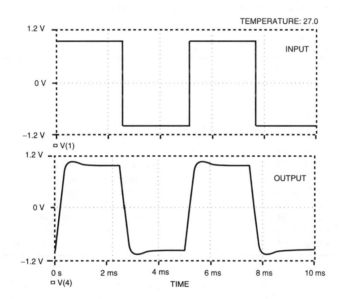

FIGURE 19.13 Simulated input and output waveforms of simulation example 3.

Fourier Analysis

Fourier analysis is a useful tool for examining the frequency spectrum of signals, and PSpice is capable of performing such an analysis using the .FOUR statement. The basic form of this statement is

.FOUR <fundamental frequency> <output variables>

where <fundamental frequency> is the frequency of the signal on which Fourier decomposition is based and <output variable> is any node voltages and/or component current. The .FOUR statement calculates the following Fourier components: DC, fundamental, and second to the ninth harmonics. The amplitude and phase of these components are tabulated in the circuit output file (.out). The program performs Fourier analysis only during a transient simulation, which means that both the .TRAN and .FOUR statements must be included in the circuit input file.

Simulation Example 4

To demonstrate Fourier analysis capability, consider the filter circuit shown in Fig. 19.9. Use PSpice to calculate the Fourier components of the input and output signals of the filter. The transient analysis of this circuit was already considered in Listing B. To perform the Fourier analysis, the following statement:

$$.\text{FOUR } 200 \text{ Hz } V(1), V(4)$$

must be added to the Listing. This statement instructs the program to calculate the Fourier component of the filter input signal, V(1), and the output signal, V(4). The decomposition of the analysis is based on a 200-Hz frequency, which is the input signal frequency. The .FOUR statement can be placed anywhere within the listing, but for clarity it should be after the .TRAN statement. Running PSpice produces the Fourier components shown in Tables 19.6(a) and 19.6(b). Table 19.6(a) provides the amplitude and phase of the DC, first, and second to the ninth Fourier components of the input signal, V(1). As expected from the Fourier series of a square waveform only the odd harmonics are present. The program outputs the harmonic values relative to the fundamental component, as well as the absolute values as shown in the table. The Fourier components of the filter output signal, V(4), are shown in Table 19.6(b). As can be seen, the fifth and higher harmonics of the input signal have been attenuated according to the filter characteristics. The -3 dB frequency point of the filter is 1 kHz.

TABLE 19.6a Fourier Components of the Input Signal

Fourier Components of Transient Response V(1)
DC Component = $-1,603199E - 03$

Harmonic No.	Frequency Hz	Fourier Component	Normalized Component	Phase	Normalized phase
1	2.000E + 02	1.273E - 00	1.000E + 00	-1.443E - 01	0.000E + 00
2	4.000E + 02	3.207E - 03	2.518E - 03	-9.045E + 01	-9.031E + 01
3	6.000E + 02	4.244E - 01	3.333E - 01	-4.329E - 01	-2.886E - 01
4	8.000E + 02	3.207E - 03	2.519E - 03	-9.090E + 01	-9.076E + 01
5	1.000E + 03	2.546E - 01	2.000E - 01	-7.215E - 01	-5.772E - 01
6	1.200E + 03	3.207E - 03	2.519E - 03	-9.135E + 01	-9.121E + 01
7	1.400E + 03	1.819E - 01	1.429E - 01	-1.010E + 00	-8.658E - 01
8	1.600E + 03	3.208E - 03	2.520E - 03	-9.180E + 01	-9.166E + 01
9	1.800E + 03	1.415E - 01	1.111E - 01	-1.299E + 00	-1.154E + 00

TABLE 19.6b Fourier Components of the Output Signal

Fourier Components of Transient Responge V(4)
DC Component = $1.231433E - 06$

Harmonic No.	Frequency Hz	Fourier Component	Normalized Component	Phase	Normalized phase
1	2.000E + 02	1.272E + 00	1.000E + 00	-1.641E + 01	0.000E + 00
2	4.000E + 02	1.130B - 05	8.881E - 06	-2.185E + 01	-5.434E + 00
3	6.000E + 02	3.994E - 01	3.139E - 01	-5.296E + 01	-3.655E + 01
4	8.000E + 02	1.905E - 05	1.497E - 05	-5.908E + 01	-4.267E + 01
5	1.000E + 03	1.801E - 01	1.416E - 01	-9.000E + 01	-7.359E + 01
6	1.200E + 03	2.211E - 05	1.738E - 05	-9.115E + 01	-7.474E + 01
7	1.400E + 03	8.268E - 02	6.499E - 02	-1.159E + 02	-9.948E + 01
8	1.600E + 03	2.052E - 05	1.613E - 05	-1.154E + 02	-9.900E + 01
9	1.800E + 03	4.171E - 02	3.279E - 02	-1.314E + 02	-1.150E + 02

Computer-Based Circuit Simulation

This example has shown capability in performing Fourier analysis using the .FOUR statement. The result of the analysis is included in the PSpice circuit output file in a tabulated format. It is also possible to obtain the Fourier components of a signal in graphic format, using the program Probe. This is achieved by first carrying out a transient analysis and plotting the required waveform(s) using Probe. Second, select X-axis from the Probe menu and then Fourier from the submenu. To illustrate this, consider the circuit shown in Fig. 19.9 and Listing B. Having run Listing B, the input and output signals, V(l) and V(4), can now be plotted (Fig. 19.13). Selecting X-axis from the main Probe menu, and then Fourier from the submenu, allows the Fourier components of both signals to be displayed as shown in Fig. 19.14. As expected they are in a good agreement with the tabulated results given in Tables 19.6.

Subcircuits

Electronic circuit designs often involve the use of similar or even identical blocks of circuitry. To allow repetitive circuitry to be described efficiently, PSpice provides the user with the **subcircuit** approach. In this approach, a repetitive or an often used circuit block can be defined once and then used each time the circuit block is required by writing a single statement.

The concept of subcircuits is similar to that of a subroutine in conventional computer programming. A typical subcircuit is made up of the following:

- .SUBCKT statement
- Circuit description statements
- .ENDS statement

The SUBCKT statement defines the subcircuit in terms of a name, nodes, and a number of optional parameters using the following format:

$$.SUBCKT\ <name>\ <nodes>\ [PARAMS:\ <<name>=<value>>]$$

where <name> is any name chosen by the user to identify the subcircuit and <nodes> are the external nodes of the subcircuit. To increase the flexibility of the subcircuit, the user is allowed to pass a number

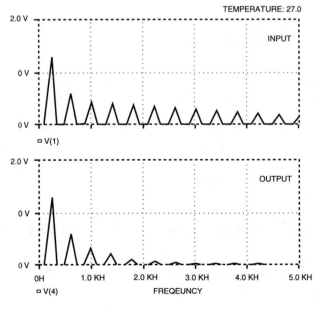

FIGURE 19.14 Fourier analysis of the signals shown in Fig.19.13.

of optional parameters into the subcircuit. The names and values of these parameters are set by <name> and <value>. Having defined the subcircuit using the .SUBCKT statement, the subcircuit components can now be described using various description statements. To mark the end of the subcircuit description, an .ENDS statement is used, which has the following format:

.ENDS [subcircuit name]

where the optional parameter [subcircuit name] is the same <name> given in the .SUBCKT statement. To call an already defined subcircuit, an X call statement is required in the main circuit input file. The subcircuit call statement has the following format:

X<name> <nodes> <subcircuit name> [PARAMS: <<name>= <value>>]

where <name> is a name chosen arbitrarily by the user to identify the subcircuit. The X statement causes the referenced subcircuit <subcircuit name> to be inserted into the main circuit with the X statement nodes matching the nodes used in the .SUBCKT statement. Note that there must be the same number of nodes and parameters in the X statement as in the .SUBCKT statement. Also it is essential that they are in the same order.

Simulation Example 5

To illustrate the development of subcircuits and their use, consider the circuit shown in Fig. 19.15. This circuit is a sixth-order low-pass active based on the cascade approach [Van Valkenburg 1982], where three identical second-order sections are connected in series. To avoid entering the same component statements describing each section, it is more efficient to define one second-order section as a subcircuit and then call it as required.

Using a text editor, the subcircuit description of the second-order low-pass section [Fig. 19.16(a)] is

.SUBCKT second_lp 100 103 PARAMS: R = 1, C1 = 1, C2 = 1
R1 100 101 {R}
R2 101 102 {R}
C1 101 103 {C1}
C2 102 0 {C2}
E 103 0 102 103 200k; ideal op-amp description
.ENDS second_lp

The subcircuit has been called second_lp with its external nodes at 100 and 103 [Fig. 19.16(b)]. The capacitor values of each second-order section of the filter are different [Van Valkenburg 1982], depending on the required filtering characteristics (Butterworth, Chebyshev, and Bessel). Therefore, we will assume that each section has equal resistors (R1=R2=R), and two different capacitors (C1 and C2). As a result, the subcircuit second_lp will have three parameters R, C1, and C2, as shown in the subcircuit description

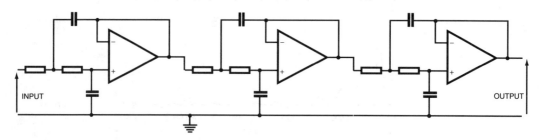

FIGURE 19.15 Sixth-order low-pass active filter.

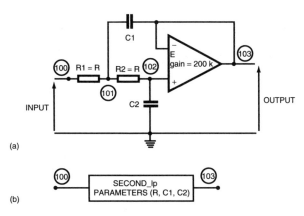

FIGURE 19.16 Second-order low-pass section in: (a) components form, (b) subcircuit form.

FIGURE 19.17 Subcircuit block-diagram representation of the circuit shown in Fig. 19.15.

and Fig. 19.16(b). These three parameters must be given initial values to satisfy the .SUBCKT statement requirement. In this case, these initial values are set arbitrarily to 1 since they will be changed to the required values when the subcircuit is called. It has been assumed that the op-amp model is ideal and described using the E component. Note that the subcircuit node numbers have been chosen such that they are different from the node numbers in the main circuit input file to improve the clarity of the subcircuit description. Now, we want to use this subcircuit to simulate a Butterworth filtering response based on the circuit in Fig. 19.15. This circuit is drawn in a subcircuit format as shown in Fig. 19.17, and its input file is given in Listing C.

The first call statement (X1) connects subcircuit second_lp nodes 100 and 103 to main circuit nodes 1 and 2. The parameters R, C1, and C2 have been given the values of 10 kΩ, 16.49 nF, and 15.36 nF, respectively (overriding the default values of 1). Similarly, X2 connects subcircuit nodes 100 and 103 to main circuit nodes 2 and 3, where R, C1, and C2 are equal to 10 kΩ, 22.5 nF, and 11.25 nF, respectively. The main circuit input file must have a reference to where the subcircuit second_lp is located, and it is good practice to collect subcircuits in a user-defined file, which is referred to using the PSpice command

.LIB <file name>

For example in Listing C, the second-order section subcircuit is included in a file called <filters>.

Simulation vs Practical

It has been shown in example 2 that there was a small difference between the simulated frequency response of a 1-kHz low-pass filter based on ideal op-amps (Fig. 19.9) and macromodel op-amps (Fig. 19.10). The op-amp macro-model filter simulation time, however, was nearly a factor of 2 higher than that of the ideal. Now, we want to build the 1-kHz filter and compare its frequency response with that obtained from the simulation to show how good or bad the simulation is. Low-pass filters are often characterized by various performance parameters such as passband ripple, -3-dB frequency point, and stopband attenuation. All of these parameters are therefore important depending on the application. To simplify the comparison, however, in this example, the -3-dB frequency point will be assumed to be the most important parameter. Table 19.7 shows the comparison, and as can be seen there is a good agreement between simulation and

Listing C Input File of Example 5

Low-pass filter (Fig.19.15) simulated	
+ using the subcircuit approach (Fig.19.17)	Title Line
*.	
.LIB A:\filters	; Second_lp subcircuit
+	location
Vin 1 0 AC 1	; AC input signal of 1 V
.AC LIN 500 1 10 k	; Frequency analysis
+	range
*	
X1 1 2 second_lp PARAMS: R = 10 k,	; Call second_lp
+C1 = 16.49 nF, C2 = 15.36 nF	subcircuit
X2 2 3 second_lp PARAMS: R = 10 k,	
+C1 = 22.5 nF, C2 = 11.25 nF	
X3 3 4 second_lp PARAMS: R = 10 k,	
+C1 = 61.49 nF, C2 = 4.12 nF	
*	
.PROBE V(4)	; Plot filter output, V(4)
*	
.END	; End of circuit netlist

measured results. The simulation (ideal and macromodel) predicted a filter with −3-dB frequency point at 1 kHz, as designed, whereas the practical filter has the −3-dB frequency point at 0.995 kHz. The op-amp macromodel used in the simulation was the uA741 selected from the evaluation copy library eval.lib. The practical filter was built using a 741 op-amp and selected and trimmed capacitors and resistors, so that the performance was mostly determined by the op-amp.

TABLE 19.7 Comparsion Between Simulation and Measured Results of a 1-kHz Low-Pass Filter

Parameter	Ideal op-amp	Macromodel op-amp	Measured
−3-dB frequency point, kHz	1	1	0.995
Simulation time, s	7	13.9	—

The Table 19.7 comparison has shown that both ideal and macromodel op-amps produce results very close to the measured in the case of a filter operating at low frequencies. Now we want to examine the frequency response of the same filter when it is operating at much higher frequency. Assume that the F-3 dB of the filter is 500 kHz. The capacitor values of the 1-kHz filter shown in Fig. 19.10 must be multiplied by the factor (1/500) so that the filter will have F-3 dB at 500 kHz, while leaving the resistor values unchanged. Table 19.8 provides the comparison that shows that there is a large discrepancy between the ideal simulation and the measured results. The ideal simulation predicted a −3-dB frequency point at 500 kHz, as designed, whereas the practical filter has the −3-dB frequency point at 370 kHz. There is a better correlation, however, between the macromodel simulation and the measured filter. The macromodel simulation predicted the −3-dB frequency point at 438 kHz.

TABLE 19.8 Comparsion Between Simulation and Measured Results of a 500 kHz Low-Pass Filter

Parameter	Ideal op-amp	Macromodel op-amp	Practical
−3-dB frequency point, kHz	500	438	370
Peaking	no peaking	0.9 dB at 239 kHz	3 dB at 220 kHz
Simulation time, s	7	13.9	—

Computer-Based Circuit Simulation

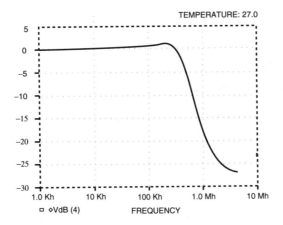

FIGURE 19.18 Frequency response simulation of a 500-kHz low-pass filter based on op-amp macromodel.

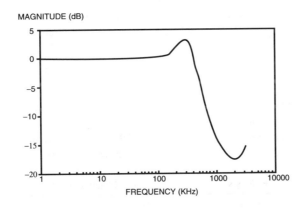

FIGURE 19.19 Measured frequency response of a 500-kHz low-pass filter.

It is known that active filters operating at high frequencies usually exhibit peaking in the passband [Van Valkenburg 1982] due to the finite bandwidth of the amplifiers. To check this effect using simulation, the macromodel simulation predicted a peaking of around 0.9 dB at 239 kHz (Fig. 19.18), whereas the measured filter has a peaking of around 3 dB at 220 kHz (Fig. 19.19). Note that the ideal op-amp filter simulation predicted no peaking at all, as shown in Fig. 19.20.

It should be mentioned that a better correlation between macromodel simulation and measured results would have been possible if more complex or full transistor op-amp models were developed and used in the simulation. This would have been achieved at the expense of higher simulation time and the possibility of simulator convergence problems. This type of simulation was not included in the comparison because the aim of the study was to compare simulation and measured results using op-amp macromodels readily available within the simulator. The conclusion of this simple example is that the user must decide on the complexity of the models used and the time needed (simulation and

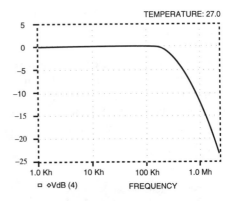

FIGURE 19.20 Frequency response simulation of a 500-kHz low-pass filter based on ideal op-amp.

development) to achieve the required simulation accuracy. For more information on the topic of simulation vs practical, the reader is referred to an excellent article by Swager [1992].

PSpice Error Messages

Two common error messages reported by PSpice when there is an error in the circuit input file are given here. The error messages usually appear in the circuit output file. The two errors are (1) *less than two connections at node* x and (2) *node* x *is floating*, where x is an arbitrarily chosen node. Both errors are rectified simply by connecting a large value resistor (such as lE9Ω) from node x to ground. The resistor value must be chosen so that it has no effect on the circuit operation.

19.3 Other Simulators

Up to this point, only analog circuit simulation has been discussed. There are other types of circuit simulators, such as digital, mixed analog–digital, and block diagram simulators. The aim of this section is to provide an introduction to these simulators and demonstrate their use through a number of simulation examples. Unlike PSpice, which is commonly used for analog simulation, there are no standard simulators for digital, mixed analog–digital, and block-diagram simulation. Because of this, the emphasis here will be placed on the principle of the simulation type, rather than discussing the syntax of a particular vendor simulator.

Digital Simulators

Digital simulators allow the user to check the function and perform a timing analysis of a digital system. The result of the simulation is a timing diagram or a truth-table; in practice, this is similar to using a logic analyzer. There are a number of commercially available digital simulators running on a variety of computers ranging from PCs to workstations. Table 19.9 shows some commonly used simulators.

TABLE 19.9 Some Commonly Used Digital Simulators

Vendor	Simulator Name
Microsim	PLogic
Mentor Graphics	Quicksim
OrCAD	OrCAD VST 386+
Viewlogic	Workview

The choice of the simulator usually depends on a number of factors including cost, hardware available, and user preference. The use of digital simulators is on the increase because they are now an integral part of the design and simulation tools for programmable logic devices (PLDs) and field programmable gate arrays (FPGAs). Most digital simulators work on similar principles and there are five basic steps involved in the simulation process. The steps are:

- Circuit entry
- Electrical check
- Netlist generation
- Test and output signals definition
- Run simulator

Circuit entry is carried out using a number of methods; examples are manual netlist entry, schematic capture, truth tables, and hardware description languages such as VHDL [Lipsett, Schaefer, and Ussery 1989]. It is often considered that the schematic capture method provides the most convenient and common approach for digital circuit entry. In this approach the circuit is drawn on the screen using a

Computer-Based Circuit Simulation

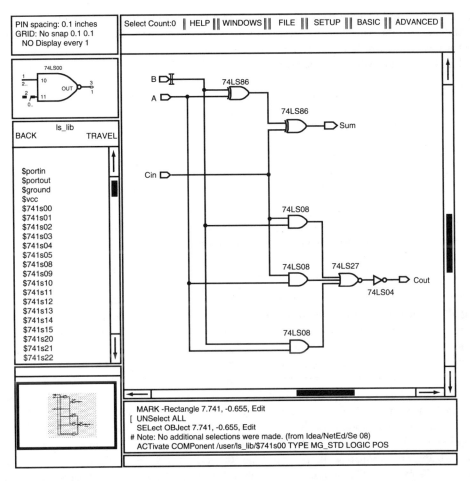

FIGURE 19.21 Schematic capture of the 1-b adder performed using the Neted package.

mouse and part symbols are picked from the simulator library. When the circuit schematic is completed, it is passed through the electrical rule check (ERC), which inspects the electrical connections on the schematic. For example, the ERC check ensures that the circuit output is not shorted or connected to the power supply. The third step in the simulation process involves generating the circuit netlist from the schematic, which is usually performed automatically by the simulation package. Once the circuit netlist is produced successfully, the next step is to define the test signals and format of the simulation results (graphical waveforms, truth tables). This is often achieved using simple description statements contained in a file commonly called *stimulus file*. Now the circuit is ready for simulation. To demonstrate how digital circuit simulation is performed, the Mentor Graphics software package will be considered.

Simulation Example 6

Consider the simulation of the simple 1-b adder circuit (Fig. 19.21), where A and B are the inputs, Cin and Cout are the carry input/output, and Sum is the adder output. The circuit was first drawn using the schematic capture program, *Neted*, which is part of the Mentor Graphics simulation package. The various symbols of the circuit digital parts were selected from the transistor-transistor-logic (TTL) library. Neted has libraries of other logic families [complementary metal oxide semiconductor (CMOS); emitter coupled logic (ECL)], memory, and microprocessor devices. To perform the simulation, the user must define on the schematic the circuit input(s) and output(s) using symbols from the library, and in the case of Mentor

Graphics, the input and output symbols are called portin and portout, respectively, as shown in Fig. 19.21. Having completed the schematic, the next two steps are performing the ERC check and producing the circuit netlist. These two tasks are easily carried out using simple commands. Now the adder circuit stimulus file needs to be defined as shown:

```
scale user time 1000                    # line 1
clock period 1                          # line 2
force B 0 0-rep                         # line 3
force B 1 0.5-rep                       # line 4
force A 0 0                             # line 5
force A 1 3                             # line 6
force A 0 5                             # line 7
force A 1 7                             # line 8
force Cin 0 0                           # line 9
force Cin 1 5                           # line 10
trace B A Cin Sum Cout                  # fine 11
list binary B A Cin Sum Cout-change     # line 12
scale trace 1                           # line 13
run 8                                   # line 14
```

The program performs timing analysis in terms of time unit, which has the default value of 1 ns. The user is allowed to change this value using the scale statement, as shown in line 1 of the stimulus file. In this case the time unit is set to 1000 ns or 1 μs. Note that the # symbol in the stimulus file indicates a comment. Lines 2–4 describe the B signal as a clock. Line 2 defines the clock period of 1 μs, line 3 sets the logic value of signal B to be 0 at 0 s, and line 4 sets the logic value to be 1 at 0.5 μs. The parameter -rep in lines 3 and 4 repeats the B signal at a clock period of 1 μs. Lines 5–8 describe the various logic and time values of signal A. The description of the carry input signal, Cin, is given in lines 9 and 10. Line 11 tells the simulator to display the input (B, A, and Cin) and the output (Sum and Cout) signals as graphic waveforms. If desired, the simulation results can be produced as a truth table; line 12 provides this option. Finally, line 13 displays the waveforms every 1 μs, and line 14 defines how long a period over which the simulation is to be perfomed—in this case it is 8 time unit or 8000 ns. To run the simulation, the program, *Quiksim*, must be invoked. Figure 19.22 shows the simulation results. As can be seen, the adder circuit functions as intended. Note that the digital devices selected from the Neted program library model the timing and delay characteristics of the devices, which means that timing problems such as glitches and races will be predicted by the simulator.

Assume that the 1-b adder circuit is to be used as part of a 4-b parallel ladder circuit; to avoid drawing repetitive designs, it is possible to define the 1-b adder circuit as a symbol and use it 4 times in this case. This type of simulation is called a *hierarchical simulation*, which is another capability of the Mentor Graphics package. This simple example has illustrated how a digital simulator can be used to predict the timing analysis of a 1-b adder circuit. The package is capable of simulating very complex digital systems; for example, the reader is referred to Hall [1992] for a microcomputer system simulation. Also, for a discussion on digital simulation using another simulator, see Al-Hashimi [1995], where the *PLogic* simulator from Microsim is used.

Mixed Analog–Digital Simulators

Thus far analog and digital simulations have been considered as two separate topics. As electronic systems increase in complexity, it is likely that the system will consist of analog and digital parts. For example, a typical signal processing system will have analog antialiasing and reconstruction filters at the front and back end of the system, whereas at the heart of the system lies a digital signal processor and memory devices. To simulate such a system, a mixed analog–digital simulator will be required. There are a number of mixed A/D simulators commercially available, as shown in Table 19.10. For a complete list of mixed A/D simulators, see Connor [1994].

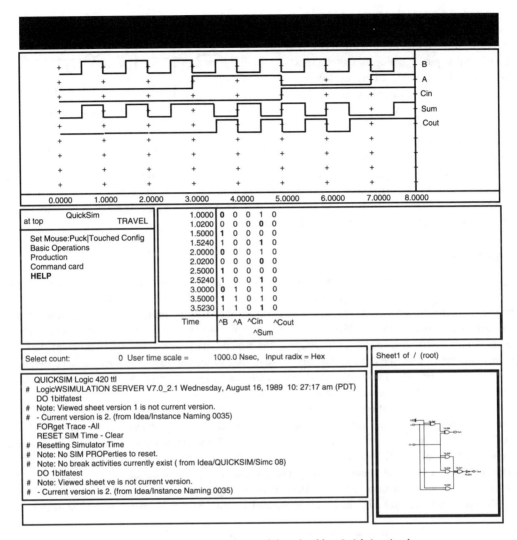

FIGURE 19.22 Simulated input and output waveforms of the 1-b adder, Quicksim simulator.

MixedA/D simulators are usually divided into two types, *native* and *glued*. A glued simulator consists of two separate simulators (analog and digital) combined through interface software, which controls data flow between the two simulators. An example of glued A/D simulator is the Endeavor from Mentor Graphics. This type of simulator is aimed primarily at systems with a large number of digital and analog parts. A native simulator on the other hand consists of one simulator that allows analog and digital simulation, examples of these simulators are the PSpice A/D from Microsim [1994] and the Saber from Analogy. When simulating systems that contain a large number of analog parts and a small number of digital devices, a native A/D simulator provides a better choice.

Analog simulators recognize only analog nodes where all components connected to a node are analog. Similarly, digital simulators deal only with digital nodes. A mixed A/D simulator recognizes three types of nodes: analog, digital, and interface. An interface node occurs when a combination of analog and digital components are connected to it, as shown in Fig. 19.23,

TABLE 19.10 Some Commerically Available Mixed Analog–Digital Simulators

Vendor	Name
Analogy	Saber
Intergraph	Apex
Intusoft	ICAP
Mentor Graphics	Endeavor
Microsim	PSpice A/D

where node 1 is an analog node and nodes 2 and 3 are interface nodes. The mixed A/D simulator must automatically translate interface nodes into purely analog or digital nodes. The translation is achieved using different techniques depending on the simulator. The Saber simulator, for example, uses special models called Hypermodels for the translation. The PSpice A/D simulator, on other hand, achieves the translation using analog/digital (A to D) and digital/analog (D to A) interface subcircuits [Microsim 1994, Al-Hashimi 1995]. The function of the Hypermodels or the A/D and D/A subcircuits is to change analog voltages and impedances to digital states (0,1, ...) and vice versa. To demonstrate how mixed analog–digital simulation is performed, the *PSpice A/D* simulator will be considered.

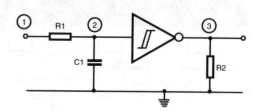

FIGURE 19.23 Circuit showing analog and interface nodes.

Simulation Example 7

Figure 19.24 shows a circuit that is often used to produce analog ramp waveforms using digital techniques [Patterson 1992]. The output of the 4-b counter determines the position of the switches. A logic 0 connects the switch to ground, whereas a 1 connects the switch to the inverting input of the amplifier, which acts as a current-to-voltage converter. Here, the simulator PSpice A/D (full production DOS Version 6.1) will be used to obtain the amplifier output voltage as the 4-b counter goes through its 16 different states. To perform the simulation, the circuit must first be described to the simulator. As mentioned earlier, there are usually two methods of achieving this task, a netlist and schematic capture. Here the netlist method will be used and Listing D gives the PSpice input file of the circuit.

First consider the description of the analog parts of the circuit. Listing D shows that the resistors have been described using passive component description statements, whereas the voltage reference (Vref) has been described using a DC independent source statement. These statements are discussed in Sec. 19.2. The op-amp has been described using the XI subcircuit statement. The op-amp is the TL082, whose model is obtained from the linear.lib library. The program is capable of simulating voltage and current controlled switches [Microsim 1994, A1-Hashimi 1995]. Figure 19.24 shows that the switches (S1, S2, ..., S8) are voltage controlled, which are recognized by the letter S. To describe such a switch, two

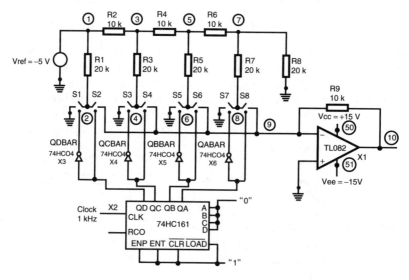

FIGURE 19.24 Circuit of simulation example 7.

statements are required. The first statement describes the input, output, control nodes, and model name of the switch. For example, the input and output of the switch S1 are connected to nodes 2 and 0, respectively, whereas the control is connected to nodes QDBAR and ground, as shown in Listing D. PSpice allows the user to express nodes as numbers or names; for example, the output node of the inverter X3 has been given the name QDBAR, which has been chosen arbitrarily. The second statement, which is a .MODEL statement, defines the optional parameters of the switch model. The model parameters are on and off resistance and the control voltage for on and off switch state. In this example, all switches have been modeled using the model name SW1, which has all its parameters set to their default values.

Now consider the description of the digital parts of the circuit. PSpice A/D has an extensive digital library containing models of commonly used digital devices, including the 74TTL and the 4000 CMOS series. These models are described in the form of subcircuits, which are readily called using X statements. For example, the counter (74HC161) has been defined using the X2 statement. To be compatible with actual digital devices, the program expresses the nodes of such devices using names; for example, the counter output nodes have been assigned the names QA, QB, QC, and QD, as shown in the X2 statement. The parameters $D_HI and $D_LO in the X2 statement represent digital 1 and 0, respectively. The inverters (74HC04) have been defined using the X3, X4, X5, and X6, statements, respectively. For example, the parameters QD and QDBAR in the X3 statement define the input and output nodes of the inverter. As mentioned earlier, the name QDBAR has been chosen arbitrarily. The digital devices models are in the library (74HC.lib), which must be included in the netlist using a .LIB statement. Clock signals are defined using the U statement [Microsim 1994, Al-Hashimi 1995]. In this example the counter clock is assumed to be 1 kHz and has been described using the U1 and loop statements as shown in Listing D. The parameters $D_DPWR and $D_DGND in the U1 statement define the digital power supply and ground nodes, respectively.

To simulate the output voltage of the amplifier as a function of time, a transient analysis or .TRAN statement is required; Listing D shows the analysis will be performed over a period of 30 ms with a print step of 0.02 ms. Transient analysis is discussed in detail in Sec.19.2. The .PROBE command generates a data file for viewing the simulation results graphically. Figure 19.25 shows the simulated digital and

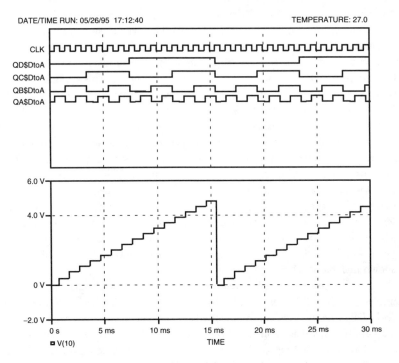

FIGURE 19.25 Simulated digital and analog waveforms of the circuit (Fig. 19.24).

analog signals of the circuit. The digital signals are the clock and the counter outputs, whereas the analog signal is amplifier output voltage. Note that the digital signals are expressed in terms of logic values (0 and 1) whereas the analog signal is described in terms of voltage levels.

There are a number of interface nodes in the circuit of Fig. 19.24; these nodes occur when the counter digital outputs (QD QC ...) and the inverter outputs (QDBAR QCBAR ...) drive the control inputs of the analog switches (S1 S2 ...). In this case the PSpice A/D simulator automatically breaks these nodes into purely analog nodes using digital/analog interface subcircuits, which are included in the library dig_io.lib (see Listing D, .LIB statement). Figure 19.26 shows the simulated waveforms of the counter outputs; note how the program has changed the outputs from digital states (0, 1) into analog voltages (0, 5 V). For more information on mixed analog–digital simulation using PSpice, see Al-Hashimi [1995].

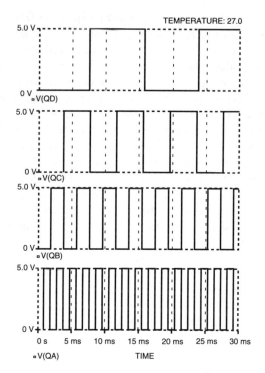

FIGURE 19.26 Counter output waveforms.

Block-Diagram Simulators

Simulators allow the prediction of system performance. To carry out the simulation, a detailed component design of the system must be available. The system is then described to the simulator using a basic set of components including resistors, transistors, and op-amps. This type of simulation is called the *primitive-level simulation*, and at the component level, clearly this type of simulation is required. When a designer is operating at the system level, however, producing component level simulation may be too detailed. At this level, a block-diagram simulation approach may be more appropriate. This approach allows the designer to simulate systems as a combination of block diagrams, each of which performs a specific function. Each function is described using a mathematical equation or a transfer function. For example, let us assume that part of the system under simulation is a low-pass filter. To optimize the system, the effect of changing the filter type (Butterworth, Chebyshev, etc.) on the system response needs to be examined. One way of evaluating the effect is to design a number of different filters and simulate them at the component level, which is a time consuming task. A more effective way of dealing with this problem is to simulate the filter as a block diagram, where the filter input/output relationship can be expressed in terms of a transfer function. This means that the filter component design is unnecessary at this stage, since the filter has been described using a mathematical expression. At a later stage the required filter can be replaced by actual circuitry. The advantage of block diagram simulation is that it enables the user to check and optimize a *system design* without the need to perform *circuit design*. There are a number of block diagram simulators commercially available, as shown in Table 19.11.

Some of these simulators have been developed mainly for communications and signal processing systems; examples are the SPW, Tesla, and Simulink simulators. These simulators have libraries of predefined block-diagram models of commonly used analog, digital, communications, and test functions. Model examples are filters, voltage controlled oscillators (VCOs), A/Ds, D/As, and AM/FM generators. Other simulators have been developed for hierarchical simulation, where system and component level simulation is possible; examples are the Saber and the PSpice simulators. Usually in these simulators, the

Listing D Input Circuit File of Example 7

Sawtooth Waveform Generator + (Fig. 19.24)	Comment Line
* analog parts	
R1 1 2 20 K	
R2 1 3 10 K	
R3 3 4 20 K	
R4 3 5 10 K	
R5 5 6 20 K	
R6 5 7 10 K	
R7 7 8 20 K	
R8 7 0 20 K	
R9 9 10 10 K	; Resistors
Vref 1 0 DC −5V	; Reference voltage source
+	
X1 0 9 50 51 10 TL082	; op-amp description
.LIB c:\msim61\lib\linear.lib	; op-amp library
Vcc 50 0 15 V	; Positive power supply
Vee 51 0 −15 V	; Negative power supply
*	
S1 2 0 QDBAR 0 SW1	; Voltage controlled switch
+	
S2 2 9 QD 0 SW1	
S3 4 0 QCBAR 0 SW1	
S4 4 9 QC 0 SW1	
S5 6 0 QBBAR 0 SW1	
S6 6 9 QB 0 SW1	
S7 8 0 QABAR 0 SW1	
S8 8 9 QA 0 SW1	
.MODEL SW1 VSWITCH	; Switch model
*	
* digital parts	
X2 CLK $D_HI $D_HI $D_HI $D_HI	
+$D_LO $D_LO $D_LO $D_LO	; 4-bit counter
+QA QB QC QD RCO 74HC161	
X3 QD QDBAR 74HC04	; Inverter
X4 QC QCBAR 74HC04	
X5 QB QBBAR 74HC04	
X6 QA QABAR 74HC04	
.LIB c:\msim61\lib\74HC.lib	; Digital devices library
.LIB c:\msim61\lib\dig_io.lib	; A to D and D to A subcircuits library
+	
.OPTIONS DIGINITSTATE = 0	; Clear digital devices
*	
* counter clock definition	
+ (period = 1 ms)	
U1 STIM (1,1) $D_DPWR $D_DGND	
+ CLK IO_STM TIMESTEP 0.5 ms	
+ 0c 0	
+ label = loop	; 1-kHz clock
+ 1c 1	
+ 2c 0	
+ 3c goto loop −1 times	
*	
.TRAN 0.02 ms 30 ms 0 m 0.02 m	; Transient analysis range
+	
.PROBE	Graphic outputs
*	
.END	; End of netlist

TABLE 19.11 Some Commercially Available Block-Diagram Simulators

Vendor	Simulator Name
Cadence	SPW
Analogy	Saber (MAST)
The Math Works	Simulink
Microsim	PSpice (analog behavioral modeling)
Tesoft	Tesla

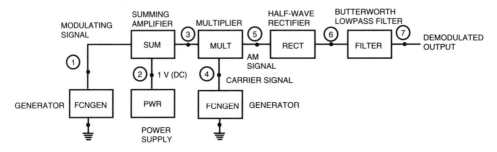

FIGURE 19.27 Block-diagram representation of an AM system.

user develops block-diagram models of the required functions using analog behavioral modeling [Al-Hashimi 1995] as the case with the PSpice simulator or hardware description language (MAST) as the case with the Saber simulator from Analogy [Paterson 1992]. Here, the simulator Tesla from Tesoft will be used to demonstrate block diagram system simulation.

Simulation Example 8

Consider the block diagram of the AM system shown in Fig. 19.27. It consists of a summing amplifier and a multiplier, which describe the modulator, and a half-wave rectifier and low-pass filter, which represent the demodulator. In this example, Tesla will be used to simulate the AM, modulating and demodulated signals. Prior to the simulation stage, the system needs to be described. Tesla allows netlist and schematic capture for system entry. The schematic capture is performed using the OrCAD schematic capture package with the appropriate Tesla library. To simplify netlist generation, Tesla uses similar syntax to SPICE in describing circuit elements and will be considered here. The basic format of a description statement is

$<ename>$ $<node1>$ $<node2>$ .... $<model\ name>$ [param1=valu1 param2=value2 .....]

where:

ename	= arbitrary 8-character element name
node1 node2...	= element nodal connections
model name	= name of the model used (Table 19.12)
param1 value1...	= optional parameters and their values

The major difference between the SPICE and the Tesla description statement is that the first letter in a SPICE statement defines the element type; for example, the letter R describes a resistor. The first letter of a Tesla statement, however, is an arbitrary element name chosen by the user. Tesla recognizes elements by their $<model\ name>$ part of the description statement. For example, the description statement defines a two-input summing amplifier connected between nodes 14 and 15. In Tesla such an amplifier model is recognized by the name, SUM, see Table 19.12. This amplifier, however, is user identified by the name Sl, which has been arbitrarily chosen. The simulator Tesla has predefined models of analog, digital, and test functions. Table 19.12 shows some of the models available.

TABLE 19.12 A Selection of Tesla Block-Diagram Models

Model Name	Function
FILTER	Butterworth, Chebyshev, Bessel, and Elliptic
INTEG	Integrator
SUM	2-input summing amplifier
MULT	Multiplier
RECT	Half-wave rectifier
VCO	Voltage controlled oscillator
ADD4	4-b binary adder with carry in/out
NAND	2-input NAND gate
SR4PS	4-b shift register (parallel in, serial out)
MUX	2:1 multiplexer
ADC8	8-b analog-to-digital converter
DAC8	8-b digital-to-analog converter
FCNGEN	Sine, square, triangle, and sawtooth waveforms generator
NOISE	White Gaussian noise source
PWR	Power supply
PSKMOD	QPSK or BPSK phase modulator

For example, the FILTER model allows the user to simulate Butterworth, Chebyshev, Bessel, and Gaussian 6-dB filters [Williams 1991]. The filter can be a low pass, high pass, band pass or band reject, with an order that can vary from 1 to 10. Using Table 19.12 and description statements, the netlist of the AM system is shown in Listing E. The system input file was created using the DOS EDLIN editor; other text editors may be used. Listing E shows that line 2 describes a function generator model, FCNGEN, which is used to define the modulating signal. The characteristics of this signal is a 1-kHz sine waveform, with 1-V amplitude and 90° phase shift. The generator is connected between nodes 1 and 0, and it is user identified by the name, g1, which has been chosen arbitrarily. The generator model, FCNGEN, is capable of producing AM and FM signals directly by assigning values to the first nodes of the generator description statement. In this example they are not used and, hence, have been set to 0 0. The parameter, FCN, of the description statement defines the type of waveform produced by the generator; in this example FCN = 1 produces a sinewave. Other waveforms can be defined by assigning different values to the parameter FCN. Similarly, line 3 defines the carrier signal (frequency = 1 MHz, amplitude = 1 V). Power supplies are described in Tesla using the PWR model; line 4 describes a power supply (ps) connected at node 2 with a value of 1V. This power supply is used to represent the DC signal shown in Fig. 19.27. Line 5 specifies a two-input summing amplifier model, SUM, with the inputs at nodes 1 and 2 and the output at node 3. The amplifier gain for the first and second input is 1 (i.e., G1 = 1 G2 = 1). Line 6 describes a multiplier function model, MULT, whereas line 7 specifies a half-wave rectifier model, RECT. Tesla has a wide choice of filters ranging from simple RC to complex elliptic models. In this example, a low-pass second-order Butterworth filter with cutoff frequency of 2 kHz is used, as shown in line 8 of Listing E.

Listing E Input File of Example 8

* AM System (Fig. 19.27) Simulation Using the + Simulator, Tesla	; Comment Line (line 1)
g1 0 0 1 0 FCNGEN FCN = 1 F = 1E3 V = 1, PHI = 90	; 1 kHz, 1 V, 90° (modulating signal)
g2 0 0 4 0 FCNGEN FCN = 1 F = 1 E6 V = 1	; 1 MHz, 1 V (carrier signal)
ps 2 PWR V = 1	; 1 V DC value
s1 1 2 3 SUM G1 = 1 G2 = 1	; Summing amplifier with gain = 1
m1 3 4 5 MULT	; Multiplier function
r1 5 6 RECT	; Half-wave rectifier
f1 6 7 FILTER BUTR LP N = 2 F = 2E3 +	; Second-order lp Butterworth filter, ; F = 2 kHz

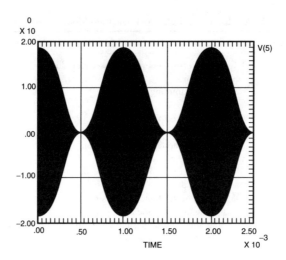

FIGURE 19.28 Simulated AM signal with 100% modulation.

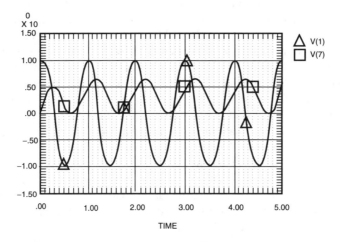

FIGURE 19.29 Simulated modulating, V(1), and demodulated signals, V (7).

Figure 19.28 shows the simulated AM signal with 100% modulation. Figure 19.29 shows the demodulated signal V(7) at the filter output, which corresponds to the input signal V(1), with a reduced amplitude due to the filtering function at the demodulator.

Defining Terms

Block-diagram simulator: A simulator that allows the user to simulate systems as a combination of block diagrams, each of which performs a specific function. Each function is described using a mathematical equation or a transfer function.

Digital simulator: A simulator that allows the user to check the function and perform a timing analysis of a digital system.

Frequency response analysis: A type of simulation that allows the user to predict the circuit response over a range of frequencies. The results of simulation are graphs of amplitude and/or phase against frequency.

Mixed A/D simulator: A simulator that is capable of simulating combined analog and digital circuitry.

Netlist: A method used for entering a circuit to the simulator using a text editor. The circuit is represented using description statements, where each statement expresses a component in terms of a name, node connections, and a value.

Op-amp macromodeling: A simulation technique used to model the function of an op-amp without describing the full internal transistor circuitry of the op amp.

Schematic capture: A graphical method of describing a circuit to the simulator using a mouse and symbols representing the circuit components.

Simulation program with integrated circuit emphasis (SPICE): A simulation package used to analyze electrical and electronic analog circuits.

Subcircuits: A simulation approach that allows an efficient description of repetitive circuitry.

Time domain analysis: A type of simulation that allows the user to predict the circuit response over a specified time range. The result of the simulation is a graph of amplitude against time.

References

Al-Hashimi, B. 1995. *The Art of Simulation Using PSpice: Analog and Digital.* CRC Press, Boca Raton, FL.

Antognetti, P. and Massobrio, G. 1988. *Semiconductor Device Modelling With SPICE.* McGraw-Hill, New York.

Biagi, H. and Stiff, M.R. 1990. Operational amplifier and instruments macromodels. *Application Bulletin.* Burr-Brown Corp.

Bogart, F.B. 1990. *Electronics Devices and Circuits,* 2nd ed. Merrill.

Boyle, G.R. et al. 1974. Macromodeling of integrated circuit operational amplifiers. *IEEE Solid-State Circuits* SC-9:353–364.

Buxton, J. 1992. Improve noise analysis with op-amp macromodel. *Electronic Design* (April):73–81.

Conner, D. 1994. Mixed analog-digital simulation. *EDN* (Sept. 1):39–44.

Hall, D.V. 1992. *Microprocessors and Interfacing: Programming and Hardware,* 2nd ed. Glencoe, pp. 380–386.

Horowitz, P. and Hill, W. 1994. *The Art of Electronics,* 2nd ed. Cambridge Univ. Press, New York.

Jung, W.G. 1990. Models can mimic behavior of real op amps. *Electronic Design* (Oct. 25): 67–78.

Lipsett, R., Schaefer, C., and Ussery, C. 1989. *VHDL: Hardware Description and Design.* Kluwer Academic, Norwell, MA.

Microsim. 1994. Circuit analysis—User's guide manual, Version 6.1, July. Microsim Corporation, CA.

Nagel, L.W. 1975. SPICE2: A computer program to simulate semiconductor circuits, ERL Memo No. ERL-M520, May. Electronics Research Lab., Univ. of California, Berkeley, CA.

Patterson, A. 1992. Evolving techniques in analogue synthesis and AHDL-based design. *Electronic Engineering* (Sept.): S51–S53.

Peic, R.V. 1991. Simple and accurate nonlinear macromodel for operational amplifiers. *IEEE J. Solid-State Circuits,* SC-26:896–899.

Rashid, M.H. 1990. *SPICE for Circuits and Electronics Using PSpice.* Prentice-Hall, Englewood Cliffs, NJ.

Robison, M.E., Sanchez-Sinencio, Y.E., E., Kao, W.K. 1993. Comparison study brings op-amp macromodel differences to light. *EDN* (Jan. 21): 95–101.

Swager, A.W. 1992. DOS-based analog simulation software. *EDN,* May 21, pp. 125–142.

Tuinenga, P.W. 1992. *SPICE: A Guide to Circuit Simulation & Analysis Using PSpice.* Prentice-Hall, Englewood Cliffs, NJ.

Van Valkenburg, M.E. 1982. *Analogue Filter Design.* Holt, Rinehart & Winston, Philadelphia, PA.

Williams, A.B. 1991. *Electronic Filter Design Handbook.* McGraw-Hill, New York.

Further Information

Additional information on the topic of circuit and system simulation is available from the following sources:

EDN, Electronic Design, and *Microwaves & RF* Magazines regularly publish articles and design ideas dealing with the application of simulation packages in solving engineering problems. Also, these magazines review the latest developments in simulation tools.

The Design Center Source, a newsletter published quarterly by MicroSim Corp. (the PSpice developer), is free to qualified subscribers. The newsletter includes PSpice technical articles and application notes.

In addition, the following books are recommended:

Al-Hashimi, B.M. 1995. *The Art of Simulation Using PSpice: Analog and Digital.* CRC Press, Boca Raton, FL.

Tuinenga, P.W. 1992. SPICE: A Guide to Circuit Simulation & Analysis Using *PSpice.* Prentice-Hall, Englewood Cliffs, NJ.

Companies producing software discussed in this chapter are as follows:

Analogy, Beaverton, OR.
Mentor Graphics, Wilsonville, OR.
PSpice A/D, MicroSim
Tesoft, Inc., 205 Crossing Creek Ct., Roswell, GA.

20
Audio Frequency Distortion Mechanisms and Analysis

	20.1	Introduction .... 20-1 Purpose of Audio Measurements
	20.2	Level Measurements .... 20-2 Root-Mean-Square Measurements • Average-Response Measurements • Peak-Response Measurements • Decibel Measurements
	20.3	Noise Measurement 20-5
	20.4	Phase Measurement 20-5
	20.5	Nonlinear Distortion Mechanisms .... 20-6 Harmonic Distortion • Notch Filter Analyzer • Intermodulation Distortion • CCIT IM • Addition and Cancellation of Distortion Components
Jerry C. Whitaker *Editor-in-Chief*	20.6	Multitone Audio Testing .... 20-10 Multitone vs. Discrete Tones • Operational Considerations • FFT Analysis
	20.7	Considerations for Digital Audio Systems 20-15

20.1 Introduction

Most measurements in the audio field involve characterizing fundamental parameters. These include signal level, phase, and frequency. Most other tests consist of measuring these fundamental parameters and displaying the results in combination by using some convenient format. For example, signal-to-noise ratio (SNR) consists of a pair of level measurements made under different conditions expressed as a logarithmic, or decibel (dB), ratio. When characterizing a device, it is common to view it as a box with input terminals and output terminals. In normal use a signal is applied to the input and the signal, modified in some way, appears at the output. Instruments are necessary to quantify these unintentional changes to the signal. Some measurements are *one-port* tests, such as impedance or noise level, and are not concerned with input/output signals, only with one or the other.

Purpose of Audio Measurements

Measurements are made on audio circuits and equipment to check performance under specified conditions and to assess suitability for use in a particular application. The measurements may be used to verify specified system performance or as a way of comparing several pieces of equipment for use in a system. Measurements may also be used to identify components in need of adjustment or repair. Whatever the application, audio measurements are a key part of audio engineering.

Many parameters are important in audio devices and merit attention in the measurement process. Some common audio measurements are frequency response, gain or loss, harmonic distortion, intermodulation distortion, noise level, phase response, and transient response.

Measurement of level is fundamental to most audio specifications. Level can be measured either in absolute terms or in relative terms. Power output is an example of an absolute level measurement; it does not require any reference. SNR and gain or loss are examples of relative measurements; the result is expressed as a ratio of two measurements. Although it may not appear so at first, frequency response is also a relative measurement. It expresses the gain of the device under test as a function of frequency, with the midband gain as a reference.

Distortion measurements are a way of quantifying the amount of unwanted components added to a signal by a piece of equipment. The most common technique is total harmonic distortion (THD), but others are often used. Distortion measurements express the amount of unwanted signal components relative to the desired signal, usually as a percentage or decibel value. This is another example of multiple level measurements that are combined to give a new measurement figure.

20.2 Level Measurements

The simplest definition of a level measurement is the alternating current (AC) amplitude at a particular place in the audio system. However, in contrast to direct current (DC) measurements, there are many ways of specifying AC voltage. The most common methods are average, root-mean-square (RMS), and peak response. Strictly speaking, the term *level* refers to a logarithmic, or decibel, measurement. However, common parlance employs the term for an AC amplitude measurement, and that convention will be followed in this chapter.

Root-Mean-Square Measurements

The RMS technique measures the effective power of the AC signal. It specifies the value of the DC equivalent that would dissipate the same power if either was applied to a load resistor. This process is illustrated in Fig. 20.1 for voltage measurements. The input signal is squared, and the average value is found. This is equivalent to finding the average power. The square root of this value is taken to translate the signal from a power value back to a voltage. For the case of a sine wave, the RMS value is 0.707 of its maximum value.

Consider the case when the signal is not a sine wave, but rather a sine wave and several of its harmonics. If the RMS amplitude of each harmonic is measured individually and added, the resulting value will be the same as an RMS measurement on the signals together. Because RMS voltages cannot be added directly, it is necessary to perform an RMS addition, as follows:

$$Vrms = \sqrt{V_{rms1}^2 + V_{rms2}^2 + V_{rms3}^2 + V_{rmsn}^2}$$

Note that the result is not dependent on the phase relationship of the signal and its harmonics. The RMS value is determined completely by the amplitude of the components. This mathematical predictability is powerful in practical applications of level measurement, enabling measurements made at different places in a system to be correlated. It is also important in correlating measurements with theoretical calculations.

Average-Response Measurements

Average-responding voltmeters were the common in audio work some years ago principally because of their low cost. Such devices measure AC voltage by rectifying and filtering the resulting waveform to its average value, which can then be read on a standard DC voltmeter. The average value of a sine wave is

Audio Frequency Distortion Mechanisms and Analysis

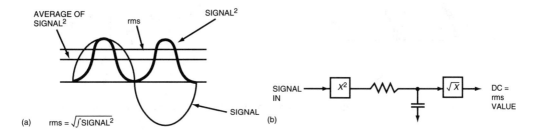

FIGURE 20.1 Root-mean-square (RMS) voltage measurements: (a) the relationship of RMS and average values, (b) RMS measurement circuit.

WAVEFORM		rms	Avg.	rms avg.	CREST FACTOR
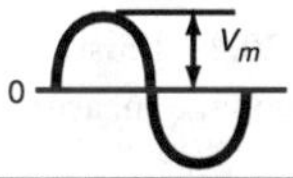 V_m	SINE WAVE	$\dfrac{V_m}{\sqrt{2}}$ $0.707\,V_m$	$\dfrac{2}{\pi}V_m$ $0.637\,V_m$	$\left|\dfrac{\pi}{2\sqrt{2}}\right| = 1.111$	$\sqrt{2} = 1.414$
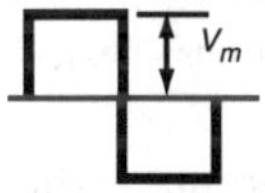 V_m	SYMMETRICAL SQUARE WAVE OR DC	V_m	V_m	1	1
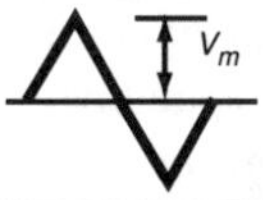 V_m	TRIANGULAR WAVE OR SAWTOOTH	$\dfrac{V_m}{\sqrt{3}}$	$\dfrac{V_m}{2}$	$\dfrac{2}{\sqrt{3}} = 1.155$	$\sqrt{3} = 1.732$

FIGURE 20.2 Comparison of RMS and average voltage characteristics.

0.637 of its maximum amplitude. Average-responding meters are usually calibrated to read the same as an RMS meter for the case of a single-sine-wave signal. This results in the measurement being scaled by a constant K of 0.707/0.637, or 1.11. Meters of this type are called average-responding, RMS calibrated. For signals other than sine waves, the response will be different and difficult to predict. If multiple sine waves are applied, the reading will depend on the phase shift between the components and will no longer match the RMS measurement. A comparison of RMS and average-response measurements is made in Fig. 20.2 for various waveforms. If the average readings are adjusted as described previously to make the average and RMS values equal for a sine wave, all the numbers in the *average* column should be increased by 11.1%, while the *RMS-average* numbers should be reduced by 11.1%.

Peak-Response Measurements

Peak-responding meters measure the maximum value that the AC signal reaches as a function of time (see Fig. 20.3). The signal is full-wave-rectified to find its absolute value and then passed through a diode to a storage capacitor. When the absolute value of the voltage rises above the value stored on the capacitor, the diode will conduct and increase the stored voltage. When the voltage decreases, the capacitor will maintain the old value. Some means for discharging the capacitor is required to allow measuring a new peak value. In a true peak detector, this is accomplished by a switch. Practical peak detectors usually include a large resistor to discharge the capacitor gradually after the user has had a chance to read the meter.

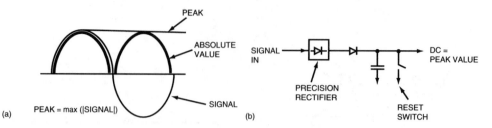

FIGURE 20.3 Peak voltage measurements: (a) illustration of peak detection, (b) peak measurement circuit.

The ratio of the true peak to the RMS value is called the *crest factor*. For any signal but an ideal square wave the crest factor will be greater than 1, as illustrated in Fig. 20.4. As the measured signal becomes more peaked, the crest factor will increase.

By introducing a controlled charge and discharge time, a *quasi-peak* detector is achieved. The charge and discharge times may be selected, for example, to simulate the ear's sensitivity to impulsive peaks. International standards define these response times and set requirements for reading accuracy on pulses and sine wave bursts of various durations. The gain of a quasi-peak detector is normally calibrated so that it reads the same as an RMS detector for sine waves.

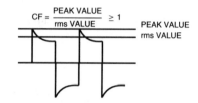

FIGURE 20.4 Illustration of the crest factor in voltage measurements.

Another method of specifying signal amplitude is the *peak-equivalent sine*, which is the RMS level of a sine wave having the same peak-to-peak amplitude as the signal under consideration. This is the peak value of the waveform scaled by the correction factor 1.414, corresponding to the peak-to-RMS ratio of a sine wave. This is useful when specifying test levels of waveforms in distortion measurements.

Decibel Measurements

Measurements in audio work are usually expressed in decibels (dB). Audio signals span a wide range of levels. The sound pressure of a live concert band performance may be one million times that of rustling leaves. This range is too wide to be accommodated on a linear scale. The logarithmic scale of the decibel compresses this wide range down to a more easily handled range. Order-of-magnitude changes result in equal increments on a decibel scale. Furthermore, the human ear perceives changes in amplitude on a logarithmic basis, making measurements with the decibel scale reflect audibility more accurately.

A decibel may be defined as the logarithmic ratio of two power measurements or as the logarithmic ratio of two voltages:

$$dB = 20 \log \frac{E_1}{E_2} \quad \text{for voltage measurements}$$

$$dB = 10 \log \frac{P_1}{P_2} \quad \text{for power measurements}$$

There is no difference between decibel values from power measurements and decibel values from voltage measurements if the impedances are equal. In both equations the denominator variable is usually a stated reference. A doubling of voltage will yield a value of 6.02 dB, while a doubling of power will yield 3.01 dB. This is true because doubling voltage results in a factor-of-4 increase in power.

Audio engineers often express the decibel value of a signal relative to some standard reference instead of another signal. The reference for decibel measurements may be predefined as a power level, as in dBm

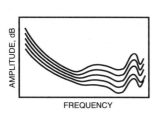

FIGURE 20.5 The Fletcher-Munson curves of hearing sensitivity vs. frequency.

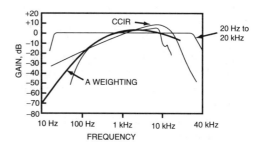

FIGURE 20.6 Response characteristics of several common weighting filters for audio measurements.

(decibels referenced to 1 mW), or it may be a voltage reference, as in dBV (decibels referenced to 1 V). When measuring dBm or any power-based decibel value, the reference impedance must be specified or understood. Often it is desirable to specify levels in terms of a reference transmission level somewhere in the system under test. These measurements are designated dBr, where the reference point or level must be separately conveyed.

20.3 Noise Measurement

Noise measurements are simply specialized level measurements. It has long been recognized that the ear's sensitivity varies with frequency, especially at low levels. This effect was studied in detail by Fletcher and Munson and later by Robinson and Dadson. The Fletcher-Munson hearing-sensitivity curve for the threshold of hearing and above is given in Fig. 20.5. The ear is most sensitive in the region of 2 to 4 kHz, with rolloffs above and below these frequencies. To predict how loud something will sound, it is necessary to use a filter that duplicates this nonflat behavior electrically. The filter *weights* the signal level on the basis of frequency, thus earning the name *weighting filter*. Various efforts have been made to do this, resulting in several standards for noise measurement. Some of the common weighting filters are shown overlaid on the hearing threshold curve in Fig. 20.6.

The most common filter used in the U.S. for weighted noise measurements is the A-weighting curve. An average-responding meter is often used for A-weighted noise measurements, although RMS meters are also used for this application.

European audio equipment is usually specified with a CCIR filter and a quasi-peak detector. The CCIR curve is significantly more peaked than the A curve and has a sharper rolloff at high frequencies. The CCIR quasi-peak standard was developed to quantify the noise in telephone systems. The quasi-peak detector more accurately represents the ear's sensitivity to impulsive sounds. When used with the CCIR filter curve, it is supposed to correlate better with the subjective level of the noise than do A-weighted average-response measurements.

Some audio equipment manufacturers specify noise with a 20 Hz to 20 kHz bandwidth filter and an RMS-responding meter. This is done to specify noise over the audio band without regard to the ear's differing sensitivity with frequency. The International Electrotechnical Commission (IEC) defines the audio band as all frequencies between 22.4 Hz and 22.4 kHz. In IEC standards, measurements over such a bandwidth are referred to as unweighted.

20.4 Phase Measurement

Phase in an audio system is typically measured and recorded as a function of frequency over the audio range. For most audio devices phase and amplitude responses are closely coupled. Any change in amplitude that varies with frequency will produce a corresponding phase shift. A fixed time delay will introduce a phase shift that is a linear function of frequency. This time delay can introduce large values of phase shift at high frequencies that are of no significance in practical applications. The time delay will not distort the waveshape of complex signals and will not be audible. There can be problems, however, with time delay when the delayed signal is used in conjunction with an undelayed signal.

When dealing with complex signals, the meaning of phase can become unclear. Viewing the signal as the sum of its components according to Fourier theory, we find a different value of phase shift at each frequency. With a different phase value on each component, the question is raised as to which one should be used as the reference. If the signal is periodic and the waveshape is unchanged passing through the device under test, a phase value may still be defined. This may be done by using the shift of the zero crossings as a fraction of the waveform period. However, if there is differential phase shift with frequency, the waveshape will be changed. It is then not possible to define any phase-shift value, and phase must be expressed as a function of frequency.

Group delay is another useful expression of the phase characteristics of an audio device. Group delay is the slope of the phase response. It expresses the relative delay of the spectral components of a complex waveform. If the group delay is flat, all components will arrive together. A peak or rise in the group delay indicates that those components will arrive later by the amount of the peak or rise. Group delay ϕ is computed by taking the derivative of the phase response vs. frequency:

$$\phi = \frac{-(\alpha_{f2} - \alpha_{f1})}{f_2 - f_1}$$

where α_{f1} = phase at f_1 and α_{f2} = phase at $f2$. This requires that phase be measured over a range of frequencies to yield a curve that can be differentiated. It also requires that the phase measurements be performed at frequencies close enough to provide a smooth and accurate derivative.

20.5 Nonlinear Distortion Mechanisms

Distortion is a measure of signal impurity. It is usually expressed as a percentage or decibel ratio of the undesired components to the desired components of a signal. There are several methods of measuring distortion, the most common being harmonic distortion and several types of intermodulation distortion.

Harmonic Distortion

The transfer characteristic of a typical amplifier is shown in Fig. 20.7. The transfer characteristic represents the output voltage at any point in the signal waveform for a given input voltage; ideally this is a straight line. The output waveform is the projection of the input sine wave on the device transfer characteristic. A change in the input produces a proportional change in the output. Because the actual transfer characteristic is nonlinear, a distorted version of the input waveshape appears at the output.

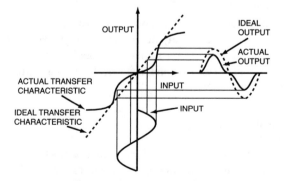

FIGURE 20.7 Illustration of total harmonic distortion (THD) measurement of an amplifier transfer characteristic.

Harmonic distortion measurements excite the device under test with a sine wave and measure the spectrum of the output. Because of the non-linearity of the transfer characteristic, the output is not sinusoidal. By using Fourier series, it can be shown that the output waveform consists of the original input sine wave plus sine waves at integer multiples (harmonics) of the input frequency. The spectrum of the distorted signal is shown in Fig. 20.8 for a 1-kHz input, and output signals consisting of 1 kHz, 2 kHz, 3 kHz, etc. The harmonic amplitudes are proportional to the amount

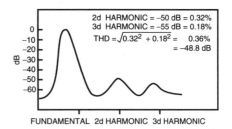

FIGURE 20.8 Example of reading THD from a spectrum analyzer.

of distortion in the device under test. The percentage harmonic distortion is the RMS sum of the harmonic amplitudes divided by the RMS amplitude of the fundamental.

Harmonic distortion may also be measured with a spectrum analyzer. As shown in Fig. 20.8, the fundamental amplitude is adjusted to the 0 dB mark on the display. The amplitudes of the harmonics are then read and converted to linear scale. The RMS sum of these values is taken, which represents the THD. This procedure is time-consuming and can be difficult for an unskilled operator.

Notch Filter Analyzer

A simpler approach to the measurement of harmonic distortion can be found in the notch-filter distortion analyzer. This device, commonly referred to as simply a distortion analyzer, removes the fundamental of the signal to be investigated and measures the remainder. A block diagram of such a unit is shown in Fig. 20.9. The fundamental is removed with a notch filter, and the output is measured with an AC voltmeter. Because distortion is normally presented as a percentage of the fundamental signal, this level must be measured or set equal to a predetermined reference value. Additional circuitry (not shown) is required to set the level to the reference value for calibrated measurements. Some analyzers use a series of step attenuators and a variable control for setting the input level to the reference value. More sophisticated units eliminate the variable control by using an electronic gain control. Others employ a second AC-to-DC converter to measure the input level and compute the percentage by using a microprocessor. Completely automatic units also provide autoranging logic to set the attenuators and ranges, and tabular or graphic display of the results.

The correct method of representing percentage distortion is to express the level of the harmonics as a fraction of the fundamental level. However, most commercial distortion analyzers use the total signal level as the reference voltage. For small amounts of distortion these two quantities are equivalent. At large values of distortion the total signal level will be greater than the fundamental level. This makes

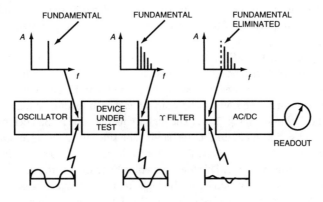

FIGURE 20.9 Simplified block diagram of a harmonic distortion analyzer.

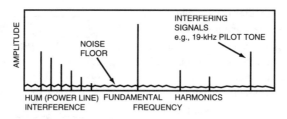

FIGURE 20.10 Example of interference sources in distortion and noise measurements.

distortion measurements on these units lower than the actual value. The errors are not significant until about 20% measured distortion.

Because of the notch-filter response, any signal other than the fundamental will influence the results, not just harmonics. Some of these interfering signals are illustrated in Fig. 20.10. Any practical signal contains some noise, and the distortion analyzer will include these in the reading. Because of these added components, the correct term for this measurement is *total harmonic distortion and noise* (THD+N). Although this fact does limit the reading of very low THD levels, it is not necessarily bad. Indeed, it can be argued that the ear hears all components present in the signal, not just the harmonics.

Additional filters are included on most distortion analyzers to reduce unwanted hum and noise. These usually consist of one or more high pass filters (400 Hz is almost universal) and several low pass filters. Common low pass filter frequencies are 22.4, 30, and 80 kHz. A selection of filters eases the tradeoff between limiting bandwidth to reduce noise and the reduction in reading accuracy that results from removing desired components of the signal. When used in conjunction with a good differential input stage on the analyzer, these filters can solve most noise problems.

Intermodulation Distortion

Many methods have been devised to measure the intermodulation (IM) of two or more signals passing through a device simultaneously. The most common of these is *SMPTE IM*, named after the Society of Motion Picture and Television Engineers, which first standardized its use. IM measurements according to the SMPTE method have been in use since the 1930s. The test signal is a low-frequency tone (usually 60 Hz) and a high-frequency tone (usually 7 kHz) mixed in a 4:1 amplitude ratio. Other amplitude ratios and frequencies are used occasionally. The signal is applied to the device under test, and the output signal is examined for modulation of the upper frequency by the low frequency tone. The amount by which the low frequency tone modulates the high frequency tone indicates the degree of nonlinearity. As with harmonic distortion measurement, this test may be done with a spectrum analyzer or with a dedicated distortion analyzer.

The modulation components of the upper signal appear as sidebands spaced at multiples of the lower-frequency tone. The RMS amplitudes of the sidebands are summed and expressed as a percentage of the upper-frequency level.

The most direct way to measure SMPTE IM distortion is to measure each component with a spectrum analyzer and add their RMS values. The spectrum analyzer approach has a drawback in that it is sensitive to frequency modulation of the carrier as well as amplitude modulation. A distortion analyzer for SMPTE testing is quite straightforward. The signal to be analyzed is passed through a high pass filter to remove the low frequency tone, as shown in Fig. 20.11. The high frequency tone is then demodulated as if it were an amplitude modulated signal to obtain the sidebands. The sidebands pass through a low pass filter to remove any remaining high frequency energy. The resulting demodulated low frequency signal will follow the envelope of the high frequency tone. This low frequency fluctuation is the distortion component and is displayed as a percentage of the amplitude of the high frequency tone. Because low pass filtering sets the measurement bandwidth, noise has little effect on SMPTE IM measurements.

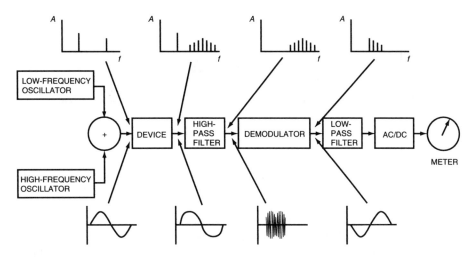

FIGURE 20.11 Simplified block diagram of an SMPTE intermodulation analyzer.

CCIT IM

Twin-tone intermodulation or CCIT difference frequency distortion is another method of measuring distortion by using two sine waves. The test signal consists of two closely spaced high frequency tones as shown in Fig. 20.12. When the tones are passed through a nonlinear device, IM products are generated at frequencies related to the difference in frequency between the original tones. For the typical case of signals at 14 kHz and 15 kHz, the IM components will be at 1 kHz, 2 kHz, 3 kHz, etc. and 13 kHz, 16 kHz, 12 kHz, 17 kHz, etc. Even-order, or asymmetrical distortions produce low frequency difference-frequency components. Odd-order, or symmetrical nonlinearities, produce components near the input signal frequencies. The most common application of this test measures only the even-order components because they may be measured with a multipole low pass filter. Measurement of the odd-order components requires a spectrum analyzer or selective voltmeter.

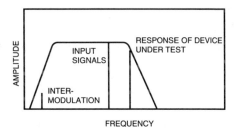

FIGURE 20.12 SMPTE intermodulation test of transfer characteristic.

The CCIT test has several advantages over either harmonic or SMPTE IM testing. The signals and distortion components may almost always be arranged to be in the passband of the device under test. This method ceases to be useful below a few hundred hertz when the required selectivity in the spectrum analyzer or selective voltmeter becomes excessive. However, FFT-based devices can extend the practical lower limit substantially below this point.

The distortion products in the CCIT IM test are usually far removed from the input signal. This positions them outside the range of the auditory system's masking effects. If a test that measures what the ear might hear is desired, the CCIT test is the most likely candidate.

Addition and Cancellation of Distortion Components

A common consideration for system-wide distortion measurements is that of distortion addition and cancellation in the devices under test. Consider the example given in Fig. 20.13. Assume that one device under test has a transfer characteristic similar to that diagrammed at the top of Fig. 20.13a and another has the characteristic diagrammed at the bottom. If the devices are cascaded, the resulting transfer-characteristic nonlinearity will be magnified as shown. The effect on sine waves in the time domain is

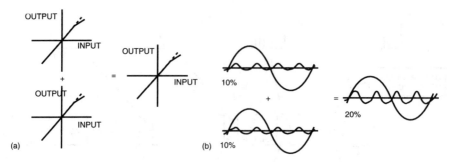

FIGURE 20.13 Illustration of the addition of distortion components: (a) addition of transfer-function nonlinearities, (b) addition of distortion components.

illustrated in Fig. 20.13b. The distortion component generated by each nonlinearity is in phase and will sum to a component of twice the original magnitude. However, if the second device under test has a complementary transfer characteristic, quite a different result is obtained. When the devices are cascaded, the effects of the two curves cancel, yielding a straight line for the transfer characteristic. The corresponding distortion products are out of phase with each other, resulting in no measured distortion components in the final output.

20.6 Multitone Audio Testing

Multitone testing operates on the premise that audio equipment can be stimulated to the same extent by a simultaneous combination of sine waves as it can be by a series of discrete tones occurring one at a time [5]. The advantage of multitone analysis is clear: it provides complete system evaluation in one step, eliminating the time-consuming process of applying a series of individual tones, allowing a settling time after each, making adjustments after each, and then repeating the process to check for interactions.

Multitone test techniques use carefully designed mixtures of tones applied simultaneously to the device under test (DUT). The individual tone elements have well-defined frequency, phase, and amplitude relationships. The frequencies of multitone components are selected to avoid mathematically predictable harmonic and intermodulation products that would fall on or near any of the fundamentals. The phase of each tone is fixed, but randomly selected relative to each of the other tones. Amplitude relationships may vary, depending on the device under test.

Multitone signals may be created with an inverse fast Fourier transform (FFT) for highly accurate and stable signal parameters. Figure 20.14 shows a spectral display of a multitone signal designed for testing wideband audio recording equipment. Because the frequency, phase, and amplitude relationships of the elements of a multitone signal are well defined, DSP techniques allow easy detection of changes in these relationships. Thus, frequencies not present in the original multitone are the result of distortion and noise. Similarly, changes in phase relationships of the multitone elements can be used to derive phase response information. It follows that deviation from the known amplitude relationships will provide frequency response information.

Although implementations of multitone analysis vary from one vendor to the next, plots of level vs. frequency, level difference between channels, phase difference between channels vs. frequency, and noise and/or distortion vs. frequency are among those typically provided. Figures 20.15 and 20.16 show two of these multitone displays, levels vs. frequency (frequency response) and level/phase differences between channels plotted against frequency. An audio spectrum display, such as the one shown in Fig. 20.14, is also a valuable tool when combined with the multitone signal.

Audio Frequency Distortion Mechanisms and Analysis

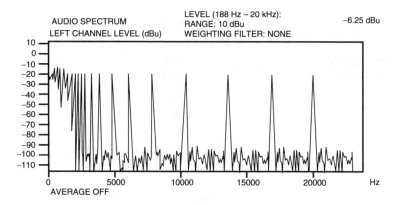

FIGURE 20.14 An audio spectrum display showing the spectral distribution of the numerous fundamental frequencies in a multitone signal. (Courtesy of Tektronix.)

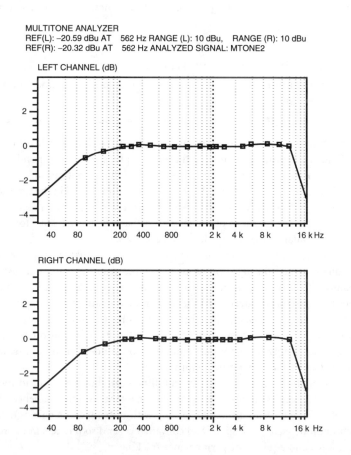

FIGURE 20.15 Level vs. frequency plots resulting from audio system measurement with a multitone signal analyzer. (Courtesy of Tektronix.)

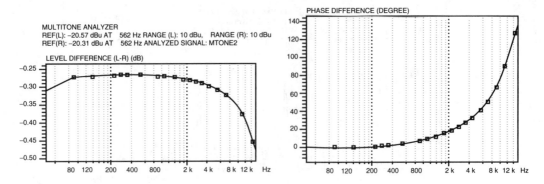

FIGURE 20.16 Level and phase difference measurements taken with a multitone signal analyzer. (Courtesy of Tektronix.)

Multitone vs. Discrete Tones

Multitone testing differs from traditional single-tone testing in several ways [5]. As previously mentioned, multitone analysis reduces testing and calibration times. This speed advantage stems from several properties:

- All tones are applied at once.
- The entire spectrum is measured/adjusted at once.
- Only one settling time is needed for the generator, DUT, and measurement device because a single signal acquisition is made.
- Level, phase, and noise and distortion calculations are made from one FFT record.

Not only does multitone eliminate the step-and-repeat process, but it can also provide immediate feedback for equipment adjustments. Watching the entire audio spectrum respond to interactive adjustments with little or no time lag further simplifies audio equipment calibration.

When compared to a single sine wave tone, looking at a multitone signal in the time domain (Fig. 20.17) illustrates how closely a multitone waveform resembles program audio. Two factors contribute to this resemblance. First, multitones fill more of the spectrum than single sine wave tones. Second, the crest factor of multitone is similar to that of music or voice signals.

With multitone's similarity to program material, it can be argued that the adjustments performed and measurement results obtained through multitone give more realistic noise and distortion values than discrete tone testing. (In this case, "more realistic" means a value better representing the noise and distortion present with program material.) In the past, noise measurements were made on audio equipment while no input signal was present. Therefore, the quantization noise and other level-related anomalies were not a part of the noise measurement result. Furthermore, signal processors often operate at different gain ranges depending on the input signal level. So, a no-input noise measurement would be even less representative of actual operating conditions.

The different type of stimulation provided by multitone can, however, lead to measurement results that do not exactly match results obtained with conventional tones or tone sequences. With this in mind, multitone testing should not be used interchangeably with tests made using single-tone input signals. The strengths of multitone testing are speed and convenience, and so it excels in the roles of calibration, production line testing, and equipment performance tracking.

Standardized tone sequence tests require settling, acquisition, and measurement times for each tone in the sequence. All of the advantages of multitone previously mentioned apply here as well. And since multitone is a DSP-based technology, it is always backed up with the computing power necessary for automation. To fully mimic standard tone sequences, a "multitone sequence" of two or three different levels of multitone can be performed in one-tenth the time of conventional sequences.

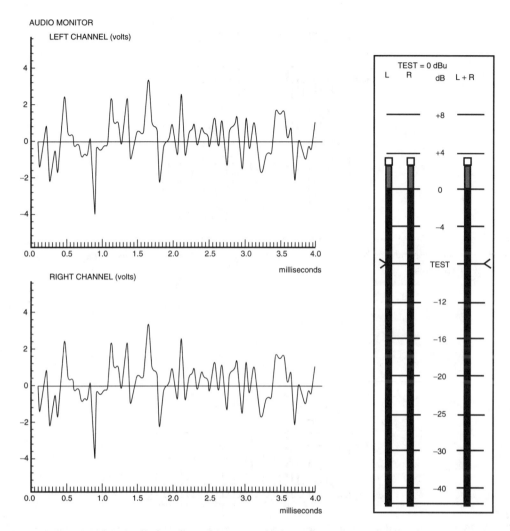

FIGURE 20.17 Time domain display of a multitone signal. The multitone bears a much stronger resemblance to program material than a single sine wave tone. (Courtesy of Tektronix.)

Operational Considerations

Multitone is a strong candidate for rapid performance tests of common carrier and broadcast audio transmission links—while on the air or otherwise in use [5]. A short burst of multitone, typically about 1 sec, is all that is needed for analyzers to do a complete audio path characterization. Although most multitones do not sound particularly symphonic, they are not abrasive either, especially in the short bursts necessary for measurement.

Given the usual distance between ends of a transmission system during split-site testing (i.e., transmitter and studio), it is not practical to lock the sampling clocks of the generator and analyzer, resulting in slight frequency shifts in the multitone elements. Split-site testing is one application where frequency shifts in the multitone can occur, and windowing prior to the FFT analysis should be employed to ensure accurate measurement results.

FFT Analysis

A closer look at multitone shows the need for caution when processing the signal [5]. Figures 20.18 and 20.19 illustrate how different processing techniques produce drastically different results, only some of

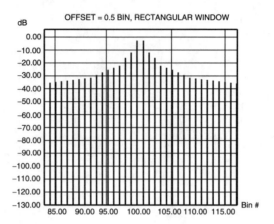

FIGURE 20.18 Computer simulation of FFT without windowing. The 588.87 Hz fundamental frequency in this model does not correspond exactly to an FFT bin. The result is leakage that provides a misleading noise floor display. (Courtesy of Tektronix.)

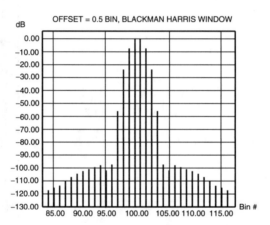

FIGURE 20.19 Computer simulation of FFT with Blackman-Harris windowing. The application of windowing prior to the FFT drastically reduces the effects of leakage. Within 23.44 Hz on either side of the fundamental, leakage is down more than 90 dB. (Courtesy of Tektronix.)

which are meaningful. Both figures are based on a simulated 8192-point (8K) FFT with a 48-kHz sampling frequency and a 588.87-Hz tone. FFT bin numbers (e.g., bin 100 out of 4096 bins) label the horizontal axes of each graph. The fundamental frequency of 588.87 Hz falls halfway between bin numbers 100 (585.94 Hz) and 101 (591.80 Hz).

Transforming time-domain information into the frequency domain with an FFT requires that the period of the time domain signal be an integer multiple of the period of the FFT. When such a periodic relationship occurs, a fundamental frequency will fall into a single FFT bin. But this requires that the same sampling clock be used for both signal generation and subsequent resampling of the signal, and further, that no shift in the tone's pitch occurs while passing through the DUT. Because shifts in pitch can occur and because some testing applications require that the generator and analyzer be at separate locations, it is not safe to assume perfect tone-to-bin correspondence.

The result of this imperfect tone-to-bin correspondence is *leakage*, which simply means energy that should fall into a single bin is spread out over a wide frequency range. To reduce the effect of leakage, the time-domain data is multiplied with a window function.

The graph in Fig. 20.18 clearly shows the effects of leakage. If the fundamental was at 585.94 Hz, a single line at bin 100 with no energy in adjacent bins would result. A rectangular window, which is equivalent to no window, was used in Fig. 20.18. The graph in Fig. 20.19 shows the effect of applying a Blackman-Harris window to the same data. After only four bins either side of the fundamental, the leakage is down more than 90 dB. In actual application, an algorithm can determine the exact frequency and amplitude of the fundamental. To compensate for significant shifts in pitch, the spectra surrounding each expected multitone fundamental may be searched ±5% of the frequency of each fundamental.

Extraneous signals introduced into the signal path will also have a serious impact on a nonwindowed analysis. Interfering carriers, hum, and sync buzz are a few types of noncoherent signals that fit into this category. These noncoherent signals do not fall within the periodicity requirements for a nonwindowed FFT and will affect the noise floor of a graphical display in the same manner as will a shift in pitch of the multitone fundamentals. With windowing, 60 Hz hum will cause a noise spike at 60 Hz. Without windowing, the same hum energy will be spread across the audio spectrum, raising the entire noise floor.

For noise and distortion measurements, the eight bins surrounding each fundamental in a multitone are set to zero. The remaining bins are plotted to represent the noise (and distortion) floor of the device under test.

20.7 Considerations for Digital Audio Systems

Perhaps the single most common operating problem in a digital audio system is *clipping*, which occurs when the input range of the A/D converter is exceeded [1]. It is important to detect this situation because it can degrade audio quality and introduce distortion artifacts. Figure 20.20 shows an audio analyzer display of input signal clipping. Because of its importance, some form of clipping analysis is necessary for audio work. The behavior of the clipping detector in an analyzer can often be modified according to the audio program material and preferences of the audio engineer. For example, if the mix and material demand that no full-scale samples be present in the media, then the trigger point for the clip indicator (and counter) can be set to its most sensitive position. At this point, just one full-scale sample will trigger the clip indicator. On the other hand, when some full-scale samples are permitted, the sensitivity of the clip detector can be reduced by specifying the number of consecutive full-scale samples that must occur before the clipping detector is triggered.

When compact discs are mastered, engineers attempt to adjust audio levels so that the clip point is just reached. This practice obtains the maximum dynamic range of the 16-bit resolution of CD media. In this situation, the engineer might choose a clip detector trigger point of two or more consecutive samples before a clip registers. A similar capability may be provided for mute detection, with the sensitivity of the trigger adjusted to indicate a single muted sample or some number of consecutive samples.

An important capability of some analyzers is the compilation of statistics associated with a digital audio program or path. A display of such statistics is shown in Fig. 20.21. The following parameters are logged in the unit shown:

- *Highest True Peak*, which retains the largest value indicated by the Peak Hold indicator.
- *Highest Bar Reading*, which retains the largest value of bar graph level. In the event that the user selects either VU or PPM behavior for the bar graphs, the Highest Bar Reading will always be less than the Highest True Peak. Conversely, when True Peak behavior is selected for the bar graphs, then Highest True Peak and Highest Bar Reading will have the same value.
- *Number of Clips*, a counter that accumulates the number of times the clip detector is triggered. Of course, the clip is only triggered when the sensitivity conditions (number of consecutive full scale samples) are satisfied.
- *Number of Mutes*, a counter that accumulates the number of times the mute detector is triggered.
- *Invalid Samples*, a counter that accumulates the number of times a sample includes a high validity bit.

- *Parity Errors and Code Violations*, counters that accumulate the number of times that each of these interface errors are detected.
- *Active Bits*, which displays the number of bits in a sample that are active or changing states.
- *DC Offset*, which displays the DC offset contained in the encoded audio. This capability enables fine-tuning of the DC offset characteristics of A/D converters.
- *Sample Rate*, which displays the sample rate measured by the analyzer.

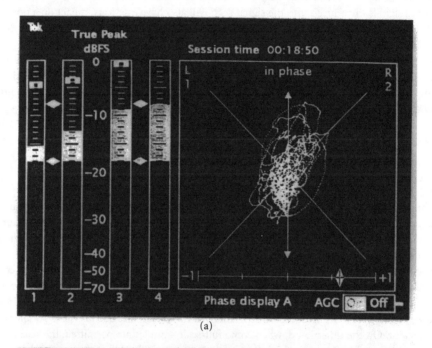

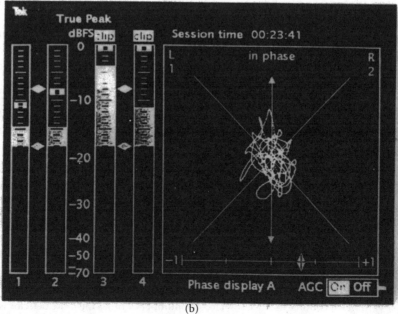

FIGURE 20.20 Detection of clipping in an A/D converter: (a) audio monitor display of normal program material below the clipping threshold, (b) display showing the effects of clipping. (Courtesy of Tektronix.)

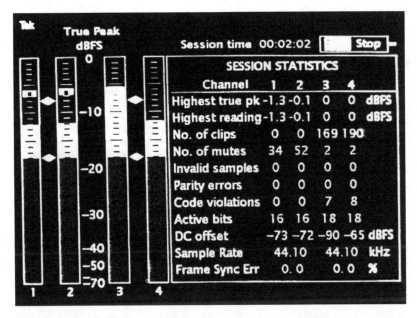

FIGURE 20.21 Audio monitor display showing statistics that characterize the performance of a digital audio system. (Courtesy of Tektronix.)

Defining Terms

Clipping: A distortion mechanism in an audio system in which the input level to one or more devices or circuits is sufficiently high in amplitude that the maximum input level of the device or circuit is exceeded, resulting in significant distortion of the output waveform.

Crest factor: In an audio system, the ratio of the true peak to the RMS value. For any signal but an ideal square wave, the crest factor will be greater than one.

Group delay: The relative delay of the spectral components of a complex waveform. If the group delay is *flat*, all components arrive together.

Multitone testing: A measurement technique whereby an audio system is characterized by the application simultaneously of a combination of sine waves to a device under test.

Peak-equivalent sine: A method of specifying signal amplitude in an audio system, equal to the RMS level of a sine wave having the same peak-to-peak amplitude as the signal under consideration.

SMPTE IM: A method of measuring intermodulation distortion in an audio system standardized by the Society of Motion Picture and Television Engineers.

Weighting filter: A standardized filter used to impart predetermined characteristics to noise measurements in an audio system.

References

[1] "764 Digital Audio Monitor," Tektronix application note 21W7269, Tektronix, Beaverton, OR, 1992.
[2] Benson, K. B., (Ed.), *Audio Engineering Handbook*, McGraw-Hill, New York, 1988.
[3] Benson, K. B., and J. C. Whitaker, *Television and Audio Handbook for Technicians and Engineers*, McGraw-Hill, New York, 1990.
[4] Fink, D. G., and D. Christiansen, (Eds.), *Electronics Engineers' Handbook*, 3rd ed., McGraw-Hill, New York, 1989.
[5] Thompson, B., "Multitone Audio Testing," Tektronix application note 21W-7182, Tektronix, Beaverton, OR, 1992.

[6] Whitaker, J. C., *Interconnecting Electronic Systems*, CRC Press, Boca Raton, FL, 1992.
[7] Whitaker, J. C., *Maintaining Electronic Systems*, CRC Press, Boca Raton, FL, 1991.

Further Information

Most of the information presented in this chapter focuses on classic audio distortion mechanisms and analysis tools. There is also, however, an emerging discipline of digital audio analysis, which was only touched on briefly in this chapter. The best source for information on developing measurement techniques is the Audio Engineering Society (New York), which conducts trade shows and highly respected technical seminars annually in the U.S. and abroad. The society also publishes a monthly journal (*Journal of the Audio Engineering Society*) and numerous specialized books on the topic of audio technology in general, and digital audio in particular.

Readers are also directed to a well-respected book in the audio field, *Audio Engineering Handbook*, edited by K. B. Benson and published by McGraw-Hill. The publication date of the Benson work is 1988, and its coverage of digital audio is minimal. However, a second edition of the work, with extensive coverage of digital audio, is underway. Another title of interest is *Principles of Digital Audio*, 3rd edition, written by Ken Pohlmann and published by McGraw-Hill (1996).

21
Video Display Distortion Mechanisms and Analysis

21.1	Video Signal Spectra	21-1
	Minimum Video Frequency • Maximum Video Frequency • Horizontal Resolution	
	Video Frequencies Arising from Scanning	
21.2	Measurement of Color Displays	21-6
21.3	Assessment of Color Reproduction	21-8
	Chromatic Adaptation and White Balance • Overall Gamma Requirements • Perception of Color Differences	
21.4	Display Resolution and Pixel Format	21-9
	Contrast Ratio	
21.5	Applications of the Zone Plate Signal	21-11
	Simple Zone Plate Patterns • Producing the Zone Plate Signal • The Time (Motion) Dimension	
21.6	CRT Measurement Techniques	21-18
	Subjective CRT Measurements • Objective CRT Measurements	

Jerry C. Whitaker
Editor-in-Chief

New applications for electronic displays are pushing design engineers to produce systems that provide higher resolution, larger screen size, and better accuracy. More stringent performance demands improved quality assessment techniques in the manufacture, installation, and maintenance of display systems.

21.1 Video Signal Spectra

The spectrum of the video signal arising from the scanning process in a television camera extends from a lower limit determined by the time rate of change of the average luminance of the scene to an upper limit determined by the time during which the scanning spots cross the sharpest vertical boundary in the scene as focused within the camera. This concept is illustrated in Fig. 21.1. The distribution of spectrum components within these limits is determined by the following:

- The distribution of energy in the camera scanning spots
- Number of lines scanned per second
- Percentage of line-scan time consumed by horizontal blanking
- Number of fields scanned per second
- Rates at which the luminances and chrominances of the scene change in size, position, and boundary sharpness

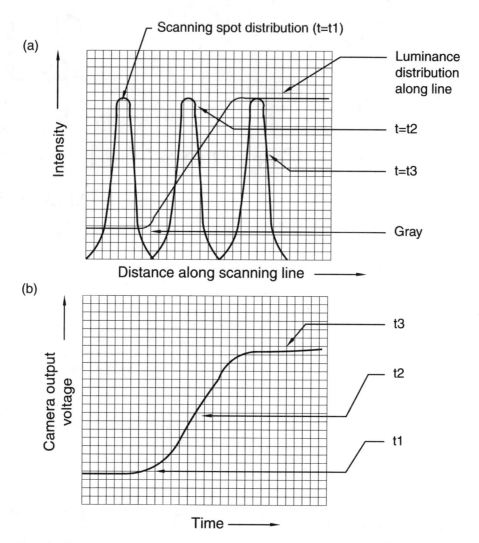

FIGURE 21.1 Video signal spectra: (a) camera scanning spot, shown with a Gaussian distribution, passing over a luminance boundary on a scanning line, (b) corresponding camera output signal resulting from convolution of the spot and luminance distributions.

To the extent that the contents and dynamic properties of the scene cannot be predicted, the spectrum limits and energy distribution are not defined. However, the spectra associated with certain static and dynamic test charts and tapes may be used as the basis for video system design. Among the configurations of interest are:

- Flat fields of uniform luminance and/or chrominance
- Fields divided into two or more segments of different luminance by sharp vertical, horizontal, or oblique boundaries

The latter case includes the horizontal and vertical wedges of test charts and the concentric circles of zone plate charts, illustrated in Fig. 21.2. The reproductions of such patterns typically display diffuse boundaries and other degradations that may be introduced by the camera scanning process, the amplitude and phase responses of the transmission system, the receiver scanning spots, and other artifacts associated with scanning.

The upper limit of the video spectrum actually employed in reproducing a particular image is most often determined by the amplitude-vs.-frequency and phase-vs.-frequency responses of the receiving

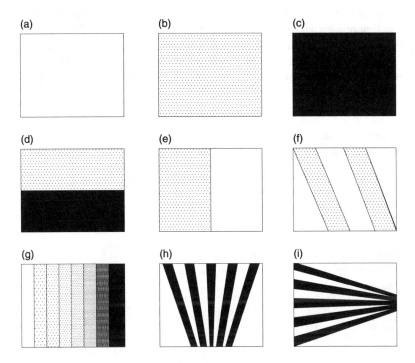

FIGURE 21.2 Scanning patterns of interest in analyzing video signals: (a), (b), (c) flat fields useful for determining color purity and transfer gradient (gamma); (d) horizontal half-field pattern for measuring low-frequency performance; (e) vertical half field for examining high-frequency transient performance; (f) display of oblique bars; (g) in monochrome, a tonal wedge for determining contrast and luminance transfer characteristics; in color, a display used for hue measurements and adjustments; (h) wedge for measuring horizontal resolution; (i) wedge for measuring vertical resolution.

system. These responses are selected as a compromise between the image sharpness demanded by typical viewers and the deleterious effects of noise, interference, and incomplete separation of the luminance and chrominance signals in the receiver.

Minimum Video Frequency

To reproduce a uniform value of luminance from top to bottom of an image scanned in the conventional interlaced fashion, the video signal spectrum must extend downward to include the field-scanning frequency. This frequency represents the lower limit of the spectrum arising from scanning an image whose luminance does not change. Changes in the average luminance are reproduced by extending the video spectrum to a lower frequency equal to the reciprocal of the duration of the luminance change. Because a given average luminance may persist for many minutes, the spectrum extends sensibly to zero frequency (DC). Various techniques of preserving or restoring the DC component are employed to extend the spectrum from the field frequency down to zero frequency.

Maximum Video Frequency

In the analysis of maximum operating frequency for a video system, three values must be distinguished:

1. Maximum output signal frequency generated by the camera or other pickup/generating device
2. Maximum modulating frequency corresponding to (1) the fully transmitted radiated sideband, or (2) the system used to convey the video signal from the source to the display
3. Maximum video frequency present at the picture tube (display) control electrodes

The maximum camera frequency is determined by the design and implementation of the imaging element. The maximum modulating frequency is determined by the extent of the video channel reserved for the fully transmitted sideband. The channel width, in turn, is chosen to provide a value of horizontal resolution approximately equal to the vertical resolution implicit in the scanning pattern. The maximum video frequency at the display is determined by the device and support circuitry of the display system.

Horizontal Resolution

The horizontal resolution factor is the proportionality factor between horizontal resolution and video frequency. It may be expressed as:

$$H_r = \frac{R_h}{\alpha} \times \iota$$

where:

H_r = horizontal resolution factor in lines per megahertz
R_h = lines of horizontal resolution per hertz of the video waveform
α = aspect ratio of the display
ι = active line period in microseconds

For NTSC, the horizontal resolution factor is:

$$78.8 = \frac{2}{4/3} \times 52.5$$

Video Frequencies Arising from Scanning

The signal spectrum arising from scanning comprises a number of discrete components at multiples of the scanning frequencies. Each spectrum component is identified by two numbers, m and n, which describe the pattern that would be produced if that component alone were present in the signal. The value of m represents the number of sinusoidal cycles of brightness measured horizontally (in the width of the picture) and the value of n represents the number of cycles measured vertically (in the picture height). The 0, 0 pattern is the DC component of the signal, the 0, 1 pattern is produced by the field-scanning frequency, and the 1, 0 pattern is produced by the line scanning frequency. Typical patterns for various values of m and n are shown in Fig. 21.3. By combining a number of such patterns (including m and n values up to several hundred), in the appropriate amplitudes and phases, any image capable of being represented by the scanning pattern may be built up. This is a two-dimensional form of the Fourier series.

The amplitudes of the spectrum components decrease as the values of m and n increase. Because m represents the order of the harmonic of the line-scanning frequency, the corresponding amplitudes are those of the left-to-right variations in brightness. A typical spectrum resulting from scanning a static scene is shown in Fig. 21.4. The components of major magnitude include:

- The DC component
- Field-frequency component
- Components of the line frequency and its harmonics

Surrounding each line-frequency harmonic is a cluster of components, each separated from the next by an interval equal to the field-scanning frequency.

It is possible for the clusters surrounding adjacent line-frequency harmonics to overlap one another. This corresponds to two patterns in Fig. 21.4 situated on adjacent vertical columns, which produce the same value of video frequency when scanned. Such "intercomponent confusion" of spectral energy is fundamental to the scanning process. Its effects are visible when a heavily striated pattern (such as that

Video Display Distortion Mechanisms and Analysis

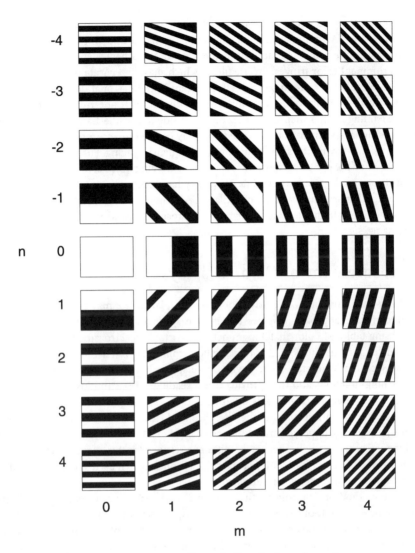

FIGURE 21.3 An array of image patterns corresponding to indicated values of m and n.

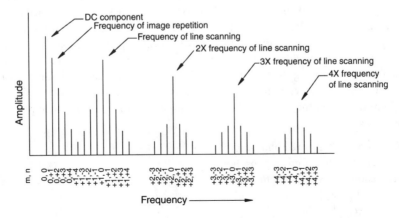

FIGURE 21.4 The typical spectrum of a video signal, showing the harmonics of the line-scanning frequency surrounded by clusters of components separated at intervals equal to the field-scanning frequency.

of a fabric having an accented weave) is scanned with the striations approximately parallel to the scanning lines. In the NTSC and PAL color systems, in which the luminance and chrominance signals occupy the same spectral region (one being interlaced in frequency with the other) such intercomponent confusion may produce prominent color fringes. Precise filters, which sharply separate the luminance and chrominance signals (comb filters), can remove this effect, except in the diagonal direction.

In static and slowly moving scenes, the clusters surrounding each line-frequency harmonic are compact, seldom extending further than 1 or 2 kHz on either side of the line-harmonic frequency. The space remaining in the signal spectrum is unoccupied and may be used to accommodate the spectral components of another signal having the same structure and frequency spacing. For scenes in which the motion is sufficiently slow for the eye to perceive the detail of the moving objects, it may be safely assumed that less than half the spectral space between line-frequency harmonics is occupied by energy of significant magnitude. It is on this principle that the NTSC and PAL compatible color television systems are based. The SECAM system uses frequency-modulated chrominance signals, which are not frequency interlaced with the luminance signal.

21.2 Measurement of Color Displays

The chromaticity and luminance of a portion of a color display device may be measured in several ways. The most fundamental approach involves a complete spectroradiometric measurement followed by computation using tables of color-matching functions. Portable spectroradiometers with built-in computers are available for this purpose. Another method, somewhat faster but less accurate, involves the use of a photoelectric colorimeter. These devices have spectral sensitivities approximately equal to the CIE color-matching functions and thus provide direct readings of tristimulus values.

For setting up the reference white, it is often simplest to use a split-field visual comparator and to adjust the display device until it matches the reference field (usually D65) of the comparator. However, because there is usually a large spectral difference (large metamerism) between the display and the reference, different observers will often make different settings by this method. Thus, settings by one observer—or a group of observers—with normal color vision are often used simply to provide a reference point for subsequent photoelectric measurements.

An alternative method of determining the luminance and chromaticity coordinates of any area of a display involves measurement of the output of each phosphor separately and combining them using the center of gravity law in which the total tristimulus output of each phosphor is considered as an equivalent weight located at the chromaticity coordinates of the phosphor.

Consider the CIE chromaticity diagram shown in Fig. 21.5 to be a uniform flat surface positioned in a horizontal plane. For the case illustrated, the center of gravity of the three weights (Tr, Tg, Tb), or the balance point, will be at the point Co. This point determines the chromaticity of the mixture color. The luminance of the color Co will be the linear sum of the luminance outputs of the red, green, and blue phosphors. The chromaticity coordinates of the display primaries may be obtained from the manufacturer. The total tristimulus output of one phosphor may be determined by turning off the other two CRT guns, measuring the luminance of the specified area, and dividing this value by the y chromaticity coordinate of the energized phosphor. This procedure is then repeated for the other two phosphors. From these data the color resulting from given excitations of the three phosphors may be calculated as follows:

Chromaticity coordinates of red phosphor = xr, yr
Chromaticity coordinates of green phosphor = xg, yg
Chromaticity coordinates of blue phosphor = xb, yb
Luminance of red phosphor = Yr
Luminance of green phosphor = Yg
Luminance of blue phosphor = Yb

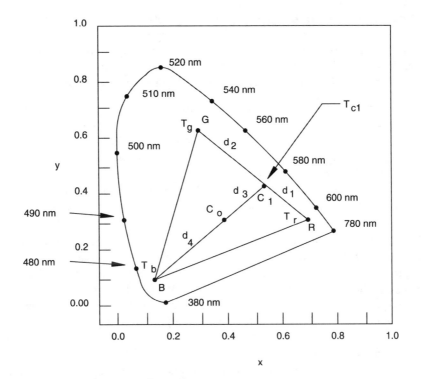

FIGURE 21.5 The CIE 1931 chromaticity diagram illustrating use of the center of gravity law ($T_r d_1 = T_g d_2$, $T_{c1} = T_r + T_g$, $T_{c1} d_3 = T_b d_4$).

$$\text{Total tristimulus value of red phosphor} = X_r + Y_r + Z_r = \frac{Y_r}{y_r} = T_r$$

$$\text{Total tristimulus value of green phosphor} = X_g + Y_g + Z_g = \frac{Y_g}{y_g} = T_g$$

$$\text{Total tristimulus value of blue phosphor} = X_b + Y_b + Z_b = \frac{Y_b}{y_b} = T_b$$

Consider T_r as a weight located at the chromaticity coordinates of the red phosphor and T_g as a weight located at the chromaticity coordinates of the green phosphor. The location of the chromaticity coordinates of color C1 (blue gun of color CRT turned off) can be determined by taking moments along line RG to determine the center of gravity of weights T_r and T_g:

$$T_r \times d_1 = T_g \times d_2$$

The total tristimulus value of C1 is equal to $T_r + T_g = T_{C1}$. Taking moments along line C1B will locate the chromaticity coordinates of the mixture color Co:

$$T_{C1} \times d_3 = T_b \times d_4$$

The luminance of the color Co is equal to $Y_r + Y_g + Y_b$.

21.3 Assessment of Color Reproduction

A number of factors may contribute to poor color rendition in a display system. To assess the effect of these factors, it is necessary to define system objectives and then establish a method of measuring departures from the objectives. Visual image display may be categorized as follows:

- *Spectral color reproduction:* the exact reproduction of the spectral power distributions of the original stimuli. Clearly this is not possible in a video system with three primaries.
- *Exact color reproduction:* the exact reproduction of tristimulus values. The reproduction is then a metameric match to the original. Exact color reproduction will result in equality of appearance only if the viewing conditions for the picture and the original scene are identical. These conditions include the angular subtense of the picture, the luminance and chromaticity of the surround, and glare. In practice, exact color reproduction often cannot be achieved because of limitations on the maximum luminance that can be produced on a color monitor.
- *Colorimetric color reproduction:* a variant of exact color reproduction in which the tristimulus values are proportional to those in the original scene. In other words, the chromaticity coordinates are reproduced exactly, but the luminances are all reduced by a constant factor. Traditionally, color video systems have been designed and evaluated for colorimetric color reproduction. If the original and the reproduced reference whites have the same chromaticity, if the viewing conditions are the same, and if the system has an overall gamma of unity, colorimetric color reproduction is indeed a useful criterion. However, these conditions often do not hold and then colorimetric color reproduction is inadequate.
- *Equivalent color reproduction:* the reproduction of the original color appearance. This might be considered as the ultimate objective but cannot be achieved because of the limited luminance that can be generated in a display system.
- *Corresponding color reproduction:* a compromise in which colors in the reproduction have the same appearance as the colors in the original would have had if they had been illuminated to produce the same average luminance level and the same reference white chromaticity as that of the reproduction. For most purposes, corresponding color reproduction is the most suitable objective of a color video system.
- *Preferred color reproduction:* a departure from the preceding categories that recognizes the preferences of the viewer. It is sometimes argued that corresponding color reproduction is not the ultimate aim for some display systems, such as color television, but that account should be taken of the fact that people prefer some colors to be different from their actual appearance. For example, sun-tanned skin color is preferred to average real skin color, and sky is preferred bluer and foliage greener than they really are.

Even if corresponding color reproduction is accepted as the target, it is important to remember that some colors are more important than others. For example, flesh tones must be acceptable—not obviously reddish, greenish, purplish, or otherwise incorrectly rendered. Similarly the sky must be blue and the clouds white, within the viewer's range of acceptance. Similar conditions apply to other well-known colors of common experience.

Chromatic Adaptation and White Balance

With properly adjusted cameras and displays, whites and neutral grays are reproduced with the chromaticity of D65. Tests have shown that such whites (and grays) appear satisfactory in home viewing situations even if the ambient light is of quite different color temperature. Problems occur, however, when the white balance is slightly different from one camera to the next or when the scene shifts from studio to daylight or vice versa. In the first case, unwanted shifts of the displayed white occur, whereas in the other, no shift occurs even though the viewer subconsciously expects a shift.

By always reproducing a white surface with the same chromaticity, the system is mimicking the human visual system, which adapts so that white surfaces always appear the same whatever the chromaticity of the illuminant (at least within the range of common light sources). The effect on other colors, however, is more complicated. In video cameras the white balance adjustment is usually made by gain controls on the R, G, and B channels. This is similar to the von Kries model of human chromatic adaptation, although the R, G, and B primaries of the model are not the same as the video primaries. It is known that the von Kries model does not accurately account for the appearance of colors after chromatic adaptation, and so it follows that simple gain changes in a video camera are not the ideal approach. Nevertheless, this approach seems to work well in practice, and the viewer does not object to the fact, for example, that the relative increase in the luminances of reddish objects in tungsten light is lost.

Overall Gamma Requirements

Colorimetric color reproduction requires that the overall gamma of the system, including the camera, the display, and any gamma-adjusting electronics, be unity. This simple criterion is the one most often used in the design of a video color rendition system. However, the more sophisticated criterion of corresponding color reproduction takes into account the effect of the viewing conditions. In particular, several studies have shown that the luminance of the surround is important. For example, a dim surround requires a gamma of about 1.2, and a dark surround requires a gamma of about 1.5 for optimum color reproduction.

Perception of Color Differences

The CIE 1931 chromaticity diagram does not map chromaticity on a uniform-perceptibility basis. A just-perceptible change of chromaticity is not represented by the same distance in different parts of the diagram. Many investigators have explored the manner in which perceptibility varies over the diagram. The most often quoted study is that of MacAdam [28], who identified a set of ellipses that are contours of equal perceptibility about a given color, as shown in Fig. 21.6.

From this and similar studies it is apparent, for example, that large distances represent relatively small perceptible changes in the green sector of the diagram. In the blue region, much smaller changes in the chromaticity coordinates are readily perceived.

Further, viewing identical images on dissimilar displays can result in observed differences in the appearance of the image [5]. There are two factors that affect the appearance of the image: (1) physical factors, including display gamut, illumination level, and black point; and (2) psychophysical factors, including chromatic induction and color constancy.

Each of these factors interacts in such a way that prediction of the appearance of an image on a given display becomes difficult. System designers have experimented with colorimetry to facilitate the sharing of image data among display devices that vary in manufacture, calibration, and location. Of particular interest is the application of colorimetry to imaging in a networked window system environment, where it is often necessary to assure that an image displayed remotely looks like the image displayed locally.

Studies have indicated that image context and image content are also factors that affect color appearance. The use of highly chromatic backgrounds in a windowed display system is popular, but will impact the appearance of the colors in the foreground.

21.4 Display Resolution and Pixel Format

The pixel represents the smallest resolvable element of a display. The size of the pixel varies from one type of display to another. In a monochrome CRT, pixel size is determined primarily by the following factors:

- Spot size of the electron beam (the current density distribution)
- Phosphor particle size
- Thickness of the phosphor layer

The term *pixel* was developed in the era of monochrome television, and the definition was—at that time—straightforward. With the advent of color triad-based CRTs and solid state display systems, the definition is not nearly so clear.

For a color CRT, a single triad of red, green, and blue phosphor dots constitutes a single pixel. This definition assumes that the mechanical and electrical parameters of the CRT will permit each triad to be addressed without illuminating other elements on the face of the tube. Most display systems, however, will not meet this criteria. Depending on the design, a number of triads may constitute a single pixel in a CRT display. A more all-inclusive definition for the pixel is the smallest spatial-information element as seen by the viewer.

Dot pitch is one of the principal mechanical criteria of a CRT that determines—to a large extent—the resolution of the display. Dot pitch is defined as the center-to-center distance between adjacent green phosphor dots of the red, green, blue triad.

The *pixel format* is the arrangement of pixels into horizontal rows and vertical columns. For example, an arrangement of 640 horizontal pixels by 480 vertical pixels results in a 640 × 480 pixel format. This description is not a resolution parameter by itself, simply the arrangement of pixel elements on the screen. *Resolution* is the measure of the ability to delineate picture detail; it is the smallest discernible and measurable detail in a visual presentation.

Pixel density is a parameter closely related to resolution, stated in terms of pixels per linear distance. Pixel density specifies how closely the pixel elements are spaced on a given display. It follows that a display with a given pixel format will not provide the same pixel density (resolution) on a large size screen—such as 19-in. diagonal—as on a small size screen—such as 12-in. diagonal.

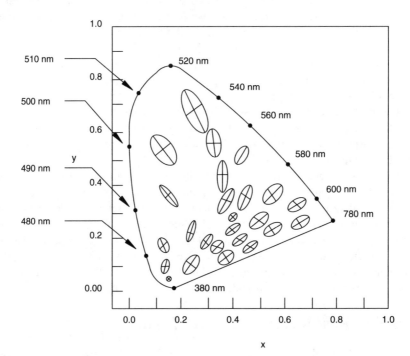

FIGURE 21.6 Ellipses of equally perceptible color differences.

Television lines is another term used to describe resolution. The term refers to the number of discernible lines on a standard test chart. As before, the specification of television lines is not by itself a description of display resolution. A 525-line display on a 17-in. monitor will appear to have greater resolution to a viewer than a 525-line display on a 30-in. monitor. Pixel density is the preferred resolution parameter.

Contrast Ratio

The purpose of a video display is to convey information by controlling the illumination of phosphor dots on a screen, or by controlling the reflectance or transmittance of a light source. The contrast ratio specifies the observable difference between a pixel that is switched on and its corresponding off state:

$$C_r = \frac{L_{on}}{L_{off}}$$

where:

C_r = contrast ratio of the display
L_{on} = luminance of a pixel in the on state
L_{off} = luminance of a pixel in the off state

The area encompassed by the contrast ratio is an important parameter when considering the performance of a display. Two contrast ratio divisions are typically specified:

- Small area: comparison of the on and off states of a pixel-sized area
- Large area: comparison of the on and off states of a group of pixels

For most display applications, the small area contrast ratio is the more critical parameter.

21.5 Applications of the Zone Plate Signal

The increased information content of advanced, high-definition display systems requires sophisticated processing to make recording and transmission practical [10]. This processing uses various forms of bandwidth compression, scan rate changes, motion detection and compensation algorithms, and other new techniques. Zone plate patterns are well suited to exercising a complex video system in the three dimensions of its signal spectrum: horizontal, vertical, and temporal. Zone plate signals, unlike most conventional test signals, can be complex and dynamic. Because of this, they are capable of simulating much of the detail and movement of actual picture video, exercising the system under test with signals representative of the intended application. These digitally generated and controlled signals also have other important characteristics needed in test signals.

A signal intended for meaningful testing of a video system must be carefully controlled, so that any departure from a known parameter of the signal is attributable to a distortion or other change in the system under test. The test signal must also be predictable, so it can be reproduced accurately at other times or places. These constraints have usually led to test signals that are electronically generated. In a few special cases, a standardized picture has been televised by a camera or monoscope—usually for a subjective, but more detailed, evaluation of system performance. A zone plate is a physical optical pattern, which was first used by televising it in this way. Now that electronic generators are capable of producing similar patterns, the label "zone plate" is applied to the wide variety of patterns created by video test instruments.

Conventional test signals, for the most part limited by the practical considerations of electronic generation, have represented relatively simple images. Each signal is capable of testing a narrow range of possible distortions; several test signals are needed for a more complete evaluation. Even with several signals, this method may not reveal all possible distortions, or allow study of all pertinent characteristics. This is especially true in video systems employing new forms of sophisticated signal processing.

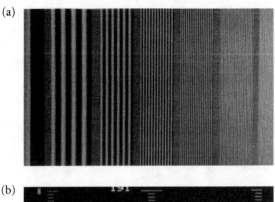

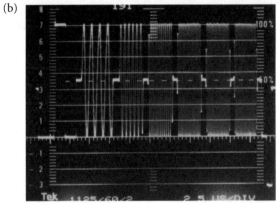

FIGURE 21.7 Multiburst video test waveform: (a) picture display, (b) multiburst signal as viewed on a waveform monitor (1H). (Courtesy of Tektronix.)

Simple Zone Plate Patterns

The basic testing of a video communication channel historically has involved the application of several single frequencies—in effect, spot checking the spectrum of interest [10]. A well known and quite practical adaptation of this idea is the multiburst signal, as shown in Fig. 21.7. This test waveform has been in use since the earliest days of video. The multiburst signal provides several discrete frequencies along a TV line.

The frequency sweep signal is an improvement on multiburst. While harder to implement in earlier generators, it was easier to use. The frequency sweep signal, illustrated in Fig. 21.8, varies the applied signal frequency continuously along the TV line.[1] In some cases, the signal is swept as a function of the vertical position (field time). Even in these cases, the signal being swept is appropriate for testing the spectrum of the horizontal dimension of the picture.

Figure 21.9 shows the output of a zone plate generator configured to produce a horizontal single frequency output. Figure 21.10 shows a zone plate generator configured to produce a frequency sweep signal. Electronic test patterns, such as these, may be used to evaluate the following system characteristics:

- Channel frequency response
- Horizontal resolution
- Moir, effects in recorders and displays
- Other impairments

---

[1] Figure 21.8(a) and other photographs in this section show the "beat" effects introduced by the screening process used for photographic printing. This is largely unavoidable. The screening process is quite similar to the scanning or sampling of a television image—the patterns are designed to identify this type of problem.

Video Display Distortion Mechanisms and Analysis

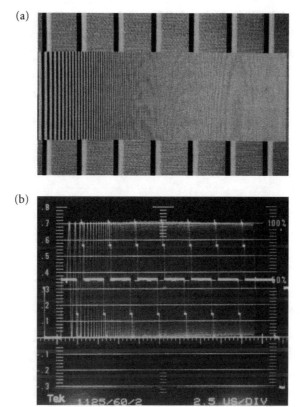

FIGURE 21.8 Conventional sweep frequency test waveform: (a) picture display, (b) waveform monitor display, with markers (1H). (Courtesy of Tektronix.)

Patterns that test vertical (field) response have—traditionally— been used less frequently. As new technologies implement conversion from interlaced to progressive scan, line doubling display techniques, vertical anti-aliasing filters, scan conversion, motion detection, or other processes that combine information from line to line, vertical testing patterns will be more in demand.

In the vertical dimension, as well as the horizontal, tests may be done at a single frequency or with a frequency sweep signal. Figure 21.11 illustrates a magnified vertical rate waveform display. Each "dash" in the photo represents one horizontal scan line. Sampling of vertical frequencies is inherent in the scanning process, and the photo shows the effects on the signal waveform. Note also that the signal voltage remains constant during each line, and changes only from line to line in accord with the vertical dimension sine function of the signal. Figure 21.12 shows a vertical frequency sweep picture display.

The horizontal and vertical sinewaves and sweeps are quite useful, but they do not use the full potential of a zone plate signal source.

Producing the Zone Plate Signal

A zone plate signal is created in real time by the test signal generator [10]. The value of the signal at any instant is represented by a number in the digital hardware. This number is incremented as the scan progresses through the three dimensions that define a point in the video image; horizontal position, vertical position, and time.

The exact method in which these dimensions alter the number is controlled by a set of coefficients. These coefficients determine the initial value of the number, and control the size of the increments as the scan progresses along each horizontal line, from line to line vertically, and from field to field

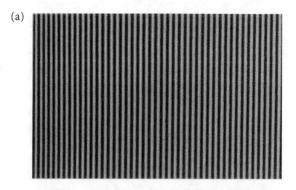
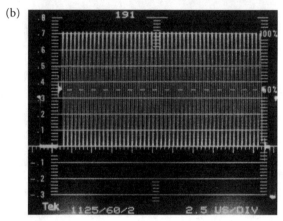

FIGURE 21.9 Single horizontal frequency test signal from a zone plate generator: (a) picture display, (b) waveform monitor display (1H). (Courtesy of Tektronix.)

temporally. A set of coefficients uniquely determines a pattern, or a sequence of patterns when the time dimension is active.

This process produces a sawtooth waveform; overflow in the accumulator holding the signal number effectively resets the value to zero at the end of each cycle of the waveform. Usually it is desirable to minimize the harmonic energy content of the output signal; in this case, the actual output is a sine function of the number generated by the incrementing process.

Complex Patterns

A pattern of sinewaves or sweeps in multiple dimensions may be produced, using the unique architecture of the zone plate generator [10]. The pattern shown in Fig. 21.13, for example, is a single signal sweeping both horizontally and vertically. Figure 21.14 shows the waveform of a single selected line (line 263 in the 1125/60/2 HDTV system). Note that the horizontal waveform is identical to Fig. 21.8(b), even though the vertical dimension sweep is active now as well. Actually, different lines will give slightly different waveforms. The horizontal frequency and sweep characteristics will be identical, but the starting phase must be different from line to line to construct the vertical signal.

Figure 21.15 shows a two-axis sweep pattern that is most often identified with zone plate generators; perhaps because it quite closely resembles the original optical pattern. In this circle pattern, both horizontal and vertical frequencies start high, sweep to zero (in the center of the screen), and sweep up again to the end of their respective scans. The concept of two-axis sweeps is actually more powerful than it might, at first, appear. In addition to purely horizontal or vertical effects, there are possible distortions or artifacts that are only apparent with simultaneous excitation in both axes. In other words, the response of a system to diagonal detail may not be predictable from information taken from the horizontal and vertical responses.

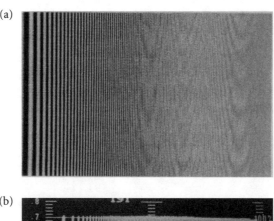

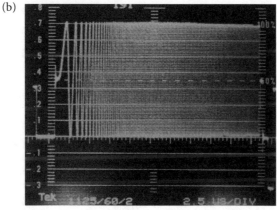

FIGURE 21.10 Horizontal frequency sweep test signal from a zone plate generator: (a) picture display, (b) waveform monitor display (1H). (Courtesy of Tektronix.)

Consider an example from NTSC. A comb filter will suppress crosscolor effects from a horizontal luminance frequency signal near the subcarrier frequency (such as the higher frequency packets in an NTSC multiburst signal). If, however, the right vertical component is added, creating a specific diagonal luminance pattern, even the most complex decoders will interpret the pattern as colored. In this case, the two-axis sweep shows very different effects than the same sweeps shown individually. A multi-dimensional sweep is a powerful tool for identifying analogous effects in other complex signal processing systems.

The Time (Motion) Dimension

Incrementing the number in the accumulator of the zone plate generator from frame to frame (or field to field in an interlaced system) creates a predictably different pattern for each vertical scan [10]. This, in turn, creates apparent motion and exercises the signal spectrum in the temporal dimension. Analogous to the single frequency and frequency sweep examples given previously, appropriate setting of the time related coefficients will create constant motion or motion sweep (acceleration).

Specific motion detection and interpolation algorithms in a system under test may be exercised by determining the coefficients of a critical sequence of patterns. These patterns may then be saved for subsequent testing during development or adjustment. In an operational environment, appropriate response to a critical sequence could ensure expected operation of the equipment or facilitate fault detection.

While motion artifacts are difficult to portray in the still image constraints of a printed book, the following example gives some idea of the potential of a versatile generator. In Fig. 21.16 the vertical sweep maximum frequency has been increased to the point where it is zero-beating with the scan at the bottom

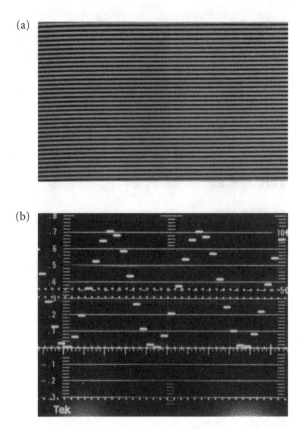

FIGURE 21.11 Single vertical frequency test signal: (a) picture display, (b) magnified vertical rate waveform, showing the effects of scan sampling. (Courtesy of Tektronix.)

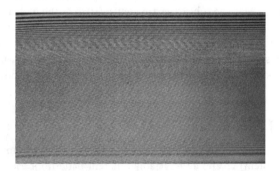

FIGURE 21.12 Vertical frequency sweep picture display. (Courtesy of Tektronix.)

of the screen. (The cycles per picture height of the pattern matches the lines per picture height per field of the scan.) Actually, in direct viewing, there is another noticeable artifact in the vertical center of the screen; it is a harmonic beat related to the gamma of the display CRT. Because of interlace, this beat flickers at the field rate. The photograph integrates the interfield flicker and thereby hides the artifact, which is readily apparent when viewed in real time.

Figure 21.17 is identical to the previous photo, except for one important difference—upward motion of $1/2$ cycle per field has been added to the pattern. Now the sweep pattern itself is integrated out, as is the first order beat at the bottom. The harmonic effects in center screen no longer flicker because the change of scan vertical position from field to field is compensated by a change in position of the image.

Video Display Distortion Mechanisms and Analysis

FIGURE 21.13 Combined horizontal and vertical frequency sweep picture display. (Courtesy of Tektronix.)

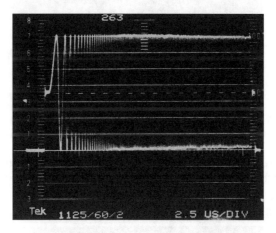

FIGURE 21.14 Combined horizontal and vertical frequency sweeps, selected line waveform display (1H). This figure shows the maintenance of horizontal structure in the presence of vertical sweep. (Courtesy of Tektronix.)

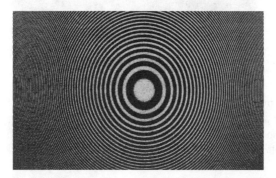

FIGURE 21.15 The best known zone plate pattern, combined horizontal and vertical frequency sweeps with zero frequency in the center screen. (Courtesy of Tektronix.)

The resulting beat pattern does not flicker and is easily photographed or, perhaps, scanned to determine depth of modulation.

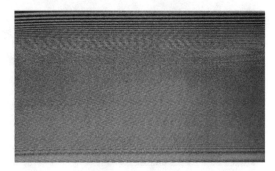

FIGURE 21.16 Vertical frequency sweep picture display. (Courtesy of Tektronix.)

FIGURE 21.17 The same vertical sweep as Fig. 21.16, except that appropriate pattern motion has been added to "freeze" the beat pattern in the center screen for photography or other analysis. (Courtesy of Tektronix.)

FIGURE 21.18 A hyperbolic variation of the two-axis zone plate frequency sweep. (Courtesy of Tektronix.)

A change in coefficients produces hyperbolic, rather than circular, two-axis patterns, as shown in Fig. 21.18. Another interesting pattern, which has been suggested for checking complex codecs, is shown in Fig. 21.19. This is also a moving pattern, which was slightly altered to freeze some aspects of the movement to take the photograph.

21.6 CRT Measurement Techniques

A number of different techniques have evolved for measuring the static performance of CRT display devices and systems [41]. Most express the measured device performance in a unique figure of merit or

Video Display Distortion Mechanisms and Analysis

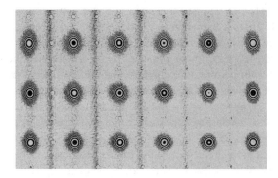

FIGURE 21.19 A two-axis frequency sweep in which the range of frequencies is swept several times in each axis. Complex patterns such as this may be created for specific test requirements. (Courtesy of Tektronix.)

metric. While each approach provides useful information, the lack of standardization in measurement techniques makes it difficult or impossible to directly compare the performance of a given class of CRT.

Regardless of the method used to measure performance, the operating parameters must be set for the anticipated operating environment. Key parameters include:

- Input signal level
- System/display line rate
- Luminance (brightness)
- Contrast
- Image size
- Aspect ratio

If the display is used in more than one environmental condition—such as under day and night conditions—a set of measurements is appropriate for each application.

Subjective CRT Measurements

Three common subjective measurement techniques are used to assess the performance of a CRT:

- Shrinking raster
- Line width
- TV limiting resolution

Predictably, subjective measurements tend to exhibit more variability than objective measurements. While not generally used for acceptance testing or quality control, subjective CRT measurements provide a fast and relatively simple means of performance assessment. Results are usually consistent when performed by the same observer. However, results for different observers are often not consistent because different observers use different visual criteria to make their judgments.

The shrinking raster and line width techniques are used to estimate the vertical dimension of the CRT beam spot size (*footprint*). There are several underlying assumptions with this approach:

- The spot is assumed to be symmetrical and Gaussian in the vertical and horizontal planes.
- The display MTF calculated from the spot size measurement results in the best performance envelope that can be expected from the CRT.
- The modulating electronics are designed with sufficient bandwidth so that spot size is the limiting performance parameter.
- The modulation contrast at low spatial frequencies approaches 100%.

Depending on the application, not all of these assumptions are valid.

Assumption 1. Verona [41] has reported that the symmetry assumption is generally not true. The vertical spot profile is only an approximation to the horizontal spot profile; most spot profiles exhibit some degree of astigmatism. However, significant deviations from the symmetry and Gaussian assumptions result in only minor deviations from the projected performance when the assumptions are correct.

Assumption 2. The optimum performance envelope assumption infers that other types of measurements will result in the same or lower modulation contrast at each spatial frequency. The MTF calculations based on a beam footprint in the vertical axis indicate the optimum performance that can be obtained from the display because finer detail (higher spatial frequency information) cannot be written onto the screen smaller than the spot size.

Assumption 3. The modulation circuit bandwidth must be sufficient to pass the full incoming video signal. Typically, the video circuit bandwidth is not a problem with current technology circuits, which are usually designed to provide significantly more bandwidth than the CRT is capable of displaying. However, in cases where this assumption is not true, the calculated MTF based purely on the vertical beam profile will be incorrect. The calculated performance will be better than the actual performance of the display.

Assumption 4. The calculated MTF is normalized to 100% modulation contrast at zero spatial frequency and ignores the light scatter and other factors that degrade the actual measured MTF. Independent modulation contrast measurements at a low spatial frequency can be used to adjust the MTF curve to correct for the normalization effects.

Shrinking Raster Method

The shrinking raster measurement procedure is relatively simple [41]. Steps include the following:

- The brightness and contrast controls are set for the desired peak luminance with an active raster background luminance (1% of peak luminance) using a stair-step video signal.
- While displaying a flat field video signal input corresponding to the peak luminance, the vertical gain/size is reduced until the raster lines are barely distinguishable.
- The raster height is measured and divided by the number of active scan lines to estimate the average height of each scan line. The number of active scan lines is typically 92% of the line rate. (For example, a 525-line display has 480 active lines, an 875-line display has 817 active lines, and a 1025-line display has 957 active lines.)

The calculated average line height is typically used as a stand-alone metric of display performance.

The most significant shortcoming of the shrinking raster method is the variability introduced through the determination of when the scan lines are *barely distinct* to the observer. Blinking and eye movements often enhance the distinctness of the scan lines; lines that were indistinct become distinct again. Further, while the shrinking raster procedure can produce acceptable results on large format CRT displays, it is less accurate for miniature devices (1-in. diameter and less). The nominal line spacing for a 1-in. CRT operating with full raster (at 875-line rate video) is approximately 15.6 to 15.8 μm. Because the half-intensity spot width is about 20 to 22 μm, there is a significant amount of spot overlap between the scan lines before the raster height is reduced.

Line Width Method

The line-width measurement technique requires a microscope with a calibrated graticule [41]. The focused raster is set to a 4:3 aspect ratio and the brightness and contrast controls are set for the desired peak luminance with an active raster background luminance (1% of peak luminance) using a stair-step video signal. A single horizontal line at the anticipated peak operating luminance is presented in the center of the display. The spot is measured by comparing its luminous profile with the graticule markings. As with the shrinking raster technique, determination of the line edge is subjective.

TV Limiting Resolution Method

This technique involves the display of two-dimensional, high contrast bar patterns or lines of various size, spacing, and angular orientation [41]. The observer subjectively determines the limiting resolution of the image by the smallest set of bars that can be resolved. There are several potential errors with this technique, including:

- A phenomenon called *spurious resolution* can occur which leads the observer to overestimate the limiting resolution. Spurious resolution occurs beyond the actual resolution limits of the display. It appears as fine structures that can be perceived as line spacings closer than the spacing at which the bar pattern first completely blurs. This situation arises when the frequency response characteristics fall to zero, then go negative, and perhaps oscillate as they die out. At the bottom of the negative trough, contrast is restored, but in reverse phase (white becomes black and black becomes white).
- The use of test charts imaged with a video camera can lead to incorrect results because of the addition of camera resolution considerations to the measurement. Electronically generated test patterns are more reliable image sources.

The proper setting of brightness and contrast is required for this measurement. Brightness and contrast controls are adjusted for the desired peak luminance with an active raster background luminance (1% of peak luminance) using a stair-step video signal. Too much contrast will result in an inflated limiting resolution measurement; too little contrast will result in a degraded limiting resolution measurement.

Electronic resolution pattern generators typically provide a variety of resolution signals from 100 to 1000 TV lines/picture height (TVL/PH) or more in a given multiple (such as 100). Figure 21.20 illustrates an electronically generated resolution test pattern for high definition video applications.

Application Considerations

The subjective techniques discussed in this section, with the exception of TV limiting resolution, measure the resolution of the *display* [41]. The TV pattern test measures *image* resolution, which is quite different.

Consider as an example a video display in which the scan lines can just be perceived—about 480 scan lines per picture height. This indicates a *display* resolution of at least 960 TV lines—counting light *and* dark lines, per the convention. If a pattern from an electronic generator is displayed, observation will show the image beginning to deteriorate at about 340 TV lines. This characteristic is the result of beats between the image pattern and the raster, with the beat frequency decreasing as the pattern spatial frequency approaches the raster spatial frequency. This ratio of 340/480 = 0.7 (approximately) is known as the *Kell factor*. Although debated at length, the factor does not change appreciably in subjective observations.

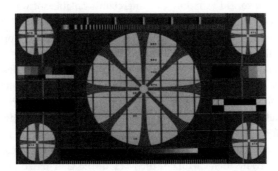

FIGURE 21.20 Wide aspect ratio resolution test chart produced by an electronic signal generator. (Courtesy of Tektronix.)

Objective CRT Measurements

Four common types of objective measurements may be performed to assess the capabilities of a CRT [41]:

1. Half power width
2. Impulse Fourier transform
3. Knife edge Fourier transform
4. Discrete frequency

While more difficult to perform than the subjective measurements discussed thus far, objective CRT measurement techniques offer greater accuracy and better repeatability. Some of the procedures require specialized hardware and/or software.

Half Power Width Method

The half power width technique is appropriate for large displays (9 in. and larger), but is not reliable when used to measure line width on a miniature CRT [41]. A single horizontal line is activated with the brightness and contrast controls set to a typical operating level (as discussed previously). The line luminance is equivalent to the highlight luminance (maximum signal level). The central portion of the line is imaged with a microscope in the plane of a variable width slit. The open slit allows all the light from the line to pass through to a photodetector. The output of the photodetector is displayed on an oscilloscope. As the slit is gradually closed, the peak amplitude of the photodetector signal decreases. When the signal drops to 50% of its initial value, the slit width is recorded. The width measurement divided by the microscope magnification represents the *half power width* of the horizontal scan line. The half power width of a miniature CRT may be measured using a scanning microphotometer and software to perform numerical integration on the luminance profile.

The half power width is defined as the distance between symmetrical integration limits, centered about the maximum intensity point, which encompasses half of the total power under the intensity curve. The half power width is not the same as the half intensity width measured between the half intensity points. The half intensity width is theoretically 1.75 times greater than the half power width for a Gaussian spot luminance distribution.

It should be noted that the half power line width technique relies on line width to predict the performance of the CRT. Many of the precautions outlined in the previous section apply here as well. The primary difference, however, is that line width is measured objectively, rather than subjectively.

Fourier Transform Methods

The impulse Fourier transform technique involves measuring the luminance profile of the spot and then taking the Fourier transform of the distribution to obtain the MTF [41]. The MTF, by definition, is the Fourier transform of the line spread function. Commercially available software may be used to perform these measurements using either an impulse or knife edge as the input waveform. Using the vertical spot profile as an approximation to the horizontal spot profile is not always appropriate, and the same reservations expressed in the previous section apply in this case.

The measurement is made by generating a single horizontal line on the display with the brightness and contrast set as discussed previously. A microphotometer with an effective slit aperture width approximately one tenth the width of the scan line is moved across the scan line (the long slit axis parallel to the scan line). The data taken is stored in an array, which represents the luminance profile of the CRT spot, distance vs luminance. The microphotometer is calibrated for luminance measures and for distance measures in the object plane. Each micron step of the microphotometer represents a known increment in the object plane. The software then calculates the MTF of the CRT based on its line spread from the calibrated luminance and distance measurements. Finite slit width corrections may also be made to the MTF curve by dividing it by a measurement system MTF curve obtained from the luminance profile of an ideal knife edge aperture or a standard source.

The knife edge Fourier transform measurement may be conducted using a low spatial frequency vertical bar pattern (5 to 10 cycles) across the display with the brightness and contrast controls set as discussed previously. The frequency response of the square wave pattern generator and video pattern generator should be greater than the frequency response of the display system (100 MHz is typical). The microphotometer scans from the center of a bright bar to the center of a dark bar (left to right), measuring the width of the boundary and comparing it to a knife edge. The microphotometer slit is oriented vertically, with its long axis parallel to the bars. The scan is usually made from a light bar to a dark bar in the direction of spot movement. This procedure is preferred because waveforms from scans in the opposite direction may contain anomalies. When the beam is turned on in a square wave pattern, it tends to overshoot and oscillate. This behavior produces artifacts in the luminance profile of the bar edge as the beam moves from an off to an on state. In the on-to-off direction, however, the effects are minimal and the measured waveform does not exhibit the same anomalies that can corrupt the MTF calculations.

The bar edge (knife edge) measurement, unlike the other techniques discussed thus far, uses the horizontal spot profile to predict display performance. All of the other techniques use the vertical profile as an approximation of the more critical horizontal spot profile. The bar edge measurement will yield a more accurate assessment of display performance because the displayed image is being generated with a spot scanned in the horizontal direction.

Discrete Frequency Method

The discrete sine wave frequency response measurement technique provides the most accurate representation of display performance [41]. With this approach there are no assumptions implied about the shape of the spot, the electronics bandwidth, or low frequency light scatter. Discrete spatial frequency sine wave patterns are used to obtain a discrete spatial frequency MTF curve that represents the signal-in to luminance-out performance of the display.

The measurement begins by setting the brightness and contrast as discussed previously, with black level luminance at 1% of the highlight luminance. A sine wave signal is produced by a function generator and fed to a pedestal generator where it is converted to an RS-170A or RS-343 signal, which is then applied to the CRT. The modulation and resulting spatial frequency pattern is measured with a scanning microphotometer. The highlight and black level measurements are used to calculate the modulation constant for each spatial frequency from 5 cycles/display width to the point that the modulation constant falls to less than 1%. The modulation constant values are then plotted as a function of spatial frequency, generating a discrete spatial frequency MTF curve.

Defining Terms

Corresponding color reproduction: The characteristics of a system in which colors in the reproduction have the same appearance that colors in the original would have if they had been illuminated to produce the same average luminance level and the same reference white chromaticity as that of the reproduction. For most purposes, corresponding color reproduction is the most suitable objective of a color video system.

Dot pitch: The center-to-center distance between adjacent green phosphor dots of the red, green, blue triad in a color display.

Pixel: The smallest spatial-information element as seen by the viewer in an imaging system.

Pixel density: A parameter that specifies how closely the pixel elements are spaced on a given display, usually stated in terms of pixels per linear distance.

Pixel format: The arrangement of pixels into horizontal rows and vertical columns.

Spectral color reproduction: The exact reproduction of the spectral power distributions of the original stimuli. Although this is the ultimate goal of any color system, it is not possible in a video system with three primaries.

Zone plate: A complement of test signals for video systems that permit the engineer to exercise a complex video system in the three dimensions of its signal spectrum: horizontal, vertical, and temporal.

References

[1] Anstey, G., and M. J. Dore, "Automatic Measurement of Cathode Ray Tube MTFs," Royal Signals and Radar Establishment, 1980.
[2] Baldwin, M. W., Jr., "Subjective Sharpness of Additive Color Pictures," *Proc. IRE*, 39, 1173-1176, October 1951.
[3] Baldwin, M. W., Jr., "The Subjective Sharpness of Simulated Television Images," *Proc. IRE*, 28, 458, October 1940.
[4] Bartleson, C. J., and E. J. Breneman, "Brightness Reproduction in the Photographic Process," *Photog. Sci. Eng.*, 11, 254-262, 1967.
[5] Bender, W., and A. Blount, "The Role of Colorimetry and Context in Color Displays," in *Human Vision, Visual Processing, and Digital Display III*, B. E. Rogowitz, Ed., *Proc. SPIE* 1666, 343-348, 1992.
[6] Benson, K. B., "Report on Sources of Variability in Color Reproduction as Viewed on the Home Television Receiver," *IEEE Trans. BTR*, 19, 269-275, 1973.
[7] Benson, K. B., and J. C. Whitaker (Eds.), *Television Engineering Handbook*, revised edition, McGraw-Hill, New York, 1991.
[8] Benson, K. B., and J. C. Whitaker, *Television and Audio Handbook for Engineers and Technicians*, McGraw-Hill, New York, 1989.
[9] Bingley, F. J., "The Application of Projective Geometry to the Theory of Color Mixture," *Proc. IRE*, 36, 709-723, 1948.
[10] "Broadening the Applications of Zone Plate Generators," Application Note 20W-7056, Tektronix, Beaverton, OR, 1992.
[11] Brodeur, R., K. R. Field, and D. H. McRae, "Measurement of Color Rendition in Color Television," in M. Pearson (Ed.), *Proc. ISCC Conf. Optimum Reproduction of Color*, Williamsburg, VA, 1971, Graphic Arts Research Center, Rochester, NY, 1971.
[12] Castellano, J. A., *Handbook of Display Technology*, Academic Press, New York, 1992.
[13] Crost, M. E., "Display Devices and the Human Observer," *Proc. Interlab. Sem. Component Technol.* Pt. 1, *R&D Tech. Rep.* ECOM-2865. U.S. Army Electronics Command, Fort Monmouth, NJ, August, 1967.
[14] DeMarsh, L. E., "Color Rendition in Television," *IEEE Trans. CE*, 23, 149-157, 1977.
[15] DeMarsh, L. E., "Colorimetric Standards in US Color Television," *J. SMPTE*, 83, 1-5, 1974.
[16] Donofrio, R., "Image Sharpness of a Color Picture Tube by Modulation Transfer Techniques," *IEEE Tran. Broadcast Television Receivers*, BTR-18(1), 16, February 1972.
[17] Epstein, D. W., "Colorimetric Analysis of RCA Color Television System," *RCA Review*, 14, 227-258, 1953.
[18] Eshbach, O. W., *Handbook of Engineering Fundamentals*, 2nd ed., Wiley, New York, 1936.
[19] Fink, D. G., *Color Television Standards*, McGraw-Hill, New York, 1955.
[20] Fink, D. G., and D. Christiansen (Eds.), *Electronics Engineers' Handbook*, 2nd ed., McGraw-Hill, New York, 1982.
[21] Fink, D. G., and H. W. Beaty (Eds.), *Standard Handbook for Electrical Engineers*, 11th ed., McGraw-Hill, New York, 1978.
[22] Green, M., "Temporal Sampling Requirements for Stereoscopic Displays," in *Human Vision, Visual Processing, and Digital Display III*, B. E. Rogowitz ed., Proc. SPIE 1666, 101-111, 1992.
[23] Herman, S., "The Design of Television Color Rendition," *J. SMPTE*, 84, 267-273, 1975.
[24] Hunt, R. W. G., *The Reproduction of Colour*, 3rd ed., Fountain Press, England, 1975.
[25] Hunter, R., *The Measurement of Appearance*, Wiley, New York, 1975.
[26] Jenkins. A. J., "Modulation Transfer Function (MTF) Measurements on Phosphor Screens," *Assessment of Imaging Systems: Visible and Infrared* (Sira). 274, 154-158, SPIE, Bellingham, WA, 1981.
[27] Kucherrov, G. V. et al., "Application of the Modulation Transfer Function Method to the Analysis of Cathode-Ray Tubes," *Radio Engineering and Electronics Physics*, 19, 150-152, February 1974.

[28] MacAdam, D. L., "Visual Sensitivities to Color Differences in Daylight," *J. Opt. Soc. Am.*, 32, 247-274, 1942.

[29] Middlebrook, B., and M. Day, "Measure CRT Spot Size to Pack More Information into High-Speed Graphic Displays: You Can Do It with the Vernier Line Method," *Electronic Design*, 15, 58-60, July 19, 1975.

[30] Naiman, A. C., and W. Makous, "Spatial Non-linearities of Grayscale CRT pixels," in *Human Vision, Visual Processing, and Digital Display III*, B. E. Rogowitz, Ed., *Proc. SPIE* 1666, 41-56, 1992.

[31] Neal, C. B., "Television Colorimetry for Receiver Engineers," *IEEE Trans. BTR*, 19, 149-162, 1973.

[32] Pearson, M. (Ed.), *Proc. ISCC Conf. on Optimum Reproduction of Color*, Williamsburg, VA, 1971, Graphic Arts Research Center, Rochester, NY, 1971.

[33] Pritchard, D. H., "U.S. Color Television Fundamentals—A Review," *IEEE Trans. CE*, 23, 467-478, 1977.

[34] Rash, C. E., and R. W. Verona, "Temporal Aspects of Electro-Optical Imaging Systems," *Imaging Sensors and Displays*, 765, 22-25, SPIE, Bellingham, WA, 1987.

[35] Reinhart, W. F., "Gray-scale Requirements for Anti-aliasing of Stereoscopic Graphic Imagery," in *Human Vision, Visual Processing, and Digital Display III*, B. E. Rogowitz, Ed., *Proc. SPIE* 1666, 90-100, 1992.

[36] "Setting Chromaticity and Luminance of White for Color Television Monitors Using Shadow Mask Picture Tubes," SMPTE Recommended Practice 71-1977.

[37] Sproson, W. N., *Colour Science in Television and Display Systems*, Adam Hilger, Bristol, England, 1983.

[38] Sullivan, J. R., and L. A. Ray, "Secondary Quantization of Gray-Level Images for Minimum Visual Distortion," in *Human Vision, Visual Processing, and Digital Display III*, B. E. Rogowitz, Ed., *Proc. SPIE* 1666, 27-40, 1992.

[39] Thomas, G. A., "An Improved Zone Plate Test Signal Generator," *Proceedings IBC*, 11th International Broadcasting Conference, Brighton, UK, 358-361, 1986.

[40] Thomas, W., Jr. (Ed.), *SPSE Handbook for Photographic Science and Engineering*, Wiley, New York, 1973.

[41] Verona, R. W., "Comparison of CRT Display Measurement Techniques, in *Helmet-Mounted Displays III*, T. M. Lippert, Ed., *Proc. SPIE* 1695, 117-127, 1992.

[42] Wentworth, J. W., *Color Television Engineering*, McGraw-Hill, New York, 1955.

[43] Weston, M., "The Zone Plate: Its Principles and Applications," *EBU Review—Technical*, No. 195, October 1982.

[44] Whitaker, J. C., *Electronic Displays: Technology, Design and Applications*, McGraw-Hill, New York, 1993.

[45] Wintringham, W. T., "Color Television and Colorimetry," *Proc. IRE*, 39, 1135-1172, 1951.

Further Information

The SPIE (the international optical technology society) offers a wide variety of publications on the topic of display systems engineering, measurement, and quality control. The organization also holds technical seminars on this and other topics several times each year. SPIE is headquartered in Bellingham, WA. Two books on the topic of display technology are also recommended:

J. C. Whitaker, *Electronic Displays: Technology, Design and Applications*, McGraw-Hill, New York, 1994.

S. Sherr, *Fundamentals of Display System Design*, 2nd Ed., Wiley-Interscience, New York, 1994.

22
Radio Frequency Distortion Mechanisms and Analysis

22.1	Introduction	22-1
22.2	Types of Distortion	22-2
	Linear Distortion • Nonlinear Distortion • Distortion Due to Time-Variant Multipath Channels	
22.3	The Wireless Radio Channel	22-6
	Ground Waves • Tropospheric Waves • Sky Waves	
22.4	Effects of Phase and Frequency Errors in Coherent Demodulation of AM Signals	22-9
	Double-Sideband Suppressed-Carrier (DSB-SC) Demodulation Errors • Single-Sideband Suppressed-Carrier (SSB-SC) Demodulation Errors	
22.5	Effects of Linear and Nonlinear Distortion in Demodulation of Angle Modulated Waves	22-13
	Amplitude Modulation Effects of Linear Distortion • Effects of Amplitude Non-linearities on Angle Modulated Waves • Phase Distortion in Angle Modulation	
22.6	Interference as a Radio Frequency Distortion Mechanism	22-18
	Interference in Amplitude Modulation • Interference in DSB-SC AM • Interference in SSB-SC • Interference in Angle Modulation	

Samuel O. Agbo
California Polytechnic State University, San Luis Obispo

22.1 Introduction

Radio frequency (RF) communication is usually understood to mean radio and television broadcasting, cellular radio communication, point-to-point radio communication, microwave radio, and other wireless radio communication. For such wireless radio communication, the communication channel is the atmosphere or free space. It is in this sense that RF distortion is treated in this section, although RF signals are also transmitted through other media such as coaxial cables.

Distortion refers to the corruption encountered by a signal during transmission or processing that mutilates the signal waveform. A simple model of the atmospheric radio channel, as shown in Fig. 22.1, will help to illustrate the sources of radio frequency distortion. The channel is modeled as a linear time-variant system with additive noise and additive interference. In the figure, $s(t)$ is the RF signal to be transmitted through the channel, $h(t, \tau)$ is the channel impulse response, $n(t)$ is the additive noise, $i(t)$ is the interference, and $r(t)$ is the received signal. Sources of signal corruption, such as signal attenuation, amplitude distortion, phase distortion, multipath effects and time-varying channel characteristics, constitute the linear time-variant channel impulse

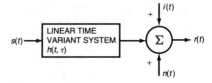

FIGURE 22.1 Model of the atmospheric radio channel as a linear time-variant system with additive noise and additive interference.

response. Noise and interference are additive sources of signal corruption. Corruption by noise may result in a roughness of the output waveform, but the original signal waveform is usually discernible. This is one sense in which noise differs from the distortion mechanisms discussed in this section.

The impulse response $h(t, \tau)$ is the response of the channel at time t due to an impulse $\delta(t-\tau)$ applied at time $t - \tau$. Let $\alpha_k(t)$ represent the time-dependent attenuation factors for the m multipath components of an atmospheric radio channel. Then a simple representation of the impulse response is

$$h(t, \tau) = \sum_{k=1}^{m} \alpha_k(t)\delta(t - \tau_k) \qquad (22.1)$$

The received signal, which is the convolution of this impulse response with the signal $s(t)$, plus the additive noise and interference acquired in the channel, is given by

$$r(t) = \sum_{k=1}^{m} \alpha_k s(t - \tau_k) + i(t) + n(t) \qquad (22.2)$$

Radio frequency signals are processed in the transmitter, prior to transmission, and in the receiver, prior to demodulation, by electronic circuits. Such processing can be regarded in a generalized way as filtering. Distortion introduced by nonideal filters and the atmospheric radio channel will be discussed. Following this, the effects on demodulation outputs of frequency and phase distortion and interference on radio frequency signals will be discussed.

22.2 Types of Distortion

Linear Distortion

The processing of radio frequency signals in transmitter or receiver by electronic circuits can be viewed as forms of filtering. Such processing will introduce distortion unless the filter is ideal or distortionless within a band of frequencies equal to or exceeding the bandwidth of the input signal. A filter is distortionless if its input and output signals have identical waveforms, save for a constant amplitude gain or attenuation and a constant time delay for all of the frequency components. Thus, if $g(t)$ with a bandwidth W is the input signal into an ideal filter of bandwidth B, which is greater than W, the output signal $y(t)$ is given by

$$y(t) = \alpha g(t - \tau) \qquad (22.3)$$

In Eq. (22.3) α is the constant gain or attenuation and τ is the time delay in passing through the system. With this definition, $H(f)$, the Fourier transforms of the ideal filter impulse response and $Y(f)$, the filter output are given by, respectively,

$$H(f) = \alpha e^{-j2\pi f \tau} \tag{22.4}$$

$$Y(f) = G(f)H(f)$$

$$= \alpha G(f)e^{-j2\pi f \tau} \tag{22.5}$$

Figure 22.2 shows the frequency-domain representation of an ideal and a nonideal filter and the time-domain representation of the input and output signals. Figures 22.2(a), 22.2(b), and 22.2(c) show, respectively, the frequency response of the ideal filter, the input signal waveform, and the output signal waveform, which is a time delayed version of the input waveform. Both the filter response and the input signal are bandlimited to B Hz.

Figure 22.2(d) shows a nonideal filter frequency response with ideal phase but nonideal amplitude response, which is given by

$$|H(f)| = \begin{cases} \alpha \cos 2\pi fT, & |f| < B \\ 0; & |f| > B \end{cases} \tag{22.6}$$

By employing the time-shifting property of the Fourier transform, the time-domain output signal is obtained as

$$y(t) = \frac{\alpha}{2}[g(t-\tau-T) + g(t-\tau+T)] \tag{22.7}$$

From Eq. 22.7, it is clear that the output signal consists of two scaled and time shifted versions of the input signal as shown in Fig. 22.2(e). The sum of these shifted versions shown in Fig. 22.2(f) indicates

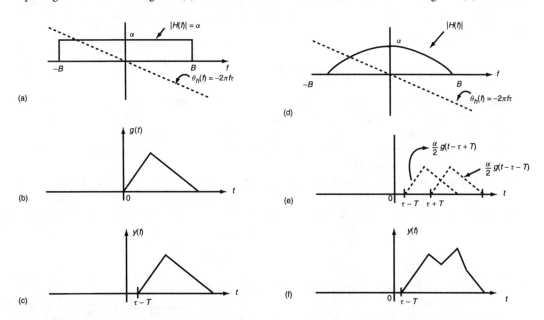

FIGURE 22.2 (a) Ideal filter frequency response, (b) waveform of input signal, (c) waveform of output signal from ideal filter, (d) nonideal filter frequency response with ideal phase but nonideal amplitude response, (e) output waveform of nonideal filter showing its components as time shifted versions of the input signal, (f) output waveform of nonideal filter: a dispersed version of the input signal.

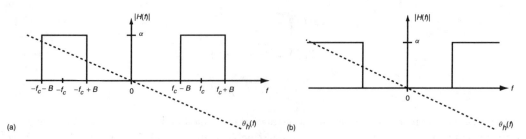

FIGURE 22.3 (a) Ideal bandpass filter, (b) high-pass filters.

that linear distortion results in **dispersion** of the output signal. It is easy to see that nonideal phase response, which implies different delays at different frequencies, will also lead to dispersion. Thus, linear distortion causes pulse spreading and interference with neighboring pulses. Consequently, linear distortion should be avoided in time-division multiplexing (TDM) systems where it causes **crosstalk** with neighboring channels. In frequency-division multiplexing (FDM), linear distortion corrupts the signal spectrum, but the pulse spreading does not result in crosstalk because the channels are adjacent in frequency, not in time.

The ideal filter of Fig. 22.2 is a low-pass filter. Bandpass and high-pass filters are more relevant for the processing of RF signals in the transmitter, prior to transmission, and in the receiver, prior to demodulation. For bandpass and high-pass filters, the condition for distortionless filtering is the same as previously indicated: the input signal bandwidth should not exceed the filter bandwidth, and the filter should have a constant gain and the same time delay for all frequencies within its bandwidth. Figure 22.3 illustrates the frequency responses for ideal bandpass and ideal high-pass filters.

Nonlinear Distortion

Nonlinearity in system response usually arises in cases involving large signal amplitudes. For a simple case of memoryless, nonlinear system or channel, with a low-pass input signal $g(t)$, of bandwidth W Hz, the output signal $y(t)$ is given by

$$y(t) = a_0 + a_1 g(t) + a_2 g^2(t) + a_3 g^3(t) + \cdots + a_k g^k(t) + \cdots \quad (22.8)$$

Consider the term $a_k g^k(t)$ in the equation. Let $G(f)$ be the Fourier transform of $g(t)$. Then the Fourier transform of $g^k(t)$ is a $k - 1$ fold convolution of $G(f)$. Thus, the bandwidth of $g^k(t)$ is $k - 1$ times the bandwidth of $g(t)$. Suppose m is the last significant term of the power series of the equation, then the bandwidth of the output signal is $(m - 1)W$ Hz. Thus, the bandwidth of the output signal will far exceed the bandwidth of the input signal.

Figure 22.4 illustrates the spectrum of the input and the output signals. If the input signal is a bandpass signal of bandwidth W and carrier frequency f_c, which would be appropriate for a radio frequency signal, the spectrum of the output signal would be much more complex. In this case, the kth term of the power

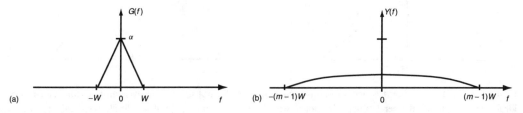

FIGURE 22.4 Input and output spectrum for a channel with nonlinear distortion: (a) amplitude spectrum of output signal, (b) amplitude spectrum of input signal.

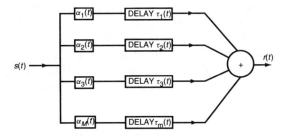

FIGURE 22.5 Model of a noiseless, time-variant, multipath channel.

series of the output signal $y(t)$ will, in general, contribute a component of bandwidth $(k-1)W$ centered at kf_c, in addition to smaller bandwidth components centered at lower harmonics of f_c.

The preceding discussion shows that nonlinear distortion causes spreading in the frequency domain. Such spreading of the spectrum causes interference in FDM systems because it results in frequency-domain overlap of neighboring channels. However, nonlinear distortion does not cause adjacent channel interference in TDM as the channels in that case are not adjacent in frequency, but in time.

Distortion Due to Time-Variant Multipath Channels

Radio waves from a transmitting antenna often reach the receiving antenna through many different paths. Examples include different reflecting layers in the **ionosphere**, numerous scattering points in the **troposphere**, reflections from the Earth's surface, and the direct line-of-sight path. Each of these paths will contribute a different attenuation and time delay. Shown in Fig. 22.5 is a model of the time-variant multipath radio channel. For the sake of simplicity, the effect of noise has been ignored.

This channel is time dispersive because the different time delays entail a spread in time of the composite received signal relative to the transmitted signal. In addition, because of variations with time and locality in weather, temperature and other atmospheric conditions, the time delay and attenuation of these paths vary with time. Because these variations in channel characteristics are generally unpredictable, a statistical characterization of the time-variant multipath channel is appropriate.

To simplify the analysis, consider the transmission of an unmodulated carrier of unit amplitude, $s(t) = \cos 2\pi f_c t$, through an m multipath channel in the absence of noise. Let $\alpha_k(t)$ be the attenuation and $\tau_k(t)$ be time delay contributed by the kth path at the receiver. The received signal is given by

$$r(t) = \sum_{k=1}^{m} \alpha_k(t)\cos[2\pi f_c(t-\tau_k(t))]$$

$$= Re\left[\sum_{k=1}^{m} \alpha_k(t)e^{j2\pi f_c t}e^{-j2\pi f_c \tau_k(t)}\right] \quad (22.9)$$

$$= [\cos 2\pi f_c t]\, Re\left[\sum_{k=1}^{m} \alpha_k e^{-j2\pi f_c \tau_k(t)}\right]$$

It is easy to see that the response of the channel to the unmodulated carrier is

$$h(t,\tau) = Re\left[\sum_{k=1}^{m} \alpha_k(t)e^{-j2\pi f_c \tau_k(t)}\right] \quad (22.10)$$

Note that although the channel input was an unmodulated carrier, the channel output contains a frequency spread around the carrier, resulting from the time-varying delays of the various paths. The bandwidth of this output signal is a measure of how rapidly the channel characteristics are changing. A large change in channel characteristics is required to produce a significant change in the time-varying attenuation $\alpha_k(t)$. Because f_c is a large number, a change of 2π in the phase of the output signal $-2\pi f_c \tau_k(t)$ corresponds to only a change of $1/f_c$ in the time delay $\tau_k(t)$. Thus, the attenuation changes slowly, while the phase (or frequency deviation) changes rapidly.

Because these changes are random in nature, the channel impulse response can be modeled as a random process. Constructive addition of the phases of the various paths result in a strong signal, whereas destructive addition of the phases results in a weak signal. Thus, the multipath propagation results in signal fading, primarily due to the time-variant phase changes. When m, the number of multiple propagation paths, is large, the central limit theorem applies and the channel response will approximate a Gaussian random process.

22.3 The Wireless Radio Channel

In wireless radio communication, electromagnetic waves are radiated into the atmosphere or free space via a transmitting antenna. For efficient radiation, the antenna length should exceed one-tenth of the wavelength of the electromagnetic wave. The wavelength λ is given by

$$\lambda = \frac{c}{f} \qquad (22.11)$$

In Eq. (22.11) $c = 3 * 10^8$ m/s, is the free-space velocity of electromagnetic waves and f is the frequency in Hz. Wireless radio communicating covers a wide range of frequencies from about 10 kHz in the very low frequency (VLF) band to above 300 GHz in the extra high frequency (EHF) band. Depending on the frequency band and the distance between the transmitter and the receiver, the transmitted wave may reach the receiving antenna by one or a combination of several propagation paths, as shown in Fig. 22.6.

Sky waves are those that reach the receiver after being reflected from the ionosphere, a band of reflecting layers ranging in altitude from 100 to 300 km. Tropospheric waves are those that reach the receiver after being reflected (scattered) by inhomogeneities within the troposphere, the region within 10 km of the Earth's surface. Other waves reach the receiving antenna via ground waves, which consists of space waves

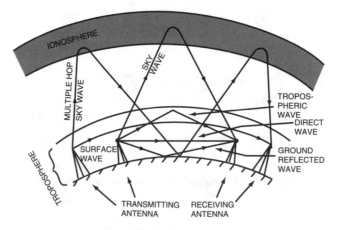

FIGURE 22.6 Wireless radio wave propagation paths.

Radio Frequency Distortion Mechanisms and Analysis

TABLE 22.1 Classification of Wireless Radio Frequency Channels by Frequency Bands, Typical Uses, and Wave Propagation Modes

Frequency Band	Name	Typical Uses	Propagation Mode
3–30 kHz	Very low frequency (VLF)	Navigation, sonar	Ground waves
30–300 kHz	Low frequency (LF)	Navigation, telephony, telegraphy	Ground waves
0.3–3 MHz	Medium frequency (MF)	AM Broadcasting, amateur, and CB radio	Sky waves and ground waves
3–30 MHz	High frequency (HF)	Mobile, amateur, CB, and military radio	Sky waves
30–300 MHz	Very high frequency (VHF)	VHF TV, FM Broadcasting police, air traffic control	Sky waves, tropospheric waves
0.3–3 GHz	Ultra high frequency (UHF)	UHF TV, radar, satellite communication	Tropospheric waves, space waves
3–30 GHz	Super high frequency (SHF)	Space and satellite, radar microwave relay	Direct waves, ionospheric penetration waves
30–300 GHz	Extra high frequency (EHF)	Experimental, radar radio astronomy	Direct waves, ionospheric penetration waves

and surface waves. Surface waves (in the LF and VLF ranges) have wavelengths comparable to the altitude of the ionosphere. The ionosphere and the ground surface act as guiding planes, so that surface waves follow the curvature of the Earth, permitting transmission over great distances. Space waves consist of direct (line-of-sight) waves and ground reflected waves, which arrive at the receiver after being refllected from the ground. A classification of radio frequency communication by frequency bands, typical uses, and electromagnetic wave propagation modes is given in Table 22.1.

Ground Waves

In the region adjacent to the Earth's surface, free-space conditions are modified by the surface topography, physical structures such as buildings, and the atmosphere. One effect of the atmosphere is to cause a refraction of the wave, which carries it beyond the optical horizon. Consequently, the **radio horizon** is longer than the optical horizon. Hills and building structures cause multiple reflections and shieldings, which contribute both attenuation and phase change to the received signal. Ground reflected waves travel a different path length and undergo different attenuation and time delay relative to direct waves and depending on the pont of reflection. In addition, at each point of reflection on the Earth]s surface, there is a phase change, which depends on the condition of the surface, but is typically a 180° phase change.

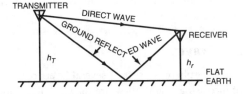

FIGURE 22.7 Simplified mdoel for ground wave propagation showing the direct wave and the ground reflected wave.

Consider the simplified model of ground wave propagation, shown in Fig. 22.7, which involves only two transmission paths: a direct wave and a ground reflected wave. The curved Earth's surface is represented by a flat Earth model after correction has been made for the longer radio horizon. The height of the transmitting antenna and the receiving antenna are h_T and h_r, respectively.

The direct wave travels a path length x from the transmitting antenna to the receiving antenna. The attenuation factor for the atmospheric medium is α. Let the corresponding distance traveled by the ground reflected wave be $x + \Delta x$ and the transmitted signal be $s(t)$. Note that a path length equal to the wavelength λ corresponds to a phase change of 2π. The received direct wave $r_d(t)$, the received ground reflected wave $r_g(t)$, and the composite received wave $r(t)$ are given, respectively, by

$$r_d(t) = e^{-\alpha x} s(t) \cos\left(\frac{2\pi x}{\lambda}\right) \tag{22.12}$$

$$r_g(t) = e^{-\alpha(x+\Delta x)} s(t) \cos\left[\frac{2\pi}{\lambda}(x+\Delta x) - \pi\right] \tag{22.13}$$

$$r(t) = r_d(t) + r_g(t) \tag{22.14}$$

From the preceding equations, it is seen that even for small differences in path lengths, the received signals from the direct and the ground reflected waves tend to have large phase differences. Consequently, the composite received signal has significant attenuation and phase error relative to the transmitted signal.

Tropospheric Waves

Waves propagating through the troposphere are only seriously attenuated by adverse atmospheric conditions at frequencies above 10 GHz. Above 10 GHz, heavy rain causes severe attenuation. However, light rain, clouds, and fog cause serious attenuation only at frequencies above 30 GHz. Attenuation due to snow is negligible at all frequencies. On the other hand, the inhomogeneities in the atmosphere due to the weather and other conditions are faborable for tropospheric scatter propagation of radio waves within the 30 MHz–4 GHz frequency band. Within this range of frequencies, the inhomogeneities within the troposphere form effective point scatters that deflect the transmitted waves downward toward the receiving antenna. By employing large transmitted power and highly directional antennas, long transmission ranges of up to 400 mi can be obtained.

The range of frequencies favored for tropospheric propagation is essentially the domain of microwave transmission, UHF, and VHF radio. Although atmospheric attenuation is higher for frequencies above 10 GHz, it is still quite significant within the 30 MHz–4 GHz range of frequencies. Hence, microwave repeater stations are spaced about 30 mi apart. This spacing is long compared to the spacing of about 3 mi for repeaters for coaxial cable systems. However, such spacing is relatively close, compared with the more recent development of optical fiber transmission in which repeater spacing of above 300 km has been achieved.

Sky Waves

The ionosphere consists of several layers containing ionized gases at altitudes of 100–300 km. In order of increasing height, these are the C, D, E, F_1, and F_2 layers. The ionization is caused by solar radiation, mainly due to ultraviolet and cosmic rays. Consequently, the electron density in the layers varies with the seasons and with the time of day and night. The gas molecules in the denser, lower layers cause more collisions and hence a higher recombination rate of electrons and ions. At night, the reduced level of radiation and the greater recombination rates cause the lower layers to disappear, so that only the F_2 layer remains.

Electromagnetic wave propagation through the ionosphere is by means of sky waves, mainly in the MF, HF, and VHF bands. The waves are successively refracted at the layers until they are eventually reflected back toward the Earth. At higher frequencies, the waves pass through the ionosphere without being reflected from it. Communication at these higher frequencies depends on a line-of-sight basis. During the daytime, ionosphere propagation is severely attenuated at frequencies below 2 MHz, due to severe absorption in the lower layers, especially the D layer. At night, after the disappearance of these lower layers, powerful AM transmission utilizing the F_2 layer is possible over the range of 100–250 mi. Of course, the maximum transmission range for sky waves depends on the frequency of transmission.

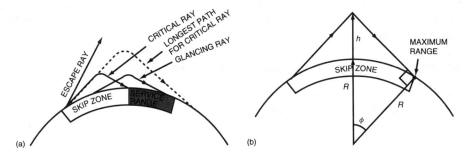

FIGURE 22.8 Geometry of ionospheric single hop: (a) raypaths, skip zone, and service range, (b) geometry for computation of maximum range.

Figure 22.8 shows the geometry relevant to the determination of the range for ionospheric propagation. Because the altitudes involved are much greater than the antenna heights, the latter have been ignored. Rays from the transmitting antenna at angles of elevation greater than the critical angle of elevation for the frequency of transmission result in escape rays, which are not reflected back from the ionosphere. The ray leaving the transmitter at the critical angle, the critical ray, which is reflected from the lowest possible layer, returns to the Earth closest to the transmitter than any other ray, giving the minimum range or the skip zone. The distance between this minimum range and the range of the glancing ray (which is tangential to the Earth at the receiver) is the service range.

The greatest transmission range is obtained when the critical ray is reflected from a layer at so high an altitude that it is also tangential to the Earth's surface at the receiver. The geometry for this case is shown in Fig. 22.8(b) in which the raypaths have been idealized as straight lines, ignoring the curvature of the paths due to refraction. The height of the virtual reflecting point above the Earth's surface is h, and R is the radius of the Earth. The angle ϕ and the maximum range d are given, respectively, by

$$\phi = \cos^{-1}\left(\frac{R}{R+h}\right) \tag{22.15}$$

$$d = 2R\cos^{-1}\left(\frac{R}{R+h}\right) \tag{22.16}$$

This computation is for a single-hop reflection. For multiple hops, the transmission range is considerably greater, although the signal strength would be correspondingly reduced. The condition of the ionosphere within a locality greatly affects the critical angle of elevation and the transmission path. Conditions vary considerably with time and locality within the ionosphere. Such irregularities within the ionosphere can be used for scatter propagation, much like tropospheric scatter propagation, thereby resulting in increased transmission range of up to 1000 mi. However, the undesirable effects of ionospheric inhomogeneities far exceed this favorable effect.

Ionospheric irregularities result from differences in electron densities in the various layers. The irregularities can be quite large in spatial dimensions, ranging from tens of meters to several hundreds of kilometers. Some of these irregularities travel through the ionosphere at speeds in excess of 3000 km/h. These time-varying irregularities are responsible for the time-variant multipath effects and the resultant fading that plagues ionospheric transmission.

22.4 Effects of Phase and Frequency Errors in Coherent Demodulation of AM Signals

RF signals acquire phase and amplitude distortion in the wireless radio channel and in processing circuits, through the many mechanisms previously discussed. For coherent (or synchronous) detection, a local carrier, which matches the transmitted signal in phase and frequency, is required at the receiver. In general, such a match is hard to obtain. These mismatch errors are additional to the phase and amplitude errors acquired in the channel.

Double-Sideband Suppressed-Carrier (DSB-SC) Demodulation Errors

The receiver for the synchronous demodulation of double-sideband suppressed-carrier (DSB-SC) waves is shown in Fig. 22.9. In general, it is easy to correct for the effect of attenuation or amplitude errors through amplification, filtering, or other processing in the receiver. It is more difficult to correct for the effect of phase or frequency errors.

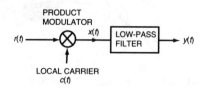

FIGURE 22.9 Receiver for synchronous demodulation of DSB-SC signals.

The received signal at the input to the receiver and the local carrier are given, respectively, by

$$r(t) = m(t) \cos 2\pi f_c t \tag{22.17}$$

$$c(t) = \cos[2\pi(f_c + \Delta f)t + \theta] \tag{22.18}$$

In the last two equations, $m(t)$ is the message signal of bandwidth W, where f_c is the carrier frequency, Δf is the frequency error, and θ is the phase error. The output of the product modulator is $x(t)$. Thus, $x(t)$ and $y(t)$ are given, respectively, by

$$\begin{aligned} x(t) &= m(t)\cos(2\pi f_c t) \cos[2\pi(f_c + \Delta f)t + \theta] \\ &= \frac{1}{2}m(t)\{\cos[2\pi(2f_c + \Delta f)t + \theta] + \cos(2\pi\Delta f t + \theta)\} \end{aligned} \tag{22.19}$$

$$y(t) = \frac{1}{2}m(t) \cos(2\pi\Delta f t + \theta) \tag{22.20}$$

It is easy to see that when both the frequency and the phase errors are zero, the demodulated output is $1/2 m(t)$, and the recovery of the message signal is perfect. Consider the case of zero phase error, but finite frequency error. The demodulated signal and its Fourier transform are given, respectively, by

$$y(t) = \frac{1}{2}m(t) \cos 2\pi\Delta f t \tag{22.21}$$

$$Y(f) = \frac{1}{4}[M(f + \Delta f) + M(f - \Delta f)] \tag{22.22}$$

Thus, the demodulated output is a replica of the message signal multiplied by a slowly varying cosine wave of frequency Δf. This is a serious type of distortion, which produces a beating effect. The human ear can tolerate at most about 30 Hz of such frequency drift. Figure 22.10 shows the frequency-domain illustration of this type of distortion. Figure 22.10(a) shows the amplitude spectrum of the original message signal. Figure 22.10(b) shows that the demodulated signal can be viewed as a DSB-SC signal whose carrier is the frequency error Δf. This error is so small that the replicas of the message signal shifted by $+\Delta f$ and $-\Delta f$ overlap in frequency and, therefore, interfere, giving rise to the beating effect.

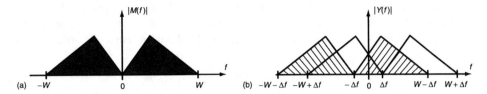

FIGURE 22.10 Frequency spectrum of message signal and demodulated output when there is a Δf frequency error between the local carrier and the carrier in the received DSB-SC signal: (a) message signal spectrum, (b) spectrum of demodulated output signal.

Next, consider the case in which the frequency error is zero, but the phase error is finite. The demodulated signal is given by

$$y(t) = \frac{1}{2} m(t) \cos\theta \tag{22.23}$$

If θ, the phase error is constant, then it contributes only a constant attenuation, which is easily corrected for through amplification. However, in the extreme case in which $\theta = \pm\pi/2$, the demodulated signal amplitude is reduced to zero. On the other hand, if θ varies randomly with time, as is the case for multipath fading channels, the attenuation induced by the phase error varies randomly and represents a serious problem.

To avoid the undesirable effects of frequency and phase errors between the local and the received carriers, various schemes are employed. One such scheme employs a **pilot carrier** at the transmitter to provide a source of matching carrier at the receiver. Some other schemes employed for providing matching carriers are the *Costas receiver* and signal squaring.

Single-Sideband Suppressed-Carrier (SSB-SC) Demodulation Errors

The receiver for the demodulation of single-sideband suppressed-carrier (SSB-SC) signals is similar to the receiver for DSB-SC signals shown in Fig. 22.9. The local carrier is the same as for the DSB-SC signal, but the received signal is given by

$$r(t) = m(t) \cos 2\pi f_c t \mp m_h(t) \sin 2\pi f_c t \tag{22.24}$$

In Eq. 22.24, the negative ($-$) sign applies to the upper-sideband (USB) case, whereas the positive sign ($+$) applies to the lower-sideband (LSB) case. The signal, $m_h(t)$, is the *Hilbert transform* of the message signal $m(t)$. The output of the product modulator is

$$\begin{aligned}x(t) &= [m(t) \cos 2\pi f_c t \mp m_h(t) \sin 2\pi f_c t] \cos[2\pi(f_c + \Delta f)t + \theta] \\ &= \frac{1}{2} m(t)\{\cos(2\pi\Delta f t + \theta) + \cos[2\pi(2f_c + \Delta f)t + \theta]\} \\ &\pm \frac{1}{2} m_h(t)\{\sin(2\pi\Delta f t + \theta) - \sin[2\pi(2f_c + \Delta f)t + \theta]\}\end{aligned} \tag{22.25}$$

The low-pass filter suppresses the bandpass components of the product modulator output. Hence, the demodulated output $y(t)$ is given by

$$y(t) = \frac{1}{2} m(t) \cos(2\pi\Delta f t + \theta) \pm \frac{1}{2} m_h(t) \sin[2\pi\Delta f t + \theta] \tag{22.26}$$

Note that when both the frequency and the phase errors are set to zero in the equation, the demodulator output signal is $1/2 m(t)$, an exact replica of the desired message signal. Consider the case of zero phase error, but nonzero frequency error. In this case, the demodulated output signal $y(t)$ can be expressed as

$$y(t) = \frac{1}{2}m(t)\cos 2\pi\Delta f t \pm \frac{1}{2}m_h(t)\sin 2\pi\Delta f t \qquad (22.27)$$

This equation shows that when only frequency error is present between the received and local carriers, the demodulated signal is essentially an SSB-SC signal in which the small frequency error Δf plays the role of the carrier. A comparison of this equation with the expression for the SSB-SC signal shows that if the modulated signal is a USB signal, the demodulated output is an LSB signal, and vice versa.

Figure 22.11 shows the spectra of the message signal, the SSB-SC modulated signal (USB case), and two cases of the demodulated signal when only frequency errors are present, Figure 22.11(c) shows the spectrum of the demodulated output of the USB signal when the frequency error is negative. Figure 22.11(e) shows the spectrum of the demodulated output of the USB signal when the frequency error is positive.

By comparing the demodulated output of an SSB-SC signal with the message signal, it is easy to see that in each case, each frequency component in the demodulated output is shifted by Δf. This does not introduce as serious a distortion as in the DSB-SC carrier case in which each frequency component is shifted by $-\Delta f$ and $+\Delta f$. As shown in Fig. 22.11, $M(f)$, the spectrum of the message signal is zero in the frequency gap from zero to f_g, as should be the case for a well-processed SSB message. By ensuring that f_g exceeds the maximum possible value of Δf, the sidebands for the demodulated output of an SSB-SC wave will never overlap in frequency or interfere with each other as in the DSB-SC case.

For the case of USB modulation in which the frequency error is zero but the phase error is nonzero, the demodulated output is obtained by setting Δf equal to zero in Eq. (22.26). This gives $y(t)$, the demodulated output signal and its frequency domain representation $Y(f)$, respectively, as

$$y(t) = \frac{1}{2}[m(t)\cos\theta + m_h(t)\sin\theta] \qquad (22.28)$$

$$Y(f) = \frac{1}{2}[M(f)\cos\theta + M(f)\sin\theta] \qquad (22.29)$$

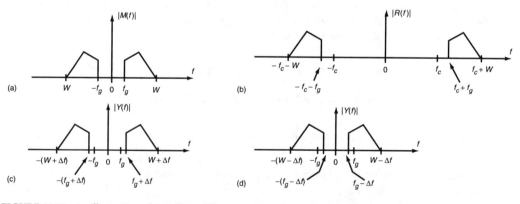

FIGURE 22.11 An illustration of the effect of frequency error in synchronous demodulation of SSB-SC signals: (a) amplitude spectrum of message signal, (b) amplitude spectrum of SSB (USB) signal, (c) amplitude spectrum of demodulated USB signal when $\Delta f < 0$, (d) amplitude spectrum of demodulated USB signal when $\Delta f > 0$.

Radio Frequency Distortion Mechanisms and Analysis

But the frequency domain representation of the Hilbert transform of the message signal, $m_h(f)$, can be expressed as

$$M_h(f) = -j\,\mathrm{sgn}(f)M(f)$$

$$= \begin{cases} -jM(f); & f>0 \\ jM(f); & f<0 \end{cases} \quad (22.30)$$

Thus, $Y(f)$ can be expressed as

$$Y(f) = \begin{cases} \frac{1}{4}\left[M(f)e^{-j\theta} - jM(f)\left(\frac{-e^{-j\theta}}{j}\right)\right]; & f>0 \\ \frac{1}{4}\left[M(f)e^{j\theta} + jM(f)\left(\frac{e^{j\theta}}{j}\right)\right]; & f<0 \end{cases}$$

$$= \begin{cases} \frac{1}{2}M(f)e^{-j\theta}; & f>0 \\ \frac{1}{2}M(f)e^{j\theta}; & f<0 \end{cases} \quad (22.31)$$

From the preceding equation, it is clear that each positive frequency component in the demodulated output signal undergoes a phase shift of $-\theta$, while each negative frequency component undergoes a phase shift of $+\theta$. Thus, phase errors in SSB-SC signals results in phase distortion in the demodulated outputs. For voice signals, the human ear is relatively insensitive to phase distortion. However, this type of phase distortion, which produces a constant phase shift at all message signal frequencies, is quite objectionable in music signals and intolerable in video signals.

22.5 Effects of Linear and Nonlinear Distortion in Demodulation of Angle Modulated Waves

The amplitude of an angle modulated wave is supposed to be constant while the information being transmitted is contained in the phase. Linear distortion can be introduced by the channel, nonideal filters and other subcircuits. In general, linear distortion results in time-variant amplitude and phase errors in the angle modulated wave. In effect, the angle modulated wave now contains both amplitude and phase modulation. Nonlinear distortion usually results from channel or power amplifier nonlinearities. Amplitude nonlinearities result in undesirable (or residual) amplitude modulation (AM), whereas phase nonlinearities result in phase distortion of the angle modulated wave. The effect of undesirable amplitude modulation resulting from linear and nonlinear distortion are discussed under separate subheadings. The effects of the time-variant phase errors resulting from linear distortion and from phase nonlinearities are discussed together under phase distortion in angle modulation.

Amplitude Modulation Effects of Linear Distortion

In an angle modulated waveform, the information resides in the phase of the modulated wave. Such an angle modulated wave can be demodulated by first passing it through a circuit whose amplitude response is proportional to the frequency of the input signal. The resulting intermediate signal has both amplitude modulation and angle modulation, and the information also resides in its amplitude. The demodulation process is then completed by passing this intermediate signal through an envelope detector.

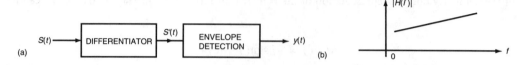

FIGURE 22.12 FM demodulator employing a differentiator and an envelope detector: (a) FM demodulator, (b) amplitude response of an ideal differentiator.

A circuit which can implement this process is shown in Fig. 22.12. The differentiator input signal is the angle modulated wave. The amplitude of its output is proportional to the signal frequency, as shown in Fig. 22.12(b). Thus, the differentiator output is an amplitude as well as a frequency modulated waveform. The envelope detector output contains a replica of the message signal. Consider the case in which the angle modulation involved is frequency modulation (FM). Suppose the FM signal contains no amplitude or phase errors, its amplitude will be constant. For simplicity, it can be assumed to be of unit amplitude. Let f_c be the carrier frequency, k_f be the frequency sensitivity of the modulator, and $m(t)$ be the message signal. Then the FM signal $s(t)$, the differentiator output $s'(t)$, and the envelope detector output $y(t)$, are given, respectively, by

$$s(t) = \cos\left[2\pi f_c t + 2\pi k_f \int_{-\infty}^{t} m(\lambda)d\lambda\right] \tag{22.32}$$

$$s'(t) = -[2\pi f_c + 2\pi k_f m(t)]\cos\left[2\pi f_c t + 2\pi k_f \int_{-\infty}^{t} m(\lambda)d\lambda\right] \tag{22.33}$$

$$y(t) = 2\pi f_c + 2\pi k_f m(t) \tag{22.34}$$

It is easy to see that an exact replica of the message signal can be recovered from the envelope detector output.

Consider the case in which the FM signal contains a time variation in its amplitude due to the channel, nonideal filter, or other system effects. Let $a(t)$ represent the time variation in the amplitude relative to a the unit carrier amplitude. Then the FM signal and the output of the differentiator are given, respectively, by

$$s(t) = a(t)\cos\left[2\pi f_c t + 2\pi k_f \int_{-\infty}^{t} m(\lambda)d\lambda\right] \tag{22.35}$$

$$s'(t) = -a(t)[2\pi f_c + 2\pi k_f m(t)]\sin\left[2\pi f_c t + 2\pi k_f \int_{-\infty}^{t} m(\lambda)d\lambda\right]$$
$$+ a'(t)\cos\left[2\pi f_c t + 2\pi k_f \int_{-\infty}^{t} m(\lambda)d\lambda\right] \tag{22.36}$$

The output of the envelope detector, $y(t)$, is the envelope of the differentiator output given in the last equation. Thus, the envelope detector output is

$$y(t) = \{a^2(t)[2\pi f_c + 2\pi k_f m(t)]^2 + [a'(t)]^2\}^{\frac{1}{2}} \tag{22.37}$$

From the last equation it is easy to see that because of the time variation in the amplitude of the FM signal, it is not possible to extract an undistorted replica of the message signal from the envelope detector

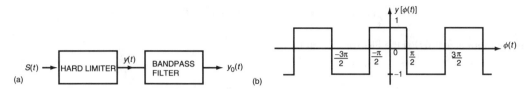

FIGURE 22.13 The bandpass limiter and an illustration of the output waveform of the hard limiter: (a) Bandpass limiter, (b) output waveform of the hard limiter.

output. Even if the second term in Eq. (22.36) is ignored, the message signal is still multiplied by the time-variant amplitude. Thus, it is necessary to rid the FM wave of variations in its amplitude prior to demodulation. This is usually accomplished by passing the FM signal through a bandpass limiter, such as is illustrated in Fig. 22.13, prior to demodulation.

The FM wave containing time-variant amplitude errors can be expressed as

$$s(t) = a(t)\cos[\phi(t)] \tag{22.38}$$

Its phase $\phi(t)$ is given by

$$\phi(t) = 2\pi f_c t + 2\pi k_f \int_{-\infty}^{t} m(\lambda)\,d\lambda \tag{22.39}$$

The hard limiter is a device with a voltage transfer characteristic such that if its input is $s(\phi)$, then its output, $y(\phi)$ is given by

$$y(\phi) = \begin{cases} 1, & \phi > 0 \\ -1, & \phi < 0 \end{cases} \tag{22.40}$$

When the input to the hard limiter is the FM signal, $a(t)\cos[\phi(t)]$, the output of the hard limiter is the square waveform shown in Fig. 22.13(b). This periodic square waveform has a Fourier series expression given by

$$y[\phi(t)] = \frac{4}{\pi}\left[\cos[\phi(t)] - \frac{1}{3}\cos[3\phi(t)] + \frac{1}{5}\cos[5\phi(t)] + \cdots\right] \tag{22.41}$$

Note that in the last equation, the terms containing $\phi(t)$, $3\phi(t)$, $5\phi(t)$, ..., correspond to FM waves with carrier frequencies f_c, $3f_c$, $5f_c$, ... Hence, the bandpass filter centered at f_c selects only the desired FM signal with carrier frequency f_c. Note that the output of the bandpass limiter, which is the output of the bandpass filter, has a constant amplitude given by

$$y_0(t) = \frac{4}{\pi}\cos\left[2\pi f_c t + 2\pi k_f \int_{-\infty}^{t} m(\lambda)\,d\lambda\right] \tag{22.42}$$

Effects of Amplitude Nonlinearities on Angle Modulated Waves

Amplitude nonlinearities can arise from nonlinearities in the channel or from nonlinearities in processing circuits such as power amplifiers. Consider a channel or system with amplitude nonlinearities, whose input signal $x(t)$ and output signal $y(t)$ are related as in the following equation:

$$y(t) = a_0 + a_1 x(t) + a_2 x^2(t) + a_3 x^3(t) + \cdots \qquad (22.43)$$

If the input signal $s(t)$ is the FM wave given by

$$s(t) = \cos[\phi(t)] = \cos\left[2\pi f_c t + 2\pi k_f \int_{-\infty}^{t} m(\lambda) d\lambda\right] \qquad (22.44)$$

then the output signal $y(t)$ is given by

$$\begin{aligned} y(t) &= a_0 + a_1 \cos[\phi(t)] + a_2 \cos^2[\phi(t)] + a_3 \cos^3[\phi(t)] + \cdots \\ &= a_0 + a_1 \cos[\phi(t)] + \frac{1}{2} a_2[1 + \cos 2\phi(t)] + \frac{1}{2} a_3 \left\{ \cos[\phi(t)] + \frac{1}{2} \cos[\phi(t)] \right. \\ &\quad \left. + \frac{1}{2} \cos[3\phi(t)] \right\} + \cdots \\ &= \left(a_0 + \frac{1}{2} a_2\right) + \left(a_1 + \frac{3}{4} a_3\right) \cos[\phi(t)] + \frac{1}{2} a_2 \cos[2\phi(t)] + \frac{1}{4} \cos[3\phi(t)] + \cdots \end{aligned} \qquad (22.45)$$

The phase angles $\phi(t)$, $2\phi(t)$, and $3\phi(t)$, correspond to FM waves with carrier frequencies f_c, $2f_c$, and $3f_c$. Thus if the nonlinear system or channel is followed by a bandpass filter centered at f_c, the filter output is the desired FM wave with a carrier f_c, given by

$$\begin{aligned} y_0(t) &= \left(a_1 + \frac{3}{4} a_3\right) \cos[\phi(t)] \\ &= \alpha \cos\left[2\pi f_c t + 2\pi k_f \int_{-\infty}^{t} m(\lambda) d\lambda\right] \end{aligned} \qquad (22.46)$$

Let B_{FM} be the transmission bandwidth of the FM signal, W the bandwidth of the message signal and Δf the frequency deviation. The FM bandwidth is given for narrowband FM (NBFM) and for wideband FM (WBFM) by

$$B_{FM} = \begin{cases} 2(\Delta f + W), & \text{for NBFM} \\ 2(\Delta f + 2W), & \text{for WBFM} \end{cases} \qquad (22.47)$$

To ensure that the FM wave with a carrier frequency of $2f_c$ does not interfere with the desired FM signal of carrier frequency f_c, the following condition must be satisfied:

$$f_c + \frac{1}{2} B_{FM} \leq 2\left(f_c - \frac{1}{2} B_{FM}\right) \qquad (22.48)$$

This condition is equivalent to ensuring that $f_c \geq 1.5 B_{FM}$. By substituting the expressions for the bandwidths for NBFM and WBFM into the last equation, it is seen that the carrier frequency must satisfy the condition

$$f_c > \begin{cases} 3(\Delta f + W) & \text{for NBFM} \\ 3(\Delta f + 2W) & \text{for WBFM} \end{cases} \quad (22.49)$$

Note that in the case discussed earlier in which the nonlinearities are not time variant, the effect of passing an FM wave through a nonlinear channel or system, followed by a bandpass filter centered at the carrier frequency, is to produce a gain or attenuation in the constant amplitude. If the nonlinearities are time varying, so that the coefficients in the expression relating the channel input to the output are $a_0(t), a_1(t), a_2(t), \ldots$, a substitution of these time-varying coefficients into the preceding equation shows that the bandpass filter output will be a time-variant amplitude version of the FM signal at the correct carrier frequency. As before, such amplitude variations can be removed with a bandpass limiter, prior to demodulation. Thus, unlike amplitude modulation, FM is fairly tolerant of channel nonlinearities. Hence, FM is widely used in microwave and satellite communication systems where the nonlinearities of the power amplifiers need to be well tolerated.

Phase Distortion in Angle Modulation

Angle modulation systems are very sensitive to phase nonlinearities because in such systems the information resides in the phase of the modulated signal. Phase distortion of the modulated wave eventually results in distortion of the demodulated output. Phase distortion can result from phase nonlinearities of channel, repeater, filter, or amplifier responses. It can also result from nonideal linear filtering of an angle modulated wave with a time-variant amplitude.

Let the input signal to a nonideal linear filter be an FM signal $s_1(t)$ with envelope $E_1(t)$ and phase $\phi_1(t)$. Let the FM wave have a residual amplitude modulation due to distortion previously acquired in the transmission channel. Let the output $s_2(t)$ of the filter have an envelope $E_2(t)$ and a phase $\phi_2(t)$. Then the input and output signal can be expressed, respectively, as

$$s_1(t) = E_1(t) \cos[\phi_1(t)] \quad (22.50)$$

$$s_2(t) = E_2(t) \cos[\phi_2(t)] \quad (22.51)$$

Note that although $s_2(t)$ is a linear function of $s_1(t)$, both the phase and the envelope of the output signal are nonlinear functions of the output signal $s_2(t)$. Hence, they are both nonlinear functions of the input signal $s_1(t)$. Thus, in general, amplitude modulation at the input to a linear filter results in phase modulation at the output. This is known as AM–PM conversion. On the other hand, phase modulation at the input to a linear filter results in residual amplitude modulation at the filter output. This is known as PM–AM conversion. PM–AM conversion does not present much of a problem in angle modulated systems, since the resulting residual amplitude modulation is easily remedied with the aid of bandpass limiters. However, AM–PM conversion presents a great problem in angle modulated systems because it is usually not possible to separate the resulting phase distortion from the demodulated signal.

In addition to the mechanism just described, phase nonlinearities of amplifiers, filters, repeaters, etc., also cause AM–PM conversion. This type of distortion is common in microwave systems where it results from the dependence of the phase characteristic of the repeaters and amplifiers on the envelope of the input signal. Typically, the phase change $\Delta\phi_2(t)$ in the output signal is proportional to the corresponding change, in decibel, of the envelope of the input signal. Let $k°/dB$ be the constant of proportionality relating the change $\Delta E_1(t)$ in the envelope of the input signal $E_1(t)$ to the phase change in the output signal. Then the phase change is given by

$$\Delta\phi_2(t) = k 20 \log \frac{\Delta E_1(t)}{E_1(t)} \quad (22.52)$$

22.6 Interference as a Radio Frequency Distortion Mechanism

Interference here refers to the corruption of the received signal in the desired channel by adjacent or nearby channels. Although each channel has an assigned bandwidth, adjacent channel interference often occurs because practical transmission filters do not provide infinite signal suppression outside the desired or assigned bandwidth. Adjacent channel interference from a strong channel can be disastrous for a weak channel. The effect of interference will be discussed for AM, DSB-SC AM, SSB-SC AM, and angle modulation. For each case, f_c will represent the carrier frequency of the desired channel, whereas $f_c + \Delta f$ will represent the carrier frequency of the interfering signal. For simplicity, the interference will be regarded as an unmodulated carrier of amplitude A_i and the phase angle between the interference and the desired modulated wave will be ignored. Thus, the interference $r_i(t)$ at the receiver input is

$$r_i(t) = A_i \cos[2\pi(f_c + \Delta f)t] \tag{22.53}$$

Interference in Amplitude Modulation

In the case of AM, demodulation is performed by passing the received signal through an envelope detector. The received signal, which consists of the desired AM signal with carrier amplitude A_c, and the interference is given by

$$\begin{aligned} r(t) &= [A_c + m(t)] \cos(2\pi f_c t) + A_i \cos 2\pi (f_c + \Delta f)t \\ &= [A_c + m(t) + A_i \cos 2\pi \Delta f t] \cos(2\pi f_c t) - A_i \sin(2\pi \Delta f t) \sin(2\pi f_c t) \\ &= E(t) \cos[2\pi f_c t + \phi(t)] \end{aligned} \tag{22.54}$$

In the last equation, $E(t)$ is the envelope and $\phi(t)$ is the phase of the received signal. The phase is not relevant to the present discussion. The envelope is given by

$$E(t) = \{[A_c + m(t) + A_i \cos(2\pi \Delta f t)]^2 + A_i^2 \sin^2(2\pi \Delta f t)\}^{\frac{1}{2}} \tag{22.55}$$

From the expression for the envelope, it is easy to see that if $A_i \ll A_c$, the envelope is approximated by

$$E(t) \approx A_c + m(t) + A_i \cos 2\pi \Delta f t \tag{22.56}$$

Note that A_c in the last equation can be removed by DC blocking to yield the demodulator output $y(t)$. Let M be the peak amplitude of the message signal. Then it is observed that M/A_i, the ratio of the peak message signal amplitude to the peak interference amplitude, is equal at the input and the output of the receiver. Thus, for the case of small interference, the interference is additive at the demodulator output, and demodulation does not increase or decrease the strength of the interference relative to the message signal. The case for small interference is illustrated in Figure. 22.14. Fig. 22.14(a) illustrates the spectrum of the modulated signal and the interference at the receiver input. Figure 22.14(b) illustrates the spectrum of the envelope detector output.

For the case in which the interference amplitude is much larger than the carrier amplitude for the desired channel, the envelope $E(t)$ can be approximated by

$$E(t) \approx [A_c + m(t)] \cos 2\pi \Delta f t + A_i \tag{22.57}$$

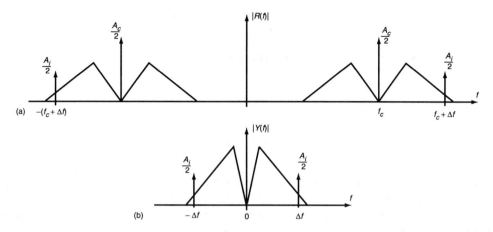

FIGURE 22.14 Illustration of interference in AM: (a) amplitude spectrum of AM wave plus interference, (b) amplitude spectrum of demodulated message signal plus interference.

In this case, after the constant A_i is blocked with a capacitor, the demodulated output consists of the message signal multiplied by the interference, $\cos 2\pi \Delta f\, t$ and an additive component of the interference, $A_c \cos 2\pi \Delta f t$. The effect of interference is much worse in this case because the multiplicative interference results in serious distortion.

Interference in DSB-SC AM

Consider the synchronous demodulation of DSB-SC signal plus the interfering sinusoid previously described. The same receiver of Fig. 22.9 is employed, but in this case, the phase and frequency errors between the received and the local carriers are ignored. The received signal plus interference $r(t)$, the product modulator output $x(t)$, and the demodulated output of the low-pass filter $y(t)$, are given, respectively, by

$$r(t) = m(t)\cos 2\pi f_c t + A_i \cos[2\pi(f_c + \Delta f)t] \tag{22.58}$$

$$x(t) = \frac{1}{2} m(t)[1 + \cos 4\pi f_c t] + \frac{1}{2} A_i [\cos 2\pi \Delta f t + \cos(4\pi f_c t + 2\pi \Delta f t)] \tag{22.59}$$

$$y(t) = \frac{1}{2} m(t) + \frac{1}{2} A_i \cos 2\pi \Delta f t \tag{22.60}$$

The last equation shows that for DSB-SC, the effect of interference is additive to the message signal at the demodulator output. Moreover, the ratio of the peak message output to the interference output M/A_i is equal at both the input and the output of the receiver.

Interference in SSB-SC

The analysis and the results in this case are very similar to those for DSB-SC. The same receiver is employed for both cases. In this case, the received signal plus interference, the product modulator output, and the demodulated output of the low-pass filter are given, respectively, by

$$r(t) = m(t) \cos 2\pi f_c t \mp m_h(t) \sin 2\pi f_c t + A_i \cos 2\pi (f_c + \Delta f) t \tag{22.61}$$

$$x(t) = \frac{1}{2} m(t)[1 + \cos 4\pi f_c t] \mp \frac{1}{2} M_h(t) \sin 4\pi f_c t +$$
$$\frac{1}{2} A_i [\cos 2\pi \Delta f t + \cos(4\pi f_c t + 2\pi \Delta f t)] \tag{22.62}$$

$$y(t) = \frac{1}{2} m(t) + \frac{1}{2} A_i(t) \cos 2\pi \Delta f t \tag{22.63}$$

The demodulated output in the last equation shows that the effect of interference on SSB-SC is the same as formerly described for DSB-SC. Comparing these results with the AM case, it is easy to see that for small interference, the effect is about the same for AM as for DSB-SC and SSB-SC. However, AM performance is quite inferior to the other two in the case of large interference.

Interference in Angle Modulation

Consider for simplicity, only the unmodulated carrier of an angle modulated wave of carrier frequency f_c given by $A_c \cos 2\pi f_c t$. For the sake of simplicity, the interference is also considered to be the unmodulated sinusoid $A_i \cos 2\pi (f_c + \Delta f) t$. The signal at the input to the receiver is

$$\begin{aligned} r(t) &= A_c \cos 2\pi f_c t + A_i \cos 2\pi (f_c + \Delta f) t \\ &= [A_c + A_i \cos 2\pi \Delta f t] \cos 2\pi f_c t - A_i \sin 2\pi \Delta f t \sin 2\pi f_c t \\ &= E(t) \cos[2\pi f_c t + \phi(t)] \end{aligned} \tag{22.64}$$

In the last equation, $E(t)$ is the envelope of the received signal, whereas $\phi(t)$ is its phase. In angle modulated waves, the phase, but not the envelope, is relevant to the demodulator output. The phase is given by

$$\begin{aligned} \phi(t) &= \tan^{-1} \left(\frac{-A_i \sin 2\pi \Delta f t}{A_c + A_i \cos 2\pi \Delta f t} \right) \\ &\approx -\frac{A_i \sin 2\pi \Delta f t}{A_c} \quad \text{for } A_i \ll A_c \end{aligned} \tag{22.65}$$

For PM, the demodulated signal is given by

$$y(t) = \phi(t) = -\frac{A_i}{A_c} \sin 2\pi \Delta f t \tag{22.66}$$

For FM, the demodulated signal is given by

$$y(t) = \phi'(t) = -\frac{2\pi \Delta f A_i}{A_c} \cos 2\pi \Delta f t \tag{22.67}$$

The last two equations show that for FM, the amplitude of the interference in the demodulated output is proportional to Δf, the difference between the carrier frequencies of the desired channel and the

Radio Frequency Distortion Mechanisms and Analysis

interference. On the other hand, the interference amplitude is constant for all values of Δf in the case of PM. These differences are illustrated in Fig. 22.15.

The last two equations also show that for both FM and PM, the interference at the demodulator output is inversely proportional to A_c, the amplitude of the carrier of the desired channel. Thus, the stronger the desired channel and the weaker the interference, the smaller is the effect of the interference. This is superior in performance to amplitude modulated systems in which, at the very best, the interference is independent of the carrier amplitude. In effect,

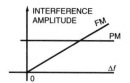

FIGURE 22.15 Effect of interference in angle modulation.

angle modulated systems suppress weak interference. This statement gives a clue to the behavior of angle modulated systems in the presence of large interference. When the interference is large relative to the desired signal, the so-called **capture effect** sets in. The interference now assumes the role of the desired channel and effectively suppresses the weaker desired channel, in the same way that a weak interference would have been suppressed.

Defining Terms

Capture effect: In angle modulation, the demodulated output of the weaker of two interfering signals is inversely proportional to the amplitude of the stronger signal. Consequently, the stronger signal or channel suppresses the weaker channel. If the stronger signal is noise or an undesired channel, it is said to have captured the desired channel.

Crosstalk: Interference between adjacent communication channels in which an interfering adjacent channel is heard or received in the desired channel.

Dispersion: The variation of wave propagation velocity (or time delay) in a medium with the frequency of the wave, which results in the spreading of the output signal over a time duration greater than the duration of the input signal.

Ionosphere: The ionosphere consists of several layers of the upper atmosphere, ranging in altitude from 100 to 300 km, containing ionized particles. The ionization is due to the effect of solar radiation, particularly ultraviolet and cosmic radiations, on gas particles. The ionosphere serves as reflecting layers for radio (sky) waves.

Pilot carrier: A means of providing a carrier at the receiver, which matches the received carrier in phase and frequency. In this method, which is employed in suppressed carrier modulation systems, a carrier of very small amplitude is inserted into the modulated signal prior to transmission, extracted and amplified in the receiver, and then employed as a matching carrier for coherent detection.

Radio horizon: The maximum range, from transmitter to receiver on the Earth's surface, of direct (line-of-sight) radio waves. This is greater than the optical horizon because the radio waves follow a curved path as a result of the continuous refraction it undergoes in the atmosphere.

Troposphere: The region of the atmosphere within about 10 km of the Earth's surface. Within this region, the wireless radio channel is modified relative to free space conditions by weather conditions, pollution, dust particles, etc. These inhomogeneities act as point scatterers which deflect radio waves downward to reach the receiving antennas, thereby providing tropospheric scatter propagation.

References

Bultitude, R.J.C., Melancon, P., Zaghloul, H., Morrison, G., and Prokiki, M. 1993. The dependence of indoor radio channel multipath characteristics on transmit receive ranges. *IEEE J. Selec. Areas Commun.* 11(7): 979–990.

Couch, L.W., II. 1990. *Digital and Analog Communication Systems*, 3rd ed. Macmillan, New York.

Fechtel, S.A. and Meyer, H. 1994. Optimal parametric feedforward estimation of frequency-selective fading radio channels. *IEEE Trans. Commun.* 42(2):1639–1650.
Gibson, J.D. 1993. *Principles of Digital and Analog Communications*, 2nd ed. Macmillan, New York.
Hashemi, H. 1993. Impulse response modeling of indoor radio propagation channels. *IEEE J. Selec. Areas Commun.* 11(7):967–978.
Haykin, S. 1994. *Communication Systems*, 3rd ed. J. Wiley, New York.
Lathi, B.P. 1989. *Modern Digital and Analog Communication Systems*, 2nd ed. Holt, Rinehart and Winston, Philadelphia, PA.
Porakis, J.G. and Salehi, M. 1994. *Communication Systems Engineering*. Prentice-Hall, Englewood Cliffs, NJ.
Roddy, D. and Coolen, J. 1981. *Electronic Communications*, 2nd ed. Reston Pub., Reston, VA.
Serboyakov, I.Y. 1989. An estimate of the effectiveness of using an amplitude-frequency equalizer to compensate selective fading on digital radio relay links. *Telecommun. Radio Eng.* 44(6):112–115.
Stremler, F.G. 1992. *Introduction to Communication Systems*, 3rd ed. Addison-Wesley, Reading, MA.
Taub, H. and Schilling, D. 1986. *Principles of Communication Systems*, 2nd ed. McGraw-Hill, New York.
Voronkov, Y.V. 1992. The generation of an AM oscillation with low nonlinear envelope distortion and low attendant angle modulation. *Telecomm. Radio Eng.* 47(7):10–12.
Ziemer, R.E. and Tranter, W.H. 1990. *Principles of Communications Systems, Modulation and Noise*, 3rd ed. Houghton Mifflin, Boston, MA.

Further Information

Electronics and Communication in Japan
IEEE Transactions on Antenna and Propagation
IEEE Transactions on Communications
IEEE Journal of Selected Areas in Communications
Telecommunications and Radio Engineering, a technical journal published by Scripta Technica, Inc., Wiley Co.

23
Digital Test Equipment and Measurement Systems

23.1	Introduction	23-1
23.2	Logic Analyzer	23-2
23.3	Signature Analyzer	23-4
23.4	Manual Probe Diagnosis	23-5
23.5	Checking Integrated Circuits	23-5
23.6	Emulative Tester	23-6
23.7	Protocol Analyzer	23-7
23.8	Automated Test Instruments	23-7
	Computer-Instrument Interface • Software Considerations • Applications	
23.9	Digital Oscilloscope	23-9
	Operating Principles • Digital Storage Oscilloscope (DSO) Features • Capturing Transient Waveforms • Triggering	

Jerry C. Whitaker
Editor-in-Chief

23.1 Introduction

As the equipment used by consumers and industry becomes more complex, the requirements for highly-skilled maintenance technicians also increases. Maintenance personnel today require advanced test equipment and must think in a *systems mode* to troubleshoot much of the hardware now in the field. New technologies and changing economic conditions have reshaped the way maintenance professionals view their jobs. As technology drives equipment design forward, maintenance difficulties will continue to increase. Such problems can be met only through improved test equipment and increased technician training.

Servicing computer-based professional equipment typically involves isolating the problem to the board level and then replacing the defective printed wiring board (PWB). Taken on a case-by-case basis, this approach seems efficient. The inefficiency in the approach, however (which is readily apparent), is the investment required to keep a stock of spare boards on hand. Further, because of the complex interrelation of circuits today, a PWB that appeared to be faulty may actually turn out to be perfect. The ideal solution is to troubleshoot down to the component level and replace the faulty device instead of swapping boards. In many cases, this approach requires sophisticated and expensive test equipment. In other cases, however, simple test instruments will do the job.

Although the cost of most professional equipment has been going up in recent years, maintenance technicians have seen a buyer's market in test instruments. The semiconductor revolution has done more than given consumers low-cost computers and disposable calculators. It has also helped to spawn a broad variety of inexpensive test instruments with impressive measurement capabilities.

23.2 Logic Analyzer

The logic analyzer is really two instruments in one: a *timing analyzer* and a *state analyzer*. The timing analyzer is analogous to an oscilloscope. It displays information in the same general form as a scope, with the horizontal axis representing time and the vertical axis representing voltage amplitude. The timing analyzer samples the input waveform to determine whether it is high or low. The instrument cares about only one voltage threshold. If the signal is above the threshold when it is sampled, it will be displayed as a 1 (or high); any signal below the threshold is displayed as a 0 (or low). From these sample points, a list of ones and zeros is generated to represent a picture of the input waveform. This data is stored in memory and used to reconstruct the input waveform, as shown in Fig. 23.1. A block diagram of a typical timing analyzer is shown in Fig. 23.2.

Sample points for the timing analyzer are developed by an internal clock. The period of the sample can be selected by the user. Because the analyzer samples asynchronously to the unit under test (UUT) under the direction of the internal clock, a long sample period results in a more accurate picture of the data bus. A sample period should be selected that permits an accurate view of data activity, while not filling up the instrument's memory with unnecessary data.

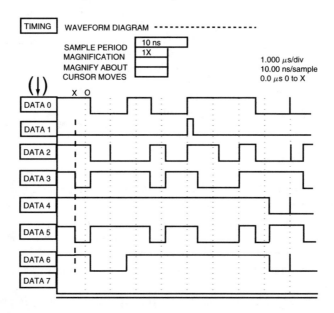

FIGURE 23.1 Typical display of a timing analyzer.

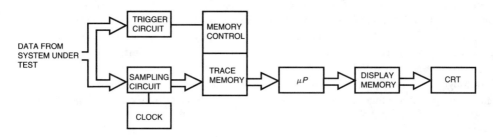

FIGURE 23.2 Block diagram of a timing analyzer.

Accurate sampling of data lines requires a trigger source to begin data acquisition. A fixed delay may be inserted between the trigger point and the *trace point* to allow for bus settling. This concept is illustrated in Fig. 23.3. A variety of trigger modes are commonly available on a timing analyzer, including:

- *Level triggering.* Data acquisition and/or display begins when a logic high or low is detected.
- *Edge triggering.* Data acquisition/disp lay begins on the rising or falling edge of a selected signal. Although many logic devices are level-dependent, clock and control signals are often edge-sensitive.
- *Bus state triggering.* Data acquisition/display is triggered when a specific code is detected (specified in binary or hexadecimal).

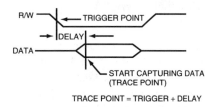

FIGURE 23.3 Use of a delay period between the trigger point and the trace point of a timing analyzer.

The timing analyzer is constantly taking data from the monitored bus. Triggering, and subsequent generation of the trace point, controls the *data window* displayed to the technician. It is possible, therefore, to configure the analyzer to display data that precedes the trace point. (See Fig. 23.4.) This feature can be a powerful troubleshooting and developmental tool.

The timing analyzer is probably the best method of detecting **glitches** in computer-based equipment. (A *glitch* being defined as any transition that crosses a logic threshold more than once between clock periods. See Fig. 23.5.) The triggering input of the analyzer is set to the bus line that is experiencing random glitches. When the analyzer detects a glitch, it displays the bus state preceding, during, or after occurrence of the disturbance.

The state analyzer is the second half of a logic analyzer. It is used most often to trace the execution of instructions through a microprocessor system. Data, address, and status codes are captured and displayed as they occur on the microprocessor bus. A **state** is a sample of the bus when the data are valid. A state is usually displayed in a tabular format of hexadecimal, binary, octal, or assembly language. Because some microprocessors multiplex data and addresses on the same lines, the analyzer must be able to clock-in information at different clock rates. The analyzer, in essence, acts as a demultiplexer to capture an address at the proper time, and then to capture data present on the same bus at a different point in time. A state analyzer also gives the operator the ability to *qualify* the data stored. Operation of the instrument may be triggered by a specific logic pattern on the bus. State analyzers usually offer a *sequence term* feature

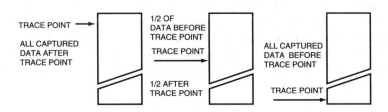

FIGURE 23.4 Data capturing options available from a timing analyzer.

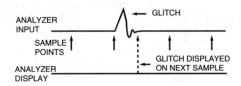

FIGURE 23.5 Use of a timing analyzer for detecting glitches on a monitored line.

that aids in triggering. A sequence term allows the operator to qualify data storage more accurately than would be possible with a single trigger point. A sequence term usually takes the following form:

   Find   xxxx
   Then find   yyyy
   Start on   zzzz

A sequence term is useful for probing a subroutine from a specific point in a program. It also makes possible selective storage of data, as shown in Fig. 23.6.

To make the acquired data easier to understand, most state analyzers include software packages that interpret the information. Such *disassemblers* (also known as *inverse assemblers*) translate hex, binary, or octal codes into assembly code (or some other format) to make them easier to read.

The logic analyzer is used routinely by design engineers to gain an in-depth look at signals within digital circuits. A logic analyzer can operate at 100 MHz and beyond, making the instrument ideal for detecting glitches resulting from timing problems. Such faults are usually associated with design flaws, not with manufacturing defects or failures in the field.

Although the logic analyzer has benefits for the service technician, its use is limited. Designers require the ability to verify hardware and software implementations with test equipment; service technicians simply need to quickly isolate a fault. It is difficult and costly to automate test procedures for a given PWB using a logic analyzer. The technician must examine a long data stream and decide if the data are good or bad. Writing programs to validate state analysis data is possible, but—again—is time-consuming.

23.3 Signature Analyzer

Signature analysis is a common troubleshooting tool based on the old analog method of signal tracing. In analog troubleshooting, the technician followed an annotated schematic that depicted waveforms and voltages that should be observed with an oscilloscope and voltmeter. The signature analyzer captures a complex data stream from a test point on the PWB and converts it into a hexadecimal signature. These hexadecimal signatures are easy to annotate on schematics, permitting their use as references for comparison against signatures obtained from a UUT. The signature by itself means nothing; the value of the signature analyzer comes from comparing a known-good signature with the UUT.

A block diagram of a simplified signature analyzer is shown in Fig. 23.7. As shown, a digital bit stream, accepted through a data probe during a specified measurement window, is marched through a 16-bit linear feedback shift register. Whatever data is left over in the register after the specified measurement window has closed is converted into a 4-bit hexadecimal readout. The feedback portion of the shift register allows just one faulty bit in a digital bit stream to create an entirely different signature than would be expected in a properly operating system. Hewlett-Packard, which developed this technique, terms it a *pseudo-random binary sequence* (PRBS) generator. This method allows the maintenance technician to identify a single bit error in a digital bit stream with 99.998% certainty, even when picking up errors that are timing-related.

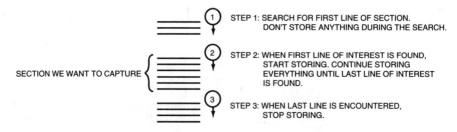

FIGURE 23.6 Illustration of the selective storage capability of state analyzer.

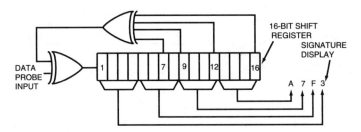

FIGURE 23.7 Simplified block diagram of a signature analyzer.

To function properly, the signature analyzer requires start and stop signals, a clock input, and the data input. The start and stop inputs are derived from the circuit being tested and are used to bracket the beginning and end of the measurement window. During this gate period, data is input through the data probe. The clock input controls the sample rate of data entering the analyzer. The clock is most often taken from the clock input pin of the microprocessor. The start, stop, and clock inputs may be configured by the technician to trigger on the rising or falling edges of the input signals.

23.4 Manual Probe Diagnosis

Manual probe diagnosis integrates the logic probe, logic analyzer, and signature analyzer to troubleshoot computer-based hardware. The manual probe technique employs a database consisting of nodal-level signature measurements. Each node of a known-good unit under test is probed and the signature information saved. This known-good data is then compared to the data from a faulty UUT. Efficient use of the manual probe technique requires a skilled operator to determine the proper order for probing circuit nodes, or a results-driven software package that directs the technician to the next logical step.

23.5 Checking Integrated Circuits

An integrated circuit tester identifies chip faults by generating test patterns that exercise all possible input-state combinations of the unit under test. For example, the NAND gate shown in Fig. 23.8 has a maximum of four input patterns that the instrument must generate to completely test the device. The input test patterns, and the resulting device outputs are arranged in a truth table. The table represents the response that should be obtained with specified input stimulus. The operator inputs the device type number, which configures the instrument to the proper pattern and truth table. As illustrated in Fig. 23.9, devices that match the table are certified as good; those failing are rejected.

Such conventional functional tests are well suited for out-of-circuit device qualification. Under out-of-circuit conditions, the instrument is free to toggle each input high or low, thereby executing all patterns of the truth table. For many in-circuit applications, however, inputs are wired in ways that prevent the instrument from executing the predetermined test pattern. Some instruments are able to circumvent the problem of nonstandard (in-circuit) configurations by including a *learning mode* in the system. Using this mode, the IC tester exercises the input of a gate that is known to be operating properly in a nonstandard configuration with the normal set of input signals. The resulting truth table is stored in memory. The instrument is then connected to an in-circuit device to be checked.

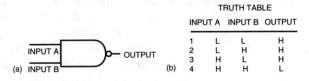

FIGURE 23.8 Truth table for a NAND gate.

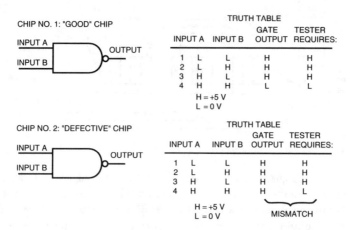

FIGURE 23.9 Component accept/reject process based on comparing observed states and an integrated circuit logic truth table.

23.6 Emulative Tester

Guided diagnostics can be used on certain types of hardware to facilitate semiautomated circuit testing. The test connector typically taps into one of the system board input/output ports and, armed with the proper software, the instrument goes through a step-by-step test routine that exercises the computer circuits. When the system encounters a fault condition, it outputs an appropriate message.

Emulative testers are a variation on the general guided diagnostics theme. Such an instrument plugs into the microprocessor socket of the computer and runs a series of diagnostic tests to identify the cause of a fault condition. An emulative tester emulates the board's microprocessor while verifying circuit operation. Testing the board *inside-out* allows easy access to all parts of the circuit; synchronization with the various data and address cycles of the PWB is automatic. Test procedures, such as read/write cycles from the microprocessor are generic, so that high quality functional tests can be quickly created for any board. Signature analysis is used to verify that circuits are operating correctly. Even with long streams of data, there is no maximum memory depth.

Several different types of emulative testers are available. Some are designed to check only one brand of computer, or only computers based on one type of microprocessor. Other instruments can check a variety of microcomputer systems using so-called *personality* modules that adapt the instrument to the system being serviced.

Guided-fault isolation (GFI) is practical with an emulative tester because the instrument maintains control over the entire system. Automated tests can isolate faults to the node level. All board information and test procedures are resident within the emulative test instrument, including prompts to the operator on what to do next. Using the microprocessor test connection combined with movable probes or clips allows a closed loop test of any part of a circuit. Input/output capabilities of emulative testers range from single point probes to well over a hundred I/O lines. These lines can provide stimulus as well as measurement capabilities.

The principal benefit of the guided probe over the manual probe is derived from the creation of a *topology database* for the UUT. The database describes devices on the UUT, their internal fault-propagating characteristics, and their interdevice connections. In this way, the tester can guide the operator down the proper logic path to isolate the fault. A guided probe accomplishes the same analysis as a conventional fault tree, but differs in that troubleshooting is based on a generic algorithm that uses the circuit model as its input. First, the system compares measurements taken from the failed UUT against the expected results. Next, it searches for a database representation of the UUT to determine the next logical place to take a measurement. The guided probe system automatically determines which nodes

Digital Test Equipment and Measurement Systems   **23**-7

impact other nodes. The algorithm tracks its way through the logic system of the board to the source of the fault. Programming of a guided probe system requires the following steps:

- Development of a stimulus routine to exercise and verify the operating condition of the UUT (go/no-go status).
- Acquisition and storage of measurements for each node affected by the stimulus routine. The programmer divides the system or board into measurement sets (MSETs) of nodes having a common timebase. In this way, the user can take maximum advantage of the time domain feature of the algorithm.
- Programming a representation of the UUT that depicts the interconnection between devices, and the manner in which signals can propagate through the system. Common *connectivity* libraries are developed and reused from one application to another.
- Programming a measurement set cross reference database. This allows the guided probe instrument to cross MSET boundaries automatically without operator intervention.
- Implementation of a test program control routine that executes the stimulus routines. When a failure is detected, the guided probe database is invoked to determine the next step in the troubleshooting process.

23.7 Protocol Analyzer

Testing a local area network (LAN)—whether copper, fiber optic, or wireless— presents special challenges to the design engineer or maintenance technician. The protocol analyzer is commonly used to service LANs and other communications systems. The protocol analyzer performs data monitoring, terminal simulation, and bit error rate tests (BERTs). Sophisticated analyzers provide *high-level decide* capability. This ability refers to the open system interconnection (OSI) network seven layer model. Typically, a sophisticated analyzer based on a personal computer can decide to level 3. PC-based devices also may incorporate features that permit performance analysis of wide area networks (WANs). Network statistics, including response time and use, are measured and displayed graphically. The characteristics of a line can be easily viewed and measured. These functions enable the engineer to observe activity of a communications link and exercise the facility to verify proper operation. Simulation capability is usually available to emulate almost any data terminal or data communications equipment.

LAN test equipment functions include monitoring of protocols and emulation of a network node, bridge, or terminal. Statistical information is provided to isolate problems to particular devices, or to high periods of activity. Statistical data includes use, packet rates, error rates, and collision rates.

23.8 Automated Test Instruments

Computers can function as powerful troubleshooting tools. Advanced technology personal computers, coupled with add-on interface cards and applications software, provide a wide range of testing capabilities. Computers can also be connected to many different instruments to provide for automated testing. There are two basic types of stand-alone automated test instruments (ATEs): *functional* and *in-circuit*.

Functional testing exercises the unit under test to identify faults. Individual PWBs or subassemblies may be checked using this approach. Functional testing provides a fast go/no-go qualification check. Dynamic faults are best discovered through this approach. Functional test instruments are well suited to high volume testing of PWB subassemblies, however, programming is a major undertaking. An intimate knowledge of the subassembly is needed to generate the required test patterns. An in-circuit emulator is one type of functional tester.

In-circuit testing is primarily a diagnostic tool. It verifies the functionality of the individual components of a subassembly. Each device is checked and failing parts are identified. In-circuit testing, while valuable for detailed component checking, does not operate at the clock rate of the subassembly. Propagation delays, race conditions, and other abnormalities may go undetected. Access to key points on the PWB may be accomplished in one of two ways:

- **Bed of nails.** The subassembly is placed on a dedicated test fixture and held in place by a vacuum or mechanical means. Probes access key electrical traces on the subassembly to check individual devices. Through a technique known as *backdriving*, the inputs of each device are isolated from the associated circuitry and the component is tested for functionality. This type of testing is expensive and is only practical for high-volume subassembly qualification testing and rework.
- **PWB clips.** Intended for lower volume applications than a bed of nails instrument, PWB clips replace the dedicated test fixture. The operator places the clips on the board as directed by the instrument. As access to all components simultaneously is not required, the tester is less hardware-intensive and programming is simplified. Many systems include a library of software routines designed to test various classes of devices. Test instruments using PWB clips tend to be slow. Test times for an average PWB may range from 8 to 20 min vs. one min or less for a bed of nails.

Computer-Instrument Interface

There are two primary nonproprietary types of computer interface systems used to connect test instruments and computers: IEEE-488 and RS-232. The IEEE-488 format is also called the *general purpose interface bus* (GPIB) or the *Hewlett Packard interface bus* (HPIB). RS-232 is the *standard serial interface* used on many computers. The two interface systems each have their own advantages. Both provide for connection of a computer to one or more measuring instruments. Both are bidirectional, which allows the computer to either send information or receive it from the outside world. Some systems provide both interfaces, but most have one or the other.

Test instruments utilizing the GPIB interface greatly outnumber those with RS-232. Several thousand test instruments are available with GPIB interfacing as an option. Some plotters and printers also accept a GPIB input.

By comparison, RS-232 is more common than GPIB in computer applications. Printers, plotters, scanners, and modems often use the standard serial interface. Test instruments incorporating RS-232 typically are those used for remote sensing, such as RF signal-strength meters or thermometers. Neither RS-232 nor GPIB is ideal for every application. Each protocol works well in some uses and marginally in others. The decision of which protocol to use for a particular application must be based on what the system needs to do. Because RS-232 is already built into personal computers, many users want to use it for automation. Yet GPIB is the preferred protocol for most test equipment applications.

Software Considerations

Most automated test instrument packages include software that permits the user to customize test instruments and procedures to meet the required task. The software, in effect, writes software. The user enters the codes needed by each automatic instrument, selects the measurements to be performed, and tells the computer where to store the test data. The program then puts the final software together after the user answers key configuration questions. The software then looks up the operational codes for each instrument and compiles the software to perform the required tests. Automatic generation of programming greatly reduces the need to have experienced programmers on staff. Once installed in the computer, programming becomes as simple as fitting graphic symbols together on the screen.

Applications

The most common applications for computer-controlled testing are data gathering, product go/no-go qualification, and troubleshooting. All depend on software to control the instruments. Acquiring data can often be accomplished with a computer and a single instrument. The computer collects dozens or hundreds of readings until the occurrence of some event. The event may be a preset elapsed time or the

Digital Test Equipment and Measurement Systems

occurrence of some condition at a test point, such as exceeding a preset voltage or dropping below a preset voltage. Readings from a variety of test points are then stored in the computer. Under computer direction, test instruments can also run checks that might be difficult or time-consuming to perform manually. The computer may control several test instruments such as power supplies, signal generators, and frequency counters. The data is stored for later analysis.

23.9 Digital Oscilloscope

The digital storage oscilloscope offers a number of significant advantages beyond the capabilities of analog instruments. A DSO can store in memory the signal being observed, permitting in-depth analysis impossible with previous technology. Because the waveform resides in memory, the data associated with the waveform can be transferred to a computer for real-time processing, or for processing at a later time.

Operating Principles

Figure 23.10 shows a block diagram of a DSO. Instead of being amplified and directly applied to the deflection plates of a cathode ray tube (CRT), the waveform is first converted into a digital form and stored in memory. To reproduce the waveform on the CRT, the data is sequentially read and converted back into an analog signal for display.

Although a DSO is specified by its maximum sampling rate, the actual rate used in acquiring a given waveform is usually dependent on the time-per-division setting of the oscilloscope. The record length (samples recorded over a given period of time) defines a finite number of sample points available for a given acquisition. The DSO must, therefore, adjust its sampling rate to fill a given record over the period set by the sweep control. To determine the sampling rate for a given sweep speed, the number of displayed points per division is divided into the sweep rate per division. Two additional features can modify the actual sampling rate:

- Use of an external clock for pacing the digitizing rate. With the internal digitizing clock disabled, the digitizer will be paced at a rate defined by the operator.
- Use of a peak detection (or glitch capture) mode. Peak detection allows the digitizer to sample at the full digitizing rate of the DSO, regardless of the time-base setting. The minimum and maximum values found between each normal sample interval are retained in memory. These minimum and maximum values are used to reconstruct the waveform display with the help of an algorithm that recreates a smooth trace along with any captured glitches. Peak detection allows the DSO to capture glitches even at its slowest sweep speed. For higher performance, a technique known as *peak-accumulation* (or *envelope mode*) may be used. With this approach, the instrument accumulates and displays the maximum and minimum excursions of a waveform for a given point in time. This builds an envelope of activity that can reveal infrequent noise spikes, long-term amplitude or time drift, and pulse jitter extremes. Figure 23.11 illustrates the advantage of peak accumulation when variations in data pulse width are monitored.

Achieving adequate samples for a high frequency waveform places stringent requirements on the sampling rate. High frequency waveforms require high sampling rates. *Equivalent-time sampling* is often used to provide high-bandwidth capture. This technique relies on sampling a repetitive waveform at a low rate to build up sample density. This concept is illustrated in Fig. 23.12. When the waveform to be

FIGURE 23.10 Simplified block diagram of a digital storage oscilloscope.

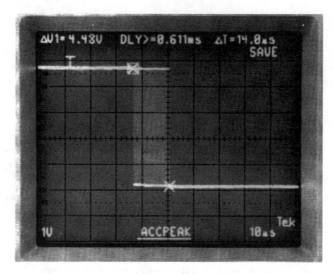

FIGURE 23.11 Use of the peak accumulation sampling mode to capture variations in pulse width. (Courtesy of Tektronix.)

acquired triggers the DSO, several samples are taken over the duration of the waveform. The next repetition of the waveform triggers the instrument again, and more samples are taken at different points on the waveform. Over many repetitions, the number of stored samples can be built up to the equivalent of a high sampling rate.

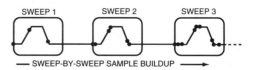

FIGURE 23.12 Increasing sample density through equivalent-time sampling.

System operating speed also has an effect on the display update rate. Update performance is critical when measuring waveform or voltage changes. If the display does not track the changes, adjustment of the circuit may be difficult.

There may be cases when, using the fastest sweep speed on a DSO, high speed sampling still does not provide as quick a display update as necessary. In this event, a conventional analog scope may provide the best performance. Some DSO instruments offer a switchable analog display path for such applications.

Digital Storage Oscilloscope (DSO) Features

The digital oscilloscope has become a valuable tool in troubleshooting both analog and computer-based products. Advanced components and construction techniques have led to lower costs for digital scopes and higher performance. Digital scopes can capture and analyze transient signals, such as race conditions, clock jitter, glitches, dropouts, and intermittent faults. Automated features reduce testing and troubleshooting costs through the use of recallable instrument setups, direct parameter readout, and unattended monitoring. Digital oscilloscopes have inherent benefits not available on most analog oscilloscopes. These benefits include:

- Increased resolution (determined by the quality of the analog-to-digital converter).
- Memory storage of digitized waveforms. Figure 23.13 shows a complex waveform captured by a DSO.
- Automatic setup for repetitive signal analysis. For complex multichannel configurations that are used often, front-panel storage/recall can save dozens of manual selections and adjustments. When multiple memory locations are available, multiple front-panel setups can be stored to save even more time.

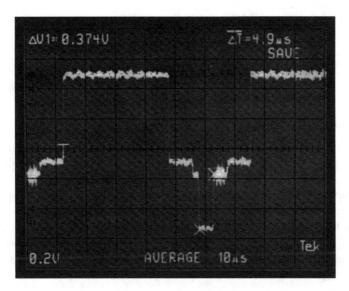

FIGURE 23.13 A digitized waveform "frozen" on the display to permit measurement and examination of detail. (Courtesy of Tektronix.)

- Auto-ranging. Many instruments will automatically adjust for optimum sweep, input sensitivity, and triggering. The instrument's microprocessor automatically configures the front panel for optimum display. Such features permit the operator to concentrate on making measurements, not on adjusting the scope.
- Instant hardcopy output from printers and plotters.
- Remote programmability via GPIB for automated test applications.
- Trigger flexibility. Single-shot digitizing oscilloscopes capture transient signals and allow the user to view the waveform which preceded the trigger point. Figure 23.14 illustrates the use of pre/post trigger for waveform analysis.
- Signal analysis. Intelligent scopes can make key measurements and comparisons. Display capabilities include voltage peak, mean voltage, RMS value, rise time, fall time, and frequency. Figure 23.15 shows the voltage measurement options available on one instrument.
- Cursor measurement. Advanced oscilloscopes permit the operator to take measurements or perform comparative operations on data appearing on the display. A measurement cursor consists of a pair of lines or dots that can be moved around the screen as needed. Figure 23.16 shows one such example. Cursors have been placed at different points on the displayed waveform. The instruments automatically determine the phase difference between the measurement points. The cursor readout follows the relative position of each cursor.
- Trace quality. Eye fatigue is reduced noticeably with a DSO when viewing low-repetition signals. For example, a 60-Hz waveform can be difficult to view for extended periods of time on a conventional scope because the display tends to flicker. A DSO overcomes this problem by writing waveforms to the screen at the same rate regardless of the input signal. (See Fig. 23.17.)

Digital memory storage offers a number of benefits, including:

- Reference memory. A previously acquired waveform can be stored in memory and compared with a sampled waveform. This feature is especially useful for repetitive testing or calibration of a device to a standard waveform pattern. Non-volatile battery-backed memory permits reference waveforms to be transported to field sites.
- Simple data transfers to a host computer for analysis or archive.

- Local data analysis through the use of a built-in microprocessor.
- Cursors capable of providing a readout of delta and absolute voltage and time.
- No CRT blooming for display of fast transients.
- Full bandwidth capture of long duration waveforms, thus storing all of the signal details. The waveform can be expanded after capture to expose the details of a particular section.

Capturing Transient Waveforms

Single-shot digitizing makes it possible to capture and clearly display transient and intermittent signals. Waveforms such as signal glitches, dropouts, logic race conditions, intermittent faults, clock jitter, and power-up sequences can be examined with the help of a digital oscilloscope. With single-shot digitizing, the waveform is captured the first time it occurs, on the first trigger. It can then be displayed immediately or held in memory for analysis at a later date. In contrast, most analog oscilloscopes can only capture repetitive signals. Hundreds of cycles of the signal are needed to construct a representation of the waveshape. Analog scopes are unable to capture transients. Figure 23.18 illustrates the benefits of digital storage in capturing transient waveforms. An analog scope will often fail to detect a transient pulse that a DSO can clearly display.

Triggering

Basic triggering modes available on a digital oscilloscope permit the user to select the desired source, its coupling, level, and slope. More advanced digital scopes contain triggering circuitry similar to that found in a logic analyzer. These powerful features let the user trigger on elusive conditions, such as pulse widths less than or greater than expected, intervals less than or greater than expected, and specified logic conditions. The logic triggering can include digital pattern, state qualified, and time/event qualified conditions. Many trigger modes are further enhanced by allowing the user to hold off the trigger by a selectable time or number of events. Hold-off is especially useful when the input signal contains bursts of data or follows a repetitive pattern.

Pulse-width triggering lets the operator quickly check for pulses narrower than expected or wider than expected. The pulse-width trigger circuit checks the time from the trigger source transition of a given

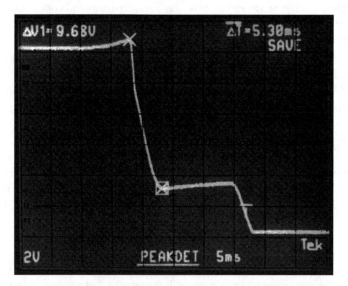

FIGURE 23.14 Use of the pre/post trigger function for waveform analysis. (Courtesy of Tektronix.)

slope (typically the rising edge) to the next transition of opposite slope (typically the falling edge). The operator can interactively set the pulse-width threshold for the trigger. For example, a glitch can be considered any signal narrower than 1/2 of a clock period. Conditions preceding the trigger can be displayed to show what events led up to the glitch.

Interval triggering lets the operator quickly check for intervals narrower than expected or wider than expected. Typical applications include monitoring for transmission phase changes in the output of a modem or for signal dropouts, such as missing bits on a computer hard disk.

Pattern triggering lets the user trigger on the logic state (high, low, or either) of several inputs. The inputs can be external triggers or the input channels themselves. The trigger can be generated either upon entering or exiting the pattern. Applications include triggering on a particular address select or

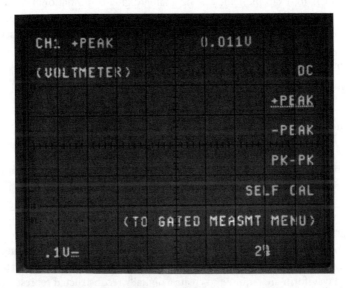

FIGURE 23.15 Menu of voltage measurement options available from a DSO. (Courtesy of Tektronix.)

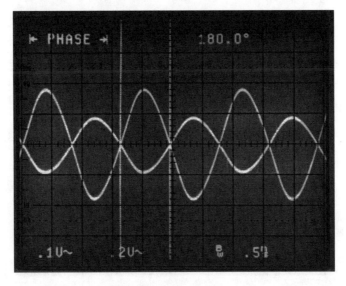

FIGURE 23.16 Use of cursors to measure the phase difference between two points on a waveform. (Courtesy of Tektronix.)

data bus condition. Once the pattern trigger is established, the operator can probe throughout the circuit, taking measurement synchronous with the trigger.

State qualified triggering enables the oscilloscope to trigger on one source, such as the input signal itself, only after the occurrence of a specified logic pattern. The pattern acts as an *enable* or *disable* for the source.

Advanced Features

Some digital oscilloscopes provide enhanced triggering modes that permit the user to select the level and slope for each input. This flexibility makes it easy to look for odd pulse shapes in the pulse-width trigger mode and for subtle dropouts in the interval trigger mode. It also simplifies testing different logic types (TTL, CMOS, and ECL, for example), and testing analog/digital combinational circuits with dual logic triggering. Additional flexibility is available on multiple channel scopes. Once a trigger has been sensed, multiple simultaneously sampled inputs permit the user to monitor conditions at several places in the unit under test, with each channel synchronized to the trigger. Additional useful trigger features for monitoring jitter or drift on a repetitive signal include:

- Enveloping
- Extremes
- Waveform delta
- Roof/floor

Each of these functions are related, and in some cases describe the same general operating mode. Various scope manufacturers use different nomenclature to describe proprietary triggering methods. Generally speaking, as the waveshape changes with respect to the trigger, the scope generates upper and lower traces. For every Nth sample point with respect to the trigger, the maximum and minimum values are saved. Thus, any jitter or drift is displayed in the envelope.

Advanced triggering features provide the greatest benefit in conjunction with single-shot sampling. Repetitive sampling scopes can only capture and display signals that repeat precisely from cycle to cycle. Several cycles of the waveform are required to create a digitally reconstructed representation of the input. If the signal varies from cycle to cycle, such scopes can be less effective than a standard analog oscilloscope for accurately viewing a waveform.

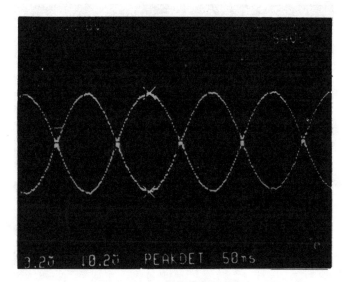

FIGURE 23.17 The DSO provides a stable, flicker-free display for viewing low-repetition signals. (Courtesy of Tektronix.)

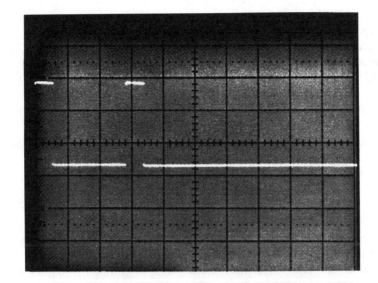

FIGURE 23.18 The benefits of a DSO in capturing transient signals: (a) analog display of a pulsed signal with transient present, but not visible; (b) DSO display of the same signal clearly showing the transient. (Courtesy of Tektronix.)

References

Albright, J. R., "Waveform Analysis with Professional-Grade Oscilloscopes," *Electronic Servicing and Technology*, Intertec Publishing, Overland Park, KS, April 1989.

Breya, M., "New Scopes Make Faster Measurements," *Mobile Radio Technology*, Intertec Publishing, Overland Park, KS, November 1988.

Carey, G. D., "Automated Test Instruments, *Broadcast Engineering*, Intertec Publishing, Overland Park, KS, November 1989.

Harris, B., "The Digital Storage Oscilloscope: Providing the Competitive Edge," *Electronic Servicing and Technology*, Intertec Publishing, Overland Park, KS, June 1988.

Harris, B., "Understanding DSO Accuracy and Measurement Performance," *Electronic Servicing and Technology*, Intertec Publishing, Overland Park, KS, April 1989.

Hoyer, M., "Bandwidth and Rise Time: Two Keys to Selecting the Right Oscilloscope," *Electronic Servicing and Technology*, Intertec Publishing, Overland Park, KS, April 1990.

Montgomery, S., "Advanced in Digital Oscilloscopes," *Broadcast Engineering*, Intertec Publishing, Overland Park, KS, November 1989.

Persson, C., "Oscilloscope Special Report," *Electronic Servicing and Technology*, Intertec Publishing, Overland Park, KS, April 1990.

Persson, C., "Oscilloscope: The Eyes of the Technician," *Electronic Servicing and Technology*, Intertec Publishing, Overland Park, KS, April 1987.

Persson, C., "Test Equipment for Personal Computers," *Electronic Servicing and Technology*, Intertec Publishing, Overland Park, KS, July 1987.

Persson, C., "The New Breed of Test Instruments," *Broadcast Engineering*, Intertec Publishing, Overland Park, KS, November 1989.

Siner, T. A., "Guided Probe Diagnosis: Affordable Automation," *Microservice Management*, Intertec Publishing, Overland Park, KS, March 1987.

Sokol, F., "Specialized Test Equipment," *Microservice Management*, Intertec Publishing, Overland Park, KS, August 1989.

Toorens, H., "Oscilloscopes: From Looking Glass to High-Tech," *Electronic Servicing and Technology*, Intertec Publishing, Overland Park, KS, April 1990.

Wickstead, M., "Signature Analyzers," *Microservice Management*, Intertec Publishing, Overland Park, KS, October, 1985.

Whitaker, J. C., *Maintaining Electronic Systems*, CRC Press, Boca Raton, FL, 1989.

Defining Terms

Bed of nails: A test fixture for automated circuit qualification in which a printed wiring board is placed in contact with a fixture that contacts the board at certain nodes required for exercising the assembly.

Bus state triggering: A data acquisition mode initiated when a specific digital code is detected.

Edge triggering: A data acquisition mode initiated by the rising or falling edge of a selected signal.

Equivalent-time sampling: An operating feature of a digital storage oscilloscope in which a repetitive waveform is sampled at a low rate to build up sample density.

Glitch: An undesired condition in a digital system in which a logic threshold transition occurs more than once between fixed or specified clock periods.

Level triggering: A data acquisition mode initiated when a logic high or low is detected.

Peak accumulation mode: An operating feature of digital storage oscilloscope in which the maximum and minimum excursions of the waveform are displayed for a given point in time.

State: In a digital system, a sample of the data bus when the data are valid.

Further Information

Most test equipment manufacturers provide detailed operational information on the use of their instruments. In addition, some companies offer applications booklets and engineering notes that explain test and measurement objectives, requirements, and procedures. Most of this information is available at little or no cost. The following book is also recommended:

Maintaining Electronic Systems, J. C. Whitaker, CRC Press, Boca Raton, FL, 1989.

24
Fourier Waveform Analysis

24.1	Introduction	24-1
24.2	The Mathematical Preliminaries for Fourier Analysis	24-3
24.3	The Fourier Series for Continuous-Time Periodic Functions	24-6
24.4	The Fourier Transform for Continuous-Time Aperiodic Functions	24-7
24.5	The Fourier Series for Discrete-Time Periodic Functions	24-7
24.6	The Fourier Transform for Discrete-Time Aperiodic Functions	24-9
24.7	Example Applications of Fourier Waveform Techniques	24-9
	Using the DFT/FFT in Fourier Analysis • Total Harmonic Distortion Measures	

Jerry C. Hamann
University of Wyoming

John W. Pierre
University of Wyoming

24.1 Introduction

Fourier waveform analysis originated in the early 1800s when Baron Jean Baptiste Joseph Fourier (1768–1830), a French mathematical physicist, developed these methods for investigating the conduction of heat in solid bodies. Fourier contended that rather complex **continuous-time waveforms** could be decomposed into a summation of simple trigonometric functions, sines and cosines of differing frequencies and magnitudes. When Fourier announced his results at a meeting of the French Academy of Scientists in 1807, his claims drew strong criticism from some of the most prominent mathematicians of the time, most notably Lagrange. Publication of his work in-full was delayed until 1822 when his seminal book, *Theorie Analytique de la Chaleur (The Analytical Theory of Heat)*, was finally released. Technical doubts regarding Fourier's methods were largely dispelled by the late 19th century through the efforts of Dirichlet, Riemann, and others.

Following this somewhat rocky beginning, the concepts of Fourier analysis have found application in many fields of science, mathematics, and engineering. Indeed, the study of Fourier techniques is now a traditional undertaking within undergraduate electrical engineering programs. With the availability of electronic computers, Fourier techniques have matured dramatically in the analysis of **discrete-time waveforms**. The now ubiquitous **fast Fourier transform (FFT)** is a somewhat general title applied to efficient algorithms for the machine computation of *discrete Fourier transforms* (DFTs). The probability is high that a given electronics technician has one or more instruments at his disposal with the capability of carrying out Fourier analysis, for example, see Fig. 24.1.

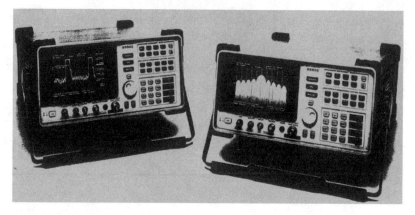

FIGURE 24.1 Fourier analyzers, spectrum analyzers, dynamic signal analyzers, and distortion analyzers are now common examples of benchtop instrumentation. Fourier capability is also popular in the new generation of digitizing oscilloscopes. (*Source:* Photograph courtesy of Hewlett-Packard Inc., Palo Alto, CA.)

24.2 The Mathematical Preliminaries for Fourier Analysis

The analysis techniques described in the following four sections can be conveniently categorized based on the variety of waveform or function to which they apply. A summary of useful properties of the analysis methods is provided in Table 24.1. We proceed herein to examine the mathematical preliminaries for the Fourier methods.

Most if not all **periodic** and **aperiodic** continuous-time **waveforms** $x(t)$, of practical interest, have Fourier series or Fourier transform counterparts. However, to aid in the existence question, the following sufficient technical constraints on the function $x(t)$, known generally as the *Dirichlet conditions*, guarantee convergence of the Fourier technique:

1. The function must be **single-valued**.
2. The function must have a finite number of discontinuities in the periodic interval, or, if aperiodic, about the entire real line.
3. The function must remain finite and have a finite number of maxima and minima in the periodic interval, or, if aperiodic, about the entire real line.
4. The function must be absolutely integrable, that is, if periodic,

$$\int_{t_0}^{t_0+T} |x(t)|\, dt$$

must be finite or if aperiodic,

$$\int_{-\infty}^{+\infty} |x(t)|\, dt$$

must be finite.

The techniques of Fourier analysis are perhaps best appreciated initially by an example. Consider the ideal continuous-time square wave of period T s, having a maximum value of A and a minimum value of 0 as shown at the top of Fig. 24.2. Because of the periodic nature of the waveform, the technique of Fourier series (FS) analysis applies, and the resulting decomposition into an infinite sum of simple trigonometric functions is given by

$$x(t) = \frac{A}{2} + \sum_{k=1,3,5,\ldots}^{\infty} \frac{2A}{k\pi}(-1)^{(k-1)/2}\cos(k\Omega_0 t) \qquad (24.1)$$

TABLE 24.1 Summary of Mathematical Properties of Fourier Analysis Methods

Fourier Series (FS)	Fourier Transform (FT)	Discrete-Time Fourier Transform (DTFT)	Discrete-Time Fourier Series (DTFS)
time domain: continuous and periodic	time domain: continuous and aperiodic	time domain: discrete and aperiodic	time domain: discrete and periodic
frequency domain: discrete and aperiodic	frequency domain: continuous and aperiodic	frequency domain: continuous and periodic	frequency domain: discrete and periodic
linearity	linearity	linearity	linearity
$r_1 x_1(t) + r_2 x_2(t) \leftrightarrow r_1 C_{1k} + r_2 C_{2k}$	$r_1 x_1(t) + r_2 x_2(t) \leftrightarrow r_1 X_1(\Omega) + r_2 X_2(\Omega)$	$r_1 x_1(n) + r_2 x_2(n) \leftrightarrow r_1 X_1(\omega) + r_2 X_2(\omega)$	$r_1 x_1(n) + r_2 x_2(n) \leftrightarrow r_1 c_{1k} + r_2 c_{2k}$
shift properties	shift properties	shift properties	shift properties
$x(t-\tau) \leftrightarrow e^{-j\Omega_0 k\tau} C_k$	$x(t-\tau) \leftrightarrow e^{-j\Omega\tau} X(\Omega)$	$x(n-m) \leftrightarrow e^{-j\omega m} X(\omega)$	$x(n-m) \leftrightarrow e^{-j\omega_0 km} c_k$
$e^{j\Omega_0 mt} x(t) \leftrightarrow C_{k-m}$	$e^{j\Omega_0 t} x(t) \leftrightarrow X(\Omega - \Omega_0)$	$e^{j\omega_0 n} x(n) \leftrightarrow X(\omega - \omega_0)$	$e^{j\omega_0 mn} x(n) \leftrightarrow c_{k-m}$
convolution	convolution	convolution	convolution
$x_1(t) * x_2(t) \leftrightarrow C_{1k} C_{2k}$	$x_1(t) * x_2(t) \leftrightarrow X_1(\Omega) X_2(\Omega)$	$x_1(n) * x_2(n) \leftrightarrow X_1(\omega) X_2(\omega)$	$x_1(n) * x_2(n) \leftrightarrow c_{1k} c_{2k}$
$x_1(t) x_2(t) \leftrightarrow C_{1k} * C_{2k}$	$x_1(t) x_2(t) \leftrightarrow 1/2\pi X_1(\Omega) * X_2(\Omega)$	$x_1(n) x_2(n) \leftrightarrow 1/2\pi X_1(\omega) * X_2(\omega)$	$x_1(n) x_2(n) \leftrightarrow c_{1k} * c_{2k}$
modulation	modulation	modulation	modulation
$x(t) \cos(\Omega_0 mt)$	$x(t) \cos(\Omega_0 t)$	$x(n) \cos(\omega_0 n)$	$x(n) \cos(\omega_0 mn)$
$\leftrightarrow \frac{1}{2}[C_{k+1} + C_{k-m}]$	$\leftrightarrow \frac{1}{2}[X(\Omega + \Omega_0) + X(\Omega - \Omega_0)]$	$\leftrightarrow \frac{1}{2}[X(\omega + \omega_0) + X(\omega - \omega_0)]$	$\leftrightarrow \frac{1}{2}[c_{k+m} + c_{k-m}]$
Parseval's Theorem	Parseval's Theorem	Parseval's Theorem	Parseval's Theorem
$\frac{1}{T}\int_{t_0}^{t_0+T} x_1(t) x_2^*(t) dt$	$\int_{-\infty}^{\infty} x_1(t) x_2^*(t) dt$	$\sum_{n=-\infty}^{\infty} x_1(n) x_2^*(n)$	$\frac{1}{N}\sum_{n=0}^{N-1} x_1(n) x_2^*(n)$
$= \sum_{k=-\infty}^{\infty} C_{1k} C_{2k}^*$	$= \frac{1}{2\pi}\int_{-\infty}^{\infty} X_1(\Omega) X_2^*(\Omega) d\Omega$	$= \frac{1}{2\pi}\int_{-\pi}^{\pi} X_1(\omega) X_2^*(\omega) d\omega$	$= \sum_{k=0}^{\infty} c_{1k} c_{2k}^*$
real signals	real signals	real signals	real signals
$C_k = C_{-k}^*$	$X(\Omega) = X^*(-\Omega)$	$X(\omega) = X^*(-\omega)$	$c_k = c_{-k}^*$
$\text{Re} C_k = \text{Re} C_{-k}$	$\text{Re} X(\Omega) = \text{Re} X(-\Omega)$	$\text{Re} X(\omega) = \text{Re} X(-\omega)$	$\text{Re} c_k = \text{Re} c_{-k}$
$\text{Im} C_k = -\text{Im} C_{-k}$	$\text{Im} X(\Omega) = -\text{Im} X(-\Omega)$	$\text{Im} X(\omega) = -\text{Im} X(-\omega)$	$\text{Im} c_k = (-\text{Im}) c_{-k}$
$|C_k| = |C_{-k}|$	$|X(\Omega)| = |X(-\Omega)|$	$|X(\omega)| = |X(-\omega)|$	$|c_k| = |c_{-k}|$
$\arg C_k = -\arg C_{-k}$	$\arg X(\Omega) = -\arg X(-\Omega)$	$\arg X(\omega) = -\arg X(-\omega)$	$\arg c_k = -\arg c_{-k}$
$x(t)$ real and even C_k real and even	$x(t)$ real and even $X(\Omega)$ real and even	$x(n)$ real and even $X(\omega)$ real and even	$x(n)$ real and even c_k real and even
$x(t)$ real and odd C_k imag. and odd	$x(t)$ real and odd $X(\Omega)$ imag. and odd	$x(t)$ real and odd $X(\omega)$ imag. and odd	$x(n)$ real and odd c_k imag. and odd

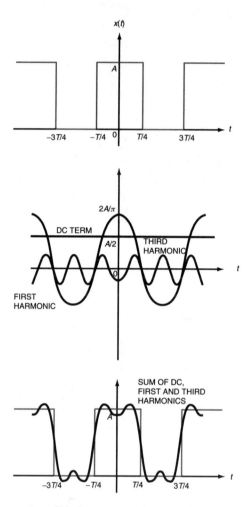

FIGURE 24.2 Example reconstruction of ideal square wave from Fourier series summation.

In Eq. (24.1), the average value or DC term of the waveform is given by $A/2$, whereas the **fundamental frequency** is described by $\Omega_0 = 2\pi/T$. Because the waveform is an **even function** only cosine terms are present. Because the waveform, less its average value, is characterized by **half-wave symmetry**, only the **odd harmonics** are present. To examine the reconstruction of the function $x(t)$ from the given decomposition, the middle plot of Fig. 24.2 displays the DC term, and the first and third harmonics. The sum of these terms is overlayed with the ideal square wave at the bottom of Fig. 24.2. This progressive reconstruction is continued in Fig. 24.3 with successive odd harmonics added as per Eq. (24.1). The ultimate goal of reconstructing the ideal square wave theoretically demands completion of the infinite sum, an unrealistic task of computation or analog signal summation. By truncating the summation at a large enough value of the harmonic number k, we are assured of having a best approximation of the ideal square wave in a least-square-error sense: that is, the truncated Fourier series is the best fit at any given truncation level $k = k_{max} < \infty$. The ringing or overshoot, which forms near the discontinuity of the ideal square wave, is described by the **Gibbs phenomenon**, in tribute to an early investigator of the effects of truncating Fourier series summations.

In practice, we are seldom presented with ideal functions such as the square wave of Fig. 24.2. Indeed, an analytical description of the waveform of interest, as is implied by the use of $x(t)$ in formulas of the next two sections, is rarely available. The discrete-time waveform counterparts of the Fourier series and Fourier transform provide viable alternatives for estimating the frequency content of signals if discrete

Fourier Waveform Analysis

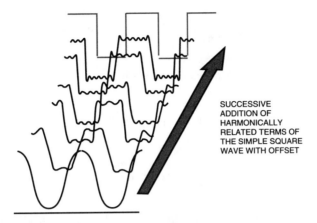

FIGURE 24.3 Continued reconstruction of ideal square wave from Fourier series summation.

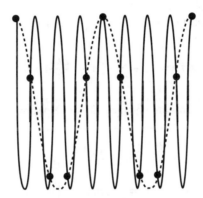

FIGURE 24.4 Example of aliasing wherein uniformly spaced samples from a high-frequency signal are indistinguishable from samples of a low-frequency signal.

measurements of the desired waveforms are available. This avenue of analysis, which is currently popular in benchtop instrumentation, is considered in further detail in Sec. 24.7.

One critical mathematical preliminary for the processing of discrete-time waveforms, which originate from sampling continous-time waveforms, is the issue of **sampling rate** and **aliasing**. The *Nyquist sampling theorem* can be summarized as follows:

If $x(t)$ is a **bandlimited** continuous-time waveform having no frequency content greater than or equal F_N HZ (the **Nyquist frequency**), then $x(t)$ is uniquely determined by a discrete-time waveform $x(n)$, which results from uniform sampling of $x(t)$ at a rate F_S, which is greater than or equal $2F_N$ (the **Nyquist rate**).

Failure to obey the Nyquist sampling theorem results in aliasing, that is, the effect of frequencies greater than the Nyquist frequency are reflected into the band of frequencies between DC and one-half the sampling rate. This situation is demonstrated in Fig. 24.4, where the samples of the high-frequency sinusoid are indistinguishable from a uniform sampling of the low-frequency sinusoid.

24.3 The Fourier Series for Continuous-Time Periodic Functions

Three forms of the FS are commonly encountered for continous-time waveforms $x(t)$ of period T. The first is the *trigonometric form*,

$$x(t) = \frac{a_0}{2} + \sum_{k=1}^{\infty} \left[a_k \cos\left(\frac{2\pi k}{T} t\right) + b_k \sin\left(\frac{2\pi k}{T} t\right) \right] \tag{24.2}$$

where the coefficients a_k and b_k are calculated via the integrals

$$a_k = \frac{2}{T} \int_{t_0}^{t_0+T} x(t) \cos\left(\frac{2\pi k}{T} t\right) dt \tag{24.3}$$

$$b_k = \frac{2}{T} \int_{t_0}^{t_0+T} x(t) \sin\left(\frac{2\pi k}{T} t\right) dt \tag{24.4}$$

with the representation for a_0 often appearing without the multiplier and divider of two [and therefore directly representing the average value of $x(t)$]. Symmetry properties of the particular function $x(t)$ can lead to simplifications of Eqs. (24.3) and (24.4).

A slight adaptation of this leads to the *alternative trigonometric form*,

$$x(t) = \frac{a_0}{2} + \sum_{k=1}^{\infty} A_k \cos\left(\frac{2\pi k}{T} t + \theta_k\right) \tag{24.5}$$

where the magnitude and phase terms are obtained via the respective identities

$$A_k = \sqrt{a_k^2 + b_k^2} \tag{24.6}$$

$$\theta_k = -\arctan\left(\frac{b_k}{a_k}\right) \tag{24.7}$$

Finally, the *complex exponential form* is given by

$$x(t) = \sum_{k=-\infty}^{\infty} C_k e^{j\frac{2\pi k}{T} t} \tag{24.8}$$

where the coefficients C_k are calculated via

$$C_k = \frac{1}{T} \int_{t_0}^{t_0+T} x(t) e^{-j\frac{2\pi k}{T} t} dt \tag{24.9}$$

It should be noted that the C_k are, in general, complex valued. They may be obtained from the trigonometric form data as follows:

$$|C_k| = |C_{-k}| = \frac{A_k}{2} \quad \text{for } k = 0, 1, 2, \ldots \tag{24.10}$$

$$\arg C_k = -\arg C_{-k} = \theta_k \quad \text{for } k = 0, 1, 2, \ldots \tag{24.11}$$

Common graphical representations of this data include plots of $|C_k|$ (or A_k) and θ_k, which are referred to as the *magnitude* and *phase spectra*, respectively. As is implied by the indexing of the spectral data by the integer k, the Fourier series of a continous-time waveform is itself discrete in character, that is, it takes on nonzero values at only discrete, isolated points of the frequency axis, namely, the frequencies $\Omega_k = 2\pi k/T$.

The average power in a periodic waveform can be found in one of two ways, with the equality noted below commonly attributed to Parseval:

$$P = \frac{1}{T}\int_{t_0}^{t_0+T} |x(t)|^2 dt = \sum_{k=-\infty}^{\infty} |C_k|^2. \tag{24.12}$$

Because of the right-hand side of Eq. (24.12), a plot of $|C_k|^2$ is commonly referred to as the *power spectral density*.

24.4 The Fourier Transform for Continuous-Time Aperiodic Functions

The Fourier transform (FT) for an aperiodic, continous-time waveform $x(t)$ is given by

$$X(\Omega) = F\{x(t)\} = \int_{-\infty}^{\infty} x(t)e^{-j\Omega t} dt \tag{24.13}$$

where the implication is that the frequency-domain counterpart $X(\Omega)$ of $x(t)$ is continuous in frequency and can be used to reconstruct $x(t)$ via

$$x(t) = F^{-1}\{X(\Omega)\} = \frac{1}{2\pi}\int_{-\infty}^{\infty} X(\Omega)e^{j\Omega t} d\Omega \tag{24.14}$$

If the integration described in Eq. (24.14) is carried out in hertz rather than radians per second, the division by the factor 2π is removed.

The transformation given by Eq. (24.13) can also be applied to some familiar periodic functions as well as functions that are not absolutely integrable. Example Fourier transform pairs are shown in Table 24.2.

The total energy in an aperiodic waveform can be found in one of two ways, with the equality noted commonly attributed to Parseval,

$$E = \int_{-\infty}^{\infty} |x(t)|^2 dt = \frac{1}{2\pi}\int_{-\infty}^{\infty} |X(\Omega)|^2 d\Omega \tag{24.15}$$

Because of the right-hand side of Eq. (24.15), a plot of $|X(\Omega)|^2$ is commonly referred to as the *energy spectral density*. One advantage to the integration vs frequency is the ability to determine the amount of energy in a signal due to frequency components within a particular range of frequencies.

24.5 The Fourier Series for Discrete-Time Periodic Functions

The discrete-time Fourier series (DTFS) for discrete-time waveforms $x(n)$ of period N can also be given in three forms; however, the *complex exponential form* is by far the most common,

$$x(n) = \sum_{k=0}^{N-1} c_k e^{j\frac{2\pi k}{N}n} \tag{24.16}$$

TABLE 24.2 Example Fourier Transform Pairs

$x(t)$	$x(t)$	$X(\Omega)$	$|X(\Omega)|$
	$\delta(t)$	1	
	1	$2\pi\delta(\Omega)$	
	$u(t)$	$\pi\delta(\Omega) + \dfrac{1}{j\Omega}$	
	$e^{-at}u(t), a > 0$	$\dfrac{1}{a+j\Omega}$	
	$te^{-at}u(t), a > 0$	$\dfrac{1}{(a+j\Omega)^2}$	
	$e^{-at}\cos(\Omega_0 t)u(t),\ a > 0$	$\dfrac{a+j\Omega}{(a+j\Omega)^2 + \Omega_0^2}$	
	$\cos(\Omega_0 t)$	$\pi[\delta(\Omega - \Omega_0) + \delta(\Omega + \Omega_0)]$	
	$1, |t| < \tau/2$; $0, |t| > \tau/2$	$\dfrac{\sin(\Omega\tau/2)}{\Omega/2}$	
	$1 - |t|/\tau, |t| < \tau$; $0, |t| > \tau$	$\dfrac{1}{\tau}\dfrac{\sin^2(\Omega\tau/2)}{(\Omega/2)^2}$	
	$\dfrac{\sin(\Omega_0 t/2)}{\pi t}$	$1, |\Omega| < \Omega_0$; $0, |\Omega| > \Omega_0$	

where the coefficients c_k are calculated via the finite summation

$$c_k = \frac{1}{N}\sum_{n=0}^{N-1} x(n) e^{-j\frac{2\pi k}{N} n} \quad (24.17)$$

Because of the periodicity of the exponential in Eq. (24.17), the DTFS coefficients c_k form a discrete, periodic sequence of period N if the index k is extended beyond the fundamental range of 0 to $N - 1$.

The average power in a periodic, discrete-time waveform can be found in one of two ways, with the equality noted commonly attributed to Parseval,

$$P = \frac{1}{N}\sum_{n=0}^{N-1}|x(n)|^2 = \sum_{k=0}^{N-1}|c_k|^2 \quad (24.18)$$

Because of the right-hand side of Eq. (24.18), a plot of $|c_k|^2$ is commonly referred to as the power spectral density.

24.6 The Fourier Transform for Discrete-Time Aperiodic Functions

The discrete-time Fourier transform (DTFT) for aperiodic, finite energy discrete-time waveforms $x(n)$ is given by

$$X(\omega) = \sum_{n=-\infty}^{\infty} x(n) e^{-j\omega n} \qquad (24.19)$$

where the implication is that the frequency-domain counterpart $X(\omega)$ of $x(n)$ is continuous in the discrete frequency variable ω and can be used to reconstruct $x(n)$ via

$$x(n) = \frac{1}{2\pi} \int_{-\pi}^{\pi} X(\omega) e^{j\omega n} d\omega \qquad (24.20)$$

If the integration described in Eq. (24.20) is carried out in cycles per sample rather than radian per sample frequency, the division by the factor 2π is removed.

The transformation given by Eq. (24.19) can also be applied to some periodic functions as well as functions that are not absolutely summable. Examples of discrete-time Fourier transform pairs are shown in Table 24.3.

The total energy in an aperiodic, discrete-time waveform can be found in one of two ways, with the equality noted commonly attributed to Parseval,

$$E = \sum_{n=-\infty}^{\infty} |x(n)|^2 = \frac{1}{2\pi} \int_{-\pi}^{\pi} |X(\omega)|^2 d\omega \qquad (24.21)$$

Because of the right-hand side of Eq. (24.21), a plot of $|X(\omega)|^2$ is commonly referred to as the energy spectral density.

24.7 Example Applications of Fourier Waveform Techniques

Using the DFT/FFT in Fourier Analysis

The FFT is arguably the most significant advancement in signal analysis in recent decades. The FFT is actually the name given to many algorithms used to efficiently compute the DFT. The DFT of $x(n)$ is defined as

$$X(k) = DFT\{x(n)\} = \sum_{n=0}^{N-1} x(n) e^{-j\frac{2\pi k}{N} n} \quad \text{for } k = 0, 1, 2, \ldots, N-1 \qquad (24.22)$$

and the inverse transform is given by

$$x(n) = IDFT\{X(k)\} = \frac{1}{N} \sum_{k=0}^{N-1} X(k) e^{j\frac{2\pi n}{N} k} \quad \text{for } n = 0, 1, 2, \ldots, N-1 \qquad (24.23)$$

TABLE 24.3 Example Discrete-Time Fourier Transform Pairs

$x(n)$	$x(n)$	$X(\omega)$	$|X(\omega)|$
(impulse at n=0)	$\delta(n)$	1	(constant 1)
(constant 1)	1	$\sum_{k=-\infty}^{\infty} 2\pi\delta(\omega + 2\pi k)$	(impulses at 2π)
(step)	$u(n)$	$\dfrac{1}{1-e^{-j\omega}} + \sum_{k=-\infty}^{\infty} \pi\delta(\omega + 2\pi k)$	
(decaying exp)	$e^{-an}u(n),\ a>0$	$\dfrac{1}{1-e^{-(a+j\omega)}}$	$(1-e^{-a})^{-1}$, $(1+e^{-a})^{-1}$
	$ne^{-an}u(n),\ a>0$	$\dfrac{e^{-(a+j\omega)}}{(1-e^{-(a+j\omega)})^2}$	$e^{-a}(1-e^{-a})^{-2}$, $e^{-a}(1+e^{-a})^{-2}$
	$e^{-an}\cos(\omega_0 n)u(n),\ a>0$	$\dfrac{1-\cos(\omega_0)e^{-(a+j\omega)}}{1-2\cos(\omega_0)e^{-(a+j\omega)} + e^{-2(a+j\omega)}}$	
	$\cos(\omega_0 n)$	$\pi \sum_{k=-\infty}^{\infty}[\delta(\omega-\omega_0+2\pi k) + \delta(\omega+\omega_0+2\pi k)]$	
	$1, |n|<L/2$; $0, |n|>(L-1)/2$; L an odd integer	$\dfrac{\sin(\omega L/2)}{\sin(\omega/2)}$	
	$1-|n|/L, |n|<L$; $0, |n|\geq L$; L an odd integer	$\dfrac{1}{L}\dfrac{\sin^2(\omega L/2)}{\sin^2(\omega/2)}$	
	$\dfrac{\sin(\omega_0 n)}{\pi n}$	$1, |\omega|<\omega_0$; $0, \omega_0<|\omega|<\pi$	

The DFT/FFT can be used to compute or approximate the four Fourier methods described in Secs. 24.3–24.6, that is the FS, FT, DTFS, and DTFT. Algorithms for the FFT are discussed extensively in the literature (see the Further Information entry for this topic).

The DTFS of a periodic, discrete-time waveform can be directly computed using the DFT. If $x(n)$ is one period of the desired signal, then the DTFS coefficients are given by

$$c_k = \frac{1}{N} DFT x(n) \quad \text{for } k = 0, 1, \ldots, N-1 \qquad (24.24)$$

The DTFT of a waveform is a continuous function of the discrete or sample frequency ω. The DFT can be used to compute or approximate the DTFT of an aperiodic waveform at uniformly separated sample frequencies $k\omega_0 = k2\pi/N$, where k is any integer between 0 and $N-1$. If N samples fully describe a finite duration $x(n)$, then

$$X(\omega)\big|_{\omega=\frac{2\pi}{N}k} = DFT x(n) \quad \text{for } k = 0, 1, \ldots, N-1 \qquad (24.25)$$

Fourier Waveform Analysis

If the $x(n)$ in Eq. (24.25) is not of finite duration, then the equality should be changed to an approximately equal. In many practical cases of interest, $x(n)$ is not of finite duration. The literature contains many approaches to truncating and **windowing** such $x(n)$ for Fourier analysis purposes.

The continuous-time FS coefficients can also be approximated using the DFT. If a continuous-time waveform $x(t)$ of period T is sampled at a rate of T_s samples per second to obtain $x(n)$, in accordance with the Nyquist sampling theorem, then the FS coefficients are approximated by

$$C_k \approx \frac{T_s}{T} DFT\{x(n)\} \quad \text{for } k = 0, 1, \ldots, N-1 \tag{24.26}$$

where N samples of $x(n)$ should represent precisely one period of the original $x(t)$. Using the DFT in this manner is equivalent to computing the integral in Eq. (24.9) with a simple rectangular or Euler approximate.

In a similar fashion, the continuous-time FT can be approximated using the DFT. If an aperiodic, continuous-time waveform $x(t)$ is sampled at a rate of T_s samples per second to obtain $x(n)$, in accordance with the Nyquist sampling theorem, then uniformly separated samples of the FT are approximated by

$$X(\Omega)\big|_{\Omega = \frac{2\pi}{NT_s}k} \approx T_s(DFT\{x(n)\}) \quad \text{for } k = 0, 1, \ldots, N-1 \tag{24.27}$$

For $x(t)$ that are not of finite duration, truncating and windowing can be applied to improve the approximation.

Total Harmonic Distortion Measures

The applications of Fourier analysis are extensive in the electronics and instrumentation industry. One typical application, the computation of total harmonic distortion (THD), is described herein. This application provides a measure of the nonlinear distortion, which is introduced to a pure sinusoidal signal when it passes through a system of interest, perhaps an amplifier. The root-mean-square (rms) total harmonic distortion (THD_{rms}) is defined as the ratio of the rms value of the sum of the harmonics, not including the fundamental, to the rms value of the fundamental.

$$THD_{rms} = \frac{\sqrt{\sum_{k=2}^{\infty} A_k^2}}{A_1} \tag{24.28}$$

As an example, consider the clipped sinusoidal waveform shown in Fig. 24.5. The Fourier series representation of this waveform is given by.

$$v(t) = A\left(\frac{1}{2} + \frac{1}{\pi}\right)\cos\left(\frac{2\pi}{T}t\right) + \frac{2\sqrt{2}A}{\pi} \sum_{k=3,5,7,\ldots}^{\infty} \frac{\sin\left(\frac{\pi}{4}k\right) - k\cos\left(\frac{\pi}{4}k\right)}{k(1-k^2)} \cos\left(\frac{2\pi}{T}kt\right) \tag{24.29}$$

The magnitude spectrum for this waveform is shown in Fig. 24.6. Because of symmetry, only the odd harmonics are present. Clearly the fundamental is the dominant component, but due to the clipping, many additional harmonics are present. The THD_{rms} is readily computed from the coefficients in Eq. (24.29), yielding 13.42% for this example. In practice, the coefficients of the Fourier series are typically approximated via the DFT/FFT, as described at the beginning of this section.

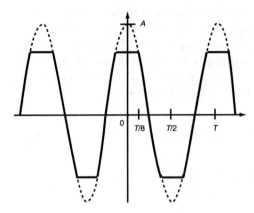

FIGURE 24.5 A clipped sinusoidal waveform for total harmonic distortion example.

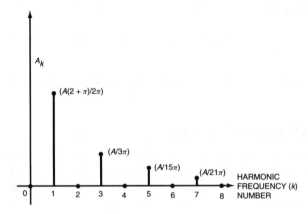

FIGURE 24.6 The magnitude spectrum, through eighth harmonic, for the chipped sinusoid total harmonic distortion example.

Defining Terms

Aliasing: Refers to an often detrimental phenomenon associated with the sampling of continuous-time waveforms at a rate below the Nyquist rate. Frequencies greater than one-half the sampling rate become indistinguishable from frequencies in the fundamental bandwidth, that is, between DC and one-half the sampling rate.

Aperiodic waveform: This phrase is used to describe a waveform that does not repeat itself in a uniform, periodic manner. Compare with periodic waveform.

Bandlimited: A waveform is described as bandlimited if the frequency content of the signal is constrained to lie within a finite band of frequencies. This band is often described by an upper limit, the Nyquist frequency, assuming frequencies from DC up to this upper limit may be present. This concept can be extended to frequency bands that do not include DC.

Continuous-time waveform: A waveform, herein represented by $x(t)$, that takes on values over the continuum of time t, the assumed independent variable. The Fourier series and Fourier transform apply to continuous-time waveforms. Compare with discrete-time waveform.

Discrete-time waveform: A waveform, herein represented by $x(n)$, that takes on values at a countable, discrete set of sample times or sample numbers n, the assumed independent variable. The discrete Fourier transform, the discrete-time Fourier series, and discrete-time Fourier transform apply to discrete-time waveforms. Compare with continuous-time waveform.

Even function: If a function $x(t)$ can be characterized as being a mirror image of itself horizontally about the origin, it is described as an even function. Mathematically, this demands that $x(t) = x(-t)$ for all t. The name arises from the standard example of an even function as a polynomial containing only even powers of the independent variable. A pure cosine is also an even function.

Fast Fourier transform (FFT): Title which is now somewhat loosely used to describe any efficient algorithm for the machine computation of the discrete Fourier transform. Perhaps the best known example of these algorithms is the seminal work by Cooley and Tukey in the early 1960s.

Fourier waveform analysis: Refers to the concept of decomposing complex waveforms into the sum of simple trigonometric or complex exponential functions.

Fundamental frequency: For a periodic waveform, the frequency corresponding to the smallest or fundamental period of repetition of the waveform is described as the fundamental frequency.

Gibbs phenomenon: Refers to an oscillatory behavior in the convergence of the Fourier transform or series in the vicinity of a discontinuity, typically observed in the reconstruction of a discontinuous $x(t)$. Formally stated, the Fourier transform does converge uniformly at a discontinuity, but rather, converges to the average value of the waveform in the neighborhood of the discontinuity.

Half-wave symmetry: If a periodic function satisfies $x(t) = -x(t - T/2)$, it is said to have half-wave symmetry.

Harmonic frequency: Refers to any frequency that is a positive integer multiple of the fundamental frequency of a periodic waveform.

Nyquist frequency: For a bandlimited waveform, the width of the band of frequencies contained within the waveform is described by the upper limit known as the Nyquist frequency.

Nyquist rate: To obey the Nyquist sampling theorem, a bandlimited waveform should be sampled at a rate which is at least twice the Nyquist frequency. This minimum sampling rate is known as the Nyquist rate. Failure to follow this restriction results in aliasing.

Odd function: If a function $x(t)$ can be characterized as being a reverse mirror image of itself horizontally about the origin, it is described as an odd function. Mathematically, this demands that $x(t) = -x(-t)$ for all t. The name arises from the standard example of an odd function as a polynomial containing only odd powers of the independent variable. A pure sine is also an odd function.

Periodic waveform: This phrase is used to describe a waveform that repeats itself in a uniform, periodic manner. Mathematically, for the case of a continuous-time waveform, this characteristic is often expressed as $x(t) = x(t \pm kT)$, which implies that the waveform described by the function $x(t)$ takes on the same value for any increment of time kT, where k is any integer and the characteristic value T, a real number greater than zero, describes the fundamental period of $x(t)$. For the case of a discrete-time waveform, we write $x(n) = x(n \pm kN)$, which implies that the waveform $x(n)$ takes on the same value for any increment of sample number kN, where k is any integer and the characteristic value N, an integer greater than zero, describes the fundamental period of $x(n)$.

Quarter-wave symmetry: If a periodic function displays half-wave symmetry and is *even* symmetric about the 1/4 and 3/4 period points (the negative and positive lobes of the function), then it is said to have quarter-wave symmetry.

Sampling rate: Refers to the frequency at which a continuous-time waveform is sampled to obtain a corresponding discrete-time waveform. Values are typically given in hertz.

Single-valued: For a function of a single variable, such as $x(t)$, single-valued refers to the quality of having one and only one value $y_0 = x(t_0)$ for any t_0. The square root is an example of a function which is not single-valued.

Windowing: A term used to describe various techniques for preconditioning a discrete-time waveform before processing by algorithms such as the discrete Fourier transform. Typical applications include extracting a finite duration approximation of an infinite duration waveform.

References

Bracewell, R.N. 1989. The Fourier transform. *Sci. Amer.* (June):86–85.
Brigham, E.O. 1974. *The Fast Fourier Transform.* Prentice-Hall, Englewood Cliffs, NJ.
Nilsson, J.W. 1993. *Electric Circuits,* 4th ed. Addison-Wesley, Reading, MA.
Oppenheim, A.V. and Schafer, R. 1989. *Discrete-Time Signal Processing.* Prentice-Hall, Englewood Cliffs, NJ.
Proakis, J.G. and Manolakis, D.G. 1992. *Digital Signal Processing: Principles, Algorithms, and Applications,* 2nd ed. Macmillan, New York.
Ramirez, R. W. 1985. *The FFT, Fundamentals and Concepts.* Prentice-Hall, Englewood Cliffs, NJ.

Further Information

For in-depth descriptions of practical instrumentation incorporating Fourier analysis capability, consult the following sources:

Witte, Robert A. *Spectrum and Network Measurements.* Prentice-Hall, Englewood Cliffs, NJ, 1993.

Witte, Robert A. *Electronic Test Instruments: Theory and Applications.* Prentice-Hall, Englewood Cliffs, NJ, 1993.

For further investigation of the history of Fourier and his transform, the following source should prove interesting:

Herviel, J. 1975. *Joseph Fourier: The Man and the Physicist.* Clarendon Press.

A thoroughly engaging introduction to Fourier concepts, accessible to the reader with only a fundamental background in trigonometry, can be found in the following publication:

Who Is Fourier? A Mathematical Adventure. Transnational College of LEX, English translation by Alan Gleason, Language Research Foundation, Boston, MA, 1995.

Perspectives on the FFT algorithms and their application can be found in the following sources:

Cooley, J.W. and Tukey, J.W. 1965. An algorithm for the machine calculation of complex Fourier series. *Mathematics of Computation* 19(April):297–301.

Cooley, J.W. 1992. How the FFT gained acceptance. *IEEE Signal Processing Magazine* (Jan.):10–13, 1992.

Heideman, M.T., Johnson, D.H., and Burrus, C.S. 1984. Gauss and the history of the fast fourier transform. *IEEE ASSP Magazine* 1:14–21.

Kraniauskas, P. 1994. A plain man's guide to the FFT. *IEEE Signal Processing Magazine* (April):24–35.

Press, W.H., Flannery, B.P., Teukolsky, S.A., and Vetterling, W.T. 1988. *Numerical Recipes in C: The Art of Scientific Computing.* Cambridge Univ. Press.

25
Computer-Based Signal Analysis

Rodger E. Ziemer
University of Colorado

25.1	Introduction... 25-1	
25.2	Signal Generation and Analysis................................. 25-1	
	Signal Generation • Curve Fitting • Statistical Data Analysis • Signal Processing	
25.3	Symbolic Mathematics 25-5	

25.1 Introduction

In recent years several mathematical packages have appeared for signal and system analysis on computers. Among these are **Mathcad**® [Math Soft 1995], **MATLAB**® [1995], and **Mathematica**® [Wolfram 1991], to name three general purpose packages. More specialized computer simulation programs include **SPW**® [Alta Group 1993] and **System View**® [Elanix 1995]. More specialized to electronic circuit analysis and design is **PSpice** [Rashid 1990] and **Electronics Workbench**® [Interactive 1993]. The purpose of this section is to discuss some of these tools and their utility for signal and system analysis by means of computer. Attention will be focused on the more general tool MATLAB. Several text books are now available that make extensive use of MATLAB in student exercises [Etter 1993, Gottling 1995, Frederick and Chow 1995].

25.2 Signal Generation and Analysis

Signal Generation

MATLAB is a vector- or array-based program. For example, if one wishes to generate and plot a sinusoid by means of MATLAB, the statements involved would be:

```
t = 0:.01:10;
x = sin(2*pi*t);
plot(t,x,'-w'), xlabel('t'), ylabel('x(t)'),grid
```

The first line generates a vector of values for the independent variable starting at 0, ending at 10, and spaced by 0.01. The second statement generates a vector of values for the dependent variable, $x = \sin(2\pi t)$, and the third statement plots the vector **x** vs the vector **t**. The resulting plot is shown in Fig. 25.1.

In Matlab, one has the options of running a program stored in an **m-file**, invoking the statements from the **command window**, or writing a **function** to perform the steps in producing the sinewave or other operation. For example, the command window option would be invoked as follows:

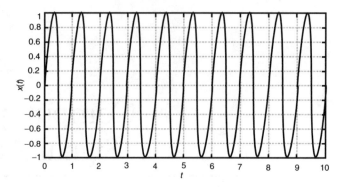

FIGURE 25.1 Plot of a sinusoid generated by MATLAB.

```
≫ t = 0:.01:10;
≫ x = sin(2*pi*t);
≫ plot(t,x,'–w'), xlabel('t'), ylabel('x(t)'), grid
```

The command window prompt is ≫ and each line is executed as it is typed and entered.

An example of a function implementation is provided by the generation of a unit step:

```
%    Function for generation of a unit step
function u = stepfn(t)
L = length(t);
u = zeros(size(t));
for i = 1:L
        if t(i) >= 0
                u(i) = 1;
        end
end
```

The command window statements for generation of a unit step starting at **t** = 2 are given next and a plot is provided in Fig. 25.2,

```
≫ t = −10:0.1:10;
≫ u = stepfn(t−2);
≫ plot(t,u, '−w'), xlabel('t'), ylabel('u(t)'), grid, title('unit step'), axis([−10 10 − 0.5 1.5])
```

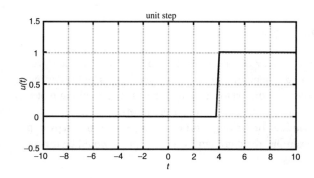

FIGURE 25.2 Unit step starting at **t** = 2 generated by the given step generation function.

Curve Fitting

MATLAB has several functions for fitting polynomials to data and a function for evaluation of a **polynomial fit** to the data. These functions include *table1, table2, spline,* and *polyfit*. The first one makes a linear fit to a set of data pairs, the second does a planar fit to data triples, and the third does a cubic fit to data pairs. The polyfit function does a **least-mean square-error** fit to data pairs. The uses of these are illustrated by the following program:

```
x = [0 1 2 3 4 5 6 7 8 9];
y = [0 20 60 68 77 110 113 120 140 135];
newx = 0:0.1:9;
newy = spline(x, y, newx);
for n = 1:5
        X = polyfit(x,y,n);
        f(:,n) = polyval(X,newx)';
end
subplot(321), plot(x,y,'w', newx,newy,'w',x,y,'ow'),axis([0 10 0 150]),grid
subplot(322), plot(newx,f(:,1),'w',x,y,'ow'),axis([0 10 0 150]),grid
subplot(323), plot(newx,f(:,2),'w',x,y,'ow'),axis([0 10 0 150]),grid
subplot(324), plot(newx,f(:,3),'w',x,y,'ow'),axis([0 10 0 150]),grid
subplot(325), plot(newx,f(:,4),'w',x,y,'ow'),axis([0 10 0 150]),grid
subplot(326), plot(newx,f(:,5),'w',x,y,'ow'),axis([0 10 0 150]),grid
```

Plots of the various fits are shown in Fig. 25.3. In the program, the linear interpolation is provided by the plotting routine itself, although the numerical value for the linear interpolation of a data point is provided by the statement *table1* (x, y, x_0) where x_0 is the x value that an interpolated y value is desired. The polyfit statement returns the coefficients of the least-squares fit polynomial of specified degree n to the data pairs. For example, the coefficients of the fifth-order polynomial returned by polyfit are -0.0150 0.3024 -1.9988 3.3400 25.0124 -1.1105 from highest to lowest degree. The **polyval** statement provides an evaluation of the polynomial at the element values of the vector *newx*.

Statistical Data Analysis

MATLAB has several **statistical data analysis functions**. Among these are random number generation, sample mean and standard deviation computation, histogram plotting, and correlation coefficient computation for pairs of random data. The next program illustrates several of these functions,

```
X = rand(1, 5000);
Y = randn(1, 5000);
mean_X = mean(X)
std_dev_X = std(X)
mean_Y = mean(Y)
std_dev_Y = std(Y)
rho = corrcoef(X, Y)
subplot(211), hist(X, [0 .1 .2 .3 .4 .5 .6 .7 .8 .9 1]), grid
subplot(212), hist(Y, 15), grid
```

The computed values returned by the program (note the semicolons left off) are mean_X = 0.5000; std_dev_X = 0.2883; mean_Y = -0.0194; std_dev_Y = 0.9958;

rho = 1.0000 0.0216
       0.0216 1.0000

The theoretical values are 0.5, 0.2887, 0, and 1, respectively, for the first four. The correlation coefficient matrix should have 1s on the main diagonal and 0s off the main diagonal. Histograms for the two cases of **uniform** and **Gaussian variates** are shown in Fig. 25.4. In the first plot statement, a vector giving the centers of the desired **histogram** bins is given. The two end values at 0 and 1 will have, on the average, half the values in the other bins since the random numbers generated are uniform in [0, 1]. In the second histogram plot statement, the number of bins is specified at 15.

Signal Processing

MATLAB has several **toolboxes** available for implementing special computations involving such areas as filter design and image processing. In this section we discuss a few of the special functions available in the **signal processing toolbox**.

A linear system, or filter, can be specified by the numerator and denominator polynomials of its **transfer function**. This is the case for both **continuous-time** and **discrete-time linear systems**. For

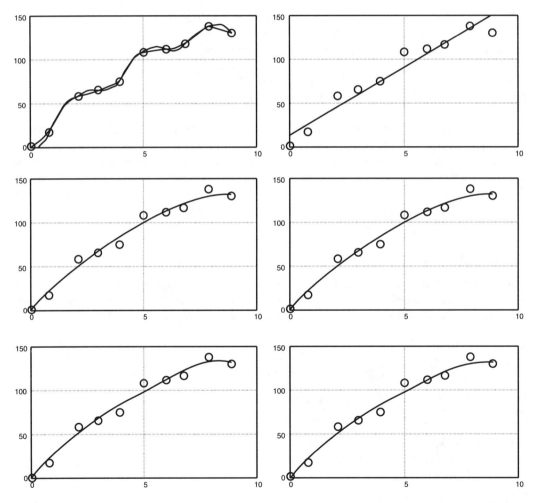

FIGURE 25.3 Various fits to the data pairs shown by the circles, from left to right and top to bottom: linear and spline fits, linear least-squares fit, quadratic least-squares fit, cubic least-squares fit, quartic least-squares fit, fifth-order least-squares fit.

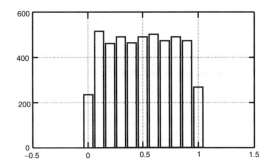 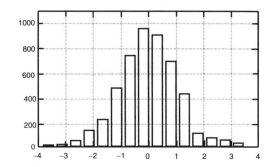

FIGURE 25.4 Histograms for 5000 pseudorandom numbers uniform in [0, 1] and 5000 Gaussian random numbers with mean zero and variance one.

example, the **amplitude, phase**, and **impulse responses** of the continuous-time linear system with transfer function

$$H(s) = \frac{s^3 + 2.5s^3 + 10s + 1}{s^3 + 10s^3 + 10s + 1} \quad (25.1)$$

can be found with the following MATLAB program:

```
%     Matlab example for creating Bode plots and step response for
%     a continuous-time linear system
%
num = [1 2.5 10 1];
den = [1 10 10 1];
[MAG,PHASE,W] = bode (num,den);
[Y,X,T] = step(num, den);
subplot(311),semilogx(W,20*log10(MAG), '—w'), xlabel('freq, rad/s'),...
      ylabel('mag. resp., dB'), grid, axis([0.1 100 − 15 5])

subplot(312),semilogx(W,PHASE,'—w'), xlabel('freq, rad/s'),...
      ylabel('phase resp.,  degrees'), grid, axis([0.1 100−40 40])
subplot(313),plot(T, Y,'−w'), grid, xlabel('time, sec'), ylabel('step resp.'),...
      axis([0 30 0 1.5])
```

Plots for these three response functions are shown in Fig. 25.5.

In addition, MATLAB has several other programs that can be used for filter design and signal analysis. The ones discussed here are meant to just give a taste of the possibilities available.

25.3 Symbolic Mathematics

MATLAB can manipulate variables symbolically. This includes algebra with scalar expressions, matrix algebra, linear algebra, calculus, differential equations, and transform calculus. For example, to enter a symbolic expression in the command window in MATLAB, one can do one of the following:

```
≫ A = 'cos(x)'
A =
cos(x)
```

```
>> B = sym('sin(x)')
B =
sin(x)
```

Once defined, it is a simple matter to perform symbolic operations on A and B. For example:

```
>> diff(A)
ans =
-sin(x)
>> int(B)
ans =
-cos(x)
```

To illustrate the Laplace transform capabilities of MATLAB, consider
```
>> F = laplace('t*exp(-3*t)')
F =
1/(s+3)^2
>> G = laplace('t^2*Heaviside(t)')
G =
2/s^2
>> h = invlaplace(symmul(F, G))
h =
2/9*t*exp(-3*t)+4/27*exp(-3*t)+2/9*t-4/27
```

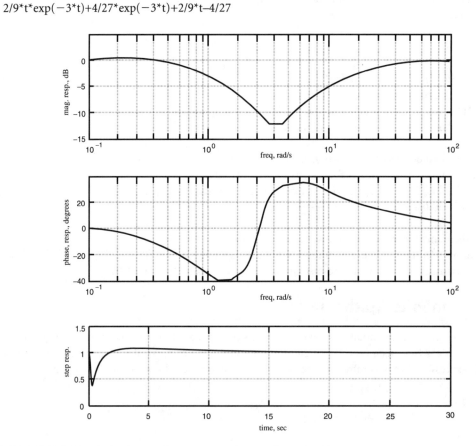

FIGURE 25.5 Magnitude, phase, and step responses for a continous-time linear system.

Alternatively, we could have carried out the symbolic multiply as a separate step:

≫ H = symmul(F, G)
H =
2/(s+3)^2/s^2
≫ h = invlaplace(H)
h =
2/9*t*exp(−3*t)+4/27*exp(−3*t)+2/9*t−4/27

Defining Terms

Amplitude (magnitude) response: The magnitude of the steady-state response of a fixed, linear system to a unit-amplitude input sinusoid.

Command window: The window in MATLAB in which the computations are done, whether with a direct command line or through an m-file.

Continuous-time fixed linear systems: A system that responds to continuous-time signals for which superposition holds and a time shift in the input results in the same output, but shifted by the amount of the time shift of the input.

Discrete-time fixed linear systems: A system that responds to discrete-time signals for which superposition holds and a time shift in the input results in the same output, but shifted by the amount of the time shift of the input.

Electronics workbench: An analysis/simulation computer program for electronic circuits, similar to Pspice.

Function: In MATLAB, a subprogram that implements a small set of statements that appear commonly enough to warrant their implementation.

Gaussian variates: Pseudorandom numbers in MATLAB, or other computational language, that obey a Gaussian, or bell-shaped, probability density function.

Histogram: A function in MATLAB, or other computational language, that provides a frequency analysis of random variates into contiguous bins.

Least-square-error fit: An algorithm or set of equations resulting from fitting a polynomial or other type curve, such as logarithmic, to data pairs such that the sum of the squared errors between data points and the curve is minimized.

Mathcad: A computer package like MATLAB that includes numerical analysis, programming, and symbolic manipulation capabilities.

Mathematica: A computer package like MATLAB that includes numerical analysis, programming, and symbolic manipulation capabilities.

MATLAB: A computer package that includes numerical analysis, programming, and symbolic manipulation capabilities.

M-file: A file in MATLAB that is the method for storing programs.

Phase response: The phase shift of the steady-state response of a fixed, linear system to a unit-amplitude input sinusoid relative to the input.

Polynomial fit: A function in MATLAB for fitting a polynomial curve to a set of data pairs. See least-squared-error fit. Another function fits a cubic spline to a set of data pairs.

Polyval: A function in MATLAB for evaluating a polynomial fit to a set of data pairs to a vector of abscissa values.

PSpice: An analysis/simulation computer program for electronic circuits, which originated at the University of California, Berkeley, as a batch processing program and was later adapted to personal computers using Windows.

Signal processing toolbox: A toolbox, or set of functions, in MATLAB for implementing signal processing and filter analysis and design.

SPW: A block-diagram oriented computer simulation package that is specifically for the analysis and design of signal processing and communications systems.
Statistical data analysis functions: Functions in MATLAB, or any other computer analysis package, that are specifically meant for analysis of random data. Functions include those for generation of pseudorandom variates, plotting histograms, computing sample mean and standard deviation, etc.
Step response: The response of a fixed, linear system to a unit step applied at time zero.
System view: A block-diagram oriented computer simulation package that is specifically for the analysis and design of signal processing and communications systems, but not as extensive as SPW.
Toolboxes: Sets of functions in MATLAB designed to facilitate the computer analysis and design of certain types of systems, such as communications, control, or image processing systems.
Transfer function: A ratio of polynomials in s that describes the input–output response characteristics of a fixed, linear system.
Uniform variates: Pseudorandom variates generated by computer that are equally likely to be any place within a fixed interval, usually [0, 1].

References

Alta Group. SPW.
Elanix. 1995. *SystemView® by Elanix: The Student Edition*. PWS Publishing, Boston, MA.
Etter, D.M. 1993. *Engineering Problem Solving Using MATLAB*. Prentice-Hall, Englewood Cliffs, NJ.
Frederick, D.K. and Chow, J.H. 1995. *Feedback Control Problems Using MATLAB and the Control System Toolbox*. PWS Publishing, Boston, MA.
Gottling, J.G. 1995. *Matrix Analysis of Circuits Using MATLAB*. Prentice-Hall, Englewood Cliffs, NJ.
Interactive. 1993. *Electronics Workbench, User's Guide*. Interactive Image Technologies, Ltd. Toronto, Ontario, Canada.
Mathsoft. 1995. *User's Guide—Mathcad®*. MathSoft, Inc. Cambridge, MA.
MATLAB. 1995. *The Student Edition of MATLAB®*. Prentice-Hall, Englewood Cliffs, NJ.
Rashid, M.H. 1990. *Spice for Circuits and Electronics Using PSpice*. Prentice-Hall, Englewood Cliffs, NJ.
Wolfram, S. 1991. *Mathematica: A System for Doing Mathematics by Computer,* 2nd ed. Addison-Wesley, New York.

Further Information

There are many books that can be referenced in regard to computer-aided analysis of signals and systems. Rather than add to the reference list, two mathematics books are suggested that give backgound pertinent to the development of many of the functions implemented in such computer analysis tools.

Kreyszig, E. 1988. *Advanced Engineering Mathematics*, 6th ed. Wiley, New York.
Smith, J.W. 1987. *Mathematical Modeling and Digital Simulation for Engineers and Scientists,* 2nd ed. Wiley, New York.

26
Systems Engineering Concepts

26.1	Introduction	26-1
26.2	Systems Theory	26-1
26.3	Systems Engineering	26-2
	Functional Analysis • Synthesis • Evaluation and Decision • Description of System Elements	
26.4	Phases of a Typical System Design Project	26-10
	Design Development • Electronic System Design • Program Management • Systems Engineer	
26.5	Conclusion	26-22

Gene DeSantis
DeSantis Associates

26.1 Introduction

Modern systems engineering emerged during World War II as, due to the degree of complexity in design, development, and deployment, weapons evolved into weapon systems. The complexities of the space program made a systems engineering approach to design and problem solving even more critical. Indeed, the Department of Defense and NASA are two of the staunchest practitioners. With the growth of digital systems, the need for systems engineering has gained increased attention. Today, most large engineering organizations utilize a systems engineering process. Much has been published about system engineering practices in the form of manuals, standards, specifications, and instruction. In 1969, MIL-STD-499 was published to help government and contractor personnel involved in support of defense acquisition programs. In 1974 this standard was updated to MIL-STD-499A, which specifies the application of system engineering principles to military development programs. The tools and techniques of this processes continue to evolve in order to do each job a little better, save time, and cut costs. This chapter will first describe systems theory in a general sense followed by its application in systems engineering and some practical examples and implementations of the process.

26.2 Systems Theory

Though there are other areas of application outside of the electronics industry, we will be concerned with systems theory as it applies to electrical systems engineering. Systems theory is applicable to engineering of control, information processing, and computing systems. These systems are made up of component elements that are interconnected and programmed to function together. Many systems theory principles are routinely applied in the aerospace, computer, telecommunications, transportation, and manufacturing industries.

For the purpose of this discussion a *system* is defined as a set of related elements that function together as a single entity.

Systems theory consists of a body of concepts and methods that guide the description, analysis and design of complex entities.

Decomposition is an essential tool of systems theory. The systems approach attempts to apply an organized methodology to completing large complex projects by breaking them down into simpler more manageable component elements. These elements are treated separately, analyzed separately, and designed separately. In the end, all of the components are recombined to build the whole.

Holism is an element of systems theory in that the end product is greater than the sum of its component elements. In systems theory, the modeling and analytical methods enable all essential effects and interactions within a system and those between a system and its surroundings to be taken into account. Errors resulting from the idealization and approximation involved in treating parts of a system in isolation, or reducing consideration to a single aspect, are thus avoided.

Another holistic aspect of system theory describes **emergent properties**. Properties that result from the interaction of system components, properties, that are not those of the components themselves, are referred to as emergent properties.

Though dealing with concrete systems, **abstraction** is an important feature of systems models. Components are described in terms of their function rather than in terms of their form. Graphical models such as block diagrams, flow diagrams, timing diagrams, and the like are commonly used.

Mathematical models may also be employed. Systems theory shows that, when modeled in abstract formal language, apparently diverse kinds of systems show significant and useful **isomorphisms** of structure and function. Similar interconnection structures occur in different types of systems. Equations that describe the behavior of electrical, thermal, fluid, and mechanical systems are essentially identical in form.

Isomorphism of structure and function implies isomorphism of behavior of a system. Different types of systems exhibit similar dynamic behavior such as response to stimulation.

The concept of **hard** and **soft systems** appears in system theory. In hard systems, the components and their interactions can be described by mathematical models. Soft systems cannot be described as easily. They are mostly human activity systems, which imply unpredictable behavior and non-uniformity They introduce difficulties and uncertainties of conceptualization, description, and measurement. The kinds of system concepts and methodology described earlier cannot be applied.

26.3 Systems Engineering

Systems engineering depends on the use of a process methodology based on systems theory. To deal with the complexity of large projects, systems theory breaks down the process into logical steps.

Even though underlying requirements differ from program to program, there is a consistent, logical process, which can best be used to accomplish system design tasks. The basic product development process is illustrated in Fig. 26.1. The systems engineering starts at the beginning of this process to describe the product to be designed. It includes four activities:

- Functional analysis
- Synthesis
- Evaluation and decision
- Description of system elements

The process is iterative, as shown in Fig. 26.2. That is, with each successive pass, the product element description becomes more detailed. At each stage in the process a decision is made whether to accept, make changes, or return to an earlier stage of the process and produce new documentation. The result of this activity is documentation that fully describes all system elements and that can be used to develop and produce the elements of the system. The systems engineering process does not produce the actual system itself.

Systems Engineering Concepts

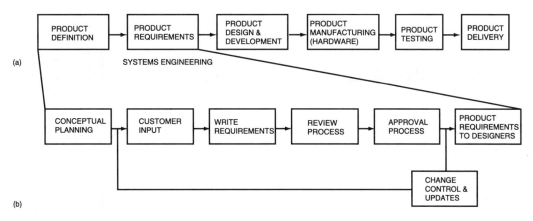

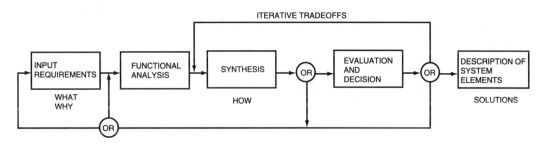

FIGURE 26.1 Systems engineering: (a) product development process, (b) requirements documentation process.

FIGURE 26.2 The systems engineering process. (After [4].)

Functional Analysis

A systematic approach to systems engineering will include elements of systems theory (see Fig. 26.3). To design a product, hardware and software engineers need to develop a vision of the product, the product requirements. These requirements are usually based on customer needs researched by a marketing department. An organized process to identify and validate customer needs will help minimize false starts. System objectives are first defined. This may take the form of a mission statement, which outlines the objectives, the constraints, the mission environment, and the means of measuring mission effectiveness. The purpose of the system is defined, and analysis is carried out to identify the requirements and what essential functions the system must perform and why. The *functional flow block diagram* is a basic tool used to identify functional needs. It shows logical sequences and relationships of operational and support functions at the system level. Other functions, such as maintenance, testing, logistics support, and productivity, may also be required in the functional analysis. The functional requirements will be used during the synthesis phase to show the allocation of the functional performance requirements to individual system elements or groups of elements. Following evaluation and decision, the functional requirements provide the functionally oriented data required in the description of the system elements.

Analysis of time critical functions is also a part of this functional analysis process when functions have to take place sequentially, or concurrently, or on a particular schedule. Time line documents are used to support the development of requirements for the operation, testing and maintenance functions.

Synthesis

Synthesis is the process by which concepts are developed to accomplish the functional requirements of a system. Performance requirements and constraints, as defined by the functional analysis, are applied

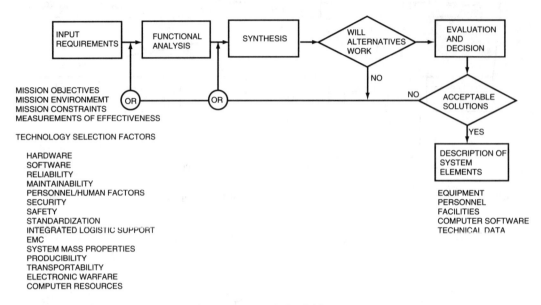

FIGURE 26.3 The systems engineering decision process. (After [4].)

to each individual element of the system, and a design approach is proposed for meeting the requirements. Conceptual schematic arrangements of system elements are developed to meet system requirements. These documents can be used to develop a description of the system elements and can be used during the acquisition phase.

Modeling

The concept of **modeling** is the starting point of synthesis. Since we must be able to weigh the effects of different design decisions in order to make choices between alternative concepts, modeling requires the determination of those quantitative features that describe the operation of the system. We would, of course, like a very detailed model with as much detail as possible describing the system. Reality and time constraints, however, dictate that the simplest possible model be selected to improve our chances of design success. The model itself is always a compromise. The model is restricted to those aspects that are important in the evaluation of system operation. A model might start off as a simple block diagram with more detail being added as the need becomes apparent.

Dynamics

Most system problems are **dynamic** in nature. The signals change over time and the components determine the dynamic response of the system. The system behavior depends on the signals at a given instant, as well as on the rates of change of the signals and their past values. The term signals can be replaced by substituting human factors such as the number of users on a computer network, for example.

Optimization

The last concept of synthesis is **optimization**. Every design project involves making a series of compromises and choices based on relative weighting of the merit of important aspects. The best candidate among several alternatives is selected. Decisions are often subjective when it comes to deciding the importance of various features.

Systems Engineering Concepts

Evaluation and Decision

Program costs are determined by the tradeoffs between operational requirements and engineering design. Throughout the design and development phase, decisions must be made based on evaluation of alternatives and their effect on cost. One approach attempts to correlate the characteristics of alternative solutions to the requirements and constraints that make up the selection criteria for a particular element. The rationale for alternative choices in the decision process are documented for review. Mathematical models or computer simulations may be employed to aid in this evaluation decision making process.

Trade Studies

A structured approach is used in the trade study process to guide the selection of alternative configurations and ensure that a logical and unbiased choice is made. Throughout development, **trade studies** are carried out to determine the best configuration that will meet the requirements of the program. In the concept exploration and demonstration phases, trade studies help define the system configuration. Trade studies are used as a detailed design analysis tool for individual system elements in the full-scale development phase. During production, trade studies are used to select alternatives when it is determined that changes need to be made. Figure 26.4 illustrates the relationship of the various types of elements that may be employed in a trade study.

Figure 26.5 is a flow diagram of the trade study process. To provide a basis for the selection criteria, the objectives of the trade study must first be defined. Functional flow diagrams and system block

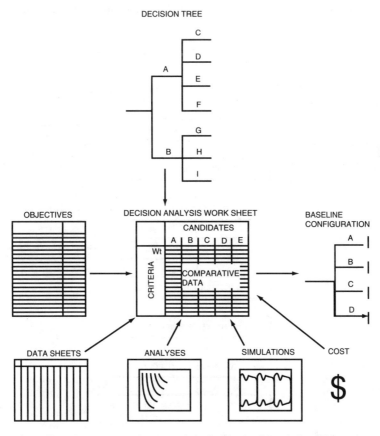

FIGURE 26.4 Trade studies using a systematic approach to decision making. (After [1].)

diagrams are used to identify trade study areas that can satisfy certain requirements. Alternative approaches to achieving the defined objectives can then be established.

Complex approaches can be broken down into several simpler areas, and a decision tree constructed to show the relationship and dependences at each level of the selection process. This *trade tree*, as it is called, is illustrated in Fig. 26.6. Several trade study areas may be identified as possible candidates for accomplishing a given function. A trade tree is constructed to show relationships and the path through selected candidate trade areas at each level to arrive at a solution.

Several alternatives may be candidates for solutions in a given area. The selected candidates are then submitted to a systematic evaluation process intended to weed out unacceptable candidates. Criteria are determined, which are intended to reflect the desirable characteristics. Undesirable characteristics may also be included to aid in the evaluation process. Weights are assigned to each criterion to reflect its value or impact on the selection process. This process is subjective. It should also take into account cost, schedule, and hardware availability restraints that may limit the selection.

The criteria data on the candidates is then collected and tabulated on a decision analysis work sheet (Fig. 26.7). The attributes and limitations are listed in the first column and the data for each candidate listed in adjacent columns to the right. The performance data is available from vendor specification sheets or may require laboratory testing and analysis to determine. Each attribute is given a relative score from 1 to 10 based on its comparative performance relative to the other candidates. Utility function graphs (Fig. 26.8) can be used to assign logical scores for each attribute. The utility curve represents the advantage rating for a particular value of an attribute. A graph is made of ratings on the y axis vs attribute value on the x axis. Specific scores can then be applied, which correspond to particular performance values.

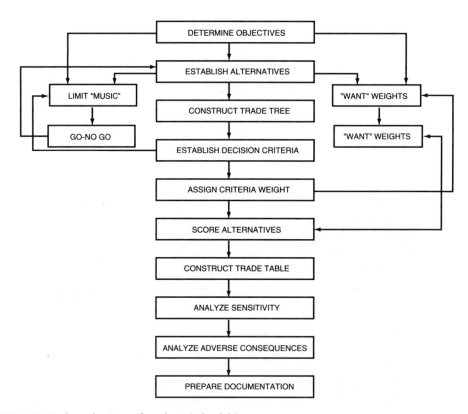

FIGURE 26.5 Trade study process flow chart. (After [1].)

Systems Engineering Concepts

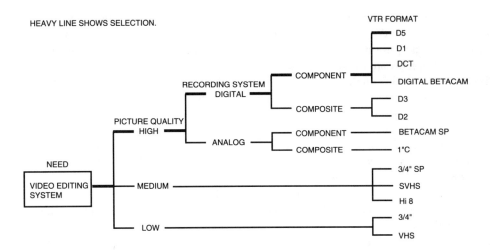

FIGURE 26.6 An example trade tree.

ALTERNATIVES:	WT	CANDIDATE 1			CANDIDATE 2			CANDIDATE 3		
WANTED			SC	WT SC		SC	WT SC		SC	WT SC
VIDEO BANDWIDTH MHz	10	5.6	10	(100)	6.0	10	(100)	5.0	9	90
SIGNAL-TO-NOISE RATIO, dB	10	60	8	80	54	6	60	62	10	(100)
10-bit QUANTIZING	10	YES	1	10	YES	1	10	YES	1	10
MAX PROGRAM LENGTH, h	10	2	2	20	3	3	(30)	1.5	1.5	15
READ BEFORE WRITE CAPABLE	5	YES	1	(5)	YES	1	(5)	NO	0	0
AUDIO PITCH CORRECTION AVAIL	5	YES	1	(5)	NO	0	0	YES	1	(5)
CAPABLE OF 16:9 ASPECT RATIO	10	NO	0	0	YES	1	(10)	YES	1	(10)
EMPLOYS COMPRESSION	–5	YES	1	–5	NO	0	(0)	YES	1	–5
SDIF (SERIAL DIGITAL INTERFACE) BUILT IN	10	YES	1	(10)	YES	1	(10)	YES	1	(10)
CURRENT INSTALLED BASE	8	MEDIUM	2	(16)	LOW	1	8	LOW	1	8
TOTAL WEIGHTED SCORE:				241			234			(243)

FIGURE 26.7 Decision analysis work sheet example. (After [1].)

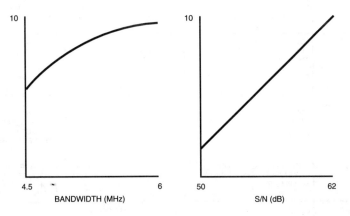

FIGURE 26.8 Attribute utility trade curve example.

The shape of the curve may take into account requirements, limitations, and any other factor that will influence its value regarding the particular criteria being evaluated. The limits to which the curves should be extended should run from the minimum value, below which no further benefit will accrue, to the maximum value, above which no further benefit will accrue.

The scores are filled in on the decision analysis work sheet and multiplied by the weights to calculate the weighted score. The total of the weighted scores for each candidate then determines their ranking. As a rule, at least a 10% difference in score is acceptable as meaningful.

Further analysis can be applied in terms of evaluating the sensitivity of the decision to changes in the value of attributes, weights, subjective estimates, and cost. Scores should be checked to see if changes in weights or scores would reverse the choice. How sensitive is the decision to changes in the system requirements or technical capabilities?

A **trade table** (Fig. 26.9) can be prepared to summarize the selection results. Pertinent criteria are listed for each alternative solution. The alternatives may be described in a quantitative manner, such as high, medium, or low.

Finally, the results of the trade study are documented in the form of a report, which discusses the reasons for the selections and may include the trade tree and the trade table.

There has to be a formal system of change control throughout the systems engineering process to prevent changes from being made without proper review and approval by all concerned parties and to keep all parties informed. Change control also ensures that all documentation is kept up to date and can help to eliminate redundant documents. Finally, change control helps to control project costs.

Description of System Elements

Five categories of interacting system elements can be defined: equipment (hardware), software, facilities, personnel, and procedural data. Performance, design, and test requirements must be specified and documented for equipment, components, and computer software elements of the system. It may be necessary to specify environmental and interface design requirements, which are necessary for proper functioning of system elements within a facility.

The documentation produced by the systems engineering process controls the evolutionary development of the system. Figure 26.10 illustrates the special purpose documentation used by one organization in each step of the systems engineering process.

CRITERIA	COOL ROOM ONLY. ONLY NORMAL CONVECTION COOLING WITHIN ENCLOSURES	FORCED COLD AIR VENTILATION THROUGH RACK THEN DIRECTLY INTO RETURN	FORCED COLD AIR VENTILATION THROUGH RACK, EXHAUSTED INTO THE ROOM, THEN RETURNED THRU THE NORMAL PLENUM
COST	LOWEST. CONVENTIONAL CENTRAL AIR CONDITIONING SYSTEM USED.	HIGH. DEDICATED DUCTING REQUIRED. SEPARATE SYSTEM REQUIRED TO COOL ROOM	MODERATE. DEDICATED DUCTING REQUIRED FOR INPUT AIR.
PERFORMANCE EQUIPMENT TEMPERATURE ROOM TEMPERATURE	POOR 80–120° F+ 65–70° F TYPICAL AS SET	VERY GOOD 55–70° F TYPICAL 65–70° F TYPICAL AS SET	VERY GOOD 55–70° F TYPICAL 65–70° F TYPICAL AS SETF
CONTROL CONTROL OF EQUIPMENT TEMPERATURE	POOR. HOT SPOTS WILL OCCUR WITHIN ENCLOSURES.	VERY GOOD.	VERY GOOD. WHEN THE THERMOSTAT IS SET TO PROVIDE A COMFORTABLE ROOM TEMPERATURE, THE ENCLOSURE WILL BE COOL INSIDE.
CONTROL OF ROOM TEMPERATURE	GOOD. HOT SPOTS MAY STILL EXIST NEAR POWER HUNGRY EQUIPMENT.	GOOD.	GOOD. IF THE ENCLOSURE EXHAUST AIR IS COMFORTABLE FOR OPERATORS, THE INTERNAL EQUIPMENT MUST BE COOL.
OPERATOR COMFORT	GOOD	GOOD. SEPARATE ROOM VENTILATION SYSTEM REQUIRED CAN BE SET FOR COMFORT.	GOOD. WHEN THE THERMOSTAT IS SET TO PROVIDE A COMFORTABLE ROOM TEMPERATURE, THE ENCLOSURE WILL BE COOL INSIDE.

FIGURE 26.9 Trade table example.

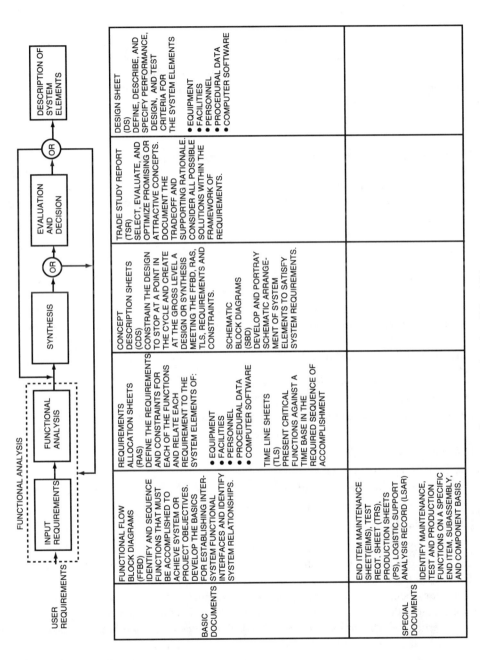

FIGURE 26.10 Basic and special purpose documentation for system engineering. (After [4].)

The requirements are formalized in written specifications. In any organization, there should be clear standards for producing specifications. This can help reduce the variability of technical content and improve product quality as a result. It is also important to make the distinction here that the product should not be overespecified to the point of describing the design or making it too costly. On the other hand, requirements should not be too general or so vague that the product would fail to meet the customer needs. In large departmentalized organizations, commitment to schedules can help assure that other members of the organization can coordinate their time.

The system engineering process does not actually design the system. The system engineering process produces the documentation necessary to define, design, develop, and test the system. The technical integrity provided by this documentation ensures that the design requirements for the system elements reflect the functional performance requirements, that all functional performance requirements are satisfied by the combined system elements, and that such requirements are optimized with respect to system performance requirements and constraints.

26.4 Phases of a Typical System Design Project

The television industry has always been a dynamic industry because of the rapid advancement of communications technology. The design of a complex modern video facility can be used to illustrate the systems engineering approach.

Design Development

System design is carried out in a series of steps that lead to an operational unit. Appropriate research and preliminary design work is completed in the first phase of the project, the design development phase. It is the intent of this phase to fully delineate all requirements of the project and to identify any constraints. Based on initial concepts and information, the design requirements are modified until all concerned parties are satisfied and approval is given for the final design work to proceed. The first objective of this phase is to answer the following questions:

- What are the functional requirements of the product of this work?
- What are the physical requirements of the product of this work?
- What are the performance requirements of the product of this work?
- Are there any constraints limiting design decisions?
- Will existing equipment be used?
- Is the existing equipment acceptable?
- Will this be a new facility or a renovation?
- Will this be a retrofit or upgrade to an existing system?
- Will this be a stand-alone system?

Working closely with the customer's representatives, the equipment and functional requirements of each of the major technical areas of the facility are identified. In the case of facility renovation, the systems engineer's first order of business is to analyze existing equipment. A visit is made to the site to gather detailed information about the existing facility. Usually confronted with a mixture of acceptable and unacceptable equipment, the systems engineer must sort out those pieces of equipment that meet current standards and determine which items should be replaced. Then, after soliciting input from the facility's technical personnel, the systems engineer develops a list of needed equipment.

One of the systems engineer's most important contributions is the ability to identify and meet the needs of the customer and do it within the project budget. Based on the customer's initial concepts and any subsequent equipment utilization research conducted by the systems engineer, the desired capabilities are identified as precisely as possible. Design parameters and objectives are defined and reviewed.

Functional efficiency is maximized to allow operation by a minimum number of personnel. Future needs are also investigated at this time. Future technical systems expansion is considered.

After the customer approves the equipment list, preliminary system plans are drawn up for review and further development. If architectural drawings of the facility are available, they can be used as a starting point for laying out an equipment floor plan. The systems engineer uses this floor plan to be certain adequate space is provided for present and future equipment, as well as adequate clearance for maintenance and convenient operation. Equipment identification is then added to the architect's drawings.

Documentation should include, but not be limited to, the generation of a list of major equipment including:

- Equipment prices
- Technical system functional block diagrams
- Custom item descriptions
- Rack and console elevations
- Equipment floor plans

The preliminary drawings and other supporting documents are prepared to record design decisions and to illustrate the design concepts to the customer. Renderings, scale models, or full-size mockups may also be needed to better illustrate, clarify, or test design ideas.

Ideas and concepts have to be exchanged and understood by all concerned parties. Good communication skills are essential. The bulk of the creative work is carried out in the design development phase. The physical layout—the look and feel—and the functionality of the facility will all have been decided and agreed upon by the completion of this phase of the project. If the design concepts appear feasible, and the cost is within the anticipated budget, management can authorize work to proceed on the final detailed design.

Electronic System Design

Performance standards and specifications have to be established up front in a technical facility project. This will determine the performance level of equipment that will be acceptable for use in the system and affect the size of the budget. Signal quality, stability, reliability, and accuracy are examples of the kinds of parameters that have to be specified. Access and processor speeds are important parameters when dealing with computer-driven products. The systems engineer has to confirm whether selected equipment conforms to the standards.

At this point it must be determined what functions each component in the system will be required to fulfill and how each will function together with other components in the system. The management and operation staff usually know what they would like the system to do and how they can best accomplish it. They have probably selected equipment that they think will do the job. With a familiarity of the capabilities of different equipment, the systems engineer should be able to contribute to this function-definition stage of the process. Questions that need to be answered include:

- What functions must be available to the operators?
- What functions are secondary and therefore not necessary?
- What level of automation should be required to perform a function?
- How accessible should the controls be?

Overengineering or overdesign must be avoided. This serious and costly mistake can be made by engineers and company staff when planning technical system requirements. A staff member may, for example, ask for a seemingly simple feature or capability without fully understanding its complexity or the additional cost burden it may impose on a project. Other portions of the system may have to be compromised to implement the additional feature. An experienced systems engineer will be able to spot this and determine if the tradeoffs and added engineering time and cost are really justified.

When existing equipment is going to be used, it will be necessary to make an inventory list. This list will be the starting point for developing a final equipment list. Usually, confronted with a mixture of acceptable and unacceptable equipment, the systems engineer must sort out what meets current standards and what should be replaced. Then, after soliciting input from facility technical personnel, the systems engineer develops a summary of equipment needs, including future acquisitions. One of the systems engineer's most important contributions is the ability to identify and meet these needs within the facility budget.

A list of major equipment is prepared. The systems engineer selects the equipment based on experience with the products, and on customer preferences. Often some existing equipment may be reused. A number of considerations are discussed with the facility customer to arrive at the best product selection. Some of the major points include:

- Budget restrictions
- Space limitations
- Performance requirements
- Ease of operation
- Flexibility of use
- Functions and features
- Past performance history
- Manufacturer support

The goal of the systems engineer is the design of equipment to meet the functional requirements of a project efficiently and economically. Simplified block diagrams for the video, audio, control, data, and communication systems are drawn. They are discussed with the customer and presented for approval.

Detailed Design

With the research and preliminary design development completed, the details of the design must now be concluded. The design engineer prepares complete detailed documentation and specifications necessary for the fabrication and installation of the technical systems, including all major and minor components. Drawings must show the final configuration and the relationship of each component to other elements of the system, as well as how they will interface with other building services, such as air conditioning and electrical power. This documentation must communicate the design requirements to the other design professionals, including the construction and installation contractors.

In this phase, the systems engineer develops final, detailed flow diagrams and schematics that show the interconnection of all equipment. Cable interconnection information for each type of signal is taken from the flow diagrams and recorded on the cable schedule. Cable paths are measured and timing calculations performed. Timed cable lengths (used for video and other special services) are entered onto the cable schedule.

The flow diagram is a schematic drawing used to show the interconnections between all equipment that will be installed. It is different from a block diagram in that it contains much more detail. Every wire and cable must be included on the drawings. A typical flow diagram for a video production facility is shown in Fig. 26.11.

The starting point for preparing a flow diagram can vary depending on the information available from the design development phase of the project and on the similarity of the project to previous projects. If a similar system has been designed in the past, the diagrams from that project can be modified to include the equipment and functionality required for the new system. New models of the equipment can be shown in place of their counterparts on the diagram, and minor wiring changes can be made to reflect the new equipment connections and changes in functional requirements. This method is efficient and easy to complete.

Systems Engineering Concepts

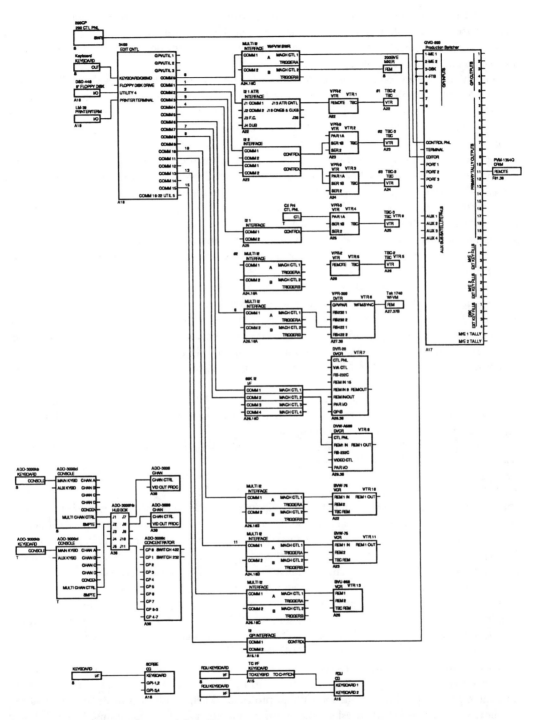

FIGURE 26.11 Video system control flow diagram.

If the facility requirements do not fit any previously completed design, the block diagram and equipment list are used as a starting point. Essentially, the block diagram is expanded and details added to show all of the equipment and their interconnections and to show any details necessary to describe the installation and wiring completely.

An additional design feature that might be desirable for specific applications is the ability to easily disconnect a rack assembly from the system and relocate it. This would be the case if the system were to be prebuilt at a systems integration facility and later moved and installed at the client's site. When this is a requirement, the interconnecting cable harnessing scheme must be well planned in advance and identified on the drawings and cable schedules.

Special custom items are defined and designed. Detailed schematics and assembly diagrams are drawn. Parts lists and specifications are finalized, and all necessary details worked out for these items. Mechanical fabrication drawings are prepared for consoles and other custom-built cabinetry.

The systems engineer provides layouts of cable runs and connections to the architect. Such detailed documentation simplifies equipment installation and facilitates future changes in the system. During preparation of final construction documents, the architect and the systems engineer can firm up the layout of the technical equipment wire ways, including access to flooring, conduits, trenches, and overhead wire ways.

Dimensioned floor plans and elevation drawings are required to show placement of equipment, lighting, electrical cable ways, duct, conduit, and heating, ventilation, and air conditioning (HVAC) ducting. Requirements for special construction, electrical, lighting, HVAC, finishes, and acoustical treatments must be prepared and submitted to the architect for inclusion in the architectural drawings and specifications. This type of information, along with cooling and electrical power requirements, also must be provided to the mechanical and electrical engineering consultants (if used on the project) so that they can begin their design calculations.

Equipment heat loads are calculated and submitted to the HVAC consultant. Steps are taken when locating equipment to avoid any excessive heat buildup within the equipment enclosures, while maintaining a comfortable environment for the operators.

Electrical power loads are calculated and submitted to the electrical consultant and steps taken to provide for sufficient power and proper phase balance.

Customer Support

The systems engineer can assist in purchasing equipment and help to coordinate the move to a new or renovated facility. This can be critical if a great deal of existing equipment is being relocated. In the case of new equipment, the customer will find the systems engineer's knowledge of prices, features, and delivery times to be an invaluable asset. A good systems engineer will see to it that equipment arrives in ample time to allow for sufficient testing and installation. A good working relationship with equipment manufacturers helps guarantee their support and speedy response to the customer's needs.

The systems engineer can also provide engineering management support during planning, construction, installation, and testing to help qualify and select contractors, resolve problems, explain design requirements, and assure quality workmanship by the contractors and the technical staff.

The procedures described in this section outline an ideal scenario. In reality, management may often try to bypass many of the foregoing steps to save money. This, the reasoning goes, will eliminate unnecessary engineering costs and allow construction to start right away. Utilizing in-house personnel, a small company may attempt to handle the job without professional help. With inadequate design detail and planning, which can result when using unqualified people, the job of setting technical standards and making the system work then defaults to the construction contractors, in-house technical staff, or the installation contractor. This can result in costly and uncoordinated work-arounds and, of course, delays and added costs during construction, installation, and testing. It makes the project less manageable and less likely to be completed successfully.

Systems Engineering Concepts

The complexity of a project can be as simple as interconnecting a few pieces of computer equipment together to designing software for the Space Shuttle. The size of a technical facility can vary from a small one-room operation to a large multimillion dollar plant or large area network. Where large amounts of money and other resources are going to be involved, management is well advised to recruit the services of qualified system engineers.

Budget Requirements Analysis

The need for a project may originate with customers, management, operations staff, technicians, or engineers. In any case, some sort of logical reasoning or a specific production requirement will justify the cost. On small projects, like the addition of a single piece of equipment, money only has to be available to make the purchase and cover installation costs. When the need may justify a large project, it is not always immediately apparent how much the project will cost to complete. The project has to be analyzed by dividing it up into its constituent elements. These elements include:

- Equipment and parts
- Materials
- Resources (including money and time needed to complete the project)

An executive summary or capital project budget request containing a detailed breakdown of these elements can provide the information needed by management to determine the return on investment and to make an informed decision on whether or not to authorize the project.

A capital project budget request containing the minimum information might consist of the following items:

- Project name. Use a name that describes the result of the project, such as control room upgrade.
- Project number (if required). A large organization that does many projects will use a project numbering system of some kind or may use a budget code assigned by the accounting department.
- Project description. A brief description of what the project will accomplish, such as design the technical system upgrade for the renovation of production control room 2.
- Initiation date. The date the request will be submitted.
- Completion date. The date the project will be completed.
- Justification. The reason the project is needed.
- Material cost breakdown. A list of equipment, parts, and materials required for construction, fabrication, and installation of the equipment.
- Total material cost.
- Labor cost breakdown. A list of personnel required to complete the project, their hourly pay rates, the number of hours they will spend on the project, and the total cost for each.
- Total project cost. The sum of material and labor costs.
- Payment schedule. Estimation of individual amounts that will have to be paid out during the course of the project and the approximate dates each will be payable.
- Preparer's name and the date prepared.
- Approval signature(s) and date(s) approved.

More detailed analysis, such as return on investment, can be carried out by an engineer, but financial analysis should be left to the accountants who have access to company financial data.

Feasibility Study and Technology Assessment

Where it is required that an attempt be made to implement new technology and where a determination must be made as to whether certain equipment can perform a desired function, it will be necessary to conduct a feasibility study. The systems engineer may be called upon to assess the state of the art to develop a new application. An executive summary or a more detailed report of evaluation test results may be required, in addition to a budget request, to help management make a decision.

Planning and Control of Scheduling and Resources

Several planning tools have been developed for planning and tracking progress toward the completion of projects and scheduling and controlling resources. The most common graphical project management tools are the Gantt chart and the *critical path method* (CPM) utilizing the *project evaluation and review* (PERT) technique. Computerized versions of these tools have greatly enhanced the ability of management to control large projects.

Project Tracking and Control

A project team member may be called upon by the project manager to report the status of the work during the course of the project. A standardized project status report form can provide consistent and complete information to the project manager. The purpose is to supply information to the project manager regarding work completed and money spent on resources and materials.

A project status report containing the minimum information might contain the following items:

- Project number (if required)
- Date prepared
- Project name
- Project description
- Start date
- Completion date (the date this part of the project was completed)
- Total material cost
- Labor cost breakdown
- Preparer's name

Change Control

After part or all of a project design has been approved and money allocated to build it, any changes may increase or decrease the cost. Factors that affect the cost include:

- Components and material
- Resources, such as labor and special tools or construction equipment
- Costs incurred because of manufacturing or construction delays

Management will want to know about such changes and will want to control them. For this reason, a method of reporting changes to management and soliciting approval to proceed with the change may have to be instituted. The best way to do this is with a *change order request* or *change order*. A change order includes a brief description of the change, the reason for the change and a summary of the effect it will have on costs and on the project schedule.

Management will exercise its authority and approve or disapprove each change based on its understanding of the cost and benefits and the perceived need for the modification of the original plan. Therefore, it is important that the systems engineer provide as much information and explanation as may be necessary to make the change clear and understandable to management.

A change order form containing the minimum information might contain the following items:

- Project number
- Date prepared
- Project name
- Labor cost breakdown
- Preparer's name
- Description of the change

- Reason for the change
- Equipment and materials to be added or deleted
- Material costs or savings
- Labor costs or savings
- Total cost of this change (increase or decrease)
- Impact on the schedule

Program Management

The Defense Systems Management College (Hoban, F.T. and Lawbaugh, W.M. 1993. *Readings in Systems Engineering Management*. NASA Science and Technical Information Program, Washington, D.C., p. 9) favors the management approach and defines systems engineering as follows:

> Systems engineering is the management function which controls the total system development effort for the purpose of achieving an optimum balance of all system elements. It is a process which transforms an operational need into a description of system parameters and integrates those parameters to optimize the overall system effectiveness.

Systems engineering is both a technical process and a management process. Both processes must be applied throughout a program if it is to be successful. The persons who plan and carry out a project constitute the project team. The makeup of a project team will vary depending on the size of the company and the complexity of the project. It is up to management to provide the necessary human resources to complete the project.

Executive Management

The executive manager is the person who can authorize that a project be undertaken. This person can allocate funds and delegate authority to others to accomplish the task. Motivation and commitment is toward the goals of the organization. The ultimate responsibility for a project's success is in the hands of the executive manager. This person's job is to get tasks completed through other people by assigning group responsibilities, coordinating activities between groups, and resolving group conflicts. The executive manager establishes policy, provides broad guidelines, approves the project master plan, resolves conflicts, and assures project compliance with commitments.

Executive management delegates the project management functions and assigns authority to qualified professionals, allocates a capital budget for the project, supports the project team, and establishes and maintains a healthy relationship with project team members.

Management has the responsibility to provide clear information and goals, up front, based on their needs and initial research. Before initiating a project, the company executive should be familiar with daily operation of the facility and analyze how the company works, how jobs are done by the staff, and what tools are needed to accomplish the work. Some points that may need to be considered by an executive before initiating a project include:

- What is the current capital budget for equipment?
- Why does the staff currently use specific equipment?
- What function of the equipment is the weakest within the organization?
- What functions are needed but cannot be accomplished with current equipment?
- Is the staff satisfied with current hardware?
- Are there any reliability problems or functional weaknesses?
- What is the maintenance budget and is it expected to remain steady?
- How soon must the changes be implemented?
- What is expected from the project team?

Only after answering the appropriate questions will the executive manager be ready to bring in expert project management and engineering assistance. Unless the manager has made a systematic effort to evaluate all of the obvious points about the facility requirements, the not so obvious points may be overlooked. Overall requirements must be broken down into their component parts. Do not try to tackle ideas that have too many branches. Keep the planning as basic as possible. If the company executive does not make a concerted effort to investigate the needs and problems of a facility thoroughly before consulting experts, the expert advice will be shallow and incomplete, no matter how good the engineer.

Engineers work with the information they are given. They put together plans, recommendations, budgets, schedules, purchases, hardware, and installation specifications based on the information they receive from interviewing management and staff. If the management and staff have failed to go through the planning, reflection, and refinement cycle before those interviews, the company will likely waste time and money.

Project Manager

Project management is an outgrowth of the need to accomplish large complex projects in the shortest possible time, within the anticipated cost, and with the required performance and reliability. Project management is based on the realization that modern organizations may be so complex as to preclude effective management using traditional organizational structures and relationships. Project management can be applied to any undertaking that has a specific end objective.

The project manager must be a competent systems engineer, accountant, and manager. As systems engineer, there must be understanding of analysis, simulation, modeling, and reliability and testing techniques. There must be awareness of state-of-the-art technologies and their limitations. As accountant, there must be awareness of the financial implications of planned decisions and knowledge of how to control them. As manager, the planning and control of schedules is an important part of controlling the costs of a project and completing it on time. Also, as manager, there must be the skills necessary to communicate clearly and convincingly with subordinates and superiors to make them aware of problems and their solutions.

The project manager is the person who has the authority to carry out a project. This person has been given the legitimate right to direct the efforts of the project team members. The manager's power comes from the acceptance and respect accorded by superiors and subordinates. The project manager has the power to act and is committed to group goals.

The project manager is responsible for getting the project completed properly, on schedule, and within budget, by utilizing whatever resources are necessary to accomplish the goal in the most efficient manner. The manager provides project schedule, financial, and technical requirement direction and evaluates and reports on project performance. This requires planning, organizing, staffing, directing, and controlling all aspects of the project.

In this leadership role, the project manager is required to perform many tasks including the following:

- Assemble the project organization.
- Develop the project plan.
- Publish the project plan.
- Set measurable and attainable project objectives.
- Set attainable performance standards.
- Determine which scheduling tools (PERT, CPM, and/or Gantt) are right for the project.
- Using the scheduling tools, develop and coordinate the project plan, which includes the budget, resources, and the project schedule.
- Develop the project schedule.
- Develop the project budget.
- Manage the budget.
- Recruit personnel for the project.
- Select subcontractors.

- Assign work, responsibility, and authority so that team members can make maximum use of their abilities.
- Estimate, allocate, coordinate, and control project resources.
- Deal with specifications and resource needs that are unrealistic.
- Decide on the right level of administrative and computer support.
- Train project members on how to fulfill their duties and responsibilities.
- Supervise project members, giving them day-to-day instructions, guidance, and discipline as required to fulfill their duties and responsibilities.
- Design and implement reporting and briefing information systems or documents that respond to project needs.
- Control the project.

Some basic project management practices can improve the chances for success. Consider the following:

- Secure the necessary commitments from top management to make the project a success.
- Set up an action plan that will be easily adopted by management.
- Use a work breakdown structure that is comprehensive and easy to use.
- Establish accounting practices that help, not hinder, successful completion of the project.
- Prepare project team job descriptions properly up front to eliminate conflict later on.
- Select project team members appropriately the first time.

After the project is under way, follow these steps:

- Manage the project, but make the oversight reasonable and predictable.
- Get team members to accept and participate in the plans.
- Motivate project team members for best performance.
- Coordinate activities so they are carried out in relation to their importance with a minimum of conflict.
- Monitor and minimize interdepartmental conflicts.
- Get the most out of project meetings without wasting the team's productive time. Develop an agenda for each meeting and start on time. Conduct one piece of business at a time. Assign responsibilities where appropriate. Agree on follow up and accountability dates. Indicate the next step for the group. Set the time and place for the next meeting. End on time.
- Spot problems and take corrective action before it is too late.
- Discover the strengths and weaknesses in project team members and manage them to get desired results.
- Help team members solve their own problems.
- Exchange information with subordinates, associates, superiors, and others about plans, progress, and problems.
- Make the best of available resources.
- Measure project performance.
- Determine, through formal and informal reports, the degree to which progress is being made.
- Determine causes of and possible ways to act upon significant deviations from planned performance.
- Take action to correct an unfavorable trend or to take advantage of an unusually favorable trend.
- Look for areas where improvements can be made.
- Develop more effective and economical methods of managing.
- Remain flexible.
- Avoid activity traps.
- Practice effective time management.

When dealing with subordinates, each person must:

- Know what is to be done, preferably in terms of an end product.
- Have a clear understanding of the authority and its limits for each individual.
- Know what the relationship with other people is.
- Know what constitutes a job well done in terms of specific results.
- Know when and what is being done exceptionally well.
- Be shown concrete evidence that there are just rewards for work well done and for work exceptionally well done.
- Know where and when expectations are not being met.
- Be made aware of what can and should be done to correct unsatisfactory results.
- Feel that the supervisor has an interest in each person as an individual.
- Feel that the supervisor both believes in each person and is anxious for individual success and progress.

By fostering a good relationship with associates, the manager will have less difficulty communicating with them. The fastest, most effective communication takes place among people with common points of view.

The competent project manager watches what is going on in great detail and can, therefore, perceive problems long before they flow through the paper system. Personal contact is faster than filling out forms. A project manager who spends most of the time in the management office instead of roaming through the places where the work is being done is headed for catastrophe.

Systems Engineer

The term *systems engineer* means different things to different people. The systems engineer is distinguished from the engineering specialist, who is concerned with only one aspect of a well-defined engineering discipline, in that the systems engineer must be able to adapt to the requirements of almost any type of system. The systems engineer provides the employer with a wealth of experience gained from many successful approaches to technical problems developed through hands-on exposure to a variety of situations. This person is a professional with knowledge and experience, possessing skills in a specialized and learned field or fields. The systems engineer is an expert in these fields; highly trained in analyzing problems and developing solutions that satisfy management objectives. The systems engineer takes data from the overall development process and, in return, provides data in the form of requirements and analysis results to the process.

Education in electronics theory is a prerequisite for designing systems that employ electronic components. As a graduate engineer, the systems engineer has the education required to design electronic systems correctly. Mathematics skill acquired in engineering school is one of the tools used by the systems engineer to formulate solutions to design problems and analyze test results. Knowledge of testing techniques and theory enables this individual to specify system components and performance and to measure the results. Drafting and writing skills are required for efficient preparation of the necessary documentation needed to communicate the design to technicians and contractors who will have to build and install the system.

A competent systems engineer has a wealth of technical information that can be used to speed up the design process and help in making cost effective decisions. If necessary information is not at hand, the systems engineer knows where to find it. The experienced systems engineer is familiar with proper fabrication, construction, installation, and wiring techniques and can spot and correct improper work.

Training in personnel relations, a part of the engineering curriculum, helps the systems engineer communicate and negotiate professionally with subordinates and management.

Small in-house projects can be completed on an informal basis and, indeed, this is probably the normal routine where the projects are simple and uncomplicated. In a large project, however, the systems engineer's involvement usually begins with preliminary planning and continues through fabrication, implementation, and testing. The degree to which program objectives are achieved is an important measure of the systems engineer's contribution.

During the design process the systems engineer:

- Concentrates on results and focuses work according to the management objectives.
- Receives input from management and staff.
- Researches the project and develops a workable design.
- Assures balanced influence of all required design specialties.
- Conducts design reviews.
- Performs tradeoff analyses.
- Assists in verifying system performance.
- Resolves technical problems related to the design, interface between system components, and integration of the system into any facility.

Aside from designing a system, the systems engineer has to answer any questions and resolve problems that may arise during fabrication and installation of the hardware. Quality and workmanship of the installation must be monitored. The hardware and software will have to be tested and calibrated upon completion. This, too, is the concern of the systems engineer. During the production or fabrication phase, systems engineering is concerned with:

- Verifying system capability
- Verifying system performance
- Maintaining the system baseline
- Forming an analytical framework for producibility analysis

Depending on the complexity of the new installation, the systems engineer may have to provide orientation and operating instruction to the users. During the operational support phase, system engineers:

- Receive input from users
- Evaluate proposed changes to the system
- Establish their effectiveness
- Facilitate the effective incorporation of changes, modifications and updates

Depending on the size of the project and the management organization, the systems engineer's duties will vary. In some cases the systems engineer may have to assume the responsibilities of planning and managing smaller projects.

Other Project Team Members

Other key members of the project team where building construction may be involved include the following:

- Architect, responsible for design of any structure.
- Electrical engineer, responsible for power system design if not handled by the systems engineer.
- Mechanical engineer, responsible for HVAC, plumbing, and related designs.
- Structural engineer, responsible for concrete and steel structures.
- Construction contractors, responsible for executing the plans developed by the architect, mechanical engineer, and structural engineer.
- Other outside contractors, responsible for certain specialized custom items which cannot be developed or fabricated internally or by any of the other contractors.

26.5 Conclusion

Systems theory is the theoretical basis of systems engineering that is an organized approach to the design of complex systems. The key components of systems theory applied in systems engineering are a holistic approach, the decomposition of problems, the exploitation of analogies, and the use of models.

A formalized technical project management technique used to define systems includes three major steps.

- Define the requirements in terms of functions to be performed and measurable and testable requirements describing how well each function must be performed.
- Synthesize a way of fulfilling the requirements.
- Study the tradeoffs and select one solution from among the possible alternative solutions.

In the final analysis, beyond any systematic approach to systems engineering, engineers have to engineer. Robert A. Frosch, a former NASA Administrator, in a speech to a group of engineers in New York, urged a common sense approach to systems engineering (Frosch, R.A. 1969. A Classic Look at System Engineering and Management. In *Readings in Systems Engineering*. ed. F.T. Hoban and W.M. Lawbaugh, pp. 1–7. NASA Science and Technical Information Program, Washington, D.C.):

> Systems, even very large systems, are not developed by the tools of systems engineering, but only by the engineers using the tools....I can best describe the spirit of what I have in mind by thinking of a music student who writes a concerto by consulting a check list of the characteristics of the concerto form, being careful to see that all of the canons of the form are observed, but having no flair for the subject, as opposed to someone who just knows roughly what a concerto is like, but has a real feeling for music. The results become obvious upon hearing them. The prescription of technique cannot be a substitute for talent and capability....

Defining Terms

Abstraction: Though dealing with concrete systems, abstraction is an important feature of systems models. Components are described in terms of their function rather than in terms of their form. Graphical models such as block diagrams, flow diagrams, timing diagrams, and the like are commonly used. Mathematical models may also be employed. Systems theory shows that, when modeled in abstract formal language, apparently diverse kinds of systems show significant and useful isomorphisms of structure and function. Similar interconnection structures occur in different types of systems. Equations that describe the behavior of electrical, thermal, fluid, and mechanical systems are essentially identical in form.

Decomposition: The systems approach attempts to apply an organized methodology to completing large complex projects by breaking them down into simpler more manageable component elements. These elements are treated separately, analyzed separately, and designed separately. In the end, all of the components are recombined to build the whole.

Dynamics: Most system problems are dynamic in nature. The signals change over time and the components determine the dynamic response of the system. The system behavior depends on the signals at a given instant, as well as on the rates of change of the signals and their past values. The term signals can be replaced by substituting human factors such as the number of users on a computer network, for example.

Emergent properties: A holistic aspect of system theory describes emergent properties. Properties which result from the interaction of system components, properties which are not those of the components themselves, are referred to as emergent properties.

Hard and soft systems: In hard systems, the components and their interactions can be described by mathematical models. Soft systems cannot be described as easily. They are mostly human activity systems, which imply unpredictable behavior and non-uniformity. They introduce difficulties and uncertainties of conceptualization, description, and measurement. Usual system concepts and methodology cannot be applied.

Holism: Holism is an element of systems theory in that the end product is greater than the sum of its component elements. In systems theory, the modeling and analytical methods enable all essential effects and interactions within a system and those between a system and its surroundings to be taken into account. Errors from the idealization and approximation involved in treating parts of a system in isolation, or reducing consideration to a single aspect are thus avoided.

Isomorphism: Similarity in elements of different kinds. Similarity of structure and function in elements implies isomorphism of behavior of a system. Different types of systems exhibit similar dynamic behavior such as response to stimulation.

Modeling: The concept of modeling is the starting point of synthesis of a system. Since we must be able to weigh the effects of different design decisions in order to make choices between alternative concepts, modeling requires the determination of those quantitative features that describe the operation of the system. We would, of course, like a very detailed model with as much detail as possible describing the system. Reality and time constraints, however, dictate that the simplest possible model be selected to improve our chances of design success. The model itself is always a compromise. The model is restricted to those aspects that are important in the evaluation of system operation. A model might start off as a simple block diagram with more detail being added as the need becomes apparent.

Optimization: Making an element as perfect, effective, or functional as possible. Every design project involves making a series of compromises and choices based on relative weighting of the merit of important aspects. The best candidate among several alternatives is selected.

Synthesis: This is the process by which concepts are developed to accomplish the functional requirements of a system. Performance requirements and constraints, as defined by the functional analysis, are applied to each individual element of the system, and a design approach is proposed for meeting the requirements. Conceptual schematic arrangements of system elements are developed to meet system requirements. These documents are used to develop a description of the system elements.

Trade Studies: A structured approach to guide the selection of alternative configurations and ensure that a logical and unbiased choice is made. Throughout development, trade studies are carried out to determine the best configuration that will meet the requirements of the program. In the concept exploration and demonstration phases, trade studies help define the system configuration. Trade studies are used as a detailed design analysis tool for individual system elements in the full-scale development phase. During production, trade studies are used to select alternatives when it is determined that changes need to be made. Alternative approaches to achieving the defined objectives can thereby be established.

Trade Table: A trade table is used to summarize selection results of a trade study. Pertinent criteria are listed for each alternative solution. The alternatives may be described in a qualitative manner, such as high, medium, or low.

References

[1] Defense Systems Management. 1983. *Systems Engineering Management Guide*. Defense Systems Management College, Fort Belvoir, VA.

[2] Delatore, J.P., Prell, E.M., and Vora, M.K. 1989. Translating customer needs into product specifications. *Quality Progress* 22(1), Jan.

[3] Finkelstein, L. 1988. Systems theory. *IEE Proceedings*, Pt. A, 135(6), pp. 401–403.

[4] Hoban, F.T. and Lawbaugh, W.M. 1993. *Readings in Systems Engineering*. NASA Science and Technical Information Program, Washington, D.C.
[5] Shinners, S.M. 1976. *A Guide to Systems Engineering and Management*. Lexington Books, Lexington, MA.
[6] Tuxal, J.G. 1972. *Introductory System Engineering*. McGraw-Hill, New York.

Further Information

Additional information on systems engineering concepts is available from the following sources:

MIL-STD-499A is available by writing to The National Council on Systems Engineering (NCOSE) at 333 Cobalt Way, Suite 107, Sunnyvale, CA 94086. NCOSE has set up a working group that specifically deals with commercial systems engineering practices.

Blanchard, B.S. and Fabrycky, W.J. 1990. *Systems Engineering and Analysis*. Prentice-Hall, Englewood Cliffs, NJ. An overview of systems engineering concepts, design methodology, and analytical tools commonly used.

Skytte, K. 1994. Engineering a small system. *IEEE Spectrum* 31(3) describes how systems engineering, once the preserve of large government projects, can benefit commercial products as well.

27
Disaster Planning and Recovery

Richard Rudman
KFWB Radio

27.1 Introduction 27-1
  A Short History of Disaster Communications
27.2 Emergency Management 101 for Designers 27-2
27.3 The Planning Process 27-4
  Starting Your Planning Process: Listing Realistic Risks • Risk Assessment and Business Resumption Planning
27.4 Workplace Safety 27-6
27.5 Outside Plant Communications Links 27-7
  Outside Plant Wire • Microwave Links • Fiber Optics Links • Satellite
27.6 Emergency Power and Batteries 27-8
27.7 Air Handling Systems 27-10
27.8 Water Hazards 27-10
27.9 Electromagnetic Pulse Protection (EMP) 27-10
27.10 Alternate Sites 27-11
27.11 Security 27-11
27.12 Workplace and Home: Hand-in-Hand Preparedness 27-12
27.13 Expectations, 9-1-1, and Emergencies 27-12
27.14 Business Rescue Planning for Dire Emergencies 27-12
27.15 Managing Fear 27-13

27.1 Introduction

Disaster-readiness must be an integral part of the design process. You cannot protect against every possible risk.

Earthquakes, hurricanes, tornados and floods come to mind when we talk about risks that can disable a facility and its staff. Risk can also be man made. Terrorist attack, arson, utility outage, or even simple accidents that get out of hand are facts of everyday life. Communications facilities, especially those that convey information from government to the public during major catastrophes, must keep their people and systems from becoming disaster victims. No major emergency has yet cut all communications to the public. We should all strive to maintain that record.

Employees all too often become crash dummies when disaster disrupts the workplace. During earthquakes, they sometimes have to dodge falling ceiling tiles, elude objects hurled off shelves, watch computers and work stations dance in different directions, breathe disturbed dust, and suffer the indignities of failed plumbing. This chapter does not have ready-to-implement plans for people, facilities or systems. It does outline major considerations for disaster planning. It is then up to you to be the champion for emergency preparedness and planning on mission critical projects.

A Short History of Disaster Communications

When man lived in caves, every day was an exercise in disaster planning and recovery. There were no "All Cave News" stations for predator reports. The National Weather Service was not around yet to warn about rain that might put out the community camp fire. We live in an increasingly interdependent information-driven society. Modern information systems managers have an absolute social, technical, and economic responsibility to get information to the public when disaster strikes. Designers of these facilities must do their part to build workplaces that will meet the special needs of the lifeline information mission during major emergencies.

The U.S. recognized the importance of getting information to the public during national emergencies in the 1950s when it devised a national warning system in case of enemy attack. This effort was and still is driven by a belief that a government that loses contact with its citizens during an emergency could fall. The emergency broadcast system (EBS), began life as control of electromagnetic radiation (CONELRAD). It was devised to prevent enemy bombers from homing in on AM broadcast stations. It was conceived amid Cold War fears of nuclear attack, and was born in 1951 to allow government to reach the public rapidly through designated radio stations. EBS replaced CONELRAD and its technical shortcomings. EBS is being replaced by emergency alert system (EAS). EAS introduces multiple alerting paths for redundancy as well as some digital technology. It strengthens the emergency communication partnerships beyond broadcasting, federal government, National Weather Service, local government, and even the Internet.

Major disasters such as Hurricane Andrew and the Northridge (CA) earthquake help us focus on assuring that information systems will work during emergencies. These events expose our weaknesses. They can also be rare windows of opportunity to learn from our mistakes and make improvements for the future.

27.2 Emergency Management 101 for Designers

There is a government office of emergency services (or management) in each state. Within each state, *operational areas* for emergency management have been defined. Each operational area has a person or persons assigned to emergency management, usually under the direction of an elected or appointed public safety official. There is usually a mechanism in place to announce when an emergency condition exists. *Declared* has a special meaning when used with the word *emergency*. An operational area normally has a formal method, usually stated in legislation, to announce to the world that it is operating under emergency conditions. That legislation places management of the emergency under a specially trained person or organization. During declared emergencies, operational areas usually open a special management office that is usually called an **emergency operations center (EOC)**. Some large businesses with a lot at stake have emergency management departments with full-time employees and rooms set aside to be their EOC. These departments usually use the same management philosophy as government emergency organizations. They also train their people to understand how government manages emergencies. The unprepared business exists in a vacuum without an understanding of how government is managing the emergency. Misunderstandings are created when people do not know the language. An emergency, when life and property are at stake, is the wrong time to have misunderstandings. Organizations that have to maintain services to the public, have outside plant responsibilities, or have significant responsibilities to the public when on their property should have their plans, people, and purpose in place when disaster strikes.

A growing number of smaller businesses which have experienced serious business disruptions during emergencies, or wish to avoid that unpleasant experience, are writing their own emergency plans and building their own emergency organizations. Broadcasters, information companies, cable systems, and other travellers on the growing information superhighway should do likewise. There are national organizations and possibly local emergency-minded groups you can contact. In Los Angeles, there is the

Business and Industry Council for Emergency Planning and Preparedness (BICEPP). A more complete listing of resource organizations follows this chapter.

Figure 27.1 shows the cyclical schematic of emergency management. Atter an emergency has been declared, *response* deals with immediate issues that have to do with saving lives and property. Finding out what has broken and who has been hurt is called *damage assessment*. Damage assessment establishes the dimensions of the response that are needed. Once you know what is damaged and who is injured, trapped, or killed, you have a much better idea of how much help you need. You

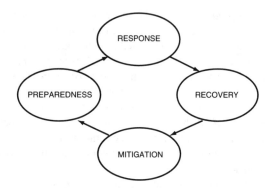

FIGURE 27.1 The response, recovery, mitigation, and preparedness cycle.

also know if you have what you need on hand to meet the challenge. Another aspect of response is to provide the basics of human existence: water, food, and shelter. An entire industry has sprung up to supply such special needs. Water is available packaged to stay drinkable for years. The same is true for food. Some emergency food is based on the old military K rations or the more modern meals ready to eat (MREs). MREs became widely known during the Gulf War.

The second phase of the emergency cycle is called *recovery*. Recovery often begins after immediate threats to life safety have been addressed. This is difficult to determine during an earthquake or flood. A series of life threats may span hours, days, or weeks. It is not uncommon for business and government emergency management organizations to be engaged in recovery and response at the same time.

In business, response covers these broad topics:

- Restoring lost capacity to your production and distribution capabilities
- Getting your people back to work
- Contacting customers and/or alternate vendors
- Contacting insurers and processing claims

A business recovery team's mission is to get the facility back to 100% capacity, or as close to it as possible. Forming a *business recovery team* should be a key action step along with conducting the *business impact analysis,* discussed later in this chapter. Major threats must be eliminated or stabilized so that employees or outside help can work safely. Salvage missions may be necessary to obtain critical records or equipment. Reassurance should also be a part of the mission. Employees may be reluctant to return to work or customers may be loathe to enter even if the building is not *red tagged*. Red tagging is a term for how inspectors identify severely damaged buildings, an instant revocation of a building's certificate of occupancy until satisfactory repairs take place. Entry, even for salvage purposes by owners or tenants, may be prohibited or severely restricted. Recovery will take a whole new turn if the facility is marked unfit for occupancy.

The third stage of the cycle is *mitigation*. Sometimes called lessons learned, mitigation covers a wide range of activities that analyze what went wrong and what went right during response and recovery. Accurate damage assessment records, along with how resources were allocated, are key elements reviewed during mitigation. At the business level, mitigation debriefings might uncover how and why the emergency generator did not start automatically. You might find that a microwave dish was knocked out of alignment by high winds or seismic activity or that a power outage disabled critical components such as security systems. On the *people* side, a review of the emergency might show that no one was available to reset circuit breakers or unlock normally locked doors. There may have been times when people on shift were not fed in a timely manner. Stress, an often overlooked factor that compounds emergencies and emergency response, may have affected performance. In the *what went right* column, you might find that certain people rose above the occasion and averted added disaster, that money spent on preparing paid off, or that you were able to resume 100% capacity much faster than expected.

Be prepared should be the watch-phrase of emergency managers and facilities designers. Lessons learned during mitigation should be implemented in the *preparedness* phase. Procedures that did not work are rewritten. Depleted supplies are replenished and augmented if indicated. New resources are ordered and stored in safe places. Training in new procedures, as well as refresher training, is done. Mock drills and exercises may be carried out. The best test of preparedness is the next actual emergency. Preparedness is the one element of the four you should adopt as a continuous process.

Emergency management, from a field incident to the state level, is based on the **incident command system (ICS)**. ICS was pioneered by fire service agencies in California as an emergency management model. Its roots go further back into military command and control. ICS has been adopted by virtually every government and government public safety emergency management organization in the country as a standard.

ICS depends on simple concepts. One is *span of control*. Span of control theory states that a manager should not directly manage more than seven people. The optimum number is five. Another ICS basic is that an emergency organization builds up in stages, from the bottom up, from the field up, in response to need. For example, a wastebasket fire in your business may call for your fire response team to activate. One trained member on the team may be all that is required to identify the source of the fire, go for a fire extinguisher on the wall, and put it out. A more serious fire may involve the entire team. The team *stages up* according to the situation.

27.3 The Planning Process

It is impossible to separate emergency planning from the facility where the plan will be put into action. Emergency planning must be integral to a functional facility. It must support the main mission and the people who must carry it out. It must work when all else fails. Designers first must obtain firm commitment and backing from top management. Commitment is always easier to get if top management has experienced first-hand a major earthquake or powerful storm. Fear is a powerful source of motivation.

Disaster planning and recovery is an art, a science, and a technology. Like entities such as the Institute of Electrical and Electronics Engineers (IEEE) or the Society of Broadcast Engineers (SBE), disaster planners have their own professional groups and certification standards. States such as California provide year-round classroom training for government disaster planners. Many planners work full time for the military or in the public safety arenas of government. Others have found homes in businesses who recognize that staying in business after a major disaster is smart business. Still others offer their skills and services as consultants to entities who need to *jump start* their disaster planning process.

The technical support group should have responsibility or supervision over the environmental infrastructure of a critical communications facility. Without oversight, electronic systems are at the mercy of whomever controls that environment. Local emergencies can be triggered by preventable failures in air supply systems, roof leaks, or uncoordinated telephone, computer, or AC wiring changes. Seemingly harmless acts such as employees plugging electric heaters into the wrong AC outlet have brought down entire facilities. Successful practitioners of systems design and support must take daily emergencies into account in the overall planning process. To do otherwise risks rapid doom, if not swift unemployment.

Starting Your Planning Process: Listing Realistic Risks

Your realistic risk list should contain specific hazards based on local conditions such as:

- Regional high water marks for the 100 and 150 year storms
- Regional social, political and governmental conditions
- Regional commercial electrical power reliability
- Regional weather conditions
- Regional geography
- Regional geology

Disaster Planning and Recovery

You should assess specific local hazards that could be triggered by:

- Threats from present or former employees who may hold grudges
- External parties who are likely to get mad at your organization
- Other factors that could make you an easy target
- Nearby man-made hazards
- Special on-site hazards
- Neighbors
- Construction of your facility
- Hazardous materials on the premises
- Communications links to the outside world
- Electrical power
- Other utilities
- Buried pipelines

For example, information from the Northridge earthquake could rewrite seismic building codes for many types of structures. The Northridge quake showed that some high rise structures thought to be earthquake safe are not. Designers should be aware that seismic building codes usually allow for safe evacuation. They do not embody design criteria to prevent major structural damage. Earthquake safe is not earthquake proof.

If possible, get help from emergency planning professionals when you finish your list. They can help you devise a well-written and comprehensive emergency plan. They can also help with detailed research on factors such as geology and hazardous materials. For instance, you may be located near a plant that uses hazardous materials. The ultimate expert advice might be to move before something bad happens! Once there is agreement on the major goals for operations under emergency conditions, you will have a clear direction for *emergency-ready* facilities planning.

Risk Assessment and Business Resumption Planning

Perform a realistic assessment of the risks that your list suggests. Do not overlook the obvious. If computers, transmitters, or telephone equipment depend on cool air, how can they continue to operate during a heat wave when your one air conditioner has malfunctioned?

What level of reliability should a designer build into a lifeline communications facility? Emergencies introduce *chaos* into the reliability equation. Most engineers are quite happy when a system achieves 99.9999% reliability. Although the glass is more than half full, *four nines reliability* still means eight minutes of outage over a one year period! Reliability is an educated prediction based on a number of factors.

Believers in Murphy's law (anything that can go wrong, will go wrong) know that the 0.0001% outage will occur at the worst possible time. Design beyond that stage of reliability so that you can have a greater chance to cheat Murphy and stay on line during major emergencies. Double, triple, and even quadruple redundancy become realistic options when 100% uptime is the goal.

The new Los Angeles County Emergency Operations Center has three diesel generators. Air handling equipment is also built in triplicate. The building rests on huge rubber isolators at the base of each supporting column. These rubber shock mounts are sometimes called *base isolators*. They are built using laminated layers of Neoprene® rubber and metal. The entire structure floats on 26 of these isolators, designed to let the earth move beneath a building during an earthquake, damping transmission of rapid and damaging acceleration. The Los Angeles County EOC is designed to allow 16 in. of lateral movement from normal, or as much as 32 in. of total movement in one axis. LA County emergency planners are hoping there will not be an earthquake that produces more than 16 in. of lateral displacement. The isolators protect the structure and its contents from most violent lateral movement during earthquakes. The design mandate for this structure came directly from the Board of Supervisors and the Sheriff for the County of Los Angeles.

Can that building or a key system still fail? Of course. It is possible, though unlikely, that all three generators will fail to start. Diminishing returns set in beyond a certain point. Designers must balance realistic redundancy with the uptime expectations for the facility.

A facility designer must have a clear understanding of how important continued operation during and after a major catastrophe will be to its future survival. Disaster planning for facilities facing a hurricane with 125-mi/h winds may well entail boarding up the windows and leaving town for the duration. Others facilities will opt for uninterrupted operation, even in the face of nature's fury. Some will do nothing, adding to the list of victims who sap the resources of those who did prepare. The facility's mission may be critical to local government emergency management. For example, the public tunes to radio and television when disaster strikes. Government emergency managers tune in too.

A **business resumption plan (BRP)** is just as important as an organization's disaster plan. Both are just as essential as having an overall strategic business plan. Some experts argue these three elements should be formulated and updated in concert. When disaster strikes, the first concern of an organization must be for the safety of its employees, customers, vendors, and visitors.

Once *life safety* issues have been addressed, the next step is to perform damage assessments, salvage operations, and, if needed, relocation of critical functions. The BRP is activated once these basic requirements have been met and may influence how they are met.

The focus of business resumption planning is maintaining or resuming core business activities following a disaster. The three major goals of the BRP are: resumption of production and delivery, customer service and notification, and cash flow. A comprehensive BRP can accelerate recovery, saving time, money, and jobs. Like any insurance policy or any other investment, a properly designed BRP will cost you both in time and money. A BRP is an insurance policy, if not a major investment. The following actions are the backbone of a BRP:

- Conduct a **business impact analysis (BIA)**.
- Promote employee buy-in and participation.
- Seek input starting at the lowest staff levels.
- Build a recovery strategy and validation process.
- Test your BRP.
- Assure a continuous update of the BRP.[1]

Most companies do not have the staff to do a proper BIA. If a BIA becomes a top management goal, retain an experienced consultant. Qualified consultants can also conduct training and design test exercises. The BIA process does work. Oklahoma City businesses that had BIAs in place recovered faster than their competitors after the terrorist bombing in 1995. First Interstate Bank was able to open for business the very next day after a fire shut down their Los Angeles highrise headquarters in 1988. On January 17, 1994, the day of the Northridge earthquake, Great Western Bank headquarters suffered major structural damage. They made an almost seamless transition to their Florida operations.

27.4 Workplace Safety

Employers must always assure safety in the workplace at all times. Some states such as California have passed legislation that mandates that most employers identity hazards and protect their workers from them. Natural emergencies create special hazards that can maim or kill. A *moment magnitude* 7.4 earthquake can hurl heavy objects such as computer monitors and daggerlike shards from plate glass windows lethally through the air. The Richter Scale is no longer used by serious seismic researchers. Moment magnitude calculates energy release based on the surface area of the planes of two adjacent rock structures (an earthquake fault) and the distance these structures will move in relation to one another. Friction across the surface of the fault holds the rocks until enough stress builds to release energy. Some of the released energy travels via low-frequency wave motion through the rock. These low-frequency

---

[1]The author thanks Mary Carrido, President, MLC & Associates of Irvine, CA, for core elements of business resumption. It is far beyond the scope of this chapter to go into more detail on the BIA process.

waves cause the shaking and sometimes violent accelerations that occur during an earthquake. For more on modern seismic research and risk, please refer to the Reference section for this chapter. A strong foundation of day-to-day safety can lessen the impact of major emergencies. For instance, assuring that plate glass in doors has a safety rating could avoid an accidental workplace injury.

Special and dangerous hazards are found in the information and communications workplace. Tall equipment racks are often not secured to floors, much less secured to load-bearing walls. Preventing equipment racks from tipping over during an earthquake may avoid crippling damage to both systems and people. Bookcases and equipment storage shelves should be secured to walls. Certain objects should be *tethered*, rather than firmly bolted. Although securing heavy objects is mostly common sense, consult experts for special cases. Do not forget seismic-rated safety chains for heavy objects like television studio lights and large speakers.

Computers and monitors are usually not secured to work surfaces. A sudden drop from work station height would ruin the day's output for most computers, video monitors, and their managers. An industry has sprung up that provides innovative fasteners for computer and office equipment. Special Velcro® quick-release anchors and fasteners can support the entire weight of a personal computer or printer, even if the work surface falls over.

Bolting work stations to the floor and securing heavy equipment with properly rated fasteners can address major seismic safety issues. G forces measured in an upper story of a high rise building during the Northridge quake were greater than 2.7 times the force of gravity; 1 g is of course equal to the force of Earth's gravity. An acceleration of 2 g doubles the effective force of a person or object in motion, and nullifies the effectiveness of restraints that worked fine just before the earthquake [Force = (mass) $\times$ (acceleration)]. Seismic accelerations cause objects to make sudden stops: 60 to zero in 1 s. A room full of unsecured work stations could do a fair imitation of a slam dance contest, even at lower accelerations. Cables can be pulled loose, monitors can implode, and delicate electronics can be smashed into scrap. Even regions to the Earth where there has not been recent seismic activity, have been given a *long overdue* rating by respected seismologists. Maybe you will be the only one on your block to take such warnings seriously. Maybe you will be the only one left operational on your block!

Maintaining safety standards is difficult in any size organization. A written safety manual that has specific practices and procedures for normal workplace hazards as well as the emergency-related hazards you identify is not only a good idea, it may lower your insurance rates. If outside workers set foot in your facility, prepare a special Safety Manual for Contractors. Include in it installation standards, compliance with *lock-out/tag-out*, and emergency contact names, and phone numbers. (Lock-out/tag-out is a set of standard safety policies that assure that energy is removed from equipment during installation and maintenance. It assures that every member of a work detail is clear before power is reapplied.) Make sure outside contractors carry proper insurance and are qualified, licensed, or certified to do the work for which you contract.

27.5 Outside Plant Communications Links

Your facility may be operational, but failure of a wire, microwave, or fiber communications link could be devastating. All outside plant links discussed next presuppose proper installation. For wire and fiber, this means adequate *service loops* (coiled slack) so quake and wind stresses will not snap taut lines. It means that the telephone company has installed terminal equipment so it will not fall over in an earthquake or be easily flooded out. A range of backup options are available.

Outside Plant Wire

Local telephone companies still use a lot of wire. If your facility is served only by wire on telephone poles or underground in flood-prone areas, you may want what the telephone industry calls *alternate routing*. Alternate routing from your location to another **central office (CO)** may be very costly since the next nearest CO is rarely close. Ask to see a map of the proposed alternate route. If it is alternate only to the next block, or duplicates your telephone pole or underground risk, the advantage you gain will be minimal.

Most telephone companies can designate as an *essential service* a limited block of telephone numbers at a given location for lifeline communications. Lines so designated are usually found at hospitals and public safety headquarters. Contact your local phone company representative to see if your facility can qualify. Many broadcasters who have close ties to local government emergency management should easily qualify.

Microwave Links

Wind and seismic activity can cause microwave dishes to go out of alignment. Earthquake-resistant towers and mounts can help prevent alignment failure, even for wind-related problems. Redundant systems should be considered part of your solution. A duplicate microwave system might lead to a false sense of security. Consider a nonmicrowave backup, such as fiber, for a primary microwave link. Smoke, heavy rain, and snow storms can cause enough path loss to disable otherwise sound wireless systems.

Fiber Optics Links

If you are not a fiber customer today, you will be tomorrow. Telephone companies will soon be joined by other providers to seek your fiber business. You may be fortunate enough to be served by separate fiber vendors with separate fiber systems and routing to enhance reliability and uptime. Special installation techniques are essential to make sure fiber links will not be bothered by earth movement, subject to vandalism, or vulnerable to single-point failure. Single-point failure can occur in any system. Single-point failure analysis and prevention is based on simple concepts: A chain is only as strong as its weakest link, but two chains, equally strong, may have the same weak link. The lesson may be make one chain much stronger, or use three chains of a material that has different stress properties.

Fiber should be installed underground in a sturdy plastic sheath, called an interliner. Interliners are usually colored bright orange to make them stand out in trenches, manholes and other places where careless digging and prodding could spell disaster. This sheath offers protection from sharp rocks or other forces that might cause a nick or break in the armor of the cable, or actually sever one or more of the bundled fibers. Cable systems that only have aerial rights-of-way on utility poles for their fiber may not prove as reliable in some areas as underground fiber. Terminal equipment for fiber should be installed in earthquake-secure equipment racks away from flooding hazards. Fiber electronics should have a minimum of two parallel DC power supplies, which are in turn paralleled with rechargeable battery backup.

Sonet® technology is a proven approach you should look for from your fiber vendor. This solution is based on topology that looks like a circle or ring. A fiber optics cable could be cut just like a wire or cable. A ringlike network will automatically provide a path in the other direction, away from the break. Caution! Fiber installations that run through unsealed access points, such as manholes, can be an easy target for terrorism or vandalism.

Satellite

Ku- or C-band satellite is a costly but effective way to link critical communications elements. C band has an added advantage over Ku during heavy rain or snow storms. Liquid or frozen water can disrupt Ku-band satellite transmission. A significant liability of satellite transmission for ultrareliable facilities is the possibility that a storm could cause a deep fade, even for C-band links. Another liability is short but deep semiannual periods of sun outage when a link is lost while the sun is focused directly into a receive dish. Although these periods are predictable and last for only a minute or two, there is nothing that can prevent their effect unless you have alternate service on another satellite with a different sun outage time, or terrestrial backup.

27.6 Emergency Power and Batteries

Uninterruptible power supplies (UPS) are common in the information workplace. From small UPS that plug into wall outlets at a personal computer work station, to giant units that can power an entire facility, they all have one thing in common, batteries. UPS batteries have a finite life span. Once exceeded, a UPS is nothing more than an expensive door stop. UPS batteries must be tested regularly. Allow the UPS to go on line to test it. Some UPS test themselves automatically. Routinely pull the UPS AC plug out of the wall for a manual test. Some UPS applications require hours of power, whereas some only need several minutes. Governing factors are:

- Availability of emergency power that can be brought on line fast
- A need to keep systems alive long enough for a graceful shutdown
- Systems so critical that they can never go down

Although UPS provide emergency power when the AC mains are dead, many are programmed with another electronic agenda: Protect the devices plugged in from what the UPS thinks is *bad power*. Many diesel generators in emergency service are not sized for the load they have to carry, or may not have proper power factor correction. Computers and other devices with switching power supplies can distort AC power wave forms; the result: bad power.

After a UPS comes on line, it should shut down after the emergency generator picks up the load and charges its batteries. If it senses the AC equivalent of poison, it stays on or cycles on and off. Its battery eventually runs down. Your best defense is to test your entire emergency power system under full load. If a UPS cycles on and off to the point that its batteries run down, you must find out why. Consult your UPS manufacturer, service provider, or service manual to see if your UPS can be adjusted to be more tolerant. Some UPS cycling cannot be avoided with engine-based emergency power, especially if heavy loads such as air conditioner compressors switch on and off line.

Technicians sometimes believe that starting an emergency generator with no equipment load is an adequate weekly test. Even a 30-min test will not get the engine up to proper operating temperature. If your generator is diesel driven, this may lead to *wet stacking*, cylinder glazing, and piston rings that can lose proper seating. Wet stacking occurs when a generator is run repeatedly with no load or a light load. When the generator is asked to come on line to power a full equipment load, deposits that build up during no-load tests prevent it from developing full power. The engine will also not develop full power if its rings are misseated and there is significant cylinder glazing. The cure is to always test with the load your diesel has to carry during an emergency. If that is not possible, obtain a resistive load bank so you can simulate a full load for an hour or two of hard running several times per year. A really hard run should burn out accumulated carbon, reseat rings, and deglaze cylinder walls.

Fuel stored in tanks gets old and old fuel is unreliable. Gum and varnish can form. Fuel begins to break down. Certain forms of algae can grow in diesel oil, especially at the boundary layer between fuel and the water that can accumulate at the bottom of most tanks. Fuel additives can extend the storage period and prevent algae growth. A good filtering system, and a planned program of cycling fuel through it, can extend storage life dramatically. Individual fuel chemical composition, fuel conditioners, and the age and type of storage tank all affect useful fuel life.

There are companies that will analyze your fuel. They can filter out dirt, water, and debris that can rob your engine of power. The cost of additives and fuel filtering is nominal compared to the cost of new fuel plus hazardous material disposal charges for old fuel. Older fuel tanks can spring leaks that either introduce water into the fuel, or introduce you to a costly hazardous materials clean up project, your tank will be out of service while it is being replaced, and fuel carrying dirt or water can stop the engine.

While you are depending on an emergency generator for your power, you would hate to see it stop. A running generator will consume fuel, crankcase oil, and possibly radiator coolant. You should know your generator's crankcase oil consumption rate so you can add oil well before the engine grinds to a screeching, nonlubricated halt. Water-cooled generators must be checked periodically to make sure there is enough coolant in the radiator. Make sure you have enough coolant and oil to get the facility through a minimum of one week of constant duty.

Most experts recommend a generator health check every six months. Generators with engine block heaters put special stress on fittings and hoses. Vibration can loosen bolts, crack fittings, and fatigue wires and connectors. If your application is supercritical, a second generator may give you a greater margin of safety. Your generator maintenance technician should take fuel and crankcase oil samples for testing at a qualified laboratory. The fuel report will let you know if your storage conditions are acceptable. The crankcase oil report might find microscopic metal particles, early warning of a major failure. How long will a generator last? Some engine experts say that a properly maintained diesel generator set can run in excess of 9,000 h before it would normally need to be replaced.

Mission dictates need. Need dictates reliability. If the design budget permits, a second or even third emergency generator is a realistic insurance policy. When you are designing a facility you know must never fail, consider redundant UPS wired in parallel. Consult the vendor for details on wiring needs for multiphase parallel UPS installations. During major overhauls and generator work, make sure you have a local source for reliable portable power. High-power diesel generators on wheels are common now to supply field power for events from rock concerts to movie shoots. Check your local telephone directory. If you are installing a new diesel, remember that engines over a certain size may have to be licensed by your local air quality management district and that permits must be obtained to construct and store fuel in an underground tank.

27.7 Air Handling Systems

Equipment crashes when it gets too hot. Clean, cool, dry, and pollutant-free air in generous quantities is critical for modern communications facilities. If you lease space in a high-rise, you may not have your own air system. Many building systems often have no backup, are not supervised nights and weekends, and may have uncertain maintenance histories. Your best protection is to get the exact terms for air conditioning nailed down in your lease. You may wish to consider adding your own backup system, a costly but essential strategy if your building air supply is unreliable or has no backup. Several rental companies specialize in emergency portable industrial-strength air conditioning. An emergency contract for **heating ventilating and air conditioning (HVAC)** that can be invoked with a phone call could save you hours or even days of downtime. Consider buying a portable HVAC unit if you are protecting a supercritical facility.

Wherever cooling air comes from, there are times when you need to make sure the system can be forced to recirculate air within the building, temporarily becoming a closed system. Smoke or toxic fumes from a fire in the neighborhood can enter an open system. Toxic air could incapacitate your people in seconds. With some advanced warning, forcing the air system to full recirculation could avoid or forestall calamity. It could buy enough time to arrange an orderly evacuation and transition to an alternate site.

27.8 Water Hazards

Water in the wrong place at the wrong time can be part of a larger emergency or be its own emergency. A simple mistake such as locating a water heater where it can flood out electrical equipment can cause short circuits when it finally wears out and begins to leak. Unsecured water heaters can tear away from gas lines, possibly causing an explosion or fire in an earthquake. The water in that heater could be lost, depriving employees of a source of emergency drinking water.

Your facility may be located near a source of water that could flood you out. Many businesses are located in flood plains that see major storms once every 100 or 150 years. If you happen to be on watch at the wrong time of the century, you may wish that you had either located elsewhere, or stocked a very large supply of sand bags.

Remember to include any wet or dry pipe fire sprinkler systems as potential water hazards.

27.9 Electromagnetic Pulse Protection (EMP)

The **electromagnetic pulse (EMP)** phenomenon associated with nuclear explosions can disable almost any component in a communications system. EMP energy can enter any component or system coupled to a wire or metal surface directly, capacitively, or inductively. Some chemical weapons can produce

Disaster Planning and Recovery

EMP, but on a smaller scale. The Federal Emergency Management Agency (FEMA) publishes a three volume set of documents on EMP. They cover the theoretical basis for EMP protection, protection applications, and protection installation. FEMA has been involved in EMP protection since 1970 and is charged at the federal level with the overall direction of the EMP program. FEMA provides detailed guidance and, in some cases, direct assistance on EMP protection to critical communications facilities in the private sector. AM, FM, and television transmitter facilities that need EMP protection should discuss EMP protection tactics with a knowledgeable consultant before installing protection devices on radio frequency (RF) circuitry. EMP devices such as gas discharge tubes can fail in the presence of high-RF voltage conditions and disable facilities through such a failure.

27.10 Alternate Sites

No matter how well you plan, something still could happen that will require you to abandon your facility for some period of time. Government emergency planners usually arrange for an alternate site for their EOCs. Communications facilities can sign mutual aid agreements. Sometimes this is the only way to access telephone lines, satellite uplink equipment, microwave, or fiber on short notice. If management shows reluctance to share, respectfully ask what they would do if their own facility is rendered useless.

27.11 Security

It is a fact of modern life that man-caused disasters must now enter into the planning and risk assessment process. Events ranging from terrorism to poor training can cause the most mighty organization to tumble. The World Trade Center and Oklahoma City bombings are a warning to us all. Your risk assessment might even prompt you to relocate if you are too close to *ground zero*.

Federal Communications Commission (FCC) rules still state that licensees of broadcast facilities must protect their facilities from hostile takeover. Breaches in basic security have often led to serious incidents at a number of places throughout the country. It has even happened at major market television stations. Here are the basics:

- Approve visits from former employees through their former supervisors.
- Escort nonemployees in critical areas.
- Assure that outside doors are never propped open.
- Secure roof hatches from the inside and have alarm contacts on the hatch.
- Use identification badges when employees will not know each other by sight.
- Check for legislation that may require a written safety and security plan.
- Use video security and card key systems where warranted.
- Repair fences, especially at unmanned sites.
- Install entry alarms at unattended sites; test weekly.
- Redesign to limit the places bombs could be planted.
- Redesign to prevent unauthorized entry.
- Redesign to limit danger from outside windows.
- Plan for fire, bomb threats, hostage situations, terrorist takeovers.
- Plan a safe way to shut the facility down in case of invasion.
- Plan guard patrol schedules to be random, not predictable.
- Plan for off-site relocation and restoration of services on short notice.

California Senate Bill 198 mandates that California businesses with more than 100 employees write an industrial health and safety plan for each facility addressing workplace safety, hazardous materials spills, employee training, and emergency response.

27.12 Workplace and Home: Hand-in-Hand Preparedness

A critical facility deprived of its staff will be paralyzed just as surely as if all of the equipment suddenly disappeared. Employees may experience guilt if they are at work when a regional emergency strikes, and they do not know what is happening at home. The first instinct is to go home. This is often the wrong move. Blocked roads, downed bridges, and flooded tunnels are dangerous traps, especially at night. People who leave work during emergencies, especially people experiencing severe stress, often become victims. Encourage employees to prepare their homes, families, and pets for the same types of risks the workplace will face. Emergency food and water and a supply of fresh batteries in the refrigerator are a start. Battery-powered radios and flashlights should be tested regularly. If employees or their families require special foods, prescription drugs, eyewear, oxygen, over-the-counter pharmaceuticals, sun block or bug repellent, remind them to have an adequate supply on hand to tide them over for a lengthy emergency.

Heavy home objects like bookcases should be secured to walls so they will not tip over. Secure or move objects mounted on walls over beds. Make sure someone in the home knows how to shut off natural gas. An extra long hose can help for emergency fire fighting or help drain flooded areas. Suggest family *hazard hunts*. Educate employees on what you are doing to make the workplace safe. The same hazards that can hurt, maim, or kill in the workplace can do the same at home. Personal and company vehicles should all have emergency kits that contain basic home or business emergency supplies. Food, water, comfortable shoes, and old clothes should be added. If their families are prepared at home or on the road, employees will have added peace of mind. It may sustain them until they can get home safely.

An excellent home family preparedness measure is to identify a distant relative or friend who can be the emergency message center. Employees may be able to call a relative from work to find out their family is safe and sound. Disasters that impair telephone communications teach us that it is often possible to make and receive long distance calls when a call across the street will not get through. Business emergency planners should not overlook this hint. A location in another city, or a key customer or supplier may make a good out-of-area emergency contact.

27.13 Expectations, 9-1-1, and Emergencies

Television shows depicting 9-1-1 saving lives over the telephone are truly inspirational. But during a major emergency, resources normally available to 9-1-1 services, including their very telephone system, may be unavailable. Emergency experts used to tell us to be prepared to be self-sufficient at the neighborhood and business level for 72 hours or more. Some now suggest a week or longer. Government will not respond to every call during a major disaster. That is a fact.

Even experienced communications professionals sometimes forget that an overloaded telephone exchange cannot supply dial tone to all customers at once. Emergency calls will often go through if callers wait patiently for dial tone. If callers do not hear dial tone after 10 or 15 minutes, it is safe to assume that there is a more significant problem.

27.14 Business Rescue Planning for Dire Emergencies

When people are trapped and professionals cannot get through, our first instinct may be to attempt a rescue. Professionals tell us that more people are injured or killed in rescue attempts during major emergencies than are actually saved. Experts in **urban search and rescue (USAR)** not only have the know-how to perform their work safely, but have special tools that make this work possible under impossible conditions. The *jaws of life*, hydraulic cutters used to free victims from wrecked automobiles, is a common USAR tool. Pneumatic jacks that look like large rubber pillows can lift heavy structural members in destroyed buildings to free trapped people.

You, as a facilities designer, may never be faced with a life-or-death decision concerning a rescue when professionals are not available. Those in the facilities you design may be faced with tough decisions. Consider that your design could make their job easier or more difficult. Also consider recommending

USAR training for those responsible for on-line management and operations of the facility as a further means to ensure readiness.

27.15 Managing Fear

Anyone who says they are not scared while experiencing a hurricane, tornado, flood, or earthquake is either lying or foolish. Normal human reactions when an emergency hits are colored by a number of factors, including fear. As the emergency unfolds, we progress from fear of the unknown, to fear of the known. While preparedness, practice, and experience may help keep fears in check, admitting fear and the normal human response to fear can help us keep calm.

Some people prepare mentally by reviewing their behavior during personal, corporate, or natural emergencies. Then they consider how they could have been better prepared to transition from *normal* human reactions like shock, denial, and panic, to abnormal reactions like grace, acceptance, and steady performance. The latter behaviors reassure those around them and encourage an effective emergency team. Grace under pressure is not a bad goal.

Another normal reaction most people experience is a rapid change of focus toward one's own personal well being. "Am I OK?" is a very normal question at such times. Even the most altruistic people have moments during calamities when they regress. They temporarily become selfish children. Once people know they do not require immediate medical assistance, they can usually start to focus again on others and on the organization.

Defining Terms

Business impact analysis (BIA): A formal study of the impact of a risk or multiple risks on a specific business. A properly conducted BIA becomes critical to the *business recovery plan.*

Business resumption planning (BRP): A blueprint to maintain or resume core business activities following a disaster. Three major goals of BRP are resumption of products and services, customer service, and cash flow.

Central office (CO): Telephone company jargon for the building where local switching is accomplished.

Electromagnetic pulse (EMP): A high burst of energy associated most commonly with nuclear explosions. EMP can instantly destroy many electronic systems and components.

Emergency operations center (EOC): A location where emergency managers receive damage assessments, allocate resources, and begin recovery.

Heating, ventilation, and air conditioning (HVAC): Architectural acronym.

Incident commander (IC): The title of the person in charge at an emergency scene from the street corner to an emergency involving an entire state or region.

Incident command system (ICS): An effective emergency management model invented by fire fighters in California.

Urban Search and Rescue (USAR): Emergency management acronym.

References

Baylus, E. 1992. *Disaster Recovery Handbook.* Chantico, NY.

Fletcher, R. 1990. Federal Response Plan. Federal Emergency Management Agency, Washington, DC.

FEMA. 1991. *Electromagnetic Pulse Protection Guidance,* Vols. 1–3. Federal Emergency Management Agency, Washington, DC.

Handmer, J. and Parker, D. 1993. *Hazard Management and Emergency Planning.* James and James Science, NY.

Rothstein Associates. 1993. *The Rothstein Catalog on Disaster Recovery and Business Resumption Planning.* Rothstein Associates.

Further Information

Associations/Groups:

The Association of Contingency Planners, 14775 Ventura Boulevard, Suite 1-885, Sherman Oaks, CA 91483.

Business and Industry Council for Emergency Planning and Preparedness (BICEPP), P.O. Box 1020, Northridge, CA 91328.

The Disaster Recovery Institute (DRT), 1810 Craig Road, Suite 125, St. Louis, MO 63146.

DRI holds national conferences, and publishes the *Disaster Recovery Journal.*

Earthquake Engineering Research Institute (EERI), 6431 Fairmont Avenue, Suite 7, El Cerritos, CA 94530.

National American Red Cross, 2025 E Street, NW, Washington, DC 20006.

National Center for Earthquake Engineering Research, State University of New York at Buffalo, Science and Engineering Library-304, Capen Hall, Buffalo, NY 14260.

National Coordination Council on Emergency Management (NCCEM), 7297 Lee Highway, Falls Church, VA 22042.

National Hazards Research & Applications Information Center, Campus Box 482, University of Colorado, Boulder, CO 80309.

Business Recovery Planning:

Harris Devlin Associates, 1285 Drummers Lane, Wayne, PA 19087.

Industrial Risk Insurers (IRI), 85 Woodland Street, Hartford, CT 06102.

MLC & Associates, Mary Carrido, President, 15398 Eiffel Circle, Irvine, CA 92714.

Price Waterhouse, Dispute Analysis and Corporate Recovery Dept., 555 California Street, Suite 3130, San Francisco, CA 94104.

Resource Referral Service, P.O. Box 2208, Arlington, VA 22202.

The Workman Group, Janet Gorman, President, P.O. Box 94236, Pasadena, CA 91109.

Life Safety/Disaster Response:

Caroline Pratt & Associates, 24104 Village #14, Camarillo, CA 93013.

Industry Training Associates, 3363 Wrightwood Drive, Suite 100, Studio City, CA 91604.

Emergency Supplies:

BEST Power Technology, P.O. Box 280, Necedah, WI 54646 (UPS).

Exide Electronics Group, Inc., 8521 Six Forks Road, Raleigh, NC 27615.

Extend-A-Life, Inc., 1010 South Arroyo, Parkway #7, Pasadena, CA 91105.

Velcro® USA, P.O. Box 2422, Capistrano Beach, CA 92624.

Worksafe Technologies, 25133 Avenue Tibbets, Building F. Valencia, CA.

Construction/Design/Seismic Bracing:

American Institute of Architects, 1735 New York Avenue, NW, Washington, DC 20006.

DATA Clean Corporation (800-328-2256).

Geotechnical/Environmental Consultants:

H.J. Degenkolb Associates, Engineers, 350 Sansome Street, San Francisco, CA 94104.

Leighton and Associates, Inc., 17781 Cowan, Irvine, CA 92714.

Miscellaneous:

Commercial Filtering, Inc., 5205 Buffalo Avenue, Sherman Oaks, CA 91423 (Fuel Filtering).

Data Processing Security, Inc., 200 East Loop 820, Forth Worth, TX 76112.

EDP Security, 7 Beaver Brook Road. Littleton, MA 01460.

ENDUR-ALL Glass Coatings, Inc., 23018 Ventura Blvd., Suite 101, Woodland Hills, CA 91464.

Mobile Home Safety Products, 28165 B Front Street, Suite 121, Temecula, CA 92390.

28
Safety and Protection Systems

Jerry C. Whitaker
Editor-in-Chief

28.1 Introduction 28-1
   Facility Safety Equipment
28.2 Electric Shock 28-3
   Effects on the Human Body • Circuit-Protection Hardware
   • Working With High Voltage • First Aid Procedures
28.3 Polychlorinated Biphenyls 28-10
   Health Risk • Governmental Action • PCB Components
   • Identifying PCB Components • Labeling PCB Components
   • Record-Keeping • Disposal • Proper Management
28.4 OSHA Safety Requirements 28-16
   Protective Covers • Identification and Marking • Extension
   Cords • Grounding • Management Responsibility

28.1 Introduction

Safety is critically important to engineering personnel who work around powered hardware, especially if they work under considerable time pressures. Safety is not something to be taken lightly. *Life safety* systems are those designed to protect life and property. Such systems include emergency lighting, fire alarms, smoke exhaust and ventilating fans, and site security.

Facility Safety Equipment

Personnel safety is the responsibility of the facility manager. Proper life safety procedures and equipment must be installed. Safety-related hardware includes the following:

- *Emergency power off* (EPO) button. EPO push buttons are required by safety code for data processing centers. One must be located at each principal exit from the data processing (DP) room. Other EPO buttons may be located near operator workstations. The EPO system, intended only for emergencies, disconnects all power to the room, except for lighting.
- Smoke detector. Two basic types of smoke detectors commonly are available. The first compares the transmission of light through air in the room with light through a sealed optical path into which smoke cannot penetrate. Smoke causes a differential or *backscattering* effect that, when detected, triggers an alarm after a preset threshold has been exceeded. The second type of smoke detector senses the ionization of combustion products rather than visible smoke. A mildly radioactive source, usually nickel, ionizes the air passing through a screened chamber. A charged probe captures ions and detects the small current that is proportional to the rate of capture. When combustion products or material other than air molecules enter the probe area, the rate of ion production changes abruptly, generating a signal that triggers the alarm.

- Flame detector. The flame sensor responds not to heated surfaces or objects, but to infrared when it flickers with the unique characteristics of a fire. Such detectors, for example, will respond to a lighted match, but not to a cigarette. The ultraviolet light from a flame also is used to distinguish between hot, glowing objects and open flame.
- Halon. The Halon fire-extinguishing agent is a low-toxicity, compressed gas that is contained in pressurized vessels. Discharge nozzles in data processing (DP) rooms and other types of equipment rooms are arranged to dispense the entire contents of a central container or of multiple smaller containers of Halon when actuated by a command from the fire control system. The discharge is sufficient to extinguish flame and stop combustion of most flammable substances. Halon is one of the more common fire-extinguishing agents used for DP applications. Halon systems usually are not practical, however, in large, open-space computer centers.
- Water sprinkler. Although water is an effective agent against a fire, activation of a sprinkler system will cause damage to the equipment it is meant to protect. Interlock systems must drop all power (except for emergency lighting) before the water system is discharged. Most water systems use a two-stage alarm. Two or more fire sensors, often of different design, must signal an alarm condition before water is discharged into the protected area. Where sprinklers are used, floor drains and EPO controls must be provided.
- Fire damper. Dampers are used to block ventilating passages in strategic parts of the system when a fire is detected. This prevents fire from spreading through the passages and keeps fresh air from fanning the flames. A fire damper system, combined with the shutdown of cooling and ventilating air, enables Halon to be retained in the protected space until the fire is extinguished.

Many life safety system functions can be automated. The decision of what to automate and what to operate manually requires considerable thought. If the life safety control panels are accessible to a large number of site employees, most functions should be automatic. Alarm-silencing controls should be maintained under lock and key. A mimic board can be used to readily identify problem areas. Figure 28.1 illustrates a well-organized life safety control system. Note that fire, HVAC (heating, ventilation, and air conditioning), security, and EPO controls all are readily accessible. Note also that operating instructions are posted for life safety equipment, and an evacuation route is shown. Important telephone numbers are posted, and a direct-line telephone (not via the building switchboard) is provided. All equipment is located adjacent to a lighted emergency exit door.

Life safety equipment must be maintained just as diligently as the computer system that it protects. Conduct regular tests and drills. It is, obviously, not necessary or advisable to discharge Halon or water during a drill.

Configure the life safety control system to monitor not only the premises for dangerous conditions, but also the equipment designed to protect the facility. Important monitoring points include HVAC machine parameters, water and/or Halon pressure, emergency battery-supply status, and other elements of the system that could compromise the ability of life safety equipment to carry out its functions. Basic guidelines for life safety systems include the following:

- Carefully analyze the primary threats to life and property within the facility. Develop contingency plans to meet each threat.
- Prepare a life safety manual, and distribute it to all employees at the facility. Require them to read it.
- Conduct drills for employees at random times without notice. Require acceptable performance from employees.
- Prepare simple, step-by-step instructions on what to do in an emergency. Post the instructions in a conspicuous place.
- Assign after-hours responsibility for emergency situations. Prepare a list of supervisors whom operators should contact if problems arise. Post the list with phone numbers. Keep the list accurate and up-to-date. Always provide the names of three individuals who can be contacted in an emergency.

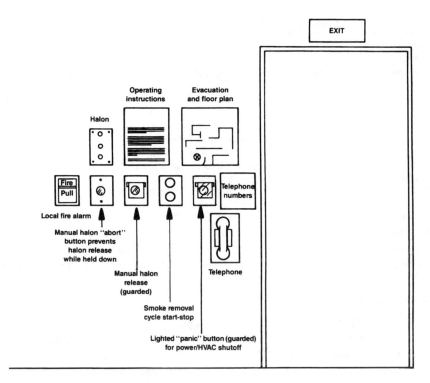

FIGURE 28.1 A well-organized life safety control station. (After [1].)

- Work with a life safety consultant to develop a coordinated control and monitoring system for the facility. Such hardware will be expensive, but it must be provided. The facility may be able to secure a reduction in insurance rates if comprehensive safety efforts can be demonstrated.
- Interface the life safety system with automatic data-logging equipment so that documentation can be assembled on any event.
- Insist upon complete, up-to-date schematic diagrams for all hardware at the facility. Insist that the diagrams include any changes made during installation or subsequent modification.
- Provide sufficient emergency lighting.
- Provide easy-access emergency exits.

The importance of providing standby power for sensitive loads at commercial and industrial facilities has been outlined previously. It is equally important to provide standby power for life safety systems. A lack of AC power must not render the life safety system inoperative. Sensors and alarm control units should include their own backup battery supplies. In a properly designed system, all life safety equipment will be fully operational despite the loss of all AC power to the facility, including backup power for sensitive loads.

Place cables linking the life safety control system with remote sensors and actuators in separate conduits containing only life safety conductors. Study the National Electrical Code and all applicable local and federal codes relating to safety. Follow them to the letter.

28.2 Electric Shock

It takes surprisingly little current to injure a person. Studies at Underwriters' Laboratories (UL) show that the electrical resistance of the human body varies with the amount of moisture on the skin, the muscular structure of the body, and the applied voltage. The typical hand-to-hand resistance ranges from 500 Ω to 600 kΩ, depending on the conditions. Higher voltages have the capability to break down the

TABLE 28.1 The Effects of Current on the Human Body

1 mA or less	No sensation, not felt
More than 3 mA	Painful shock
More than 10 mA	Local muscle contractions, sufficient to cause "freezing" to the circuit for 2.5% of the population
More than 15 mA	Local muscle contractions, sufficient to cause "freezing" to the circuit for 50% of the population
More than 30 mA	Breathing is difficult, can cause unconsciousness
50 mA to 100 mA	Possible ventricular fibrillation of the heart
100 mA to 200 mA	Certain ventricular fibrillation of the heart
More than 200 mA	Severe burns and muscular contractions; heart more apt to stop than to go into fibrillation
More than a few amperes	Irreparable damage to body tissues

outer layers of the skin, which can reduce the overall resistance value. UL uses the lower value, 500 Ω, as the standard resistance between major extremities, such as from the hand to the foot. This value generally is considered the minimum that would be encountered. In fact, it may not be unusual because wet conditions or a cut or other break in the skin significantly reduce human body resistance.

Effects on the Human Body

Table 28.1 lists some effects that typically result when a person is connected across a current source with a hand-to-hand resistance of 2.4 kΩ. The table shows that a current of 50 mA will flow between the hands, if one hand is in contact with a 120 V AC source and the other hand is grounded. The table also indicates that even the relatively small current of 50 mA can produce *ventricular fibrillation* of the heart, and maybe even cause death. Medical literature describes ventricular fibrillation as very rapid, uncoordinated contractions of the ventricles of the heart resulting in loss of synchronization between heartbeat and pulse beat. The electrocardiograms shown in Fig. 28.2 compare a healthy heart rhythm with one in ventricular fibrillation. Unfortunately, once ventricular fibrillation occurs, it will continue. Barring resuscitation techniques, death will ensue within a few minutes.

The route taken by the current through the body greatly affects the degree of injury. Even a small current, passing from one extremity through the heart to another extremity, is dangerous and capable of causing severe injury or electrocution. There are cases in which a person has contacted extremely high current levels and lived to tell about it. However, when this happens, it is usually because the current

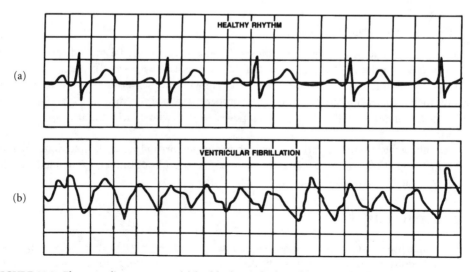

FIGURE 28.2 Electrocardiogram traces: (a) healthy heart rhythm, (b) ventricular fibrillation of the heart.

Safety and Protection Systems

passes only through a single limb and not through the entire body. In these instances, the limb is often lost but the person survives.

Current is not the only factor in electrocution. Figure 28.3 summarizes the relationship between current and time on the human body. The graph shows that 100 mA flowing through an adult human body for 2 s will cause death by electrocution. An important factor in electrocution, the *let-go range*, also is shown on the graph. This point marks the amount of current that causes *freezing*, or the inability to let go of a conductor. At 10 mA, 2.5% of the population would be unable to let go of a live conductor; at 15 mA, 50% of the population would be unable to let go of an energized conductor. It is apparent from the graph that even a small amount of current can freeze someone to a conductor. The objective for those who must work around electric equipment is to protect themselves from electric shock. Table 28.2 lists required precautions for maintenance personnel working near high voltages.

Circuit-Protection Hardware

A common primary panel or equipment circuit breaker or fuse will not protect an individual from electrocution. However, the *ground-fault current interrupter* (GFCI), used properly, can help prevent electrocution. Shown in Fig. 28.4, the GFCI works by monitoring the current being applied to the load. It uses a differential transformer that senses an imbalance in load current. If a current (typically 5 mA, ± 1 mA on a low-current 120 V AC line) begins flowing between the neutral and ground or between the hot and ground leads, the differential transformer detects the leakage and opens the primary circuit (typically within 2.5 ms).

OSHA (Occupational Safety and Health Administration) rules specify that temporary receptacles (those not permanently wired) and receptacles used on construction sites be equipped with GFCI protection. Receptacles on two-wire, single-phase portable and vehicle-mounted generators of not more than 5 kW, where the generator circuit conductors are insulated from the generator frame and all other grounded surfaces, need not be equipped with GFCI outlets.

GFCIs will not protect a person from every type of electrocution. If you become connected to both the neutral and the hot wire, the GFCI will treat you as if you are merely a part of the load and will not open the primary circuit.

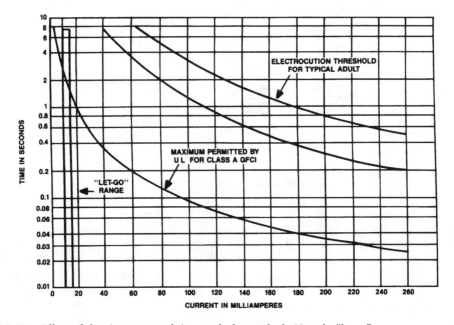

FIGURE 28.3 Effects of electric current and time on the human body. Note the "let-go" range.

TABLE 28.2 Required Safety Practices for Engineers Working Around High-Voltage Equipment

- ✓ Remove all AC power from the equipment. Do not rely on internal contactors or SCRs to remove dangerous AC.
- ✓ Trip the appropriate power-distribution circuit breakers at the main breaker panel.
- ✓ Place signs as needed to indicate that the circuit is being serviced.
- ✓ Switch the equipment being serviced to the *local control* mode as provided.
- ✓ Discharge all capacitors using the discharge stick provided by the manufacturer.
- ✓ Do not remove, short-circuit, or tamper with interlock switches on access covers, doors, enclosures, gates, panels, or shields.
- ✓ Keep away from live circuits.
- ✓ Allow any component to cool completely before attempting to replace it.
- ✓ If a leak or bulge is found on the case of an oil-filled or electrolytic capacitor, do not attempt to service the part until it has cooled completely.
- ✓ Know which parts in the system contain PCBs. Handle them appropriately.
- ✓ Minimize exposure to RF radiation.
- ✓ Avoid contact with hot surfaces within the system.
- ✓ Do not take chances.

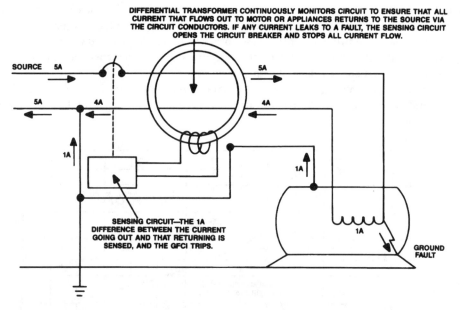

FIGURE 28.4 Basic design of a ground-fault current interrupter (GFCI).

For large, three-phase loads, detecting ground currents and interrupting the circuit before injury or damage can occur is a more complicated proposition. The classic method of protection involves the use of a zero-sequence current transformer (CT). Such devices are basically an extension of the single-phase GFCI circuit shown in Fig. 28.4. Three-phase CTs have been developed to fit over bus ducts, switchboard buses, and circuit-breaker studs. Rectangular core-balanced CTs are able to detect leakage currents as small as several milliamperes when the system carries as much as 4 kA. "Doughnut-type" toroidal zero-sequence CTs also are available in varying diameters.

The zero-sequence current transformer is designed to detect the magnetic field surrounding a group of conductors. As shown in Fig. 28.5, in a properly operating three-phase system, the current flowing

Safety and Protection Systems

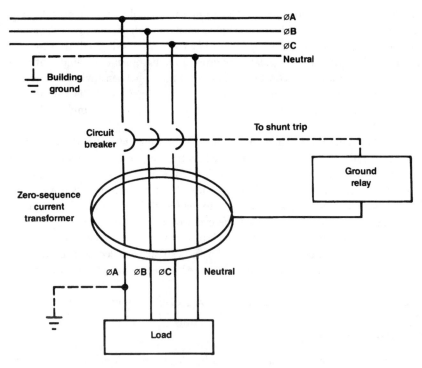

FIGURE 28.5 Ground-fault detection in a three-phase AC system.

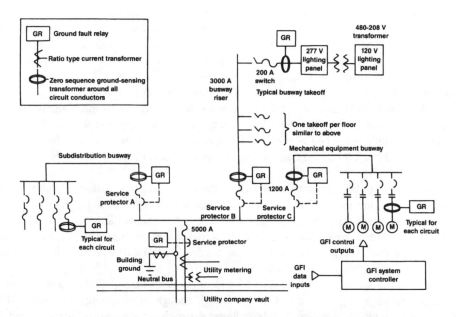

FIGURE 28.6 Ground-fault protection system for a large, multistory building.

through the conductors of the system, including the neutral, goes out and returns along those same conductors. The net magnetic flux detected by the CT is zero. No signal is generated in the transformer winding, regardless of current magnitudes—symmetrical or asymmetrical. If one phase conductor is faulted to ground, however, the current balance will be upset. The ground-fault-detection circuit then will trip the breaker and open the line.

For optimum protection in a large facility, GFCI units are placed at natural branch points of the AC power system. It is, obviously, preferable to lose only a small portion of a facility in the event of a ground fault than it is to have the entire plant dropped. Figure 28.6 illustrates such a distributed system. Sensors are placed at major branch points to isolate any ground fault from the remainder of the distribution network. In this way, the individual GFCI units can be set for higher sensitivity and shorter time delays than would be practical with a large, distributed load. The technology of GFCI devices has improved significantly within the past few years. New integrated circuit devices and improved CT designs have provided improved protection components at a lower cost.

Sophisticated GFCI monitoring systems are available which analyze ground-fault currents and isolate the faulty branch circuit. This feature prevents needless tripping of GFCI units up the line toward the utility service entrance. For example, if a ground fault is sensed in a fourth-level branch circuit, the GFCI system controller automatically locks out first-, second-, and third-level devices from operating to clear the fault. The problem, therefore, is safely confined to the fourth-level branch. The GFCI control system is designed to operate in a fail-safe mode. In the event of a control-system shutdown, the individual GFCI trip relays would operate independently to clear whatever fault currents may exist.

Any facility manager would be well-advised to hire an experienced electrical contractor to conduct a full ground-fault protection study. Direct the contractor to identify possible failure points, and to recommend corrective actions.

An extensive discussion of GFCI principals and practices can be found in Reference [2].

Working with High Voltage

Rubber gloves are a common safety measure used by engineers working on high-voltage equipment. These gloves are designed to provide protection from hazardous voltages when the wearer is working on "hot" circuits. Although the gloves may provide some protection from these hazards, placing too much reliance on them poses the potential for disastrous consequences. There are several reasons why gloves should be used only with a great deal of caution and respect. A common mistake made by engineers is to assume that the gloves always provide complete protection. The gloves found in some facilities may be old and untested. Some may even have been "repaired" by users, perhaps with electrical tape. Few tools could be more hazardous than such a pair of gloves.

Know the voltage rating of the gloves. Gloves are rated differently for AC and DC voltages. For instance, a *class 0* glove has a minimum DC breakdown voltage of 35 kV; the minimum AC breakdown voltage, however, is only 6 kV. Furthermore, high-voltage rubber gloves are not tested at RF frequencies, and RF can burn a hole in the best of them. Working on live circuits involves much more than simply wearing a pair of gloves. It involves a frame of mind—an awareness of everything in the area, especially ground points.

Gloves alone may not be enough to protect an individual in certain situations. Recall the axiom of keeping one hand in your pocket while working on a device with current flowing? The axiom actually is based on simple electricity. It is not the hot connection that causes the problem; it is the ground connection that permits current flow. Studies have showed that more than 90% of electric equipment fatalities occurred when the grounded person contacted a live conductor. Line-to-line electrocution accounted for less than 10% of the deaths.

When working around high voltages, always look for grounded surfaces—and keep away from them. Even concrete can act as a ground if the voltage is high enough. If work must be conducted in live cabinets, consider using—in addition to rubber gloves—a rubber floor mat, rubber vest, and rubber sleeves. Although this may seem to be a lot of trouble, consider the consequences of making a mistake. Of course, the best troubleshooting methodology is never to work on any circuit unless you are sure no hazardous voltages are present. In addition, any circuits or contactors that normally contain hazardous voltages should be grounded firmly before work begins.

Another important safety rule is to never work alone. Even if a trained assistant is not available when maintenance is performed, someone should accompany you and be available to help in an emergency.

First Aid Procedures

Be familiar with first aid treatment for electric shock and burns. Always keep a first aid kit on hand at the facility. Figure 28.7 illustrates the basic treatment for electric shock victims. Copy the information, and post it in a prominent location. Better yet, obtain more detailed information from your local Heart Association or Red Cross chapter. Personalized instruction on first aid usually is available locally. Table 28.3 lists basic first aid procedures for burns.

For electric shock, the best first aid is prevention. In the event that an individual has sustained or is sustaining an electric shock at the work place, several guidelines are suggested, as detailed next.

Shock in Progress

For the case when a co-worker is receiving an electric shock and cannot let go of the electrical source, the safest action is to trip the circuit breaker that energizes the circuit involved, or to pull the power-line plug on the equipment involved if the latter can be accomplished safely [2]. Under no circumstances

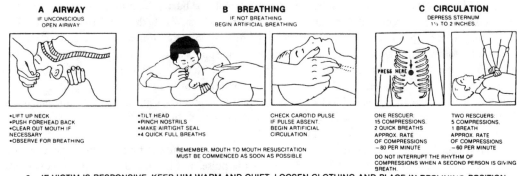

FIGURE 28.7 Basic first aid treatment for electric shock.

TABLE 28.3 Basic First Aid Procedures

For extensively burned and broken skin:
- ✓ Cover affected area with a clean sheet or cloth.
- ✓ Do not break blisters, remove tissue, remove adhered particles of clothing, or apply any salve or ointment.
- ✓ Treat victim for shock as required.
- ✓ Arrange for transportation to a hospital as quickly as possible.
- ✓ If victim's arms or legs are affected, keep them elevated.
- ✓ If medical help will not be available within 1 hour and the victim is conscious and not vomiting, prepare a weak solution of salt and soda. Mix 1 teaspoon of salt and 1/2-teaspoon of baking soda to each quart of tepid water. Allow the victim to sip slowly about 4 oz (half a glass) over a period of 15 min. Discontinue fluid intake if vomiting occurs. (Do not allow alcohol consumption.)

For less severe burns (first- and second-degree):
- ✓ Apply cool (not ice-cold) compresses using the cleanest available cloth article.
- ✓ Do not break blisters, remove tissue, remove adhered particles of clothing, or apply salve or ointment.
- ✓ Apply clean, dry dressing if necessary.
- ✓ Treat victim for shock as required.
- ✓ Arrange for transportation to a hospital as quickly as possible.
- ✓ If victim's arms or legs are affected, keep them elevated.

should the rescuer touch the individual who is being shocked, because the rescuer's body may then also be in the dangerous current path. If the circuit breaker or equipment plug cannot be located, then an attempt can be made to separate the victim from the electrical source through the use of a nonconducting object such as a wooden stool or a wooden broom handle. Use only an *insulating* object and nothing that contains metal or other electrically conductive material. The rescuer must be very careful not to touch the victim or the electrical source and thus become a second victim.

If such equipment is available, hot sticks *used in conjunction with lineman's gloves* may be applied to push or pull the victim away from the electrical source. Pulling the hot stick normally provides the greatest control over the victim's motion and is the safest action for the rescuer. After the electrical source has been turned off, or the victim can be reached safely, immediate first aid procedures should be implemented.

Shock No Longer in Progress

If the victim is conscious and moving about, have the victim sit down or lie down. Sometimes there is a delayed reaction to an electrical shock that causes the victim to collapse. Call 911 or the appropriate plant-site paramedic team immediately. If there is a delay in the arrival of medical personnel, check for electrical burns. In the case of severe shock, there will normally be burns at a minimum of two sites: the entry point for the current and the exit point(s). Cover the burns with dry (and sterile, preferably) dressings. Check for possible bone fractures if the victim was violently thrown away from the electrical source and possibly impacted objects in the vicinity. Apply splints as required if suitable materials are available and you have appropriate training. Cover the victim with a coat or blanket if the environmental temperature is below room temperature, or the victim complains of feeling cold.

If the victim is unconscious, call 911 or the appropriate plant-site paramedic team immediately. In the interim, check to see if the victim is breathing and if a pulse can be felt at either the inside of a wrist above the thumb joint (radial pulse) or in the neck above and to either side of the Adam's apple (carotid pulse). It is usually easier to feel the pulse in the neck as opposed to the wrist pulse, which may be weak. The index and middle finger should be used to sense the pulse, and not the thumb. Many individuals have an apparent thumb pulse that can be mistaken for the victim's pulse. If a pulse can be detected but the victim is not breathing, begin mouth-to-mouth respiration if you know how to do so. If no pulse can be detected (presumably the victim will not be breathing), carefully move the victim to a firm surface and begin cardiopulmonary resuscitation if you have been trained in the use of CPR. Respiratory arrest and cardiac arrest are crisis situations. Because of loss of the oxygen supply to the brain, permanent brain damage can occur after several minutes even if the victim is successfully resuscitated.

Ironically, the treatment for cardiac arrest induced by an electric shock is a massive counter shock, which causes the entire heart muscle to contract. The random and uncoordinated ventricular fibrillation contractions (if present) are thus stilled. Under ideal conditions, normal heart rhythm is restored once the shock current ceases. The counter shock is generated by a cardiac defibrillator, various portable models of which are available for use by emergency medical technicians and other *trained* personnel. Although portable defibrillators may be available at industrial sites where there is a high risk of electrical shock to plant personnel, they should be used only by *trained* personnel. Application of a defibrillator to an unconscious subject whose heart is beating can induce cardiac standstill or ventricular fibrillation; just the conditions that the defibrillator was designed to correct.

28.3 Polychlorinated Biphenyls

Polychlorinated biphenyls (PCBs) belong to a family of organic compounds known as *chlorinated hydrocarbons*. Virtually all PCBs in existence today have been synthetically manufactured. PCBs are of a heavy, oil-like consistency and have a high boiling point, a high degree of chemical stability, low flammability, and low electrical conductivity. These characteristics led to the past widespread use of PCBs in high-voltage capacitors and transformers. Commercial products containing PCBs were distributed widely from 1957 to 1977 under several trade names including:

- Aroclor
- Pyroclor
- Sanotherm
- Pyranol
- Askarel

Askarel also is a generic name used for nonflammable dielectric fluids containing PCBs. Table 28.4 lists some common trade names for Askarel. These trade names typically are listed on the nameplate of a PCB transformer or capacitor.

TABLE 28.4 Commonly Used Names for PCB Insulating Material

Apirolio	Abestol	Askarel	Aroclor B	Chlorextol	Chlophen
Chlorinol	Clorphon	Diaclor	DK	Dykanol	EEC-18
Elemex	Eucarel	Fenclor	Hyvol	Inclor	Inerteen
Kanechlor	No-Flamol	Phenodlor	Pydraul	Pyralene	Pyranol
Pyroclor	Sal-T-Kuhl	Santothern FR	Santovac	Solvol	Therminal

Health Risk

PCBs are harmful because, once they are released into the environment, they tend not to break apart into other substances. Instead, PCBs persist, taking several decades to slowly decompose. By remaining in the environment, they can be taken up and stored in the fatty tissues of all organisms, from which they are released slowly into the bloodstream. Therefore, because of the storage in fat, the concentration of PCBs in body tissues can increase with time, even though PCB exposure levels may be quite low. This process is called *bioaccumulation*. Furthermore, as PCBs accumulate in the tissues of simple organisms, which are consumed by progressively higher organisms, the concentration increases. This process is referred to as *biomagnification*. These two factors are especially significant because PCBs are harmful even at low levels. Specifically, PCBs have been shown to cause chronic (long-term) toxic effects in some species of animals and aquatic life. Well-documented tests on laboratory animals show that various levels of PCBs can cause reproductive effects, gastric disorders, skin lesions, and cancerous tumors.

PCBs can enter the body through the lungs, the gastrointestinal tract, and the skin. After absorption, PCBs are circulated in the blood throughout the body and stored in fatty tissues and skin, as well as in a variety of organs, including the liver, kidneys, lungs, adrenal glands, brain, and heart.

The health risk lies not only in the PCB itself, but also in the chemicals developed when PCBs are heated. Laboratory studies have confirmed that PCB by-products, including *polychlorinated dibenzofurans* (PCDFs) and *polychlorinated dibenzo-p-dioxins* (PCDDs), are formed when PCBs or chlorobenzenes are heated to temperatures ranging from approximately 900°F to 1300°F. Unfortunately, these products are more toxic than PCBs themselves.

The problem for the owner of PCB equipment is that the liability from a PCB spill or fire contamination can be tremendous. A fire involving a PCB large transformer in Binghamton, NY, resulted in $20 million in cleanup expenses. The consequences of being responsible for a fire-related incident with a PCB transformer can be monumental.

Governmental Action

The U.S. Congress took action to control PCBs in October 1975 by passing the Toxic Substances Control Act (TSCA). A section of this law specifically directed the EPA to regulate PCBs. Three years later, the EPA issued regulations to implement a congressional ban on the manufacture, processing, distribution, and disposal of PCBs. Since that time, several revisions and updates have been issued by the EPA. One of these revisions, issued in 1982, specifically addressed the type of equipment used in industrial plants.

Failure to properly follow the rules regarding the use and disposal of PCBs has resulted in high fines and some jail sentences.

Although PCBs no longer are being produced for electric products in the U.S., significant numbers still exist. The threat of widespread contamination from PCB fire-related incidents is one reason behind the EPA's efforts to reduce the number of PCB products in the environment. The users of high-power equipment are affected by the regulations primarily because of the widespread use of PCB transformers and capacitors. These components usually are located in older (pre-1979) systems, so this is the first place to look for them. However, some facilities also maintain their own primary power transformers. Unless these transformers are of recent vintage, it is quite likely that they too contain a PCB dielectric. Table 28.5 lists the primary classifications of PCB devices.

TABLE 28.5 Definition of PCB Terms as Identified by the EPA

Term	Definition	Examples
PCB	Any chemical substance that is limited to the biphenyl molecule that has been chlorinated to varying degrees, or any combination of substances that contain such substances.	PCB dielectric fluids, PCB heat-transfer fluids, PCB hydraulic fluids, 2,2',4-trichlorobiphenyl
PCB article	Any manufactured article, other than a PCB container, that contains PCBs and whose surface has been in direct contact with PCBs.	Capacitors, transformers, electric motors, pumps, pipes
PCB container	A device used to contain PCBs or PCB articles, and whose surface has been in direct contact with PCBs.	Packages, cans, bottles, bags, barrels, drums, tanks
PCB article container	A device used to contain PCB articles or equipment, and whose surface has not been in direct contact with PCBs.	Packages, cans, bottles, bags, barrels, drums, tanks
PCB equipment	Any manufactured item, other than a PCB container or PCB article container, which contains a PCB article or other PCB equipment.	Microwave ovens, fluorescent light ballasts, electronic equipment
PCB item	Any PCB article, PCB article container, PCB container, or PCB equipment that deliberately or unintentionally contains, or has as a part of it, any PCBs.	See PCB article, PCB article container, PCB container, and PCB equipment
PCB transformer	Any transformer that contains PCBs in concentrations of 500 ppm or greater.	High-power transformers
PCB contaminated	Any electric equipment that contains more than 50 ppm, but less than 500 ppm, of PCBs. (Oil-filled electric equipment other than circuit breakers, reclosers, and cable whose PCB concentration is unknown must be assumed to be PCB-contaminated electric equipment.)	Transformers, capacitors, circuit breakers, reclosers, voltage regulators, switches, cable, electromagnets

PCB Components

The two most common PCB components are transformers and capacitors. A PCB transformer is one containing at least 500 ppm (parts per million) PCBs in the dielectric fluid. An Askarel transformer generally has 600,000 ppm or more. A PCB transformer can be converted to a *PCB-contaminated device* (50 to 500 ppm) or a *non-PCB device* (less than 50 ppm) by being drained, refilled, and tested. The testing must not take place until the transformer has been in service for a minimum of 90 days. Note that this is *not* something that a maintenance technician can do. It is the exclusive domain of specialized remanufacturing companies.

PCB transformers must be inspected quarterly for leaks. However, if an impervious dike (sufficient to contain all the liquid material) is built around the transformer, the inspections can be conducted yearly. Similarly, if the transformer is tested and found to contain less than 60,000 ppm, a yearly inspection is sufficient. Failed PCB transformers cannot be repaired; they must be disposed of properly.

If a leak develops, it must be contained and daily inspections must begin. A cleanup must be initiated as soon as possible, but no later than 48 hours after the leak is discovered. Adequate records must be kept of all inspections, leaks, and actions taken for 3 years after disposal of the component. Combustible materials must be kept a minimum of 5 m from a PCB transformer and its enclosure.

As of October 1, 1990, the use of PCB transformers (500 ppm or greater) was prohibited in or near commercial buildings with secondary voltages of 480 Vac or higher. The use of radial PCB transformers was allowed if certain electrical protection was provided.

The EPA regulations also require that the operator notify others of the possible dangers. All PCB transformers (including those in storage for reuse) must be registered with the local fire department. Supply the following information:

- The location of the PCB transformer(s).
- Address(es) of the building(s). For outdoor PCB transformers, provide the outdoor location.
- Principal constituent of the dielectric fluid in the transformer(s).
- Name and telephone number of the contact person in the event of a fire involving the equipment.

Any PCB transformers used in a commercial building must be registered with the building owner. All owners of buildings within 30 m of such PCB transformers also must be notified. In the event of a fire-related incident involving the release of PCBs, immediately notify the Coast Guard National Spill Response Center at 1-800-424-8802. Also take appropriate measures to contain and control any possible PCB release into water.

Capacitors are divided into two size classes: *large* and *small*. The following are guidelines for classification:

- A PCB small capacitor contains less than 1.36 kg (3 lb) dielectric fluid. A capacitor having less than 100 in.^{3} also is considered to contain less than 3 lb dielectric fluid.
- A PCB large capacitor has a volume of more than 200 in.^{3} and is considered to contain more than 3 lb dielectric fluid. Any capacitor having a volume from 100 to 200 in.^{3} is considered to contain 3 lb dielectric, provided the total weight is less than 9 lb.
- A PCB large low-voltage capacitor contains 3 lb or more dielectric fluid and operates below 2 kV.
- A PCB large high-voltage capacitor contains 3 lb or more dielectric fluid and operates at 2 kV or greater voltages.

The use, servicing, and disposal of PCB small capacitors is not restricted by the EPA unless there is a leak. In that event, the leak must be repaired or the capacitor disposed of. Disposal can be performed by an approved incineration facility, or the component can be placed in a specified container and buried in an approved chemical waste landfill. Currently, chemical waste landfills are only for disposal of liquids containing 50 to 500 ppm PCBs and for solid PCB debris. Items such as capacitors that are leaking oil containing greater than 500 ppm PCBs should be taken to an EPA-approved PCB disposal facility.

Identifying PCB Components

The first task for the facility manager is to identify any PCB items on the premises. Equipment built after 1979 probably does not contain any PCB-filled devices. Even so, inspect all capacitors, transformers, and power switches to be sure. A call to the manufacturer also may help. Older equipment (pre-1979) is more likely to contain PCB transformers and capacitors. A liquid-filled transformer usually has cooling fins, and the nameplate may provide useful information about its contents. If the transformer is unlabeled or the fluid is not identified, it must be treated as a PCB transformer. Untested (not analyzed) mineral-oil-filled transformers are assumed to contain at least 50 ppm, but less than 500 ppm PCBs. This places them in the category of PCB-contaminated electric equipment, which has different requirements than PCB transformers. Older high-voltage systems are likely to include both large and small PCB capacitors. Equipment rectifier panels, exciter/modulators, and power-amplifier cabinets may contain a significant number of small capacitors. In older equipment, these capacitors often are Askarel-filled. Unless leaking, these devices pose no particular hazard. If a leak does develop, follow proper disposal techniques. Also, liquid-cooled rectifiers may contain Askarel. Even though their use is not regulated, treat them as a PCB article, as if they contain at least 50 ppm PCBs. Never make assumptions about PCB contamination; check with the manufacturer to be sure.

Any PCB article or container being stored for disposal must be date-tagged when removed, and inspected for leaks every 30 days. It must be removed from storage and disposed of within 1 year from the date it was placed in storage. Items being stored for disposal must be kept in a storage facility meeting the requirements of 40 CFR (Code of Federal Regulations), Part 761.65(b)(1) unless they fall under alternative regulation provisions. There is a difference between PCB items stored for disposal and those stored for reuse. Once an item has been removed from service and tagged for disposal, it cannot be returned to service.

Labeling PCB Components

After identifying PCB devices, proper labeling is the second step that must be taken by the facility manager. PCB article containers, PCB transformers, and large high-voltage capacitors must be marked with a standard 6-in. × 6-in. large marking label (ML) as shown in Fig. 28.8. Equipment containing these transformers or capacitors also should be marked. PCB large low-voltage (less than 2 kV) capacitors need not be labeled until removed from service. If the capacitor or transformer is too small to hold the large label, a smaller 1-in. × 2-in. label is approved for use. Labeling each PCB small capacitor is not required. However, any equipment containing PCB small capacitors should be labeled on the outside of the cabinet or on access panels. Properly label any spare capacitors and transformers that fall under the regulations. Identify with the large label any doors, cabinet panels, or other means of access to PCB transformers. The label must be placed so that it can be read easily by firefighters. All areas used to store PCBs and PCB items for disposal must be marked with the large (6-in. × 6-in.) PCB label.

FIGURE 28.8 Marking label (ML) used to identify PCB transformers and PCB large capacitors.

Record-Keeping

Inspections are a critical component in the management of PCBs. EPA regulations specify a number of steps that must be taken and the information that must recorded. Table 28.6 summarizes the schedule requirement, and Table 28.7 can be used as a checklist for each transformer inspection. This record must be retained for 3 years. In addition to the inspection records, some facilities may need to maintain an annual report. This report details the number of PCB capacitors, transformers, and other PCB items on the premises. The report must contain the dates when the items were removed from service, their disposition, and detailed information regarding their characteristics. Such a report must be prepared if the facility uses or stores at least one PCB transformer containing greater than 500 ppm PCBs, 50 or more PCB large capacitors, or at least 45 kg of PCBs in PCB containers. Retain the report for 5 years after the facility ceases using or storing PCBs and PCB items in the prescribed quantities. Table 28.8 lists the information required in the annual PCB report.

Disposal

Disposing of PCBs is not a minor consideration. Before contracting with a company for PCB disposal, verify its license with the area EPA office. That office also can supply background information on the company's compliance and enforcement history.

TABLE 28.6 The Inspection Schedule Required for PCB Transformers and Other Contaminated Devices

PCB Transformers	Standard PCB transformer	Quarterly
	If full-capacity impervious dike is added	Annually
	If retrofitted to < 60,000 ppm PCB	Annually
	If leak is discovered, clean up ASAP (retain these records for 3 years)	Daily
PCB article or container stored for disposal (remove and dispose of within 1 year)		Monthly
Retain all records for 3 years after disposing of transformers.		

TABLE 28.7 Inspection Checklist for PCB Components

Transformer location:
Date of visual inspection:
Leak discovered? (Yes/No):
If yes, date discovered (if different from inspection date):
Location of leak:
Person performing inspection:
Estimate of the amount of dielectric fluid released from leak:
Date of cleanup, containment, repair, or replacement:
Description of cleanup, containment, or repair performed:
Results of any containment and daily inspection required for uncorrected active leaks:

TABLE 28.8 Required Information for PCB Annual Report

I. **PCB device background information:**
   a. Dates when PCBs and PCB items are removed from service.
   b. Dates when PCBs and PCB items are placed into storage for disposal, and are placed into transport for disposal.
   c. The quantities of the items removed from service, stored, and placed into transport are to be indicated using the following breakdown:
      (1) Total weight, in kilograms, of any PCB and PCB items in PCB containers, including identification of container contents (such as liquids and capacitors).
      (2) Total number of PCB transformers and total weight, in kilograms, of any PCBs contained in the transformers.
      (3) Total number of PCB large high- or low-voltage capacitors.

II. **The location of the initial disposal or storage facility for PCBs and PCB items removed from service, and the name of the facility owner or operator.**

III. **Total quantities of PCBs and PCB items remaining in service at the end of calendar year per the following breakdown:**
   a. Total weight, in kilograms, of any PCB and PCB items in PCB containers, including the identification of container contents (such as liquids and capacitors).
   b. Total number of PCB transformers and total weight, in kilograms, of any PCBs contained in the transformers.
   c. Total number of PCB large high- or low-voltage capacitors.

The fines levied for improper disposal are not mandated by federal regulations. Rather, the local EPA administrator, usually in consultation with local authorities, determines the cleanup procedures and costs. Civil penalties for administrative complaints issued for violations of the PCB regulations are determined according to a matrix provided in the PCB penalty policy. This policy, published in the Federal Register, considers the amount of PCBs involved and the potential for harm posed by the violation.

Proper Management

Properly managing the PCB risk is not difficult. The keys are to understand the regulations and to follow them carefully. A PCB management program should include the following steps:

- Locate and identify all PCB devices. Check all stored or spare devices.
- Properly label PCB transformers and capacitors according to EPA requirements.
- Perform the required inspections, and maintain an accurate log of PCB items, their location, inspection results, and actions taken. These records must be maintained for 3 years after disposal of the PCB component.
- Complete the annual report of PCBs and PCB items by July 1 of each year. This report must be retained for 5 years.
- Arrange for any necessary disposal through a company licensed to handle PCBs. If there are any doubts about the company's license, contact the EPA.
- Report the location of all PCB transformers to the local fire department and owners of any nearby buildings.

The importance of following the EPA regulations cannot be overstated.

28.4 OSHA Safety Requirements

The federal government has taken a number of steps to help improve safety within the workplace. OSHA, for example, helps industries to monitor and correct safety practices. The agency's records show that electrical standards are among the most frequently violated of all safety standards. Table 28.9 lists 16 of the most common electrical violations, which include these areas:

- Protective covers
- Identification and marking
- Extension cords
- Grounding

TABLE 28.9 Sixteen Common OSHA Violations

Fact Sheet No.	Subject	NEC Ref.
1	Guarding of live parts	110-17
2	Identification	110-22
3	Uses allowed for flexible cord	400-7
4	Prohibited uses of flexible cord	400-8
5	Pull at joints and terminals must be prevented	400-10
6-1	Effective grounding, Part 1	250-51
6-2	Effective grounding, Part 2	250-51
7	Grounding of fixed equipment, general	250-42
8	Grounding of fixed equipment, specific	250-43
9	Grounding of equipment connected by cord and plug	250-45
10	Methods of grounding, cord and plug-connected equipment	250-59
11	AC circuits and systems to be grounded	250-5
12	Location of overcurrent devices	240-24
13	Splices in flexible cords	400-9
14	Electrical connections	110-14
15	Marking equipment	110-21
16	Working clearances about electric equipment	110-16

After [3].

Protective Covers

Exposure of live conductors is a common safety violation. All potentially dangerous electric conductors should be covered with protective panels. The danger is that someone can come into contact with the exposed, current-carrying conductors. It also is possible for metallic objects such as ladders, cable, or tools to contact a hazardous voltage, creating a life-threatening condition. Open panels also present a fire hazard.

Identification and Marking

Properly identify and label all circuit breakers and switch panels. The labels for breakers and equipment switches may be years old, and may no longer describe the equipment that is actually in use. This confusion poses a safety hazard. Improper labeling of the circuit panel can lead to unnecessary damage—or worse, casualties—if the only person who understands the system is unavailable in an emergency. If there are a number of devices connected to a single disconnect switch or breaker, provide a diagram or drawing for clarification. Label with brief phrases, and use clear, permanent, and legible markings.

Equipment marking is a closely related area of concern. This is not the same thing as equipment identification. Marking equipment means labeling the equipment breaker panels and AC disconnect switches according to device rating. Breaker boxes should contain a nameplate showing the manufacturer name, rating, and other pertinent electrical factors. The intent of this rule is to prevent devices from being subjected to excessive loads or voltages.

Extension Cords

Extension (flexible) cords often are misused. Although it may be easy to connect a new piece of equipment with a flexible cord, be careful. The National Electrical Code lists only eight approved uses for flexible cords.

The use of a flexible cord where the cable passes through a hole in the wall, ceiling, or floor is an often-violated rule. Running the cord through doorways, windows, or similar openings also is prohibited. A flexible cord should not be attached to building surfaces or concealed behind building walls or ceilings. These common violations are illustrated in Fig. 28.9.

Along with improper use of flexible cords, failure to provide adequate strain relief on connectors is a common problem. Whenever possible, use manufactured cable connections.

Grounding

OSHA regulations describe two types of grounding: *system grounding* and *equipment grounding*. System grounding actually connects one of the current-carrying conductors (such as the terminals of a supply transformer) to ground. (See Fig. 28.10.) Equipment grounding connects all the noncurrent-carrying metal surfaces together and to ground. From a grounding standpoint, the only difference between a

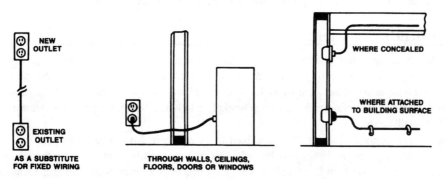

FIGURE 28.9 Flexible cord uses prohibited under NEC rules.

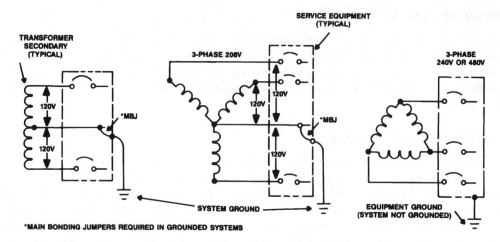

FIGURE 28.10 Even though regulations have been in place for many years, OSHA inspections still uncover violations in the grounding of primary electrical service systems.

grounded electrical system and an ungrounded electrical system is that the *main-bonding jumper* from the service equipment ground to a current-carrying conductor is omitted in the ungrounded system.

The system ground performs two tasks:

- It provides the final connection from equipment-grounding conductors to the grounded circuit conductor, thus completing the ground-fault loop.
- It solidly ties the electrical system and its enclosures to their surroundings (usually earth, structural steel, and plumbing). This prevents voltages at any source from rising to harmfully high voltage-to-ground levels.

It should be noted that equipment grounding—bonding all electric equipment to ground—is required whether or not the system is grounded. System grounding should be handled by the electrical contractor installing the power feeds.

Equipment grounding serves two important functions:

- It bonds all surfaces together so that there can be no voltage differences among them.
- It provides a ground-fault current path from a fault location back to the electrical source, so that if a fault current develops, it will operate the breaker, fuse, or GFCI.

The National Electrical Code is complex, and it contains numerous requirements concerning electrical safety. If the facility electric wiring system has gone through many changes over the years, have the entire system inspected by a qualified consultant. The fact sheets listed in Table 28.9 provide a good starting point for a self-evaluation. The fact sheets are available from any local OSHA office.

Management Responsibility

The key to operating a safe facility is diligent management. A carefully thought-out plan ensures a coordinated approach to protecting staff members from injury and the facility from potential litigation. Facilities that have effective accident-prevention programs follow seven basic guidelines. Although the details and overall organization may vary from workplace to workplace, these practices—summarized in Table 28.10—still apply.

If managers are concerned about safety, it is likely that employees also will be. Display safety pamphlets, and recruit employee help in identifying hazards. Reward workers for good safety performance. Often, an incentive program will help to encourage safe work practices. Eliminate any hazards identified, and obtain OSHA forms and any first aid supplies that would be needed in an emergency. The OSHA *Handbook for Small Business* outlines the legal requirements imposed by the Occupational Safety and

TABLE 28.10 Major Points to Consider When Developing a Facility Safety Program

- ✓ Management assumes the leadership role regarding safety policies.
- ✓ Responsibility for safety- and health-related activities is clearly assigned.
- ✓ Hazards are identified, and steps are taken to eliminate them.
- ✓ Employees at all levels are trained in proper safety procedures.
- ✓ Thorough accident/injury records are maintained.
- ✓ Medical attention and first aid is readily available.
- ✓ Employee awareness and participation is fostered through incentives and an ongoing, high-profile approach to workplace safety.

TABLE 28.11 Sample Checklist of Important Safety Items

Refer regularly to this checklist to maintain a safe facility. For each category shown, be sure that:

Electrical Safety

- ✓ Fuses of the proper size have been installed.
- ✓ All AC switches are mounted in clean, tightly closed metal boxes.
- ✓ Each electrical switch is marked to show its purpose.
- ✓ Motors are clean and free of excessive grease and oil.
- ✓ Motors are maintained properly and provided with adequate overcurrent protection.
- ✓ Bearings are in good condition.
- ✓ Portable lights are equipped with proper guards.
- ✓ All portable equipment is double-insulated or properly grounded.
- ✓ The facility electrical system is checked periodically by a contractor competent in the NEC.
- ✓ The equipment-grounding conductor or separate ground wire has been carried all the way back to the supply ground connection.
- ✓ All extension cords are in good condition, and the grounding pin is not missing or bent.
- ✓ Ground-fault interrupters are installed as required.

Exits and Access

- ✓ All exits are visible and unobstructed.
- ✓ All exits are marked with a readily visible, properly illuminated sign.
- ✓ There are sufficient exits to ensure prompt escape in the event of an emergency.

Fire Protection

- ✓ Portable fire extinguishers of the appropriate type are provided in adequate numbers.
- ✓ All remote vehicles have proper fire extinguishers.
- ✓ Fire extinguishers are inspected monthly for general condition and operability, which is noted on the inspection tag.
- ✓ Fire extinguishers are mounted in readily accessible locations.
- ✓ The fire alarm system is tested annually.

Health Act of 1970. The handbook, which is available from OSHA, also suggests ways in which a company can develop an effective safety program.

Free on-site consultations also are available from OSHA. A consultant will tour the facility and offer practical advice about safety. These consultants do not issue citations, propose penalties, or routinely provide information about workplace conditions to the federal inspection staff. Contact the nearest OSHA office for additional information. Table 28.11 provides a basic checklist of safety points for consideration.

Maintaining safety standards is difficult in any size organization. A written safety manual that has specific practices and procedures for normal workplace hazards as well as the emergency-related hazards you identify is a good idea, and may lower your insurance rates [4]. If outside workers set foot in your facility, prepare a special Safety Manual for Contractors. Include in it installation standards, compliance with *Lock-Out/Tag-Out*, and emergency contact names and phone numbers. *Lock-Out/Tag-Out* is a set of standard safety policies that assures that energy is removed from equipment during installation and maintenance. It assures that every member of a work detail is clear before power is reapplied. Make sure outside contractors carry proper insurance, and are qualified, licensed, or certified to do the work for which you contract.

References

1. Federal Information Processing Standards Publication No. 94, *Guideline on Electrical Power for ADP Installations*, U.S. Department of Commerce, National Bureau of Standards, Washington, D.C., 1983.
2. *Practical Guide to Ground Fault Protection*, PRIMEDIA Intertec, Overland Park, KS, 1995.
3. National Electrical Code, NFPA no. 70.
4. Rudman, R., "Disaster Planning and Recovery," in *The Electronics Handbook*, J. C. Whitaker (Ed.), CRC Press, Boca Raton, FL, pp. 2266–2267, 1996.

Further Information

Code of Federal Regulations, 40, Part 761.
"Current Intelligence Bulletin #45," National Institute for Occupational Safety and Health Division of Standards Development and Technology Transfer, February 24, 1986.
"Electrical Standards Reference Manual," U.S. Department of Labor, Washington, D.C.
Hammar, W., *Occupational Safety Management and Engineering*, Prentice-Hall, New York.
Lawrie, R., *Electrical Systems for Computer Installations*, McGraw-Hill, New York, 1988.
Pfrimmer, J., "Identifying and Managing PCBs in Broadcast Facilities," *NAB Engineering Conference Proceedings*, National Association of Broadcasters, Washington, D.C., 1987.
"Occupational Injuries and Illnesses in the United States by Industry," OSHA Bulletin 2278, U.S. Department of Labor, Washington, D.C., 1985.
OSHA, "Handbook for Small Business," U.S. Department of Labor, Washington, D.C.
OSHA, "Electrical Hazard Fact Sheets," U.S. Department of Labor, Washington, D.C., January 1987.

29
Conversion Tables

TABLE 29.1 Standard Units

Name	Symbol	Quantity
Ampere	A	Electric current
Ampere per meter	A/m	Magnetic field strength
Ampere per square meter	A/m^2	Current density
Becquerel	Bg	Activity (of a radionuclide)
Candela	cd	Luminous intensity
Coulomb	C	Electric charge
Coulomb per kilogram	C/kg	Exposure (x and gamma rays)
Coulomb per sq. meter	C/m^2	Electric flux density
Cubic meter	m^3	Volume
Cubic meter per kilogram	m^3/kg	Specific volume
Degree Celsius	°C	Celsius temperature
Farad	F	Capacitance
Farad per meter	F/m	Permittivity
Henry	H	Inductance
Henry per meter	H/m	Permeability
Hertz	Hz	Frequency
Joule	J	Energy, work, quantity of heat
Joule per cubic meter	J/m^3	Energy density
Joule per kelvin	J/K	Heat capacity
Joule per kilogram K	J/(kg-K)	Specific heat capacity
Joule per mole	J/mol	Molar energy
Kelvin	K	Thermodynamic temperature
Kilogram	kg	Mass
Kilogram per cubic meter	kg/m^3	Density, mass density
Lumen	lm	Luminous flux
Lux	lx	Luminance
Meter	m	Length
Meter per second	m/s	Speed, velocity
Meter per second sq.	m/s^2	Acceleration
Mole	mol	Amount of substance
Newton	N	Force
Newton per meter	N/m	Surface tension
Ohm	Ω	Electrical resistance
Pascal	Pa	Pressure, stress
Pascal second	Pa-s	Dynamic viscosity
Radian	rad	Plane angle
Radian per second	rad/s	Angular velocity
Radian per second sq.	rad/s^2	Angular acceleration
Second	s	Time
Siemens	S	Electrical conductance
Square meter	m^2	Area
Steradian	sr	Solid angle
Tesla	T	Magnetic flux density
Volt	V	Electrical potential
Volt per meter	V/m	Electric field strength
Watt	W	Power, radiant flux
Watt per meter kelvin	W/(m-K)	Thermal conductivity
Watt per square meter	W/m^2	Heat (power) flux density
Weber	Wb	Magnetic flux

TABLE 29.2 Standard Prefixes

Multiple	Prefix	Symbol
10^{18}	exa	E
10^{15}	peta	P
10^{12}	tera	T
10^{9}	giga	G
10^{6}	mega	M
10^{3}	kilo	k
10^{2}	hecto	h
10	deka	da
10^{-1}	deci	d
10^{-2}	centi	c
10^{-3}	milli	m
10^{-6}	micro	m
10^{-9}	nano	n
10^{-12}	pico	p
10^{-15}	femto	f
10^{-18}	atto	a

TABLE 29.3 Standard Units for Electrical Work

Unit	Symbol
Centimeter	cm
Cubic centimeter	cm^3
Cubic meter per second	m^3/s
Gigahertz	GHz
Gram	g
Kilohertz	kHz
Kilohm	kΩ
Kilojoule	kJ
Kilometer	km
Kilovolt	kV
Kilovoltampere	kVA
Kilowatt	kW
Megahertz	MHz
Megavolt	MV
Megawatt	MW
Megohm	MΩ
Microampere	µA
Microfarad	µF
Microgram	µg
Microhenry	µH
Microsecond	µs
Microwatt	µW
Milliampere	mA
Milligram	mg
Millihenry	mH
Millimeter	mm
Millisecond	ms
Millivolt	mV
Milliwatt	mW
Nanoampere	nA
Nanofarad	nF
Nanometer	nm
Nanosecond	ns
Nanowatt	nW
Picoampere	pA
Picofarad	pF
Picosecond	ps
Picowatt	pW

TABLE 29.4 Specifications of Standard Copper Wire

Wire Size AWG	Dia. in Mils	Cir. Mil Area	Turns per Linear Inch[a]			Ohms per 100ft[b]	Current Carrying Capacity[c]	Dia. in mm
			Enamel	S.C.E	D.C.C			
1	289.3	83810	—	—	—	0.1239	119.6	7.348
2	257.6	05370	—	—	—	0.1563	94.8	6.544
3	229.4	62640	—	—	—	0.1970	75.2	5.827
4	204.3	41740	—	—	—	0.2485	59.6	5.189
5	181.9	33100	—	—	—	0.3133	47.3	4.621
6	162.0	26250	—	—	—	0.3951	37.5	4.115
7	144.3	20820	—	—	—	0.4982	29.7	3.665
8	128.5	16510	7.6	—	7.1	0.6282	23.6	3.264
9	114.4	13090	8.6	—	7.8	0.7921	18.7	2.906
10	101.9	10380	9.6	9.1	8.9	0.9989	14.8	2.588
11	90.7	8234	10.7	—	9.8	1.26	11.8	2.305
12	80.8	6530	12.0	11.3	10.9	1.588	9.33	2.063
13	72.0	5178	13.5	—	12.8	2.003	7.40	1.828
14	64.1	4107	15.0	14.0	13.8	2.525	5.87	1.628
15	57.1	3257	16.8	—	14.7	3.184	4.65	1.450
16	50.8	2583	18.9	17.3	16.4	4.016	3.69	1.291
17	45.3	2048	21.2	—	18.1	5.064	2.93	1.150
18	40.3	1624	23.6	21.2	19.8	6.386	2.32	1.024
19	35.9	1288	26.4	—	21.8	8.051	1.84	0.912
20	32.0	1022	29.4	25.8	23.8	10.15	1.46	0.812
21	28.5	810	33.1	—	26.0	12.8	1.16	0.723
22	25.3	642	37.0	31.3	30.0	16.14	0.918	0.644
23	22.6	510	41.3	—	37.6	20.36	0.728	0.573
24	20.1	404	46.3	37.6	35.6	25.67	0.577	0.511
25	17.9	320	51.7	—	38.6	32.37	0.458	0.455
26	15.9	254	58.0	46.1	41.8	40.81	0.363	0.406
27	14.2	202	64.9	—	45.0	51.47	0.288	0.361
28	12.6	160	72.7	54.6	48.5	64.9	0.228	0.321
29	11.3	127	81.6	—	51.8	81.83	0.181	0.286
30	10.0	101	90.5	64.1	55.5	103.2	0.144	0.255
31	8.9	50	101	—	59.2	130.1	0.114	0.227
32	8.0	63	113	74.1	61.6	164.1	0.090	0.202
33	7.1	50	127	—	66.3	206.9	0.072	0.180
34	6.3	40	143	86.2	70.0	260.9	0.057	0.160
35	5.6	32	158	—	73.5	329.0	0.045	0.143
36	5.0	25	175	103.1	77.0	414.8	0.036	0.127
37	4.5	20	198	—	80.3	523.1	0.028	0.113
38	4.0	16	224	116.3	83.6	659.6	0.022	0.101
39	3.5	12	248	—	86.6	831.8	0.018	0.090

[a] Based on 25.4 mm.
[b] Ohms per 1000 ft measured at 20°C.
[c] Current carrying capacity at 700 C.M./A.

TABLE 29.5 Celsius-to-Fahrenheit Conversion

°Celsius	°Fahrenheit	°Celsius	°Fahrenheit
−50	−58	125	257
−45	−49	130	266
−40	−40	135	275
−35	−31	140	284
−30	−22	145	293
−25	−13	150	302
−20	4	155	311
−15	5	160	320
−10	14	165	329
−5	23	170	338
0	32	175	347
5	41	180	356
10	50	185	365
15	59	190	374
20	68	195	383
25	77	200	392
30	86	205	401
35	95	210	410
40	104	215	419
45	113	220	428
50	122	225	437
55	131	230	446
60	140	235	455
65	149	240	464
70	158	245	473
75	167	250	482
80	176	255	491
85	185	260	500
90	194	265	509
95	203	270	518
100	212	275	527
105	221	280	536
110	230	285	545
115	239	290	554
120	248	295	563

TABLE 29.6 Inch-to-Millimeter Conversion

In.	0	1/8	1/4	3/8	1/2	5/8	3/4	7/8	In.
0	0.0	3.18	6.35	9.52	12.70	15.88	19.05	22.22	0
1	25.40	28.58	31.75	34.92	38.10	41.28	44.45	47.62	1
2	50.80	53.98	57.15	60.32	63.50	66.68	69.85	73.02	2
3	76.20	79.38	82.55	85.72	88.90	92.08	95.25	98.42	3
4	101.6	104.8	108.0	111.1	114.3	117.5	120.6	123.8	4
5	127.0	130.2	133.4	136.5	139.7	142.9	146.0	149.2	5
6	152.4	155.6	158.8	161.9	165.1	168.3	171.4	174.6	6
7	177.8	181.0	184.2	187.3	190.5	193.7	196.8	200.0	7
8	203.2	206.4	209.6	212.7	215.9	219.1	222.2	225.4	8
9	228.6	231.8	235.0	238.1	241.3	244.5	247.6	250.8	9
10	254.0	257.2	260.4	263.5	266.7	269.9	273.0	276.2	10
11	279	283	286	289	292	295	298	302	11
12	305	308	311	314	317	321	324	327	12
13	330	333	337	340	343	346	349	352	13
14	356	359	362	365	368	371	375	378	14
15	381	384	387	391	394	397	400	403	15
16	406	410	413	416	419	422	425	429	16
17	432	435	438	441	445	448	451	454	17
18	457	460	464	467	470	473	476	479	18
19	483	486	489	492	495	498	502	505	19
20	508	511	514	518	521	524	527	530	20

TABLE 29.7 Millimeters-to-Decimal Inches Conversion

mm	In.	mm	In.	mm	In.	mm	In.	mm	In.
1	0.039370	31	1.220470	61	2.401570	91	3.582670	210	8.267700
2	0.078740	32	1.259840	62	2.440940	92	3.622040	220	8.661400
3	0.118110	33	1.299210	63	2.480310	93	3.661410	230	9.055100
4	0.157480	34	1.338580	64	2.519680	94	3.700780	240	9.448800
5	0.196850	35	1.377949	65	2.559050	95	3.740150	250	9.842500
6	0.236220	36	1.417319	66	2.598420	96	3.779520	260	10.236200
7	0.275590	37	1.456689	67	2.637790	97	3.818890	270	10.629900
8	0.314960	38	1.496050	68	2.677160	98	3.858260	280	11.032600
9	0.354330	39	1.535430	69	2.716530	99	3.897630	290	11.417300
10	0.393700	40	1.574800	70	2.755900	100	3.937000	300	11.811000
11	0.433070	41	1.614170	71	2.795270	105	4.133848	310	12.204700
12	0.472440	42	1.653540	72	2.834640	110	4.330700	320	12.598400
13	0.511810	43	1.692910	73	2.874010	115	4.527550	330	12.992100
14	0.551180	44	1.732280	74	2.913380	120	4.724400	340	13.385800
15	0.590550	45	1.771650	75	2.952750	125	4.921250	350	13.779500
16	0.629920	46	1.811020	76	2.992120	130	5.118100	360	14.173200
17	0.669290	47	1.850390	77	3.031490	135	5.314950	370	14.566900
18	0.708660	48	1.889760	78	3.070860	140	5.511800	380	14.960600
19	0.748030	49	1.929130	79	3.110230	145	5.708650	390	15.354300
20	0.787400	50	1.968500	80	3.149600	150	5.905500	400	15.748000
21	0.826770	51	2.007870	81	3.188970	155	6.102350	500	19.685000
22	0.866140	52	2.047240	82	3.228340	160	6.299200	600	23.622000
23	0.905510	53	2.086610	83	3.267710	165	6.496050	700	27.559000
24	0.944880	54	2.125980	84	3.307080	170	6.692900	800	31.496000
25	0.984250	55	2.165350	85	3.346450	175	6.889750	900	35.433000
26	1.023620	56	2.204720	86	3.385820	180	7.086600	1000	39.370000
27	1.062990	57	2.244090	87	3.425190	185	7.283450	2000	78.740000
28	1.102360	58	2.283460	88	3.464560	190	7.480300	3000	118.110000
29	1.141730	59	2.322830	89	3.503903	195	7.677150	4000	157.480000
30	1.181100	60	2.362200	90	3.543300	200	7.874000	5000	196.850000

TABLE 29.8 Common Fractions to Decimal and Millimeter Units

Common Fractions	Decimal Fractions	mm (approx)	Common Fractions	Decimal Fractions	mm (approx)	Common Fractions	Decimal Fractions	mm (approx)
1/128	0.008	0.20	11/32	0.344	8.73	43/64	0.672	17.07
1/64	0.016	0.40	23/64	0.359	9.13	11/16	0.688	17.46
1/32	0.031	0.79	3/8	0.375	9.53	45/64	0.703	17.86
3/64	0.047	1.19	25/64	0.391	9.92	23/32	0.719	18.26
1/16	0.063	1.59	13/32	0.406	10.32	47/64	0.734	18.65
5/64	0.078	1.98	27/64	0.422	10.72	3/4	0.750	19.05
3/32	0.094	2.38	7/16	0.438	11.11	49/64	0.766	19.45
7/64	0.109	2.78	29/64	0.453	11.51	25/32	0.781	19.84
1/8	0.125	3.18	15/32	0.469	11.91	51/64	0.797	20.24
9/64	0.141	3.57	31/64	0.484	12.30	13/16	0.813	20.64
5/32	0.156	3.97	1/2	0.500	12.70	53/64	0.828	21.03
11/64	0.172	4.37	33/64	0.516	13.10	27/32	0.844	21.43
3/16	0.188	4.76	17/32	0.531	13.49	55/64	0.859	21.83
13/64	0.203	5.16	35/64	0.547	13.89	7/8	0.875	22.23
7/32	0.219	5.56	9/16	0.563	14.29	57/64	0.891	22.62
15/64	0.234	5.95	37/64	0.578	14.68	29/32	0.906	23.02
1/4	0.250	6.35	19/32	0.594	15.08	59/64	0.922	23.42
17/64	0.266	6.75	39/64	0.609	15.48	15/16	0.938	23.81
9/32	0.281	7.14	5/8	0.625	15.88	61/64	0.953	24.21
19/64	0.297	7.54	41/64	0.641	16.27	31/32	0.969	24.61
5/16	0.313	7.94	21/32	0.656	16.67	63/64	0.984	25.00
21/64	0.328	8.33						

TABLE 29.9 Conversion Ratios for Length

Known Quantity	Multiply By	Quantity to Find
Inches (in)	2.54	Centimeters (cm)
Feet (ft)	30	Centimeters (cm)
Yards (yd)	0.9	Meters (m)
Miles (mi)	1.6	Kilometers (km)
Millimeters (mm)	0.04	Inches (in)
Centimeters (cm)	0.4	Inches (in)
Meters (m)	3.3	Feet (ft)
Meters (m)	1.1	Yards (yd)
Kilometers (km)	0.6	Miles (mi)
Centimeters (cm)	10	Millimeters (mm)
Decimeters (dm)	10	Centimeters (cm)
Decimeters (dm)	100	Millimeters (mm)
Meters (m)	10	Decimeters (dm)
Meters (m)	1000	Millimeters (mm)
Dekameters (dam)	10	Meters (m)
Hectometers (hm)	10	Dekameters (dam)
Hectometers (hm)	100	Meters (m)
Kilometers (km)	10	Hectometers (hm)
Kilometers (km)	1000	Meters (m)

TABLE 29.10 Conversion Ratios for Area

Known Quantity	Multiply By	Quantity to Find
Square inches (in^2)	6.5	Square centimeters (cm^2)
Square feet (ft^2)	0.09	Square meters (m^2)
Square yards (yd^2)	0.8	Square meters (m^2)
Square miles (mi^2)	2.6	Square kilometers (km^2)
Acres	0.4	Hectares (ha)
Square centimeters (cm^2)	0.16	Square inches (in^2)
Square meters (m^2)	1.2	Square yards (yd^2)
Square kilometers (km^2)	0.4	Square miles (mi^2)
Hectares (ha)	2.5	Acres
Square centimeters (cm^2)	100	Square millimeters (mm^2)
Square meters (m^2)	10,000	Square centimeters (cm^2)
Square meters (m^2)	1,000,000	Square millimeters (mm^2)
Ares (a)	100	Square meters (m^2)
Hectares (ha)	100	Ares (a)
Hectares (ha)	10,000	Square meters (m^2)
Square kilometers (km^2)	100	Hectares (ha)
Square kilometers (km^2)	1,000	Square meters (m^2)

TABLE 29.11 Conversion Ratios for Mass

Known Quantity	Multiply By	Quantity to Find
Ounces (oz)	28	Grams (g)
Pounds (lb)	0.45	Kilograms (kg)
Tons	0.9	Tonnes (t)
Grams (g)	0.035	Ounces (oz)
Kilograms (kg)	2.2	Pounds (lb)
Tonnes (t)	100	Kilograms (kg)
Tonnes (t)	1.1	Tons
Centigrams (cg)	10	Milligrams (mg)
Decigrams (dg)	10	Centigrams (cg)
Decigrams (dg)	100	Milligrams (mg)
Grams (g)	10	Decigrams (dg)
Grams (g)	1000	Milligrams (mg)
Dekagram (dag)	10	Grams (g)
Hectogram (hg)	10	Dekagrams (dag)
Hectogram (hg)	100	Grams (g)
Kilograms (kg)	10	Hectograms (hg)
Kilograms (kg)	1000	Grams (g)
Metric tons (t)	1000	Kilograms (kg)

TABLE 29.12 Conversion Ratios for Cubic Measure

Known Quantity	Multiply By	Quantity to Find
Cubic meters (m³)	35	Cubic feet (ft³)
Cubic meters (m³)	1.3	Cubic yards (yd³)
Cubic yards (yd³)	0.76	Cubic meters (m³)
Cubic feet (ft³)	0.028	Cubic meters (m³)
Cubic centimeters (cm³)	1000	Cubic millimeters (mm³)
Cubic decimeters (dm³)	1000	Cubic centimeters (cm³)
Cubic decimeters (dm³)	1,000,000	Cubic millimeters (mm³)
Cubic meters (m³)	1000	Cubic decimeters (dm³)
Cubic meters (m³)	1	Steres
Cubic feet (ft³)	1728	Cubic inches (in³)
Cubic feet (ft³)	28.32	Liters (L)
Cubic inches (in³)	16.39	Cubic centimeters (cm³)
Cubic meters (m³)	264	Gallons (gal)
Cubic yards (yd³)	27	Cubic feet (ft³)
Cubic yards (yd³)	202	Gallons (gal)
Gallons (gal)	231	Cubic inches (in³)

TABLE 29.13 Conversion Ratios for Electrical Quantities

Known Quantity	Multiply By	Quantity to Find
Btu per minute	0.024	Horsepower (hp)
Btu per minute	17.57	Watts (W)
Horsepower (hp)	33,000	Foot-pounds per min (ft-lb/min)
Horsepower (hp)	746	Watts (W)
Kilowatts (kW)	57	Btu per minute
Kilowatts (kW)	1.34	Horsepower (hp)
Watts (W)	44.3	Foot-pounds per min (ft-lb/min)

Index

A

AC system grounding. see also Facility grounding
 and building codes, 15-38
 and conductor size, 15-43 to 15-45
 equipment racks, 15-48 to 15-50
 facility ground systems and, 15-41 to 15-46
 high-frequency effects, 15-46
 isolated grounding and, 15-39
 isolation transformers and, 15-47 to 15-48
 power-center grounding and, 15-46 to 15-48
 separately derived systems and, 15-39
 single-point ground and, 15-38 to 15-39
 terminology, 15-40
Active power line conditioners, 14-40 to 14-41
Adhesives, 12-10 to 12-12
Adjustable resistors, 7-2 to 7-3
Air-core inductors, 9-3
Air handling systems, 13-25 to 13-28, 27-10
Aliasing, 18-7 to 18-9
Alpha multiplication, 13-17
Aluminum electroytic capacitors, 8-6
Analog-digital simulators, 19-26 to 19-30
Analog input architecture, 18-4 to 18-6
Analog-to-digital converters (ADCs), 18-2 to 18-4
Angle modulated waves, 22-13 to 22-17, 22-20 to 22-21
Aperture-leakage control, 6-6 to 6-12
Area conversion ratios, 29-7 (table)

Asynchronous transfer mode (ATM)
 networks, 16-1 to 16-3, 16-5 to 16-7
 configuration and operations, 16-9 to 16-10
 versus SONET (STM) networks, 16-10
Attenuators, 7-3
Audio frequency measurements
 addition and cancellation of distortion components in, 20-9 to 20-10
 CCIT difference frequency distortion and, 20-9
 decibel, 20-4 to 20-5
 digital audio systems and, 20-15 to 20-17
 discrete tones and, 20-12
 distortion mechanisms and, 20-6 to 20-10
 FFT analysis and, 20-13 to 20-15
 harmonic distortion and, 20-6 to 20-8
 intermodulation distortion and, 20-8
 level, 20-2 to 20-5
 multitone audio testing and, 20-10 to 20-15
 noise measurement and, 20-5
 peak-response, 20-3 to 20-4
 phase measurements and, 20-5 to 20-6
 purpose of, 20-1 to 20-2
 twin-tone intermodulation distortion and, 20-9
Autoincrementing, 18-15
Avalanche breakdown, 13-16 to 13-17
Average-response measurements, 20-2 to 20-3

B

Battery supply, uninterruptible power systems, 14-29 to 14-30
Binomial sampling, 1-13 to 1-15
Block-diagram simulators, 19-30 to 19-34
Bond pad structure, 13-8 to 13-9
Building codes, 15-38
Bulkhead panels, 15-28 to 15-33

C

Cables and connectors, 6-1
 bulkhead panels and, 15-28 to 15-33
 data acquisition and, 18-24
 grounding signal-carrying, 15-50 to 15-59
 routing, 15-58 to 15-59
 transmission-system grounding and, 15-24 to 15-25
Capability maturity model (CMM), 3-1 to 3-2
Capacitors
 aluminum electrolytic, 8-6
 ceramic, 8-5
 electrolyte failures and, 8-7 to 8-8
 electrolytic, 8-5 to 8-6
 failure modes, 8-6 to 8-7
 film, 8-3 to 8-5
 foil, 8-5
 introduction to, 8-1 to 8-2
 life spans, 8-8
 nonpolarized, 8-2
 operating losses in, 8-2 to 8-3
 parameters and characteristics of, 8-4 (table)
 polarized, 8-2, 8-5
 practical, 8-2 to 8-8

I-1

tantalum electrolytic, 8-6
  temperature cycling and, 8-7
Carbon composition resistors, 7-2
Carbon film resistors, 7-2
Card-cage shielding, 6-4
CCIT difference frequency distortion, 20-9
Central limit theorem, 1-11 to 1-12
Ceramics
  capacitors, 8-5
  hybrid microcircuits and, 11-2 to 11-3
  printed wiring boards and, 10-4
  resistivity of, 7-4 (table)
Chemical ground rods, 15-10 to 15-12
Chip-and-wire process, 11-17 to 11-19
Chromatic adaptation and white balance, 21-8 to 21-9
Circuit analysis, 4-7
  digital testing and, 23-5
Circuit-protection hardware, 28-5 to 28-8
Clean room methodology, 3-9
Coin tosses, 1-1
  probability space and, 1-3 to 1-4
Combinatorial and fault trees, 1-17 to 1-18
Components
  basics of, 1-15 to 1-16
  shielding, 6-4
Composites, shielding, 6-11 to 6-12
Conditional probability, 1-5 to 1-7
Conductive-impregnated elastomers, 6-10
Conductivity, thermal, 4-2 to 4-3, 5-19
  conductive coating processes and, 6-5 to 6-6
Conductors, 4-5
  conductive coating processes and, 6-5 to 6-6
  grounding, 15-43 to 15-45
  practical capacitors and, 8-2 to 8-8
  printed wiring boards and, 10-3
  thick-film, 11-4 to 11-6
  thin-film, 11-15
Confidence intervals, 1-12 to 1-13
Control and limiting resistors, 7-2
Conversion tables, 29-1 to 29-8
Cooling techniques, 5-18 to 5-19, 13-25 to 13-28
Copper, 11-6, 29-3 (table)
Cratering, 13-9
CRT measurement techniques, 21-18 to 21-23
Cubic measure conversion, 29-8 (table)

D

Data acquisition
  accuracy, 18-16 to 18-22
  aliasing and, 18-7 to 18-9
  analog input architecture and, 18-4 to 18-6
  analog-to-data converters (ADCs) and, 18-2 to 18-4
  boards, 18-1 to 18-6
  continuous scanning and, 18-10
  differential nonlinearity and, 18-17 to 18-19
  digital sampling and, 18-6 to 18-14
  equivalent-time sampling and, 18-14 to 18-16
  external sampling and, 18-9
  fundamentals of, 18-1 to 18-6
  grounding and, 18-27
  integral nonlinearity and, 18-19 to 18-20
  interval scanning and, 18-11 to 18-12
  linearity and, 18-17
  multirate scanning and, 18-10
  noise and, 18-22 to 18-24, 18-25 to 18-26
  PC-based, 18-22 to 18-31
  plug-in data acquisition boards and, 18-2 to 18-4
  seamless changing of sampling rates and, 18-12 to 18-14
  settling time, 18-20 to 18-22
  signals and, 18-2, 18-24 to 18-31
  simultaneous sampling and, 18-11
  software, 18-6
  software polling and, 18-9
  types of measurement inputs, 18-27 to 18-30
Decibel measurements, 20-4 to 20-5
Decomposition, 26-2
Dedicated protection systems
  active power line conditioners and, 14-40 to 14-41
  ferroresonant transformer, 14-31 to 14-32
  isolation transformer, 14-33 to 14-34
  line conditioners and, 14-39 to 14-40
  magnetic-coupling-controlled voltage regulator, 14-32 to 14-33
  tap-changing regulator, 14-34 to 14-38
  variable ratio regulators and, 14-35 to 14-38
  variable voltage transformers and, 14-38 to 14-39
Delta magnetic inverters, 14-24

Density functions and integrals
  estimators and, 1-10 to 1-11
  probability and, 1-4 to 1-5, 1-7 to 1-8
Deposition technology, 11-12 to 11-13
Derating, 2-6 to 2-7, 13-14
Design guidelines and techniques
  motor-generator set, 14-12 to 14-15
  printed wiring boards and, 10-6 to 10-8
  product, 2-4 to 2-7
  systems, 26-10 to 26-21
Dielectric absorption, 8-2 to 8-3
  hybrid microcircuits and, 11-9 to 11-10
Dielectric constant, 4-4, 10-10
Differential equations
  for Markov models, 1-22
Digital audio systems, 20-15 to 20-17
Digital sampling
  real-time, 18-7
  simulators, 19-24 to 19-26
Digital test equipment
  automated test instruments, 23-7 to 23-9
  checking integrated circuits and, 23-5
  computer, 23-7 to 23-9
  emulative testers and, 23-6 to 23-7
  introduction to, 23-1 to 23-2
  logic analyzer, 23-2 to 23-4
  manual probe diagnosis and, 23-5
  oscilloscope, 23-9 to 23-15
  protocol analyzer, 23-7
  signature analyzer, 23-4 to 23-5
Disaster readiness. see also Safety
  air handling systems and, 27-10
  business rescue planning and, 27-12 to 27-13
  business resumption planning and, 27-5 to 27-6
  electromagnetic pulse protection (EMP) and, 27-10 to 27-11
  emergency management design and, 27-2 to 27-4
  emergency power and batteries and, 27-8 to 27-10
  introduction to, 27-1 to 27-2
  managing fear and, 27-13
  outside plant communication links and, 27-7 to 27-8
  planning process, 27-4 to 27-5
  risk assessment and, 27-5 to 27-6
  security and, 27-11
  telephones and, 27-12
  water hazards and, 27-10
  workplace and home, 27-12
  workplace safety and, 27-6 to 27-7

Index

Discrete tones, 20-12
Discrete transistor failure modes, 13-16 to 13-17
Dissipation factor, 8-2
Distortion
 addition and cancellation of, 20-9 to 20-10
 due to time-variant multipath channels, 22-5 to 22-6
 harmonic, 20-6 to 20-8, 24-11 to 24-12
 intermodulation, 20-8
 linear, 22-2 to 22-4, 22-13 to 22-15
 nonlinear, 22-4 to 22-5
 phase, 22-17
 radio frequency (RF), 22-2 to 22-6
Distribution functions, 1-4 to 1-5, 1-7 to 1-8
Double-sideband suppressed-carrier (DSB-SC) demodulation errors, 22-10 to 22-11

E

Earth electrode, 15-6
Earth grounding. see also Facility grounding
 on bare rock, 15-23 to 15-24
 bonding ground-system elements and, 15-14 to 15-15
 chemical ground rods and, 15-10 to 15-12
 exothermic bonding and, 15-15
 ground electrode testing and, 15-9 to 15-10
 ground-system inductance and, 15-15 to 15-17
 ground-wire dressing and, 15-19 to 15-20
 interconnection, 15-20 to 15-23
 interface, 15-7 to 15-9
 personnel protection and, 15-21 to 15-23
 tower elements, 15-17 to 15-19
 the Ufer ground system and, 15-12 to 15-14
Electrical gaskets, 6-10
Electrical overstress (EOS), 2-3 to 2-4
Electrode, grounding, 15-5 to 15-6
Electrolyte failures, 8-7 to 8-8
Electromagnetic compatibility (EMC), 6-1
 printed wiring boards and, 10-14 to 10-16
Electromagnetic interface (EMI) radiated emission, 6-1
Electromagnetic pulse protection (EMP), 27-10 to 27-11
Electromagnetism, 9-1
Electronic hardware reliability

 characterization of materials, parts and processe ands, 2-3
 derating and, 2-6 to 2-7
 design guidelines and techniques and, 2-4 to 2-5
 failure mechanisms and, 2-3 to 2-4
 life cycle usage environment and, 2-2 to 2-3
 manufacturability and, 2-9
 manufacturing issues and, 2-8 to 2-11
 preferred parts and, 2-5
 process qualification and, 2-8 to 2-9
 process verification testing and, 2-10 to 2-11
 product performance and, 2-1 to 2-2
 protective architectures and, 2-5 to 2-6
 qualification and accelerated testing, 2-7 to 2-8
 redundancy and, 2-5
 stress margins and, 2-6
Electroplating, 11-13
Electrostatic discharge (ESD), 2-3 to 2-4
Electrolytic capacitors, 8-5
Emulative testers, 23-6 to 23-7
Engineering data, 4-6 to 4-8
Environment, life cycle usage, 2-2 to 2-3
Equipment grounding, 15-2
Equivalent series resistance, 8-2
Equivalent-time sampling, 18-14 to 18-16
Estimators, 1-9 to 1-10
Evaporation, 11-12 to 11-13
Exothermic bonding, 15-15
Expansion, thermal, 4-3

F

Facility grounding
 AC system grounding and, 15-38 to 15-50
 on bare rock, 15-23 to 15-24
 bonding ground-system elements and, 15-14 to 15-15
 bulkhead panels and, 15-28 to 15-33
 cable routing, 15-58 to 15-59
 checklist for, 15-35 to 15-37
 chemical ground rods and, 15-10 to 15-12
 data acquisition and, 18-27
 designing a system for, 15-26 to 15-37
 earth electrodes and, 15-6
 earth grounding and, 15-6 to 15-24

 equipment grounding and, 15-2
 equipment racks, 15-48 to 15-50
 exothermic bonding and, 15-15
 ground electrode testing and, 15-9 to 15-10
 the grounding electrode and, 15-5 to 15-6
 grounding interface and, 15-7 to 15-9
 ground-system inductance and, 15-15 to 15-17
 ground-wire dressing and, 15-19 to 15-20
 input/output circuits and, 15-56 to 15-58
 interconnection, 15-20 to 15-23
 introduction to, 15-1 to 15-6
 lightning protectors and, 15-33 to 15-35
 noise currents and, 15-50 to 15-54
 patch-bay grounding and, 15-55 to 15-56
 personnel protection and, 15-21 to 15-23
 power-center grounding and, 15-46 to 15-48
 reasons for, 15-2
 safety, 28-17 to 28-18
 satellite antenna grounding and, 15-25 to 15-26
 signal-carrying cables, 15-50 to 15-59
 systems grounding and, 15-3 to 15-5
 terms and codes, 15-2 to 15-5, 15-40
 tower elements, 15-17 to 15-19
 transmission-system grounding and, 15-24 to 15-26
 the Ufer ground system and, 15-12 to 15-14
Failure mechanisms
 alpha multiplication and, 13-17
 avalanche breakdown and, 13-16 to 13-17
 breakdown effects and, 13-17
 capacitors and, 8-6 to 8-7
 computers and, 14-4
 discrete transistor failure modes and, 13-16 to 13-17
 electronic hardware, 2-3 to 2-4, 2-6 to 2-7
 package, 13-18 to 13-19
 semiconductors and, 13-12 to 13-15, 13-16 to 13-21
 software, 3-18 to 3-21
 surface-mounted device, 13-20 to 13-21
 temperature and, 5-2 to 5-3
 thermal runaway and, 13-17

thermal stress-induced failures and, 13-17 to 13-18
Failure rates, 1-15 to 1-16, 1-17
  derating and, 2-6 to 2-7
  Markov models and, 1-23 to 1-24
  product performance and, 2-1 to 2-2
Fastening techniques, semiconductor, 13-24 to 13-25
Federal Information Processing Standards (FIPS) Publication 94, 14-5
Ferromagnetic cores, 9-3, 9-8 (table), 9-9 (table)
Ferroresonant inverters, 14-22 to 14-23
Ferroresonant transformers, 14-31 to 14-32
FFT analysis, 20-13 to 20-15
Film capacitors, 8-3 to 8-5
FIPS Publication 94, 14-5
First aid procedures, 28-9 to 28-10
Flexible data acquisition signal routing, 18-15
Flip-chip bonding, 11-19 to 11-21
Flux, heat, 5-5
Flywheel considerations, 14-14 to 14-15
Foil capacitors, 8-5
Forward bias safe operating area, 13-12 to 13-13
Fourier waveform analysis
  applications of, 24-9 to 24-12
  for continuous-time aperiodic functions, 24-7
  for continuous-time periodic functions, 24-5 to 24-7
  for discrete-time aperiodic functions, 24-9
  for discrete-time periodic functions, 24-7 to 24-8
  introduction to, 24-1
  mathematical preliminaries for, 24-2 to 24-5
  total harmonic distortion measures and, 24-11 to 24-12

G

Gases, 4-4
Gaskets, electrical, 6-10
Gaussian distributions, 1-1 to 1-2
Gibbs phenomenon, 24-4
Glass transition temperature, 12-8
Gold, 11-5
Grounding. see Facility grounding

H

Harmonic distortion, 20-6 to 20-8, 24-11 to 24-12
Heat. see also Temperature; Thermal properties
  air handling systems and, 13-25 to 13-28
  capacity, 4-2
  cooling techniques and, 5-18 to 5-19
  experimental characterization, 5-16 to 5-18
  flow relations, 5-4 to 5-11
  flux, 5-5
  fundamentals of, 4-1 to 4-4
  internal thermal resistance and, 5-13
  laminar flow and, 5-6
  management, 5-1 to 5-2
  modeling and simulation, 5-13 to 5-16
  radiation heat transfer and, 5-11
  removal, 5-18 to 5-19
  sink considerations, 13-21 to 13-28
  thermal gradients and, 5-17 to 5-18
  transfer fundamentals, 5-3 to 5-11
  turbulent flow and, 5-6
Holism, 26-2
Hybrid microelectronics technology
  assembly process, 11-16 to 11-21, 13-5
  chip-and-wire process and, 11-17 to 11-19
  copper and, 11-6
  deposition technology and, 11-12 to 11-13
  dielectric materials and, 11-9 to 11-10
  electroplating and, 11-13
  flip-chip bonding and, 11-19 to 11-21
  gold and, 11-5
  high-temperature drift and, 11-8 to 11-9
  introduction to, 11-1 to 11-2
  overglaze materials, 11-10
  palladium and, 11-5
  photolithographic processes and, 11-13 to 11-14
  platinum and, 11-5
  power handling capability and, 11-8 to 11-9
  process considerations, 11-9
  processing thick-film circuits and, 11-10 to 11-11
  resistor noise and, 11-8
  resistor trimming and, 11-15 to 11-16
  silver and, 11-5
  substrates for, 11-2 to 11-3
  tape automated bonding and, 11-19 to 11-21
  temperature coefficient of resistance and, 11-7 to 11-8
  thick-film conductor materials and, 11-4 to 11-6
  thick-film materials and, 11-3 to 11-11
  thick-film resistor material and, 11-6 to 11-7
  thin-film resistors and, 11-14 to 11-15
  thin-film technology and, 11-11 to 11-15
  voltage coefficient of resistance (VCR) and, 11-8
Hybrid transient suppressor, 14-40
Hysteresis loops, 9-1

I

Inch-to-millimeter conversion, 29-4 (table)
Independence and probability, 1-5 to 1-7
Inductors, 9-2 to 9-4, 14-39
  ground-system, 15-15 to 15-17
Input/output circuits, 15-56 to 15-58
Insulators, 4-4
Interconnection models, printed wiring board, 10-8 to 10-14
Interference of radio frequency (RF) communication, 22-18 to 22-21
Internal thermal resistance, 5-13
Interval scanning, 18-11 to 18-12
Inverter-fed L/C tank, 14-24
Isolated grounding, 15-39, 18-29 to 18-30
Isolation transformers, 14-33 to 14-34, 15-47 to 15-48

J

Jelinski and Moranda's Model, 3-10 to 3-12

K

Key tolerance envelope, 14-2
Kinetic battery storage systems, 14-19
Knitted-wire meshes and screens, 6-8

Index

Knowledge
  incomplete versus complete lack of, 1-2 to 1-3

L

Laminar flow, 5-6
Lead coplanarity, 12-13 to 12-14
Legislation and polychlorinated biphenyls (PCBs), 28-11 to 28-12
Length conversion ratios, 29-6 (table)
Life cycle usage environment, product, 2-2 to 2-3
  spiral model of, 3-3 to 3-5
  waterfall model of, 3-2 to 3-3
Light bulbs, 1-1 to 1-2
Lightning hazards, 14-2 to 14-4
  facility grounding and, 15-33 to 15-35
Linearity, 18-17
Line conditioners, 14-39 to 14-40
Liquids, 4-3
Local area networks. see Open system interconnections (OSI) networks
Logic analyzer, 23-2 to 23-4
Low-noise voltimeter standard, 18-22 to 18-24

M

Magnetic-coupling-controlled voltage regulators, 14-32 to 14-33
Magnetic fields
  air-core inductors and, 9-3
  characteristics of high-permeability materials and, 9-6 (table)
  characteristics of permanent magnet alloys and, 9-7 (table)
  ferromagnetic cores and, 9-3, 9-8 (table), 9-9 (table)
  hysteresis loops and, 9-1
  losses in inductors and transformers and, 9-2 to 9-3
  magnetic flux lines and, 9-1
  properties of antiferromagnetic compounds and, 9-8 (table)
  properties of magnetic materials and, 9-5 (table)
  remnance and, 9-1
  saturation constants for ferromagnetic elements and, 9-9 (table)
  saturation constants for magnetic substances and, 9-8 (table)
  shielding and, 9-1, 9-3

Manual probe diagnosis, 23-5
Manufacturing processes
  design guidelines and techniques and, 2-4 to 2-7
  failure mechanisms and, 2-3 to 2-4
  hybrid microcircuit assembly, 11-16 to 11-21
  manufacturability and, 2-9
  preferred parts and, 2-5
  printed wiring boards and, 10-2 to 10-6
  process qualification and, 2-8 to 2-9
  product performance and reliability and, 2-3
  protective architectures and, 2-5 to 2-6
  qualification and accelerated testing and, 2-7 to 2-8
  redundancy and, 2-5
  semiconductor, 13-3 to 13-6
  stress margins and, 2-6
  verification testing, 2-10 to 2-11
Markov models, 1-18
  basic constructions, 1-19 to 1-20
  combinatorial failure of components and, 1-23 to 1-24
  correlated faults, 1-22
  differential equations for, 1-22
  for a reconfigurable fourplex, 1-20 to 1-21, 1-24
Mass conversion ratios, 29-7 (table)
Melting points, 4-5
Memoryless events, 1-7 to 1-8
Metal film resistors, 7-2
Millimeter-to-decimal inches conversion, 29-4 (table)
Mills Fault Seeding Model, 3-13
Modeling and computation methods, 1-17 to 1-18
  data collection, 3-15 to 3-18
  heat, 5-13 to 5-16
  model verification and, 1-18
  software, 3-5 to 3-21
  software reliability and, 3-8 to 3-18
Moments and probability, 1-8
Monte Carlo computations, 1-13 to 1-15, 1-18
Monte Hall problem, the, 1-6 to 1-7
MOSFET devices, 13-14 to 13-15
  failure modes, 13-19 to 13-20
Motor-generator sets
  description of, 14-10 to 14-11
  design considerations, 14-12 to 14-15
  flywheel considerations, 14-14 to 14-15

kinetic battery storage systems, 14-19
  maintenance considerations, 14-15 to 14-16
  single-shaft systems, 14-14
  system configuration, 14-11 to 14-12
  UPS, 14-16 to 14-19
Mounting power semiconductors, 13-23 to 13-24
Multiplicative probabilty, 1-5 to 1-7
Multirate scanning, 18-10
Multitone audio testing, 20-10 to 20-15
Musa Basic Execution Time Model (BETM), 3-12
Musa-Okumoto Logarithmic Poisson Execution Time Model (LPETM), 3-12 to 3-13

N

Nelson's Model, 3-13 to 3-14
Noise, 15-50 to 15-54, 18-22 to 18-24, 18-25 to 18-26
  measurement, 20-5
Nonlinearity, 18-17 to 18-20
Nonpolarized capacitors, 8-2
Notch filter analyzer, 20-7 to 20-8

O

Open system interconnections (OSI) networks
  application layer, 17-4 to 17-5
  classifications, 17-5
  data link layer, 17-2 to 17-3
  introduction to, 17-1
  model, 17-1 to 17-5
  network layer, 17-3 to 17-4
  physical layer, 17-2
  presentation layer, 17-4
  session layer, 17-4
  transport layer, 17-4
Opinion surveys, 1-13 to 1-15
Organic binders, 11-4
Oscilloscope, digital, 23-9 to 23-15
Output transfer switching, 14-28 to 14-29
Overglaze materials, 11-10
Overstress failures, 2-3 to 2-4

P

Packaging systems
  cooling techniques in, 5-18 to 5-19
  failure mechanisms, 13-18 to 13-19
  thermal effects in, 5-12 to 5-18

Palladium, 11-5
Patch-bay grounding, 15-55 to 15-56
Phase measurement, 20-5 to 20-6
Phase modulation inverters, 14-26
Photolithographic processes, 11-13 to 11-14
Pixel formats, 21-9 to 21-11
Planarity, 12-13 to 12-14
Plastic
  capacitors and, 8-3 to 8-5
  leaded chip carriers (PLCCs), 12-4 to 12-5
Platinum, 11-5
Plug-in data acquisition boards, 18-2 to 18-4
Polarized capacitors, 8-2, 8-5
Polychlorinated biphenyls (PCBs)
  components, identifying, 28-13 to 28-14
  components of, 28-12 to 28-13
  disposal of, 28-14 to 28-15
  health risk from, 28-11
  labeling components of, 28-14
  legislation affecting, 28-11 to 28-12
  proper management of, 28-16
  record keeping and, 28-14
Power-conversion methods, 14-22 to 14-26
Power systems. see also Uninterruptible power systems
  dedicated protection systems and, 14-31 to 14-43
  emergency power off (EPO) buttons and, 28-1
  FIPS Publication 94 and, 14-5
  introduction to, 14-1 to 14-9
  the key tolerance envelope and, 14-2
  lightning hazards and, 14-2 to 14-4
  line conditioners, 14-39 to 14-41
  motor-generator sets and, 14-10 to 14-19
  power-center grounding and, 15-46 to 15-48
  protection hardware, 14-6 to 14-9
  transient protection alternatives and, 14-6
  uninterruptible, 14-19 to 14-30
Predictive models, software, 3-6 to 3-8
Preferred parts and manufacturing, 2-5
Printed-circuit board shielding, 6-4
Printed wiring boards (PWB)
  design of, 10-6 to 10-8
  electromagnetic compatibility (EMC) and, 10-14 to 10-16
  interconnection models, 10-8 to 10-14

introduction to, 10-1
signal-integrity and, 10-14 to 10-16
types, materials, and fabrication, 10-2 to 10-6
Probability
  the central limit theorem and, 1-11 to 1-12
  components and, 1-15 to 1-16
  conditional, 1-5 to 1-7
  confidence intervals and, 1-12 to 1-13
  from density functions and integrals, 1-4 to 1-5
  general introduction to, 1-1 to 1-15
  incomplete versus complete lack of knowledge and, 1-2 to 1-3
  independence and, 1-5 to 1-7
  memoryless events and, 1-7 to 1-8
  modeling and computation methods and, 1-17 to 1-18
  moments and, 1-8
  multiplicative, 1-5 to 1-7
  space, 1-3 to 1-4
Process qualification, 2-8 to 2-9
Process verification testing, 2-10 to 2-11
Product performance and reliability life cycle usage environment, 2-2 to 2-3
Product performance and reliability, 2-1 to 2-2
  derating and, 2-6 to 2-7
  design guidelines and techniques, 2-4 to 2-7
  failure mechanisms and, 2-3 to 2-4
  manufacturability and, 2-9
  manufacturing processes and, 2-3
  preferred parts and, 2-5
  process qualification and, 2-8 to 2-9
  process verification testing, 2-10 to 2-11
  protective architectures and, 2-5 to 2-6
  qualification and accelerated testing, 2-7 to 2-8
  redundancy and, 2-5
  stress margins and, 2-6
Project managers, 26-17 to 26-20
Protection switching, 16-10 to 16-12
Protective architectures, 2-5 to 2-6. see also Dedicated protection systems
Protocol analyzers, 23-7
Prototype systems, 12-17 to 12-18
Pulse-width modulation inverter, 14-26
Punch-through, 13-17

Q

Quad flat packs (QFPs), 12-4 to 12-5
Qualification and accelerated testing, 2-7 to 2-8
Quasi-square wave inverters, 14-25

R

Radiation heat transfer, 5-11
Radio frequency (RF) communication
  angle modulated waves and, 22-13 to 22-17
  coherent demodulation of AM signals and, 22-9 to 22-13
  distortion types, 22-2 to 22-6
  ground waves and, 22-7 to 22-8
  interference and, 22-18 to 22-21
  introduction to, 22-1 to 22-2
  linear distortion and, 22-2 to 22-4, 22-13 to 22-15
  nonlinear distortion and, 22-4 to 22-5
  phase distortion and, 22-17
  sky waves and, 22-8 to 22-9
  time-variant multipath channels and, 22-5 to 22-6
  tropospheric waves and, 22-8
  the wireless radio channel and, 22-6 to 22-9
Ramamoorthy and Bastani model, 3-14 to 3-15
Real-time digital sampling, 18-6 to 18-14
Reconfigurable fourplex models, 1-20 to 1-21, 1-24
Redundancy, systems, 1-16 to 1-17, 2-5
  uninterruptible power, 14-26 to 14-28
Reflow soldering, 12-14 to 12-17
Reliability. see Electronic hardware reliability; Software reliability
Remnance, 9-1
Resistivity, 4-4, 5-4, 5-12 to 5-13
  electrical, 7-4 to 7-6 (table)
  of semiconducting minerals, 7-8 (table)
Resistors
  adjustable, 7-2 to 7-3
  attenuators and, 7-3
  carbon composition, 7-2
  carbon film, 7-2
  control and limiting, 7-2
  metal film, 7-2
  networks, 7-2
  noise, 11-8
  thick-film, 11-6 to 11-8

Index

thin-film, 11-14 to 11-15
trimming, 11-15 to 11-16
types, 7-1 to 7-8
wire-wound, 7-1
Reverse bias safe operating area, 13-13
Risk assessment, 27-5 to 27-6
Rock-based radial elements, 15-23 to 15-24
Rome Air Development Center Reliability Metric (RADC), 3-6 to 3-8
Root-mean-square measurements, 20-2

S

Safety. see also Disaster readiness
  circuit-protection hardware and, 28-5 to 28-8
  from electric shock, 28-3 to 28-10
  equipment, facility, 28-1 to 28-3
  extension cords and, 28-17
  first aid procedures and, 28-9 to 28-10
  grounding, 28-17 to 28-18
  high voltage equipment and, 28-8
  identification and marking for, 28-17
  management responsibility for, 28-18 to 28-20
  OSHA requirements for workplace, 28-16 to 28-20
  polychlorinated biphenyls (PCBs) and, 28-10 to 28-16
  protective covers for, 28-17
  workplace, 27-6 to 27-7
Sampling, statistical, 1-8 to 1-9
  accuracy, 18-16 to 18-22
  binomial, 1-13 to 1-15
  digital, 18-6 to 18-14
  equivalent-time, 18-14 to 18-16
  rates, 18-12 to 18-14
  settling time, 18-20 to 18-22
  simultaneous, 18-11
Satellite antenna grounding, 15-25 to 15-26
Screen printing, 11-10
Self-inductance, capacitor, 8-3
Semiconductors, 4-4 to 4-5
  air handling systems and, 13-25 to 13-28
  bond pad structure, 13-8 to 13-9
  burn-in, 13-8
  case histories, 13-10 to 13-11
  computer-based simulation and, 19-4 to 19-5
  cooling techniques, 13-25 to 13-28
  cratering and, 13-9
  derating, 13-14
  device assembly, 13-3 to 13-6
  device ruggedness, 13-12 to 13-15
  failure mode, 13-16 to 13-21
  fastening techniques, 13-24 to 13-25
  forward bias safe operating area and, 13-12 to 13-13
  heat sink considerations and, 13-21 to 13-28
  hybrid assembly process, 11-16 to 11-21
  hybrid devices and, 13-5
  MOSFET devices and, 13-14 to 13-15, 13-19 to 13-20
  mounting power, 13-23 to 13-24
  packaging, 13-4
  power handling capability, 13-13 to 13-14
  reliability, 13-6 to 13-11
  reliability of VLSI devices and, 13-9 to 13-10
  reverse bias safe operating area and, 13-13
  terminology, 13-2 to 13-3
  VLSI and VHSIC devices and, 13-5 to 13-6, 13-9 to 13-10
Semi-Markov models, 1-18
Settling time, 18-20 to 18-22
Shielding
  aperture-leakage control and, 6-6 to 6-12
  at box-housing level, 6-5 to 6-6
  cables and connectors and, 6-1
  card-cage, 6-4
  component, 6-4
  composites, 6-11 to 6-12
  effectiveness, 6-2 to 6-3
  electromagnetic interface (EMI) radiated emission and, 6-1
  levels of, 6-3 to 6-6
  magnetic, 9-1, 9-3
  printed-circuit board, 6-4
  system design approach, 6-12 to 6-13
  using electrical gaskets, 6-10
  using knitted-wire meshes and screens, 6-8
  using thin films, 6-8
  using ventilating holes, 6-8 to 6-10
Shock, electric
  circuit-protection hardware and, 28-5 to 28-8
  effects on the human body, 28-4 to 28-5
  first aid for, 28-9 to 28-10
  high voltage equipment and, 28-8
Signal analysis, computer-based
  introduction to, 25-1
  signal generation and, 25-1 to 25-5
  signal processing and, 25-4 to 25-5
  symbolic mathematics and, 25-5 to 25-7
Signal conditioning architectures, 18-30 to 18-31
Signal-integrity and printed wiring boards, 10-14 to 10-16
Signature analyzer, 23-4 to 23-5
Silver, 11-5
Simulation, computer-based
  AC independent voltage source and, 19-6 to 19-7
  block-diagram, 19-30 to 19-34
  circuit entry, 19-2
  control shell, 19-7 to 19-9
  DC independent voltage source and, 19-5 to 19-6
  digital, 19-24 to 19-26
  Fourier analysis and, 19-17 to 19-19
  graphics outputs, 19-7
  introduction to, 19-1
  measured results and, 19-21 to 19-24
  mixed analog-digital, 19-26 to 19-30
  Op-Amp description, 19-9 to 19-13
  passive components description, 19-2 to 19-3
  program with integrated circuit emphasis (SPICE), 19-1 to 19-24
  pulse independent source (PULSE) and, 19-14 to 19-15
  semiconductor component description, 19-4 to 19-5
  sinusoidal independent source (SIN) and, 19-15
  subcircuits and, 19-19 to 19-21
  time-domain analysis, 19-14 to 19-16
  the .TRAN statement and, 19-15 to 19-16
Single-point ground, 15-38 to 15-39
Single-shaft systems, 14-14
Single-sideband suppressed-carrier (SSB-SC) demodulation errors, 22-11 to 22-13
Software polling, 18-9
Software reliability
  assessment, 3-8 to 3-18
  the capability maturity model and, 3-1 to 3-2
  data collection, 3-15 to 3-18
  derived models, 3-15
  fault tree, 3-22f
  Jelinski and Moranda's Model and, 3-10 to 3-12
  life cycle models and, 3-2 to 3-5

Mills Fault Seeding Model and, 3-13
model limitations, 3-18 to 3-21
models, 3-5 to 3-21
Musa Basic Execution Time Model (BETM) and, 3-12
Musa-Okumoto Logarithmic Poisson Execution Time Model and, 3-12 to 3-13
Nelson's Model and, 3-13 to 3-14
predictive models, 3-6 to 3-8
Ramamoorthy and Bastani model and, 3-14 to 3-15
the Rome Air Development Center Reliability Metric (RADC) and, 3-6 to 3-8
the spiral life cycle model and, 3-3 to 3-5
the waterfall life cycle model and, 3-2 to 3-3
Solder paste and joint formation, 12-12 to 12-13
Solids, 4-3
SONET networks
versus asynchronous transfer mode (ATM) networks, 16-10
broadband switching layer model and formats, 16-3 to 16-7
system configuration and operations, 16-8
Specific heat, 4-2
Spiral model, 3-3 to 3-5
Spring-finger stock, 6-10
Sputtering, 11-12 to 11-13
Standard units conversion tables, 29-1 (table)
Statistics, 1-13 to 1-15
binomial sampling and, 1-13 to 1-15
estimators and, 1-9 to 1-10
failure rates and, 1-15 to 1-16
general introduction to, 1-1 to 1-15
incomplete versus complete lack of knowledge and, 1-2 to 1-3
Markov and semi-Markov models and, 1-18, 1-19 to 1-24
Monte Carlo computations and, 1-13 to 1-15, 1-18
opinion surveys and, 1-13 to 1-15
sampling, 1-8 to 1-9
signal analysis and, 25-3 to 25-4
Step wave inverters, 14-25
Stress margins, 2-6
Substrates
for hybrid circuits, 11-2 to 11-3
for surface mount technology (SMT), 12-5 to 12-7
Surface mount technology (SMT)
adhesives and, 12-10 to 12-12
coefficients of thermal expansion (CTE or TCE) and, 12-9
components, 12-4 to 12-5
definition of, 12-1 to 12-4
failure modes, 13-20 to 13-21
joint formation, 12-12 to 12-13
parts inspection and placement, 12-13 to 12-14
prototype systems, 12-17 to 12-18
reflow soldering, 12-14 to 12-17
solder paste, 12-12 to 12-13
substrate design guidelines, 12-5 to 12-7
thermal design considerations for, 12-7 to 12-9
through-hole technology (THT) and, 12-1 to 12-4
Switching, network
asynchronous transfer mode (ATM) networks and, 16-1 to 16-3, 16-9 to 16-10
broadband layer model and formats, 16-3 to 16-7
channel formats, 16-5 to 16-7
introduction to, 16-1 to 16-3
multilayer transport model, 16-3 to 16-4
physical path versus virtual path and, 16-4 to 16-5
protection, 16-10 to 16-12
system configuration and operations, 16-8 to 16-10
Systems
budget requirement analysis, 26-15
change control, 26-16 to 26-17
customer support, 26-14 to 26-15
decomposition and, 26-2
design development, 26-10 to 26-11
design projects, 26-10 to 26-21
detailed design, 26-12 to 26-14
dynamics, 26-4
electronic, 26-11 to 26-12
elements, 26-8 to 26-10
engineering, 26-2 to 26-10
engineers, 26-20 to 26-21
evaluation and decision, 26-5 to 26-8
feasibility study and technology assessment, 26-15 to 26-16
functional analysis, 26-3 to 26-4
grounding, 15-3 to 15-5
and holism, 26-2
modeling, 26-4
modeling and computation methods and, 1-17 to 1-18
optimization, 26-4
program management, 26-17 to 26-20
project tracking and control, 26-16
protection hardware, power, 14-6 to 14-9
prototype, 12-17 to 12-18
redundancy, 1-16 to 1-17, 2-5
scheduling and resources planning, 26-16
shielding and, 6-12 to 6-13
synthesis, 26-3 to 26-4
theory, 26-1 to 26-2
trade studies and, 26-5 to 26-8

T

Tantalum electrolytic capacitors, 8-6
Tap-changing regulators, 14-34 to 14-38
Tape automated bonding, 11-19 to 11-21
Temperature, 4-2. see also Heat
coefficient of capacitance, 4-6
coefficient of resistance, 4-6, 11-7 to 11-8
compensation, 4-7
conversion table, 29-4 (table)
cycling and capacitors, 8-7
glass transition, 12-8
thermistors and, 4-7
Thermal properties. see also Heat
circuit analysis and, 4-7
coefficients of thermal expansion (CTE or TCE) and, 12-9
conductors and, 4-5, 5-19
cooling techniques and, 5-18 to 5-19
dielectric constan andt, 4-4
engineering data and, 4-6 to 4-8
gases and, 4-4
insulators and, 4-4
introduction to, 4-1
liquids and, 4-3
management, 5-1 to 5-2
melting point and, 4-5
modeling and simulation, 5-13 to 5-16
mounting power semiconductors and, 13-23 to 13-24
nomenclature, 5-20 to 5-22
other material properties and, 4-4 to 4-5
packaging systems and, 5-12 to 5-18
resistivity and, 4-4, 5-4, 5-12 to 5-13
semiconductors and, 4-4 to 4-5
solids and, 4-3
surface mount technology (SMT) and, 12-7 to 12-9
thermal analysis and, 4-8 to 4-9
thermal runaway and, 13-17

Index

thermal stress-induced failures and, 13-17 to 13-18
Thermal runaway, 13-17
Thermistors, 4-7
Thick-film technology
  conductor materials, 11-4 to 11-6
  dielectric materials, 11-9 to 11-10
  materials, 11-3 to 11-4
  overglaze materials, 11-10
  processing circuits for, 11-10 to 11-11
  resistor material, 11-6 to 11-8
  temperature of coefficient of resistance and, 11-7 to 11-8
Thin-film technology, 11-11 to 11-15
Through-hole technology (THT), 12-1 to 12-4
Time-domain analyses, 19-14 to 19-16
Trade studies, 26-5 to 26-8
Transformers, 9-2 to 9-4
  ferroresonant, 14-31 to 14-32
  isolation, 14-33 to 14-34, 15-47 to 15-48
  tap-changing regulator, 14-34 to 14-38
  variable voltage, 14-38 to 14-39, 14-42 to 14-43
Transient protection alternatives, 14-6
Transmission-system grounding, 15-24 to 15-26
Triggering, 23-12 to 23-14
Turbulent flow, 5-6
Twin-tone intermodulation distortion, 20-9

U

Ufer ground system, 15-12 to 15-14
Uninterruptible power systems. see also Power systems
  battery supply, 14-29 to 14-30
  configuration, 14-21 to 14-20
  delta magnetic inverter and, 14-24
  description of, 14-19 to 14-21
  ferroresonant inverters and, 14-22 to 14-23
  inverter-fed L/C tanks and, 14-24
  output transfer switching, 14-28 to 14-29
  and phase modulation inverters, 14-26
  power-conversion methods, 14-22 to 14-26
  pulse-width modulation inverters and, 14-26
  quasi-square wave inverters and, 14-25
  redundant operation, 14-26 to 14-28
  step wave inverters and, 14-25
Ups configuration, 14-21 to 14-22

V

Variable ratio regulators, 14-35 to 14-38
Variable voltage transformers, 14-38 to 14-39
Venn diagrams and coin tosses, 1-3
Ventilating holes and shielding, 6-8 to 6-10
VHSIC devices, 13-5 to 13-6
Video displays
  assessment of color reproduction in, 21-8 to 21-9
  CRT measurement techniques and, 21-18 to 21-23
  horizontal resolution and, 21-4
  maximum video frequency and, 21-3 to 21-4
  measurement of color, 21-6 to 21-7
  minimum video frequency and, 21-3
  overall gamma requirements and, 21-9
  perception of color differences and, 21-9
  resolution and pixel format, 21-9 to 21-11
  the time (motion) dimension and, 21-15 to 21-18
  video frequencies arising from scanning and, 21-4 to 21-6
  video signal spectra and, 21-1 to 21-6
  zone plate signal and, 21-11 to 21-18
VLSI devices, 13-5 to 13-6, 13-9 to 13-10
Voltage
  capacitors and, 8-2
  coefficient of resistance (VCR), 11-8
  computer-based simulation and, 19-5 to 19-7
  ferroresonant transformers and, 14-31 to 14-32
  line conditioners and, 14-39 to 14-41
  printed wiring boards and, 10-10 to 10-11
  regulator, magnetic-coupling-controlled, 14-32 to 14-33
  safety and, 28-8
  tap-changing regulators and, 14-34 to 14-38
  transformers, variable, 14-38 to 14-39, 14-42 to 14-43

W

Waterfall life cycle model, 3-2 to 3-3
Water hazards, 27-10
Wearout, 2-3 to 2-4
Wireless radio channels, 22-6 to 22-9
Wire-mesh gaskets, 6-10
Wire-wound resistors, 7-1

Z

Zone plate signal, 21-11 to 21-18